Wang / Morris

**Rubber Reinforcement with Particulate Fillers**

Meng-Jiao Wang
Michael Morris

# Rubber Reinforcement with Particulate Fillers

Chemical Industry Press Co., Ltd.

HANSER
Hanser Publications, Cincinnati
Hanser Publishers, Munich

The Authors:

*Meng-Jiao Wang,* EVE Rubber Institute, Zhengzhou Road, Qingdao, China

*Michael Morris,* Cabot Corporation, Billerica, MA 01821, USA

Distributed in the Americas by:
Hanser Publications
414 Walnut Street, Cincinnati, OH 45202 USA
Phone: (800) 950-8977
www.hanserpublications.com

Distributed in all other countries by:
Carl Hanser Verlag
Postfach 86 04 20, 81631 Munich, Germany
Fax: +49 (89) 98 48 09
www.hanser-fachbuch.de



Library of Congress Control Number: 2020943840



Editor: Dr. Mark Smith
Coverconcept: Marc Müller-Bremer, www.rebranding.de, Munich
Coverdesign: Max Kostopoulos
Typesetting: Beijing Keyin Company
Printed and bound by Beijing Hucais Culture Communication Co., Ltd
Printed in Beijing, China

ISBN: 978-1-56990-719-1
E-Book ISBN: 978-1-56990-720-7

# Preface

Soon after rubber's discovery as a remarkable material in the 18$^{th}$ century, the application of particulate fillers – alongside vulcanization – became the most important factor in the manufacture of rubber products, with the consumption of these particulate fillers second only to rubber itself. Fillers have held this important position not only as a cost savings measure by increasing volume, but more importantly, due to their unique ability to enhance the physical properties of rubber, a well-documented phenomenon termed "reinforcement." In fact, the term filler is misleading because for a large portion of rubber products, tires in particular, the cost of filler per unit volume is even higher than that of the polymer. This is especially true for the reinforcement of elastomers by extremely fine fillers such as carbon black and silica. This subject has been comprehensively reviewed in the monographs "Reinforcement of Elastomers," edited by G. Kraus (1964), "Carbon Black: Physics, Chemistry, and Elastomer Reinforcement," written by J.-B. Donnet and A. Voet (1975), and "Carbon Black: Science and Technology," edited by J.-B. Donnet, R. C. Bansal, and M.-J. Wang (1993). There has since been much progress in the fundamental understanding of rubber reinforcement, the application of conventional fillers, and the development of new products to improve the performance of rubber products.

While all agree that fillers as one of the main components of a filled-rubber composite have the most important bearing on improving the performance of rubber products, many new ideas, theories, practices, phenomena, and observations have been presented about how and especially why they alter the processability of filled compounds and the mechanical properties of filled vulcanizates.

This suggests that the real world of filled rubber is so complex and sophisticated that multiple mechanisms must be involved. It is possible to explain the effect of all fillers on rubber properties ultimately in similar and relatively nonspecific terms, i.e., the phenomena related to all filler parameters should follow general rules or principles. It is the authors' belief that, regarding the impact of filler on all aspects of rubber reinforcement, filler properties, such as microstructure, morphology, and surface characteristics, play a dominant role in determining the properties of filled rubbers, hence the performance of rubber products, via their effects in rubber. These effects, which include hydrodynamic, interfacial, occlusion, and agglomeration of fillers, determine the structure of this book.

The first part of the book is dedicated to the basic properties of fillers and their characterization, followed by a chapter dealing with the effect of fillers in rubber. Based on these two parts, the processing of the filled compounds and the properties of the filled vulcanizates are discussed in detail. The last few chapters cover some special applications of fillers in tires, the new development of filler-related materials for tire applications, and application of fumed silica in silicone rubber. All chapters emphasize an internal logic and consistency, giving a full picture about rubber reinforcement by particulate fillers. As such, this work is intended for those working academically and industrially in the areas of rubber and filler.

We would like to express our heartfelt thanks to Wang's colleagues at the EVE Rubber Institute Mr. Weijie Jia, Mr. Fujin He, Dr. Bin Wang, Dr. Wenrong Zhao, Dr. Hao Zhang, Dr. Mingxiu Xie, Dr. Yudian Song, Dr. Feng Liu, Dr. Liang Zhong, Dr. Bing Yao, Dr. Dan Zhang, Dr. Kai Fu, and Mr. Shuai Lu for their assistance in preparing this book. Special thanks are due to the EVE Rubber Institute, Qingdao, China and Cabot Corporation, USA. Without their firm backing and continuous understanding, this effort could not have been accomplished.

Meng-Jiao Wang, Sc. D., Professor
EVE Rubber Institute, Qingdao, China

Michael Morris, Ph. D., Cabot Corporation
Billerica, Massachusetts USA

# About the Authors

Meng-Jiao Wang began his career in rubber research in 1964 after graduating from Shandong University and joining the Beijing Research and Design Institute of the Rubber Industry, China, where his last position was a professor and chief engineer. In 1982, he moved to France to work at the CNRS in Mulhouse as an associate researcher. He was awarded a doctoral degree (Docteur d'Etat es Sciences) in 1984. He then spent a year and half as a visiting scientist working at the University of Akron. From 1988 to 1989 he worked at the German Institute for Rubber Technology (DIK) as a visiting scientist and later he joined Degussa as a Senior Scientist. In 1993, Dr. Wang moved again to the United States to join Cabot, and soon became a scientific fellow of the company. From 2011 onwards, he has been the scientific fellow of National Engineering Research Center for Rubber and Tire and the director of the EVE Rubber Institute, China.

Over his 56-year career in rubber research, Meng-Jiao Wang has published over 140 scientific papers and 10 book chapters, and he has 55 different US and Chinese patents and 24 equivalent international patents to his name. He co-edited the book "Carbon Black: Science and Technology" and co-authored 6 other books. He was previously a member of the Editorial Board of the journal *Rubber Chemistry and Technology* (USA).

Michael Morris is currently a Principal Scientist at Cabot Corporation in Massachusetts, USA. He was awarded a Ph.D. from the University of Southampton, UK, in 1985 and began his career in rubber research at the Malaysian Rubber Producers Research Association (MRPRA; now TARRC) the same year. In 1987, he was seconded to the Rubber Research Institute of Malaysia in Kuala Lumpur to work on various aspects of natural rubber research. After four years, in 1991, he returned to the MRPRA and continued research, mainly on NR latex. In 1996, Dr. Morris moved to the USA to join the R&D organization of Cabot Corporation. His research and applications development were initially focused on fumed silica, as part of the Cab-O-Sil division. Since 2005, he has been part of the rubber reinforcement group at Cabot, focused on carbon black reinforcement of rubber.

During his career, Michael Morris has published 18 journal papers, 2 book chapters, and 14 conference presentations. In addition, he is an inventor on 12 granted US patents and numerous other international patents.

# Contents

# 1 Manufacture of Fillers

The history of particulate fillers used in rubber is almost as long as that of rubber itself[1–3]. One aspect of filler addition has been improvement of rubber properties. Another aspect was extension of the rubber with less expensive materials. After Hancock developed the earliest device using rollers to crumb natural rubber (NR) in 1820, and the two-roll mill for NR mastication and compounding was patented by Chaffee in 1836 and 1841, incorporation of inert fillers in finely divided particulate form became standard practice. Fillers such as ground limestone, barites, clay, kaolin, etc. were used in order to extend and cheapen the compounds since it was found that in natural rubber, quite a bit of filler could be added without detracting too much from the final vulcanizate properties. Zinc oxide was originally used for its whiteness, and later was found to have some reinforcing effect, becoming known as an "active" filler. Carbon black, which was known as a black pigment, was also found to be able to improve the rubber properties significantly at low concentrations, especially the stiffness. Systematic studies of the effect of fillers had been reported by Heinzerling and Pahl in Germany in 1891. Part of this effect may be due to its activating effect on many vulcanization accelerators for which zinc oxide is still utilized. In 1904, Mote in England, discovered the reinforcing effect of carbon black. He reported that the tensile strength of the filled NR increased drastically, compared with the values obtained with the techniques of that time. Although automobiles had been around and running on rubber tires for more than a decade, the importance of this discovery was recognized and developed when black tires were demonstrated to have better wear resistance than white ones, which contained mainly zinc oxide as a filler. Carbon black is now the most important filler used in rubber. In the last century, the production techniques and designation of types of carbon black have developed rapidly.

In the meantime, non-black fillers have also developed. Among these non-black ones, the first reinforcing filler, calcium silicate, was introduced in 1939. It was prepared by wet precipitation from sodium silicate solution with calcium chloride. In further development of the process, the calcium was leached out by hydrochloric acid to yield a reinforcing silica pigment of comparable particle size. About 10 years later, direct precipitation of silica from sodium silicate solution had developed to a commercial process and this is still a major process today. In 1950, a different type of anhydrous silica appeared, which was made by reacting silicon tetrachloride or silica chloroform (trichlorosilane) with water vapor in a hydrogen-oxygen flame. Because of the high temperature at the formation (about 1400°C), this pyrogenic silica has a lower concentration of silanol groups on the surface than the precipitated silicas. The latter contain about 88%–92% $SiO_2$ and have ignition losses of 10%–14%, whereas pyrogenic silica contains 99.8% silica. Because of its lower surface concentration of silanols, ultra-high purity with total impurities in many cases below 100 ppm (parts per million), and much higher price, pyrogenic silica is mainly used as a filler for high cost compounds such as silicone rubber.

In contrast, since the beginning of the industrial-scale production of fine-particle silicas and silicates in 1948, precipitated silica manufacturers have always desired to find their products used in tires as well. Whereas silicas were rapidly able to replace up to 100 percent of carbon black in shoe sole materials and also made their way into the mechanical goods sector, mostly as blends with carbon blacks, their use in tires in any quantities worth mentioning has long been limited to two types of compounds: off-the-road tread compounds containing 10 phr to 15 phr of silica blended with carbon black in order to improve tear properties, and textile and steel cord bonding compounds containing 15 phr of silica, again blended with carbon black, in combination with resorcinol/formaldehyde systems[4].

During the two oil crises in the 1970s, which had led to a steep rise in the price of carbon black, the question arose whether silica in tires could be an alternative to carbon black. When the price of oil fell and the fear of a lack of availability of carbon blacks subsided, this question was soon forgotten, especially since the price of silica was always higher than that of carbon black, at least in Japan and the USA. Experience has shown that silicas only have a chance to be used in tires if they offer technological advantages which are superior to those of carbon blacks.

Two developments have created a new opportunity for silicas to be used in tires: the increased awareness of the pollution from industry and the necessity of protecting the environment have given rise to a call for tires combining a long service life with driving safety and low fuel consumption. The introduction of bifunctional organosilanes as coupling agents now permits the reinforcing mechanism of silicas to be controlled by chemical means[4,5]. Based on systematic studies of surface characteristics, polymer-filler interactions, and better understanding of compounding and processing, silica was successfully used to replace carbon black as the principal filler in the tread compound of the "green tire" patented in 1992[6]. Since then, the application of precipitated silica in tire has been continuously growing, not only in tread compounds, but also in other parts of tires.

In the last two decades, research on rubber reinforcement with particulate fillers and the development of new fillers have been very hot. Since the main fillers used in the rubber industry are still carbon blacks and silicas, the topics of this book will focus on these two materials and their derivatives.

# 1.1 Manufacture of Carbon Black

The history of carbon black manufacture is very long, such as in China, about 3000 B.C., carbon black for pigment use was made by burning vegetable oils in small lamps and collecting the carbon on a ceramic lid; in Egypt, carbon black was used as a pigment for paints and lacquers. Starting in 1870, natural gas began to be used as the feedstock for carbon black manufacture. Over a couple of decades, the channel process was developed in which small gas flames burning in restricted air supply impinged on iron channels. In 1976, the last channel black plant was closed in the USA, due to the pollution of smoke plumes.

A critical event in the development of the carbon black industry was the discovery of the benefits of carbon black as a reinforcing agent for rubber in 1904[1]. As the automobile became ubiquitous during the 1920s, the application of pneumatic tires grew rapidly and soon by-passed other applications, causing rapid growth in consumption of carbon black. Also in the 1920s, two other processes concerning carbon black production were introduced, both using natural gas as feedstock, but having better yields and lower emissions than the channel process. One was the thermal black process, in which a brick checker is employed and works alternately by absorbing heat from a natural gas air flame, and then giving up heat to crack natural gas to carbon and hydrogen. The other process was the gas furnace process, which is no longer practiced.

The oil furnace process was first introduced by Phillips Petroleum at its plant in Borger, Texas, in 1943. This process rapidly replaced all others for the production of carbon black for use in rubber. In a modern version of the oil furnace process, carbon black yields range from 65% downward depending on the surface area of the product. Product recovery is essentially 100% as a result of high efficiency bag filters. The overwhelming majority of carbon black reactors today are based on the oil furnace process.

## 1.1.1 Mechanisms of Carbon Black Formation

The formation of particulate carbon involves either pyrolysis or incomplete combustion of hydrocarbon materials. Enormous literature has been published to describe the mechanism of carbon black formation, from a series of lectures by Michael Faraday at the Royal Institution in London in the 1840s[2], to a more recent intensive review[7]. Since Faraday's time, many theories have been proposed to account for carbon formation, but controversy still exists regarding the mechanism.

Mechanisms of carbon black formation must account for the experimental observations of the unique morphology and microstructure of carbon black. These include the presence of nodules, or particles, multiple growth centers within some nodules, the fusion of nodules into large aggregates, and the paracrystalline or concentric layer plane structure of the aggregates. It is generally accepted that the mechanism of formation involves a series of stages as follows:

**Formation of gaseous carbon black precursors at high temperature** – This involves dehydrogenation of primary hydrocarbon molecular species to atomic carbon or primary free radical and ions which condense to semi-solid carbon precursors (or poly-nuclear-aromatic sheet) and/or formation of large hydrocarbon molecules by polymerization which then is dehydrogenated to particle precursors[3]. Taking production of furnace black with high aromatic feedstock as an example, Figure 1.1 represents several of the possible paths that feedstock can take as it is mixed with the primary fire in the reactor at the early stage. The primary fire has excess oxygen, carbon dioxide, and water, all of which act as combustion (or oxidation) reactants to the feedstock molecules. These molecules can react with and break up any feedstock molecules into small combustion species; any feedstock that goes this route is lost for carbon black production. As there is a limit to oxidative species, the remaining feedstock can either be broken down by pyrolysis or survive and become directly involved in carbon black formation reactions. Typical pyrolysis species are hydrogen, acetylene, and polyynes, which are essentially chained acetylenes. At least two formation paths are thought to occur. The first one is ring growth from acetylene, polyyne,or polycyclic aromatic hydrocarbon (PAH) collision with PAH molecules. When the number of rings reaches five or six, the molecules become thermally stable and will only be attacked by remaining oxidant molecules. These PAH molecules will

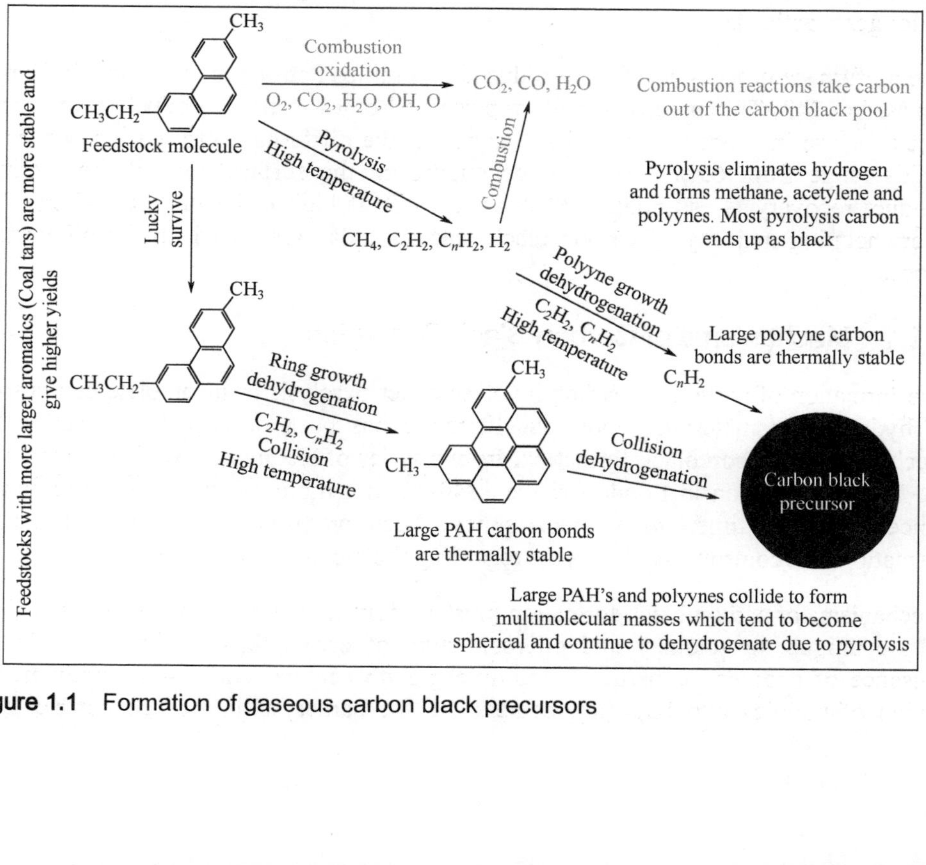


**Figure 1.1** Formation of gaseous carbon black precursors

eventually stack up after collisions and then begin to form the crystallites which are found in the finished primary particles. The other route to carbon black occurs as the acetylene polymerizes (polyacetylene) to form long chain polyynes which also reach a thermally stable size and begin to collide with the large PAH molecules. These polyynes can then restructure themselves to increase the crystallite size or provide the amorphous portion of the carbon black particles. Once these particles grow to about one to two nanometers they tend to become spherical and are referred to as carbon black precursors.

**Nucleation** – Because of increasing mass of the carbon particle precursors through collision, the larger fragments are no longer stable and condense out of the vapor phase to form nuclei or growth centers.

**Particle growth and aggregation** – In this period, three processes go on simultaneously as shown in Figure 1.2: condensation of more carbon precursors on the existing nuclei, coalescence of small particles into larger ones, and formation of new nuclei. Coalescence and growth seem to predominate. The products of this stage are "proto-nodules".

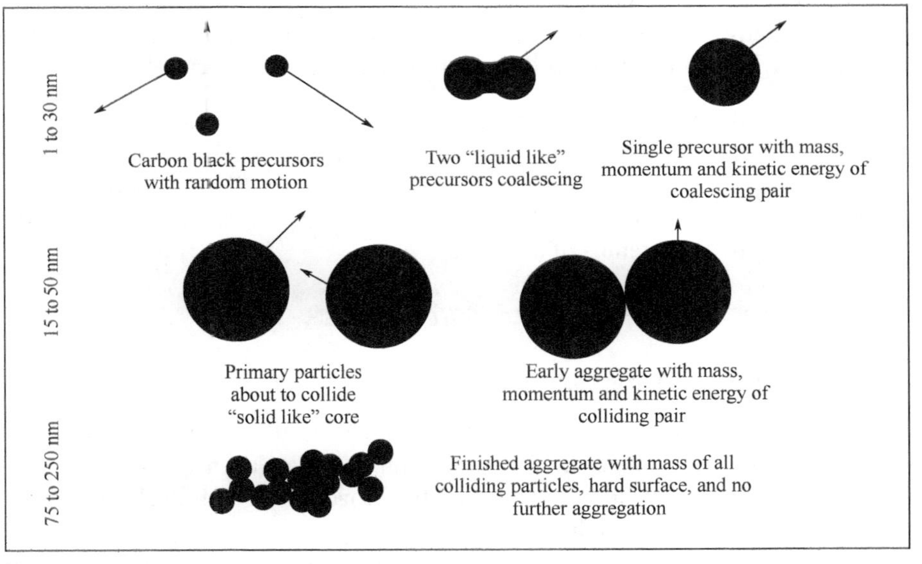


**Figure 1.2** Particle growth and aggregation

**Surface growth** – Surface growth includes the processes in which the small species attach to or deposit on the surfaces of existing particles or aggregates, forming the nodules and aggregates with their characteristic onion micro-structure (note: aggregates are permanent structures cemented by carbon). The surface growth represents about 90% of total carbon yield. It is responsible for the stability of the aggregates because of the continuous carbon network formation. Aggregates are formed and cemented in this stage.

**Agglomeration** – Once no more carbon is forming and aggregation ceases, aggregates collide and adhere by van der Waals forces but there is no material to cement them together, hence they form temporary structures (agglomerates).

**Aggregate gasification** – After its formation and growth, the carbon black surface undergoes reaction with the gas phase, resulting in an etched surface. Species such as $CO_2$, $H_2O$, and of course any residual oxygen attack the carbon surface. The oxidation is determined by gas phase conditions, such as temperature, oxidant concentration, and flow rates.

Practically, the carbon black morphology and surface chemistry can be well-controlled by changing the reaction parameters. For furnace carbon blacks, the reaction temperature is the key variable that governs the surface area. The higher the temperature, the higher is the pyrolysis rate and the more nuclei are formed, resulting in an earlier stop of the growth of the particles and aggregates due to the limitation of starting materials. Therefore, with higher reaction temperature, achieved by adjusting air rate, fuel rate, and feedstock rate, the surface area of carbon black can be increased. Addition of alkali metal salts into the reactor can modify the aggregation process, influencing carbon black structure. At the reactor temperature, the salts of alkali metals, such as potassium, are ionized. The positive ions adsorb on the forming carbon black nodules and provide some electrostatic barrier to internodule collisions, resulting in lower structure[8].

The time scale of carbon black formation varies substantially across the range of particle sizes found in commercial furnace blacks. For blacks with surface areas around 120 $m^2/g$, the carbon black formation process from oil atomization to quench takes less than 10 milliseconds. For blacks with surface areas around 30 $m^2/g$, formation times are a few tenths of a second.

## 1.1.2 Manufacturing Process of Carbon Black

### 1.1.2.1 Oil-Furnace Process

The oil-furnace process accounts for over 95 percent of all carbon black produced in the world. It was developed in 1943 and rapidly displaced previous gas-based technologies because of its higher yields and the broader range of blacks that could be produced. It also provides highly effective capture of particulates and has greatly improved the environment around carbon black plants. As indicated in the mechanism discussion, it is based on the partial combustion of residual aromatic oils. Because residual oils are ubiquitous and are easily transported, the process can be practiced with little geographic limitation. This has allowed construction of carbon black plants all over the world. Plants are typically located in areas of tire and rubber goods manufacture. Because carbon black is of relatively low density, it is far less expensive to transport feedstock oil than to transport the black.

For nearly 80 years since its invention, the oil-furnace process has undergone several cycles of improvement. These improvements have resulted in increased yields, larger process trains, better energy economy, and enhanced product performance. A simplified flow diagram of a modern furnace black production line is shown in Figure 1.3[9]. This is intended to be a generic diagram and contains elements from several operators' processes. The principal pieces of equipment are the air blower, process air and oil preheaters, reactors, quench tower, bag filter, pelletizer, and rotary dryer. The basic process consists of atomizing the preheated oil in a combustion gas stream formed by burning fuel in preheated air. The atomization is carried out in a region of intense turbulent mixing. Some of the atomized feedstock is combusted with excess oxidant in the combustion gas. Temperatures in the region of carbon black formation range from 1400°C to over 1800°C. The details of reactor construction vary from manufacturer to manufacturer and are confidential to each manufacturer. Leaving the formation zone, the carbon black containing gases are quenched by spraying water into the stream. The partially cooled smoke is then passed through a heat exchanger where incoming air is preheated. Additional quench water is used to cool the smoke to a temperature consistent with the life of the bag material used in the bag filter. The bag filter separates the unagglomerated carbon black from the by-product tail gas which contains nitrogen, hydrogen, carbon monoxide, carbon dioxide, and water vapor. It is mainly nitrogen and water vapor. The tail gas is frequently used to fuel the dryers in the plant, to provide other process heat, or sometimes is burned to manufacture steam and electric power either for internal plant use or for sale.

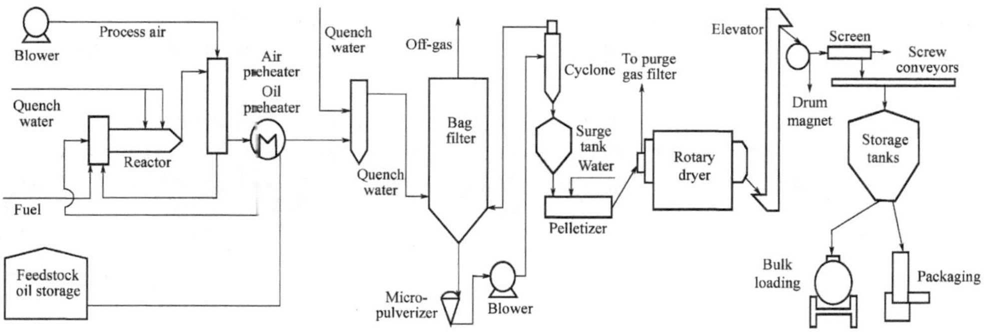


**Figure 1.3** Flow diagram of a modern furnace black process

The fluffy black from the bag filter is mixed with water, typically in a pin mixer, to form wet granules. These are dried in a rotary dryer, and the dried product is conveyed to bulk storage tanks. For special purposes, dry pelletization in rotary drums is also practiced. Most carbon black is shipped by rail or in bulk trucks. Various semi-bulk containers are also used including IBC's and large semi-bulk bags. Some special purpose blacks are packed in paper or plastic bags.

While the reactor and its associated air-moving and heat-exchange equipment are where the properties of the black are determined, they tend to be dwarfed by the bag collectors, the dryers, and particularly the storage tanks.

### Feedstocks

Feedstocks for the oil-furnace process are heavy fuel oils. Preferred oils have high aromaticity, are free of suspended solids, and have a minimum of asphaltenes. Suitable oils are catalytic cracker residue (once residual catalyst has been removed), ethylene cracker residues, and distilled heavy coal tar fractions. Other specifications of importance are freedom from solid materials, moderate to low sulfur, and low alkali metals. The ability to handle such oils in tanks, pumps, transfer lines, and spray nozzles is also a primary requirement.

### Reactor

The heart of a furnace black plant is the furnace or reactor where carbon black formation takes place under high temperature, partial combustion conditions. The reactors are designed and constructed to be as trouble-free as possible over long periods of operation under extremely aggressive conditions. They are monitored constantly for signs of deterioration in order to ensure constant product quality. The wide variety of furnace black grades for rubber and pigment applications requires different reactor designs and sizes to cover the complete range, though closely related grades can be made in the same reactor by adjusting input variables. Reactors for higher surface area and reinforcing grades operate under high gas velocities, temperatures, and turbulence to ensure rapid mixing of reactant gases and feedstock. Lower surface area and less reinforcing grades are produced in larger reactors at lower temperatures, lower velocities, and longer residence time. Table 1.1 lists carbon formation temperatures, and residence times for the various grades of rubber blacks.

**Table 1.1** Reactor conditions for various grades of carbon blacks

| Black | Surface area/($m^2$/g) | Temperature/°C | Residence time/s | Maximum velocity/(m/s) |
|---|---|---|---|---|
| N100 series | 145 | 1800 | 0.008 | |
| N200 series | 120 | | 0.010 | 180–400 |
| N300 series | 80 | 1550 | 0.031 | |
| N500 series | 42 | | 1 | 30–80 |
| N700 series | 25 | 1400 | 1.5 | 0.5–1.5 |
| N990 thermal | 8 | 1200–1350 | 10 | 10 |

A key development in the carbon black reactor technology was the development of the zoned axial flow reactor for reinforcing blacks in the early 1960s[8]. The reactor consists of three zones. The first zone is a combustion zone in which fuel and air are completely burned to produce combustion gases with excess oxygen. This gas flow is accelerated to high velocity in a throat zone with intense turbulent mixing. The feedstock is injected either into this throat zone or just ahead; therefore, the reacting gases issue from the throat into a second cylindrical zone as a turbulent diffusion jet.

Depending on the desired black, the jet may be allowed to expand freely, or may be confined by bricking. Downstream of the reaction zone is a water quench zone. The throughput of a single reactor train varies from manufacturer to manufacturer and with grade of black. The largest reactors in operation have capacities of over 30,000 metric tons per year. Many producers operate smaller reactors in parallel. Reactors are typically designed to make a series of related blacks. Air and gas may be introduced to the primary combustion zone either axially, tangentially, or radially. The feedstock can be introduced into the primary fire either axially or radially in the high velocity section of the mixing zone. The high velocity section may be Venturi-shaped or consist of a narrow diameter choke. Plants may have from one to several operating trains.

Carbon black reactors are made of carbon steel shells lined with several courses of refractory. The most severe services are in the combustor and in the throat zone. Different manufacturers take different approaches to these elements, some using exotic materials or selected water cooled metal surfaces, others using conventional materials and limiting temperatures to what their materials can stand. Most manufacturers achieve refractory life of one to several years. For the rubber grade carbon blacks, at least three different reactor designs must be used to make this range of furnace blacks. Figure 1.4 and Figure 1.5 show the designs of commercial reactors found in patents.

The quality and yield of carbon black depend on the quality and carbon content of the feedstock, the reactor design, and the input variables. Surface area in particular is controlled by adjusting the temperature in the reaction zone. Structure is adjusted by introducing potassium into the combustion gas. This may be done in any of a variety of ways.

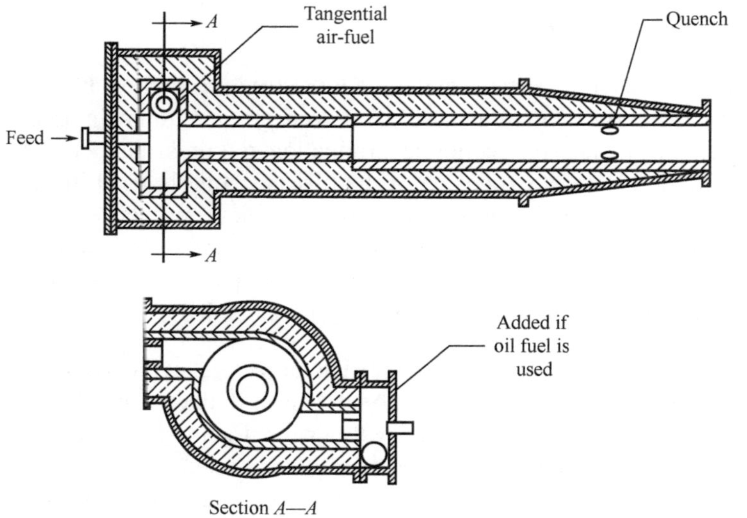


**Figure 1.4** Reactor for N300–N200 carbon blacks[10]

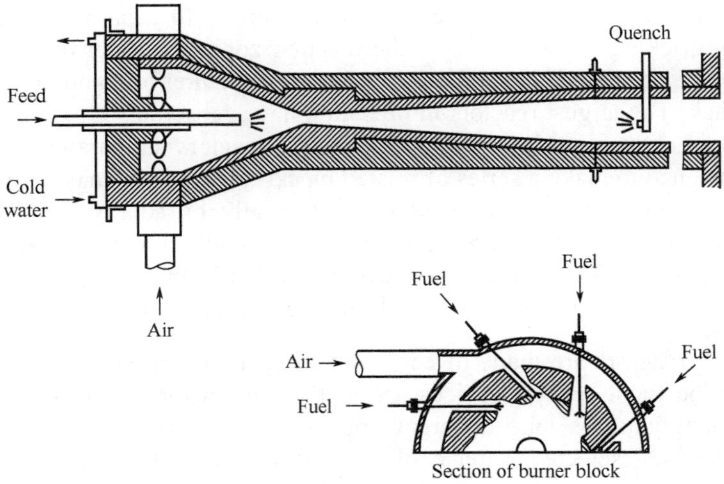


**Figure 1.5** Reactor for tread blacks[11]

The energy utilization in the production of one kilogram of oil-furnace carbon black is in the range of (9–16)×$10^7$ J, and the yields are 300–660 kg/$m^3$ depending on the grade. The energy inputs to the reactor are the heat of combustion of the preheated feedstock, heat of combustion of natural gas, and the thermal energy of the preheated air. The energy output consists of the heat of combustion of the carbon black product, the heat of combustion and the sensible heat of the tail gas, the heat loss from the water quench, the heat loss by radiation to atmosphere, and the heat transferred to preheat the primary combustion air.

### 1.1.2.2 The Thermal Black Process

Thermal black is a large particle size and low structure carbon black made by the thermal decomposition of natural gas, coke oven gas, or liquid hydrocarbons in the absence of air or flames. Its economic production requires inexpensive natural gas. Today it is among the most expensive of the blacks regularly used in rubber goods. It is used in rubber and plastics applications for its unique properties of low hardness, high extensibility, low compression set, low hysteresis, and excellent processability. Its main uses are in O-rings and seals, hose, tire inner liners, V-belts, other mechanical goods, and in cross-linked polyethylene for electrical cables.

The thermal black process dates from 1922. The process is cyclic using two refractory-lined cylindrical furnaces or generators about 4 m in diameter and 10 m high. During operation, one generator is being heated with a near stoichiometric ratio of air and off-gas from the making generator whereas the other generator, heated to an average temperature of 1300°C, is fed with natural gas. The cycle between black production and heating is five minutes alternating between generators, resulting in a reasonably continuous flow of product and off-gases to downstream equipment. The effluent gas from the make cycle, which is about 90% hydrogen, carries the black to a quench tower where water sprays lower the temperature before entering the bag filter. The effluent gas is cooled and dehumidified in a water scrubber for use as fuel in the heating cycle. The collected black from the filters is conveyed to a magnetic separator, screened, and hammer-milled. It is then bagged or pelletized. The pelletized form is bagged or sent to bulk loading facilities.

#### 1.1.2.3 Acetylene Black Process

The high carbon content of acetylene (92%) and its property of decomposing exothermically to carbon and hydrogen make it an attractive raw material for conversion to carbon black. Acetylene black is made by a continuous decomposition process at atmospheric pressure and temperatures of 800–1000°C in water-cooled metal retorts lined with refractory. The process consists of feeding acetylene into the hot reactors. The exothermic reaction is self-sustaining and requires water cooling to maintain a constant reaction temperature. The carbon black-laden hydrogen stream is then cooled, followed by separation of the carbon from the hydrogen tail gas. The tail gas is either flared or used as fuel. After separation from the gas stream, acetylene black is very fluffy with a bulk density of only 19 $kg/m^3$. It is difficult to compact and resists pelletization. Commercial grades are compressed to various bulk densities up to 200 $kg/m^3$.

Acetylene black is very pure with a carbon content of 99.7%. It has a surface area of about 65 $m^2/g$, an average particle diameter of 40 nm, and a very high but rather weak structure with a DBP number of 250 mL/100 g. It is the most crystalline or graphitic of the commercial blacks. These unique features result in high electrical and thermal conductivity, low moisture adsorption, and high liquid absorption.

A significant use of acetylene black is in dry cell batteries where it contributes low electrical resistance and high capacity. In rubber it gives electrically conductive properties to heater pads, tapes, antistatic belt drives, conveyor belts, and shoe soles. It is also useful in electrically conductive plastics such as electrical magnetic interference (EMI) shielding enclosures. Its contribution to thermal conductivity has been useful in rubber curing bags for tire manufacture.

#### 1.1.2.4 Lampblack Process

The lampblack process has the distinction of being the oldest and most primitive carbon black process still being practiced. The ancient Egyptians and Chinese employed techniques similar to modern methods collecting the lampblack by deposition on cool surfaces. Basically, the process consists of burning various liquid or molten raw materials in large, open, shallow pans 0.5–2 m in diameter and 16 cm deep under brick-lined flue enclosures with a restricted air supply. The smoke from the burning pans passes through low velocity settling chambers from which the carbon black is cleared by motor-driven ploughs. In more modern installations, the black is separated by cyclones and filters. By varying the size of the burner pans and the amount of combustion air, the particle size and surface area can be controlled within narrow limits. Lampblacks have similar properties to the low surface area oil-furnace blacks. A typical lampblack has an average particle diameter of 65 nm, a surface area of 22 $m^2/g$, and a DBP number of 130 mL/100 g. Its main use is in paints, as a tinting pigment where blue tone is desired. In the rubber industry lampblack finds some special applications.

#### 1.1.2.5 Impingement (Channel, Roller) Black Process

From World War I to World War II the channel black process made most of the carbon black used worldwide for rubber and pigment applications. The last channel black plant in the United States was closed in 1976. The demise of channel black was caused by environmental problems, cost, smoke pollution, and the rapid development of oil-furnace process grades that were equal or superior to channel black products particularly for use in synthetic rubber tires.

The name channel black came from the steel channel irons used to collect carbon black deposited by small natural gas flames impinging on their surface iron channels. Today tar fractions are used as raw material in addition to natural gas. In modern installations channels have been replaced by water cooled rollers. The black is scraped off the rollers, and the off-gases from the steel box enclosed rollers are passed through bag filters where additional black is collected. The purified exhaust gases are vented to the atmosphere. The oils used in this process must be vaporized and conveyed to the large number of small burners by means of a combustible carrier gas. Yield of rubber-grade black is 60%, and 10%–30% for high quality color grades.

The characteristics of roller process impingement blacks are basically similar to those of channel blacks. They have an acidic pH, a volatile content of about 5%, surface area of about 100 $m^2/g$, and an average particle diameter of 10–30 nm. The smaller particle size grades are used as color (pigment) blacks, and the 30 nm grade is used in rubber.

#### 1.1.2.6 Recycle Blacks

The pyrolysis of carbon black-containing rubber goods has been promoted as a solution to the accumulation of waste tires. In the processes in question, tires are pyrolyzed in the absence of oxygen, usually in indirect fired rotary kiln type units. The rubber and extender oils are cracked to hydrocarbons which are collected and sold as fuels or petrochemical feedstocks. The gaseous pyrolysis products are burned as fuel for the process. Steel tire cord is removed magnetically and the remainder of the residue is milled into a "pyrolysis black". This contains the carbon black, silica, and other metal oxides from the rubber and some newly created char. Typically these materials have 8% to 10% ash, and contain a lot of coarse residue. Most are difficult to pelletize. They have, on average, the reinforcing properties of a N300 black. But because they are a mixture of N600 and N700 blacks with N100 and N200 blacks, they are not particularly suitable for either reinforcing or semi-reinforcing applications. To-date they find application in relatively non-demanding uses such as playground and floor mats.

### 1.1.2.7 Surface Modification of Carbon Blacks

For the most of its long history, the carbon black industry had concentrated on morphology as the key factor controlling product performance and grade differentiation. Scientists have only recently recognized the importance of interface composition between the carbon blacks and the medium in the composite where the carbon black is used.

The early stages of surface modification can be traced back to the 1940s and 1950s. The approaches include physical adsorption of some chemicals on the carbon black surface, heat treatment, and frequently oxidation. During the 1980s and 1990s, some work on plasma treatment was reported. For chemical and polymer grafting modifications, a great deal of academic work was done in the 1950s and 1960s in France, US, and Japan, using surface oxygen groups as functional groups. However, because of rapid development of applications of carbon black in different areas and the challenge from other reinforcing particles in its traditional applications, surface modification technologies for carbon black have been developing very rapidly over the last decades. These include surfactant treated surfaces, chemically modified surfaces, and deposition of other phases during or after black formation. Today there is active commercial development and new product introduction in all areas.

#### 1.1.2.7.1 Attachments of the Aromatic Ring Nucleus to Carbon Black

Two approaches characterize this area. A number of patents have been issued to Cabot Corporation[12,13], which describe that the decomposition of a diazonium compound derived from a substituted aromatic or aliphatic amine results in the attachment of a substituted aromatic ring or chain onto the surface of the carbon black. This results in a stable attachment which is not sensitive to moisture. Examples show attachment of amines, anionic and cationic moieties, polysulfide moieties that can be attached into an elastomer network, and alkyl, polyethoxyl, and vinyl groups. Practically, the surface chemistry and physical chemistry can be tailored according to the applications of carbon blacks. Some applications are claimed in aqueous media for dispersion[14], in oil based coatings and inks for dispersion[15], and in rubbers for reduction of hysteresis and wear resistance improvement[16]. The initially attached groups can also function as sites for further chemical substitution. Another approach has been developed by Xerox Corporation in which oligomers of polymer are prepared using stable free radical polymerization and these are attached to the carbon black surface by reaction of the stable radical[17].

#### 1.1.2.7.2 Attachments to the Aromatic Ring Structure through Oxidized Groups

The acidic surface groups that result from surface oxidation of carbon black are natural synthons for the attachment of functionality. Generally, chemistry is done through either phenolic or carboxylic acid groups on the surface. Some of these groups are present in most blacks, but their density can be increased by treating with various oxidants such as ozone, nitric acid, or hypochlorite[18–20]. Compared to the previous class, these C–O attachments are somewhat more labile, being particularly susceptible to hydrolysis. The concept of using phenolic groups as points of attachment for conventional silane treating agents has been described in several patents with the particular aim of attaching polysulfide moieties that can be vulcanized into elastomer networks for hysteresis reduction[21]. Some patents have been issued on using the acidic sites on carbon black surfaces as points of reaction of amines. In the particular case in point, the attachment was used to improve compound stability and dispersion in conductive plastics applications[22].

#### 1.1.2.7.3 Metal Oxide Treatment

The carbon black industry has worked on ways to respond to the challenge of silica in tire treads for low rolling resistance (replacing all or some of the carbon black). Cabot has filed on, and widely published, a class of dual phase fillers in which silica or other metal oxides and carbon are co-formed in a carbon black like reactor[23–26]. In the particular product they describe, the carbon black and silica are intimately intermixed on a scale that is about the same size as the carbon black crystallite. In more recent variants, materials where the silica location is more on the exterior of the particle are described[27]. In these materials, the silica is the minor constituent. The main characteristics of these carbon blacks are their lower filler-filler interactions. Filler-polymer interactions are also increased, but by incorporating coupling agents these interactions can be adjusted as required. These materials are used as fillers for low rolling resistance, higher wet skid resistance, and improved wear resistance in tire treads when used with conventional sulfide-silane coupling agents[28,29], or as fillers for silicone rubber when used with alkyl silane and vinyl silane agents[30]. The patent literature suggests that other applications have been considered as well[31]. Patents have also been issued on coated carbon black made by depositing silica on the black surface in an aqueous solution of sodium silicate by adjusting pH with acid[32–34].

## 1.2 Manufacture of Silica

There are two families of silicas: natural and synthetic. The former are generally crystallized, such as quartz or diatoms. They have contorted and large particles, even after milling, and do not find their application in rubber industry. By synthesis, it is possible to obtain better-defined particles in shape and especially of submicron size. There are two synthetic processes: pyrohydrolysis and precipitation. Both processes produce amorphous products. The first is performed in a high temperature vapor phase from silicon compounds, principally silicon tetrachloride, air, and hydrogen. The products are called “anhydrous” or fumed silica. They have less than 1.5% each of bound water and adsorbed water, which is defined as volatiles removed at, respectively, 105°C and in vacuum. With the precipitation process, the silicas, prepared by precipitation from water-soluble silicates, have approximately 5% each of bound and adsorbed water.

### 1.2.1 Mechanisms of Precipitated Silica Formation

The formation of precipitated silica is governed by a complex process in which the soluble silicic species which are present in the solution polymerize and form nuclei which grow to form spherical particles of various sizes, depending on the precipitation conditions, mainly temperature and concentration (Figure 1.6).

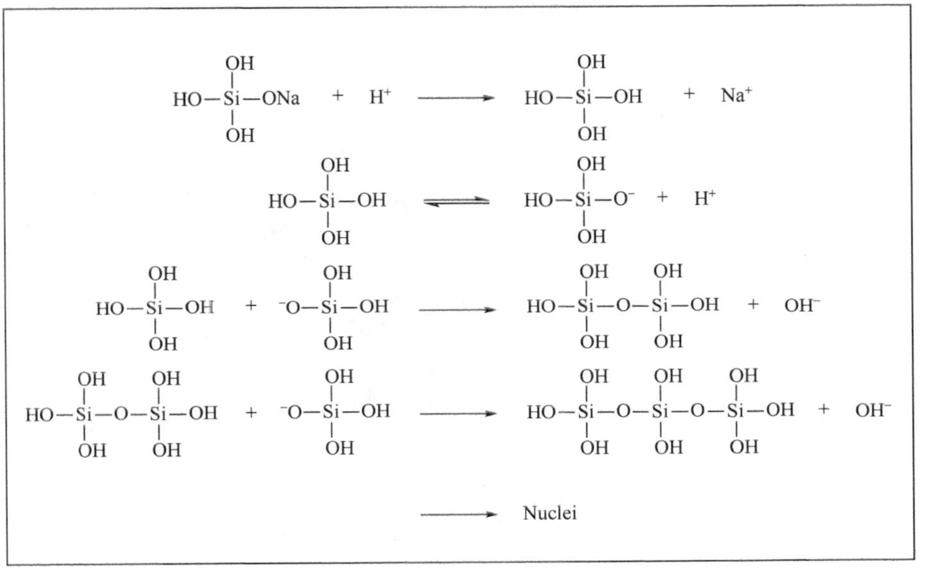


**Figure 1.6** Polymerization of silicate

The polymerization, which leads to the formation and the coagulation of colloids, produces silica aggregates. Depending on the reaction conditions, the particles collide with each other forming a gel or aggregates which are then consolidated by further deposition of $SiO_2$ on their surface as shown in Figure 1.7[35]. Under basic pH conditions, the colloidal particles are negatively charged. They repel each other in an electrostatic way. The electrostatic repulsions are suppressed under the action of an electrolyte such as sodium cation, resulting in the coagulation of the colloidal species. The floc formed is a relatively fragile structure. An addition of water regenerates the colloids. The floc can be made rigid by polymerization of soluble silica on its surface. This is how the silica aggregates are obtained. The morphological characteristics of silica, namely its primary particle size and surface area, determine the shape and size of its aggregates and their distribution. The most important parameters governing the silica morphologies include the proportion of $SiO_2$ in the alkaline silicate solution, the concentration of the soluble alkali metal salt (such as sodium chloride) in the silicate solution, the reaction temperature, the speed of addition of the acid to the solution, and the other components found in the acid used[36].

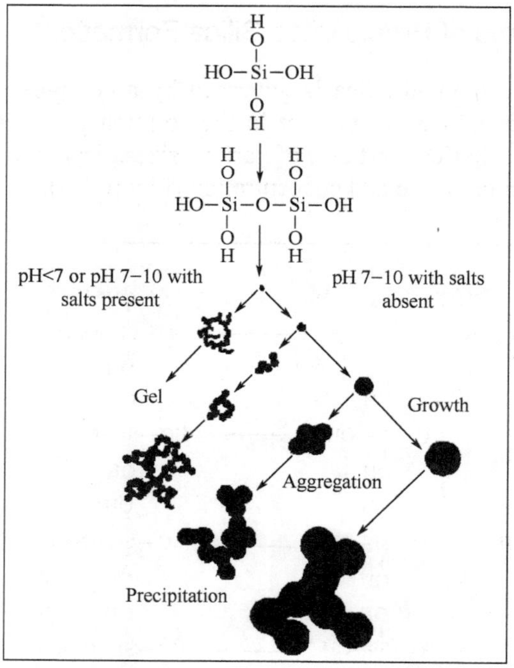


**Figure 1.7** Silica formation, growing, and aggregation during precipitation

## 1.2.2 Manufacturing Process of Precipitated Silica

Fine particle silicas for rubber reinforcement are precipitated by the controlled neutralization of sodium silicate solution by either sulfuric or carbonic acids[36,37]. Their production process is schematically shown in Figure 1.8. The basic raw materials are those required for sodium silicate: quartz sand, sodium carbonate, and water. Sodium silicate (water glass) can be produced in furnace or digester operations, but in either case the ratio of $SiO_2$ to $Na_2O$ is generally within a range of 2.5–3.5. Practically, the precipitation reaction is carried out in agitated reactors in which the silicate is dissolved, forming a diluted solution. Between pH 3 and 6–7, a silica gel is formed, but the dried silica gel is very difficult to disperse in elastomer systems. The usual pH range for commercial precipitated silica used for elastomer applications is 7.5–9.5 with the optimum being around 8.0–8.5. Above pH 9 there is a high proportion of soluble silica (as silicate) and the remaining precipitate is also extremely fine, which makes filtration extremely difficult. In this case, the filtrate has to be washed to remove the soluble silicate, which is returned as feedstock for subsequent precipitations.

Precipitation produces a low solids-content slurry of silica and sodium sulfate or sodium carbonate from which the salts are removed by washing, either in a counter-current decantation system or during a filter press concentration step. Washing generally reduces the salt content to 1% to 2%. Further concentration in rotary or plate and frame filters produces a solid filtration cake which still contains only 15% to 25%

silica. Because of this high water content the drying step is a large consumer of energy, whether the process involves rotary, tray, belt, flash dryers or spray dryers.

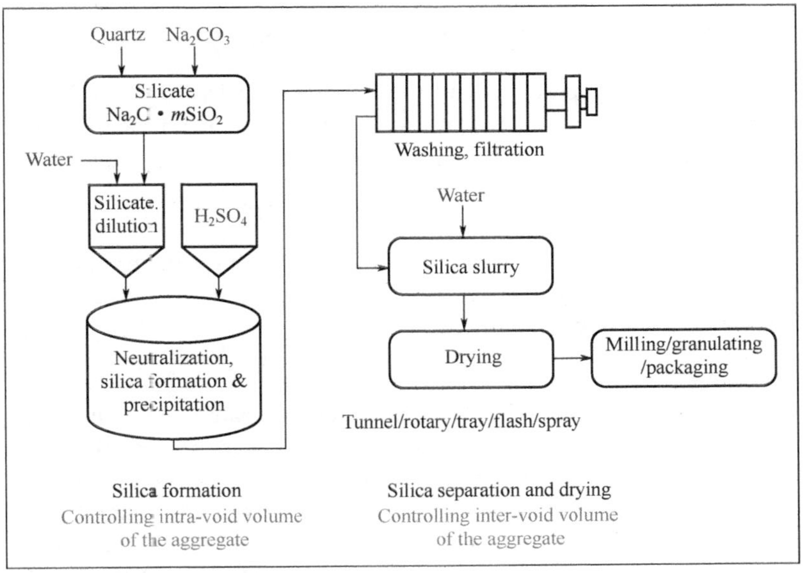


**Figure 1.8** Process of precipitated silica production

Generally, the last two dryers are commonly used for rubber silica. With the conventional dryers, the drying is a very slow process at lower temperature. In this case, the aggregates form very tight agglomerates that are very difficult to be dispersed in rubbers.

In the flash drying process, the temperature is usually much higher but the residence time of silica in the dryer is much shorter. During the filtration and drying processes, the silica aggregates bind to each other to form agglomerates. The main distinction between aggregates and agglomerates is related to the nature of the forces between the entities: in aggregates, the particles interact through chemicals bonds whereas in agglomerates, the interactions between aggregates are physical in nature. As will be discussed later, spray drying is the important process for highly dispersible silica.

The process involves producing small hydrated particles which partially explode during the rapid vaporization of water stage. These weakly associated particles are easier to disperse via attrition and infiltration of the dispersing medium. It is essential that during the spray drying process the silica (dry material) is not allowed to reach 180°C, at this temperature condensation of surface silanols starts to occur (moderate rate) which will cause particle-particle fusion and reduced dispersion. The optimum exposure temperature of the dry silica particles is 140–150°C. The air temperature is significantly above this temperature; however, residence time inhibits absorption of heat by the dry particles.

Dry silica is frequently milled and compacted or granulated to attain an optimum balance between dispersibility and dusting.

### 1.2.3 Mechanisms of Fumed Silica Formation

Fumed silica is generally produced by the vapor phase hydrolysis of chlorosilanes, such as silicon tetrachloride, in a hydrogen-oxygen flame. Such processes are generally referred to as pyrogenic processes. For silicon tetrachloride, the overall reaction is:

$$SiCl_4 + 2H_2 + O_2 \longrightarrow SiO_2 + 4HCl$$

Organosilanes also have been used in pyrogenic processes for the production of fumed silica. In the vapor phase hydrolysis of organosilanes, the carbon-bearing fragments undergo oxidation to form carbon dioxide as a byproduct along with hydrochloric acid.

Ulrich[38] gave a description of the flame process based on the immediate formation of protoparticles directly related to the chemical reaction – rather than surface deposition. At high flame temperatures collision and coalescence of protoparticles lead to the formation of primary particles. The rate of coalescence depends on the viscosity of the molten oxide, which is exceedingly high for silicon dioxide at a flame temperature of about 1500 K. Then the fumed silica primary particles collide and fuse to form three-dimensional, branched, chain-like aggregates. The properties of the silica formed are strongly related to the flame temperature. At lower temperature, collision and sticking of primary particles only results in partial fusion and stable particle aggregates are formed. The silica aggregates leave the flame and cool, but they still collide. As their surfaces are now solid, agglomerates of aggregates are formed that are held together by physico-chemical surface interactions (Figure 1.9).

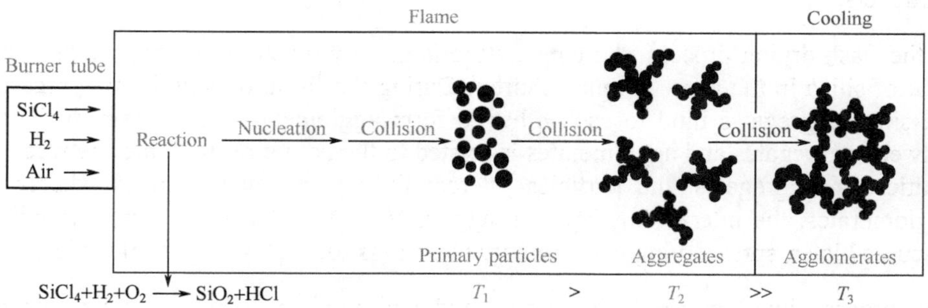


**Figure 1.9** Fumed silica formation, growing, and aggregation during production process[39]

### 1.2.4 Manufacture Process of Fumed Silica

Numerous methods have been developed to produce fumed silica via pyrogenic processes[39–41]. Figure 1.10 describes a process for the pyrogenic production of fumed silica. In accordance with this process, a gaseous feedstock consists of a fuel, such as hydrogen or methane, oxygen, or air, and a volatile silicon compound, such as silicon tetrachloride, wherein the oxygen is present in a stoichiometric or hyper-stoichiometric proportion. The feedstock is fed to a flame supported by a burner at various flow rates to produce fumed silica. The volume ratios of the individual gas components are reported not to be of critical importance. The molar ratio of the organosilane to the water-forming gases generally is said to be in the range from 1:0 to 1:12, preferably from 1:3 to 1:4.5. Water formed by the combustion of the fuel in oxygen reacts with the silicon tetrachloride to produce silicon dioxide particles, which coalesce and aggregate to form fumed silica. The effluent from the burner is cooled, and the fumed silica is then collected.

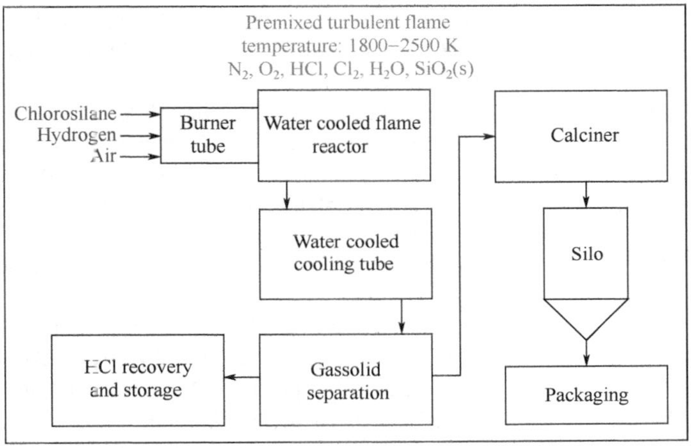


**Figure 1.10** Production process of fumed silica

## References


[1] Dannenberg E M. The Carbon Black Industry: Over a Century of Progress, Rubber World Mag. Spec. Pub.-Rubber Div. 75th Anniv. (1907-1984).

[2] Faraday M. The Chemical History of a Candle. New York: Viking Press, 1960.

[3] Wang M-J, Gray C A, Reznek S A, et al."Carbon Black" Encyclopedia of Chemical Technology, vol. 4, p. 761, 2004.

[4] Wolff S, Görl U, Wang M-J, et al. Silica-based Tread Compounds: Background and Performance. Paper presented at *TyreTech'93 Conference*, Basel, Switzerland, October 28-29, 1993.

[5] Wolff S, Görl U, Wang M-J, et al. Silane modified silicas. European Rubber Journal, 1994, 16.

[6] Roland R. Rubber Compound and Tires Based on such a Compound. EP. Patent, 0501227A1, 1992.

[7] Bansal R C, Donnet J-B, Wang M-J. Chapter 2. Carbon Black, Science and Technology. New York: Marcel Dekker, Inc., 1993.

[8] Frianf G F, Thorley B. Carbon Black Process. U.S. Patent, 3010794, 1961.

[9] Rivin D, Smith R G. Environmental Health Aspects of Carbon Black. *Rubber. Chem. Technol.*, 1982, 55: 707.

[10] Krejci J C. Production of Carbon Black. U.S. Patent, 2564700, 1951.

[11] Heller G L. Vortex Reactor for Carbon Black Manufacture. U.S. Patent, 3490869, 1970.
[12] Belmont J A. Process for Preparing Carbon Materials with Diazonium Salts and Resultant Carbon Products. U.S. Patent, 5554739, 1996.
[13] Belmont J A, Amici R M, Galloway C P. Reaction of Carbon Black with Diazonium Salts, Resultant Carbon Black Products and Their Uses. U.S. Patent, 5851280, 1998.
[14] Belmont J A. Aqueous Inks and Coatings Containing Modified Carbon Products. U.S. Patent, 5672198, 1997.
[15] Belmont J A, Adams C E. Non-aqueous Inks and Coatings Containing Modified Carbon Products. U.S. Patent, 5713988, 1998.
[16] Belmont J A, Amici R M, Galloway C P. Reaction of Carbon Black with Diazonium Salts, Resultant Carbon Black Products and Their Uses. U.S. Patent, 6494946, 2002.
[17] Keoshkerian B, Georges M K, Drappel S V. Ink Jettable Toner Compositions and Processes for Making and Using. U.S. Patent, 5545504, 1996.
[18] Bansal R C, Donnet J-B. Chapter 4. Donnet J-B, Bansal R C, Wang M-J. Carbon Black, Science and Technology. New York: Marcel Dekker, Inc., 1993.
[19] Eisenmenger E, Engel R, Kuehner G, et al. Carbon Black Useful for Pigment for Black Lacquers. U.S. Patent, 4366138, 1982.
[20] Amon F H, Thornhill F S. Process of Making Hydrophilic Carbon Black. U.S. Patent, 2439442, 1948.
[21] Wolff S, Görl U. Carbon Blacks Modified with Organosilicon Compounds, Method of Their Production and Their Use in Rubber Mixtures. U.S. Patent, 5159009, 1992.
[22] Joyce G A, Little E L. Thermoplastic Composition Comprising Chemically Modified Carbon Black and Their Applications. U.S. Patent, 5708055, 1998.
[23] Mahmud K, Wang M-J, Francis R A. Elastomeric Compounds Incorporating Silicon-treated Carbon Blacks. U.S. Patent, 5830930, 1998.
[24] Mahmud K, Wang M-J, Francis R A. Elastomeric Compounds Incorporating Silicon-treated Carbon Blacks and Coupling Agents. U.S. Patent, 5877238, 1999.
[25] Mahmud K, Wang M-J. Method of Making a Multi-phase Aggregate Using a Multi-stage Process. U.S. Patent, 5904762, 1999.
[26] Mahmud K, Wang M-J. Method of Making a Multi-phase Aggregate Using a Multi-stage Process. U.S. Patent, 6211279, 2001.
[27] Mahmud K, Wang M-J, Kutsovsky Y. Method of Making a Multi-phase Aggregate Using a Multi-stage Process. U.S. Patent, 6364944, 2002.
[28] Wang M-J, Mahmud K, Murphy L J, et al. Carbon-silica Dual Phase Filler, a New Generation Reinforcing Agent for Rubber-Part I. Characterization. *Kautsch. Gummi Kunstst.,* 1998, 51: 348.
[29] Wang M-J, Kutsovsky Y, Zhang P, et al. New Generation Carbon-Silica Dual Phase Filler Part I. Characterization and Application to Passenger Tire. *Rubber Chem. Technol.,* 2002, 75: 247.
[30] Anand J N, Mills J E, Reznek S R. Silicone Rubber Compositions Incorporating Silicon-treated Carbon Blacks. U.S. Patent, 6020402, 2000.
[31] Reed T, Mahmud K. Use of Modified Carbon Black in Gas-phase Polymerizations. U.S. Patent, 5919855, 1999.
[32] Kawazura T, Kaido H, Ikai K, et al. Surface-treated Carbon Black and Rubber Composition Containing Same. U.S. Patent, 5679728, 1997.
[33] Mahmud K, Wang M-J, Reznek S R, et al. Elastomeric Compounds Incorporating Partially Coated Carbon Blacks. U.S. Patent, 5916934, 1999.

[34] Mahmud K, Wang M-J, Belmont J A, et al. Silica Coated Carbon Blacks. U.S. Patent, 6197274, 2001.

[35] Iler R K. The Chemistry of Silica. New York: Wiley, 1979.

[36] Chevallier Y, Morawski J C. Precipitated Silica Having Improved Morphological Characteristics and Process for the Production Thereof. U.S. Patent, 4590052, 1986.

[37] Thornhill F S. Method of Preparing Silica Pigments. U.S. Patent, 2940830, 1960.

[38] Ulrich G D. Aggregation and Growth of Submicron Oxide Particles in Flames. *J. Colloid Interf. Sci.,* 1982, 87: 257.

[39] Barthel H, Rösch L, Weis J. Organosilicon Chemistry II: From Molecules to Materials. Weinheim: VCH, 1996.

[40] Pratsinis S E. Flame Aerosol Synthesis of Ceramic Powders. *Prog. Energy Combust. Sci.,* 1998, 47: 197.

[41] Kratel G. Process for the Manufacture of Silicon Dioxide. U.S. Patent, 4108964, 1978.

# 2 Characterization of Fillers

Fillers are characterized by their chemical composition, micro-structure, morphology, and physical chemistry of the surface. Figure 2.1 and Figure 2.2 schematically show the basic structures of carbon black and synthetic silica, respectively. The primary dispersible unit of fillers is referred to as an "aggregate", that is, a discrete, rigid colloidal entity. It is the functional unit in well-dispersed systems. For most carbon blacks and synthetic silicas, the aggregate is composed of spheres that are fused together. These spheres are generally termed as primary "particles" or "nodules". While primary particles of silicas are composed of amorphous silicon dioxide, in carbon black aggregates, these nodules are composed of many tiny graphite-like stacks. Within the nodule of carbon blacks, the stacks are oriented so that their *c*-axis is normal to the sphere surface, at least near the nodule surface.

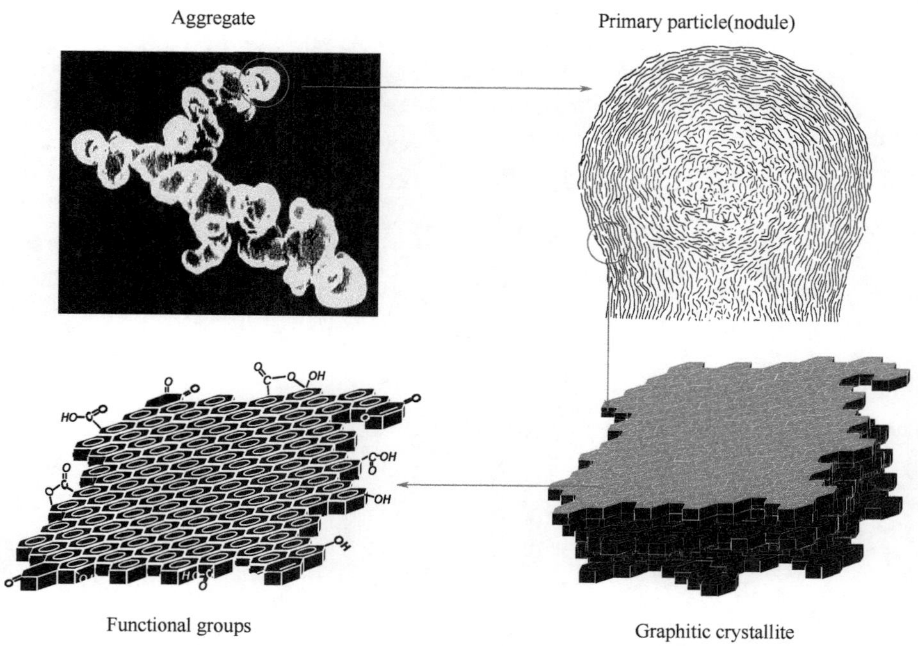


**Figure 2.1** Structure of carbon black

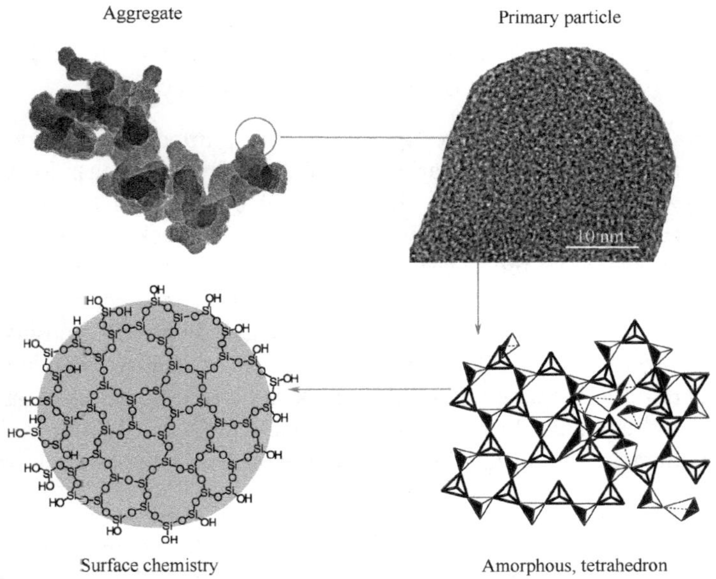


**Figure 2.2** Structure of silica

# ■ 2.1 Chemical Composition

## 2.1.1 Carbon Black

Oil-furnace blacks used by the rubber industry contain over 97% elemental carbon. Thermal and acetylene blacks consist of over 99% carbon. Table 2.1 shows the chemical composition of some carbon blacks. Other elements included in furnace black, apart from carbon, are hydrogen, oxygen, sulfur, and nitrogen. In addition, there are mineral oxides, salts, and traces of adsorbed hydrocarbons. Hydrogen and sulfur are distributed on the surface and the interior of the aggregates. The oxygen content is located on the surface of the aggregates as $C_xO_y$ complexes.

**Table 2.1** Chemical composition of carbon blacks

| Type | Carbon/ % | Hydrogen/ % | Oxygen/ % | Sulfur/ % | Nitrogen/ % | Ash/ % | Volatile/ % |
|---|---|---|---|---|---|---|---|
| Furnace rubber grade | 97.3–99 3 | 0.20–0.80 | 0.20–1.50 | 0.20–1.20 | 0.05–0.30 | 0.10–1.00 | 0.60–1.50 |
| Medium thermal | 99.4 | 0.30–0.50 | 0.00–0.12 | 0.00–0.25 | NA | 0.20–0.38 | - |
| Acetylene black | 99.8 | 0.05–0.10 | 0.10–0.15 | 0.02–0.05 | NA | 0.00 | <0.40 |

NA – not available.

Hydrocarbon materials produce carbon blacks; therefore, it is mostly hydrogen that saturates the dangling bonds at the edges of the basal planes of the graphitic layers. The graphitic layers are large, polycyclic aromatic ring systems.

Oxygen-containing complexes are by far the most important surface groups. The oxygen content of carbon blacks varies from 0.2% to 1.5% in mass for furnace blacks and 3% to 4% for channel blacks. Some special blacks used for pigment purposes contain larger quantities of oxygen than normal furnace blacks. These blacks are made by oxidation in a separate process step using nitric acid, ozone, air, or other oxidizing agents. They may contain from 2% to 12% oxygen. The oxygen-containing groups influence the physico-chemical properties, such as chemical reactivity, wettability, catalytic properties, electrical properties, and adsorbability. Oxidation improves dispersion and flow characteristics in pigment vehicle systems such as lithographic inks, paints, and enamels. In rubber-grade blacks, surface oxidation reduces pH and changes the kinetics of vulcanization, making the rubber compounds slower curing.

A convenient method for assessing the extent of surface oxidation is the measurement of volatile content. This standard method measures the weight loss of the evolved gases on heating up from 120°C to 950°C in an inert atmosphere. The composition of these gases consists of three principal components: hydrogen, carbon monoxide, and carbon dioxide. The volatile content of normal furnace blacks is under 1.5%, and the volatile content of oxidized special grades is 2% to 22%.

The origin of the volatile gases is the functional groups attached to carbon black, especially those on the surface. Surface oxides bound to the edges of the carbon layers are phenols, hydroquinones, quinones, neutral groups with one oxygen, carboxylic acids, lactones, and neutral groups containing two oxygens[1,2]. Figure 2.1 shows an idealized graphite-surface-layer plane with the various functional groups located at the periphery of the plane. Carbon blacks with few oxygen groups show basic surface properties and anion exchange behavior[3,4].

In addition to combined hydrogen and oxygen, carbon blacks may contain as much as 1.2% combined sulfur resulting from the sulfur content of the aromatic feedstock that contains thiophenes, mercaptans, and sulfides. The majority of the sulfur is not potentially reactive as it is inaccessibly bound in the interior of carbon black particles and does not contribute to sulfur cross-linking during the vulcanization of rubber compounds.

The nitrogen in carbon blacks is the residue of nitrogen heterocycles in the feedstocks. Thus, carbon blacks derived from coal tars have far more nitrogen than petroleum-derived blacks.

The ash content of furnace blacks is normally a few tenths of a percent but in some products may be as high as one percent. The chief source of ash is the water used to quench the hot black from the reactors during manufacture and for wet pelletizing the black.

### 2.1.2 Silica

Chemically, it is not the "ideal silica" with which we are concerned in reinforcing rubber. Wagner[5] in 1976 reviewed the chemistry and the micro-structure of silica. His review remains the best summary of silica chemistry.

Regarding silica composition, the purity of silicon dioxide is one parameter influencing rubber reinforcement, and such purities vary among production processes and post treatments. The purity of fumed silica is quite high, with a silica content of 98% or more, while precipitated silica is about 88%–92%. Impurities include silicate and Ca or Al elements among others; however, the main impurities are oxygen and hydrogen in the form of silanols with adsorbed and bonded water.

It has been found[6–10] that in an ideal state the surface silanol content is 4.6 OH groups per $nm^2$. This corresponds to the available surface silicon atoms permitted by the lattice configuration of silica. Lattice imperfections at the surface, and even within the bulky particle, permit values greater than 4.6 to occur. Incomplete silanol condensation and the presence of geminal hydroxyls (i.e., two hydroxyls on one silicon atom) are observed on freshly prepared, nonannealed silicas. Pyrogenic silicas approach the ideal state more closely than precipitated silicas. By the process of high-temperature dehydration and annealing (~450°C), followed by rehydration, both types approach this ideal state of 4.6 $OH/nm^2$.

Reports indicate[11] a considerable difference between the virgin surfaces of the two types of silicas. To begin with, precipitated silica strongly holds physically adsorbed water, causing high hydration. Whereas outgassing (in vacuum) at room temperature was sufficient at removing physically adsorbed water from fumed silica, a significant quantity of more strongly adsorbed water on precipitated silica was lost during heating from 25°C to 150°C. Readsorption of water indicates a hydroxyl concentration of 12.5 $OH/nm^2$, considerably higher than the ideal of 4.6. Similarly, readsorption of water on fumed silica (after vacuum outgassing) gives a value of 2.19 $OH/nm^2$, increased by rehydration to 3.31, still less than the "fully hydrated" condition proposed[8–10]. Furthermore, comparative isopropanol adsorption indicates that the hydroxyls on the precipitated silica surface are sufficiently close together that steric hindrance prevents a 1:1 interaction between silanol and isopropanol. The hydroxyls in fumed silica, even after rehydration, are sufficiently isolated to permit essentially equivalent site adsorption by water and isopropanol. Wang's findings show the hydroxyl concentration on precipitated silica reaching 19 $OH/nm^2$ when testing an alkylation of silica with methanol at 200°C[12]. This value decreases to 4 $OH/nm^2$ for fumed silica. Such high concentrations of OH groups on both silicas are attributed to the imperfection of silica lattice and micro-pore on the surface as the methanol treatments are performed under a very high pressure.

There is some evidence that adsorbed water may exist on silicas up to 300°C[13]. This may contribute to the very high hydroxyl concentration[11], since adsorbed water provides additional sites for further adsorption.

The silanol concentration at the surface depends on the number of silicon atoms per unit area at the surface and the number of hydroxyl groups presenting on each silicon atom. Amorphous silicas bear a close resemblance to silica glass, with a short-range order but with a random arrangement of more distant neighbors. For convenience, a resemblance to one or more faces of $\beta$-cristobalite or $\beta$-tridymite is postulated. The differences in surface silicon atoms between these two are small, 3.95 per $nm^2$ and 4.6 per $nm^2$, but the former can accommodate two hydroxyl groups to give 7.9 $OH/nm^2$. On the surface, silicon atoms in the other form can hold only one hydroxyl group, but the spatial arrangement could permit the second-layer silicons to participate, yielding either 4.6 $OH/nm^2$ or 13.8 $OH/nm^2$. As will be seen later, the evidence points to the value of 7.9 $OH/nm^2$ for fully hydrated amorphous silicas.

The number of surface hydroxyls does not fully characterize the silica surface. Their distribution and, particularly, the close proximity of hydroxyls have an influence on adsorption of polar molecules and, to some extent, their reactivity.

Three types of surface hydroxyls have been identified: isolated, vicinal (on adjacent silicon atoms), and geminal (two hydroxyls on the same silicon atom), which are schematically shown in Figure 2.3[6,14–16]. Even additional "anomalous adsorption sites" have been claimed[17].

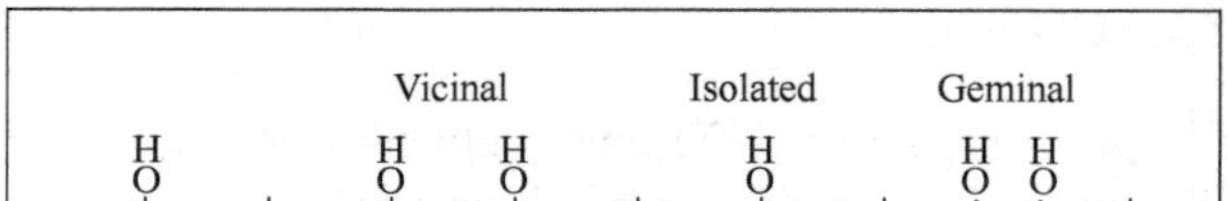


**Figure 2.3** Type of silanols on silica surface

There is reason to believe that there are several types of adsorption sites attributed to hydroxyl configurations on the silica surface. Hockey and Pethica[18] present evidence from infrared spectroscopy and thermogravimetric analysis that geminal hydroxyls exist on fully hydrated, nonannealed surfaces. Incomplete silanol condensation between first- and second-layer surface silicon atoms forced the first-layer silicon to carry two hydroxyls. After annealing (above 400°C) geminal hydroxyls are no longer formed on rehydration.

Isolated hydroxyls, i.e., not interacting with near-neighbor hydroxyls, exist predominantly on extensively dehydrated surfaces. Fumed silicas and, to a much lesser extent, precipitated silicas have isolated hydroxyls. As the degree of hydration increases, the existence of isolated hydroxyls decreases, and vicinal hydroxyls increases[14].

This is significant for adsorption of water and other polar adsorbates. Adjacent silanols are more powerful adsorption sites for water than isolated silanols[13,14]. The same can be said for other polar chemicals. The heat of adsorption of methanol on silica decreases significantly as the silica is heated from 105°C to 900°C[19], while that of the nonspecific adsorbate, carbon tetrachloride, does not. The favorable disposition of adjacent silanols apparently favors much stronger adsorption of polar species. This will be discussed in the characterization of filler surface with inverse gas chromatography.

# ■ 2.2 Micro-Structure of Fillers

## 2.2.1 Carbon Black

The arrangement of carbon atoms in carbon black has been well-established by X-ray diffraction methods[20,21]. The diffraction patterns show diffuse rings at the same positions as diffraction rings from pure graphite. Heating carbon black to 3000°C further emphasizes the suggested relation to graphite. The diffuse reflections sharpen, but the pattern never approaches that of true graphite. Carbon black has a degenerated graphitic crystalline structure as defined above. Whereas graphite has a three-dimensional order, as seen in the model structures of Figure 2.4, carbon black has a two-dimensional order. The X-ray data indicate that carbon black consists of well-developed graphite platelets stacked roughly parallel to one another but random in orientation with respect to adjacent layers. As shown in Figure 2.4 the carbon atoms in the graphitic structure of carbon black form large sheets of condensed aromatic ring systems with an interatomic spacing of 0.142 nm within the sheet identical to that found in graphite. However, the interplanar distances are quite different. While graphite interplanar distance is 0.335 nm which results in a relative density of 2.26, the interplanar distance of carbon black is larger, in the range of 0.350–0.365 nm, as a consequence of the random planar orientations or so-called turbostratic arrangement. The relative density of commercial carbon blacks is 1.76–1.90 depending on the grade. About half of the decrease in density is attributed to stacking height, $L_c$, in the crystallites. X-ray diffraction data provide estimates of crystallite size. For a typical carbon black, the average crystallite diameter, $L_a$, is about 1.7 nm and average $L_c$ is 1.5 nm, which corresponds to an average of four planal layers per crystallite containing ~375 carbon atoms.

It was originally suggested that these discrete crystallites were in random orientation within the particle. This view was later abandoned when electron microscopy of graphitized and oxidized carbon blacks indicated more of a concentric layer plane arrangement that can be described by a paracrystalline model. High resolution phase-contrast electron microscopy, which made possible the direct imaging of graphitic layers in carbon black, has confirmed this structure[22]. Figure 2.5 shows transmission electron micrographs of carbon black and graphitized blacks at high resolution. For carbon black, the image displays the marked concentric arrangement of the planal layers at the surface and around what appear to be growth centers. Upon heat treatment of the carbon black in nitrogen atmosphere at high temperature, it seems that almost all of the carbon black atoms are graphitized.

The micro-structure of the carbon black surface has been investigated by means of scanning tunneling microscopy (STM)[23,24]. Figure 2.6 shows the STM images obtained in the current mode for graphite, graphitized carbon black and normal carbon black N234. Compared to graphite, the structure of carbon blacks graphitized for 24 hours at a temperature of 2700°C in an inert atmosphere still remains in a certain imperfect state, shown by different tunneling current patterns in the organized domains.

The surface structure of carbon black can be classified as two types: organized domains and unorganized domains. The organized domains occupy the majority of the carbon black surface, and their size generally decreases with decreasing particle size.

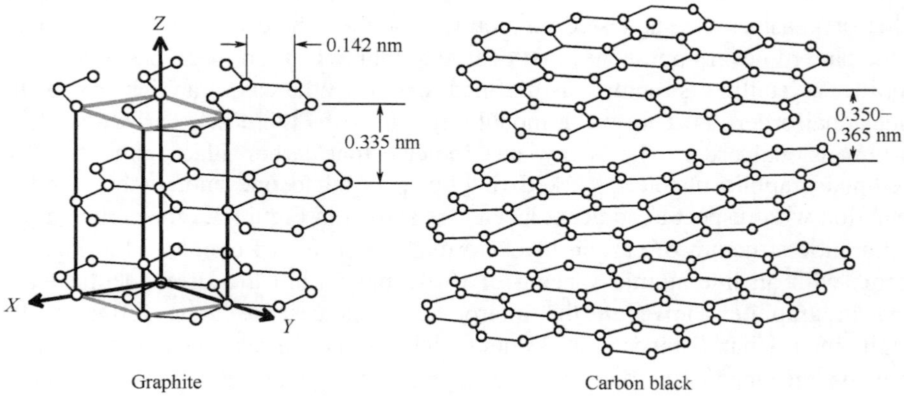


**Figure 2.4** Atomic structural models of graphite and carbon black

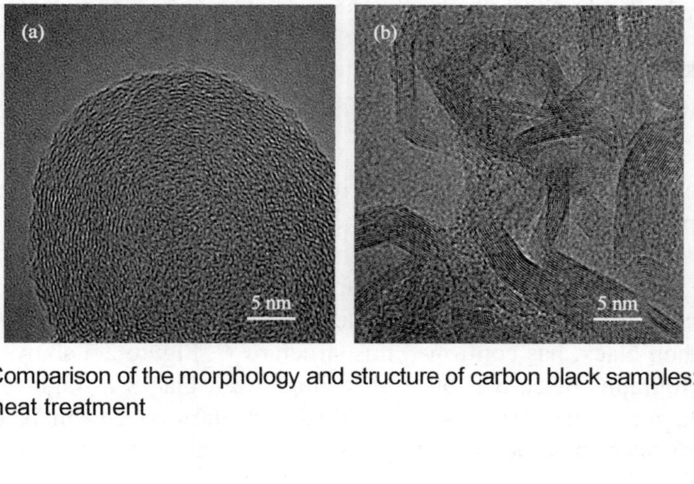


**Figure 2.5** Comparison of the morphology and structure of carbon black samples: (a) before and (b) after heat treatment

**Figure 2.6** STM images of graphite, graphitized carbon black, and carbon black N234

### 2.2.2 Silica

Both precipitated and fumed silicas are amorphous materials that consist of silicon and oxygen tetrahedrally bound into an imperfect three-dimensional structure (Figure 2.2). The degree of lattice imperfection, which depends upon the conditions of synthesis, has several consequences. First, there is no long-range crystal order, only short-range, with a random arrangement of more distant neighbors. In this regard, it resembles silica glass. Secondly, imperfections within the lattice structure leave free silanol groups. These are evident in infrared and thermogravimetric studies, but are inaccessible to adsorption or reaction with external agents. Finally, the surface contains uncondensed silanols on a siloxane lattice, which has features of both the 100 face of $\beta$-cristobalite and 001 face of $\beta$-tridymite. The number of silanols, their distribution, and even the conformation of the surface siloxane lattice also depend on the synthetic method and thermal treatment.

The structure of amorphous silicas has been extensively studied with infrared spectroscopy, thermogravimetry, chemical reactivity, and specific adsorptivity. It has helped to semiquantitatively describe the silica surface structure, which is important to rubber reinforcement.

## ■ 2.3 Filler Morphologies

Besides their composition and micro-structure, fillers differ in their primary "particle" or nodule size, surface area, aggregate size, aggregate shape, and in the distribution of each of these. Morphology is a set of properties related to the average magnitude and frequency distribution of their sizes and the way primary particles are connected in the aggregates.

### 2.3.1 Primary Particles – Surface Area

Although the smallest discrete entity of filler is the aggregate, the "particle" size, and its distribution is one of the most important morphological parameters with regard to its end-use applications, even though the particles do not exist as discrete entities except for thermal black and some special non-black fillers. The particle size is directly related to the specific surface area as the latter increases exponentially with the decrease of the former. It is an extensive parameter in rubber compounds and determines the interfacial area between rubber and filler, because rubber compounding is always based on the weight of materials. It is therefore of critical importance to the specific surface area of the fillers and has been taken as the principal parameter for grade classification of rubber fillers. In almost all types of carbon blacks and silicas, the primary particles within a single aggregate are similar. However, the types of fillers can differ in the uniformity of the primary particles of different aggregates. While many types show a quite narrow range of primary particles, others are clearly quite broad mixtures of aggregates of different primary particle size or specific surface area. The electron microscope is the universally accepted instrument for measuring particle size, surface area, aggregate size, and aggregate morphology, but the adsorption methods, both gas- and liquid-phase adsorptions, are convenient and accurate measurements for specific surface area.

#### 2.3.1.1 Transmission Electron Microscope (TEM)

Almost all filler aggregates used in the rubber industry, particularly the tire industry, are nano-scale sized and the classification of the various grades of carbon blacks is based on the size of the primary particles, which determines the surface area of the filler. TEM examination alone has made it possible to measure the particle size and aggregate morphology, as it can reach a nanometric resolution and, in combination with image analytical procedures, allows a two-dimensional statistical analysis.

Figure 2.7 and Figure 2.8 show typical electron micrographs of rubber-grade carbon blacks and precipitated silicas.

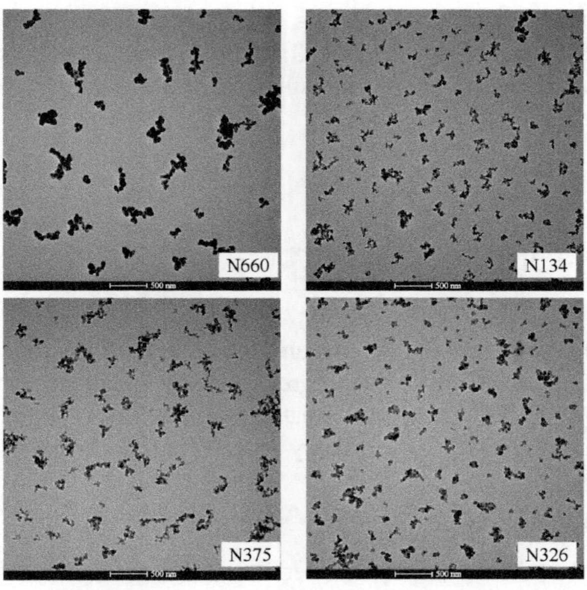


**Figure 2.7** Electron micrographs of rubber-grade carbon blacks

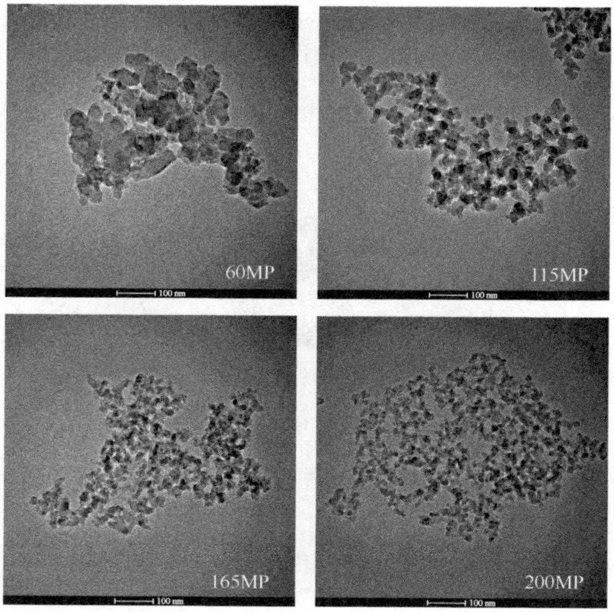


**Figure 2.8** Electron micrographs of rubber-grade precipitated silicas

The major limitations to the morphological analysis of fillers are related to the low speed of operation and to the complexity of aggregates under investigation. Therefore, many attempts have been made to speed up the data processing from the morphological images.

Nowadays, the potential of TEM has been greatly enhanced by the possibility of using automated image analysis (AIA), which allows acquisition of structural details and quantitative information on the filler through the use of descriptors based on geometrical parameters, such as size, volume of the primary particles and aggregates, and morphological parameters, describing the shape of the aggregates such as asymmetry and bulkiness.

Hess, Ban, and McDonald, in 1969, fully automated the analysis of carbon black particle and aggregate size in a dry state[24]. A Quantimet (QTM) performed the analyses, which were based on a television scanning device able to directly analyze a microscopic image. Later, in 1974, they reported a more practical automated size analysis method for carbon black[25], which is part of the ASTM procedure D3849.

Since then, many new imaging and analyzing systems have been developed commercially, and the techniques of image analysis have rapidly evolved due to progress in image acquisition and development of algorithms and software, either for general or specific applications. For these reasons more reliable and quantitative results on the morphology of fillers can be obtained which was reviewed by Conzatti, Costa, Falqui, and Turturro[26]. TEM micrographs provide digitalized images; the automated image analyzers provide measurements of the area and perimeter of an individual aggregate. Based on the two parameters, a variety of parameters characterizing the size and morphology of the aggregate and particle can be derived.

Shown in Figure 2.9 is the image of a carbon black aggregate, its area, *A*, that is, the projected area, and perimeter, *P*.

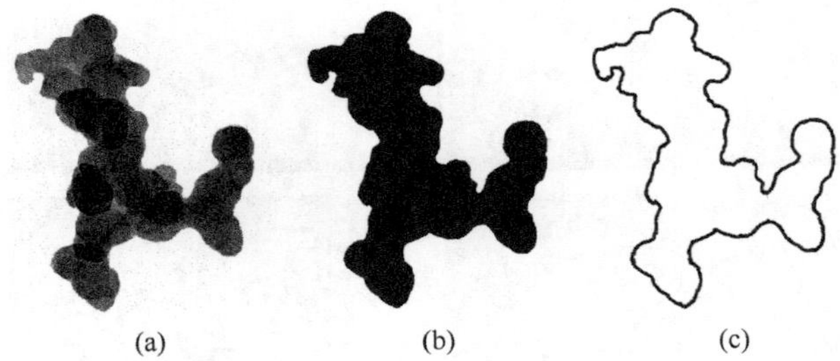

**Figure 2.9** Image of (a) a carbon black aggregate, (b) its area, and (c) its perimeter

According to Hess and McDonald[27], an average particle size, $d_p$ (nm) determines each individual aggregate using the mean chord length, $\pi A/P$:

$$d_p = \alpha \pi A/P \tag{2.1}$$

where $\alpha$ is an aggregation factor to correct for the fact that the mean chord is usually a measure of the average distance across multiple particles. To achieve more accurate average particle size measurements within each pigment aggregate, an equation for self-determination of $\alpha$ was developed. This is expressed as:

$$\lg \alpha = C_1 \lg\left(P^2/A\right) + C_2 \tag{2.2}$$

where $P^2/A$ is a nondimensional aggregate irregularity factor, and $C_1$ and $C_2$ are constants based on the slope and intercept of the relationship. Means of correlations such as that shown in Figure 2.10 empirically established the general form of the equation. This plots $P^2/A$ against calculated values for $\alpha$. Eq. 2.3 is fine-tuned for $C_1$ and $C_2$ using a series of non-porous carbon blacks with well-established size and surface area data. $C_1$ and $C_2$ are varied systematically to give the best overall data fit, but with the restriction that all equations must give the correct value of $\alpha$ for a perfect sphere ($4/\pi$). The value for a sphere is greater than one because the diameter actually represents the longest chord. The $P^2/A$ value for the projection of a sphere is equal to $4\pi$. The values established for the $C_1$ and $C_2$ constants are −0.92 and 1.117, respectively. With these constants, Eq. 2.2 can be converted to:

$$\alpha = 13.092\left(P^2/A\right)^{-0.92} \tag{2.3}$$

With this equation, no size cutoffs are required using Eq. 2.1 to calculate *d*. However, it was found to be desirable to limit the lower value of $\alpha$ to 0.45. It can be seen that the $P^2/A$ curve (Figure 2.10) flattens out in this general region. The 0.45 lower limit was determined in the same manner as $C_1$ and $C_2$. With no limitations on $\alpha$, the higher moments of the particle size distribution are relatively unaffected, but the number average values tend to be on the low side. The use of the 0.45 limit also permits application of the equation to the screening of dry blacks and blacks extracted from rubber compounds. When the calculated $\alpha$ is lower than 0.4, use $\alpha = 0.4$.

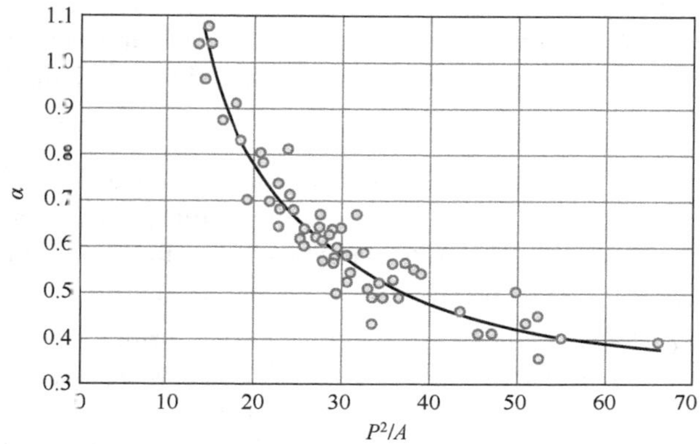


**Figure 2.10** Calculated $\alpha$ *vs.* $P^2/A$[27]

For one filler, the particle size mean diameter, $d_{sm}$ (nm), can be calculated based on the $d_p$ of several thousand aggregate images of known magnification as:

$$d_{sm} = \Sigma(nd_p^3)/(nd_p^2) \qquad (2.4)$$

The electron microscope surface area, EMSA ($m^2/g$), can be obtained by:

$$\text{EMSA} = 6000/(\rho d_{sm}) \qquad (2.5)$$

where $\rho$ is the filler density, which is assumed to be 1.8 $g/cm^3$ for carbon black.

Generally, for rubber-grade carbon blacks, the surface areas determined by TEM are in reasonable agreement with surface areas determined by nitrogen adsorption measurements. However, for those carbon blacks that have highly developed micro-pores, particularly special pigment blacks and blacks used for electrical conductivity, the surface areas calculated from their particle diameters are smaller than those calculated from gas absorption, as the internal surface area in the micro-pore is excluded.

Therefore, compared with BET surface area (NSA), the EMSA value is closer to STSA, the external surface area measured with nitrogen adsorption, because in the TEM measurement, the microscope is set at a magnification to obtain surface values that are similar to STSA (see Section 2.3.1.2.2).

It should be pointed out that the TEM method is the only way to obtain the particle size distribution which may be related to certain what in rubber properties of fillers.

#### 2.3.1.2 Gas Phase Adsorptions

While TEM images contain very detailed statistical information, using this method to obtain the surface area of the filler is expensive and time consuming. Direct measurement of specific surface area is much faster and less expensive. Gas- and liquid-phase adsorption, which depend on the amount of adsorbate required to form a surface monolayer, are simple techniques to make such measurements. If the area occupied by a single-adsorbate molecule is known, a simple calculation will yield the surface area.

For gas adsorption, the amount of gas adsorbate on solid adsorbent is a function of the mass of the solid, temperature, gas pressure, and nature of both gas and solid. The adsorption isotherms are classified into five types according to the theory of Brunauer, Emmett, and Teller[28], as shown in Figure 2.11. The isotherm type Ⅰ, also known as Langmuir isotherm, has a monolayer adsorption, i.e., further adsorption cannot take place on the adsorbed molecules. However, for most adsorbate-adsorbent systems, especially for the fillers used in elastomers, at temperatures not far from their condensation points, the adsorption isotherms of gases show two regions for most adsorbents: at low pressures, the isotherms are concave, at higher pressures convex toward the pressure axis (type Ⅱ). Such isotherms are believed to indicate the formation of multimolecular adsorbed layers. The BET equation, based on Brunauer et al.'s multimolecular adsorption, is the only theory able to explain all types of adsorption.

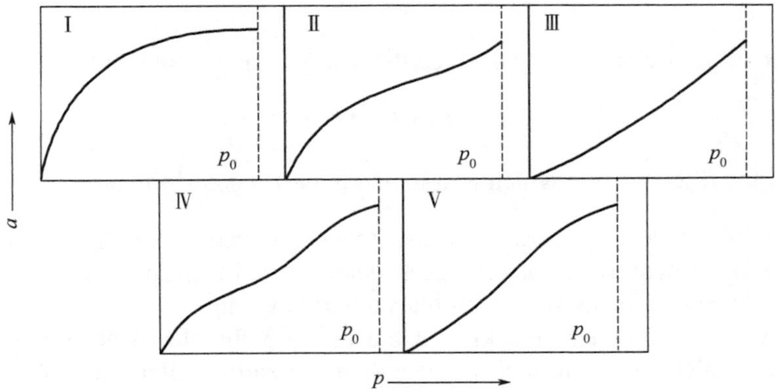


**Figure 2.11** Multilayer adsorption of gas molecules[28]

ASTM has adopted as standard method D6556, a low-temperature nitrogen-adsorption process, based on the original method of Brunauer, Emmett, and Teller (BET)[29]. This method is not susceptible to changes in the surface chemistry of carbon black such as those resulting from surface oxidation and the presence of trace amounts of tarry materials. With a molecular diameter of less than 0.5 nm, nitrogen is small enough to enter the micro-pore space so that the surface measured by BET is the total area, including micro-pores. For some applications, such as rubber reinforcement, the internal surface area in micro-pores with less than 2 nm diameter is inaccessible to large rubber molecules, thus it has no effect or has a negative effect on rubber reinforcement. The specific surface area that is accessible to rubber is defined as "external" surface area. This is conveniently measured by a multilayer nitrogen adsorption, also defined in ASTM D6556 and known as the statistical thickness surface area (STSA)[30].

#### 2.3.1.2.1 Total Surface Area Measured by Nitrogen Adsorption – BET/NSA

The BET theory assumes that not only does the adsorbate-adsorbent interaction determine the adsorption of gases on the adsorbent, but also the interaction between adsorbate molecules. This implies that when the adsorbate molecules in a gas state collide with those adsorbed on the adsorbent they may also be adsorbed on the solid surface, resulting in multimolecular adsorption. This is schematically shown in Figure 2.12.

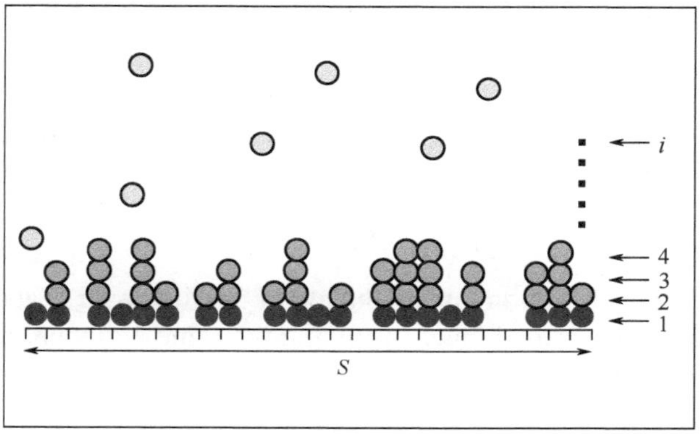


**Figure 2.12** Multilayer adsorption of gas molecules

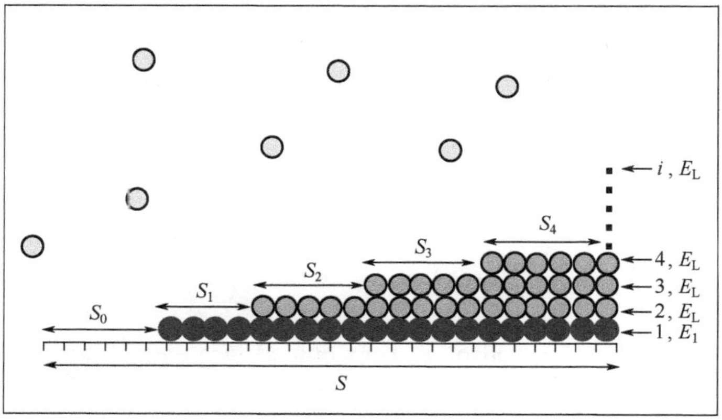


**Figure 2.13** Schematic multilayer adsorption and interaction energy of different layers

As shown in Figure 2.13, at a given pressure, $S_0$, $S_1$, $S_2$, ..., $S_i$ represent the surface area covered by 0, 1, 2, ..., $i$ layers of adsorbate. When the adsorption is at equilibrium, $S_0$ must remain constant. In this case, the adsorption rate of gas molecules on the bare surface is equal to the rate of desorption of the adsorbed adsorbate from the first layer. While the adsorption rate of the gas is proportional to the pressure, $p$, and the bare surface area, $S_0$, the desorption rate of the adsorbed adsorbate is in proportion with the number of molecules that are able to evaporate from the monolayer. According to Boltzmann's law, if the heat of adsorption of the first layer is $E_1$, this number is directly proportional to $\exp(-E_1/RT)$, in which $R$ is the gas constant and $T$ is the temperature. Thus, one has:

$$a_1 p S_0 = b_1 S_1 \mathrm{e}^{-E_1/RT} \tag{2.6}$$

where $a_1$ and $b_1$ are constants that are assumed to be independent of the number of molecules adsorbed already in the first layer.

$S_1$ can change in four different ways: by adsorption on the bare surface, by desorption from the first layer, by condensation on the first layer, and by evaporation from the second layer. At equilibrium adsorption, $S_1$ must also remain constant. Thus obtained is:

$$a_2 p S_1 + b_1 S_1 \mathrm{e}^{-E_1/RT} = b_2 S_2 \mathrm{e}^{-E_2/RT} + a_1 p S_0 \tag{2.7}$$

where the constants $a_2$, $b_2$, and $E_2$ are similarly defined to $a_1$, $b_1$, and $E_1$. From Eqs. 2.6 and 2.7, it follows that:

$$a_2 p S_1 = b_2 S_2 \mathrm{e}^{-E_2/RT} \tag{2.8}$$

This suggests that the adsorption rate on the top of the first layer is also equal to that of desorption rate from the second layer. Applying the same argument to the second and consecutive layers, it follows:

$$a_3 p S_2 = b_3 S_3 \mathrm{e}^{-E_3/RT} \ldots \tag{2.9}$$

and

$$a_i p S_{i-1} = b_i S_i \mathrm{e}^{-E_i/RT} \tag{2.10}$$

The total surface area of the solid, $S$, and the total volume adsorbed, $V$, are given by:

$$S = S_0 + S_1 + S_2 + \cdots + S_i = \sum_{i=0}^{\infty} S_i \tag{2.11}$$

and

$$V = V_0 \left(S_1 + 2S_2 + \cdots + iS_i\right) = V_0 \sum_{i=1}^{\infty} iS_i \tag{2.12}$$

where $V_0$ is the volume of gas adsorbed on a unit surface area of the adsorbent when it is covered with a complete mono-molecular layer of adsorbate. Therefore,

$$\frac{V}{SV_0} = \frac{V}{V_\mathrm{m}} = \frac{\sum_{i=1}^{\infty} iS_i}{\sum_{i=0}^{\infty} S_i} \tag{2.13}$$

where $V_m$ is the volume of gas adsorbed when the entire adsorbent surface is adsorbed with a complete mono-molecular layer.

The summations indicated in Eq. 2.13 are performed if two simplifying assumptions are made as:

$$E_2 = E_3 = \cdots = E_i = E_L \tag{2.14}$$

$E_L$ being the heat of liquefaction of the adsorbate, and

$$\frac{b_2}{a_2} = \frac{b_3}{a_3} = \cdots = \frac{b_i}{a_i} = K \tag{2.15}$$

where $K$ is an appropriate constant.

This suggests that the adsorption-desorption properties of the molecules in the second and higher adsorbed layers are the same as evaporation-condensation properties, which are the same as those of the liquid state of adsorbate. This is reasonable since the only adsorbed molecules that directly contact the solid surface are those in the first layer. From the second layer, all adsorbed molecules are in contact with the same type of molecules and they are not or are much less influenced by the solid surface. Supposing

$$x = \frac{p}{K} e^{E_L/RT} \tag{2.16}$$

and

$$y = \frac{a_1}{b_1} p e^{E_1/RT} \tag{2.17}$$

it was obtained:

$$S_1 = yS_0 \tag{2.18}$$

$$S_2 = xS_1 \tag{2.19}$$

$$S_3 = xS_2 = x^2 S_1 \tag{2.20}$$

and

$$S_i = xS_{i-1} = x^{i-1} S_1 = yx^{i-1} S_0 \tag{2.21}$$

Therefore, one has:

$$S_i = Cx^i S_0 \tag{2.22}$$

where

$$C \equiv \frac{y}{x} = \frac{a_1 K}{b_1} e^{(E_1 - E_L)/RT} \tag{2.23}$$

Substituting Eq. 2.22 into Eq. 2.13, one has:

$$\frac{V}{V_{\mathrm{m}}}=\frac{CS_0\sum_{i=1}^{\infty}ix^i}{S_0\left(1+C\sum_{i=1}^{\infty}x^i\right)} \tag{2.24}$$

The summation represented in the denominator is merely the sum of an infinite geometric progression:

$$\sum_{i=1}^{\infty}x^i=\frac{x}{1-x} \tag{2.25}$$

Concerning the summation in the numerator, it can be expressed by:

$$\sum_{i=1}^{\infty}ix^i=x\frac{\mathrm{d}}{\mathrm{d}x}\sum_{i=1}^{\infty}x^i=\frac{x}{(1-x)^2} \tag{2.26}$$

It is the case, therefore, that

$$\frac{V}{V_{\mathrm{m}}}=\frac{Cx}{(1-x)(1-x+Cx)} \tag{2.27}$$

When the adsorption takes place on a free surface, at the saturation pressure of the gas, $p_0$, an infinite number of layers can build up on the adsorbent, i.e., $V\rightarrow\infty$. According to Eq. 2.27, to make $V=\infty$ at $p=p_0$, $x$ must be equal to unity, i.e.,

$$\frac{p_0}{K}\mathrm{e}^{E_1/RT}=1 \tag{2.28}$$

and

$$x=\frac{p}{p_0} \tag{2.29}$$

where $x$ is relative pressure. Substituting it into Eq. 2.27, the adsorption isotherm can be expressed as:

$$V=\frac{V_{\mathrm{m}}Cp}{(p_0-p)\left[1+(C-1)\dfrac{p}{p_0}\right]} \tag{2.30}$$

which is the famous two-constant BET equation.

If the thickness of the adsorption layers cannot exceed some finite number, $n$, such as on a porous surface, then the summation of the two series in Eq. 2.24 is limited to $n$ terms only, and not to infinity. In this case, Eq. 2.27 becomes:

$$\frac{V}{V_{\mathrm{m}}}=\frac{Cx}{(1-x)}\left[\frac{1-(n+1)x^n+nx^{n+1}}{1+(C-1)x-Cx^{n+1}}\right] \tag{2.31}$$

This is the three-constant BET equation. For $n = 1$, i.e., monolayer adsorption, Eq. 2.31 can be simplified to:

$$V = V_{\mathrm{m}} \frac{C}{p_0} p \Big/ \left(1 + \frac{C}{p_0} p\right) \tag{2.32}$$

which is the Langmuir equation of monolayer adsorption.

Eq. 2.31 represents the first three types of adsorption isotherms reported by Brunauer et al.[28]. If $n = 1$, it is a type 1 isotherm. When $E_1 > E_L$, it is type 2. Isotherm type 3 represents the case of $E_1 < E_L$. Types 4 and 5 indicate, besides multilayer adsorption, the involvement of capillary condensation, which, as discussed later, does not influence the BET equation to the surface area measurement of powders.

### Application of BET Equation to Surface Area Measurements

Although there are some arguments about the difference between the assumptions of BET theory and actual situations, it does not seem to show any significant effect on its application in surface area measurement. Such arguments are mainly related to the adsorption energies, such as the energetic heterogeneity and the interaction between the second layer with the adsorbent surface, even though the contribution of the latter to the adsorption energy of the second layer is very small, since it reduces exponentially with distance. Generally, distribution of the surface energies of the adsorbent is quite heterogeneous, and the adsorption energy of the second layer should be influenced by the surface energies of the solid. This is especially true for adsorption at very low pressure. At relatively higher coverage of the adsorbent, at least for the surface coverage between 0.5–1.5 that is calculated by $V/V_{\mathrm{m}}$, the intermolecular interaction of the adsorbed molecules can significantly compensate the effects of the surface heterogeneity. When Eq. 2.30 is used, this may be the reason why the applicable range of relative pressure, $p/p_0$, is 0.05–0.35 for surface measurement, which corresponds to the surface coverage of 0.5–1.5.

If $V_{\mathrm{m}}$ at standard temperature and pressure (STP) and the effective cross-sectional area of the adsorbate, $\sigma_a$, at the adsorption temperature are obtained, the surface area of solid (NSA) in $m^2/g$ can be calculated according to the equation:

$$\mathrm{NSA} = 4.35 V_{\mathrm{m}} \tag{2.33}$$

where 4.35 is the area occupied by 1 $cm^3$ of nitrogen in STP.

### Measurement of $V_m$

For the purpose of testing, Eq. 2.30 can be written in the linearity form:

$$\frac{p}{V(p_0 - p)} = \frac{1}{V_{\mathrm{m}} C} + \frac{C - 1}{V_{\mathrm{m}} C} \times \frac{p}{p_0} \tag{2.34}$$

Plotting $p/V(p_0 - p)$ against $p/p_0$ should give a straight line whose slope, $M$, is $(C - 1)/V_{\mathrm{m}}C$ and intercept, $B$, is $1/V_{\mathrm{m}}C$. From $M$ and $B$, the $V_{\mathrm{m}}$ and $C$ can be calculated from:

$$V_m = \frac{1}{M+B} \tag{2.35}$$

and

$$C = \frac{M}{B} + 1 \tag{2.36}$$

It has been demonstrated that between relative pressures, $p/p_0$, of 0.05 and 0.35, the experimental data fits Eq. 2.34 quite well.

The test can be done with different adsorbates, such as argon[31], krypton[32], xenon[33], propane[34], and butane[35]. For the fillers used in rubbers, the unique and reliable measurement has been performed with adsorption of nitrogen at its boiling point (−196°C).

Generally, multiple points are taken from the range of relative pressure from 0.05 to 0.35 for calculation. In fact, for nitrogen adsorption, a set of data having $p/p_0$ in the range of 0.05 to 0.3 are used to calculate the $V_m$ and $C$ value. However, for most adsorbates, the intercepts of the straight-line plots of Eq. 2.34, $1/V_mC$, are small.

For simplification, a so-called "single point" method may be used. In this case, the slopes of the straight line connecting one pressure point to the origin of the straight-line plots can determine $V_m$. This standard test method measures the surface area based on a single partial pressure of 0.30 kPa ± 0.01 kPa. The time required to run a single point test is less than that of the multi-point test, but can be less accurate due to the effect of the surface properties. However, because the equipment and test are simpler, it may be adequate for quality control depending on the requirements of the producer and the customer.

Using a classical glass vacuum apparatus to measure the adsorption isotherms of nitrogen adsorption is a time-consuming and expensive operation. Another method for building up isotherms is based on selective adsorption of nitrogen from a mixture with other gases, helium in particular. This technique does not require a vacuum system and is time saving as the adsorption equilibrium is established rapidly[36]. Nelsen and Eggertsen[37–39] further developed this method using a so-called continuous flow chromatography. In this technique, a nitrogen/helium mixture passes through the filler in a sample cell that is immersed in liquid nitrogen. When the adsorption is complete, the liquid nitrogen reservoir is quickly removed and the sample cell is placed in a heating mantle. A thermal conductivity detector disrobes and records the change in the adsorbed nitrogen. From the resulting peak area, the volume of nitrogen adsorption is calculated. Varying the concentration of the mixture determines the adsorption isotherm point by point.

### Calculation of $\sigma_a$

After the value of $V_m$ is obtained, the average area by each molecule on the surface, $\sigma_a$, is required to calculate the surface area of the adsorbent. $\sigma_a$ may not be the actual molecular size of the adsorbate. Even for the same type of adsorbate, this value may somewhat vary due to changes in adsorption states, such as the adsorption temperature and the surface properties of adsorbents.

Generally, the adsorbed layers can be treated as a liquid. Therefore, the molecular area, $\sigma_a$, from liquid density is established based on the spherical model in the most dense face-centered packing[40], e.g.,

$$\sigma_a = 1.091 \times 10^{16} \left( \frac{M_W}{\rho N_A} \right)^{2/3} \tag{2.37}$$

where $M_W$ is the molecular weight of the adsorbate, $\rho$ the density of the adsorbate at experimental temperature, and $N_A$ Avogadro's number.

According to Eq. 2.37 the molecular area of nitrogen, $\sigma_a$, used for surface area calculation is 16.2 Å$^2$ (1 Å = $10^{-10}$ m) as the density of nitrogen at the adsorption temperature (−196°C) is 0.808 g/cm$^3$. This value is in agreement with those obtained by the same adsorbents with known surface areas.

It should be pointed out that before the surface area can be determined, it is necessary to remove any material that may already be on the surface. Removal of this material (usually by heating under a vacuum) is essential; otherwise two potential errors are possible. The first error can affect the mass, and the second error can interfere with the nitrogen's access to the adsorbent surface.

#### 2.3.1.2.2 External Surface Area Measured by Nitrogen Adsorption – STSA

Specific surface area is an important property for filler grade classification, production control, and prediction of rubber reinforcement characteristics. It is generally accepted that some internal area arises from micro-pores that are available to nitrogen which is commonly used in evaluating surface area, but they do not appear to be accessible to elastomer molecules. Therefore, this part of the area does not contribute to application performance. The NSA or total surface area alone is insufficient for estimating the reinforcing properties of microporous fillers. Thus, in predicting performance from area measurements, one should consider the external area.

Nitrogen adsorption isotherms at −196°C have been used extensively to calculate pore sizes and size distribution in carbon blacks[41–46]. These analyses are based on comparisons of the isotherms of the samples under investigation with those from standard planes of non-porous surfaces.

Originally, the comparison was made by plotting the nitrogen isotherms on a reduced basis, i.e., $V_a/V_m$, where $V_a$ is the equilibrium volume and $V_m$ the monolayer volume (all in cm$^3$ at normal temperature and pressure), as a function of relative nitrogen pressure $p/p_0$. If the reduced isotherm of the standard non-porous adsorbent superposes on that of the sample, one may then conclude that the sample is also non-porous. Departure from the standard isotherm was interpreted in terms of micro-pore filling[44,45].

Later, a so-called $t$ procedure was proposed in which the volume, $V_a$, of $N_2$ adsorbed by the sample at relative pressure $p/p_0$ can be transformed to functions of $t$, the statistical thickness of the adsorbed nitrogen layer at the same $p/p_0$ read from the reduced isotherm of the standard. The term, $t$, is called the statistical layer thickness due to the fact that the adsorbed gas molecules do not actually cover the entire surface of a pore wall in monolayer increments. These data are known as $t$ values, following a particularly elegant procedure for adsorption isotherm analysis developed by de Boer and Lippens[47].

As discussed before (see BET), for the adsorption of nitrogen at −196°C on a non-porous solid surface, the multi-molecular layer of adsorbed nitrogen could be formed freely on all parts of the surface. Assuming the density of the adsorbed layer of nitrogen is equal to that of bulk liquid, the statistical thickness $t$ in Å could be calculated according to the equation:

$$t = (V_L / \text{NSA}) \times 10^4 \tag{2.38}$$

where $V_L$ is the adsorbed volume in liquid (cm$^3$) and NSA the surface area of the adsorbent (m$^2$/g). Assuming at −196°C the adsorbed nitrogen molecules to be spherical in hexagonal close packing, a density to be equal to that of bulk liquid, i.e., 0.808 g/cm$^3$, and the effective cross-sectional area of the molecule to be the widely accepted value of 16.2 Å$^2$, for an adsorption volume of nitrogen, $V_a$, in STP, one has:

$$t = 3.54 V_a / V_m \tag{2.39}$$

where $V_a/V_m$ is the statistical thickness of adsorbed layers and 3.54 the thickness of a single layer in Å.

Based on the NSA calculated from $V_m$ (Eq. 2.33), the statistical thickness, $t$, in Å of nitrogen adsorption becomes:

$$t = 15.47 V_a / \text{NSA} \tag{2.40}$$

where 15.47 is a constant for conversion of nitrogen gas to liquid volume. Equation 2.40 may also be written in the form:

$$V_a = 0.0646 \text{NSA} t \tag{2.41}$$

This suggests that for non-porous adsorbents, when plotting the adsorbed volume of nitrogen against statistical thickness, a straight-line passes the origin. On the other hand, for the same adsorbent, the adsorption volume of nitrogen, $V_a$ is only determined by $p/p_0$. In other words, the statistical thickness $t$ is also a single function of $p/p_0$.

With several non-porous powders, de Boer et al. were able to obtain the data of $t$ as a function of relative pressure, from which a so-called universal $t$ master curve was built up[48]. With this master curve, the measured $V_a$ at given $p/p_0$ can be easily converted to $t$. The plot of $V_a$ as a function of the value of $t$ read from the standard isotherm provides the so-called $t$ plot. This procedure has the advantage over that described earlier as it can provide a linear relation between $V_a$ and $t$, departure from which is much easier to detect than comparison of the more complex $S$ shaped isotherms. The slope of the $t$ plot provides a means of evaluating external surface, referred to as statistical thickness surface area (STSA), from the relation:

$$\text{STSA} = 15.47 V_a / t \ \left(\text{m}^2/\text{g}\right) \tag{2.42}$$

If the $t$ plot is linear and passes through the origin, the sample may be assumed to be non-porous (Figure 2.14). The area derived from the slope, STSA, will then correspond to the BET area, NSA. If the $t$ plot is linear but upon extrapolation provides a positive intercept, porosity is indicated and will be evidenced by a value for STSA less than that of the total or NSA. A $V_a$–$t$ curve can form if there are multiple-size pores on the filler surfaces. In this case, the internal surface area in different size pores can be calculated from their slopes, which is schematically shown in Figure 2.15.

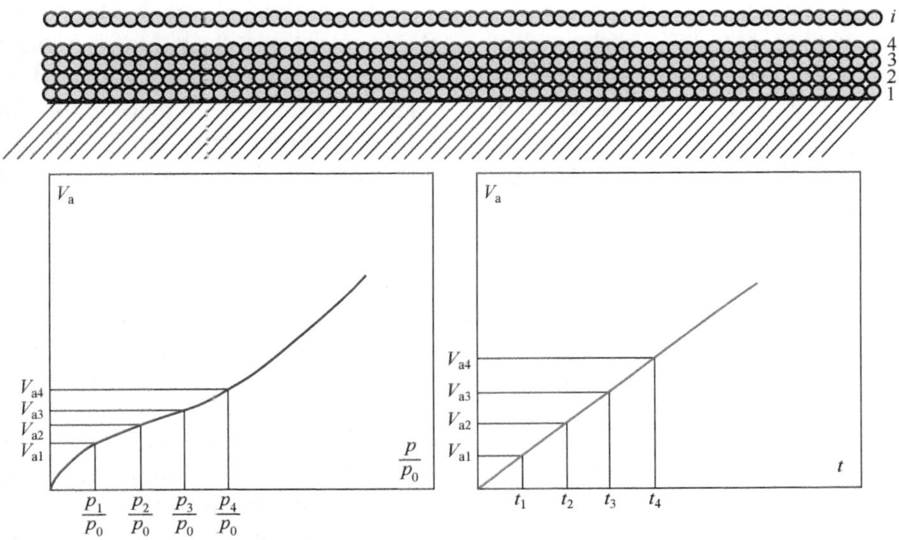


**Figure 2.14** Schematic $t$ plot of non-porous adsorbents

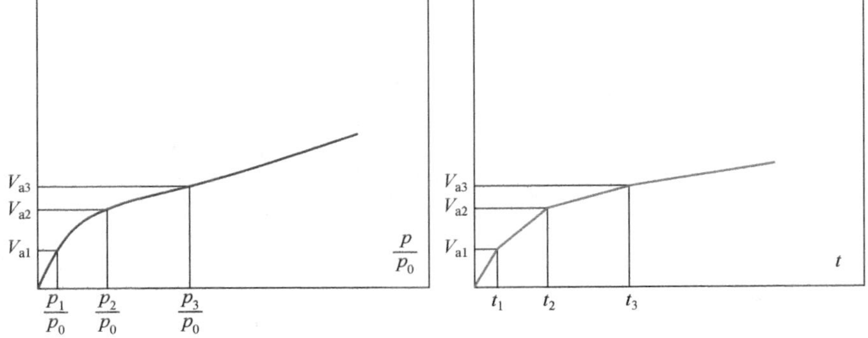


**Figure 2.15** Schematic $t$ plot of porous adsorbents

It should be noted that using the $t$-method as a comparative procedure for detecting porosity in fillers can provide useful information for their reinforcing abilities in rubbers. However, the $t$-method is empirical as the STSA values are highly dependent upon the $V_a$–$p/p_0$ or $V_a$–$t$ functions used. There are some uncertainties that require a non-porous adsorbent to serve as a standard with which other samples may be compared. For the de Boer standard, the universal master curve is constructed, and it is based on a calcined alumina. In fact, there are some differences in the surfaces of various substances[49,50]. For example, in porosity studies on carbon blacks, it was found that the low polar carbon black as the standard adsorbent is preferable to a highly polar surface, such as oxide[51]. Accordingly, Smith and Kasten selected a fine thermal black as a $t$ curve standard. Then, Magee took an N762 furnace carbon black as the $t$ layer thickness standard because of its low surface area and low degree of aggregation (low structure)[52]. The $t$ values as a function of $p/p_0$ are approximated by the equation as:

$$t_{CB} = 0.88(p/p_0)^2 + 6.45(p/p_0) + 2.98(\text{Å}) \tag{2.43}$$

Magee's research claims $V_a$–$t$ plots of furnace carbon blacks exhibit a classical shape and nonnegative intercepts (Figure 2.15). The linear portion of the furnace black $t$ plot is over the same $p/p_0$ range as the de Boer plot.

Figure 2.16 shows the STSA values obtained from the regression of the $V_a$–$t$ plots over the relative pressure range of 0.2 to 0.5, which corresponds to the statistical thickness of 4.31 Å and 6.43 Å or 1.22 and 1.82 times the monolayer adsorption, respectively. The positive intercept represents the "micro-pore volume", expressed as the volume of the nitrogen gas at STP adsorbed. Although plotted in terms of gas volume adsorbed, it is visualized as the resulting filling of liquid-like nitrogen of a certain average thickness.

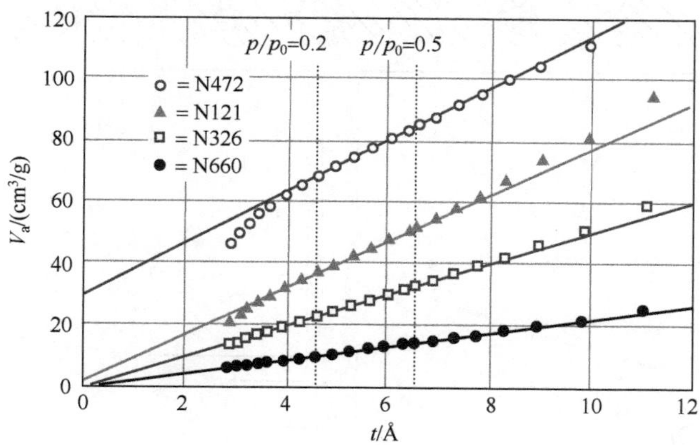


**Figure 2.16** $t$ plots of a variety of furnace carbon blacks[52]

In practice, one may also see some deviations of the adsorption values from the regression lines obtained at relative pressures higher than 0.5. For carbon blacks, such as N472, the lower adsorption values may be due to the fact that there are still some micro-pores which cannot be filled at relative pressure of 0.5 (see Figure 2.16). For some blacks like N121 and N326, the upward deviation from a straight line when the relative pressure is higher than 0.5 may reflect the increased uptake due to capillary condensation of nitrogen between particles of the aggregates (see Figure 2.17) or in micro-pores with large diameters. If the condensation occurs in the region below relative pressure 0.5, negative intercepts should be obtained, resulting in STSA higher than NSA. One may easily observe this phenomenon for lower structure and higher surface area carbon blacks such as N326, as in these cases more contact points are formed (see Chapter 4 "Filler Dispersion").

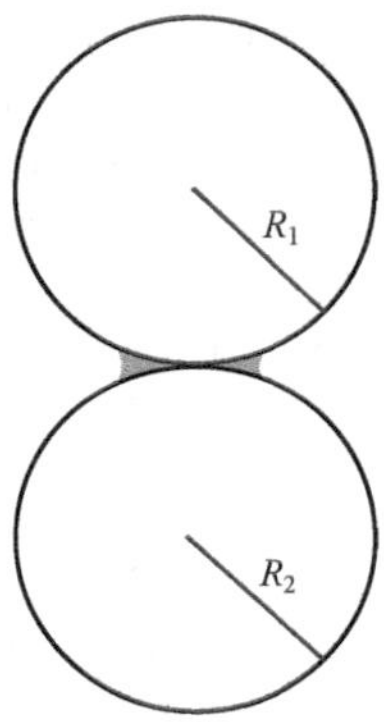


**Figure 2.17** Gas condensation between two particles

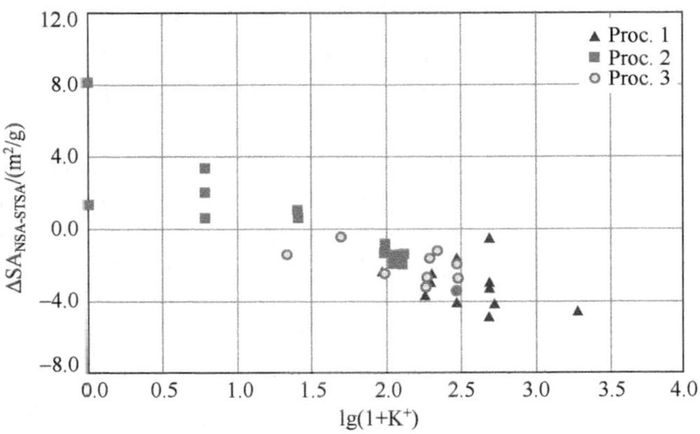


**Figure 2.18** The porosity of carbon black as a function of $K^+$ added in carbon black production processes[53]

Adding a potassium salt as a structure-adjusting chemical demonstrates the argument that structure affects the numbers of contact points on nitrogen condensation at relatively higher pressure. With increasing potassium salt, the structure is reduced, which should lead to more condensation (see Figure 2.18). On the other hand, the condensation is also related to the contact angle of liquid nitrogen, $\theta$, as:

$$F = 4\pi\left(1/R_1 + 1/R_2\right)^{-1} \gamma \cos\theta \tag{2.44}$$

where $F$ is the attractive force between the two particles due to the capillary phenomenon that is the driving force of condensation, $R_1$ and $R_2$ the radiuses of particles 1 and 2, respectively, and $\gamma$ the surface energy of liquid nitrogen. From this equation, the high concentration of potassium may also influence nitrogen condensation due to a change in surface characteristics of the black.

For carbon blacks, the external surface area derived from the *t*-method has been compared to the CTAB (cetyl trimethyl ammonium bromide) surface area for a number of rubber carbon blacks[52]. The CTAB surface area test, based on adsorption of a molecule too large to enter the micro-pore, measures the external surface area available to rubber (see Section 2.3.1.3.2). Over the range of the carbon blacks used, the linear relationship observed between the two methods indicates that they are measuring primarily the same surface area property. The slope is slightly greater than one which may suggest that the external surface area of carbon black is somewhat overestimated by CTAB method. However, different results have been reported, which will be further discussed in the section on CTAB adsorption.

Current rubber industry practices favor the *t*-method over the CTAB method for measuring the external surface area of carbon black, as the former provides the following advantages: 1) precise results that are less affected by surface oxidation; 2) less test time; 3) measurement carried out simultaneously with NSA test; and 4) no reagent preparation needed.

#### 2.3.1.2.3 Micro-Pore Size Distribution Measured by Nitrogen Adsorption

Several methods for calculating the pore size distribution from adsorption data of gas, nitrogen in general, on porous solids have been developed[42,54–56]. Generally, the desorption loop of an adsorption isotherm is used to relate the amount of adsorbate lost in a theoretical desorption step to the average size of pore emptied in the step. Dollimore and Heal[57] reviewed the fundamental principle of the methods before they proposed their new approach that is more exact and less tedious to use. In this section both their review and the improved method are introduced, even though some new treatments have been reported[58,59].

**Early Methods**

Adsorption isotherms of vapors on porous solids commonly exhibit the phenomenon of condensation of adsorbate within the confines of the pore structure at pressures considerably below the prevailing bulk vapor pressure of the adsorbate. The well-known Kelvin equation[60] derived from classical thermodynamic treatments can be used to relate the lowering of vapor pressure to the size of a pore of simple geometry (on the assumption of cylindrical pores, which is generally true for the rubber fillers, carbon black and silica in particular):

$$r = -\frac{2\gamma V}{RT \ln p/p_0} \cos\theta \tag{2.45}$$

where $p$ is the pore filling pressure of the pore having radius of $r$, $p_0$ is the saturated vapor pressure of the adsorbate at the temperature $T$, $\gamma$ is the surface tension of the liquid adsorbate, $V$ is the molar volume of the adsorbate, $R$ is the gas constant, and $\theta$ is the contact angle of adsorbate and adsorbent (taken to be zero when the adsorbent is wetted).

According to Dollimore and Heal[57], if the amount of adsorbed material, at any point on the isotherm, is converted to liquid volume, and the amount lost in a desorption step is $\Delta V$; then $\Delta V$ is the sum of capillary desorption $\Delta V_c$, and multi-layer desorption $\Delta V_m$, i.e.,

$$\Delta V = \Delta V_c + \Delta V_m \tag{2.46}$$

The number of molecules in the multi-layer varies over the surface, but for any relative pressure $p/p_0$, the average thickness $t$ may be determined, and also the decrease in multi-layer thickness $\Delta t$ for the desorption step. If it is assumed for the moment that $\Delta V_m$ may be calculated, then $\Delta V_c$ for the desorption step is obtained. This volume, $\Delta V_c$, is the capillary liquid lost between the beginning and the end of a step between two relative pressures $p_1/p_0$ and $p_2/p_0$. When the capillary liquid is lost, multi-layers are left on the walls, and thus the radius of the meniscus of the capillary condensate is less than the actual radius of the pores by the amount $t$. The capillary condensate radii $r_{k1}$ and $r_{k2}$ are related to $p_1/p_0$ and $p_2/p_0$ by Eq. 2.45:

$$\ln p/p_0 = -\frac{2\gamma V}{RTr_k}\cos\theta \tag{2.47}$$

The two radii $r_{k1}$ and $r_{k2}$ are related to actual pore radii $r_1$, and $r_2$ by:

$$r_{k1} = r_1 - t_1 \tag{2.48}$$

and

$$r_{k2} = r_2 - t_2 \tag{2.49}$$

where $t_1$ and $t_2$ are multi-layer thickness at the given pressures. The value of $\Delta V_c$, may be assigned to pores of radius of the mean of $r_1$ and $r_2$, namely $r_p$. Similarly $\overline{r_k}$ is the mean of $r_{k1}$ and $r_{k2}$.

The quantity $\Delta V_c$ must be multiplied by a factor $[r/(r-t)]^2$ to give the value of $\Delta V_p$, the actual volume of the pores, because multi-layers of adsorbate of thickness $t$ are left on the walls of pores. At the beginning of a desorption step, this factor, called $R_n$, becomes $[r_1/(r_{k1} + \Delta t)]^2$, where $r_{k1} + \Delta t$ is the radius of the space left inside the multi-layers in a pore of radius $r_1$ at the end of a step. Since this factor will change for pores in the range $r_1$ to $r_2$, the average value, $\left[r_p/(\overline{r_k} + \Delta t)\right]^2$, should be used in an actual desorption step. Thus,

$$\Delta V_p = R_n \Delta V_c \tag{2.50}$$

The surface area $S_p$ of the pores involved in the step may be found from:

$$S_p = 2\Delta V_p / r_p \tag{2.51}$$

The areas of these pores may also be summed as the calculation proceeds for several desorption steps, giving $\Sigma S_p$. The value of $\Delta V_m$, for a particular step, would then appear to be $\Delta t\,\Sigma S_p$, where $\Delta t$ is the multi-layer thinning for the same step, and the value of $\Sigma S_p$ is the one current for the preceding step. However, this would assume the area $\Sigma S_p$ to be planar, whereas it is actually made up of the curving walls of pores. Thus, $\Delta t\,\Sigma S_p$ overestimates the value of $\Delta V_m$. A correction factor for the curvature of the pore walls is then required. Unfortunately, when the calculation has proceeded over several steps, the range of pore sizes is considerable, and the factor required varies slightly for all these pores.

Barret et al[55] used a factor $C$, equal to $(r_p - t)/r_p$ and assigned a choice of constant values to it, namely 0.9, 0.85, 0.80, or 0.75 (to be chosen according to the approximate range of sizes of pores expected). This procedure gives reasonably accurate results for pores down to radii of 3.5 nm. However, for pores of 0.7 nm (the lower limit of the method[55]) a value of $C$ of 0.55 is required, and the value changes with increasing pore size too rapidly to be correctly assigned a constant value. The final equation becomes:

$$\Delta V_p = R_n \left( \Delta V_p - C \Delta t \Sigma S_p \right) \tag{2.52}$$

Cranston and Inkley[56] overcame the error involved by using, in effect, a value of $C$ variable, both for the step in desorption being calculated, and each of the pore sizes involved in the previous desorption steps. They multiplied $S_p$ for each pore size range by its individual value of $C$ instead of summing $S_p$.

According to Dollimore and his coworker, as calculation proceeds down to small pore sizes, this method results in as many as 14 multiplications and additions to find the value of $\Delta V_m$ for one line of calculation and thus is much too laborious.

### Dollimore Improved Method

Alternatively, Dollimore and Heal[57] proposed a new method for the calculation of pore size distribution based on the work of Wheeler[61]. However, some tables are necessary for use with this method, which were published in their paper[57], together with the equations required for calculation. The values of $t$ used by previous workers were obtained from adsorption data on a series of non-porous adsorbates[54]. Wheeler proposed the use of the Halsey equation[62] to obtain values of $t$:

$$t = 4.3 \left( \frac{5}{\ln p/p_0} \right)^{1/3} \tag{2.53}$$

for nitrogen adsorption at −196°C.

These values coincide with the measured ones for $p/p_0 > 0.5$, but are higher for lower pressures. Wheeler believed this to be the case for adsorption in narrow pores, where the surrounding walls enhance adsorption. To produce a general equation for desorption, he used the following terms for total volume $\Delta V_c$, area $S$, and length of pores $L$, $n$ being the number of the desorption steps completed:

$$V_c\left[<r_{pn}\right]=\int_0^{r_{pn}}\pi r_p^2 L\left(r_p\right)\mathrm{d}r_p \text{ (for pores in future steps)} \tag{2.54}$$

$$S\left[>r_{pn}\right]=\int_{r_{pn}}^{\infty}2\pi r_p L\left(r_p\right)\mathrm{d}r_p \text{ (for pores in previous steps)} \tag{2.55}$$

and

$$L\left[>r_{pn}\right]=\int_{r_{pn}}^{\infty}L\left(r_p\right)\mathrm{d}r_p \text{ (for pores in previous steps).} \tag{2.56}$$

The bracketed terms on the left signify the range of calculation of the terms. The length of the pores is a continuous function of $r_p$, namely $L(r_p)$, which is the distribution required.

Similar to Eq. 2.46, one has:

$$\Delta V_c=\Delta V-\Delta V_m \tag{2.57}$$

or for the step $n$:

$$\Delta V_c=\Delta V_n-\Delta V_m \tag{2.58}$$

In a pore of radius $r_p$ ($> r_{pn}$) the volume of adsorbate on the walls is given by the geometry of a cylinder as:

$$\text{Wall adsorption in one pore}=\pi\left[r_p^2-\left(r_p-t_n\right)^2\right]L\left(r_p\right)=\pi\left(2r_p t_n-t_n^2\right)L\left(r_p\right) \tag{2.59}$$

the value of $t_n$ being for the step $n$.

The total wall adsorption in all pores of radii from $r_{pn}$ to $\infty$ is:

$$\begin{aligned}V_m&=\int_{r_{pn}}^{\infty}\pi\left(2r_p t_n-t_n^2\right)L\left(r_p\right)\mathrm{d}r_p=t_n\int_{r_{pn}}^{\infty}2\pi r_p L\left(r_p\right)\mathrm{d}r_p\\&-\pi t_n^2\int_{r_{pn}}^{\infty}L\left(r_p\right)\mathrm{d}r_p=t_n S\left[>r_{pn}\right]-\pi t_n^2 L\left[>r_{pn}\right]\end{aligned} \tag{2.60}$$

The change on desorption may be found by differentiation:

$$\mathrm{d}V_m=\mathrm{d}t_n S\left[>r_{pn}\right]-2\pi t_n \mathrm{d}t_n L\left[>r_{pn}\right] \tag{2.61}$$

Changing to $\Delta$ terms for finite steps:

$$\Delta V_m=\Delta t_n S\left[>r_{pn}\right]-2\pi t_n \Delta t_n L\left[>r_{pn}\right] \tag{2.62}$$

The length and area functions may be put in finite terms also, as the summations of length and area of pores involved in previous steps:

$$\Delta V_{\mathrm{m}} = \Delta t_{\mathrm{n}} \Sigma S_{\mathrm{p}} - 2\pi t_{\mathrm{n}} \Delta t_{\mathrm{n}} \Sigma L_{\mathrm{p}}, \tag{2.63}$$

where $\Sigma L_p$ is the sum of individual lengths of pores in each previous desorption step. Then using Eq. 2.58, one has:

$$\Delta V_{\mathrm{c}} = \Delta V_{\mathrm{n}} - \Delta t_{\mathrm{n}} \Sigma S_{\mathrm{p}} + 2\pi t_{\mathrm{n}} \Delta t_{\mathrm{n}} \Sigma L_{\mathrm{p}}, \tag{2.64}$$

and since $\Delta V_p = R_n \Delta V_c$ (Eq. 2.50), then:

$$\Delta V_{\mathrm{p}} = R_{\mathrm{n}} \left( \Delta V_{\mathrm{n}} - \Delta t_{\mathrm{n}} \Sigma S_{\mathrm{p}} + 2\pi t_{\mathrm{n}} \Delta t_{\mathrm{n}} \Sigma L_{\mathrm{p}} \right). \tag{2.65}$$

Therefore, the value of $\Delta V_p$ may be found for a particular step.

As shown in Eq. 2.51, $S_p = 2\Delta V_p/r_p$, similarly,

$$L_{\mathrm{p}} = S_{\mathrm{p}} / 2\pi r_{\mathrm{p}} \tag{2.66}$$

These two terms may then be summed, line by line. Actually, there is no need to find $L_p$; instead, since the value of $2\pi \Sigma L_p$ is required, $2\pi L_p$ is summed, given by:

$$2\pi L_{\mathrm{p}} = S_{\mathrm{p}} / r_{\mathrm{p}} \tag{2.67}$$

As pointed out by Dollimore, the use of Eq. 2.65 does not require any approximation for multi-layer desorption, nor does it involve the tedious calculations.

### Application of Dollimore Method

To make use of the method for calculation of pore size distribution, a master sheet (Table 2.2) showing the method has to be worked out. The pore radius at the end of each step is $r$, but the actual radius related to $p/p_0$ by the Kelvin equation is $r_k$, given by $r_k = r - t$ as pointed out above. In this Table the Kelvin equation is shown with the required constants inserted. The values of $p/p_0$ are chosen to give convenient values of $r$. The value of $\Delta t$ gives the change in $t$ between steps, and $r_p$ is the average value of $r$ between steps. The amount adsorbed at any point is converted to liquid volume $V_{liq}$.

From the measurements of the adsorption isotherm and desorption of nitrogen at −196°C, the $\Delta V_p$ in each interval of $r$, i.e., $\Delta V_p/\Delta r$, can be calculated.

**Table 2.2** The master sheet for calculation of pore size distribution

| $p/p_0$ | $t$/nm | $r_k$/nm | $r$/nm | $\Delta t$/nm | $r_p$/nm |
|---|---|---|---|---|---|
| 0.894 | 1.523 | 8.477 | 10 | – | – |
| 0.881 | 1.465 | 7.535 | 9.0 | 0.058 | 9.5 |
| 0.866 | 1.401 | 6.599 | 8.0 | 0.064 | 8.5 |
| 0.854 | 1.332 | 5.668 | 7.0 | 0.069 | 7.5 |
| 0.818 | 1.256 | 4.744 | 6.0 | 0.076 | 6.5 |
| 0.780 | 1.169 | 3.831 | 5.0 | 0.087 | 5.5 |

Continued

| $p/p_0$ | $t$/nm | $r_k$/nm | $r$/nm | $\Delta t$/nm | $r_p$/nm |
|---|---|---|---|---|---|
| 0.754 | 1.121 | 3.379 | 4.5 | 0.048 | 4.75 |
| 0.722 | 1.069 | 2.931 | 4.0 | 0.052 | 4.25 |
| 0.682 | 1.012 | 2.488 | 3.5 | 0.057 | 3.75 |
| 0.628 | 0.949 | 2.051 | 3.0 | 0.063 | 3.25 |
| 0.556 | 0.878 | 1.622 | 2.5 | 0.071 | 2.75 |
| 0.538 | 0.862 | 1.538 | 2.4 | 0.016 | 2.45 |
| 0.519 | 0.846 | 1.454 | 2.3 | 0.016 | 2.35 |
| 0.499 | 0.830 | 1.370 | 2.2 | 0.016 | 2.25 |
| 0.477 | 0.813 | 1.287 | 2.1 | 0.017 | 2.15 |
| 0.453 | 0.795 | 1.205 | 2.0 | 0.018 | 2.05 |
| 0.428 | 0.777 | 1.123 | 1.9 | 0.018 | 1.95 |
| 0.401 | 0.758 | 1.042 | 1.8 | 0.019 | 1.85 |
| 0.371 | 0.738 | 0.962 | 1.7 | 0.020 | 1.75 |
| 0.340 | 0.717 | 0.883 | 1.6 | 0.021 | 1.65 |
| 0.306 | 0.695 | 0.805 | 1.5 | 0.022 | 1.55 |
| 0.270 | 0.672 | 0.728 | 1.4 | 0.023 | 1.45 |
| 0.232 | 0.648 | 0.652 | 1.3 | 0.024 | 1.35 |
| 0.192 | 0.622 | 0.578 | 1.2 | 0.026 | 1.25 |
| 0.152 | 0.595 | 0.505 | 1.1 | 0.027 | 1.15 |
| 0.111 | 0.566 | 0.434 | 1.0 | 0.029 | 1.05 |
| 0.074 | 0.534 | 0.366 | 0.9 | 0.032 | 0.95 |
| 0.042 | 0.500 | 0.300 | 0.8 | 0.034 | 0.85 |
| 0.018 | 0.462 | 0.238 | 0.7 | 0.038 | 0.75 |

#### 2.3.1.3 Liquid Phase Adsorptions

The fundamental principle of surface area measurement of fillers by liquid-phase adsorptions methods is similar to that of gas phase adsorptions. The surface area of the fillers is calculated by the adsorption of adsorbates in liquid solutions. The adsorbates used for studying the relationships between surface area of solid powder and adsorption include phenol, fatty acid, and benzoic acid. However, for rubber fillers, especially carbon black, iodine and cetyl trimethyl ammonium bromide (CTAB) provide the most useful information on surface area.

#### 2.3.1.3.1 Iodine Adsorptions

The adsorption of iodine on carbon black surfaces in an aqueous solution has been studied since the beginning of the last century[63]. In 1921, Carson and Sebrell[64] investigated correlations between iodine adsorption of carbon blacks and filled rubber properties. The iodine adsorption was performed using an iodine solution with small amounts of potassium iodide. Later, Smith, Thornhill, and Bray[65] reported that a good linear relation exists between iodine adsorption and surface area measured with nitrogen adsorption. This is for carbon blacks whose iodine adsorption is independent of the specific nature of the carbon surface, with the exception of the two blacks having a volatile content over 10%. However, when the volatile is removed by calcination, the iodine adsorption and surface areas are in much better agreement. From their data, 130 mg of iodine is associated with about 100 $m^2$ of surface area, corresponding to a covering power of 32 $Å^2$ per adsorbed iodine molecule and a diameter of 6.4 Å. The collision diameter of iodine from diffusion measurements[66] is found to be 4.6 Å, in sufficient agreement with the approximate value obtained from the present measurements to indicate the possibility of formation of a monolayer of adsorbed iodine.

Kendall's[67] systematic study on the adsorption of iodine from an aqueous solution demonstrated that the mechanism was more complex than what was previously assumed. Present in an aqueous solution of iodine-potassium iodide are potassium iodide, potassium triiodide, and iodine molecules because of the equilibrium:

$$KI + I_2 \rightleftharpoons KI_3$$

Hence, it is necessary to consider the identity of the adsorbed species. Kendall's blank experiments revealed that potassium iodide was not adsorbed. The isotherms of iodine adsorption for a number of carbon blacks are only independent of the amount of potassium iodide in the solution when the adsorption was expressed as a function of the free iodine concentration. The analysis of the iodine solution, before and after adsorption, demonstrated that the total concentration of iodide ($KI + KI_3$) decreased very little. These facts prove that only neutral iodine molecules are adsorbed.

The adsorption data of iodine with several carbon blacks satisfactorily corresponds to the Langmuir adsorption equation as:

$$m = m_1 K(c/c_0)/\left[1 + K(c/c_0)\right] \tag{2.68}$$

where $m$ is solute adsorption in mg/g, $m_1$ the mass required to form a monolayer, $c$ the concentration of the solute, $c_0$ the solubility of the adsorbate in the solvent, and $K$ a constant. $K$ is given approximately by:

$$K = e^{\Delta H/RT} \tag{2.69}$$

where $\Delta H$ is the net heat of adsorption.

Equation 2.68 shows that adsorption from the solution depends on the solubility of the adsorbate in addition to the concentration. Therefore, when measuring the iodine adsorption, the temperature and solvent have to be considered.

Watson and Parkinson[68] found the net heats of adsorption for furnace and graphitized carbon blacks were all about 2000 cal/mol, showing the reversible adsorption for these blacks to be physical in nature. The deviation from the Langmuir equation shown by channel blacks suggested that their surfaces might be heterogeneous. Based on fitting their data to different models, including the BET equation, they proposed a dual surface model for the iodine adsorption of channel black. The Langmuir equation for adsorption on a dual surface may be written as:

$$m = \frac{m_1^1 K_1 \left( \frac{c}{c_0} \right)}{1 + K_1 \left( \frac{c}{c_0} \right)} + \frac{m_2^1 K_2 \left( \frac{c}{c_0} \right)}{1 + K_2 \left( \frac{c}{c_0} \right)} \tag{2.70}$$

where the subscripts refer to the two types of surfaces which are considered to be of equal extent, so $m_1^1 = m_2^1$, although the heats of adsorption are different, corresponding to two different $K$ values, i.e., the heat of adsorption for one part of surface is higher than that of the other.

For channel carbon blacks, whose volatile contents are all about 5% by weight, and whose specific surface areas cover the range from 227 $m^2/g$ to 100 $m^2/g$, the volatile content per unit surface area increases with decreasing surface area. From the experimental data, the adsorption of iodine on unit surface also decreases. Obviously, the volatile content is responsible for the less active part of the surface. Thus, the progressive decrease in adsorption per unit area of these blacks is in agreement with the dual surface theory.

The logical extension of the theory to furnace blacks shows their Langmuir-type isotherms correspond to the case when the less active part of the surface becomes very small (volatile content < 1%). It was also confirmed that the graphitized carbon black (Graphon) corresponds to the case when volatile content (less active part of the surface) becomes zero. The adsorption in this case is even greater than that for furnace blacks. These findings are in line with the observation of Smith et al[65]. In their study, the carbon black with a higher volatile content shows a much lower iodine adsorption than that of its counterpart with a lower volatile. When the volatile is removed by calcination, the iodine adsorption increases drastically and follows the same line with the surface area as that obtained with the carbon black having a low volatile. In other words, the devolatilization makes the adsorption isotherm of iodine correspond to the Langmuir equation without changing the surface area.

For channel carbon blacks it is generally accepted that the volatile is mainly, if not only, determined by oxygen-containing groups. The process of devolatilization under high temperature in an inert atmosphere is related to the decomposition of oxygen groups. This suggests that the surface covered by oxygen groups has a lower surface activity for iodine adsorption. This can be further confirmed by the observations of Sweitzer, Venuto, and Estelow[69] who claimed that the decrease in adsorption of iodine follows the oxidation of a thermal black. The decrease in adsorption can be well interpreted as a consequence of the low heat of adsorption of iodine by oxygen groups introduced onto the carbon black surface rather than a chemisorption taking place on the unoxidized black surface.

From the discussion above, for the carbon blacks with a low volatile content, there is a linear relationship between iodine number and NSA. Therefore, it is possible to take the iodine number as a measure of surface area.

Snow[70], after analyzing an enormous amount of data on the adsorption of iodine by various carbon blacks, developed a fast and simple laboratory method for determining the surface areas of basic carbon blacks, which is suitable for control work at furnace black and rubber factories, based on the observations that follow.

In addition to the surface characteristics, the amount of iodine adsorbed increases as total iodine available increases, and the rate of increasing decreases rapidly. For the high concentration of iodine in the solution at equilibrium, the rate of change is negligible. In this case, everything is constant except total iodine present ($a + L$ in Figure 2.19). The ratio of iodide ions to iodine and the concentrations of the solutions are very important for the iodine adsorption.

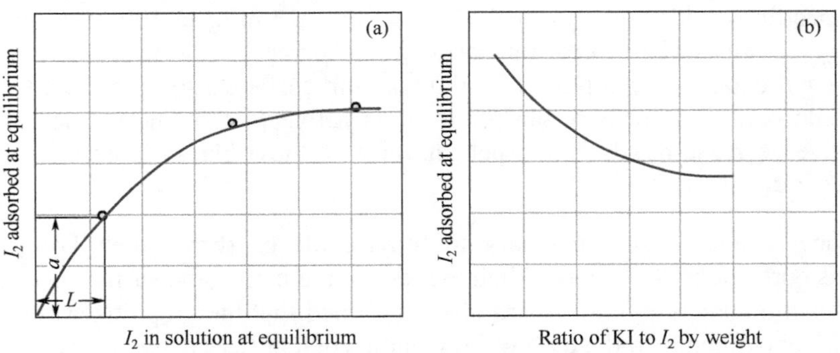


**Figure 2.19** Schematic drawing of the iodine adsorbed as a function of the iodine concentration (a) and the ratio of iodide to iodine (b)[70]

Furthermore, the surface of the sample should not be so large as to adsorb more than about 90 mg to 100 mg of iodine from the solution[71]. Under the conditions specified by Snow, such as adsorbate concentration and the amount of carbon black sample used, the values of iodine numbers turn out to be numerically about the same as the values for surface areas in $m^2/g$ that are measured by nitrogen adsorption for non-porous and non-oxidized furnace carbon blacks.

To demonstrate the utility of this test, iodine adsorptions were run on a variety of carbon blacks, and the iodine numbers obtained are plotted against the NSA in Figure 2.20[71]. The correlation is good for all definitely basic carbons. The presence of high levels of surface oxygen functionality also causes a severe reduction in the iodine number. The oxygen groups on the surface of the carbon black actually react with the potassium iodide in the test solution, thereby causing a release of free iodine, which makes the surface area appear to be lower[72]. The two carbon blacks that show the greatest deviation of the basic and acidic carbon blacks are, respectively, Vulcan SC, with very high surface area and very feebly basic (pH 7.22), and Shawinigan-acetylene black, with strong acidity (pH 4.82)[71].

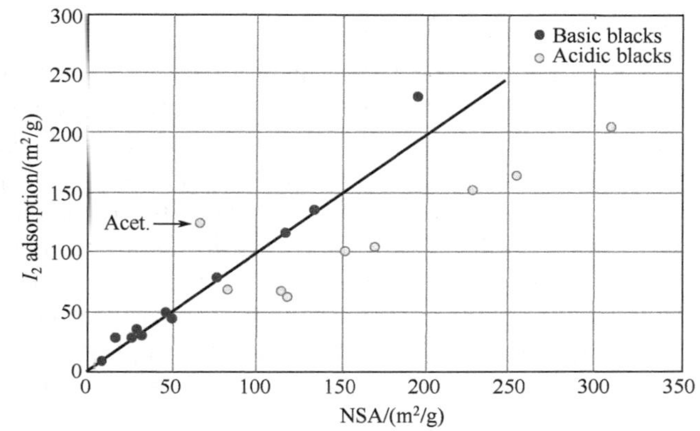


**Figure 2.20** Comparison between iodine numbers and NSA[71]

Iodine adsorption can be increased by the presence of porosity and depressed by surface impurities such as residual oil or oil used for beading[72]. The effects of tar content are shown in Figure 2.21. For a tar-free furnace black approaching a discoloration reading (percent transmittance of a benzene or chloroform extract) of 100 there is once more good agreement, but the iodine number falls below the true surface area as tarriness increases (transmittance decreases). Tar-free microporous blacks again show a discrepancy between the two surface areas, this time because iodine and nitrogen penetrate the pore space to different extents. The iodine number does not generally yield a true specific surface area and will not accurately reflect the state of subdivision of carbon black due to its sensitivity to both porosity and the surface chemical constitution of carbon blacks.

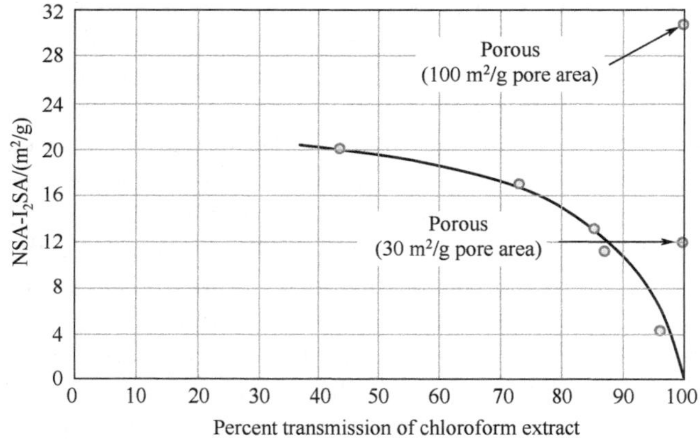


**Figure 2.21** Effect of incomplete de-tarring on the apparent surface area by iodine adsorption[72]

Although, some limitations of the iodine adsorption method have been recognized for some time, which were discussed by Kipling[73], it has been the standard ASTM method D1510 since 1957 and adopted by ISO and GB. It is still the most easily measured surface area estimate, and is used extensively, especially for process control at furnace black production and rubber factories.

#### 2.3.1.3.2 Adsorption of Large Molecules

Another method for surface area measurement is based on the adsorption of large molecules from an aqueous solution. For some adsorbate whose molecular size is too large to enter the micro-pore of filler, the adsorption should be an effective means for eliminating the microporosity problem to measure an external surface representative of the area accessible to rubber. According to Janzen and Kraus[72], with adsorption of large molecules, other criteria must be met. Beyond the requirement of chemical inertness toward the filler surface, which severely limits the choice of a suitable adsorbate molecule, it is necessary to meet the following conditions: 1) the adsorption isotherm exhibits a definite plateau corresponding to monolayer coverage, 2) the adsorption is sufficiently great to assure good precision, and 3) there is a reliable and convenient method of analysis. All these requirements are met by the adsorption of certain surfactants from an aqueous solution, such as cetyl trimethyl ammonium bromide (CTAB) and dioctyl sodium sulfosuccinate (Aerosol OT):

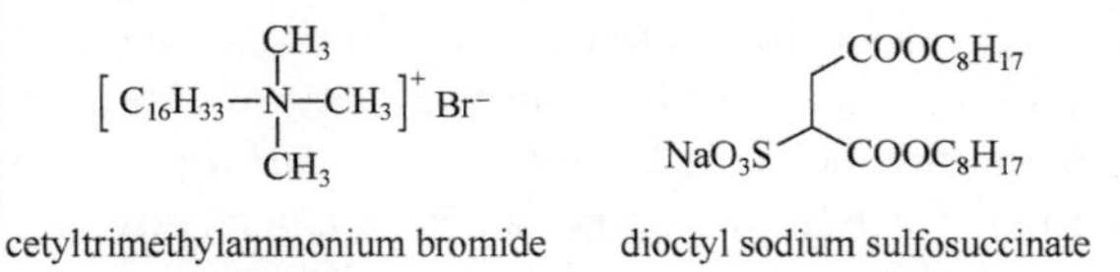


cetyltrimethylammonium bromide CTAB

dioctyl sodium sulfosuccinate Aerosol OT

Saleeb and Kitchener[74] were the first to report the adsorption isotherms of OT (short for Aerosol OT), an anionic surfactant, and CTAB, a cationic surfactant from aqueous solutions as adsorbates on six carbon blacks. Three of the blacks were original, and the others were their graphitized counterparts. From the adsorption isotherms of the surfactants, it was found that the adsorption is reversible over the experimental range. All the isotherms are of the general "Langmuir" form, with a long plateau. The data conforms approximately to the Langmuir isotherm within experimental accuracy. However, the high coverage is reached only at concentrations above the critical micellar concentration (c.m.c.). Figure 2.22 shows typical isotherms of CTAB on channel black Spheron 6, thermal black Sterling FT, and their graphitized counterparts. It is suggested that for both types of carbons, saturation from surfactant solution corresponds to a monolayer physically adsorbed with polar groups outward. The basal planes of graphite are non-polar, and adsorption of surfactant is due to escape of the non-polar group from the water phase. The presence of polar groups (the oxygen complexes) on the filler surface accounts for the decrease in adsorption on the original carbons.

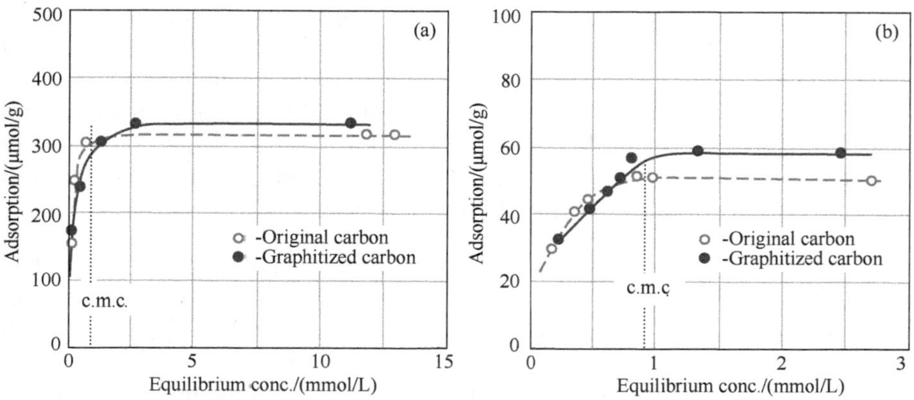


**Figure 2.22** Adsorption isotherms for CTAB from water: (a) Spheron 6; (b) Sterling FT[74]

### The State of CTAB and OT in Monolayer Adsorption

Dividing the NSA by the amount of CTAB on the graphitized carbons at plateau gives a surface area of approximately 40 Å$^2$ per molecule. This corresponds to the bulky trimethyl ammonium headgroup, rather than the cetyl paraffin chain. On the original channel black Spheron, CTAB adsorption is less packed. This suggests that interaction with pre-existing oxygen-polar groups restricts the freedom of packing of the surfactant. On the other hand, the affinity of the original carbons for this cationic surfactant is not less than that of the graphitized carbons, even somewhat greater. Therefore, some degree of salt formation occurs between the cationic surfactant and the anionic oxygen complexes. It may be that a small proportion of the cationic molecules are anchored firmly in "reverse orientation" (i.e., chemisorbed with chains outermost), while the majority then fill in as best they can by physical adsorption in the spaces left.

In the Aerosol OT molecule, these chains are double ethyl-hexyl groups and therefore appreciably compressible. The level of the plateau is therefore dependent on the thermodynamic activity of the solution and the activity coefficient of the adsorbed monolayer.

### Surface Area of CTAB and OT

From a surface area measurement view, two parameters are important with large molecule adsorption: $m_1$, the mass required to form a monolayer, and $\sigma_a$, the average area occupied by each molecule on the surface. The former is easy to obtain from the adsorption at isotherm plateau, and to calculate the latter, divide the NSA by the amount of surfactant adsorbed at "saturation".

Based on his work with six carbon blacks, Saleeb calculated a surface area of about 70 Å$^2$ per molecule of Aerosol OT on two graphitized thermal blacks. This is consistent with Abram's[75] value derived from a molecular-scale model in an unstrained configuration with the molecule oriented vertically. The apparently different areas occupied by OT on Spheron (channel) and Graphon (graphitized) are likely due to roughness or porosity of these surfaces, as nitrogen is able to penetrate where Aerosol cannot. This slightly higher area for the original blacks may be due to the slightly looser packing of the adsorbate molecules on their surface compared with their graphitized ones. This is especially true for Graphon black which may be related to its higher volatile or oxygen-containing groups and more developed surface porosity, resulting in less-than-perfect crystallization by heating than the Sterlings (furnace black).

With a series of carbon blacks, Abram and Bennett[76] calibrated two surfactant molecules for use in assessing the activity of carbon adsorbents. Careful examination of various measurements of the surface area of the blacks has allowed the determination of the surface areas occupied by OT (71 Å$^2$/molecule) and CTAB (44 Å$^2$/molecule), respectively. According to the analysis of the *t* plots, the smallest pore width that a CTAB molecule can penetrate must lie between 11 Å and 15 Å, and for OT, this value is 15 Å. On the basis of the area/molecule values, the effective molecular diameters of CTAB and OT are 7.5 Å and 9.5 Å, respectively.

In practice, both CTAB and OT have been used for routine characterization of carbon blacks. In laboratories, CTAB adsorption has been used successfully as a method for determining external specific surface areas[77]. In either case, a sample of carbon black is equilibrated with an aqueous solution of the surfactant, and then the black is separated by centrifugation or ultrafiltration. After the equilibrium liquid is analyzed for unabsorbed surfactant, the adsorption is determined from the difference in initial and final concentrations, provided the analysis is carried out in the monolayer plateau region. The amount of adsorption is simply proportional to the surface area of the carbon black. For analysis, CTAB and OT may be titrated against each other[72,78], so that the analysis is the same, no matter which is used as the adsorbate and which as the titrant.

It has been found that CTAB and OT are very nearly equivalent as adsorbates (both discriminate against micro-pores). The reasons for preferring CTAB as the adsorbate are that carbon black is more easily removed from its solutions by centrifugation, and that the titration of CTAB with OT is more convenient than the reverse, because better visual warning of the approaching end point can be obtained. There is also some indication that the CTAB molecule may be large enough for purposes of steric exclusion from micro-pores.

### Measurement of CTAB Surface Area of Carbon Black

In practice, the isotherm of CTAB adsorption on carbon black has a long plateau corresponding to a monolayer coverage of the filler surface from which the adsorbate is not sterically excluded. The CTAB adsorption by carbon black is independent of functional hydrogen- and oxygen-containing groups and residual tars, etc. Rapid equilibrium is achieved by using mechanical or ultrasonic vibration. After removal of the carbon black, measuring the unabsorbed CTAB in the solution with OT titration reveals the amount of CTAB adsorbed. The results are scaled by using the Industry Tint Reference Black (ITRB) as a primary standard and taking its accessible surface area to be exactly 83.0 m$^2$/g[79].

Comparing CTAB adsorption for external surface area with other techniques known to discriminate against microporosity provides further evidence for the validity of the CTAB method. One of these is, of course, the electron microscope method. Figure 2.23[72] shows that for the more easily counted large particle blacks, the electron microscope surface area is somewhat larger than the CTAB area, which is the expected result in view of the neglect in the electron microscope counting technique of area lost at the fusion points. This effect is more than compensated for in very small particle blacks, where difficulty of resolving individual particles from clusters is likely to lead to too small an electron microscope surface area.

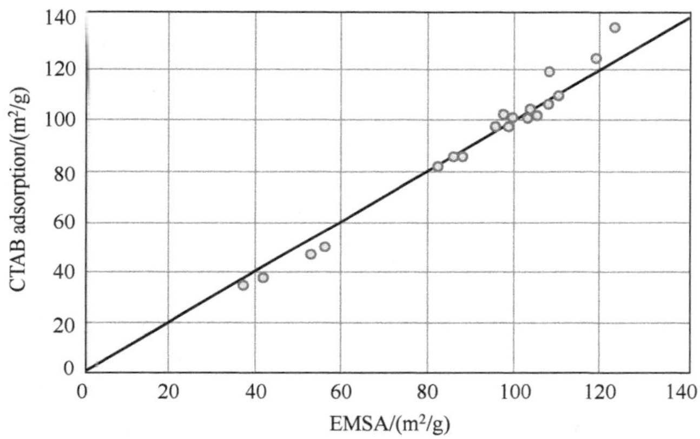


**Figure 2.23** CTAB surface area *vs.* electron microscope area[72]

More convincing is a comparison of the CTAB test with the absolute surface area method of Harkins and Jura[80]. This technique, in the form described by Kraus and Rollmann[81] and others[82], measures the heat of immersion of the black in *n*-hexane after it has been covered with a multilayer of *n*-hexane, the pores having been filled in the process. The heat liberated is the product of the surface area and the known surface enthalpy of *n*-hexane, provided the multilayer is thick enough to blanket the force field of the carbon black surface. Figure 2.24 shows Kraus's results of the correlation between CTAB and the Harkins-Jura areas[72]. The points divide themselves naturally into three groups: 1) the thermal and ordinary furnace blacks, for which agreement between the two methods is excellent; 2) the acetylene and channel blacks, which appear as a group to have high Harkins-Jura areas; 3) the conductive (SCF and CC) carbons, which tend to be slightly on the high side in CTAB area. It is not unreasonable to suspect that the Harkins-Jura experiment might give results that are too high for channel and other oxidized blacks. This would be possible if interactions between filler surface (especially polar groups) and immersion medium were incompletely shielded by the pre-adsorbed multilayer. That the heat of immersion per unit area of (hydrocarbon) adsorbate film on a pre-covered channel black might be greater than on a furnace black is not inconsistent with Wade and Deviney's calorimetric data[83]. The CTAB results for the very highly porous blacks of Group 3 are probably high because of the presence of pores larger than micro-pores. For Group

1 it is clear that the H.-J. and CTAB methods measure the same area. Because of this and its inherently superior precision, Kraus regards the CTAB technique as the most reliable direct method for accessible surface area.

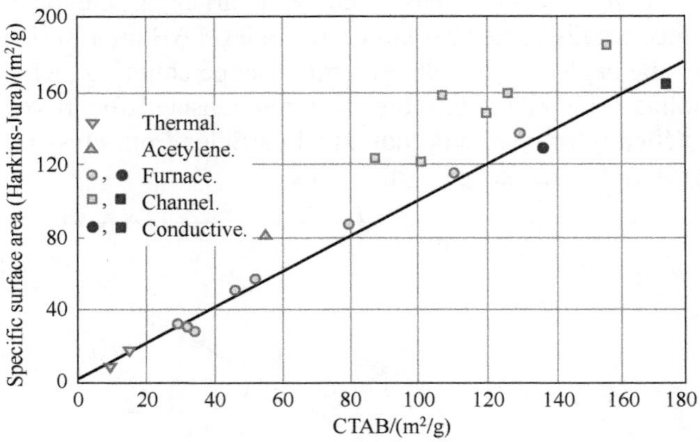


**Figure 2.24** Harkins-Jura surface area *vs.* CTAB surface area[72]

From the discussion of surface area measurement with gas adsorption, one can see that the *t*-area test measures a surface smoothed over the initial nitrogen adsorption below 0.5 relative pressure, whereas CTAB cannot penetrate pores smaller than 10−15 Å. Therefore, a question remains about which method should be used to measure the external area.

With 13 tire blacks spanning the range of 18 $m^2/g$ to 150 $m^2/g$ (CTAB), including IRB 3, Sanders[84] conducted a systematic study on the external surface area measurements, using Auto CTAB system for CTAB tests and the Gemini Surface Area Instrument 2370 (Micromeritics) for the *t*-area.

Figure 2.25 shows the correlation between CTAB and *t*-area (average of triplicates), with $R^2$ of 0.9995, slope of 0.988, intercept of 1.56, and RMS deviation from the regression line of 0.9 $m^2/g$.

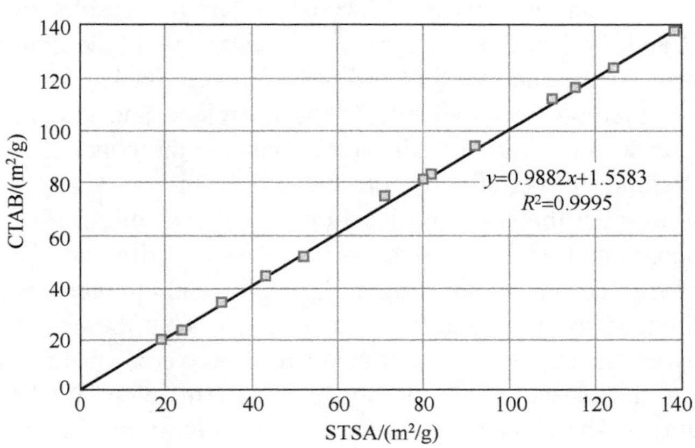


**Figure 2.25** CTAB surface area *vs.* STSA[84]

These data show many instances of higher CTAB surface area than STSA, which is more evident for low surface area blacks. This is consistent with the data presented by Magee[52], showing a somewhat overestimation of external surface area by the CTAB method, even though both authors used a different $t$–$(p/p_0)$ transform for $t$-area calculations. While Sanders used a de Boer $t$ transform generated on an alumina surface, the $t$ transform used by Magee is based the nitrogen adsorption on low surface area furnace black N762.

On the other hand, the CTAB test is based on normalization of all results to the standard black, and the ITRB (N330) with an assigned CTAB value of 83 $m^2/g$ are assumed to have zero microporosity[79]. However, the STSA value of the standard black shows some degree of microporosity[52]. The presence of this small amount of microporosity probably results in the differences between the STSA and CTAB surface areas.

In summary, both CTAB area and $t$-area are about the same as the measure of external surface area of carbon blacks. However, from the practical consideration, including test throughput, operator time, simplicity of use, and operating cost, the $t$-method has certain advantages.

#### Application of CTAB Adsorption to Measurement of Silica Surface Area

Voet, Morawski, and Donnet[85] were the first to use CTAB adsorption as a measure of external surface area of silica. Like carbon black, the isotherm for adsorption of an aqueous solution of CTAB on silica also has a long plateau, indicating a monolayer adsorption in its nature. This adsorption is independent of the surface chemistry. The test procedure is similar to that used for carbon black, but test conditions, such as pH of the buffer solution, are adjusted[86]. The surface area of silica is calculated by multiplying the amount of CTAB adsorbed by the area occupied by a CTAB molecule, which is 35 $Å^2$.

### 2.3.2 Structure – Aggregate Size and Shape

Aggregate morphology is another important characteristic that influences performance. The term “structure”, a somewhat ill-defined property, is widely used in the carbon black and rubber industries to describe the aggregate morphology. It was originally introduced in 1944[87] to describe the ensemble of aggregates, which is a stochastic distribution of the number and arrangement of the nodules that make up the aggregates.

Structure comparisons of grades with different surface areas cannot be made. It is known that the structure properties are associated principally with the bulkiness of individual aggregates. Aggregates of the same mass, surface area, and number of nodules have high structure in the open bulky and filamentous arrangement and a low structure in a more clustered compact arrangement. Therefore, the structure is now used to describe the relative void volume that characterizes grades of black having the same surface area. Structure is determined by aggregate size and shape, and their distribution. These geometrical factors affect aggregate packing and the volume of voids in the bulk material. Therefore, in composite systems, structure is also a principal feature that determines the performance of carbon black as a reinforcing agent and as a filler[20]. In liquid media, structure affects rheological properties such as viscosity and yield point. Also, in rubber, the filler structure affects viscosity, extrusion die swell of uncured compounds, modulus, abrasion resistance, dynamic properties, and electrical conductivity.

The structure of fillers can be measured with transmission electron microscopy and characterized by disc centrifuge photosedimentometry. Void volume measurement and tinting strength are measures that are only applicable to carbon black.

### 2.3.2.1 Transmission Electron Microscopy

The direct method for structure measurement of carbon black is TEM (ASTM D3849). This method is unique in furnishing information about the aggregate size, shape, and their distribution. Typical electron micrographs of rubber-grade carbon blacks and silicas are shown in Figure 2.7 and Figure 2.8, respectively (see Section 2.3.1.1). There is an enormous range in aggregate size.

The size of the aggregates is generally related to the size of the primary particles. The shapes of the aggregates have infinite variety from tight grape-like clusters to open dendritic or branched arrangements to fibrous configurations.

According to Hess, McDonald, and Urban[88], the aggregates are generally classified into four shape categories, as schematically shown in Figure 2.26.

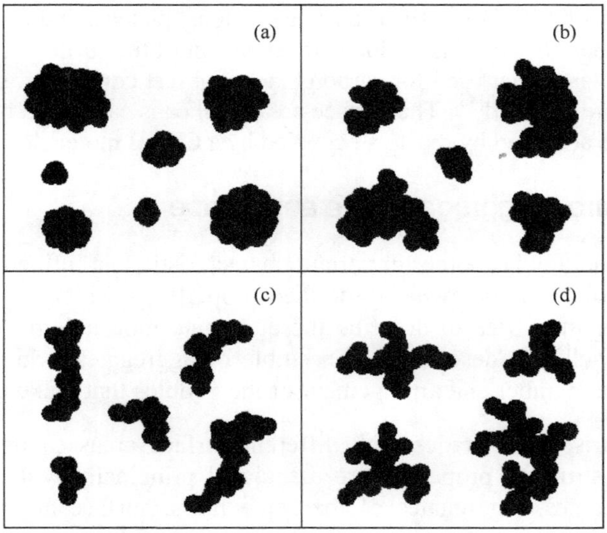


**Figure 2.26** Shape categories of carbon black aggregates: (a) spheroidal, (b) ellipsoidal, (c) linear, and (d) branched, as shown[88]

Although the parameters listed in the ASTM method are very useful for characterizing the filler morphology, the following parameters more fully depict the image of individual aggregates: area ($A$), perimeter ($P$), maximum diameter ($D_{max}$) and minimum diameter ($D_{min}$), $F_{shape}$ ($D_{min}/D_{max}$), convex perimeter ($C_{perim}$), and structure ($C_{perim}/P$).

Taking the two-dimensional image of a carbon black aggregate as an example (Figure 2.9, see Section 2.3.1.1), other than the definitions and the measurement of the aggregate area ($A$) and perimeter ($P$) that were described before, we have the definitions of the other parameters as follows:

$D_A$ – equivalent circle diameter as:

$$D_A = 2(A/\pi)^{1/2} \tag{2.71}$$

$D_{max}$ and $D_{min}$ – the longest diameter ($D_{max}$) and shortest diameter ($D_{min}$) of an object, obtained by selecting the longest and shortest of the Feret diameters measured in 32 directions (i.e., at an angular resolution of 5.7°), and $C_{perim}$ – convex perimeter of an object which is defined as the mean object diameter, multiplied by 3.1416 ($\pi$), as:

$$C_{perim} = \frac{1}{32}\sum_{1}^{32} (\text{Feret diameter})\,\pi \tag{2.72}$$

$C_{perim}$ can also be viewed as the length of a regular polygon with 64 corners, stretched around an object. The following figure shows this polygon, using only 6 corners for the purpose of drawing. Obviously, the actual $C_{perim}$ with 32 directions fits the shape much closer. It will always be smaller than the perimeter (perim). The ratio $P/C_{perim}$ is commonly used to assess the surface bulkiness of objects.

Accordingly, the parameters listed in ASTM 3849 can be calculated, such as the following:

$V_A$ – aggregate volume:

$$V_A = (8/3)A^2 / P \tag{2.73}$$

$V_P$ – particle volume, given by:

$$V_P = \pi d_p^3 / 6 \tag{2.74}$$

and $n$ – number of particles in the aggregate, which can be calculated by:

$$n = V_A / V_P \tag{2.75}$$

In addition, with images of several thousand aggregates, the linear dimensions can be derived as:

$\overline{P}$ – number average aggregate perimeter:

$$\overline{P} = \frac{\Sigma P}{N} \times \frac{k}{M} \tag{2.76}$$

where $P$ is aggregate perimeter (nm), $N$ = total aggregate count, $k$ = factor required to convert the instrumental aggregate dimensions to nanometers, and $M$ = linear magnification factor.

$\overline{A}$ – number average aggregate areas follows:

$$\overline{A} = \frac{\Sigma A}{N} \times \left(\frac{k}{M}\right)^2 \tag{2.77}$$

$\overline{V}$ – number average volumetric functions (for example, aggregate volume) as:

$$\overline{V} = \frac{\Sigma V}{N} \times \left(\frac{k}{M}\right)^3 \tag{2.78}$$

The following distributional parameters for particles and aggregates can also provide:

$n_t$ – for all of the aggregates measured, or total number of all particles, i.e., $\Sigma n$.

$m$ – mean particle size (nm)

$$m = \left[\Sigma\left(n d_p\right)\right] / n_t \tag{2.79}$$

$sd$ – particle size standard deviation (nm)

$$sd = \left[\Sigma n (d_p - m)^2 / (n_t - 1)\right]^{1/2} \tag{2.80}$$

$w_m$ – weight mean particle size (nm)

$$w_{\mathrm{m}} = \left[\Sigma\left(nd_{\mathrm{p}}^{4}\right)\right] / \left[\Sigma\left(nd_{\mathrm{p}}^{3}\right)\right] \tag{2.81}$$

$h_i$ – particle size heterogeneity index

$$h_{\mathrm{i}} = w_{\mathrm{m}} / m \tag{2.82}$$

$N_t$ – number of aggregates measured

$M$ – mean aggregate size (nm)

$$M = \Sigma D_{\mathrm{A}} / N_{\mathrm{t}} \tag{2.83}$$

SD – aggregate size standard deviation (nm)

$$\mathrm{SD} = \left[\Sigma\left(D_{\mathrm{A}} - M\right)^{2} / \left(N_{\mathrm{t}} - 1\right)\right]^{1/2} \tag{2.84}$$

$W_M$ – weight mean aggregate size (nm)

$$W_{\mathrm{M}} = \Sigma D_{\mathrm{A}}^{4} / \Sigma D_{\mathrm{A}}^{3} \tag{2.85}$$

And HI – aggregate size heterogeneity index

$$\mathrm{HI} = W_{\mathrm{M}} / M \tag{2.86}$$

For example, the morphologies of carbon black CRX 1436 and fumed silica LM150D from Cabot Corporation are characterized by the mean parameters derived from analyzing 2018 and 2013 aggregate images, respectively. The data are shown in Tables 2.3 and 2.4.

**Table 2.3** Aggregate parameters of TEM for carbon black CRX 1436

| Carbon black CRX1436 | Units | Arithmetic number | | Arithmetic mass | | Geometric number | | Geometric mass | |
|---|---|---|---|---|---|---|---|---|---|
| | | Mean | St. Dev. | Mean | St. Dev. | Mean | ln(St. Dev.)[a] | Mean | ln(St. Dev.)[a] |
| $A$ | nm$^2$ | 9221 | 13,740 | 37,240 | 31,330 | 4613 | 1.15 | 25,130 | 0.98 |
| $P$ | nm | 659 | 644 | 1798 | 1241 | 475 | 0.78 | 1417 | 0.72 |
| $D_A$ | nm | 91 | 59 | 199 | 88 | 77 | 0.57 | 179 | 0.49 |
| $D_{max}$ | nm | 156 | 112 | 354 | 180 | 127 | 0.62 | 308 | 0.55 |
| $D_{min}$ | nm | 95.6 | 68.2 | 216.8 | 107.2 | 78.0 | 0.62 | 189.5 | 0.55 |
| $F_{shape}$ | – | 0.63 | 0.13 | 0.63 | 0.13 | 0.61 | 0.21 | 0.62 | 0.21 |
| $C_{perim}$ | nm | 406 | 287 | 921 | 453 | 333 | 0.61 | 806 | 0.54 |
| Structure | – | 0.71 | 0.13 | 0.58 | 0.12 | 0.70 | 0.20 | 0.57 | 0.22 |

a. Geometric standard deviations reported in natural log units.

Calculation based on 2018 aggregates.

**Table 2.4** Aggregate parameters of TEM for fumed silica LM150D

| Fumed silica LM150D | Units | Arithmetic number | | Arithmetic mass | | Geometric number | | Geometric mass | |
|---|---|---|---|---|---|---|---|---|---|
| | | Mean | St. Dev. | Mean | St. Dev. | Mean | ln(St. Dev.)[a] | Mean | ln(St. Dev.)[a] |
| $A$ | $nm^2$ | 9139 | 8123 | 18,563 | 13,129 | 6168 | 0.97 | 14,598 | 0.73 |
| $P$ | nm | 907 | 645 | 1511 | 823 | 704 | 0.75 | 1299 | 0.57 |
| $D_A$ | nm | 99 | 44 | 145 | 51 | 89 | 0.48 | 136 | 0.36 |
| $D_{max}$ | nm | 185 | 97 | 280 | 119 | 160 | 0.56 | 256 | 0.43 |
| $D_{min}$ | nm | 111 | 56 | 165 | 64 | 97 | 0.55 | 153 | 0.41 |
| $F_{shape}$ | – | 0.62 | 0.13 | 0.61 | 0.13 | 0.60 | 0.22 | 0.60 | 0.22 |
| $C_{perim}$ | nm | 479 | 243 | 723 | 287 | 417 | 0.55 | 666 | 0.42 |
| Structure | – | 0.61 | 0.14 | 0.52 | 0.11 | 0.59 | 0.23 | 0.51 | 0.21 |

a. Geometric standard deviations reported in natural log units.

Calculation based on 2013 aggregates.

### 2.3.2.2 Disc Centrifuge Photosedimentometer

The disc centrifuge photosedimentometer (DCP), i.e., a disc centrifuge[89,90] with an optical detector, is an excellent instrument for the measurement of the particle size and the breadth of a typical distribution, covering more than one decade in size. DCP has been used for measuring the size distribution of carbon black aggregates for a long time[91]. The theory of the technique is described in the literature[92,93] reviewed by Weiner, Tscharnuter, and Bernt[94].

There are two techniques used with a DCP: one is line start, the other is homogeneous start. In the measurement of rubber fillers, the former is generally applied. In this technique, a small volume of the dispersion is injected onto the meniscus of a rotating, hollow disc filled with liquid as shown in Figure 2.27. The liquid is called the spin fluid. Typically, it contains a gradient to prevent hydrodynamic instabilities, surfactants to match those in the dispersion, and a very thin layer of a low vapor pressure liquid to prevent evaporative cooling and thermal inversions. After injection, assuming the particle density is larger than the liquid density, all the particles of various sizes are forced radially outward through a spin fluid under high centrifugal force. The particles settle at rates determined by their sizes and densities. At a specific radial distance, the particles interrupt a light beam and the particle size and relative concentrations of the particles are calculated from known parameters. Particle settling is described by Stokes' Law for centrifugation. For a sphere, the time, $t$, that it takes for the particle to travel from the meniscus, with radial distance $R_0$, to the detector, at a radial distance $R$, is given by:

$$t = \frac{18\eta_f \ln(R / R_0)}{\omega^2 \Delta\rho D_p^2} \tag{2.87}$$

where $\eta_f$ is the viscosity of spin fluid, $\omega$ the angular velocity of the disc, $D_p$ the diameter of spherical particle, and $\Delta\rho$ the difference in density between the particle and the fluid, which is:

$$\Delta\rho = \rho_p - \rho_f \tag{2.88}$$

where $\rho_p$ and $\rho_f$ are the densities of particle and fluid, respectively. The $R_0$ is easily determined by the volume of spin fluid used in the disc.

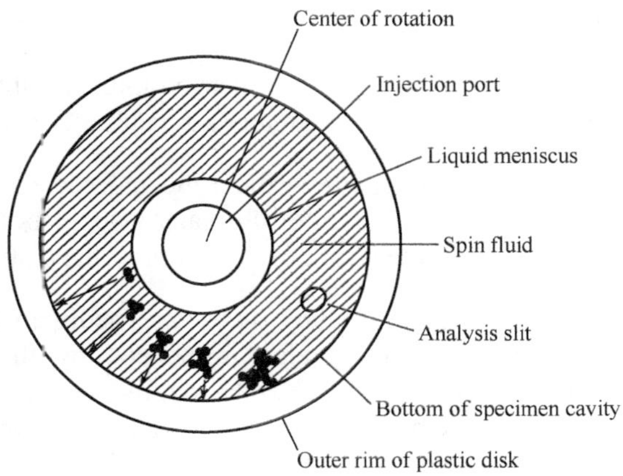


**Figure 2.27** Disc of DCP, Schematic

For a sphere whose density is higher than that of the spin fluid, the particles sediment radially outward and the detector is placed near the internal perimeter of the disc. The diameter of the sphere can be derived from the sedimentation rates in the gravitational field according to the Stokes' law (Eq. 2.87). For a non-spherical aggregate of carbon black in water, as $\Delta\rho > 0$, an equivalent Stokes diameter, $D_{st}$ is determined as the diameter of a carbon black sphere with the same settling behavior.

However, to complete the particle size distribution, the amounts of the material at different sedimentation times, $t$, are required. This is calculated from the turbidity, $\tau$, which is obtained from the measured output of the detector with and without particles.

The differential volume distribution (equivalent to the differential weight distribution assuming all particles have the same density) is given by:

$$\frac{dV}{dD_p} = \frac{C\tau}{Q_{ext}} \tag{2.89}$$

where $C$ is a constant for a fixed detector position. $Q_{ext}$ is the extinction efficiency that is calculated from Mie scattering theory, given the wavelength of the light, the particle size (determined by Stokes' law above), and the refractive index of the particle and liquid. The calculation assumes a sphere. Even though carbon black aggregates are not spheres, it is assumed that strong absorption is more important than shape in determining the extinction correction.

The weight cumulative undersize and differential aggregate size distributions of a carbon black are shown in Figure 2.28. For characterizing the structure of carbon blacks, some parameters can be derived from the measurement and defined as follows:

$D_{mode}$ – most frequently occurring aggregate size – the highest point on the weight distribution curve;

$\Delta D_{50}$ – width of the distribution at 50% of the maximum peak height;

$D_{st}$ – Stokes diameter – median value of the distribution, i.e., the aggregate size at which 50% of the total sample weight is of greater size and 50% is of lesser size;

L.Q. and U.Q. – lower quartile and upper quartile – the aggregate size at which 25% of total sample weight is of lesser size and 25% of greater size, respectively, and quartile ratio – the ratio of U.Q. to L.Q. – a measure of the width of the distribution.

It should be pointed out that although the large aggregates sediment at a faster rate than smaller ones, the sedimentation rate is also influenced by the bulkiness of the aggregates. At constant volume or mass, a bulky aggregate sediments more slowly than a compact aggregate because of frictional drag. The DCP curve is characteristic of the black structure but the measured diameters need to be viewed carefully with a different method such as TEM analysis.

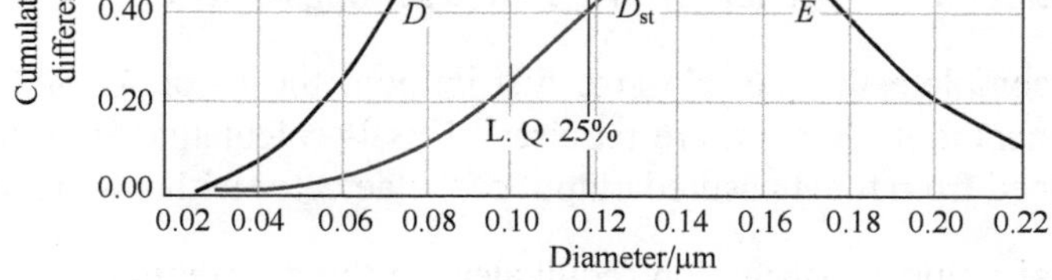


**Figure 2.28** Mass cumulative undersize and differential aggregate size distribution

### 2.3.2.3 Void Volume Measurement

Filler aggregates vary in shape from the individual spheroidal particles. The presence of these more complex shapes creates internal voids within any given bulk sampling of fillers, which are much greater than those that occur in a simple packing of spheres. It is not surprising then that the most commonly used techniques for measuring carbon black "structure" have been based on internal void volume using absorptive measurements or volumetric measurements under specific pressure conditions. In fact, the amount of void at the maximum packing has almost become synonymous with filler structure. At least two approaches are widely used. The first determines the amount of liquid that is needed to just fill all the spaces between aggregates when the aggregates are pulled together by the surface tension forces of that liquid. Compressed volume is a second procedure for determining the void volume within the filler aggregate structure, which is based on the volume of a given weight of carbon black at a specified level of pressure.

#### 2.3.2.3.1 Oil Absorption

The void volume or the bulkiness is generally measured by a vehicle absorption test using an oil absorptometer equipped with a constant rate burette. This is based on the change in torque during mixing of filler and the liquid. The liquid is gradually added to the dry powder, which remains dry as it soaks up the liquid, until the liquid fills the void volume within the aggregates and most of the void volume between the aggregates. At this point, the crumbly mass starts to cohere, and there is a sharp increase in viscosity of the mixture as it changes from free-flowing powder to a semi-plastic continuous paste.

The viscosity, hence the torque, will drop as the liquid is continually added, due to the lubrication effect. The volume of the liquid needed for a unit mass of fillers to reach a predetermined level of torque is termed the oil number (Figure 2.29).

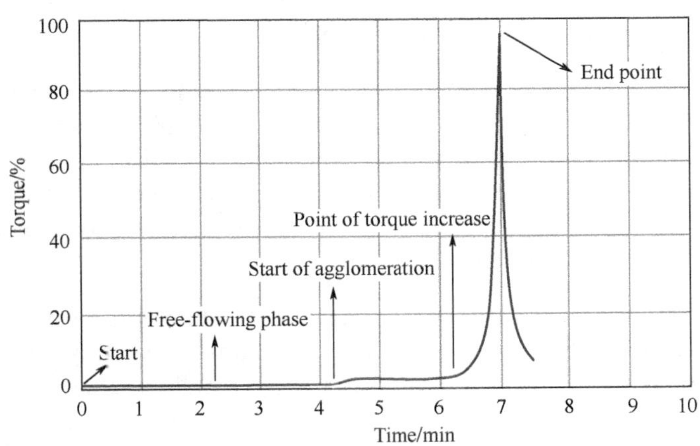


**Figure 2.29** The change in torque of the absorptometer during addition of oil[95]

#### Oil Absorption Number of Compressed Sample

The oil absorption numbers are somewhat influenced by the post treatment of the filler production processes. In addition, the tests per se are sensitive to shear conditions and to previous densification. Therefore, a cumbersome procedure of multiple compressing is recommended to bring the filler to a more standardized state of densification prior to running the oil absorption test[96].

In this test the carbon black is compressed using a mechanical press that crushes a carbon black sample in a cylinder at 165 MPa. The compression process is repeated four times. Each compression is broken up prior to the next compression. The oil absorption number is then measured. The results for the compressed samples are more related to the carbon black structure in the rubber compounds as a considerable number of filler processing and filled-compound processing steps commonly used, such as pelletization of carbon black and mixing of filled compound, are very dependent on the amount of mechanical work done on the materials. However, they vary from company to company, from plant to plant, and sometimes even from hour to hour. This test simulates the typical breakdown a carbon black may encounter during filler pelletization and mixing with rubber. Upon compression, the high severity of packing conditions will cause a further increase in packing density, owing to improved fit of the tentacles and increased interpenetration of the aggregates, both of which may be regarded as a diminution of the effective void value, and also owing to fracture of the aggregates. Therefore, the difference between conventional oil absorption (OAN) and compressed absorption (COAN), which is referred to as ΔOAN, is an indicator of the structure's stability. Typically, the higher structure fillers show a greater reduction in oil absorption after compression than do lower structure fillers. Another factor that can affect the amount of breakdown is the linearity of the carbon black structures. More linear structures typically do not break down as much as those exhibiting branching.

The ΔOAN is also an indicator of the void volume between the sample as received and the sample compressed. For the same grade of carbon black, dispersing the filler in rubber compounds is easier and superior with higher ΔOAN. This will be discussed later in Chapter 4 in relation to filler dispersion.

In the carbon black industry, while dibutylphthalate (DBP) is commonly used as the vehicle, paraffin oil has also been suggested as the liquid to replace DBP due to environmental protection issues. The experimental results show some different absorption numbers between these two oils[97]. The average bulkiness of individual aggregates is calculated by making suitable allowances for the volume of liquid between the aggregates at the end-point.

With other reinforcing fillers, the measurement of structure is less satisfactory. Silicas, both fumed and precipitated, give very high endpoints, apparently because the inter-aggregate agglomeration is so strong that it is not fully broken down in the test procedure. In fact, these fillers cannot be tested with the compressed samples under the pressure applied, as the aggregates are severely damaged, and the filler is highly densified, resulting in very hard discs.

### Calculation of DBP Absorption from TEM

According to Medalia[98], the DBP absorption can be calculated based on the TEM image analysis. In other words, the DBP number should have some physical meaning related to the size and shape of carbon black. The following paragraphs will detail Medalia's theoretical considerations and practical treatments of this issue.

### Equivalent Sphere Model for a Single Aggregate

Assuming that the TEM projected area $A$ of an aggregate as measured by TEM has diameter $D$ of an equivalent sphere, such that:

$$D = (4A/\pi)^{1/2} \tag{2.90}$$

The volume of the equivalent sphere, $V_{es}$, is:

$$V_{es} = \frac{\pi D^3}{6} = \frac{\pi}{6}\left(\frac{4A}{\pi}\right)^{3/2} = \frac{4A^{3/2}}{3\sqrt{\pi}} \tag{2.91}$$

The volume of solid carbon within an aggregate (or equivalent sphere) $V_a$ is:

$$V_a = N_p V_p \tag{2.92}$$

where $N_p$ is the number of (primary) particles in the aggregate and $V_p$ is the volume of each particle. The theoretical treatment of the TEM data is based on the assumption that the relation between the measured (two-dimensional) and calculated (three-dimensional) properties of an aggregate is the same as for a randomly oriented simulated floc – at least on a statistical basis. For a simulated floc, Medalia et al.[99,100] determined that:

$$N_p = (A/A_p)^a \tag{2.93}$$

or

$$A = A_p N_p^{1/a} \tag{2.94}$$

where $A_p$ is the projected area of a single particle and $a$ is an exponent which was found[99] to be equal to 1.15. Since:

$$V_p = \frac{4A_p^{3/2}}{3\sqrt{\pi}} \tag{2.95}$$

$$V_{es} = V_p N_p^{1.5/a} = V_p N_p^{1.305} \tag{2.96}$$

Calculation of DBP absorption from electron microscopy involves the calculation of void ratios, as discussed below. The void ratio, $e_{floc}$, of a single aggregate, or floc, is the void space within the equivalent sphere divided by the volume of solid carbon:

$$e_{floc} = \frac{V_{es} - V_a}{V_a} = N_p^{(1.5/a)-1} - 1 = N_p^{0.305} - 1$$

$$= \left(\frac{A}{A_p}\right)^{1.5-a} - 1 = \left(\frac{A}{A_p}\right)^{0.35} - 1 \tag{2.97}$$

As illustrated above, it is important to determine the particle size of each aggregate. Though at first all the aggregates were assumed to have the same particle size[100], it was later observed that aggregates with small projected areas tend to be made up of small particles, whereas large aggregates tend to be made up of large particles. Within each aggregate size range, there is a broad range of particle sizes.

### Void Ratio of a Bulky Sample

Determining the critical filler volume concentration, or end point, of an oil or DBP absorption titration involves filling the void spaces with vehicle, bringing about good packing by shear, and wetting of filler surface with vehicle. At the "soft ball" end point the equivalent spheres of carbon black are close-packed and essentially all void spaces are filled with vehicle. The total quantity of vehicle is made up of two things: (a) the amount of vehicle required to fill the void spaces within the equivalent spheres; and (b) the amount of vehicle required to fill the void spaces between the equivalent spheres.

The void ratio of the bulk sample, $\epsilon$, is defined as:

$$\epsilon = \frac{\text{Volume of Voids between Equivalent Spheres}}{\text{Volume of Carbon}} \tag{2.98}$$

This ratio of carbon black derived from packing of spheres which exists at the "soft ball" end point, is shown in Eqs. 2.99–2.101:

$$e = \frac{\begin{array}{c}\text{Volume of Voids within Equivalent Spheres}\\ +\text{Volume of Voids between Equivalent Spheres}\end{array}}{\text{Volume of Solid Carbon}} \tag{2.99}$$

$$e = \frac{\Sigma(V_{\text{es}} - V_{\text{a}})}{\Sigma V} + \frac{\text{Volume of Voids between Equivalent Spheres}}{\text{Volume of Equivalent Spheres}} \times \frac{\text{Volume of Equivalent Spheres}}{\text{Volume of Solid Carbon}} \tag{2.100}$$

$$e = \frac{\Sigma V_{\text{es}}}{\Sigma V_{\text{a}}} - 1 + \epsilon \frac{\Sigma V_{\text{es}}}{\Sigma V_{\text{a}}} = \frac{\Sigma V_{\text{es}}}{\Sigma V_{\text{a}}}(1+\epsilon) - 1 \tag{2.101}$$

In their TEM studies, Medalia et al. measured the projected area of individual aggregates, and also estimated the mean diameter of the primary particles within each aggregate. In a given run of over 200 aggregates, the computer output gives the value of $\Sigma (A^{3/2})$, from which they calculate:

$$\Sigma V_{\text{es}} = \frac{4}{3\sqrt{\pi}} \Sigma\left(A^{3/2}\right) \tag{2.102}$$

By combining Eqs. 2.92 and 2.93, the volume of carbon in each aggregate is also calculated as:

$$V_{\text{a}} = \left(\frac{4A}{\pi d_p^2}\right)^a \left(\frac{\pi d_p^3}{6}\right) \tag{2.103}$$

and the $\sum V_a$ can also be obtained. Thus, the experimental data required for calculation of Eq. 2.101 is available from the image analysis of TEM.

The theoretical treatment given so far is based on flocs made up of spherical primary particles in point contact. In actual carbon black aggregates, there is extensive interparticle fusion, so that most particles have the appearance of nodules rather than spherical particles (Figure 2.30).

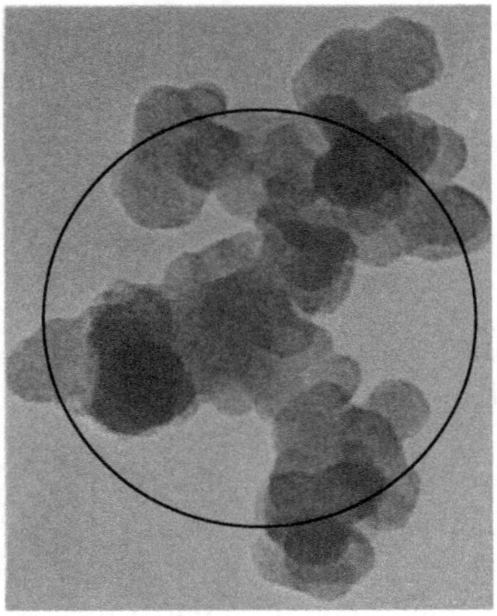

**Figure 2.30** Electron micrograph of aggregate of Vulcan 3 (N330), with circles of same projected area[98]

Based on the correction for interparticle fusion[99,100], size and packing of equivalent spheres[101,102] and anisometry of aggregates, when the projected areas of randomly oriented aggregates are taken, the void ratio of unit volume of carbon black, $e_c^*$, will be

$$e_c^* = \left(\frac{\sum V_{es}}{\sum V_a}\right)\left(\frac{1+\epsilon}{C}\right)g - 1 \tag{2.104}$$

where $C$ is correction factor for interparticle fusion and $g$ is anisometry correction factor.

### Calculation of DBP Number from TEM Image

From TEM image analysis, the area of each aggregate, $A$, is easily measured, and an average particle size is determined for each aggregate by measuring the discernible particles within it. Accordingly, the values of $V_a$ and $A^{3/2}$ for each aggregate, and the totals for the entire test run, $\sum V_a$ and $\sum A^{3/2}$, are calculated from which, the $\sum V_{es}$ can be derived by Eq. 2.102. The justification for this procedure is that individual aggregates are generally made up of nearly same-size particles, or at least of a narrower size distribution than that of the material as a whole. Table 2.5 gives the experimental values of the ratio $\sum V_{es}/\sum V_a$ as measured for eight rubber blacks varying in structure and particle size. From this ratio, the void ratio, $e_c^*$, is calculated in accordance with Eq. 2.104, with $\epsilon = 0.46$, $C = 1.4$, and $g = 0.94$.

**Table 2.5** Calculation of DBP absorption from TEM images[98]

| Carbon black | ASTM classification | $\left(\frac{\Sigma V_{es}}{\Sigma V_a}\right)$ | $e_c^*$ | $DBP_{Abs}$ | $e_{DBP}$ | $DBP_{EM}$ | $\left\{\frac{DBP_{EM}}{DBP_{Abs}}\right\}$ |
|---|---|---|---|---|---|---|---|
| Regal 660 | N219 | 3.233 | 2.17 | 72.4 | 1.55 | 66.6 | 0.92 |
| Regal 300 | N326 | 3.92 | 2.84 | 73.1 | 1.56 | 90.6 | 1.24 |
| Spheron 6 | S301 | 3.293 | 2.23 | 97.5 | 2.09 | 68.7 | 0.70 |
| Vulcan 3 | N330 | 4.385 | 3.30 | 97.9 | 2.09 | 106.9 | 1.09 |
| Sterling SO-1 | N539 | 4.100 | 3.02 | 97.0 | 2.07 | 97.0 | 1.00 |
| Vulcan 6 | N220 | 4.959 | 3.86 | 117.4 | 2.51 | 127.1 | 1.08 |
| Vulcan 3H | N347 | 5.039 | 3.94 | 128.9 | 2.76 | 129.9 | 1.01 |
| Vulcan 6H | N242 | 5.602 | 4.49 | 139.5 | 2.98 | 149.6 | 1.07 |

In Table 2.5, the void ratio, $e_{DBP}$, is calculated with the experimentally measured DBP absorption. As mentioned before, at the end point of a DBP titration with the absorptiometer, the black-DBP mixture is a crumbly mass containing an indeterminate amount of air. However, if somewhat more DBP is added, the mass can be rolled into a soft ball containing few visible air pockets. It was found that the air content of such balls formed from DBP and rubber blacks is approximately 2% of the volume of the DBP. The "soft ball" end point requires approximately 13% more DBP than the instrumental end point[95] for rubber blacks. Adding these two corrections, it is concluded that the volume of DBP required to reach the instrumental end point is about 15% less than the volume of voids. To convert to a volume basis, the density of carbon black (1.86 g/cm$^3$) must also be factored in. Thus,

$$e_{DBP} = (DBP_{Abs})(1.15)(1.86)/100 = DBP_{Abs} \times 0.02139 \tag{2.105}$$

where $DBP_{Abs}$ is the DBP absorption (absorptometer) in cubic centimeters/100 g.

As shown in Table 2.5, the values of $e_{DBP}$ are all lower than those of $e_c^*$. In order to bring about better agreement, an empirical correction factor must be introduced in Eq. 2.104. To calculate this factor, it is noted that the mean ratio of $(e_{DBP} + 1)/(e_c^* + 1)$ in Table 2.5 is 0.765. Thus, the data can be brought into best agreement by multiplying $\Sigma V_{es}$ by 0.765 in Eq. 2.104. This correction leads to an empirically corrected void volume of:

$$e_c' = \frac{0.765 \Sigma V_{es}}{\Sigma V_a} \times \frac{1.46}{1.4}(0.94) - 1 \tag{2.106}$$

and a calculated DBP absorption from TEM of:

$$DBP_{EM} = \frac{e_c' \times 100}{1.86 \times 1.15} = \frac{e_c'}{0.02139} \tag{2.107}$$

As expected, this correction gives values of $DBP_{EM}$ which are in close agreement with the measured values. The mean value of $DBP_{EM}/DBP_{Abs}$ is 1.015 and the standard deviation is 0.16.

The results of Table 2.5 are based on eight runs, covering blacks of low, normal, and high structure and of an approximately twofold range in DBP absorption and average particle size. In order to bring the electron microscope results into close agreement with the DBP titration data, it was necessary to multiply $\Sigma V_{es}$ by an empirical correction factor of 0.765. This factor may be accounted for by modifying the original assumption that the equivalent sphere is of exactly the same projected area as the aggregate, and assuming instead that the volume of the spheres is 0.765 times the volume originally assumed. This corresponds to a diameter of 0.915 (= $0.765^{1/3}$) times the original diameter, or in other words an 8.5% reduction in $D$. Such a reduction does not seem unreasonable in view of the uncertainties in the floc simulation work and in the analysis of packing.

An alternative explanation for the correction factor could be in the value of the exponent $a$ in Eq. 2.93. It is clear from Eq. 2.97 that the void ratio calculated for a single floc is quite sensitive to the value of $a$. For a given floc, the calculated value of $(e_{floc} + 1)$ is proportional to $A^{1.5-a}$; thus, an increase in $a$ will lead to a smaller calculated value of $e_{floc}$.

The calculation of DBP absorption from TEM image analysis which was developed by Medalia is based on detailed knowledge of carbon black morphology, and has further improved our understanding on carbon black properties in its application in rubber reinforcement.

#### 2.3.2.3.2 Compressed Volume

Another procedure for determining the void volume within the carbon black aggregate structure is based on the volume of a given weight of carbon black at a specified level of pressure. It is an objective method, as the actual density of a carbon black is known, the void volume at a given compression may be calculated from the bulk density of the powder by taking into account the constant black volume. This calculated void volume is a true value and, unlike other methods, is not affected by particle size[103].

The equipment and procedure of test is quite simple. The main parts of the compression system are a cylinder and two plugs, which fit loosely and so permit air to escape and give relatively little friction. The lower plug rests on a steel pad; the upper plug is driven by a piston of the same diameter attached to a pressure generator. The pressure applied to the piston can be read during the test. In a typical compression test, a few grams' sample of black is placed in the cell and pressure is applied on the piston. The height of carbon black between the two plugs quickly becomes constant. The compressed volume of carbon black is calculated by:

$$V_A = h \times 3.1416 D^2 / 4000 \qquad (2.108)$$

where $V_A$ is the (actual) compressed volume of the weighted black sample (cm$^3$), $h$ the height of the compressed carbon black in cylinder (mm), and $D$ the diameter of the cylinder (mm).

The theoretical volume of the carbon black is evaluated by:

$$V_T = m / d_{CB} \tag{2.109}$$

where $V_T$ is the theoretical volume of the weighed sample (cm$^3$), $d_{CB}$ the accepted density of carbon black, which is 1.90 g/cm$^3$, and $m$ the mass of weighed candidate black sample (g).

The void volume of the candidate carbon black per unit mass (100 g) is given by:

$$V_{(v)} = V_A - V_T \tag{2.110}$$

where $V_{(v)}$ is the void volume of the carbon black in cm$^3$/100 g.

The void ratio $e$ of the carbon black is:

$$e = (V_A - V_T) / V_T = V_{(v)} / V_T \tag{2.111}$$

Although this method was originally adopted in ASTM 6086, several studies have been carried out since then. In 1946, Benson, Gluck, and Kaufmann[104] demonstrated a relationship between compressibility and structure as part of their study on "the electrical resistivity" of dry blacks. Later, Studebaker[71] measured the specific volume of rubber-grade carbon blacks at a fixed pressure of 5.06 MPa, and the values showed a good correlation with mineral oil absorption. Mrozowski, Chaberski, Loebner, and Pinnick[105] carried out a detailed study about the effects of test conditions on experimental results at much higher pressure. At about the same time, Medalia and Sawyer's comprehensive report[103] outlined the mechanism of carbon black compression with different test conditions and various types of carbon blacks. Voet and Whitten[106,107] broadened the range of rubber-grade and other types of carbon blacks using a wide range of pressures.

##### 2.3.2.3.2.1 Relation between Compressibility and Pressure

With a stainless steel cylinder, Medalia and Sawyer carried out multipoint measurements at pressures up to 51.7 MPa on carbon blacks, giving the typical plots shown in Figure 2.31. The data are plotted semi-logarithmically, giving good straight lines that afford a convenient representation of the data. The line for acetylene black with an extremely high oil absorption number of 337 mL/100 g crosses the lines for three other blacks; thus, comparison of results with other carbon blacks at a single pressure would give different interpretation depending on whether high or low pressure was chosen. For furnace and thermal blacks of high, normal and low structure, the semi-logarithmic plot gives an excellent straight line from 5.7 MPa to 207 MPa. These data were obtained with pelletized blacks. Fluffy blacks deviate at low pressures but give almost the same results as pellets at high pressures.

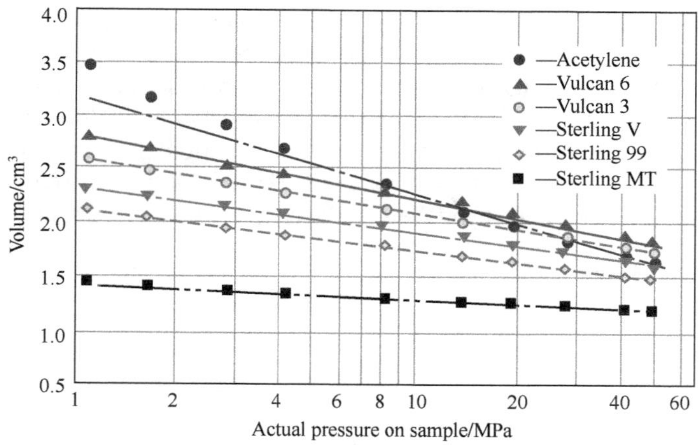


**Figure 2.31** Cartesian coordinate semi-logarithmic plot of compressibility data[103]

Voet and Whitten[106,107] applied the Mrozowski technique to a broad range of rubber-grade and other types of carbon blacks using a wide variety of pressures. Specific volume exhibited a linear relationship to the logarithm of pressure except at very low pressures below 0.07 MPa or at very high pressures above 69 MPa. This relationship is illustrated in Figure 2.32 for eight different rubber-grade carbon blacks ranging from the N200 to N700 particle size levels. The higher structure carbon blacks clearly show the highest specific volume at any given pressure, and also the sharpest drop-off in volume with increasing pressure. The plots could be extrapolated to a common point at about 90 MPa. At this point, the void volumes are approaching the calculated value for a random packing of contacting spheres of the same size. Voet et al. concluded that a pressure of about 9.79 MPa was a good single point measurement for comparing specific volumes among different grades of carbon blacks.

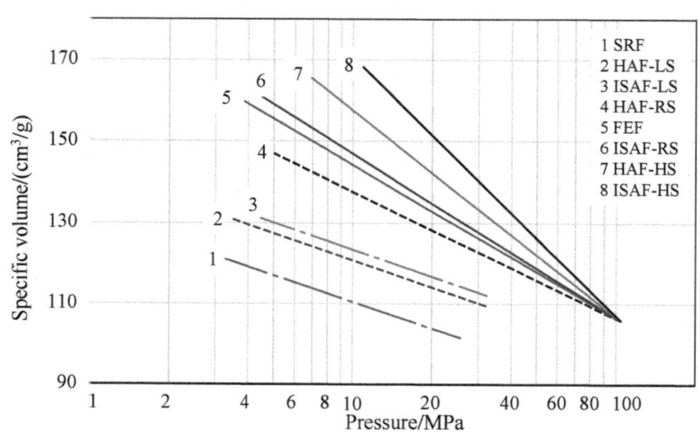


**Figure 2.32** Specific volume-pressure relationship for carbon blacks[106,107]

Since the relationship between compressed volume and the logarithm of pressure is linear over a limited pressure range, the void ratio, *e*, for each carbon black can also be expressed as:

$$e = A - B \lg p \tag{2.112}$$

where $p$ is the pressure and $A$ and $B$ are constants representing the intercept and slope, respectively, of the linear plot. Medalia et al. found that for a wide range of furnace and thermal blacks, there is a fairly good correlation between the oil absorption and the compressibility parameter, $B$ (Figure 2.33). This relationship holds for blacks of the same particle size but widely different structures [the squares of Figure 2.33 represent various experimental blacks of the particle size for ISAF (N220) black], as well as for blacks of comparable structure but different particle sizes [such as HAF-LS (N326) and SAF (N660)].

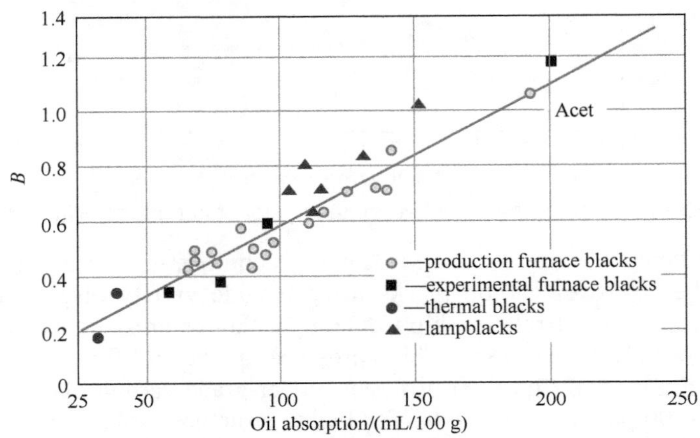


**Figure 2.33** Relation between compressibility slope parameter $B$ and oil absorption for furnace and thermal black and lampblacks[103]

From the straight lines of the semi-logarithmic compressibility plots (extrapolated if necessary), void ratios can be calculated at various pressures. As shown in Figure 2.34, the void ratios calculated from compressibility at 0.69 MPa (extrapolated) are fairly close to those calculated from oil absorption (dashed line of Figure 2.34) for all blacks shown except the two of highest structure. As it goes to higher pressures, the void ratio from compressibility becomes less than that from oil absorption for blacks of progressively lower structure[71].

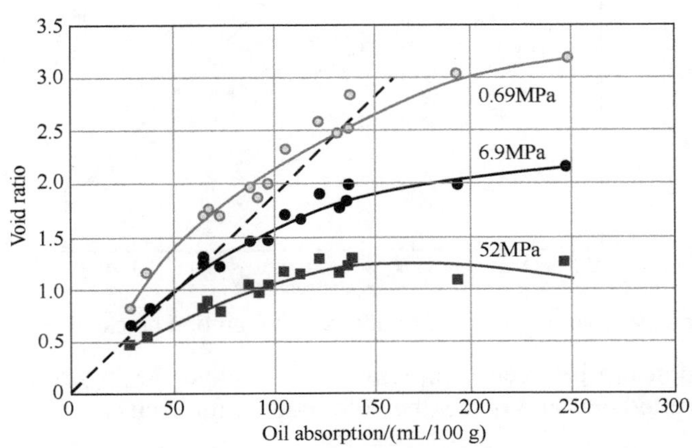


**Figure 2.34** Void ratio ($e$) at indicated pressures *vs.* oil absorption for production and thermal blacks[103]

Evidently, the void ratio at a single pressure can be used as an alternative to *B* as a measure of structure, but in order to be useful over the entire structure range of blacks, the void ratio would have to be taken at a fairly low pressure, no higher than 7 MPa, and possibly around 2–4 MPa. At these low pressures, the measured void ratio is somewhat dependent on initial bulk density and possibly also on pellet hardness, so using *B* as a measure of structure is considered better than the void ratio at a single pressure.

#### 2.3.2.3.2.2 Mechanism of Compression

Based on Medalia's work, it is believed that compression leads to a breakdown of filler agglomerates, followed by packing of the fragments thus formed. By simply releasing the pressure after compressing to 52 MPa, the compressed samples expand slightly, but even on complete removal of the applied pressure the volume of the black is much less than before compression (Figure 2.35). The elastic recovery is greatest for the blacks of highest structure, demonstrating that it is due to spring-like action of aggregates rather than elastic compression of primary particles, as shown also by Mrozowski et al[105].

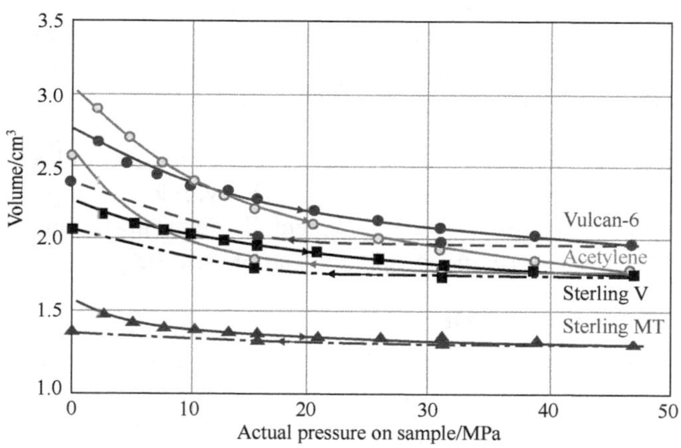


**Figure 2.35** Recovery on release of pressure[103]

Packing of a powdered material can be facilitated by vibrating or tapping. Figure 2.36 shows typical tapping results during compression. After each increase in pressure, a volume reading was taken without tapping; then the cylinder was struck, either by hand or using an electric masonry hammer, until no further change in volume occurred. Evidently, tapping under a given pressure brings about an additional compaction or compression of the black beyond the given static pressure alone.

In Figure 2.36, straight (dotted) lines are drawn through the equilibrium tapped points; from these lines, values of *A* and *B* can be calculated. For five blacks studied in this way, the values of *B* are 10%–15% lower with tapping than without. This is because tapping brings about more compaction at low pressures than at high pressures.

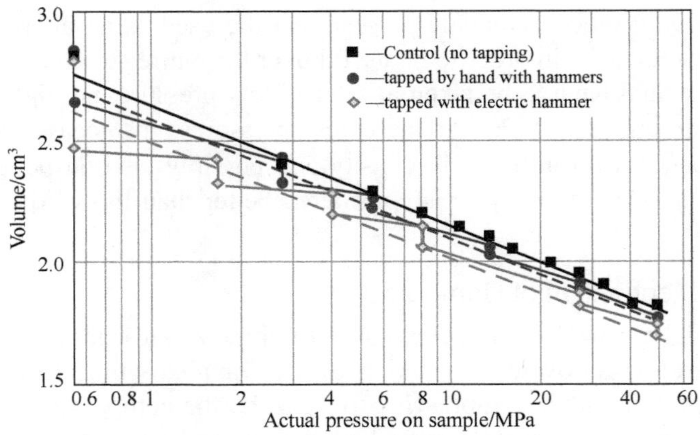


**Figure 2.36** Effect of tapping on compressibility of Vulcan 3[103]

Friction between the steel plug and the inside wall of the cylinder was studied by placing a pressure gauge under the center plug, while supporting the cylinder independently, and measuring the difference between the applied force and the force transmitted to the bottom plug. It was found that the frictional loss for furnace and thermal blacks is a function only of the height of the sample, independent of structure, particle size, initial bulk density, or weight of sample.

With a 5 g sample of pelletized black, the average loss amounts to less than 10% at the higher pressures when the inside diameter of the cylinder measures 28.7 mm. The frictional loss is small enough that the experimentally determined compressibility curves (on a per gram basis) are nearly independent of sample size between 3 g and 7.5 g.

### 2.3.2.3.3 Mercury Porosimetry

The most commonly used techniques for measuring carbon black "structure" have been based on measuring internal void volume using absorptive measurements or volumetric measurements under specific pressure conditions. However, both methods are based on the determination of the total pore volume only. Mercury porosimetry is used to obtain detail-defined structure of fillers in terms of pore size distribution.

The principle of mercury porosimetry is based on the Washburn theory[108], i.e., the pressure, $p$, required to force mercury into a capillary pore of diameter, $d$, is:

$$p = -4\gamma \cos\theta / d \tag{2.113}$$

where $\gamma$ is the surface tension of the mercury and $\theta$ the contact angle between the liquid and the material surface. For mercury porosimetry, the liquid is mercury, taking $\gamma = 480$ dyne/cm (1 dyne/cm = $10^{-3}$ N/m) and $\theta = 140°$, which transforms Eq. 2.113 into:

$$d = 1.5 / p \tag{2.114}$$

where $d$ is in μm and $p$ in kgf/cm$^2$(1 kgf/cm$^2$ $\approx$ 0.1 MPa).

Mercury porosimetry measures the pore size distribution by pressing mercury into the pores of the evacuated sample and calculating a series of pressures and volumes at various pressures.

In practice, the test procedure is as follows: the granular sample of the thoroughly outgassed material is weighed and placed in a steel pressure cylindrical tube that is then evacuated until all adsorbed gases are removed. Pure mercury is then admitted to fill the tube, and a series of pressure and volume measurements are made at various pressures up to the highest desired pressure to employ.

The decrease in volume, $\Delta V$, accompanying a small pressure increase, $\Delta p$, in any part of the range must evidently be due to the filling of pores whose effective diameters lie between the limits $d$ and $d - \Delta d$.

Moscou, Lub, and Bussemaker have applied this method to carbon black characterization[109]. Figure 2.37 shows a general shape of the recorded porosity curves. The shape of this cumulative curve is explained in terms of the different types of pores present in the carbon black agglomerates. The very steep left part of the curve represents the volume of the pores in the aggregates (volume *A*), this being the real "structural pore volume". Volume *B* is the pore volume between aggregates. The total volume *C* is comparable with the void volume in the filler agglomerates.

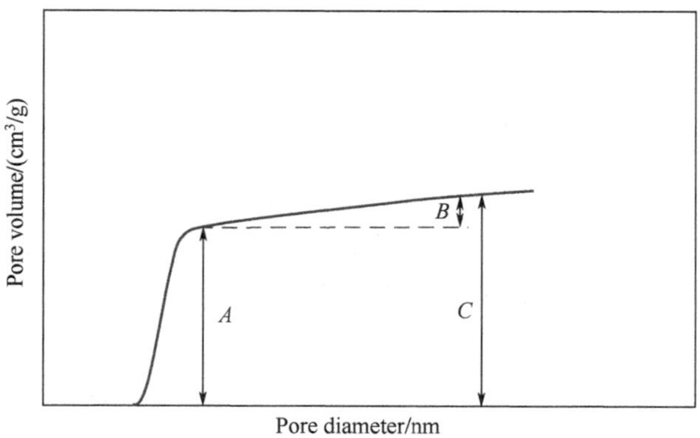


**Figure 2.37** General porosity curve from mercury porosimetry[109]

Figure 2.38 shows the different types of pore volume. The void volume of the fillers consists of three parts: intra-aggregate, inter-aggregate, and inter-agglomerate void volume.

Figure 2.39 shows the cumulative curves of the pore volume blending two carbon blacks, namely N234 and N550. The former has a surface area of 119 $m^2/g$ and a compressed DBP number of 102 mL/100 g; the latter has a 40 $m^2/g$ surface area and 85 mL/100 g compressed DBP number. All the single blacks have the same shape; however, for N550 the pore diameter is larger where the pore volume increases very rapidly, and its cumulative pore volume is significantly smaller. This can also be seen in Figure 2.40 for the pore size distribution. In contrast with N550, carbon black N234 seems to have more void volume, and its most frequently occurring pore diameter is much smaller. The pore size distribution is also much narrow for N234. When the two blacks are blended, an additive function seems to arise, at least in first approximation.

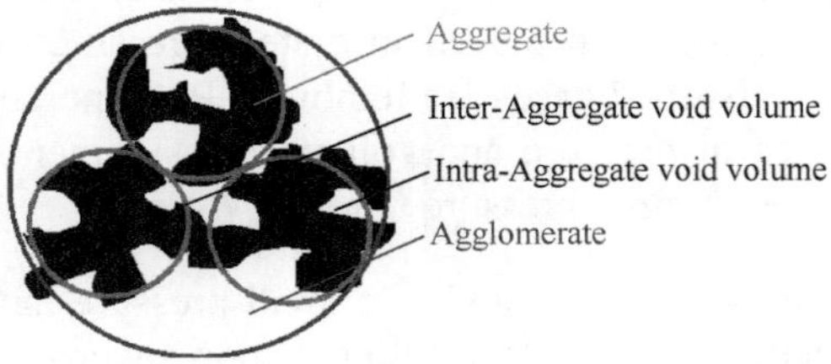


**Figure 2.38** Schematic view of different types of void volume of filler

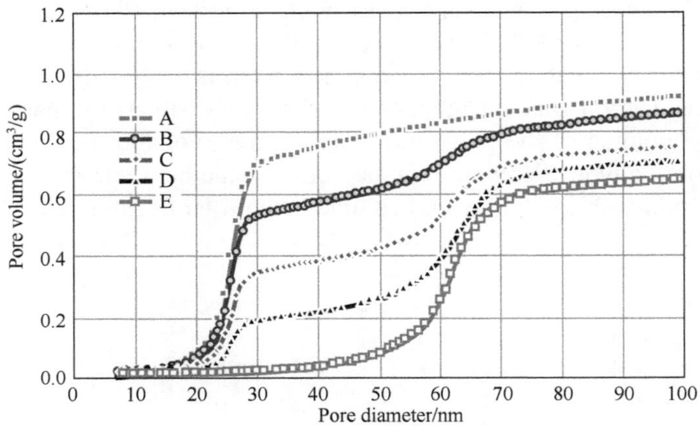


**Figure 2.39** Cumulative pore volume as a function of pore diameter of blend of carbon blacks N550 and N234. N234/N550: A—100/0; B—75/25; C—50/50; D—25/75; E—0/100

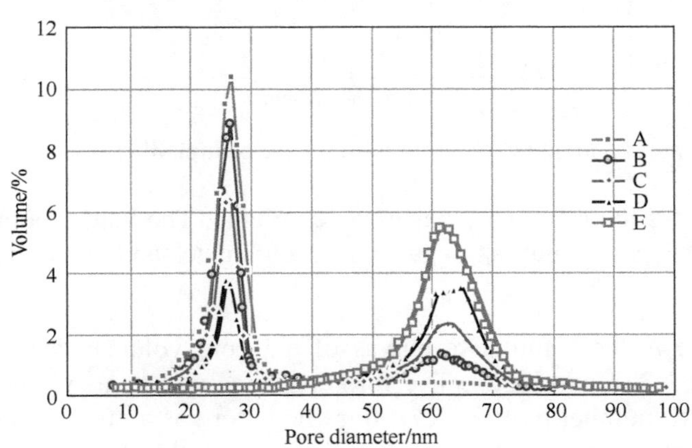


**Figure 2.40** Pore size distribution of blend of carbon blacks N550 and N234. N234/N550: A—100/0; B—75/25; C—50/50; D—25/75; E—0/100

Mercury porosimetry is now the most useful tool for characterizing silica. It is well accepted that silica has some important advantages over carbon black as a principal filler in tire tread compound. It provides improved temperature dependences of filled rubber hysteresis, imparting substantially lower rolling resistance and better traction for tires. However, due to its high surface polarity, this material is very difficult to disperse in low polar rubbers, which gives poor abrasion resistance. A new type of silica, called highly dispersible silica, has been developed to improve its dispersability by increasing the void volume between aggregates as well as between agglomerates. As will be discussed later in Chapter 4 "Filler Dispersion", the void volume in the filler pellets plays a very important role in filler dispersion. Moreover, the approach of determining the distribution of pore sizes in the filler becomes an indispensable method for identifying highly dispersible silica.

Shown in Figure 2.41 are the pore distributions of conventional silica 165 and highly dispersible silica 165MP, measured with mercury porosimetry. The two silicas have similar surface area around 165 $m^2/g$. Obviously, while the pore size and volume related to the intra-aggregate void, which is the real "structure" nature of the filler, is similar for the two silicas, the inter-aggregate and inter-agglomerate void volumes are much higher for the highly dispersible silica. This is certainly the root cause of the better dispersion of silica 165MP in compounds.

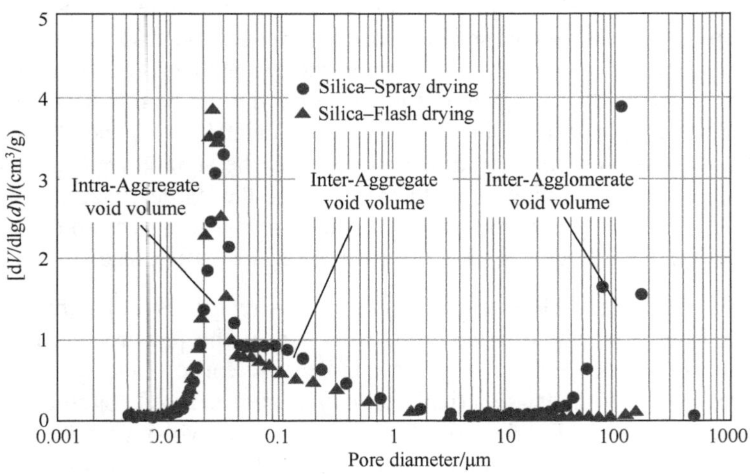


**Figure 2.41** Mercury porosimetry of highly dispersed silica and conventional silica

## 2.3.3 Tinting Strength

The tinting strength of carbon black is a measure of its ability to darken a mixture of black and white pigment, i.e., to decrease the amount of light scattered or reflected by the mixture over the entire visible spectrum. Tinting strength, or tint, of carbon black has been used for a long time for quality control of grades of carbon blacks. While the basic principles of tinting strength are well understood, the relationship between tint and carbon black structural properties is not simple.

It is generally accepted that the tinting strength of a carbon black is determined mainly by its particle size[87,110], and this relation is the chief rationale for the use of the tinting strength test for quality control of carbon black. However, it is also generally accepted that the carbon black structure measured by the oil-absorption or dibutylphthalate (DBP) absorption tests, has a significant influence on tinting strength as well. At a given particle size, higher "structure" or DBP absorption generally leads to a lower tinting strength. The relation between tint and these two inherent morphological properties of the carbon black has been established empirically.

In a tinting strength test, carbon black is mixed with a white pigment, such as zinc oxide, in a suitable vehicle. The gray paste is then spread out as a thin but opaque film, and its reflectance is measured. The reflectance depends not only on the optical properties of the black and white pigments but also on the test conditions such as the concentration and degree of dispersion of the pigments, the refractive index of the vehicle, the wavelength of the light, and the optics of the reflectance measurement. Specific test methods are given in ISO and various national standards.

### Theory of Tinting Strength

It was assumed by Medalia and Richards[111] that in the test mixture of black and white pigments, virtually all of the scattering is due to the white pigment, while virtually all of the absorption is due to the black. In addition, the tests are carried out in such a way that the scattering properties of the white pigment remain the same from test to test. Therefore, the Kubelka-Munk[112] scattering coefficient $S$ is a constant, and the absorption coefficient $K$ of the paste is determined by the tinting strength of the carbon black. Because the reflectance $R$ of the gray paste used in a typical tint test is normally less than 0.08, the carbon black tinting strength is proportional to the absorption coefficient $K$. The general effect of carbon black morphological properties on tinting strength can be summarized as follows: a) for large black particles, the tinting strength increases as the particle size decreases; b) for small particles, the tinting strength is less sensitive to particle size; and c) for the very smallest particles, the tinting strength is independent of the particle size.

The carbon-sphere size below which the tinting strength is independent of size is considerably larger (by about 4-fold) than the size of the primary particles at which this break is observed experimentally. This is because carbon blacks exist in the form of aggregates of many quasi-spherical primary particles fused together. Figure 2.42 shows a model of a typical carbon black aggregate along with two spheres. The "solid sphere" on the left contains the same volume of solid carbon as the aggregate, and the "equivalent sphere" on the right has the same projected area as the aggregate.

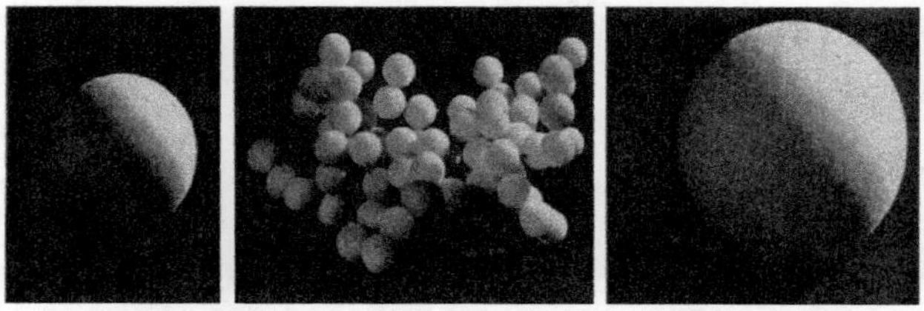


**Figure 2.42** Model aggregate (floc) with "solid" and "equivalent" sphere[111]

In a dilute dispersion of carbon black, the aggregates are the independently mobile "working units" that interact individually with the light. When the size of the solid sphere used in the Mie calculations is chosen so that it contains the same amount of carbon as the aggregates, it is found that the Mie theory gives a good estimate of the carbon black aggregate size below which the tint is independent of the aggregate size. The amount of carbon in each aggregate depends on the mean diameter of the individual particles of which it is composed, $\bar{d}$, and on the number of particles, $N_p$, that have been identified with the "structure" of the black[98]. This accounts for the influence of particle size and "structure" on the diameter of the solid sphere, and thus on the absorption coefficient and tinting strength.

Medalia derived a relation giving the diameter $D_a$ of the solid sphere representing an aggregate in terms of its particle size and DBP absorption. This treatment is strictly for a mono-disperse carbon black, i.e., one in which all particles are the same size, $d$, that is diameter of the particle and all aggregates have the same number of particles, $N_p$. He showed[98] that the DBP absorption can be calculated from the value of $N_p$ determined by electron microscopy.

For all practical purposes, Eq. 2.115 may be applied:

$$D_a = \left(1.321 + 0.0370\langle DBP\rangle\right)d \qquad (2.115)$$

that gives values of $D_a$ within about 1% error over the range of practical values of DBP absorption (30–200 mL/100 g).

The tinting strength of solid carbon spheres can be calculated using Mie theory from their diameter $D$ and the relative refractive index $m' = m/n_s$, where $m$ is the complex refractive index of the carbon, and $n_s$ is the real refractive index of the suspending medium.

### Comparison of Theory and Experiment

Because of the many assumptions and simplifications in the theory, it is important to compare the theoretical predictions with experimental data for a wide range of carbon blacks.

In the manufacture of carbon blacks for use in rubber, tinting strength is one of the commonly used quality control tests. A satisfactory quality control test for these blacks is with a procedure involving a standard dispersion procedure and standard amounts of black and white pigments and vehicle is described. Photometric comparison establishes the tinting strength using a fresh paste prepared with the closest of a series of standard carbon blacks. The tinting strength is also established by stepwise comparison with a reference carbon black of SRF grade.

Since the rubber-grade blacks are generally non-porous[49], it is legitimate to use the specific surface area, $S$, as a measure of particle size. For spherical particles, $S$ is inversely proportional to the volume-surface-average particle diameter, $d_s$:

$$d_s = 6\times10^3 / \rho S = 3226 / S \qquad (2.116)$$

where $d_s$ is in nm and $S$ is in m$^2$/g. $d_s$ can be converted to the number average particle diameter, $d_n$, by dividing by an empirical factor of 1.68. This factor is the average value of $d_s/d_n$ for various non-porous blacks listed in Ref.[113], and includes, in principle, a contribution from the effect of particle fusion as well as a contribution from the particle size distribution. Thus,

$$d_n = d_s / 1.68 = 1920 / S \tag{2.117}$$

Which, on substitution in Eq. 2.115 gives:

$$D_a = \left(2540 + 71\langle DBP\rangle\right) / S \tag{2.118}$$

where $D_a$ is in nanometers.

While the values $d_s$ in Ref. [113] are calculated from surface areas determined by nitrogen adsorption, experience has shown that the ASTM iodine number[114] is generally within 10% of the nitrogen surface area, and can thus be used for $S$ in Eq. 2.118. This equation thus enables one to calculate $D_a$ from the results of two commonly used ASTM procedures: DBP absorption and iodine number.

Figure 2.43 shows tinting strengths plotted against the solid-sphere diameter calculated from Eq. 2.118. The data were measured by a standard test method for carbon blacks representing all the grades supplied to the rubber industry, with the exception of the ultrahigh structure conductive blacks. The theoretical curve has been placed on Figure 2.43, with the values of $D_a$ calculated from the values of $\alpha$ for green light in linseed oil, and the vertical position adjusted to give the best fit with the data in the leveling-off region.

In Figure 2.43, the points are designated as to structure level, in arbitrary ranges of "extra-low" ($\langle$DBP$\rangle$ < 65), "low" (65 ⩽ $\langle$DBP$\rangle$ < 90), "normal" (90 ⩽ $\langle$DBP$\rangle$ ⩽120) and "high" ($\langle$DBP$\rangle$ > 120). All structure levels give the same tinting strength in the leveling-off portion of the curve, as expected from theory. In the screening region, the extra low and low structure blacks are generally below (or to the left of) the curve, while the high structure blacks are generally above (or to the right of) the curve. This suggests that the screening efficiency of the high structure blacks is greater than that of the low structure blacks. This effect to some extent counteracts the primary effect of high structure in raising $D_a$ (Eq. 2.115) and thus lowering the tinting strength at a given particle size.

Further comparison of the effects of structure and particle size (or surface area) is shown in Table 2.6. The first four carbon blacks listed are all of approximately the same surface area or particle size, but cover a range of over two-fold in DBP absorption. The effect of these differences in structure on tinting strength is significant over the entire range, although it is somewhat less than predicted from the solid sphere model, as just discussed. The last three blacks are of nearly the same structure level as the SRF-HS black, and (with this black) cover nearly a five-fold range in surface area.

The theory gives generally good agreement with the experimental values, and thus appears to account satisfactorily for the effect of particle size on tinting strength. The incorporation of structure in the theoretical treatment is essential to obtain the right order of magnitude of $D_a$, and does account for the direction of the observed effect of structure on tinting strength, if not exactly for its magnitude.

It should be mentioned that certain high color channel blacks, for which this tinting strength procedure is not normally employed, as well as certain experimental furnace blacks, give tinting strengths of up to 25% higher than the theoretical values (i.e., the curve of Figure 2.43). These high values are found in the leveling-off region as well as in the other regions of the tinting strength curve. The high values found with the experimental furnace blacks appear to be due, at least in part, to a narrow aggregate size distribution. The reason for the high values found for the channel blacks is not clear but may be associated with the effect of surface oxygen on the adsorptive behavior of these blacks, thus possibly affecting the dispersion as well as the tendency to float and flood.

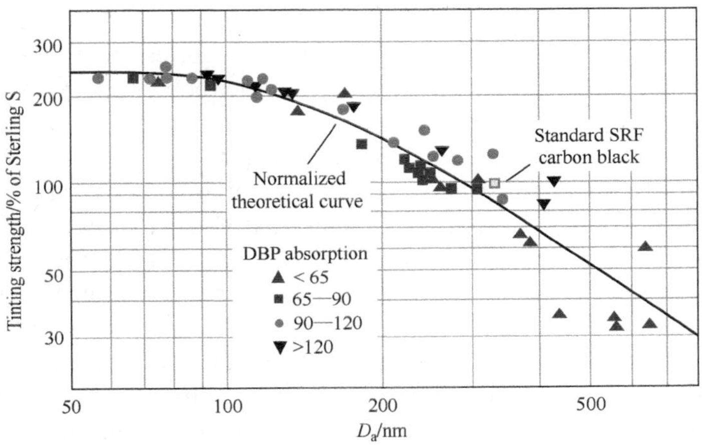


**Figure 2.43** Relative tinting strength of rubber-grade carbon black as percentage of Sterling S[111]

**Table 2.6** Effect of structure and surface area on tinting strength

| Grade of black | $S^a$/(m$^2$/g) | DBP Abs. /(mL/100 g) | $D_a^b$/nm | Tinting strength | |
|---|---|---|---|---|---|
| | | | | From $D_a^c$ | Measured procedure B |
| SRF-LS | 30 | 66 | 242 | 129 | 109 |
| SRF-MS | 29 | 87 | 300 | 102 | 97 |
| SRF-HS | 31 | 109 | 335 | 91 | 90 |
| SRF-VHS | 30 | 137 | 409 | 74 | 85 |
| FEF-LS | 46 | 98 | 205 | 150 | 143 |
| HAF | 87 | 102 | 113 | 231 | 203 |
| SAF | 145 | 118 | 75 | 262 | 254 |

a. ASTM iodine number, in milligrams per gram, taken as equal to the surface area.

b. Calculated by Eq. 2.118.

c. Based on the curve of Figure 2.43.

### Effect of Aggregate Size Distribution

In the previous discussion, it has been assumed that carbon blacks are mono-disperse, as defined under "Size of Solid Sphere". Actual carbon blacks, however, exhibit a broad distribution of particle size[115,116] and of even broader distribution of $N_p$[100,115]. Both of these properties appear to follow a log normal distribution. The distributions of $d$ and $N_p$ are not independent[115]. The largest particles are predominantly in aggregates of lower-than-average $N_p$, and thus the distribution of $D_a$ is not log normal. In order to calculate the effect of the distribution of $D_a$ on the tinting strength of a carbon black, it is necessary to determine the distribution of $D_a$ experimentally for that black.

In the developed TEM comparison chord method of characterizing carbon blacks[115], individual aggregates are assigned to categories according to their mean particle size and aggregate size. Analysis of a carbon black by this procedure directly gives the number of aggregates at the mean value of $D_a$ of each category. Figure 2.44 shows a histogram of $D_a$ by grouping these values in suitable ranges for the N220 black described above. The nominal $D_a$ of this sample (84.6 nm), based on iodine number and ⟨DBP⟩, is somewhat lower than the number-average $D_a$ calculated from the histogram data (105 nm), and considerably lower than the weight average $D_a$ (160 nm). However, in view of the nonlinear dependence of tinting strength on $D_a$, no simple average value is "proper" for calculating the tinting strength of the poly-disperse black. Instead, the data for the individual categories must be summed up, as described below. First, however, it is necessary to consider whether the tinting strength of the relatively mono-disperse black in a given category can be calculated from the same general relation used for the hetero-disperse blacks (Figure 2.43).

The vertical adjustment of the theoretical curve of Figure 2.43 is made to fit the tinting strength of blacks of $D_a$ in the leveling-off region. As the histogram shows, a typical black contains aggregates covering nearly a 10-fold range of $D_a$. A large fraction of the carbon is in aggregates of considerably larger size and lower tinting strength than the nominal $D_a$; because of the leveling off of the tinting strength curve at low $D_a$, the aggregates of smaller size do not have a higher tinting strength than those of the nominal $D_a$. This is shown in the theoretical curve of Figure 2.43, which is reproduced in Figure 2.44 with vertical adjustment as described below. Since there are no high-tint aggregates to compensate for the low-tint aggregates, the tinting strength of the (poly-disperse) N220 carbon black must be significantly lower than that of a mono-disperse black of $D_a$ = 84.6 nm. In fact, this is probably the case with all commercial channel and furnace blacks, since all such blacks that have been studied are of broad enough distribution in $d_n$ and $N_p$ to span the region of the theoretical curve in which the dependence of tinting strength on $D_a$ is nonlinear. Therefore, it appears that for mono-disperse blacks the leveling-off region (and thus the entire tinting strength curve) must be at a somewhat higher level than those shown in Figure 2.43, which was based on poly-disperse blacks.

Based on these concepts, Medalia et al. were able to calculate the tinting strength of a carbon black directly from the aggregate size data obtained by the comparison chord method[115]. The weight fraction of carbon black in each category is multiplied by the tinting strength of a mono-disperse black of the corresponding $D_a$, and these tinting strength contributions are summed up to give the tinting strength of the poly-disperse sample. Since mono-disperse blacks are not available for adjustment of the tinting strength curve, the tinting strengths are multiplied by an empirical factor. These results are presently insufficient to establish the value of this factor accurately, but for tinting strengths determined by Procedure B, it is in the neighborhood of 1.25. This factor is included in the normalization of the theoretical curve of Figure 2.44; i.e., the leveling off is taken at a tinting strength of 300 rather than at 240 as in Figure 2.43. As expected, they have found that carbon blacks of unusually narrow distribution of $D_a$ are of higher tinting strength than would be predicted from Eq. 2.118 and Figure 2.43. A correction for poly-dispersity should also be made in calculating the absorption.

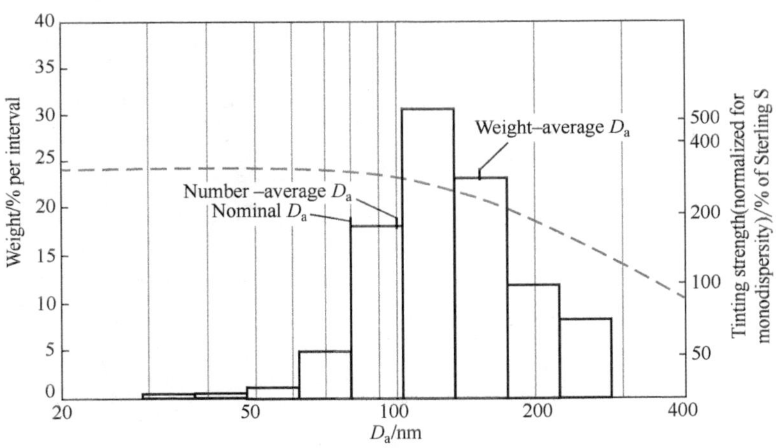


**Figure 2.44** Aggregate size distribution of N220 carbon black, with normalized theoretical tinting strength curve[111]

So far, the basic principles of some test methods for characterizing the filler morphologies have been discussed, and the specific experimental procedures are given in ISO and national standards in various countries. In conclusion, the concrete values of a variety of carbon blacks, precipitated silicas, and fumed silicas are summarized in Tables 2.7, 2.8, and 2.9, taking some commonly used products in the rubber industry as examples.

**Table 2.7** Carbon black properties

| ASTM classification | Target value | | Typical descriptive values | | | |
|---|---|---|---|---|---|---|
| | Iodine adsorption No. D1510 /(g/kg) | Oil absorption No. D2414 /($10^{-5}m^3$/kg) | Oil absorption compressed sample No. D3493 /($10^{-5}m^3$/kg) | NSA multipoint D6556 /[$10^3m^2$/kg ($m^2$/g)] | STSA D6556 /[$10^3m^2$/kg ($m^2$/g)] | Tint strength D3265 |
| N110 | 145 | 113 | 97 | 127 | 115 | 123 |
| N115 | 160 | 113 | 97 | 137 | 124 | 123 |
| N120 | 122 | 114 | 99 | 126 | 113 | 129 |
| N121 | 121 | 132 | 111 | 122 | 114 | 119 |
| N125 | 117 | 104 | 89 | 122 | 121 | 125 |
| N134 | 142 | 127 | 103 | 143 | 137 | 131 |
| N135 | 151 | 135 | 117 | 141 | – | 119 |
| S212 | – | 85 | 82 | 120 | 107 | 115 |
| N219 | 118 | 78 | 75 | – | – | 123 |
| N220 | 121 | 114 | 98 | 114 | 106 | 116 |
| N231 | 121 | 92 | 86 | 111 | 107 | 120 |
| N234 | 120 | 125 | 102 | 119 | 112 | 123 |
| N293 | 145 | 100 | 88 | 122 | 111 | 120 |
| N299 | 108 | 124 | 104 | 104 | 97 | 113 |
| S315 | – | 79 | 77 | 89 | 86 | 117 |
| N326 | 82 | 72 | 68 | 78 | 76 | 111 |
| N330 | 82 | 102 | 88 | 78 | 75 | 104 |
| N335 | 92 | 110 | 94 | 85 | 85 | 110 |
| N339 | 90 | 120 | 99 | 91 | 88 | 111 |
| N343 | 92 | 130 | 104 | 96 | 92 | 112 |
| N347 | 90 | 124 | 99 | 85 | 83 | 105 |
| N351 | 68 | 120 | 95 | 71 | 70 | 100 |
| N356 | 92 | 154 | 112 | 91 | 87 | 106 |
| N358 | 84 | 150 | 108 | 80 | 78 | 98 |
| N375 | 90 | 114 | 96 | 93 | 91 | 114 |
| N539 | 43 | 111 | 81 | 39 | 38 | – |
| N550 | 43 | 121 | 85 | 40 | 39 | – |
| N582 | 100 | 180 | 114 | 80 | – | 67 |
| N630 | 36 | 78 | 62 | 32 | 32 | – |
| N642 | 36 | 64 | 62 | 39 | – | – |
| N650 | 36 | 122 | 84 | 36 | 35 | – |
| N660 | 36 | 90 | 74 | 35 | 34 | – |

Continued

| ASTM classification | Target value | | Typical descriptive values | | | |
|---|---|---|---|---|---|---|
| | Iodine adsorption No. D1510 /(g/kg) | Oil absorption No. D2414 /($10^{-5}$ $m^3$/kg) | Oil absorption compressed sample No. D3493 /($10^{-5}$ $m^3$/kg) | NSA multipoint D6556 /[$10^3$ $m^2$/kg ($m^2$/g)] | STSA D6556 /[$10^3$ $m^2$/kg ($m^2$/g)] | Tint strength D3265 |
| N683 | 35 | 133 | 85 | 36 | 34 | – |
| N754 | 24 | 58 | 57 | 25 | 24 | – |
| N762 | 27 | 65 | 59 | 29 | 28 | – |
| N765 | 31 | 115 | 81 | 34 | 32 | – |
| N772 | 30 | 65 | 59 | 32 | 30 | – |
| N774 | 29 | 72 | 63 | 30 | 29 | – |
| N787 | 30 | 80 | 70 | 32 | 32 | – |
| N907 | – | 34 | – | 9 | 9 | – |
| N908 | – | 34 | – | 9 | 9 | – |
| N990 | – | 43 | 37 | 8 | 8 | – |
| N991 | – | 35 | 37 | 8 | 8 | – |

**Table 2.8** Precipitated silicas properties

| Type | NSA/($m^2$/g) | Type | NSA/($m^2$/g) |
|---|---|---|---|
| Ultrasil® 9100 GR | 235 | Newsil® 115 | 100–130 |
| Ultrasil® 7005 | 190 | Newsil® 125 | 115–135 |
| Ultrasil® 7000 GR | 175 | Newsil® 155 | 140–165 |
| Ultrasil® 5000 GR | 115 | Newsil® 175 | 165–185 |
| Ultrasil® VN 3 | 180 | Newsil® 195 | 185–205 |
| Ultrasil® VN 2 | 130 | Newsil® HD90MP | 80–100 |
| Ultrasil® 360 | 55 | Newsil® HD115MP | 100–130 |
| Zeosil® 1085 GR | 90 | Newsil® HD165MP | 150–180 |
| Zeosil® 1115MP | 115 | Newsil® HD175MP | 160–190 |
| Zeosil® 1165MP | 165 | Newsil® HD200MP | 200–230 |
| Zeosil® Premium 200MP | 215 | Newsil® HD250MP | 220–270 |
| ZHRS® 1200MP | 200 | | |

**Table 2.9** Fumed silicas properties

| Type | NSA/($m^2$/g) | Type | NSA/($m^2$/g) |
|---|---|---|---|
| Cab-O-Sil-L90 | 89.8 | Cab-O-Sil-HP60 | 216.0 |
| Cab-O-Sil-LM130 | 122.3 | Cab-O-Sil-MS75D | 258.4 |
| Cab-O-Sil-LM150 | 167.2 | Cab-O-Sil-HS5 | 285.5 |
| Cab-O-Sil-LM150D-T (Tuscola) | 154.5 | Cab-O-Sil-S17D | 406.8 |
| Cab-O-Sil-LM150D-B | 183.4 | Cab-O-Sil-EH5 | 417.3 |
| Cab-O-Sil-M7D | 196.3 | Fumed silica A | 193.5 |
| Cab-O-Sil-M5 | 208.7 | | |

## 2.4 Filler Surface Characteristics

It has long been recognized that, in addition to the morphology of the filler, namely surface area and structure, the third important parameter is the surface activity of the filler, which also plays a vital role in terms of rubber reinforcement[1,2,50,117-119].

The surface activity is an intensive parameter that dominates polymer-filler interaction, filler aggregate-aggregate interaction, and filler-ingredient interaction. It has a considerable effect on the efficiencies of the first two parameters. For example, without polymer-filler interaction, a filler will have no reinforcing ability whatsoever, irrespective of the size of the interfacial area. The reinforcing ability of the filler should, therefore, be the product of surface area and surface activity. On the other hand, if the polymer molecules are unable to be effectively anchored on the filler surface, the rubber in the internal voids of the filler aggregate cannot be occluded when stress is applied, and the filler would be unable to act as a buffer against strain. This would reduce the stress (or strain) amplification of the rubber matrix.

It is known that the surface activity of the filler depends mainly on the surface chemistry (i.e., chemical functional groups) of the filler which is related to the chemical reactivity with other chemicals, and physical chemistry of the surface, surface energy in particular, which is determined by the physical interaction with the filler surface.

### 2.4.1 Characterization of Surface Chemistry of Filler-Surface Groups

The chemical groups on filler surfaces are well identified (see Section 2.1) and play an important role in surface activity when the filler interacts with other chemicals, materials, and even with its own surface. While the surface chemistry has a strong effect on its chemical reactivity, which maybe used to change the filler properties through surface modification with other chemicals, the functional groups per se show a strong effect on the physical interaction with other substances. Since the functional groups are quite different for carbon black and silica, their characterizations will be discussed separately in the different sections.

### 2.4.2 Characterization of Physical Chemistry of Filler Surface-Surface Energy

While the surface chemistry aspect of the filler surface activity is today sufficiently characterized, a satisfactory description of the physical chemistry of surface characteristics is still lacking. Present knowledge of the relationships between surface activity and rubber properties is far from complete, although its importance in rubber reinforcement has been recognized long ago. One of the reasons is that there is only a limited number of tools available that can be used to characterize the filler surface energetically, and most of these methods are not accurate enough.

It is known that a certain amount of energy input is needed during mixing to breakdown the agglomerates and to disperse the aggregates in the polymer matrix. The energy input is of course, related to the filler-filler interactions and polymer-filler interactions. As will be discussed later in Chapter 3, even when the filler is well dispersed in the polymer, the aggregates trend to flocculate forming filler agglomerates. Evidently, the attractive force between filler particles or aggregates, the interaction between polymer molecules, as well as the interaction between filler and polymer, determine the filler agglomerate formation in the polymer matrix. The driving force originates from intermolecular forces and these are generally expressed by the potential of intermolecular interactions.

#### A. Intermolecular Interactions

It is well accepted that the nature and strength of molecular interactions vary in different materials, depending on the composition and structure of the molecules. Generally, the sum of contributions from various interactions comprise the total interaction among molecules, which include:

- dispersion,
- induction (induced dipole-dipole),
- orientation (dipole-dipole),
- hydrogen bonding,
- acid-base interaction,
- chemical bonding, and
- repulsion.

Only the first five types of interactions are involved in the filler interactions with other molecules. These five types of interactions are classified into two groups: non-specific or dispersive interactions, and specific or polar interactions.

### Dispersion Interaction

The dispersion interaction between molecules is universal and especially related to the concordance of the electron displacement and the resulting instantaneous dipole in the molecule. This instantaneous dipole generates an electric field that polarizes a nearby neutral molecule, inducing a dipole moment in it. The resulting interaction between the two dipoles gives rise to an instantaneous attractive force between the two molecules. The attractive potential force from this dispersion interaction between two molecules, denoted by 1 and 2 respectively, is related to the polarisability $\alpha$ and potential of ionization $I$ of the molecules 1 and 2 by[120]:

$$U_{1,2}^{d} = -\frac{3}{2} \times \frac{I_1 I_2}{I_1 + I_2} \alpha_1 \alpha_2 \frac{1}{r_{1,2}^{6}} \tag{2.119}$$

where $r_{1,2}$ is the distance between the two molecules.

### Orientation (Dipole-Dipole) Interaction

The interaction between two dipoles is dependent on their dipole moment, $\mu$, their relative position, and the temperature. The mean attraction potential is given by[121,122]:

$$U_{1,2}^{p} = -\frac{2\mu_1^2 \mu_2^2}{3kTr_{1,2}^{6}} \tag{2.120}$$

where $k$ is Boltzmann constant and $T$ is the temperature in K.

### Induction (Induced Dipole-Dipole) Interaction

This component is sometimes called Debye force or induction force[123,124]. It is related to the presence of polarizable molecules in which a dipole moment, $\mu$, may be induced by an electric field $E(\boldsymbol{\mu} = \alpha \boldsymbol{E})$. If two molecules are within a distance, $r_{1,2}$, the total induced potential energy can be expressed by:

$$U_{1,2}^{i} = -\frac{\alpha_2 \mu_1^2 + \alpha_1 \mu_2^2}{r_{1,2}^{6}} \tag{2.121}$$

For a given molecule, the induction interaction is generally smaller than those of dipole-dipole and dispersion.

### Hydrogen Bonding Interaction[125]

The force between two molecules would be very strong if one of them contains hydrogen atoms and the other possesses heteroatoms such as oxygen, nitrogen, or fluorine which are very electronegative. Pauling considers the hydrogen bonding interaction, which is different from other types of force, as a combination of dipole-dipole interaction and covalent nature. Each portion of these contributions varies from one bond to another. On the other hand, the hydrogen bonding is very "directional" and acts only between molecules which are very close to each other.

### Acid-Base Interaction

The acid-base interaction is referred to as an electron donor-acceptor interaction, which was proposed by Fowkes[126,127]. It plays an important role in the specific interaction. Hydrogen bonding is a special case of acid-base interaction.

### Repulsion

The repulsive force between two molecules in close proximity is given by:

$$U_{1,2}^{r} = +\frac{B}{r_{1,2}^{n}} \tag{2.122}$$

where $B$ is an empirical constant and $n$ is between 10 and 16, generally 12 for most molecules[128].

### General Expression of Attractive Interaction Potential between Two Molecules

The general expression of the attractive potential between two molecules, without considering hydrogen bonding and acid-base interaction, can be obtained by summing all types of interactions (from Eqs. 2.119 to 2.122):

$$U_{1,2} = -\frac{1}{r_{1,2}^{6}}\left(\frac{3}{2}\times\frac{I_1 I_2}{I_1+I_2}\alpha_1\alpha_2 + \frac{2\mu_1^2\mu_2^2}{3kT} + \alpha_2\mu_1^2 + \alpha_1\mu_2^2\right) + \frac{B}{r_{1,2}^{12}} \tag{2.123}$$

or

$$U_{1,2} = -\frac{\lambda}{r_{1,2}^{6}} + \frac{B}{r_{1,2}^{12}} \tag{2.124}$$

with

$$\lambda = \lambda^{\mathrm{d}} + \lambda^{\mathrm{sp}} \tag{2.125}$$

where $\lambda$, $\lambda^{\mathrm{d}}$, and $\lambda^{\mathrm{sp}}$ are attractive constants, its dispersive component and specific component, respectively. They are, as shown in Eq. 2.123, dependent on the nature of the molecules involved and temperature. This interaction is also known as the Lennard-Jones potential or 6–12 potential as a function of the distance between two molecules[129].

### B. Attractive Force between Two Particles

For a better understanding of the attractive force between two filler particles or aggregates, we may refer to the attractive force between two spheres made of like molecules. Based on the attractive potential from the dispersive interaction, induced dipole-dipole interaction, and dipole-dipole interaction discussed above, the attractive potential energy between two spherical particles, $U_{\mathrm{A}}$, is given by[130]:

$$U_{\mathrm{A}} = -\frac{A}{12}\times\frac{r}{H_0} \tag{2.126}$$

where $A$ is the so-called Hamaker constant, which is determined by attractive constants between molecules (see Eq. 2.123) and their number density, $r$ is the radius of the particles, and $H_0$ the shortest distance between the two sphere surfaces. This equation is only applied to the spherical particles in a vacuum and within a short distance. If the particles are surrounded by medium, such as for a filled polymer, the attractive force between filler aggregates would be substantially reduced. In this case the filler agglomerate formation may be estimated from the surface energies of the filler and polymer which also originate from intermolecular interaction. Such consideration is related not only to the interaction between filler particles but also to the interaction between polymer and filler.

### C. Surface Energy of Solids and Interactions between Materials

According to the types of interaction existing between molecules close to each other, i.e., dispersive, induced dipole-dipole, orientation (dipole-dipole), hydrogen bonding, acid-base, etc., there are different types of cohesion forces in a material. In the mass of a substance, the molecule interacts with all its neighboring molecules in the same way. Therefore, the resultant would be zero. On the surface, however, the resultant is not zero, but is directed towards the interior, as schematically shown in Figure 2.45. In the case of liquids, there is a tendency to reduce the surface to a minimum. The surface free energy, $\gamma$, (also called surface tension) is therefore defined as the work, $W$, necessary to increase a unit surface:

$$\gamma = \left( \frac{\partial W}{\partial A} \right)_{T,p} \tag{2.127}$$

where $A$ is the surface area, $T$ the temperature, and $p$ the pressure. This definition cannot be used for solids, because the molecules lack mobility. In this case, the surface free energy of the solid, $\gamma_s$, can be defined as half of the energy necessary to cleave a unit plane parallel to the surface reversibly, $W_{\text{cleavage}}$. Thus, for a unit surface of a solid one has:

$$\gamma_s = \frac{W_{\text{cleavage}}}{2} \tag{2.128}$$

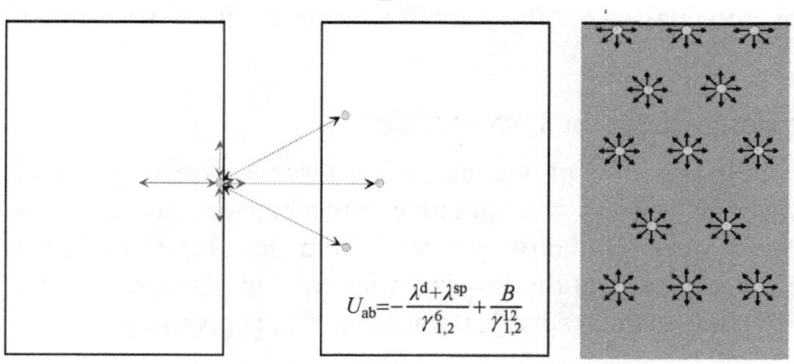


**Figure 2.45** Intermolecular interaction and surface energy of material

In the case of all cohesion forces involved in independent ways, the surface free energy can be expressed as the sum of several components, each corresponding to a specific type of interaction (dispersive, polar, hydrogen bonding, etc.). Since the effect of dispersive forces is universal, the dispersive component of the surface free energy, $\gamma_s^d$, is particularly important. If a substance is only able to exchange dispersion interaction with its environment, its surface free energy would be:

$$\gamma_s = \gamma_s^d \tag{2.129}$$

For most substances one has:

$$\gamma_s = \gamma_s^d + \gamma_s^{sp} \tag{2.130}$$

where $\gamma_s^{sp}$ is the sum of the other components of the surface free energy.

It is known that the possible interaction between two materials 1 and 2 is determined by their surface energies. When only dispersive forces are responsible for the interaction, according to Fowkes' model[131], the energy of adhesion between these two materials would correspond to the geometric mean value of their $\gamma_s^d$:

$$W_a^d = 2\left(\gamma_1^d \gamma_2^d\right)^{1/2} \tag{2.131}$$

where $W_a^d$ is the dispersive component of the adhesive energy. Similarly, the polar component of the adhesive energy, $W_a^p$, can be described by the polar components of their surface free energy[132,133]:

$$W_a^p = 2\left(\gamma_1^p \gamma_2^p\right)^{1/2} \tag{2.132}$$

Hence, the total adhesive energy, $W_a$, can be given by:

$$W_a = W_a^d + W_a^p + W_a^h + W_a^{ab} \tag{2.133}$$

or

$$W_a = 2\left(\gamma_1^d \gamma_2^d\right)^{1/2} + 2\left(\gamma_1^p \gamma_2^p\right)^{1/2} + W_a^h + W_a^{ab} \tag{2.134}$$

where $W_a^h$ is the adhesive energy due to hydrogen bonding and $W_a^{ab}$ the adhesive energy due to acid-base interactions. Therefore, the filler-polymer interaction and filler-filler interaction, the most important parameters for rubber reinforcement, can be represented by their corresponding adhesion energies, $W_a^{pf}$ and $W_a^{ff}$, respectively, as:

$$W_a^{pf} = 2\left(\gamma_p^d \gamma_f^d\right)^{1/2} + 2\left(\gamma_p^p \gamma_f^p\right)^{1/2} + W_{pf}^h + W_{pf}^{ab} \tag{2.135}$$

and

$$W_a^{ff} = 2\gamma_f^d + 2\gamma_f^p + W_{ff}^h + W_{ff}^{ab} \tag{2.136}$$

in which $\gamma_f^d$ and $\gamma_f^p$ are the dispersive and polar components of the surface energy of the filler; $\gamma_p^d$ and $\gamma_p^p$ the dispersive and polar components of the surface energy of the polymer; and $W_f^h$, $W_p^h$, and $W_{fp}^h$ the hydrogen bonding work, and $W_f^{ab}$, $W_p^{ab}$, and $W_{fp}^{ab}$ the work from acid-base interactions between filler surface, polymer surface, and between filler surface and polymer surface, respectively.

It was therefore concluded that polymer-filler interaction and filler-filler interaction in a given polymer system which is related to the filler are determined by filler surface energy and chemical nature, particularly when physical interaction is concerned.

Several techniques have been used to estimate filler surface energy, e.g., different contact angle measurements[134–136], or calorimetry for measuring heat of immersion[137–139]. Surface energy can also be evaluated from the adsorption behavior of fillers, especially from the thermodynamic parameters of gas adsorption. In this regard, IGC has obvious advantages since it is easy to operate, requires little time, is highly accurate, and provides sample information regarding filler surface characteristics[3,140,141].

### 2.4.2.1 Contact Angle

#### 2.4.2.1.1 Single Liquid Phase

When a drop of liquid is placed on a flat, horizontal solid surface it may remain as a drop of finite area, or it may spread indefinitely over the surface. The condition for spreading is that the energy gained in forming unit area of the solid-liquid interface should exceed that required to form unit area of the liquid-vapor surface, which is[142]:

$$\gamma_{sv} - \gamma_{sl} > \gamma_{lv} \tag{2.137}$$

where $\gamma_{sv}$ is the solid-vapor interfacial energy, $\gamma_{sl}$ is the solid-liquid interfacial energy, and $\gamma_{lv}$ is the liquid-vapor interfacial energy.

When this inequality is not fulfilled, the drop remains finite in size and an equilibrium contact angle exists. According to Young[143], the condition for equilibrium requires that, in the absence of gravity effects, the contact angle $\theta$ between the liquid and the solid surfaces is given by:

$$\cos\theta = \frac{\gamma_{sv} - \gamma_{sl}}{\gamma_{lv}} \tag{2.138}$$

Equation 2.138 shows that the cosine of the contact angle gives the ratio of the energy gained in forming unit area of the solid-liquid interface to that required to form unit area of the liquid-vapor interface. The finite contact angle, $\theta$, is dependent on the relative magnitude of the forces operating on the liquid (Figure 2.46). In fact, there are three forces operating on the periphery of the drop to determine the $\theta$.

On the other hand, according to Dupre[144], the adhesion energy between a liquid and a solid, $W_{sl}$, is given by:

$$W_{sl} = \gamma_s + \gamma_l - \gamma_{sl} \quad (2.139)$$

where $\gamma_s$ and $\gamma_l$ are the surface energies of the solid and liquid, respectively.

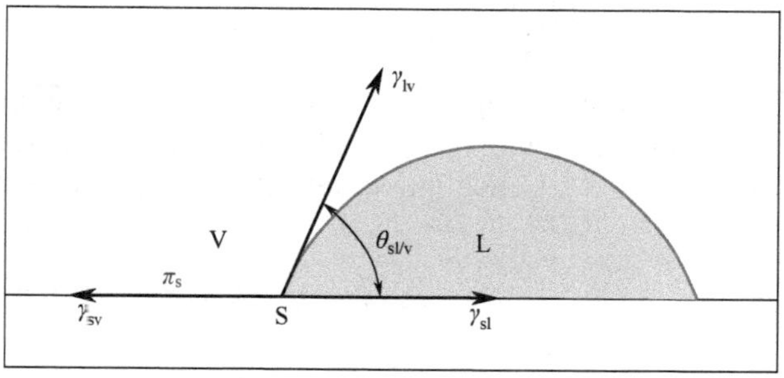


**Figure 2.46** The contact angle between liquid and solid surface in a vapor

Combining Eqs. 2.138 and 2.139, the relationship between the adhesion energy and the contact angle of liquid on solid surface can be expressed as:

$$W_{sl} = \gamma_l(1+\cos\theta) + \gamma_s - \gamma_{sv} \quad (2.140)$$

The difference $\gamma_s - \gamma_{sv}$ presents the diminution of surface energy due to vapor adsorption, which is defined as spreading pressure of the liquid, $\pi_s$, on the solid. Therefore, one has:

$$W_{sl} = \gamma_l(1+\cos\theta) + \pi_s \quad (2.141)$$

Based on Eq. 2.134, the adhesion energy, $W_{sl}$ between liquid and solid can be expressed by:

$$W_{sl} = 2\left(\gamma_s^d \gamma_l^d\right)^{1/2} + 2\left(\gamma_s^p \gamma_l^p\right)^{1/2} + W_{sl}^h + W_{sl}^{ab} \quad (2.142)$$

or

$$W_{sl} = I_{sl}^d + I_{sl}^p \quad (2.143)$$

with

$$I_{sl}^d = 2\left(\gamma_s^d \gamma_l^d\right)^{1/2} \quad (2.144)$$

and

$$I_{sl}^p = 2\left(\gamma_s^p \gamma_l^p\right)^{1/2} + W_{sl}^h + W_{sl}^{ab} \quad (2.145)$$

Combining Eqs. 2.141, 2.143, and 2.144 gives[145]:

$$W_{sl} = 2\left(\gamma_s^d \gamma_l^d\right)^{1/2} + I_{sl}^p = \gamma_l\left(1+\cos\theta\right) + \pi_s \quad (2.146)$$

or

$$\cos\theta = 2\left(\gamma_s^d\right)^{1/2}\frac{\left(\gamma_l^d\right)^{1/2}}{\gamma_l}+\frac{I_{sl}^p}{\gamma_l}-\frac{\pi_s}{\gamma_l}-1 \quad (2.147)$$

Generally speaking, the contact angle changes with surface energy and spreading pressure. However, for the solid with relatively lower surface energy, such as polymer, the term of spreading pressure is negligible, and the above equation can be written as[134]:

$$\cos\theta = 2\left(\gamma_s^d\right)^{1/2}\frac{\left(\gamma_l^d\right)^{1/2}}{\gamma_l}+\frac{I_{sl}^p}{\gamma_l}-1 \quad (2.148)$$

If the liquid used is non-polar in its nature, then $I_{sl}^p = 0$. Therefore, one has:

$$\cos\theta = 2\left(\gamma_s^d\right)^{1/2}\frac{\left(\gamma_l^d\right)^{1/2}}{\gamma_l}-1 \quad (2.149)$$

In this case, a plot of cos $\theta$ *vs.* $(\gamma_l^d)^{1/2}/\gamma_l$ should give a straight line with origin at cos $\theta=-1$ and with slope $2(\gamma_s^d)^{1/2}$ (Figure 2.47). Moreover, since the origin is fixed, one contact angle measurement is sufficient to determine the dispersion force component of the surface energy of the solid ($\gamma_s^d$).

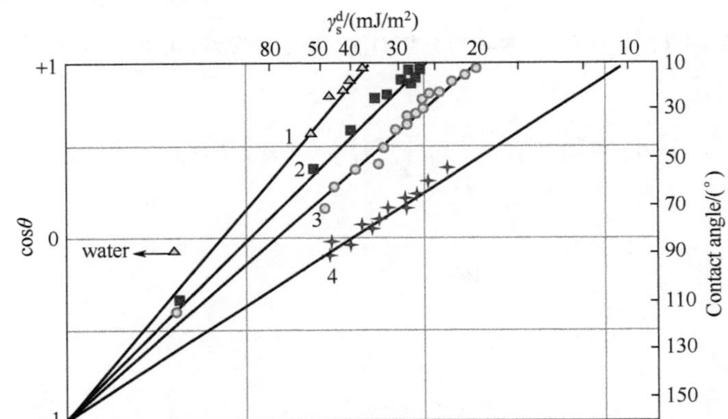


**Figure 2.47** Contact angles of a number of liquids on four low energy surfaces of (1) polyethylene, (2) paraffin wax, (3) $C_{36}H_{74}$, and (4) fluorododecanoic acid monolayers on platinum. All points below arrow are contact angles with water

For liquids with different polarities, when their cos $\theta$ are plotted as a function of $(\gamma_l^d)^{1/2}/\gamma_l$, and compared with the straight line obtained with the non-polar liquid, the difference in the value of cos $\theta$ can be taken as a measure of the polar component of interaction energy between the solid and the liquid, $I_{sl}^{p}$. From Eqs. 2.148 and 2.149, one obtains:

$$I_{sl}^{p} = \gamma_l \left(\cos\theta_p - \cos\theta_{np}\right) \tag{2.150}$$

where $\theta_p$ is the contact angle of a polar liquid with a solid, and $\theta_{np}$ the contact angle of a non-polar liquid with the same solid[12] (see Figure 2.48).

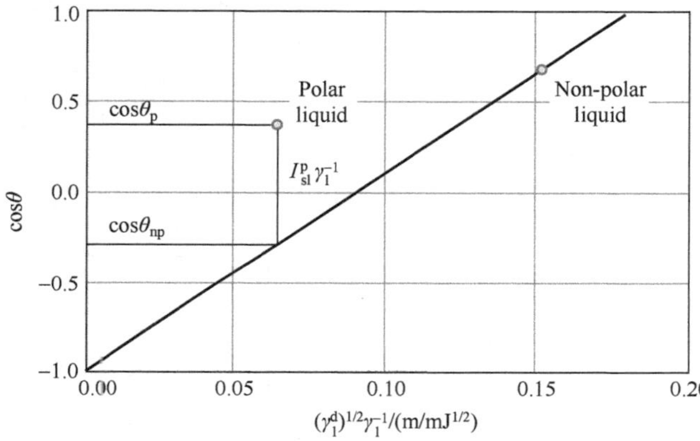


**Figure 2.48** Estimation of polar interaction energies of solid surfaces

Experimentally, the liquids used for measuring the surface energies of solids are listed in Table 2.10 with their surface energies. As examples, the plots of cos $\theta$ *vs.* $(\gamma_l^d)^{1/2} / \gamma_l$ of rubbers NR and NBR are presented in Figure 2.49 and the $I_{sl}^{p}$ of the rubbers NR, SBR, and NBR with different liquids are given in Table 2.11.

**Table 2.10** Surface energies of some liquids

| Liquids | $\gamma_l$/(mJ/m$^2$) | $\gamma_l^d$/(mJ/m$^2$) | $\gamma_l^p$/(mJ/m$^2$) |
|---|---|---|---|
| Water | 72.6 | 21.6 | 51.0 |
| Glycerol | 63.4 | 37.0 | 26.4 |
| Formamide | 58.2 | 39.5 | 18.7 |
| Ethylene glycol | 48.3 | 29.3 | 19.0 |
| Diiodomethane | 50.8 | 48.5 | 2.3 |
| Tribenzylphosphate | 40.9 | 39.2 | 1.7 |
| $\alpha$-Bromonaphthalene | 44.6 | 44.6 | 0 |

**Table 2.11** Contact angles of some liquids on rubbers; polar interactions

| Liquids | NR | | SBR | | NBR | |
|---|---|---|---|---|---|---|
| | $\theta$ ±2° | $\gamma_{sl}^{p}$ /(mJ/m$^2$) | $\theta$ ±2° | $\gamma_{sl}^{p}$ /(mJ/m$^2$) | $\theta$ ±2° | $\gamma_{sl}^{p}$ /(mJ/m$^2$) |
| Water | 97.2 | 13.0 | 96.9 | 14.0 | 68.8 | 47.3 |
| Glycerol | 86.4 | 1.3 | 82.0 | 6.9 | 64.0 | 23.6 |
| Formamide | 80.8 | −2.3 | 75.6 | 4.0 | 64.7 | 12.0 |
| Ethylene glycol | 77.9 | 0 | 76.0 | 1.9 | 62.7 | 10.4 |
| Diiodomethane | 62.6 | −1.5 | 59.3 | 2.1 | 58.0 | 0.4 |
| Tribenzylphosphate | 47.2 | 0.65 | 44.8 | 2.8 | 49.8 | −2.2 |
| $\alpha$-Bromonaphthalene | 50.9 | 0 | 52.4 | 0 | 48.1 | 0 |

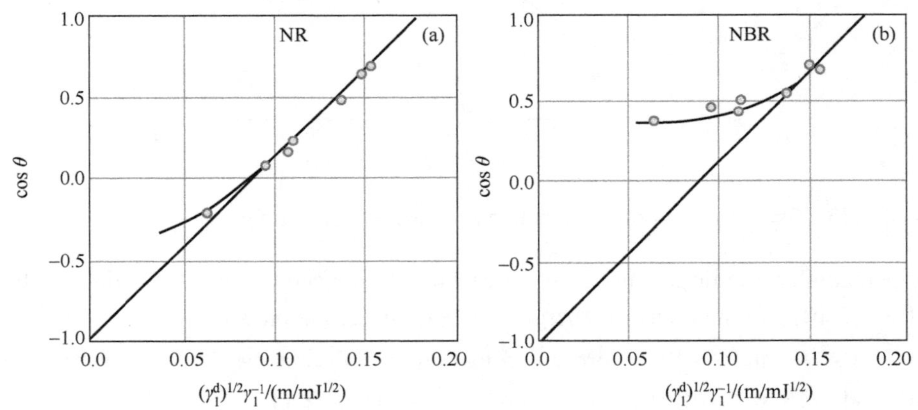


**Figure 2.49** cos $\theta$ as a function of $(\gamma_1^d)^{1/2}/\gamma_1$ for (a) NR and (b) NBR

There are many cases in contact angle measurement with high energy solid surface where $\pi_s$ is not negligible. For the case of water on graphite, for example, $\pi_s$ was measured by Harkins[139] to be 19 mJ/m$^2$ at saturation. In such a case, one has:

$$\cos\theta = -1 + \left[2\left(\gamma_s^d \gamma_l^d\right)^{1/2} - \pi_s\right] / \gamma_{lv} \tag{2.151}$$

When Eq. 2.151 is used for water at 20°C on graphite ($\gamma_l^d$ = 21.8 mJ/m$^2$, $\pi_s$ = 19 mJ/m$^2$, $\gamma_{lv}$ = 72.8 mJ/m$^2$, and $\theta$ = 85.7°), one obtains $\gamma_s^d$ = 109 mJ/m$^2$.

### 2.4.2.1.2 Dual Liquid Phases

Later, Tamai[146] and Schultz[147,148] developed a new method to measure surface energies of solids, using contact angle measurement with dual liquid phases. When a drop of liquid is placed on a flat, horizontal solid surface it may remain as a drop of finite area, or it may spread indefinitely over the surface. The contact angle measurement of a liquid drop on the solid surface is carried out when another non-miscible liquid has been presented instead of air (Figure 2.50).

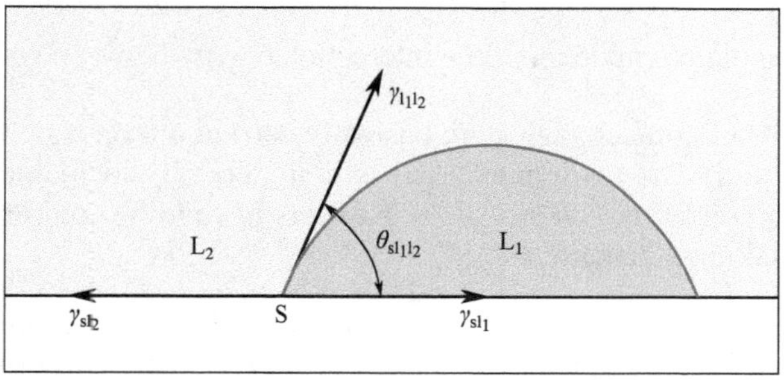


**Figure 2.50** The contact angle of liquid $L_1$ on a solid surface when liquid $L_2$ is present

For this dual liquid system, at equilibrium, the Young's equation can be written as follows:

$$\gamma_{sl_2} = \gamma_{sl_1} + \gamma_{l_1l_2} \cos\theta_{l_1/l_2} \tag{2.152}$$

where $\gamma_{sl_1}$, $\gamma_{sl_2}$, and $\gamma_{l_1l_2}$ are the interfacial energies between the different phases and $\cos\theta_{l_1/l_2}$ is the contact angle of liquid $l_1$ on the solid in liquid $l_2$, respectively.

According to Eq. 2.134, if the sum of $2(\gamma_1^p\gamma_2^p)^{1/2}$, $W_a^h$, and $W_a^{ab}$ is taken as the polar component of the adhesion energies, $I_{12}^p$, between the phase 1 and 2, one has:

$$W_a = 2\left(\gamma_1^d\gamma_2^d\right)^{1/2} + I_{12}^p \tag{2.153}$$

Consequently, the interfacial energy between solid and liquid can be written as:

$$\gamma_{sl_1} = \gamma_s + \gamma_{l_1} - W_{a-sl_1} = \gamma_s + \gamma_{l_1} - 2\left(\gamma_s^d\gamma_{l_1}^d\right)^{1/2} + I_{sl_1}^p \tag{2.154}$$

And

$$\gamma_{sl_2} = \gamma_s + \gamma_{l_2} - W_{a-sl_2} = \gamma_s + \gamma_{l_2} - 2\left(\gamma_s^d\gamma_{l_2}^d\right)^{1/2} + I_{sl_2}^p \tag{2.155}$$

Combining Eqs. 2.152, 2.154, and 2.155, the following is obtained:

$$\gamma_{l_1} - \gamma_{l_2} + \gamma_{l_1l_2} \cos\theta_{sl_1/l_2} = 2\left(\gamma_s^d\right)^{\frac{1}{2}}\left[\left(\gamma_w^d\right)^{\frac{1}{2}} - \left(\gamma_h^d\right)^{\frac{1}{2}}\right] + I_{sl_1}^p - I_{sl_2}^p \tag{2.156}$$

Generally, $l_1$ is water (expressed by *w*) and $l_2$ is a normal alkane (expressed with *h*) which is a non-polar chemical. In this case, the term $I_{sl_2}^p$ can be considered as zero. Then Eq. 2.156 can be rewritten as:

$$\gamma_w - \gamma_h + \gamma_{hw} \cos\theta_{sw/h} = 2\left(\gamma_s^d\right)^{\frac{1}{2}}\left[\left(\gamma_w^d\right)^{\frac{1}{2}} - \left(\gamma_h^d\right)^{\frac{1}{2}}\right] + I_{sw}^p \tag{2.157}$$

When the contact angles of water on a silica surface are measured in a series of normal alkanes, a plot of $\gamma_w - \gamma_h + \gamma_{hw} \cos\theta_{sw/h}$ *vs.* $(\gamma_w^d)^{\frac{1}{2}} - (\gamma_h^d)^{\frac{1}{2}}$ should give a straight line with slope to obtain $\gamma_s^d$, the dispersive component of the solid surface energy, and with intercept $I_{sw}^p$, the polar component of the interaction energy between water and the solid.

The measurements of contact angles are basically carried out on a smooth solid surface with adequate liquid. The equipment used is commercially available. For particulate fillers, such as carbon black and silica, the powders have to be compressed with a high pressure press, making a diskette.

Practically, for estimating the different components of the surface energies of a filler with the dual liquid method, a series of contact angles of water on the filler diskettes are measured in the presence of several alkanes, such as *n*-hexane, *n*-octane, *n*-decane, and *n*-hexadecane. With the known data of the alkanes and water listed in Table 2.12, the plot of $\gamma_w - \gamma_h + \gamma_{hw} \cos\theta_{sw/h}$ as a function of $(\gamma_w^d)^{\frac{1}{2}} - (\gamma_h^d)^{\frac{1}{2}}$ can be constructed from which $\gamma_f^d$, the dispersive component of filler surface energy, and $I_{fw}^p$, the polar component of the adhesion energy between filler and water, can be obtained. The plots of an alkylated precipitated silica P and Aerosil 130 fumed silica are shown in Figure 2.51[12].

**Table 2.12** In surface energies and interfacial energies of some alkanes and water

| *n*-alkane | $\gamma_h$/(mJ/m$^2$) | $\gamma_{hw}$/(mJ/m$^2$) |
|---|---|---|
| *n*-hexane | 16.2 | 51.4 |
| *n*-octane | 21.3 | 51.0 |
| *n*-decane | 23.4 | 51.0 |
| *n*-hexadecane | 27.1 | 51.3 |
| water | $\gamma_w$ = 72.6 mJ/m$^2$<br>$\gamma_w^d$ = 21.6 mJ/m$^2$<br>$\gamma_w^p$ = 51.0 mJ/m$^2$ | |

Until recently, there were no theoretical approaches to calculate the polar component of surface energy of solids based on $I_{sw}^p$. However, based on a similar consideration about the dispersive component of the surface energy, Owens and Wendt[132], and Kaelble and Uy[133] proposed that the $I_{sl}^p$ is the geometric mean of the polar components of the liquid and solid, i.e.,

$$I_{sl}^p = 2\left(\gamma_s^p \gamma_l^p\right)^{\frac{1}{2}} \tag{2.158}$$

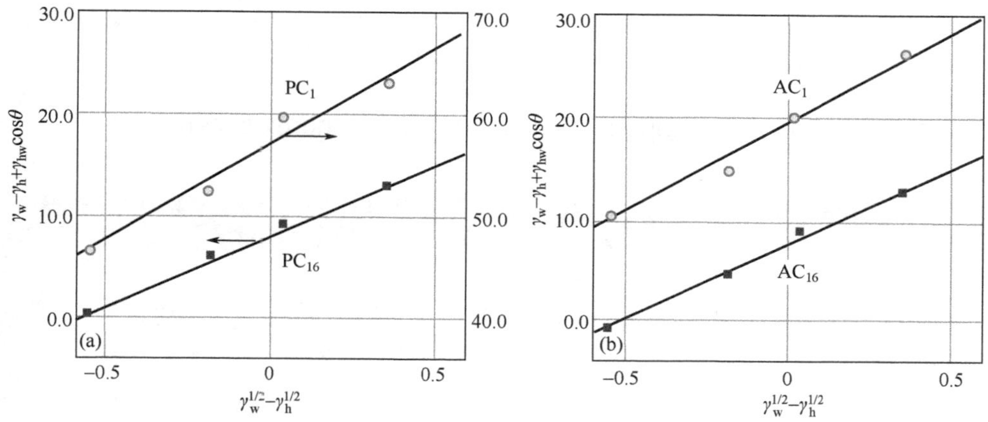


**Figure 2.51** Determination of surface energies of silicas: (a) $PC_1$ and $PC_{16}$, (b) $AC_1$ and $AC_{16}$[12]

This consideration may not be reasonable as the polar component of the interaction between solid and liquid may not originate from dipole-dipole or induced dipole interactions.

In the case where the interaction is between one phase having high surface energy and another phase having weak surface energies, such as polymers, Wu[149] proposed the following equation:

$$I_{sl}^{p} = \frac{4\gamma_s^p \gamma_l^p}{\gamma_s^p + \gamma_l^p} \tag{2.159}$$

This equation has been used for calculating the $\gamma_s^p$ of solids for which the surface energies are not very high. When water is used as the liquid for measuring contact angle, the polar component of the solid surface energy, $\gamma_s^p$ can be obtained by:

$$\gamma_s^p = \frac{I_{sw}^p \gamma_w^p}{4\gamma_w^p - I_{sw}^p} \tag{2.160}$$

Consequently, the surface energy of the solid is given by:

$$\gamma_s = \gamma_s^d + \gamma_s^p \tag{2.161}$$

As an example, the surface energies of four modified silicas measured with a dual liquid method are listed in Table 2.13. The silica $PC_1$ and $PC_{16}$ are precipitated silica (surface area: 130 $m^2$/g) modified by esterification reaction with methanol and hexadecanol, and $AC_1$ and $AC_{16}$ are the fumed silica Aerosil 130 (surface area: 137 $m^2$/g) modified with methanol and hexadecanol, respectively.

It should be pointed out that in the last few decades, the test methods and the equipment have been improved significantly, and the effects of sample preparation and test conditions on the contact angle measurements have been studied intensively. These have been reviewed in several excellent reports[150,151].

**Table 2.13** Different components of silica surface energies measured with contact angles

| Silica | $\gamma_s^d$/(mJ/m$^2$) | $\gamma_{sw}^P$/(mJ/m$^2$) | $\gamma_s^P$/(mJ/m$^2$) | $\gamma_s$/(mJ/m$^2$) |
|---|---|---|---|---|
| $PC_1$ | 87.2±3 | 57.1±2 | 19.8±2 | 107.0±5 |
| $PC_{16}$ | 47.3±3 | 8.3±2 | 2.2±2 | 49.5±5 |
| $AC_1$ | 76.0±3 | 19.3±2 | 5.3±2 | 81.3±5 |
| $AC_{16}$ | 49.6±3 | 7.6±2 | 2.0±2 | 51.6±5 |

#### 2.4.2.2 Heat of Immersion

As discussed by Chessick and Zettlemoyer[152], when a drop of liquid is placed on a solid surface, it may be wetted. The condition for wetting of the solid with the liquid is that the energy gained in forming unit area of the solid-liquid interface should be equal to or smaller than that required to form unit area of the liquid-vapor surface, which is:

$$\gamma_{sv} - \gamma_{sl} \leqslant \gamma_{lv} \tag{2.162}$$

Theoretically, the most suitable parameters for describing the wettability of a solid by a liquid are the surface free energies of the substances involved. Actually, the initial spreading coefficient,

$$S_{lv^0/s^0} = \gamma_{s^0} - \left(\gamma_{sl} + \gamma_{lv^0}\right) \tag{2.163}$$

where $\gamma_{s^0}$, $\gamma_{lv^0}$, and $\gamma_{sl}$ are the surface free energies of the solid in vacuum, of the liquid and of the solid-liquid interface, respectively, is just such a measure, since it is the free energy change for extending a film a cerain surface area over the solid. When a duplex film forms, the spreading coefficient is equal to the film pressure at saturation, $\pi_e$, according to the expression:

$$S_{lv^0/s^0} = \pi_e = \gamma_{s^0} - \gamma_{sv^0} \tag{2.164}$$

where $\pi_e$ is equal to the difference in the surface free energy of the solid, $\gamma_{s^0}$, and of the solid covered with an adsorbed film, $\gamma_{sv^0}$, in equilibrium with its saturated vapor. This equation is not valid for nonduplex spreading, since in this case $\gamma_{sv^0}$ for the monolayer is not equal to the term ($\gamma_{sl} + \gamma_{lv^0}$). Nevertheless, $\pi_e$ is a good measure of wettability for both systems and can be calculated from adsorption data by the Gibbs equation. Unfortunately, accurate values for $\pi_e$ are difficult to obtain because of the need for accurate low-pressure adsorption data and because capillary condensation, particularly between particles of small diameter, cannot be avoided at high-equilibrium pressures unless special precautions are taken[152,153].

For nonspreading systems, the final spreading coefficient is related to the equilibrium contact angle by the equation:

$$S_{\mathrm{lv}^0/\mathrm{sv}^0} = \gamma_{\mathrm{lv}^0} \cos\theta - \gamma_{\mathrm{lv}^0} \tag{2.165}$$

Therefore, contact-angle measurements are suitable for rating the wettability of low-energy solids.

In principle, heats of immersion have been obtained by immersing an evacuated (clean) solid into a carefully purified liquid. Some informative data are obtained by comparing the immersion of samples of the powder possessing known increasing amounts of preadsorbed wetting liquid. The small exothermic heat values are generally reported on a per unit area basis except for porous solids, swelling systems which have large internal areas, or agglomerates of small particles where capillary condensation can occur between primary particles. In such cases, it is difficult to establish the amount of the available surfaces remaining at any time as the amount preadsorbed increases[152]. For a solid with known specific surface area, $S$, the heat of immersion per unit surface area is:

$$\Delta H_i / S = h_{i(\mathrm{sl})} \cong e_{i(\mathrm{sl})} = e_{\mathrm{sl}} - e_{\mathrm{s}^0} \tag{2.166}$$

where $e_{\mathrm{s}^0}$ and $e_{\mathrm{sl}}$ are energies of the solid surface and the solid-liquid interface, respectively. The energy changes are essentially the same as changes in enthalpy because volume changes during the immersion process are usually negligible. The energy of adhesion of the liquid to the solid is defined as:

$$e_{\mathrm{a(sl)}} = e_{\mathrm{lv}^0} + \left(e_{\mathrm{s}^0} - e_{\mathrm{sl}}\right) \tag{2.167}$$

or

$$e_{\mathrm{a(sl)}} = e_{\mathrm{lv}^0} - h_{i(\mathrm{sl})} \tag{2.168}$$

Therefore, besides the heat of immersion, only the surface energy of the liquid is required to obtain the adhesion energy.

Adsorption and immersion processes can also be related through the heat effects. The integral heat of adsorption of $N_{\mathrm{A}}$ molecules of adsorbate in the vapor state at equilibrium pressure $p$ and temperature $T$ is:

$$h_{\mathrm{ads}} = \left[h_{i(\mathrm{sl})} - h_{i(\mathrm{s/l})}\right] + \Gamma \Delta H_{\mathrm{l}} \tag{2.169}$$

where $h_{i(\mathrm{s/l})}$ is the heat liberated on immersion of a solid precovered with $N_{\mathrm{A}}$ adsorbed molecules at a surface concentration $\Gamma = N_{\mathrm{A}}/S$ and $\Delta H_{\mathrm{l}}$ is the molar heat of liquefaction.

The net heat of adsorption $[h_{i(\mathrm{sl})} - h_{i(\mathrm{s/l})}]$ is equal to the difference in the integral molar energy of the liquid $E_\mathrm{l}$ and the energy $E'_\mathrm{a}$ of the adsorbate-solid system. Then,

$$\left[h_{i(\mathrm{sl})} - h_{i(\mathrm{s/l})}\right] = \Delta h_{\mathrm{ads}} - \Gamma \Delta H_\mathrm{l} = \Gamma\left(E'_\mathrm{a} - E_\mathrm{l}\right) \tag{2.170}$$

The term $h_{i(\mathrm{s/l})}$ can be less than, equal to, or greater than the energy of the liquid surface $\gamma_{\mathrm{lv}^0}$ at surface coverage beyond the monolayer. It is very often that the values of the heat of immersion $h_{i(\mathrm{sl})}$ of a solid into a variety of liquids are compared directly when the additional knowledge of the value of $h_{i(\mathrm{s/l})}$ at a specified surface coverage is necessary for proper interpretation of wetting phenomena.

The relationship,

$$\left[h_{i(\mathrm{sl})} - h_{i(\mathrm{s/l})}\right] = \Gamma\left(E'_\mathrm{a} - E_\mathrm{l}\right) = \int_0^\Gamma q_{\mathrm{sl}} \mathrm{d}\Gamma - \Gamma \Delta H_\mathrm{l} \tag{2.171}$$

can be used to obtain the isosteric heat of adsorption, $q_{\mathrm{sl}}$, from heat of immersion data. These isosteric heats should agree with those obtained from the Clausius-Clapeyron equation from adsorption equilibrium pressure data at constant coverage at different temperatures.

Based on Hill[154], Chessick and Zettlemoyer[152] assumed that changes in the solid surface being covered by physical adsorption can be neglected, then the molar energy $E_\mathrm{a}$ or other pertinent thermodynamic functions of the adsorbate itself can be calculated from adsorption and calorimetric data.

In chemisorption, where severe surface perturbations can occur, the Clausius-Clapeyron equation cannot be applied, since equilibrium pressures are low and often unobtainable. Nonetheless, a differential heat analogous to the isosteric heat can be obtained from heats of immersion without recourse to pressure data where the amounts adsorbed prior to immersion can be measured gravimetrically.

The state of the adsorbed film on solids could be elucidated most simply from entropy data. If the entire entropy change can be attributed to the adsorbate, Jura and Hill[155] showed that the difference in entropy of the adsorbed film, $S_\mathrm{a}$, and the entropy of the bulk liquid, $S_\mathrm{l}$, is given by the expression:

$$T\left(S_\mathrm{a} - S_\mathrm{l}\right) = \frac{\left[h_{i(\mathrm{sl})} - h_{i(\mathrm{s/l})}\right]}{\Gamma} + \frac{\pi_\mathrm{e}}{\Gamma} - kT \ln X \tag{2.172}$$

Here, $X$ is the relative equilibrium pressure. Again, this equation for the change in $\Delta S$ is exact if the solid perturbation is not negligible, although the necessity for precise adsorption data over the entire range of equilibrium relative pressures severely limits its use.

### Measurement of Surface Energy of Solids

Surface free energy is one of the most important parameters determining the adsorptive and wetting characteristics of a solid in the presence of a vapor or liquid. These in turn influence flocculation aggregation, crystal growth, and indeed most other colloidal behavior exhibited by solid particles. Despite the importance of these energy values, only scanty data exist in the literature. Lack of knowledge of the nature of real surfaces precludes accurate calculations of these energy values from intermolecular potentials. Surface-energy values can be obtained from measured differences in the heats of solution of finely divided particles and large crystals of a substance.

Girifalco and co-workers[156], derived the following expression which can be used to estimate the surface energies of solids from the available heats of immersion:

$$e_{s^0} = \frac{\left[e_{lv^0} - h_{i(sl)}\right]^2}{4e_{lv^0}\Phi^2} \tag{2.173}$$

Here, $e_{lv^0}$, the surface energy of the wetting liquid, and the heat of immersion, $h_{i(sl)}$, are the measurable quantities. The quantity $\Phi$ is given by Eq. 2.174 and may be assumed equal to unity for certain solid-liquid systems.

$$\Phi = \frac{\gamma_{s^0} + \gamma_{lv^0} - \gamma_{sl}}{2\sqrt{\gamma_{s^0}\gamma_{lv^0}}} \tag{2.174}$$

### Homogeneous and Heterogeneous Polar Solid

Water is the most important substance in immersional calorimetry whether as the wetting liquid itself or as an impurity. Therefore, more comprehensive wetting studies have been conducted with solid-water systems. The various wetting curves for different solids are illustrated in Figure 2.52, where heat values are plotted against either the volume preadsorbed or the relative pressure of water at which the solids were equilibrated at 25°C before immersion into water.

Figure 2.52(a) and (b) are typical curves found for the heat of immersion of polar solids in water. An example of Figure 2.52(a) is found in the immersion of chrysotile asbestos having known and increasing amounts of physically adsorbed water on its surface[157]. The linear relationship between the heat of wetting and the volume adsorbed up to about a monolayer is significant and indicates surface homogeneity since the heat evolved is proportional to the amount of bare surface present.

The more frequently encountered exponential decrease in the heat of wetting with increasing surface coverage illustrated in Figure 2.52(b) was first observed by Harkins and Jura[158] for the system water-anatase ($TiO_2$). Such curves are indicative of a heterogeneous surface structure.

Figure 2.52(c) differs markedly from those obtained for the immersion of polar solids in water; initially the heat values are small but increase with increasing amounts of preadsorbed water. This is the case for the system Graphon-water[159]. Graphon is a graphitized carbon black which has an essentially homogeneous, homopolar surface[160]. Nevertheless, a small fraction of heterogeneous sites is responsible for the limited adsorption of water on the surface of this solid. Similar curves can be expected for other hydrophobic solids.

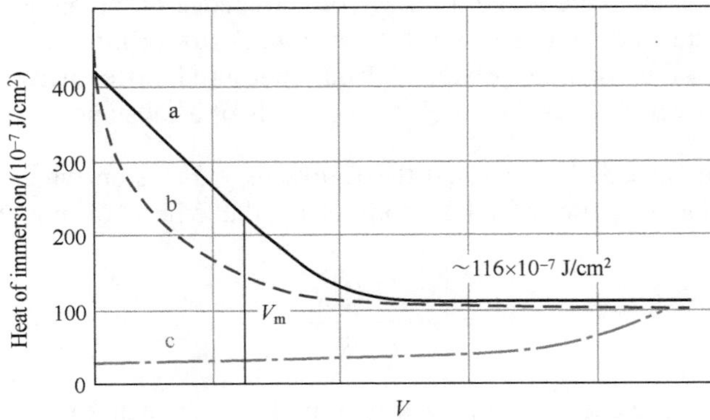


**Figure 2.52** Types of heat of immersion curves obtained for different solid as a function of pre-adsorbed wetting liquid: (a) homogeneous surface, e.g., water on some hydrophilic surfaces (chrysotile asbestos); (b) heterogeneous surfaces, e.g., water on $TiO_2$; (c) lyophobic surfaces, e.g., water on graphite surface[159]

In the comprehensive review of heat of immersion of solids, Chessick and Zettlemoyer[152] also described the application of heat immersion to determination of energetic heterogeneity and polarity of the solid surface with different types of liquids on different solids.

Some investigators have used heat of immersion data in determination of surface areas of carbon black[81,161], in studies of mechanism of carbon black oxidation[162] and energetic heterogeneity of the black surfaces in combination with gas adsorption[83].

Using three isoprenoid olefins and four liquid elastomers of polybutadiene, polyisoprene, and SBR types as the liquids, the immersion heats of a series carbon blacks, including furnace and channel blacks, were determined. It is surprising that the heats of immersion of a series of *n*-alkanes on carbon black N330 increase with the number of the carbon atoms of the liquid, and the differences in the heats of immersion of the blacks having different morphologies, surface area in particular, are not significant[163]. Similar results were also reported by Kraus[164].

Patrick[165] measured the heats of wetting of a fumed silica (high surface area Aerosil) in water after dehydrations up to 900°C and found a nearly linear decrease in the heat values as the surface hydroxyl content decreased. From these data, Iler[166] estimated the heat of wetting of a siloxane and silanol surface as −130 mJ/m$^2$ and −190 mJ/m$^2$, respectively.

A great number of authors have investigated the heats of immersion of different types of silicas and surface areas in both water and organic liquids with different polarities. Generally, the heat of immersion in water increases drastically with the surface concentration of silanols[167–172], and the dependence is less with surface area of silica[173–175].

### 2.4.2.3 Inverse Gas Chromatograph

Although several methods can be used to measure the solid surface energy of fillers, inverse gas chromatography (IGC) is one of the most sensitive and convenient methods for measuring filler surface energy. In IGC, the filler to be characterized is used as the stationary phase and the solute injected is called a probe. When the probe is operated at infinite dilution, the net retention volume determines the adsorption energy of the probe on the filler surface, and hence the surface energy of the filler. On the other hand, if the surface is energetically heterogeneous, the values of parameters obtained from IGC measurement are mean values over the whole surface of the fillers. However, they are "energy-weighted", i.e., the high-energy sites play a very important role in determining the adsorption parameters measured[176]. Operating the probe at finite concentration allows the determination of the pressure dependence of the retention volume to generate the probes' adsorption isotherms on the filler surface. These isotherms yield the distribution of the probe chemicals' free energy of adsorption[176].

#### 2.4.2.3.1 Principle of Measuring Filler Surface Energy with IGC

The thermodynamic parameters of the adsorption of the probe (solute) on the filler surface can be calculated from the retention data of the chromatogram and the filler surface energy can thus be determined[177–179].

In chromatography, the net retention volume is calculated from:

$$V_N = Dj\left(t_\mathrm{r} - t_\mathrm{m}\right)\left(1 - \frac{p_\mathrm{w}}{p_0}\right)\frac{T_\mathrm{c}}{T_\mathrm{f}} \tag{2.175}$$

where $t_\mathrm{r}$ is the retention time of the given probe, $t_\mathrm{m}$ is the zero retention time measured with a nonadsorbing probe (for example methane), $D$ is the uncorrected flow rate determined by timing a soap film, $p_0$ is the pressure at the flowmeter, $p_\mathrm{w}$ is the vapor pressure of the pure water at the flowmeter temperature, $T_\mathrm{c}$ is the column temperature, $T_\mathrm{f}$ is the retention flowmeter temperature, and $j$ is the James-Martin factor for the correction of gas compressibility if there is a pressure difference between column inlet ($p_\mathrm{in}$) and outlet ($p_\mathrm{out}$). It is obtained by:

$$j = \frac{3}{2}\frac{\left(p_\mathrm{in} / p_\mathrm{out}\right)^2 - 1}{\left(p_\mathrm{in} / p_\mathrm{out}\right)^3 - 1} \tag{2.176}$$

On the other hand, the general retention volume of the adsorbate (probe) is related to the gradient $\mathrm{d}\Gamma/\mathrm{d}c$ of the partition isotherm as[180]:

$$V_N = A\left(1 - jY_0\right)\left(\frac{\partial \Gamma}{\partial c}\right)_\mathrm{p} \tag{2.177}$$

where $c$ is the concentration of the probe in the gas phase, $\Gamma$ is the concentration of the probe on the solid surface, $p$ is the equilibrium pressure, $A$ is the total surface area of the solid adsorbent in the column, and $Y_0$ is the mole fraction of the probe in the gas phase at the column outlet. The term $jY_0$ in Eq. 2.177 is a correction factor for the sorption effect.

#### 2.4.2.3.2 Adsorption at Infinite Dilution

**Thermodynamic Parameters of Adsorption**

At very low pressure, the sorption effect is so small that it can be neglected. Hence, Eq. 2.177 can be written as:

$$V_N = A\left(\frac{\partial \Gamma}{\partial c}\right)_p \tag{2.178}$$

When adsorption takes place in the Henry's law region, i.e., at infinite dilution, one has:

$$\frac{V_N}{A} = \left(\frac{\partial \Gamma}{\partial c}\right)_{c\to 0} = \left(\frac{\pi}{p}\right)_{p\to 0} = K_s \tag{2.179}$$

with $p = cRT$ and $\pi = \Gamma RT$, where $\pi$ is the two-dimensional spreading pressure of the adsorbed probe on the filler surface, $K_s$ is the the surface partition coefficient of the given probe between the adsorbed and gaseous state, $R$ is the gas constant, and $T$ is the temperature.

The variation of the standard free energy in isothermal transfer of one mole of probe from reference pressure $p^0$ to the adsorbed state at equilibrium with a pressure $p$ of the gas phase is given by:

$$\Delta G^0 = -RT\ln\left(\frac{p^0}{p}\right) \tag{2.180}$$

When a surface with a spreading pressure $\pi^0$ is taken as the reference state of the surface, the change in standard free energy according to Eq. 2.180 for transferring one mole of probe from the gas phase at pressure $p^0$ to the solid surface at a spreading pressure $\pi^0$ is:

$$\Delta G^0 = -RT\ln\left(\frac{V_N p^0}{Sg\pi^0}\right) \tag{2.181}$$

Where $S$ is the specific surface area of the solid, and $g$ is the mass of the solid in the column.

If de Boer's surface state in which the average distance between adsorbed molecules is identical with the separation distance between molecules in the gas phase at 1 atm ($1.013 \times 10^5$ Pa) and 0°C is adopted[181], $\pi^0$ is $3.38 \times 10^{-4}$ N/m. Thus,

$$\Delta G^0 = -RT\ln\left(\frac{V_N}{Sg}2.99\times 10^8\right) \tag{2.182}$$

For practical reasons, $V_N$ is usually expressed in cm$^3$ so that:

$$\Delta G^0 = -RT\ln\left(299\frac{V_N}{Sg}\right) \tag{2.183}$$

At zero surface coverage, the enthalpy of adsorption, $\Delta H$, can be identified with the differential heat of adsorption. According to the Gibbs-Helmholtz equation, it can be calculated from the temperature dependence of the retention volume as:

$$-\frac{\partial\left(\Delta G^0 / T\right)}{\partial T} = \frac{\Delta H}{T^2} = R\frac{\mathrm{d}\left(\ln V_N\right)}{\mathrm{d}T} \tag{2.184}$$

or

$$\frac{\Delta H}{R} = \frac{\mathrm{d}\left(\ln V_N\right)}{\mathrm{d}\left(1/T\right)} \tag{2.185}$$

Thus, $\Delta H$ can be obtained from the slope of $\ln V_N$ *vs.* $1/T$. This parameter is independent of the choice of reference state of the filler surface. Consequently, the entropy of adsorption, $\Delta S^0$, is given by:

$$\Delta S^0 = \frac{\Delta H - \Delta G^0}{T} \tag{2.186}$$

### Evaluation of the Dispersive Component of Filler Surface Energy

Numerous studies have shown that the free energies of adsorption of a series of homologous *n*-alkanes on rubber-grade fillers and other solid surfaces at very low pressure vary linearly with the number of carbon atoms. The free energy of adsorption corresponding to one methylene group, $\Delta G_{\mathrm{CH_2}}$, can be obtained from the slope of the plot of $\Delta G^0$ of a series of *n*-alkanes *vs.* their carbon numbers and can also be calculated from Eq. 2.183 by:

$$\Delta G_{\mathrm{CH_2}} = -RT\ln\frac{V_{N(n)}}{V_{N(n+1)}} \tag{2.187}$$

where $V_{N(n)}$ and $V_{N(n+1)}$ are the retention volumes of *n*-alkanes with ($n$) and ($n + 1$) carbon atoms, respectively. $\Delta G_{\mathrm{CH_2}}$ provides an estimate of the dispersive interaction between a -$CH_2$- group and an adsorbent, since no polar interaction takes place between alkanes and a solid surface. According to Fowkes[131,182,183] (and Eq. 2.131), the adhesive work, $W_a$, between a nonpolar liquid and a solid surface is given by:

$$W_{\mathrm{a}} = 2\left(\gamma_{\mathrm{l}}^{\mathrm{d}}\gamma_{\mathrm{s}}^{\mathrm{d}}\right)^{1/2} \tag{2.188}$$

where $\gamma_{\mathrm{l}}^{\mathrm{d}}$ and $\gamma_{\mathrm{s}}^{\mathrm{d}}$ are the dispersive surface energies of the liquid and the solid, respectively.

In the case of an alkane, $\gamma_1^d$ is identical with its surface tension. Dorris and Gray proposed that, to a first approximation, the adhesive work is related to the increment in the free energy of adsorption associated with a $-CH_2-$ group, $\Delta G_{CH_2}$ :

$$\Delta G_{CH_2} = N_A a W_a \tag{2.189}$$

and hence,

$$W_a = \frac{\Delta G_{CH_2}}{N_A a} = 2\left(\gamma_{CH_2} \gamma_s^d\right)^{1/2} \tag{2.190}$$

where $N_A$ is Avogadro's number, $a$ is the area covered by a $-CH_2-$ group (0.06 $nm^2$), and $\gamma_{CH_2}$ is the surface tension of a surface composed of closely packed $-CH_2-$ groups analogous to polyethylene, which is given by[184] as:

$$\gamma_{CH_2} = 35.6 + 0.058(20 - T)(\text{mJ/m}^2) \tag{2.191}$$

where $T$ is the experimental temperature in °C. Therefore, by injecting a series of $n$-alkanes as probes, the free energy of adsorption generates the dispersive component of the filler surface energy.

#### Evaluation of the Specific Component of Filler Surface Energy

The specific component of filler surface energy can be estimated from the specific interaction that the filler is able to exchange with polar probes. Fowkes proposes[131] that the adhesive energy can be divided into several terms as expressed by Eq. 2.133. To a first approximation, the adhesive energy, $W_a$, can simply be expressed as the sum of two terms, the interaction due to dispersive forces per unit surface area, $W_a^d$, and the interaction due to forces called specific which represents the sum of all nondispersive interaction, $W_a^{sp}$, Hence,

$$W_a = W_a^d + W_a^{sp} \tag{2.192}$$

In the case of the adsorption of a probe on the stationary phase in the chromatography column, one can write:

$$-\Delta G^0 = N_A a W_a^d + N_A a W_a^{sp} \tag{2.193}$$

If the probe is able to exchange only dispersive interaction with the solid surface, this equation reduces to Eq. 2.189, which is the case for alkane adsorption.

Several approaches were proposed to distinguish dispersive interaction from specific interaction[185–190]. Each approach has been successful in estimating the specific interaction of given surfaces with given probes[1–4,50,117–119,140,178–180]. A number of practical problems arise with different surfaces or when a variety of probes are used. The method proposed by Wang and co-workers[191] which is described below can easily be used for estimating the specific interaction between a probe and rubber-grade fillers, such as carbon blacks, silicas, clays, $CaCO_3$, etc.

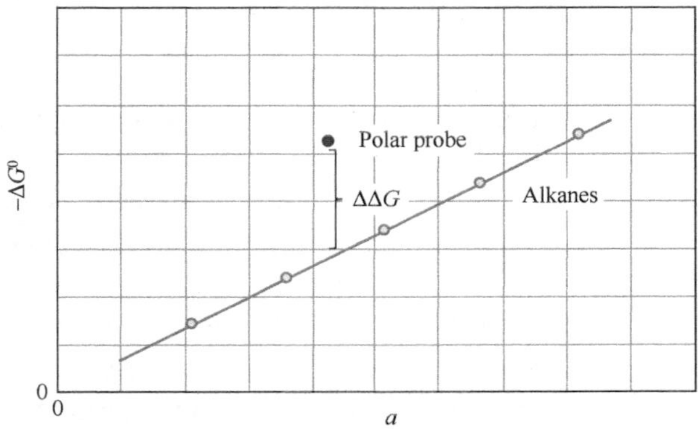


**Figure 2.53** Schematic principle of estimating specific interaction

In practice, a linear relationship between the free energy of adsorption of a series of alkanes and the surface area of their molecules is obtained, and this straight line provides a reference. The experimental points corresponding to polar probes able to exchange specific interaction with a filler surface always lie well above the reference line, as shown schematically in Figure 2.53. Thus, Eq. 2.193 can be rewritten as:

$$-\Delta G^0 = N_A a W_a^{d-ref} + N_A a W_a^{sp} \tag{2.194}$$

or as

$$-\Delta G^0 = -\Delta G^{0-ref} + N_A a W_a^{sp} \tag{2.195}$$

Therefore,

$$-\Delta\Delta G = -\left(\Delta G^0 - \Delta G^{0-ref}\right) = N_A a W_a^{sp} \tag{2.196}$$

where the $-\Delta G^{0-ref}$ is the free energy of adsorption of an alkane (real or hypothetical) with a surface area identical with that of the given polar probe. Consequently, at a given surface area of a polar molecule, the difference in ordinate between the point corresponding to the specific probe and the reference line leads to the value of specific free energy of adsorption, expressed by the term $\Delta\Delta G$ in Figure 2.53. The specific interaction per unit surface, $I^{sp}$, can thus be calculated from:

$$I^{sp} = W_a^{sp} = \frac{-\Delta\Delta G}{N_A a} = f\left(\gamma_s^{sp}\right) \tag{2.197}$$

The surface area, $a$, occupied by the probe molecule may vary with the adsorbent and measurement conditions. For the same probe, variations in the values reported in the literature are sometimes as high as 60%[176,192–194]. This surface area may also be calculated on the basis of the closest packing, either from the cylinder model for linear hydrocarbons and their derivatives[195], or, in practice, from the liquid density and the molecular weight[196].

Although this treatment is empirical, it allows a comparison of the surface polarities of different fillers and the specific interaction between a filler and an adsorbate by using a unified scale.

### $S_f$ Factor

Generally speaking, while fillers have polar surfaces, hydrocarbon rubbers generally possess no or only low polarity. The higher adsorption energy of a nonpolar probe such as an alkane, resulting from a higher $\gamma_s^d$ of the filler surface, would be indicative of stronger interaction between the filler and hydrocarbon rubbers. On the other hand, the higher adsorption energy of a highly polar probe such as acetonitrile, which is associated with a high $\gamma_s^{sp}$ of the filler, would be representative of pronounced filler aggregate-aggregate interaction. If $\Delta G^0$ of a nonpolar hydrocarbon probe on the filler is constant, increases in $\Delta G^0$ for a highly polar probe should promote incompatibility of the filler with hydrocarbon rubbers, thus enhancing aggregate agglomeration. By the same token, at the same level of $\Delta G^0$ of a highly polar probe on the filler, the higher $\Delta G^0$ of a nonpolar hydrocarbon probe would indicate greater filler-polymer interaction and a lesser degree of association of the aggregates would be expected. Based on this consideration, a specific interaction factor $S_f$ was defined as the adsorption energy of a given probe, $\Delta G^0$, divided by that of an alkane (real or hypothetical), $\Delta G^0_{alk}$, the surface area of which is identical with that of the given adsorbent (Figure 2.54):

$$S_f = \frac{\Delta G^0}{\Delta G^0_{alk}} = f\left(\gamma_s^d, \gamma_s^{sp}\right) \tag{2.198}$$

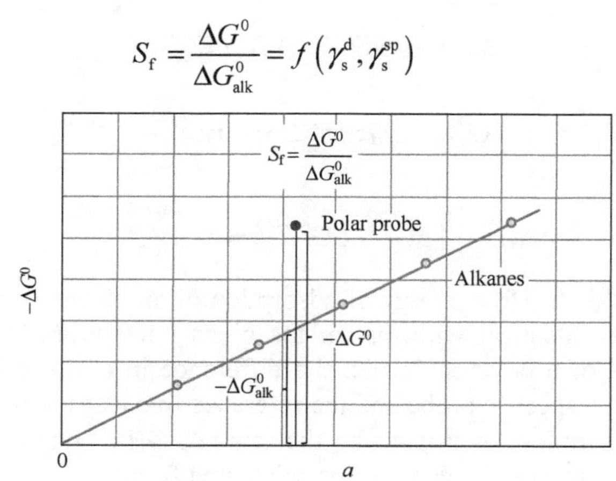


**Figure 2.54** Definition of the factor $S_f$

In this definition, the reference state for the calculation of $\Delta G^0$ is the $\Delta G^0$ of a hypothetical alkane with zero surface area which is extrapolated from $\Delta G^0$ of a series of *n*-alkanes[22]. The factor $S_f$ is related to both the dispersive and specific components of filler surface energy. The higher the value of $S_f$ for the adsorption of a highly polar probe, the stronger is the inter-aggregate interaction and the more developed is the agglomeration of filler aggregates in hydrocarbon rubbers.

Table 2.14 gives the $\gamma_s^d$ data of 26 commercial carbon blacks and 4 silicas along with adsorption parameters.

**Table 2.14** Surface energies and adsorption parameters of some probes on commercial fillers at 180°C

| Filler | $\gamma_s^d$/(mJ/m²) | $I^{sp}_{benzene}$/(mJ/m²) | $I^{sp}_{MeCN}$/(mJ/m²) | $S_f$/(benzene) | $S_f$/(MeCN) |
|---|---|---|---|---|---|
| N115 | 411.8 | 105 | 169 | 1.27 | 1.44 |
| Black 1 | 429.0 | 105 | 173 | 1.27 | 1.44 |
| N121 | 361.6 | 102 | 207 | 1.28 | 1.56 |
| N220-1 | 378.0 | 107 | 186 | 1.28 | 1.49 |
| N220-2 | 395.8 | 113 | 190 | 1.29 | 1.49 |
| N234 | 403.1 | 105 | 197 | 1.27 | 1.51 |
| N339 | 286.6 | 99 | 181 | 1.30 | 1.55 |
| N326-1 | 260.2 | 100 | 175 | 1.32 | 1.56 |
| N326-2 | 271.2 | 97 | 178 | 1.30 | 1.55 |
| N332 | 275.4 | 100 | 183 | 1.31 | 1.57 |
| N347 | 274.3 | 97 | 177 | 1.30 | 1.55 |
| N375 | 267.8 | 98 | 172 | 1.31 | 1.54 |
| N330-1 | 272.7 | 98 | 183 | 1.30 | 1.57 |
| N330-2 | 276.2 | 99 | 186 | 1.31 | 1.57 |
| Black 2 | 238.1 | 99 | 187 | 1.33 | 1.62 |
| Black 3 | 271.5 | 103 | 187 | 1.32 | 1.58 |
| Black 4 | 270.3 | 98 | 194 | 1.30 | 1.60 |
| Black 5 | 285.8 | 105 | 180 | 1.32 | 1.52 |
| N550 | 189.5 | 88 | 163 | 1.32 | 1.60 |
| N660 | 132.8 | 67 | 137 | 1.29 | 1.60 |
| N683 | 153.8 | 76 | 167 | 1.31 | 1.69 |
| N772 | 129.0 | 66 | 150 | 1.29 | 1.67 |
| N765 | 137.8 | 69 | 146 | 1.30 | 1.62 |
| Silica P1 [a] | 22.9 | 64 | 252 | 1.66 | 3.58 |
| Silica P2 [a] | 34.3 | 71.9 | 285 | 1.60 | 3.37 |
| Silica A1 [a] | 30.7 | 45.4 | 156 | 1.40 | 2.38 |
| Silica A2 [a] | 44.3 | 55.1 | 173 | 1.40 | 2.26 |

a. at 150°C, Silica P1: Ultrasil VN2; Silica P2: Ultrasil V+; Silica A1: Acrosil 130; Silica A2: Acrosil 200.

### 2.4.2.3.3 Adsorption at Finite Concentration

#### Adsorption Isotherm

For IGC operated at finite concentration, i.e., by injecting a specified amount of a liquid probe, the adsorption isotherm can be constructed from the pressure dependence of the retention volume. Considering the adsorption effect and gas compressibility and ignoring gas imperfection, Eq. 2.177 can be written in integral form as:

$$\Gamma = \frac{1}{A}\int_0^p \frac{V_N}{1 - jY_0}\,\mathrm{d}c \tag{2.199}$$

or

$$q = \frac{1}{gRT}\int_0^p \frac{V_N}{1 - jY_0}\,\mathrm{d}p \tag{2.200}$$

where $q$ is the amount adsorbed in mol/g.

The equilibrium pressure can be obtained by[3]:

$$p = \frac{VURTh}{\overline{V}D_s S_p} \tag{2.201}$$

where $V$ is the volume of the liquid probe injected, $\overline{V}$ is the molar volume of the probe, $U$ is the recorder chart feed, $D_s$ is the average carrier gas flow at column temperature $T$ (in $K$), $S_p$ is the peak area for the injected volume of the probe, and $h$ is the height of the chromatographic peak. Thus, a series of chromatograms corresponding to discrete sample injection of known volumes determine the adsorption isotherm.

#### Spreading Pressure

When gas is adsorbed on a solid surface, it leads to a spreading pressure, $\pi$, which is defined as:

$$\pi = \gamma_s + \gamma_{sv} \tag{2.202}$$

where $\gamma_s$ and $\gamma_{sv}$ are the surface free energies of the solid at the solid-vacuum interface and the solid-vapor interface, respectively. The spreading pressure can be calculated from an adsorption isotherm using the integrated form of Gibbs' adsorption equation[197,198]:

$$\pi = \frac{RT}{A}\int_0^p q\frac{\mathrm{d}p}{p} \tag{2.203}$$

The pressure dependence of $q$ can be obtained from Eq. 2.200. Investigating the spreading pressure could provide information concerning the change in the surface free energy of a solid upon adsorption.

### Difference in Chemical Potential between Polar and Nonpolar Probes

A comparison of the chemical potentials of adsorption of a polar probe (p) and a nonpolar probe (np) serving as a reference, could yield additional information concerning the specific interaction between the filler surface and a polar probe[193,199]. At equilibrium conditions, the chemical potential of a specific amount of probe adsorbed on a solid surface, $\mu_a^{ads}$, is equal to the chemical potential of the same probe in the gas phase at the related pressure $p$, $\mu_p^{gas}$. Thus,

$$\mu_a^{ads} = \mu_p^{gas} = \mu^0 + RT\ln\left(\frac{p}{p^0}\right) \tag{2.204}$$

where $\mu^0$ and $p^0$ are the chemical potential and the pressure at reference state, respectively. Taking the same reference state for these two probes (identical pressure, temperature, and adsorbed quantity for probe p and np) and assuming that they behave as ideal gases, one has:

$$\mu_{(p,a)}^{ads} - \mu_{(np,a)}^{ads} = \Delta\mu_{(p/np,a)}^{ads} \tag{2.205}$$

and

$$\Delta\mu_{(p/np,a)}^{ads} = \Delta\mu_{(p/np)}^{0} + RT\ln\left(\frac{p_{(p,a)}}{p_{(np,a)}}\right) \tag{2.206}$$

with $\Delta\mu_{(p/np)}^{0} = \mu_{(p)}^{0} - \mu_{(np)}^{0}$, a constant which is independent of the solid surface. Thus,

$$\Delta\Delta\mu^{ads} = \Delta\mu_{(p/np,a)}^{ads} - \Delta\mu_{(p/np)}^{0} = RT\ln\left(\frac{p_{(p,a)}}{p_{(np,a)}}\right) \tag{2.207}$$

Therefore, $\Delta\Delta\mu^{ads}$ can be used for the evaluation of $\Delta\mu_{(p/np,a)}^{ads}$ for different solids at different amounts adsorbed and may thus serve as a measure of the surface polarity of silicas.

### Energy Distribution Function of Adsorption

When inverse gas-solid chromatography is operated at infinite dilution (zero surface coverage), the thermodynamic functions of adsorption are only dependent on adsorbate-filler surface interaction, and the thermodynamic parameters, and hence the surface energies, can be calculated from the retention volumes, $V_N$[200–202]. As shown in Eq. 2.179, this volume is related to the total surface area of the filler in the column, $A$, and the surface partition coefficient, $K_s$, of the adsorbate between the adsorbed and the gaseous phase by:

$$V_N = AK_s \tag{2.208}$$

If the surface is energetically heterogeneous, the statistical treatment of the chromatographic process leads to the same result as that obtained by assuming an additive function of the elementary processes taking place in small patches within which adsorption is homogeneous but between which adsorption varies. Each patch is characterized by its surface partition coefficient $K_{s,i}$. The corresponding retention volume, $V_{N,i}$, is:

$$V_{N,i} = A_i K_{s,i} \tag{2.209}$$

where $A_i$ is the surface area of the $i$th surface patch. For $n$ patches, the total retention volume $V_N$ can thus be given by[23]:

$$V_N = AK_s = \sum_{i=1}^{n} A_i K_{s,i} = A\left(\sum_{i=1}^{n} A_i K_{s,i} / \sum_{i=1}^{n} A_i\right) \tag{2.210}$$

with

$$A = \sum_{i=1}^{n} A_i \tag{2.211}$$

It is obvious from Eq. 2.210 that the surface partition coefficient $K_s$ is a mean value, and so are the thermodynamic parameters of adsorption and the surface energies. However, these values are "energy-weighted" since $K_s$ is related to the adsorption energy, i.e., the high-energy patches (or high-energy sites) play a very important role in the determination of adsorption properties and surface energies of the fillers[23].

The precise energetic heterogeneity of the filler surface can be estimated from the energy distribution function of adsorption of a given probe. For an adsorbate-adsorbent system, the adsorption energy distribution function, $\chi(\varepsilon)$, can be defined as:

$$\chi(\varepsilon) = \frac{\partial N}{\partial \varepsilon} \tag{2.212}$$

where $\varepsilon$ is the energy of adsorption, and $N$ is the volume or number of adsorption sites.

As mentioned earlier, for energetically heterogeneous surfaces, the entire chromatographic process can be treated as an additive function of elementary processes taking place in small patches. Each patch possesses an adsorption energy $\varepsilon_i$. According to Eq. 2.178, where the sorption effect can be neglected, for $n$ patches, the total retention volume $V_N$ can be given by[202]:

$$V_N = \sum_{i=1}^{n} \frac{A_i \partial \Gamma_i (c, \varepsilon_i, T)}{\partial c} = \sum_{i=1}^{n} \frac{\partial \theta_1 (c, \varepsilon_i, T)}{\partial c} N_{m,i} \tag{2.213}$$

where $\Gamma_i (c, \varepsilon_i, T)$ is the concentration of the adsorbate on the $i$th surface patch, $N_{m,i}$ is the capacity of the $i$th surface patch, and $\theta_1 (c, \varepsilon_i, T)$ is the adsorption isotherm function expressed by the coverage of $i$th patch.

Equation 2.213 can be written in integral form:

$$V_N = \int_{\Omega} \frac{\partial \theta_1(c,\varepsilon_i,T)}{\partial c} \chi(\varepsilon)\,\mathrm{d}\varepsilon \tag{2.214}$$

where $\Omega$ is the range of possible variation of $\varepsilon$.

In Eq. 2.214, the energy distribution function, $\chi(\varepsilon)$, requires the following condition for normalization:

$$\int_{\Omega} \chi(\varepsilon)\,\mathrm{d}\varepsilon = N_{\mathrm{m}} \tag{2.215}$$

where $N_{\mathrm{m}}$ is the total capacity of the surface phase.

For an ideal gas, one has:

$$V_N = f(p) = RT \int_{\Omega} \frac{\partial \theta_1(p,\varepsilon,T)}{\partial p} \chi(\varepsilon)\,\mathrm{d}\varepsilon \tag{2.216}$$

In Eq. 2.216, the adsorption model in an energetically homogeneous patch of the surface determines the function $\partial\theta_1(p,\varepsilon,T)/\partial p$.

Among the various adsorption models proposed, the model suggested by Hobson[203] for the local adsorption isotherm on each patch with an energy $\varepsilon$ follows the relation:

$$\theta_1(p,\varepsilon,T) = \frac{p}{K} = \exp\left(\frac{\varepsilon}{RT}\right) \quad \text{for } p < p' \tag{2.217}$$

$$\theta_1(p,\varepsilon,T) = 1 \quad \text{for } p \geqslant p' \tag{2.218}$$

$$p' = K\exp\left[\frac{-(\varepsilon - \varepsilon_0)}{RT}\right] \tag{2.219}$$

Since, at a pressure higher than $p'$, condensation occurs and the isotherm rises vertically, $\varepsilon_0$ would represent the interaction energy between adsorbate molecules on the surface. $K$ is the pre-exponential factor of Henry's constant. The numerical value of $K$ depends only upon the nature of the adsorbate and can be obtained from the saturated pressure and heat of evaporation[204], or from the pressure and temperature dependence of the chromatographic retention volume[205]. In Hobson's model, this factor is determined by the molecular weight of the adsorbate, $M$, and temperature as:

$$K = 1.76\times 10^4 (MT)^{1/2} \tag{2.220}$$

It has been noted that the shape of the energy distribution function, $\chi(\varepsilon)$, is independent of the parameter $K$, and its different values result only in a shift of the function $\chi(\varepsilon)$, along the energy axis $\varepsilon$.

According to Hobson and Rudzinski et al.[201,203], the choice of $\varepsilon_0$ has only a minor influence on the path of $\chi(\varepsilon)$. At normal conditions, $\varepsilon_0$ is assumed to be zero and $\varepsilon$ can thus be defined as:

$$\varepsilon = -RT \ln \frac{p}{K} \tag{2.221}$$

With this adsorption model, the term $\partial\theta_1/\partial p$ can be expressed by:

$$\frac{\partial\theta_1(p,\varepsilon,T)}{\partial p} = \frac{1}{K}\exp\left(\frac{\varepsilon}{RT}\right) \tag{2.222}$$

Then from Eqs. 2.216 and 2.222 we obtain:

$$V_N = RT\int_{\Omega}\frac{1}{K}\exp\left(\frac{\varepsilon}{RT}\right)\chi(\varepsilon)\,\mathrm{d}\varepsilon \tag{2.223}$$

Differentiating Eq. 2.223 with respect to $\varepsilon$, one has:

$$\frac{\partial V_N(\varepsilon,T)}{\partial\varepsilon} = \frac{RT}{K}\exp\left(\frac{\varepsilon}{RT}\right)\chi(\varepsilon) \tag{2.224}$$

Thus the energy distribution function can be written as:

$$\chi(\varepsilon) = \frac{K}{RT}\exp\left(\frac{-\varepsilon}{RT}\right)\frac{\partial V_N(\varepsilon,T)}{\partial\varepsilon} \tag{2.225}$$

and from Eq. 2.221 one obtains:

$$\chi(\varepsilon) = \frac{p}{RT}\times\frac{\partial V_N(\varepsilon,T)}{\partial\varepsilon} \tag{2.226}$$

and

$$\mathrm{d}\varepsilon = -\frac{RT}{p}\mathrm{d}p \tag{2.227}$$

Thus, we have:

$$\chi(\varepsilon) = -\left(\frac{p}{RT}\right)^2\frac{\partial V_N(p,T)}{\partial p} \tag{2.228}$$

Equation 2.228 is a general equation allowing the retention data of chromatographic measurements to directly reveal the energetic heterogeneity of the adsorbent's surface. Differentiating the retention volume presented as a function of pressure determines the energy distribution function, $\chi(\varepsilon)$.

### 2.4.2.3.4 Surface Energy of the Fillers

#### 2.4.2.3.4.1 Dispersive Component of Filler Surface Energy

**Silicas**

Figure 2.55 presents the values for $\gamma_s^d$ of silicas as a function of temperature. P1 and P2 are precipitated silicas, Ultrasil VN2 and VN3; and A1 and A2 are fumed silicas, Aerosil 130 and 200. BET surface areas of these silicas are 134 $m^2/g$, 181 $m^2/g$, 128 $m^2/g$, and 189 $m^2/g$, respectively. The temperature dependence of the $\gamma_s^d$ of the silicas, determined at temperatures ranging between 70°C and 150°C, and the values at 20°C obtained by extrapolation, are shown in Table 2.15. Since $d\gamma_s^d / dT$ is higher for precipitated silicas, it appears that, regardless of the temperature over the range investigated, fumed silicas are associated with a higher $\gamma_s^d$ than precipitated silicas. Moreover, the $\gamma_s^d$ values also show a dependence on particle size. The increase in particle size (or decrease in specific surface area[5,206]) points towards an attenuation of the $\gamma_s^d$ for silicas, irrespective of their production process.

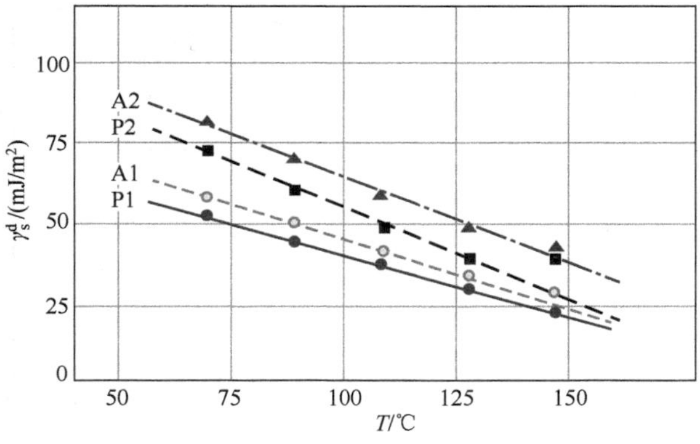


**Figure 2.55** Variation in $\gamma_s^d$ of silicas with temperature

**Table 2.15** $\gamma_s^d$ of silicas and their temperature dependence

| Silica | $\gamma_s^{d*}$(20°C)/(mJ/m²) | $(-d\gamma_s^d/dT)$/[mJ/(m²·°C)] |
|---|---|---|
| P1 | 72.3 | 0.392 |
| P2 | 100.6 | 0.562 |
| A1 | 74.4 | 0.362 |
| A2 | 104.6 | 0.490 |

* Extrapolated values.

The difference in $\gamma_s^d$ between precipitated and fumed silicas may be related to their surface complexity. It seems that the concentration of hydroxyl groups, which have been considered as high-energy sites, is not the dominant factor for $\gamma_s^d$, since these groups are much more frequent on the precipitated silica surfaces (generally 2 OH/nm$^2$ to 3 OH/nm$^2$ for fumed silicas and 4 OH/nm$^2$ to 8 OH/nm$^2$ for precipitated silicas). One of the reasons for the higher $\gamma_s^d$ of fumed silicas may be their flatter surface, which allows the probe molecules to fix more easily on them. This results in a higher free energy of adsorption of the -$CH_2$- groups from which $\gamma_s^d$ is calculated. The development of the crystal lattice in short range order may be another reason for the different $\gamma_s^d$ among silicas, especially for the dependence on particle size. The less organized micro-structure of the surfaces of finer silicas may be responsible for their higher surface energies.

### Carbon Blacks

Figure 2.56 shows $\gamma_s^d$ of dry pelletized carbon blacks extracted with toluene at room temperature, plotted against nitrogen surface area; and Figure 2.57 shows $\gamma_s^d$ *vs.* surface area for wet pelletized carbon blacks extracted in a Soxhlet extractor with toluene (hot extraction)[22]. In both cases, $\gamma_s^d$ increases with surface area in a more or less linear fashion.

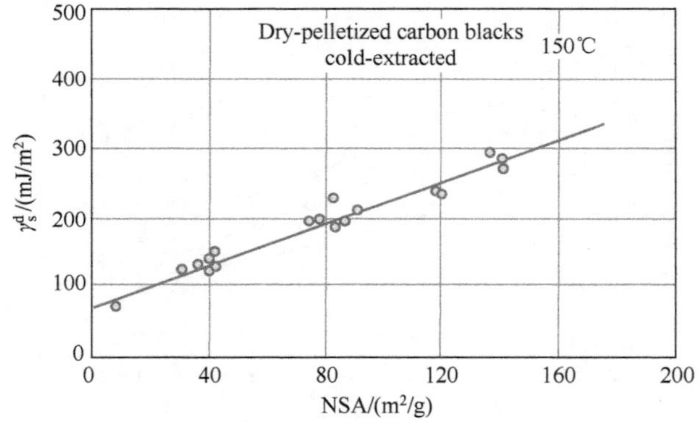


**Figure 2.56** $\gamma_s^d$ *vs.* nitrogen surface area for dry-pelletized carbon blacks at 150°C

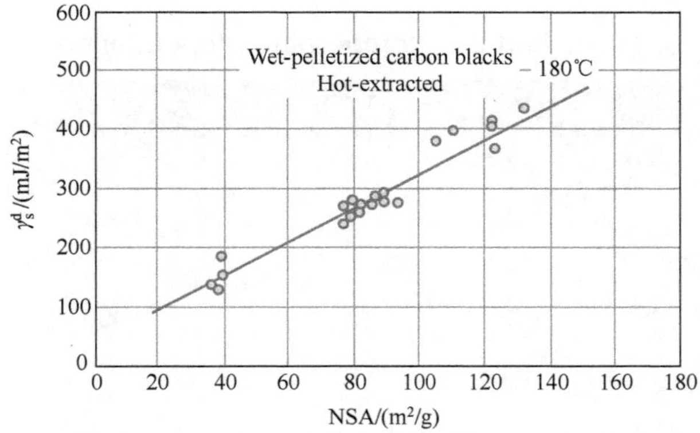


**Figure 2.57** $\gamma_s^d$ *vs.* nitrogen surface area for wet-pelletized carbon blacks at 180°C

The higher values of $\gamma_s^d$ for wet pelletized carbon blacks may be associated with high efficiency of the Soxhlet extraction at high temperature, resulting in more exposed active centers. The dependence of $\gamma_s^d$ on carbon black grade can be interpreted in terms of their micro-structure. It is known that in carbon black the carbon atoms are largely considered to be parts of quasi-graphitic crystallites, consisting of several turbostratic layers at an interplanar distance, *d*, of 0.35 nm to 0.38 nm. The crystallite dimension is characterized by the average stacking height of the parallel planes in the *c* direction, $L_c$, and the average diameter of the parallel layers in *ab* plane, $L_a$. The difference in $\gamma_s^d$ between carbon blacks may be associated with differences in their micro-structure rather than their surface area or particle size. The apparent surface area dependence of $\gamma_s^d$ may reflect the particle size dependence of the micro-structure of the blacks. Significant variations in the crystallographic parameters of carbon blacks, measured with X-ray diffraction, have been reported[22,207,208], generally showing a decrease of the crystallite dimensions with increasing surface area. The investigation of carbon black surface with scanning tunneling microscopy (STM) has also demonstrated this. It is evident from the STM images (Figure 2.58) that the organized portion of the carbon atoms on the surface is more developed for large-particle size carbon blacks[209].

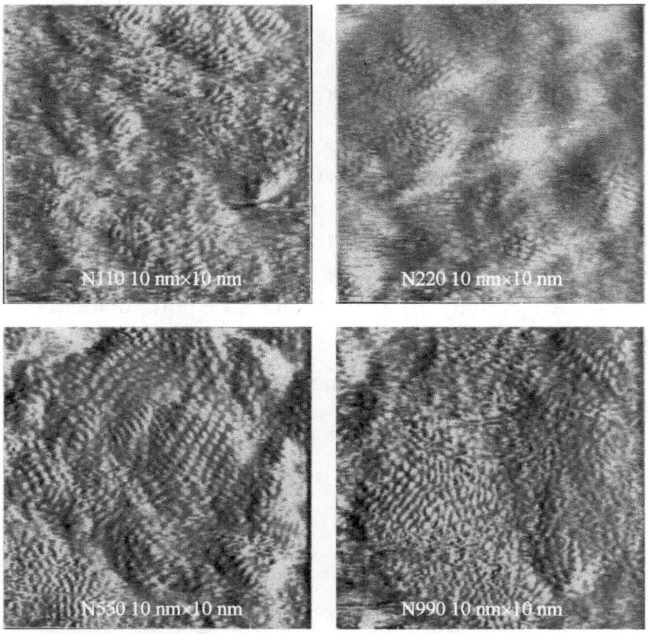


**Figure 2.58** Top-view STM images of carbon black N110, N220, N550, and N990

In Figure 2.59, $\gamma_s^d$ is presented as a function of $L_c$, indicating that surface energy decreases with increasing crystallite dimensions. This is understandable since more crystal edges and a greater unsaturated charge, which can be considered as high-energy sites, would be expected in smaller crystallites and less developed organization of the carbon atoms, corresponding to smaller particle size products. It is now almost axiomatic that the surface area dependence or particle size dependence of carbon black surface energy is more representative of the effect of micro-structure.

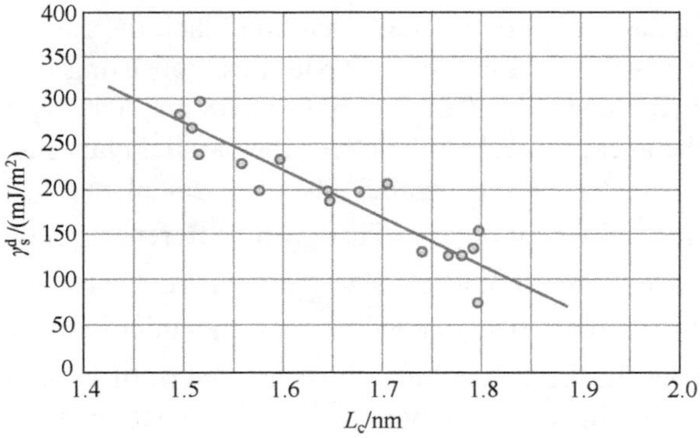


**Figure 2.59** $\gamma_s^d$ *vs.* $L_c$ for dry-pelletized carbon blacks at 150°C

The graphitization of carbon black also confirms the micro-structure dependence of the surface free energy. A drastic decrease in $\gamma_s^d$ of N330 from 276 mJ/m$^2$ to 171 mJ/m$^2$ at 180°C upon graphitization is obviously due to the growth of crystallites and the atom organization. This can be seen from Figure 2.60 where the surface organization imaged by STM is compared. The fact that the $\gamma_s^d$ of the graphitized black is still higher than that of graphite (see below) may be caused by the incomplete graphitization and/or limited growth of the crystallites. As shown in Figure 2.61, after graphitization of the carbon black at 2700°C for 48 hours, there is still certain number of defects remaining in the graphitic layers that may be considered as high energetic sites. Plasma treatment provides further evidence demonstrating the role of crystallographic defects in determining the surface activity of carbon black. Upon treatment of a graphitized carbon black N550 with air plasma, $\gamma_s^d$ of the black increases drastically from 150 mJ/m$^2$ to 420 mJ/m$^2$, while STM images show numerous crystal defects on the graphitic layer (Figure 2.62)[210,211].

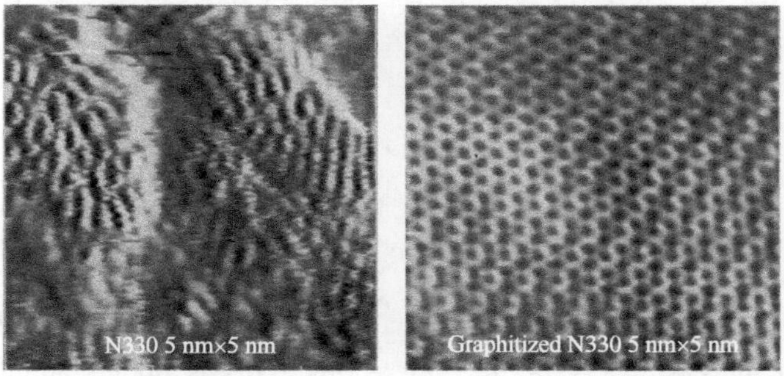


**Figure 2.60** Top-view STM images of carbon black N330 and its graphitized counterpart

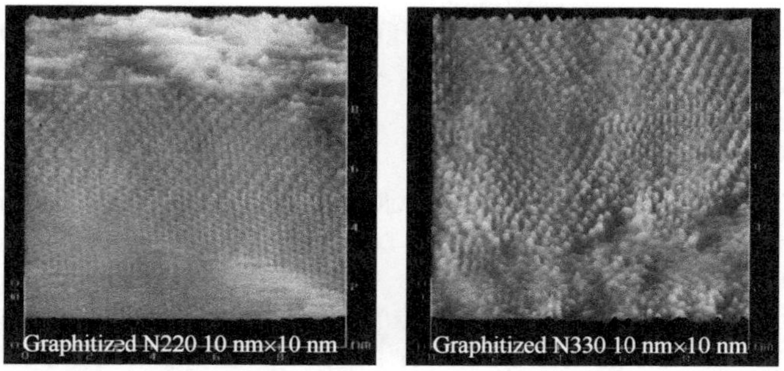


**Figure 2.61** Perspective-view STM images of carbon black N220 and N330, showing the defects of surface graphitic layers

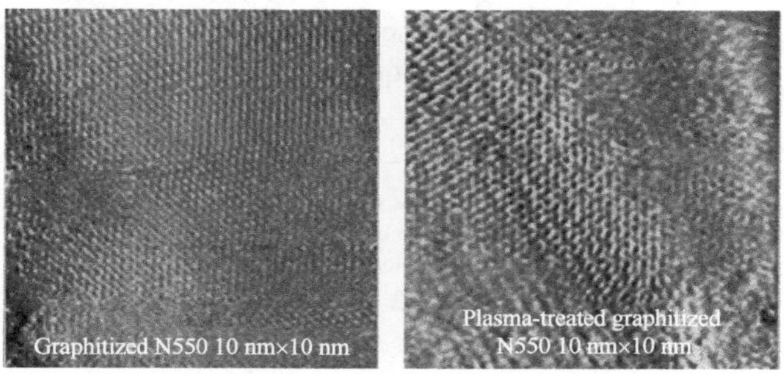


**Figure 2.62** Top-view STM images of graphitized N550 and its air plasma treated counterpart

At present, all test results seem to indicate that the high-energy centers on the carbon black surface originate mainly from defects and edges of the graphitic layers and unorganized carbon atoms, at least as far as the dispersive component of the surface energy is concerned. The coverage of these centers by adsorbed substances or/and by some functional groups, such as oxygen-containing groups, would reduce surface activity, particularly the ability for physical adsorption.

#### 2.4.2.3.4.2 Specific Component of Filler Surface Energy

**Silicas**

In contrast to the dispersive component of surface energy of silicas, which show higher values for fumed silicas than for precipitated ones, the specific component of surface energy, estimated from the specific interaction free energy, $I^{sp}$, of benzene and acetonitrile, was found to be higher for precipitated silicas (Table 2.14)[191]. Undoubtedly the high concentration of hydroxyl groups on precipitated silicas plays a very important role with regard to their higher $I^{sp}$. These groups possess not only a strong ability to polarize the benzene molecule, forming induced polar interaction, but also a strong ability to exchange dipole-dipole interaction with the polar probe acetonitrile. In the case of acetonitrile, a specific interaction due to hydrogen bonding between nitrogen atoms and the silanols may be involved. On the other hand, as in the case of $\gamma_s^d$, a small but definitely higher polarity is observed for both types of silicas when dealing with the finer silicas.

### Carbon Blacks

A comparison of dry pelletized carbon blacks is shown in Figure 2.63 and Figure 2.64, where the $I^{sp}$ of benzene and acetonitrile respectively are plotted against the surface areas of the carbon black[22]. The results suggest that the smaller particle products exhibit higher specific interaction with polar probes even though the data are somewhat scattered. This pattern is similar for wet-pelletized carbon blacks. However, similar to the dispersive components of surface energies, the surface area dependence of the specific interaction is more representative of surface state rather than surface area itself. One reason is that the specific interaction of the channel black with the polar probes is slightly but definitely higher than that expected from the surface area dependence of furnace blacks. This is undoubtedly related to its high concentration of oxygen groups, which are polar sites by nature.

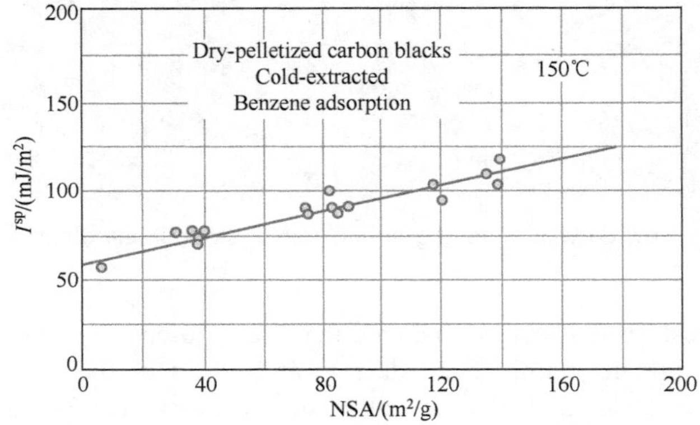


**Figure 2.63** $I^{sp}$ of benzene *vs.* nitrogen surface area for dry-pelletized carbon blacks at 150°C

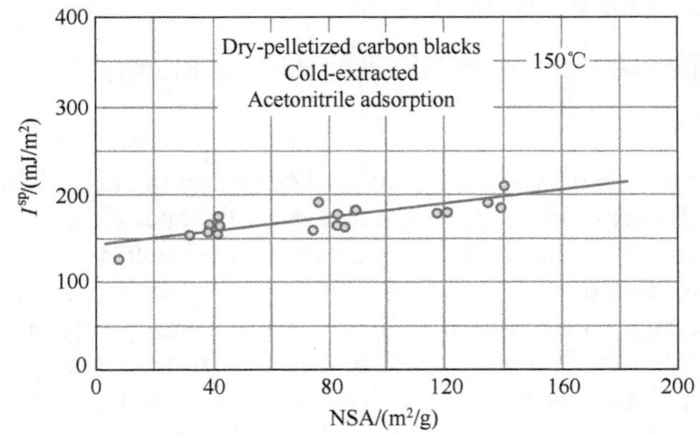


**Figure 2.64** $I^{sp}$ of acetonitrile *vs.* nitrogen surface area for dry-pelletized carbon blacks at 150°C

It is also found that the graphitization of carbon black produces a considerable decrease in the specific component of the surface energy. Taking N330 as an example, $I^{sp}$ of 186 mJ/m$^2$ for acetonitrile adsorption on original carbon black diminished to 95 mJ/m$^2$ for its graphitized counterpart. This is probably due to the crystallites' growth mechanism (decrease of crystal structure defect) as well as a reduced number of oxygen groups that are completely decomposed below 800°C.

#### 2.4.2.3.4.3 $S_f$ Value

The $S_f$ of benzene and acetonitrile on carbon blacks and silicas are illustrated in Figure 2.65 and Figure 2.66, respectively[22]. In the case of benzene adsorption, silicas, especially precipitated silicas, show higher relative polarity than carbon blacks. The values of $S_f$ seem to be constant for all carbon blacks, only very slightly higher for the large particle blacks. The same can be seen from acetonitrile adsorption, but the difference in $S_f$ between carbon blacks and silicas is much greater, due to the strong polar interaction as well as hydrogen bonding between silica surface and the probe, as mentioned earlier.

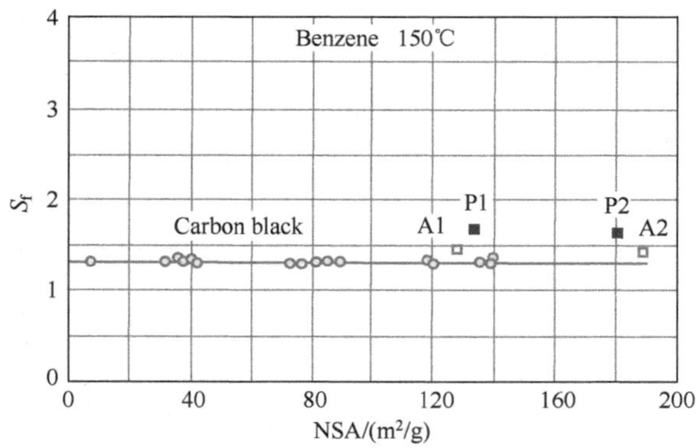


**Figure 2.65** $S_f$ of benzene *vs.* nitrogen surface area for dry-pelletized carbon blacks and silicas at 150°C

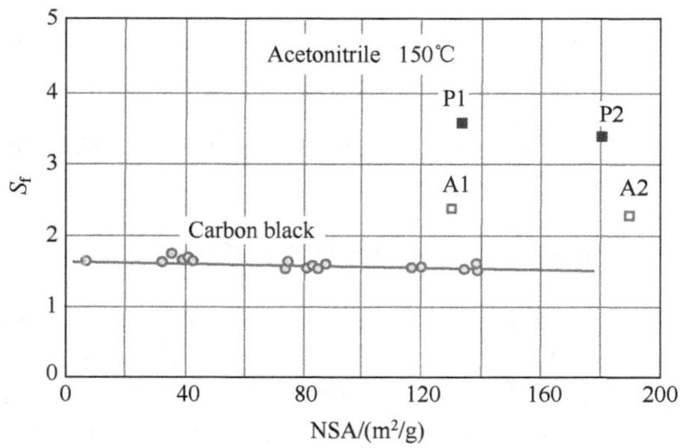


**Figure 2.66** $S_f$ of acetonitrile *vs.* nitrogen surface area for dry-pelletized carbon blacks and silicas at 150°C

Acid-base properties of filler surfaces can also be assessed from the $S_f$ value. In Figure 2.67 and Figure 2.68, the $S_f$ of THF and $CHCl_3$ are plotted *vs.* the surface area of the fillers. The former is a basic substance and the latter is taken as the acid probe. It can be seen that contrary to the acid probe, $CHCl_3$, where no significant difference between carbon blacks and silicas was found, the $S_f$ value of THF is much higher in the case of silicas. This is even more pronounced in precipitated silicas. The very high $S_f$ of THF on silica suggests a strong acid-base interaction between them.

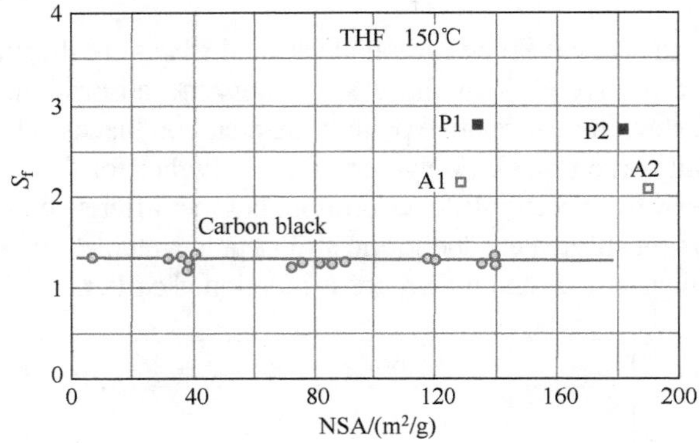


**Figure 2.67** $S_f$ of THF *vs.* nitrogen surface area for dry-pelletized carbon blacks and silicas at 150°C

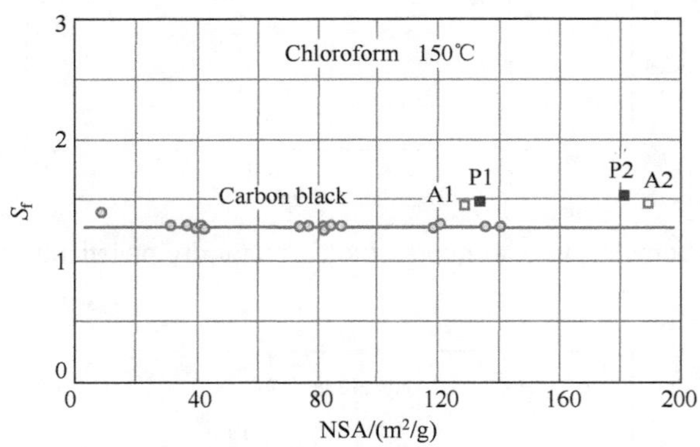


**Figure 2.68** $S_f$ of chloroform *vs.* nitrogen surface area for dry-pelletized carbon blacks and silicas at 150°C

### 2.4.2.3.4.4 Energy Heterogeneity of Filler Surfaces

As discussed in Section 2.4.2.3.3, the surface energy measured by adsorption is a mean value over the total filler surface, even when combined with other techniques such as wettability measurements[23,200]. It has been recognized that surface energy is not distributed equally over the surface. Measurements of adsorption heat have shown that the adsorptive activity of carbon blacks and silicas is concentrated on a small percentage of the surface, which exhibits much greater activity than the rest of the surface[193,212]. This small part of the surface plays a very important role in rubber reinforcement, which has been confirmed by heat treatment and chemical treatment evaluations[213,214]. Losing the high-energy sites through such treatment (e.g., the graphitization of the carbon blacks or the modification of silicas by grafting alkyl chains on their surfaces) profoundly changes the properties of the rubber compounds and vulcanizates containing these fillers. Therefore, investigation of this energetic heterogeneity of fillers could provide supplementary information concerning the nature of filler surface energies.

The energy distribution functions of adsorption, i.e., $\chi(\varepsilon)$ *vs.* $\varepsilon$, obtained by differentiation of the retention volume with respect to equilibrium pressure (Eq. 2.228) are illustrated in Figure 2.69 and Figure 2.70 for *n*-hexane and benzene adsorbed on precipitated silica and fumed silica, respectively.

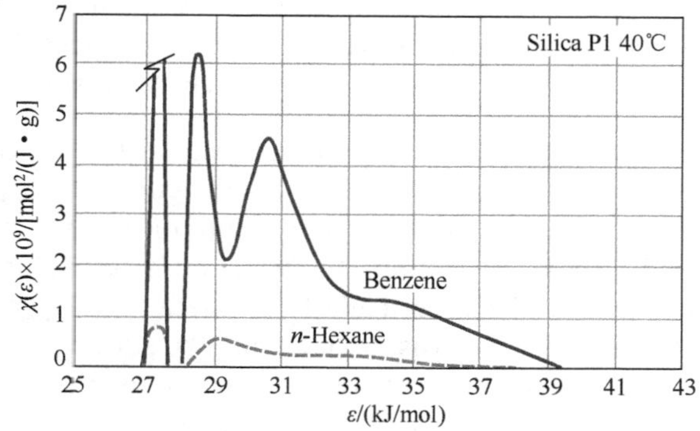


**Figure 2.69** Energy distribution function of benzene and *n*-hexane adsorption on silica P1 at 40°C

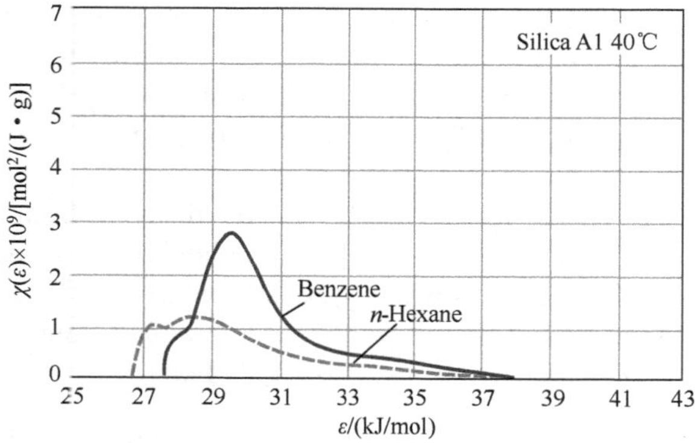


**Figure 2.70** Energy distribution function of benzene and *n*-hexane adsorption on silica A1 at 40°C

In the case of *n*-hexane adsorption on P1, there are two distinct maxima and one less distinct peak on the high-energy side of the energy function. The fumed silica, A1, exhibits less heterogeneity towards the adsorption of hexane than the precipitated counterpart P1. Only two peaks on the energy distribution functions are observed. It seems that there are a great number of medium and low energetic sites on fumed silicas, whereas the high-energy adsorption centers are more frequent on precipitated silicas. The difference between the functions of different silicas would be related to the nature of their surface chemistry, i.e., the concentration and distribution of different types of silanols which may, according to their environment, be isolated, vicinal, or geminal types. In some cases, depending on the preparation of the silicas, the silanol groups may crowd together in certain regions, forming clusters. The siloxane bridges present on the silica surface may be another type of adsorption center, even though the energy is lower in relation to that of silanols. On the other hand, silica is an amorphous product which consists of silicon and oxygen tetrahedrally bound into an imperfect three-dimensional structure in which only short-range crystal order exists. The degree of lattice imperfection depends on the methods and conditions of silica synthesis. The "broken bond" or "edge" of the crystal lattice could result in uncompensated charge leading to the high-energy sites as compared to the lattice faces. All these may be responsible for the different pattern of the adsorption energy distribution.

Since benzene is also capable of specific interactions with the silica surface via its electronic localization of a $\pi$-bond system (besides the universal dispersive interaction between adsorbent and adsorbate), the comparison of energy distribution curves between benzene and hexane adsorptions further reveals the nature of the energetic heterogeneity of the silica surface, particularly the heterogeneity of surface polarity. From Figure 2.69 and Figure 2.70, it is obvious that the surfaces of silicas, especially precipitated silicas, exhibit greater heterogeneity towards benzene molecules than towards *n*-hexane, and the intensities of the peaks are much more pronounced for precipitated silicas. This is further evidence of the high polarity of the silica surface, and its active centers in particular.

With regard to carbon black, the picture of surface energetic heterogeneity is quite different. The energy distribution curves of a variety of rubber-grade carbon blacks are illustrated in Figure 2.71 and Figure 2.72 for benzene and cyclohexane adsorption at 50°C, respectively. It is immediately apparent that the shape of the distribution curves is the same for all carbon blacks, i.e., they are skewed Gaussian-type distributions with very broad tails on the high-energy sides. This is quite different from what was observed for silicas. The energy distribution curves of benzene and *n*-hexane adsorption on silica at 40°C show several distinct peaks which were attributed to different types of silanols, siloxanes, and impurities[200]. The very broad distribution functions of the carbon blacks could be an indication of surfaces with highly heterogeneous energies.

On the other hand, a comparison of the distribution functions of the different carbon blacks suggests that the peaks of the curves, i.e., the greatest concentrations of energy sites, are approximately the same for all carbon blacks, with the exception of the thermal black N990 whose peak is shifted slightly towards the low-energy side. The relatively large difference in the distribution curves between different blacks is observed on the high-energy side. Evidently, there is a consistent increase in the concentration of high-energy centers with an increase in the surface area of the carbon black. It is therefore reasonable to conclude that the higher surface energy measured by chromatography at infinite dilution for hard carbon blacks is mainly caused by the greater concentration of active centers.

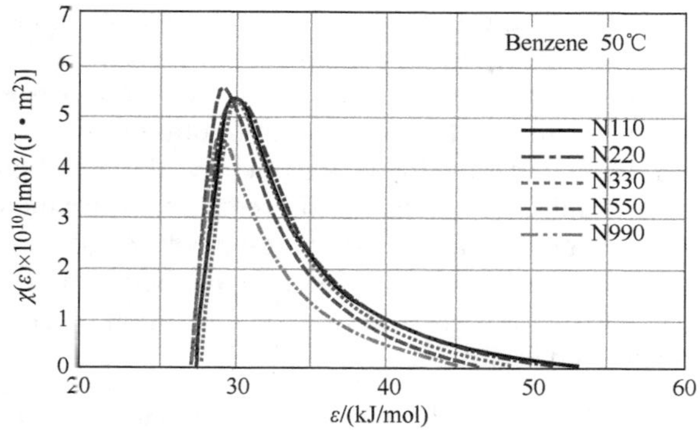


**Figure 2.71** Energy distribution function of benzene adsorption on a variety of carbon blacks at 50°C

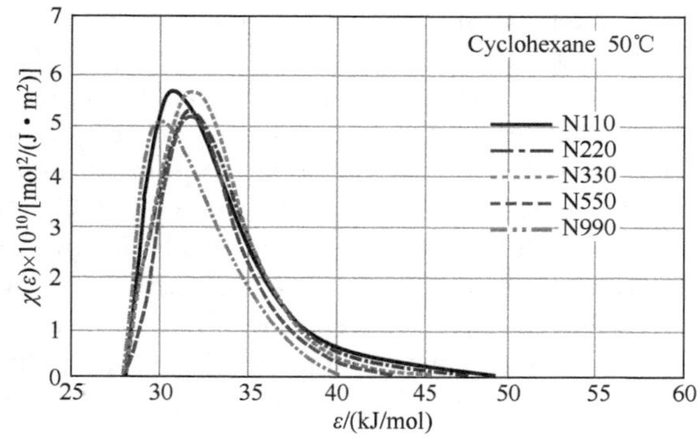


**Figure 2.72** Energy distribution function of cyclohexane adsorption on a variety of carbon blacks at 50°C

Given that the energies of the most probable sites are comparable for furnace carbon blacks, it is assumed that the sites dominating the black surface are related to the basal planes of graphitic crystallites. If this is so, then the high-energy centers should be the edges and defects which occupy only a small percentage of the surface. It is therefore understandable that the disordered part of the surface is more developed on the small-particle blacks, in accordance with their smaller crystallites, as demonstrated by X-ray diffraction and STM investigation of carbon black surfaces[209]. On the other hand, if the surface is composed of these two parts with different energies, the distribution curve should have two distinct response peaks. The fact that only one broad peak is observed is not yet fully understood. One reason for this phenomenon may be that the chemical groups located on the edges of the graphitic basal planes, such as hydrogen, oxygen-containing groups, and possibly heteroatoms, as well as foreign matter adsorbed on the carbon black surface which cannot be removed by extraction or degassing, could change the energy of these two parts of the surface, thus broadening the distribution curves. Moreover, the turbostratic structure under the first layer of the crystal lattice plane may also play a role in changing the surface energy.

It was also found that there is a difference in the energetic site concentration found at high-energy sites between benzene and cyclohexane adsorptions, but the difference is much smaller compared to that found on silica between benzene and *n*-hexane adsorption. Regardless of the difference in configuration between these two C-6 molecules, this observation confirms the lower polarity of the carbon blacks compared to silicas. On the other hand, no significant difference in the frequency of energy sites on the low-energy side, which dominate the carbon black surface and are related to the basal planes of the graphitic crystallites, was found between hexane and benzene adsorption. This suggests that the greatest part of the carbon black surface, and of the energy sites, is characterized by little or no polarity.

The growth of the crystallographic dimension upon graphitization of the carbon black would also change the energy distribution function of adsorption. Figure 2.73 compares the distribution curves of the adsorption energies of benzene and cyclohexane on carbon black N110 and its graphitized counterpart. It is noticeable that both probes' energy distribution for adsorptions narrows after graphitization. Both the high and low-energy sites have disappeared. While the loss of the high-energy centers can be attributed to the disappearance of the crystal edges, the decrease in the number of low-energy sites may be due to the removal of certain chemical groups and of substances which have a lower adsorption activity than the graphitic planes during the graphitization at temperatures as high as 2700°C. This is in line with the result obtained at infinite dilution for carbon blacks treated in an inert atmosphere (helium stream) at relatively high temperatures (e.g., 400°C). The surface energy increases considerably after heat treatment, which has been assumed to be the result of the removal of certain chemical groups from the surface. On the other hand, there is no reason to believe that any change in the crystallographic structure occurs at this temperature. In the case of cyclohexane adsorption, instead of the right-skewed type of distribution encountered for the adsorption of benzene and cyclohexane on nongraphitized blacks and for benzene on graphitized blacks, symmetric Gaussian curves appear for graphitized carbon black.

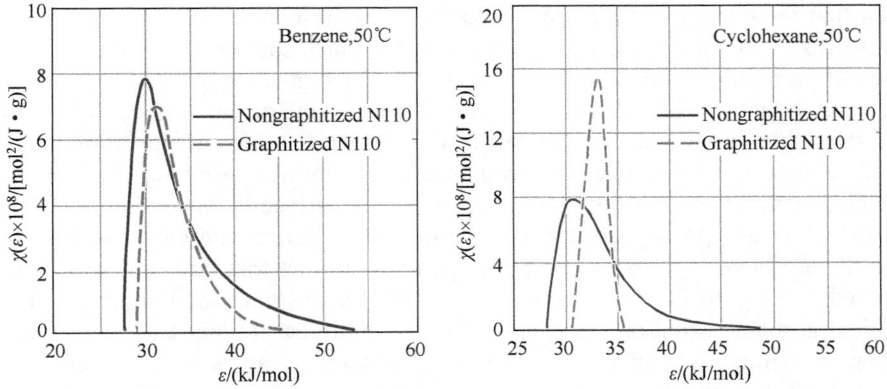


**Figure 2.73** Energy distribution function of benzene and cyclohexane adsorption on carbon black N110 and its graphitized counterpart at 50°C

It is interesting to note on the enlarged-scale plots (Figure 2.74), in contrast to normal carbon blacks (for example N330), that only a very small number of active centers still remain on the surface after graphitization, which are indicated by distinct, separate peaks to the left of the main peak. The concentration of these residual centers, leading to substep isotherms at very low pressure, is so low that they cannot be detected on a normal scale. As explained above for $\gamma_s^d$, these centers may be due to the remaining crystal defects due to the imperfect graphitization, as "visualized" by STM (Figure 2.61).

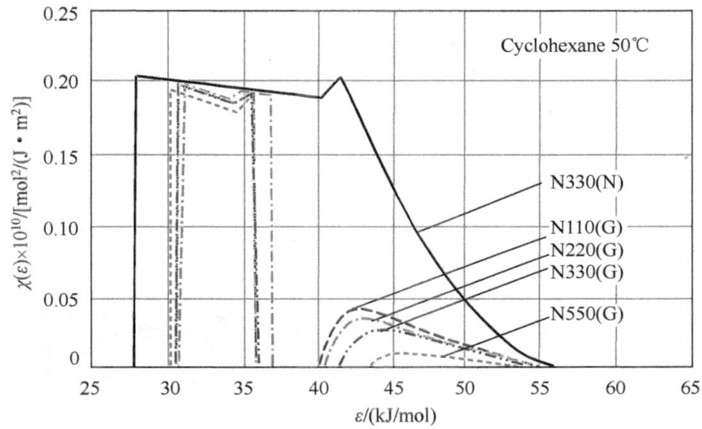


**Figure 2.74** Energy distribution function of cyclohexane adsorption on carbon black N330 and a variety of graphitized carbon blacks at 50°C (with enlarged scale)

### 2.4.2.3.4.5 Surface Energy of Silane-Modified Silicas

The test results described above yield a picture of the silica surface with regard to surface energy characteristics which distinguishes it from carbon black as follows[215]:

- The dispersive component of the surface energy, $\gamma_s^d$, is very low.
- The specific or polar component of the surface energy, $\gamma_s^{sp}$, is relatively high.

Such surface characteristics are related to certain disadvantages in rubber reinforcement with respect to carbon black. The low $\gamma_s^d$ of silica is related to the lack of filler-polymer interaction, resulting in poor strength properties of the filled vulcanizates, while the high $\gamma_s^{sp}$, as suggested by specific interaction energy and the $S_f$ value of highly polar probes, such as acetonitrile, indicates strong aggregate-aggregate interaction, resulting in developed agglomeration of the filler in the polymer matrix. This is responsible for differences in dispersion of the fillers and processability of the compounds, as well as high hardness and high Young's modulus of the vulcanizates. Enhancing polymer-filler interaction and reducing the polarity of the silica surface are the only means to overcome such shortcomings. Indeed, researchers have made considerable improvements to the silica reinforcement of elastomers, consciously or unconsciously implementing these practices. A successful example is the application of organosilanes, particularly bifunctional silanes.

The first function of silane modification is to reduce surface polarity of the silica, enhancing its wettability with polymer, namely nonpolar or low polar hydrocarbon polymer. In Table 2.16, the free energies of benzene adsorption and its specific component are listed for nonmodified silica, P1, and its counterparts modified with hexadecyltrimethoxy silane (HDTMS) and bis(3-triethoxysilylpropyl)-tetrasulfide (TESPT). A degree of modification of 100% is the highest graft ratio which corresponds to 2.6 molecules/nm$^2$ and 1.3 molecules/nm$^2$ for TESPT and HDTMS, respectively. In this study, the highly polar probe acetonitrile was not used since its molecules are too small that they can penetrate the graft layer adsorbed on filler surface. It can be seen that the specific interaction energy decreases drastically upon silanization. Likewise, the dispersive component of the silica surface energy also slightly diminishes simultaneously.

**Table 2.16** Effect of silane modification of silica on free energy of benzene adsorption

| Silane modification of silica | P1 | P1-HDTMS | P1-TESPT |
|---|---|---|---|
| $G^0$ of benzene at 130°C/(kJ/mol) | 16.6 | 8.6 | 10.9 |
| $I^{sp}$ of benzene at 130°C/(mJ/m$^2$) | 88.0 | 1.6 | 43.7 |

The change in surface energetic characteristics after silanization of silica can be verified by using adsorption isotherms of different probes on silicas modified 100% with octadecyltrimethoxy silane (ODTMS), a homolog of HDTMS, and bis(3-trimethoxysilylpropyl)-tetrasulfide (TMSPT), a homolog of TESPT. The adsorption isotherms of *n*-hexane on the silicas at 40°C, determined according to Eq. 2.200 from the retention volumes of a series of chromatograms, are shown in Figure 2.75. The adsorption isotherms are expressed as the amount of probe adsorbed *vs.* relative pressure. It appears that the isotherms of all silicas, especially of the initial (nonmodified) one, can be considered as type Ⅱ isotherms according to the classification by Brunauer, Emmett, and Teller[216], indicating strong adsorbate-adsorbent interaction. This interaction is of a dispersive type due to the inability of *n*-hexane to exchange any specific interaction with silica surfaces. As expected from the surface energy measurement by IGC at infinite dilution, a small reduction in adsorption is found after silane modification, particularly for ODTMS modification. Figure 2.76 immediately shows that silica-benzene interaction is drastically reduced upon silane modification. This confirms the function of the silane grafts in lowering the specific component of silica surface energy. Again, it is found that the layer of long octadecyl chains provides an effective shield and changes the surface from hydrophilic to hydrophobic.

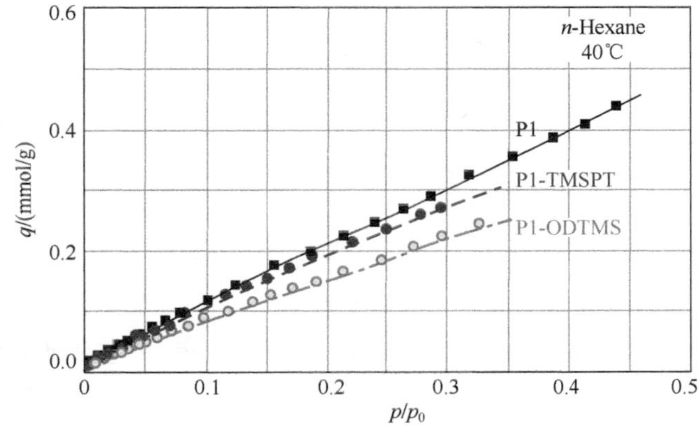


**Figure 2.75** Adsorption isotherms of *n*-hexane on silicas at 40°C

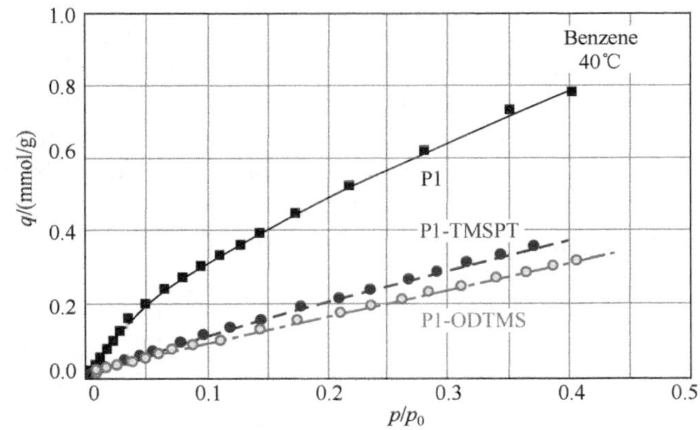


**Figure 2.76** Adsorption isotherms of benzene on silicas at 40°C

As mentioned before in Section 2.4.2.3.3, more information concerning the polarity of the silica surface can be obtained from comparison of the chemical potentials of benzene (polar probe) and hexane (nonpolar probe). This is shown in Figure 2.77, where the evolution of $\Delta\Delta\mu^{ads}$ as a function of the amount of probe adsorbed is illustrated. Besides the reduction in surface polarity due to silane modification, which is in agreement with the results shown above, the specific interaction also varies significantly with increasing adsorption (surface coverage). This also reflects the energetic heterogeneity of the filler surface. Obviously, modification is able to reduce the heterogeneity. This is further confirmed by the investigation of the energy distribution function of adsorption.

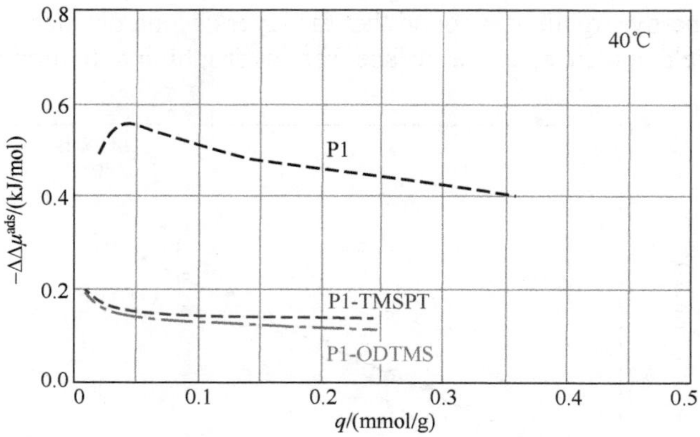


**Figure 2.77** $\Delta\Delta\mu^{ads}$ for benzene and *n*-hexane adsorption *vs.* amount of adsorption

Figure 2.78 shows the curves of $\chi(\varepsilon)$ versus $\varepsilon$ for the adsorption of *n*-hexane on silicas. The energy distribution function of hexane adsorption changes considerably upon grafting with TMSPT. Besides the fact that the strongest adsorption centers are shielded by the grafts, the greater number of adsorption centers on the low-energy side may be a reflection of the chemical complexity of the TMSPT graft. The energy distribution function of ODTMS-modified silicas yields quite a different picture of the surface heterogeneity. This function clearly demonstrates that the long alkyl chains, $C_{18}$, give a homogeneous surface, showing only one predominantly active center with a rather low frequency. This result is most readily interpreted on the basis of long alkyl chains being able to form a hydrocarbon layer and effectively to screen the interaction between the adsorbate molecules and the high-energy sites.

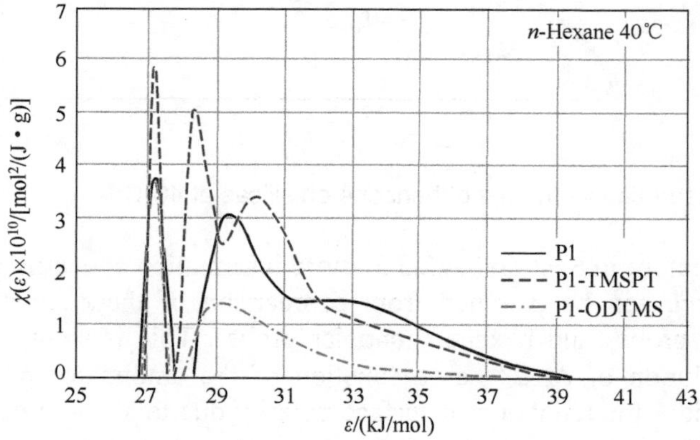


**Figure 2.78** Energy distribution function of *n*-hexane adsorption on silica P1 and its silane-modified counterparts at 40°C

Figure 2.79 seems to suggest that the silica modified with TMSPT is less heterogeneous towards benzene adsorption than towards hexane adsorption, showing less distinct maxima on the $\chi(\varepsilon)$ versus $\varepsilon$ curve in relation to *n*-hexane adsorption. The disappearance of the adsorption centers on the low-energy side implies that these sites exhibit a high affinity towards nonpolar adsorbates. It follows that the low-energy centers for *n*-hexane adsorption on TMSPT-modified silicas differ from those on the initial silica for which the low-energy sites show a high affinity towards benzene. Therefore, the natural conclusion would be that the low-energy centers for hexane adsorption come from the grafts themselves and that the low-energy sites on the initial silica surface, which show very high polarity, are essentially screened with silane chains, resulting in an energetically less heterogeneous surface for benzene adsorption. Furthermore, besides the disappearance of low-energy sites on the initial product after modification, the number of adsorption centers is drastically reduced. This effect is even more pronounced with grafts having long alkyl chains.

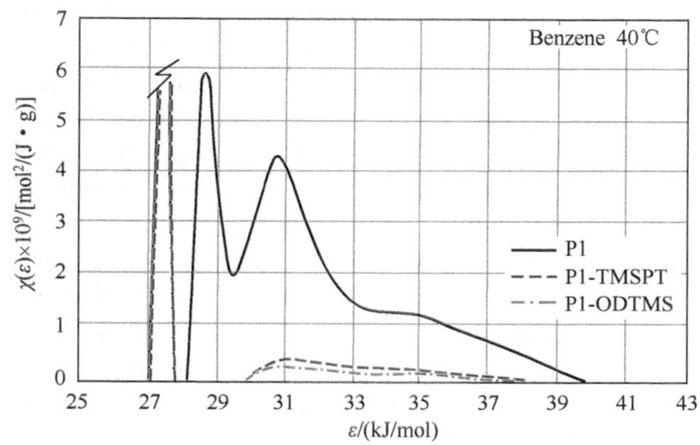


**Figure 2.79** Energy distribution function of benzene adsorption on silica P1 and its silane-modified counterparts at 40°C

To conclude, silane modification can essentially lower the specific component of surface energy. This could increase the compatibility of silicas with hydrocarbon elastomers, hence improve filler dispersion, processability and certain vulcanizate properties. Introducing covalent linkages during vulcanization by employing bifunctional silanes compensates for the lower dispersive component of the surface energy of modified silicas.

#### 2.4.2.3.5 Estimation of Rubber-Filler Interaction from Adsorption Energy of Elastomer Analogs

The free energies of adsorption of low molecular weight analogs of rubbers measured with IGC at infinite dilution could provide information concerning the ability of the filler surface to interact with polymers[22,191]. For this purpose, olefins were used as model compounds of unsaturated rubbers such as natural rubber and polybutadiene, and alkanes as models for saturated rubbers such as EPR and butyl rubber. A series of alkylated benzenes and homologous nitriles were used to evaluate the contribution of an aromatic ring to the interactions of SBR with carbon blacks and the contribution of the -CN group to the interactions of NBR with blacks, respectively.

As an example, Figure 2.80 shows the free energies of adsorption of a variety of rubber analogs on silica P1 as a function of the surface area covered by the probe molecule. The plots of all homologous probes are straight lines which can be treated as family plots. The difference in the ordinate between the given family plot and the plot of *n*-alkanes is related to the contribution of the functional groups. The slopes of the family plots of the homologous probes are representative of the contribution of $\Delta G^0$ of the $-CH_2-$ group per unit surface area if the interaction between filler and functional group is kept constant. It is obvious that the presence of double bonds in the alkene probes leads to specific interaction with the silica surfaces. This is attributed to induced dipole-dipole interaction since the π-bonds are polarized by the polar surfaces of the silica. Such an effect is particularly pronounced for precipitated silicas which seem able to interact more strongly with olefins than fumed silicas. Moreover, a comparison of the olefin isomers shows higher $\Delta G^0$ for trans-isomers than for 1-alkenes. This may be interpreted in terms of higher polarizability of the trans-isomers, due to the electron density of the π-bond induced from the methylene groups. When a side methyl group isolated from the double bond is introduced into the olefin, corresponding to 3-methyl-1-alkenes (i.e., the analog compound of high-vinyl polybutadiene), $\Delta G^0$ is even lower. This phenomenon may be related to a "steric effect" of packing, since the side group in this molecule may cause a greater distance of the master center of the probe from a flat surface of the silica, thus weakening the interaction between probe molecules and the filler surface. Such a "steric effect" is also observed when dealing with two 2-dimethyl-alkanes, the analogs of butyl rubber. These derivatives show much lower interaction free energy with silica than their *n*-alkane counterparts.

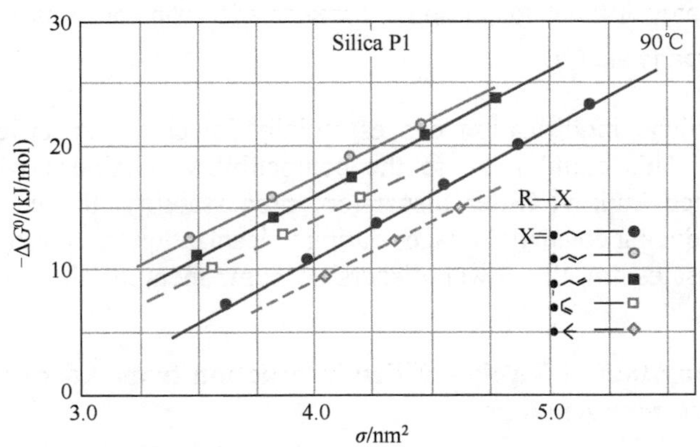


**Figure 2.80** $\Delta G^0$ of rubber analogs on silica P1 *vs.* their molecular surface areas at 90°C

The contribution of phenyl groups in SBR and -CN groups in NBR may be evaluated from Figure 2.81 in which the experimental data of 1-alkenes and *n*-alkanes are also included as a reference for comparison. It can be easily concluded that the aromatic hydrocarbons exhibit stronger interaction with silica surface than olefins, and nitriles show the highest interactions with silicas. The same conclusion about olefins, phenyl and nitrile groups contributions to the rubber-filler interaction can be drawn for carbon blacks as shown in Figure 2.82, taking N330 carbon black as an example. In comparison with silicas, the specific interactions of these functional groups with carbon black are much lower due to the lower surface polarity. All carbon blacks show the same pattern regardless of their production process and post-treatment, but, as mentioned previously, the level of the interaction energies for all rubber analogs decreases with increasing particle size, in accordance with their dispersive and specific components of surface energy.

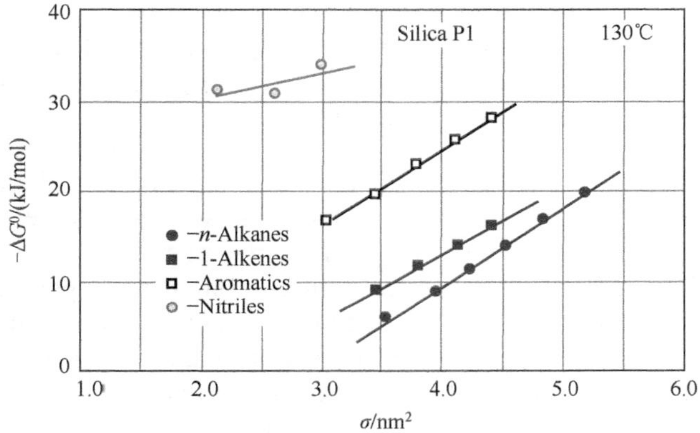


**Figure 2.81** $\Delta G^0$ of rubber analogs on silica P1 *vs.* their molecular surface areas at 130°C

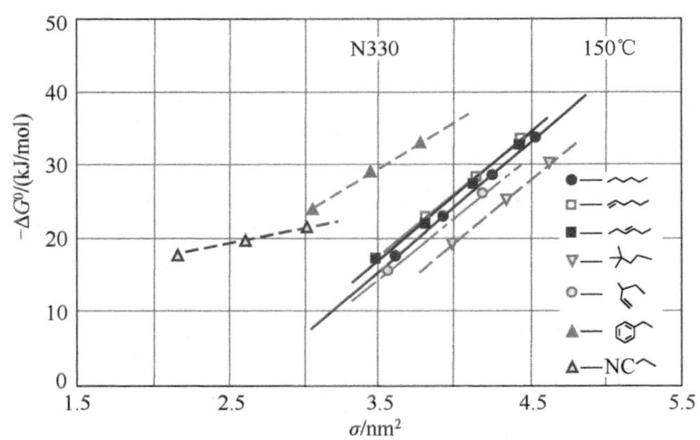


**Figure 2.82** $\Delta G^0$ of rubber analogs on carbon black N330 *vs.* their molecular surface areas at 150°C

From the above-mentioned results of filler-rubber analog interactions, the elastomers can be classified in the same order as that obtained with silicas, i.e., NBR and SBR exhibit strong interactions with carbon blacks, followed by unsaturated rubbers, such as NR and BR. The saturated rubbers, particularly branched polymers, e.g., butyl rubber, exhibit weak interaction with fillers.

This observation is in good agreement with the results of bound rubber filled with a variety of carbon blacks, indicating a higher activity for SBR, followed by polybutadiene.

Butyl rubber gives the lowest bound rubber content[217]. In an electron microscope observation of stretched vulcanizates of NR, BR, and SBR filled with carbon blacks, Todani and Sagaye[218] were able to show that carbon black particles agglomerate during elongation, forming a very heterogeneous structure in the polymer matrix for NR and BR, but this phenomenon was much less developed in SBR. We believe that one of the reasons for this phenomenon is the stronger interaction between SBR and carbon black.

#### 2.4.2.4 Bound Rubber Measurement

The bound rubber is generally considered to be the portion of rubber in an unvulcanized, filled rubber compound that is not extracted by a good solvent at room temperature. It is an important part for rubber reinforcement.

The normal procedure for measuring bound rubber is to take a small amount (<1 g) of rubber compound, cut into small pieces, and immerse it in a large amount of good solvent for the polymer for up to one week at room temperature. The solvent is usually replaced with fresh solvent at least once during the immersion period. At the end of immersion, the solution is separated from the bound rubber-filler mixture by decantation if a coherent gel is present, or by centrifugation if it is incoherent. After some washing with fresh solvent, the remaining solvent is removed by evaporation and the amount of rubber remaining on the filler is determined by thermogravimetric analysis (TGA), or simply by weight difference from the starting compound. Bound rubber is usually expressed as the fraction or percentage of the original rubber in the compound that remains unextracted. It can be calculated from the following equation:

$$BR\% = \frac{W_{\mathrm{fg}} - Wm_{\mathrm{f}} / (m_{\mathrm{f}} + m_{\mathrm{p}})}{Wm_{\mathrm{p}} / (m_{\mathrm{f}} + m_{\mathrm{p}})} \times 100 \tag{2.229}$$

where $W_{\mathrm{fg}}$ is the weight of filler and gel, $m_{\mathrm{f}}$ the loading of filler in the compound (in phr), $m_{\mathrm{p}}$ the loading of the polymer in the compound (in phr), and $W$ the weight of the dry specimen before extraction.

For common hydrocarbon rubbers such as SBR, BR, and butyl rubber, the solvents typically used include toluene, xylene, and *n*-heptane. A study on the effect of different solvents on the bound rubber in butyl compounds showed that benzene, cyclohexane, cyclohexene, ethyl benzene, xylene, chloroform, and carbon tetrachloride gave results with differences similar to the precision of the test with a single solvent[194]. Generally, the choice of solvent is not critical for bound rubber measurement in hydrocarbon rubbers; however, the choice of solvent for bound rubber determination in silicone rubber compounds, seems to be more important.

Bound rubber determination in the silicone rubber-silica system seems to be more complicated. Extraction of a silicone-silica compound for periods up to four years showed that unextracted polymer was linearly related to the square root of extraction time in toluene.

Bound rubber in the silicone-silica combination is dependent on the solvent, and the type and strength of the interactions between polymer and filler are quite different from the carbon black-hydrocarbon rubber system.

## References


[1] Studebaker M L, Beatty J R. Chapter 10. In: Eirich F R (Editor). Science and Technology of Rubber. New York: Academic Press, 1978.

[2] Wolff S, Wang M-J. Chapter 3. In: Donnet J-B, Bansal R C, Wang M-J (Editors). Carbon Black, Second Edition, Science and Technology. New York: Marcel Dekker Inc., 1993.

[3] Kiselev A V, Yashin Y I. Gas Adsorption Chromatography. New York: Plenum Press, 1969.

[4] Dorris G M, Gray D G. Adsorption, Spreading Pressure, and London Force Interactions of Hydrocarbons on Cellulose and Wood Fiber Surfaces. *J. Colloid Interface Sci.*, 1979, 71: 93.

[5] Wagner M P. Reinforcing Silicas and Silicates. *Rubber Chem. Technol.*, 1976, 49: 703.

[6] Hockey J A. *Chem. Ind.( London)*, 1965, 57.

[7] Hair M L. Chapter 4. In: Infrared Spectroscopy in Surface Chemistry. New York: Marcel Dekker Inc., 1967.

[8] de Boer J H, Hermans M E A, Vleeskens J M. The Chemisorption and Physical Adsorption of Water on Silica. I. *Proc. K. Ned. Akad. Wet., Ser. B*, 1957, 60: 45, 54.

[9] de Boer J H, Vleeskens J M. *Proc. K. Ned. Akad. Wet., Ser. B*, 1957, 60: 23.

[10] de Boer J H, Vleeskens J M. Chemisorption and Physical Adsorption of Water on Silica. Ⅳ. Nature of the Surface. *Proc. K. Ned. Akad. Wet., Ser. B*, 1958, 61: 2.

[11] Bassett D R, Boucher E A, Zettlemoyer A C. Adsorption Studies on Hydrated and Dehydrated Silicas. *J. Colloid Interface Sci.*, 1968, 27: 649.

[12] Wang M-J. Etude du Renforcement des Élastomères par les Charges: Effet Exercé par l'emploi de Silices Modifiées par Greffage de Chaines Hydrocarbonées. Doctor's thesis, Alsace, Haute-Alsace University, 1984.

[13] Fripiat J J, Uytterhoeven J. Hydroxyl Content in Silica Gel "Aerosil". *J. Phys. Chem.*, 1962, 66: 800.

[14] Fripiat J J, Gastuche M C, Brichard R. Surface Heterogeneity in Silica Gel from Kinetics of Isotopic Exchange OH-OD. *J. Phys. Chem.*, 1962, 66: 805.

[15] Armistead C G, Tyler A J, Hambleton F H, et al. Surface Hydroxylation of Silica. *J. Phys. Chem.*, 1969, 73: 3947.

[16] Van Cauwelaert F H, Jacobs P A, Uytterhoeven J B. Identification of the A-Type Hydroxyls on Silica Surfaces. *J. Phys. Chem.*, 1972, 76: 1434.

[17] Clark-Monks C, Ellis B. The Characterization of Anomalous Adsorption Sites on Silica Surfaces. *J. Colloid Interface Sci.*, 1973, 44: 37.

[18] Hockey J A, Pethica B A. Surface Hydration of Silicas. *Trans. Faraday Soc.*, 1961, 57: 2247.

[19] Tul'bovich B I, Priimak E I. Heats of Adsorption of the Vapours of Certain Organic Substances on a Silica. *Russ, J. Phys. Chem.*, 1969, 43: 195.

[20] Medalia A I. Chapter 1. In: Sichel E K (Editor). Carbon Black-Polymer Composites. New York: Marcel Dekker Inc., 1982.

[21] Redman E, Heckman F A, Connolly J E. Paper No. 14, Presented at a meeting of the Rubber Division, ACS, Chicago, Ill., 1967.
[22] Wang M-J, Wolff S, Donnet J -B. Filler-Elastomer Interactions. Part III. Carbon-Black-Surface Energies and Interactions with Elastomer Analogs. *Rubber Chem. Technol.*, 1991, 64: 714.
[23] Wang M-J, Wolff S. Filler-Elastomer Interactions. Part VI. Characterization of Carbon Blacks by Inverse Gas Chromatography at Finite Concentration. *Rubber Chem. Technol.*, 1992, 65: 890.
[24] Hess W M, Ban L L, McDonald G C. Carbon Black Morphology: I. Particle Microstructure. II. Automated EM Analysis of Aggregate Size and Shape. *Rubber Chem. Technol.*, 1969, 42: 1209.
[25] Hess W M, McDonald G C. Measuring Dynamic properties of Vulcanisates, Rubber and Related Products: New Methods for Testing and Analysing, ASTM, STP 553, ASTM, West Conshohocken, USA, 1974.
[26] Conzatti L, Costa G, Falqui L, et al. Rubber Technologist's Handbook, Volume 2, Microscopic Imaging of Rubber Compounds, Smithers Rapra Publishing; 2001.
[27] Hess W M, McDonald G C. Improved Particle Size Measurements on Pigments for Rubber. *Rubber Chem. Technol.*, 1983, 56: 892.
[28] Brunauer S, Deming L S, Deming W E, et al. On a Theory of the Van Der Waals Adsorption of Gases. *J. Am. Chem. Soc.*, 1940, 62: 1723.
[29] Fisher C, Cole M. *The Microscope*, 1968, 16: 81.
[30] Gibbard D W, Smith D J, Wells A. Area Sizing and Pattern Recognition on the Quantimet 720. *The Microscope*, 1972, 20: 37.
[31] Kraus G. Applied Polymer Symposium. *J. Appl. Polym. Sci., Appl. Polymer Symp.*, 1984, 39: 75.
[32] Medalia A I. Effect of Carbon Black on Dynamic Properties of Rubber Vulcanizates. *Rubber Chem. Technol.*, 1978, 51: 437.
[33] Rigbi Z, Boonstra B B. Presented at a meeting of the Rubber Division, ACS, Chicago, Ill., Sept. 13-15, 1967.
[34] Clint J H. Adsorption of n-Alkane Vapours on Graphon. *J. Chem. Soc. Faraday Trans.* 1972, *1*, 68: 2239.
[35] Krejci J C, Roland C H. Presented at a meeting of the Rubber Division, ACS, Fall, 1965.
[36] Loebenstein W V, Deitz V R. Surface-Area Determination by Adsorption of Nitrogen from Nitrogen-Helium Mixtures. *J. Research Natl. Bur. Standards.*, 1951, 46: 51.
[37] Nelsen F M, Eggertsen F T. Determination of Surface Area. Adsorption Measurements by Continuous Flow Method. *Anal. Chem.*, 1958, 30: 1387.
[38] Ettie L S. *11$^{th}$ Conf. Anal. Chem. Pittsburgh*, Mar., 1960.
[39] Maryasin I L, Pishchulina S L, Rafal'kes I S, et al. *Zavod. Lab.*, 1971, 37: 41.
[40] Young D M, Crowell A D. p. 226. In: Physical Adsorption of Gases. Butterworths, London, 1962.
[41] Pierce C, Smith R N. Adsorption in Capillaries. *J. Phys. Chem.*, 1953, 57: 64.
[42] Pierce C. Computation of Pore Sizes from Physical Adsorption Data. *J. Phys. Chem.*, 1953, 57: 149.
[43] Voet A. *Rubber World*, 1958, 139: 63, 232.
[44] de Boer J H, Linsen B G, Osinga Th J. Studies on Pore Systems in Catalysts: VI. The Universal *t* Curve. *J. Catalysis*, 1965, 4: 643.

[45] Atkins J H. Porosity and Surface Area of Carbon Black. *Carbon*, 1965, 3: 299.

[46] Voet A, Lamond T G, Sweigart D. Surface Area and Porosity of Carbon Blacks. *Carbon*, 1968, 6: 707.

[47] Lippens B C, de Boer J H. Studies on Pore Systems in Catalysts: V. The *t* Method. *J. Catalysis*, 1965, 4 319.

[48] Lippens B C, Linsen B G, de Boer J H. Studies on Pore Systems in Catalysts I. The Adsorption of Nitrogen; Apparatus and Calculation. *J. Catalysis*, 1964, 3: 32.

[49] Smith W R, Kasten G A. Porosity Studies on Some Oil Furnace Blacks. *Rubber Chem. Technol.*, 1970, 43: 960.

[50] Donnet J-B, Voet A. p. 66. In: Carbon Black, Physics, Chemistry, and Elastomer Reinforcement. New York: Marcel Dekker Inc., 1976.

[51] Smith W R, Ford D G. Adsorption Studies on Heterogeneous Titania and Homogeneous Carbon Surfaces. *J. Phys. Chem.*, 1965, 69: 3587.

[52] Magee R W. Evaluation of the External Surface Area of Carbon Black by Nitrogen Adsorption. *Rubber Chem. Technol.*, 1995, 68: 590.

[53] Cabot Corporation, Boston, MA, USA, unpublished.

[54] Shull C G. The Determination of Pore Size Distribution from Gas Adsorption Data. *J. Am. Chem. Soc.*, 1948, 70: 1405.

[55] Barrett E P, Joyner L G, Halenda P P. The Determination of Pore Volume and Area Distributions in Porous Substances. I. Computations from Nitrogen Isotherms. *J. Am. Chem. Soc.*, 1951, 73: 373.

[56] Cranston R W, Inkley F A. Vol. 9, p. 143. In: Advances in Catalysis and Related Subjects. New York: Academic Press, 1957.

[57] Dollimore D, Hea G R. An Improved Method for the Calculation of Pore Size Distribution from Adsorption Data. *J. Appl. Chem.*, 1964, 14: 109.

[58] Olivier J P, Conklin W B, Szombathely M V. Determination of Pore Size Distribution from Density Functional Theory: A Comparison of Nitrogen and Argon Results. *Studies in Surface Science and Catalysis*, 1994, 87: 81.

[59] Kowalczyk P, Terzyk A P, Gauden P A, et al. Estimation of the pore-size distribution function from the nitrogen adsorption isotherm. Comparison of density functional theory and the method of Do and co-workers. *Carbon*, 2003, 41: 1113.

[60] Kelvin J. *Phil. Mag.*, 1948, 47: 448.

[61] Wheeler A. Vol. 2, p. 116. In: Emmet P H (Editor). Catalysis. New York: Reinhold, 1955.

[62] Halsey G. Physical Adsorption on Non-uniform Surfaces. *J. Chem. Phys.*, 1948, 16: 931.

[63] Davis O C M. The adsorption of Iodine by Carbon. *J. Chem. Soc., Trans.*, 1907, 91: 1666.

[64] Carson G M, Sebrell L B. Some Observations on Carbon Black. *Ind. Eng. Chem.*, 1929, 21: 911.

[65] Smith W R, Thornhill F S, Bray R I. Surface Area and Properties of Carbon Black. *Ind. Eng. Chem.*, 1941, 33: 1303.

[66] Mack Jr E. Average Cross-Sectional Areas of Molecules by Gaseous Diffusion Methods. *J. Am. Chem. Soc.*, 1925, 47: 2468.

[67] Kendall C E, Dunlop Research Center, Birmingham, England, Internal Report, C. R. 1103, 1947.

[68] Watson J W, Parkinson D. Adsorption of Iodine and Bromine by Carbon Black. *Ind. Eng. Chem.*, 1955, 47: 1053.
[69] Sweitzer C W, Venuto L J, Estelow R K. *Paint Oil Chem. Rev.*, 1952, 115: 22.
[70] Snow C W. Use Iodine Number for Effective, Low Cost Evaluation of Carbon Blacks. *Rubber Age*, 1963, 93: 547.
[71] Studebaker M L. The Chemistry of Carbon Black and Reinforcement. *Rubber Chem. Technol.*, 1957, 30: 1400.
[72] Janzen J, Kraus G. Specific Surface Area Measurements on Carbon Black. *Rubber Chem. Technol.*, 1971, 44: 1287.
[73] Kipling J J. pp. 126-128, 293-294. In: Adsorption from Solutions of Non-Electrolytes. New York: Academic Press, 1965.
[74] Saleeb F Z, Kitchener J A. The Effect Of Graphitization On The Adsorption Of Surfactants By Carbon Blacks. *J. Chem. Soc.*, 1965, 911.
[75] Saleeb F Z. Doctor's thesis, London, University of London, 1962.
[76] Abram J C, Bennett M C. Carbon Blacks as Model Porous Adsorbents. *J. Colloid Interface Sci.*, 1968, 27: 1.
[77] Lamond T G, Price C R. The Adsorption of Aerosol OT by Carbon Blacks-A Simple Method of External Surface Area Measurement. Presented at a meeting of the Rubber Chemistry Division, ACS, Buffalo, New York, Oct., 1969.
[78] Barr T, Oliver J, Stubbings W V. The Determination of Surface Active Agents in Solution. *J. Soc. Chem. Ind.*, 1948, 67: 45.
[79] ASTM D 3765-02.
[80] Harkins W D, Jura G. Surface of Solids. X. Extension of the Attractive Energy of a Solid into an Adjacent Liquid or Film, the Decrease of Energy with Distance, and the Thickness of Films. *J. Am. Chem. Soc.*, 1944, 66: 919.
[81] Kraus G, Rollmann K W. Carbon Black Surface Areas by the Harkins and Jura Absolute Method. *Rubber Chem. Technol.*, 1967, 40: 1305.
[82] Wang M-J. Determination of particle size and surface area of carbon black III, *Rubber Industry*, 1978, 58.
[83] Wade W H, Deviney Jr M L. Adsorption and Calorimetric Investigations on Carbon Black Surfaces I. Immersional Energetics of Heat Treated Channel and Furnace Blacks. *Rubber Chem. Technol.*, 1971, 44: 218.
[84] Sanders D R. Surface Area Measurement of Tire Grade Carbon Blacks. CIM-95-06, 1995, unpublished.
[85] Voet A, Morawski J C, Donnet J-B. Reinforcement of Elastomers by Silica. *Rubber Chem. Technol.*, 1977, 50: 342.
[86] ASTM D 6845-12.
[87] Sweitzer C W, Goodrich W C. The Carbon Spectrum for the Rubber Compounder. *Rubber Age*, 1944, 55: 469.
[88] Hess W M, McDonald G C, Urban E. Specific Shape Characterization of Carbon Black Primary Units. *Rubber Chem. Technol.*, 1973, 46: 204.
[89] Kaye B H. Improvements in or Relating to Centrifuges. British Patent, 895222, 1962.
[90] Hildreth J D, Patterson D. Colour Differences in Azo Pigments: II—Measurements of Particle Size and of Colour. *J. Soc. Dyers Colourists*, 1964, 80: 474.
[91] E. Redman, F. A. Heckman and J. E. Connolly, *Rubber Chem. Technol.*, 51, 1000 (1977).

[92] Weiner B B, Fairhurst D, Tscharnuter W W. Chapter 12, pp. 184-195. In: Provder T (Editor). Particle Size Analysis with a Disc Centrifuge: Importance of the Extinction Correction. In Particle Size Distribution II: Assessment and Characterization. American Chemical Society: Washington D.C., *ACS Symposium Series*, 1991, 472.

[93] Devon M J, Provder T, Rudin A. Chapter 9, pp. 134-153. In: Provder T (Editor). Measurement of Particle Size Distributions with a Disc Centrifuge. In Particle Size Distribution II: Assessment and Characterization. American Chemical Society: Washington D.C., *ACS Symposium Series*, 1991, 472.

[94] Weiner B B, Tscharnuter W W, Bernt W. Characterizing ASTM Carbon Black Reference Materials Using a Disc Centrifuge Photosedimentometer. *J. Disper. Sci. Technol.*, 2002, 23: 671.

[95] Eaton E R, Middleton J S. *Rubber World*, 1965, 152: 94.

[96] Dollinger R E, Kallenberger R H, Studebaker M L. Effect of Carbon Black Densification on Structure Measurements. *Rubber Chem. Technol.*, 1967, 40: 1311.

[97] Mongardi M. A Critical Review of DBP and Paraffin Oils for OAN Testing. Cabot Corporation, Boston, MA, USA, unpublished.

[98] Medalia A I. Morphology of Aggregates: VI. Effective Volume of Aggregates of Carbon Black from Electron Microscopy; Application to Vehicle Absorption and to Die Swell of Filled Rubber. *J. Colloid Interface Sci.*, 1970, 32: 115.

[99] Medalia A I. Morphology of Aggregates: I. Calculation of Shape and Bulkiness Factors; Application to Computer-Simulated Random Flocs. *J. Colloid Interface Sci.*, 1967, 24: 393.

[100] Medalia A I, Heckman F A. Morphology of Aggregates: II. Size and Shape Factors of Carbon Black Aggregates from Electron Microscopy. *Carbon*, 1969, 7: 567.

[101] Heckman F A, Medalia A I, *J. Inst. Rubber Ind.*,1969, 3, 66.

[102] Furnas C C. Grading Aggregates-I.-Mathematical Relations for Beds of Broken Solids of Maximum Density. *Ind. Eng. Chem.*, 1931, 23: 1052.

[103] Medalia A I, Sawyer R L. Compressibility of Carbon Black. *Proc. 5th Carbon Conf.*, New York: Pergammon Press, 1963, 563.

[104] Benson G, Gluck J, Kaufmann C. Electrical Conductivity Measurements of Carbon Blacks. *Trans. Electrochem. Soc.*, 1946, 90: 441.

[105] Mrozowski S, Chaberski A, Loebner E E, et al. Electronic Properties of Heat-Treated Carbon Blacks. *Proc. 3rd Carbon Conf.*, London: Pergamon Press, 1959, 211.

[106] Voet A, Whitten Jr W N. *Rubber World*, 1962, 146: 77.

[107] Voet A, Whitten W N. *Rubber World*, 1963, 148: 33.

[108] Washburn E W. Note on a Method of Determining the Distribution of Pore Sizes in a Porous Material. *Physics*, 1921, 7: 115.

[109] Moscou L, Lub S, Bussemaker O K F. Characterization of Carbon Black Structure by Mercury Penetration. *Rubber Chem. Technol.*, 1971, 44: 805.

[110] Medalia A I, Eaton E R. *Kautsch Gummi Kunstst.*, 1967, 20: 61.

[111] Medalia A I, Richards L W. Tinting Strength of Carbon Black. *J. Colloid Interface Sci.*, 1972, 40: 233.

[112] Kubelka P, Munk F. An Article on Optics of Paint Layers. *Z. Tech. Phys.*, 1931, 12: 593.

[113] Cabot Corporation, Special Blacks Division, "Cabot Carbon Black Pigments."

[114] Iodine Adsorption Number of Carbon Black, ASTM Procedure D1510-70.

[115] Medalia A I, Heckman F A. Morphology of Aggregates: VII. Comparison Chart Method for Electron Microscopic Determination of Carbon Black Aggregate Morphology. *J. Colloid Interface Sci.*, 1971, 36: 173.
[116] "Cabot Carbon Blacks under the Electron Microscope," Cabot Corporation, Boston, Mass., 1953.
[117] Kraus G (Editor). Reinforcement of Elastomers", New York: Wiley, 1965.
[118] Boonstra B B. Chapter 7. In: Blow C M, Hepburn G (Editors). Rubber Technology and Manufacture, $2^{nd}$ ed. London: Butterworth Scientific, 1982.
[119] Kraus G. Chapter 10. In: Eirich F R (Editor). Science and Technology of Rubber. New York: Academic Press, 1978.
[120] London F. The General Theory of Molecular Forces. *Trans. Faraday Soc.*, 1937, 33: 8.
[121] Keesom W M. Van Der Waals Attractive Force. *Phys. Z.*, 1921, 22: 129.
[122] Keesom W M. *Trans. Faraday Soc.*, 1922, 23: 225.
[123] Debye P. Die Van Der Waalsschen Kohasion-skrafte. *Phys. Z.*, 1920, 21: 178;
[124] Debye P. *Trans. Faraday Soc.*, 1921, 22: 302.
[125] Pimentel C C, McClellan A L. The Hydrogen Bond. Freemann and Co., San Francisco, 1960.
[126] Fowkes F M, Mostafa M A. Acid-base Interactions in Polymer Adsorption. *Ind. Eng. Chem. Prod. Res. Dev.*, 1978, 17: 3.
[127] Fowkes F M. Presented at a meeting on organic coating and plastics chemistry, ACS, Honolulu, Apr. 1-6, 1979.
[128] Payne A R. The Role of Hysteresis in Polymers. *Rubber J.*, 1964, 146: 36.
[129] Lennard-Jones J E. The Equation of State of Gases and Critical Phenomena. *Physica*, 1937, 4: 941.
[130] Chen Z Q, Dai M G. Colloid Chemistry. Beijing: High Education Publishing House, 1984.
[131] Fowkes F M. Determination of Interfacial Tensions, Contact Angles, and Dispersion Forces in Surfaces by Assuming Additivity of Intermolecular Interactions in Surfaces. *J. Phys. Chem.*, 1962, 66: 382.
[132] Owens D K, Wendt R C. Estimation of the Surface Free Energy of Polymers. *J. Appl. Polymer Sci.*, 1969, 13: 1741.
[133] Kaelble D H, Uy K C. A Reinterpretation of Organic Liquid-polytetrafluoroethylene Surface Interactions. *J. Adhesion*, 1970, 2: 50.
[134] Fowkes F M. Vol. 1. In: Patrick R L (Editor). Treatise on Adhesion and Adhesives. New York: Marcel. Dekker, 1967.
[135] Bernett M K, Zisman W A. Effect of Adsorbed Water on Wetting Properties of Borosilicate Glass, Quartz, and Sapphire. *J. Colloid Interface Sci.*, 1969, 29: 413.
[136] Kessaissia Z, Papirer E, Donnet J B. The Surface Energy of Silicas, Grafted with Alkyl Chains of Increasing Lengths, as Measured by Contact Angle Techniques. *J. Colloid Interface Sci.*, 1981, 82: 526.
[137] Dubinin M M. Effects of Surface and Structural Properties of Carbons on the Behavior of Carbon-Supported Catalysts. *Prog. Surf. Membr. Sci.*, 1975, 9: 1.
[138] Dubun B V, Kiselev V F, Aleksandrov T I. *DAN. SSSR*, 1955, 102: 1155.
[139] Harkins W D. p. 255. In: The Physical Chemistry of Surface Films. New York: Reinhold, 1952.

[140] Conder J R, Young C L. Physical Measurement by Gas Chromatography. New York: Wiley, 1979.

[141] Laub R J, Pecsok R L. Physicochemical Applications of Gas Chromatography. New York: Wiley, 1978.

[142] Cassie A B D. Contact Angles. *Discuss. Faraday Soc.*, 1948, 3: 11.

[143] Young T. An Essay on the Cohesion of Fluids. *Philos. Trans. R. Soc. London*, 1805, 95: 65.

[144] Dupre A. Théorie mécanique de la chaleur. Paris, Gauthier-Villars, 1869.

[145] Fowkes F M. Wetting. *Soc. Chem. Ind. Monograph*, London, 1967, 25: 3.

[146] Tamai Y, Makuuchi K, Suzuki M. Experimental Analysis of Interfacial Forces at the Plane Surface of Solids. *J. Phys. Chem.*, 1967, 71: 4176.

[147] Schultz J, Tsutsumi K, Donnet J B. Surface Properties of High-Energy Solids: I. Determination of the Dispersive Component of the Surface Free Energy of Mica and Its Energy of Adhesion to Water and n-Alkanes. *J. Colloid Interface Sci.*, 1977, 59: 272.

[148] Schultz J, Tsutsumi K, Donnet J-B. Surface Properties of High-Energy Solids: II. Determination of the Nondispersive Component of the Surface Free Energy of Mica and Its Energy of Adhesion to Polar Liquids. *J. Colloid Interface Sci.*, 1977, 59: 277.

[149] Wu S. Surface and Interfacial Tensions of Polymer Melts. II. Poly(methyl methacrylate), Poly(n-butyl methacrylate), and Polystyrene. *J. Phys. Chem.*, 1970, 74: 632.

[150] Yuan Y, Lee T R. *Surface Science Techniques*, 2013, 51, 3-34.

[151] Chibowski E, Perea-Carpio R. Problems of Contact Angle and Solid Surface Free Energy Determination. *Advances in Colloid and Interface Sci.*, 2002, 98: 245.

[152] Chessick J J, Zettlemoyer A C. Immersional Heats and the Nature of Solid Surfaces. *Advances in Catalysis*, Academic Press, 1959, 11: 263.

[153] Craig R G, Van Voorhis J J, Bartell F E. Free Energy of Immersion of Compressed Powders with Different Liquids. I. Graphite Powders. *J. Phys. Chem.*, 1956, 60: 1225.

[154] Hill T L. Statistical Mechanics of Adsorption. V. Thermodynamics and Heat of Adsorption. *J. Chem. Phys.*, 1949, 17: 520.

[155] Jura G, Hill T L. Thermodynamic Functions of Adsorbed Molecules from Heats of Immersion. *J. Am. Chem. Soc.*, 1952, 74: 1598.

[156] Good R J, Girifalco L A, Kraus G. A Theory for Estimation of Interfacial Energies. II. Application to Surface Thermodynamics of Teflon and Graphite. *J. Phys. Chem.*, 1958, 62: 1418.

[157] Pierce C, Smith R N. The Adsorption−Desorption Hysteresis in Relation to Capillarity of Adsorbents. *J. Phys. Chem.*, 1950, 54: 784.

[158] Harkins W D, Jura G. p. 256. In: Hakins W D (Editor). The Physical Chemistry of Surface Films. New York: Reinhold, 1952.

[159] Young G J, Chessick J J, Healey F H, et al. Thermodynamics of the Adsorption of Water on Graphon from Heats of Immersion and Adsorption Data. *J. Phys. Chem.*, 1954, 58: 313.

[160] Hill T L, Emmett P H, Joyner L G. Calculation of Thermodynamic Functions of Adsorbed Molecules from Adsorption Isotherm Measurements: Nitrogen on Graphon. *J. Am. Chem. Soc.*, 1951, 73: 5102.

[161] Broadbent K A, Dollimore D, Dollimore J. The Surface Area of Graphite Calculated from Adsorption Isotherms and Heats of Wetting Experiments. *Carbon*, 1966, 4: 281.

[162] Brusset H, Martin J J P, Mendelbaum H G. *Bull. Soc. Chim. Fr.*, 1967, 2346.
[163] Wade W H, Deviney Jr M L, Brown W A, et al. Adsorption and Calorimetric Investigations on Carbon Black Surfaces. III. Immersion Heats in Model Elastomers and Isosteric Heats of Adsorption of C-4 Hydrocarbons. *Rubber. Chem. Technol.*, 1972, 45: 117.
[164] Kraus G. The Heat of Immersion of Carbon Black in Water, Methanol and n-Hexane. *J. Phys. Chem.*, 1955, 59: 343.
[165] Patrick W A. p. 241. In: Iler R K (Editor). The Colloid Chemistry of Silica and Silicates. Ithaca, New York: Cornell University Press, 1955.
[166] p. 234. In: Iler R K (Editor). The Colloid Chemistry of Silica and Silicates. Ithaca, New York: Cornell University Press, 1955.
[167] Young G J, Bursh T P. Immersion Calorimetry Studies of the Interaction of Water with Silica Surfaces. *J. Colloid Sci.*, 1960, 15: 361.
[168] Ganichenko L G, Kiselev V F, Krasilnikov K G, et al. Adsorption Properties of Silica Gel and Quartz as Function of the Nature of Their Surfaces. 4. Adsorption and Heat of Adsorption of Aliphatic Alcohols on Aerosil. *Zh. Fiz. Khim.*, 1961, 35: 1718.
[169] Egorova T S, Zarifyants Y A, Kiselev V F, et al. Effect of the Nature of Silica Gel and Quartz Surfaces on Their Adsorption Properties. 8. Differential Heats of Water Vapor Adsorption on the Silica Surface. *Zh. Fiz. Khim.*, 1962, 36: 1458.
[170] Mikhail R Sh, Khalil A M, Nashed S. Effects of Thermal Treatment on the Surface Properties of a Wide Pore Silica Gel. *Thermochimica Acta*, 1978, 24: 383.
[171] Kiselev V F. Poverkhnostnye yavleniya v poluprovodnikakh i dielektrikakh. *Surface Phenomena in Semiconductors and Dielectrics*, Nauka, Moscow, 1970.
[172] Milonjić S K. Heat of Immersion of Silica in Water. *Thermochimica Acta*, 1984, 78: 341.
[173] Makrides A C, Hackerman N. Heats of Immersion. I. The System Silica–Water. *J. Phys. Chem.*, 1959, 63: 594.
[174] Wade W H, Cole H, Meyer D E, et al. pp. 35-41. In: Gould R F (Editor). Solid Surfaces, the Gas-Solid Interface. Washington, DC: Advanced Chemistry Series, No. 33, American Chemical Society, 1961.
[175] Wade W H, Hackerman N. pp. 222-231. In: Gould R F (Editor). Contact Angle, Wettability, Adhesion. Washington, DC: Advanced Chemistry Series, No. 43, American Chemical Society, 1964.
[176] Kunath D, Schulz D. Correlations Between the Infrared Absorption of the Vibration of Surface Hydroxyl Groups on Aerosil and Specific Adsorption Enthalpies. *J. Colloid Interface Sci.*, 1978, 66: 379.
[177] Dorris G M, Gray D G. Adsorption of n-Alkanes at Zero Surface Coverage on Cellulose Paper and Wood Fibers *J. Colloid Interface Sci.*, 1980, 77: 353.
[178] Anhang J, Gray D G. Surface Characterization of Poly(Ethylene Terephthalate) Film by Inverse Gas Chromatography. *J. Appl. Polymer Sci.*, 1982, 27: 71.
[179] Meyer E F. On Thermodynamics of Adsorption Using Gas-Solid Chromatography. *J. Chem. Educ.*, 1980, 57: 120.
[180] Conder J R. Thermodynamic Measurement by Gas Chromatography at Finite Solute Concentration. *Chromatographia*, 1974, 7: 387.
[181] de Boer J H. The Dynamical Character of Adsorption. London: Oxford University Press, 1953.
[182] Fowkes F M. Attractive Forces at Interfaces. *Ind. Eng. Chem.*, 1964, 56: 40.

[183] Fowkes F M. Calculation of Work of Adhesion by Pair Potential Suummation. *J. Colloid Interface Sci.*, 1968, 28: 493.

[184] Gaines Jr G L. Surface and Interfacial Tension of Polymer Liquids−a Review. *Polym. Eng. Sci.*, 1972, 12: 1.

[185] Saint-Flour C, Papirer E. Gas-Solid Chromatography: a Quick Method of Estimating Surface Free Energy Variations Induced by the Treatment of Short Glass Fibers. *J. Colloid Interface Sci.*, 1983, 91: 69.

[186] Schultz J, Lavielle L, Martin C. Propriétés de Surface des Fibres de Carbone Déterminées par Chromatographie Gazeuse Inverse. *J. Chim. Phys.*, 1987, 84: 231.

[187] Dong S, Brendle M, Donnet J B. Study of Solid Surface Polarity by Inverse Gas Chromatography at Infinite Dilution. *Chromatographia*, 1989, 28: 469.

[188] Donnet J-B, Park S J, Balard H. Evaluation of Specific Interactions of Solid Surfaces by Inverse Gas Chromatography. *Chromatographia*, 1991, 31: 434.

[189] Donnet J-B, Park S J. Surface Characteristics of Pitch-Based Carbon Fibers by Inverse Gas Chromatography Method. *Carbon*, 1991, 29: 955.

[190] Donnet J-B, Park S J, Brendle M. The Effect of Microwave Plasma Treatment on the Surface Energy of Graphite as Measured by Inverse Gas Chromatography. *Carbon*, 1992, 30: 263.

[191] Wang M-J, Wolff S, Donnet J B. Filler-Elastomer Interactions. Part I: Silica Surface Energies and Interactions with Model Compounds. *Rubber Chem, Technol.*, 1991, 64: 559.

[192] Snyder L R. Principle of Adsorption Chromatography. New York: Marcel Dekker, 1986.

[193] Papirer E, Vidal A, Wang M J, et al. Modification of Silica Surfaces by Grafting of Alkyl Chains. II-Characterization of Silica Surfaces by Inverse Gas-Solid Chromatography at Finite Concentration. *Chromatographia*, 1987, 23: 279.

[194] Bilinski B, Chibowski E. The Determination of the Dispersion and Polar Free Surface Energy of Quartz by the Elution Gas Chromatography Method. *Powder Technol.*, 1983, 35: 39.

[195] Donnet J-B, R Qin, Wang M-J. A New Approach for Estimating the Molecular Areas of Linear Hydrocarbons and Their Derivatives. *J. Colloid Interface Sci.*, 1992, 153: 572.

[196] Emmett P. H, Brunauer S. The Use of Low Temperature Van Der Waals Adsorption Isotherms in Determining the Surface Area of Iron Synthetic Ammonia Catalysts. *J. Am. Chem. Soc.*, 1937, 59: 1553.

[197] Bangham D H. *Trans. Faraday Soc.*, 1938, 60: 309.

[198] Van Voorhis J J, Craig R G, Bartell F E. Free Energy of Immersion of Compressed Powders with Different Liquids. II. Silica Powder. *J. Phys. Chem.*, 1957, 61: 1513.

[199] Hillerová E, Jirátová K, Zdražil M. Determination of the Surface Polarity of Peptized Aluminas by Gas Chromatography. *Applied Catalysis*, 1981, 1: 343.

[200] Wang M J, Wolff S, Donnet J B. Filler-Elastomer Interactions. II: Investigation of the Energetic Heterogeneity of Silica Surfaces. *Kautsch Gummi Kunstst.*, 1992, 45: 11.

[201] Rudzinski W, Waksmundzki A, Leboda R, et al. Investigations of Adsorbent Heterogeneity by Gas Chromatography : II. Evaluation of the Energy Distribution Function. *J. Chromatogr.*, 1974, 92: 25.

[202] Waksmundzki A, Sokolowski S, Rayss J, et al. Application of Gas-Adsorption Chromatography Data to Investigation of the Adsorptive Properties of Adsorbents. *Separ. Sci.*, 1976, 11: 29.
[203] Hobson J P. A New Method for Finding Hetergeneous Energy Distributions from Physical Adsorption Isotherms. *Can. J. Phys.*, 1965, 43: 1934.
[204] Jaroniec M. Adsorption on Heterogeneous Surfaces: The Exponential Equation for the Overall Adsorption Isotherm. *Surf. Sci.*, 1975, 50: 553.
[205] Gawdzik J, Suprynowicz Z, Jaroniec M. Determination of the Pre-exponential Factor of Henry's Constant by Gas Adsorption Chromatography. *J. Chromatogr.*, 1976, 121: 185.
[206] Iler R K. The Chemistry of Silica. Wiley., New York, (1979).
[207] Austin A E. *Proc. 3rd Conf. on Carbon*, University of Buffalo, New York, 1958, 389.
[208] Gerspacher M, Lansinger C M. Presented at a meeting of the Rubber Division, ACS, Dallas, Texas, , Apr. 19-22, 1988.
[209] Wang M -J, Wolff S, Freund B. Filler-Elastomer Interactions. Part XI. Investigation of the Carbon-Black Surface by Scanning Tunneling Microscopy, *Rubber Chem. Technol.*, 1994, 67: 27.
[210] Wang W. Doctor's thesis, Alsace, Haute-Alsace University, 1992.
[211] Donnet J-B, Wang W, Vidal A, et al. Cabot Corporation, Boston, MA, USA, unpublished.
[212] Taylor G L, Atkins J H. Adsorption of Propane on Carbon Black. *J. Phys. Chem.*, 1966, 70: 1678.
[213] Schaeffer W D, Smith W R. Effect of Heat Treatment on Reinforcing Properties of Carbon Black. *Ind. Eng. Chem.*, 1995, 47: 1286.
[214] Donnet J-B, Papirer E, Vidal A, et al. *Rubbercon*, 1988, 1: 113.
[215] Wang M-J, Wolff S. Filler-Elastomer Interactions. Part V. Investigation of the Surface Energies of Silane-Modified Silicas. *Rubber Chem. Technol.*, 1992, 65: 715.
[216] Brunauer S, Emmett P H. The Use of Low Temperature Van Der Waals Adsorption Isotherms in Determining the Surface Area of Iron Synthetic Ammonia Catalysts. *J. Am. Chem. Soc.*, 1937, 59: 1553.
[217] Dannenberg E M. Bound Rubber and Carbon Black Reinforcement, *Rubber Chem. Technol.*, 1986, 59: 512.
[218] Todani Y, Sagaye S. *Nippon Gomu Kyokaishi*, 1973, 46: 1031.

# 3 Effect of Fillers in Rubber

During the last several decades, a vast amount of information on elastomer reinforcement by carbon blacks and silica has been published. It is recognized that the main parameters of the filler which govern their reinforcing ability in rubber are the following:

- The size and distribution of primary particles, sometime referred to as "nodules", which are joined by fusion into aggregates arranged at random. The particle size and its distribution directly determine the surface area of the filler;
- The size, shape and their distribution of aggregates (aggregate complexity), i.e., the degree of irregularity of the filler units or the development of branches due to the aggregation of primary particles and the asymmetry of the aggregates. These parameters are generally termed filler "structure"; and
- Surface activity, which, in a chemical sense, is related to the reactivity of the chemical groups on the filler surface, and, in terms of physical chemistry, is referred to as adsorption capacity. This capacity is determined by filler surface energy, both its dispersive and specific components, and the energy distribution on the filler surface.

All these parameters play a role in rubber reinforcement through different mechanisms, such as hydrodynamic effect-strain amplification, interfacial interaction between filler and rubber, occlusion of the polymer in the internal voids of the aggregate, and the agglomeration of filler aggregates in the polymer matrix.

## 3.1 Hydrodynamic Effect–Strain Amplification

The presence of rigid particles has important geometric and dynamic consequences, since these particles will not be deformed under stress applied on the material[1]. As a result, the elastomeric matrix between the filler particles will experience larger strains than for the given overall stretches. This non-affine response is called the "strain amplification" of the elastomeric matrix in the presence of filler[2]. The deformation curves of filled samples thus enter on the steep branch of the stress-strain curves at lower nominal (overall) strains. Note that the stress maximum of Figure 3.1 lies a small distance away from the particle surface.

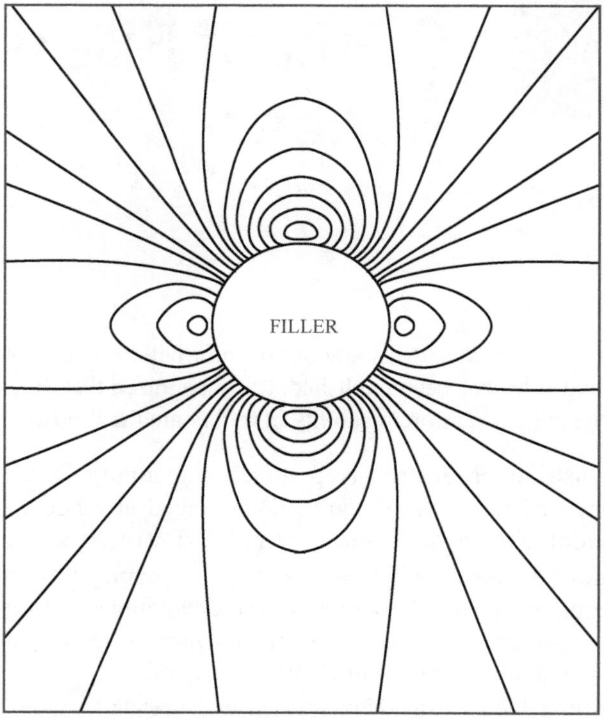


**Figure 3.1** Binder stresses around filler particle (loading vertical)[3]

Strain amplification, while being the primary factor of the changes in mechanical responses which the filler causes, such as the accelerated rise in moduli, will be modified by the extent of strain transmission through the interfaces. If the deformation of the interface is small, the stress transmission is almost complete and the interface can be counted as part of the solid particle. If the interfacial bond strength is larger than the cavitation stress of the matrix, the failure process begins by the formation of hole away from the interface. If, on the other hand, the interface becomes substantially deformed, its own flow or compliance and its integrity must be considered. Whenever the interface ruptures or dewets, an entirely new situation is created in that now some parts of the matrix become more strained while others can relax so that the stress levels off and a substantial portion of the further work of deformation is invested in continuing dewetting processes and in the enlargement of the cavities.

It can be understood that, while stored (elastic) free energy is usually the mechanically valuable state function, it is also the potentially destructive one since the extra free energy accelerates scission and crack growth processes. Thus, apart from applications in which energy dissipation or abrasion resistance are the main concern, conversion of the stored free energy into heat by friction and relaxation processes, enhanced by filler, will in general improve the mechanical performance and life span of elastomers in most constructions or goods.

Due to the existence of the filler, hence the strain amplification, several factors influencing the rubber reinforcement need to be considered:

- the translation of the nominal external macroscopic stress field into the micro stress fields throughout the matrix (especially near the particle surfaces) taking into account filler loading, particle size, shape, and orientation distribution;
- the deformation stability of the interphase, establishing again stress field and distribution as well as the limits of the stresses that can be tolerated without dewetting or rupture;
- the course of the stress-strain response of the matrix as critical strains are approached and reached, i.e., taking into account degrees of energy dissipation, work-hardening, crystallization, and hysteresis;
- the extent and severity of dewetting, crack propagation, and molecular rupture processes in a given case have to be determined.

In all these aspects, the role played by the filler is related more or less to the strain amplification[1].

## 3.2 Interfacial Interaction between Filler and Polymer

One of the consequences of the incorporation of fillers into a polymer is the creation of an interface between a rigid solid phase and a soft solid phase. For rubber-grade fillers, whose surfaces exhibit very little porosity, the total area of the interface depends on both filler loading and the specific surface area of the filler. In a unit volume of compound, the interfacial area, $\psi$, is given by:

$$\psi = \phi \rho S \tag{3.1}$$

where $\phi$ is the volume fraction of the filler in the compound, and $\rho$ and $S$ are the density and specific surface area of the filler, respectively (Eq. 3.1).

Due to the interaction between rubber and filler, the polymer molecules can be adsorbed onto the filler surface either chemically or physically. This adsorption leads to two phenomena which are well documented, the formation of bound rubber and a rubber shell on the filler surface. Both are related to the restriction of the segmental movement of polymer molecules.

### 3.2.1 Bound Rubber

Bound rubber is defined as the rubber portion in an uncured compound which cannot be extracted by a good solvent due to the adsorption of the rubber molecules onto the filler surface. This phenomenon has been studied extensively and is recognized as a typical feature of filler surface activity. Several reviews[4–9] have dealt with the mechanism and the factors affecting the formation of the bound rubber and its influence on rubber properties. A comprehensive survey on this subject was given by Dannenberg[10], and more detailed description of bound rubber will be discussed in Section 5.1.

Generally speaking, bound rubber is a parameter which is simple to measure, but the factors which influence the test results are highly complicated. It was reported that the filler-polymer interaction leading to the formation of bound rubber involves physical adsorption, chemisorption, and mechanical interaction. As far as the filler is concerned, bound rubber is not only affected by the physicochemical characteristics of the filler surface, but also by filler morphology. With regard to the polymer, both the chemical structure of the molecules (saturated vs unsaturated, polar vs nonpolar) and their microstructure (configuration, molecular weight, and its distribution) influence the level of bound rubber. Moreover, bound rubber also shows a strong dependence on the processing conditions of the compound, such as mixing and storage time. Furthermore, there are important factors during the measurement itself which determine the bound rubber content, e.g., the nature of the solvent and the temperature of extraction. Results reported in publications are, therefore, sometimes contradictory, since they are affected by the compound composition, compound preparation, as well as by the test procedures. Bound rubber is thus a phenomenon which, although well-known and the subject of numerous publications, is still a matter of controversy.

Two arguments concerning bound rubber have been widely debated in the literature for many years. One is the nature of the polymer-filler interaction involved in bound rubber formation, and the other is the validity of bound rubber being considered as a measure of filler surface activity.

For carbon black, it has been proposed that the chemisorptive mechanism of bound rubber formation is a free radical reaction between filler and polymer. The existence of unpaired spin electrons on carbon blacks has been demonstrated by electron spin resonance[11] and they were assumed to be free radicals[11,12]. If free polymer radicals can be generated by chain scission during processing, the stabilization of the polymeric radical by carbon black may result in the grafting of rubber molecules onto the carbon black surface by means of covalent carbon–carbon bonds. Gessler[13,14] further postulated that the free radicals of carbon black, originating from the mechanical breakage of the carbon black aggregate structure during milling of rubber and carbon black, is a new source of free radicals, and responsible for the higher bound rubber content when a high-structure carbon black, which is easily broken up, is used instead of a low-structure black. Although this conclusion was supported by the suppression of bound rubber formation by radical chain stoppers such as thiophenols, observations indicating the contrary were also reported. Donnet and coworkers[15–17], with model free radicals, were able to show that the free radical activity of carbon blacks does not play a significant role with regard to chemical reactivity and that the reinforcement of natural rubber is not related to the free radical content of carbon blacks.

On the other hand, while the oxygen-containing functional groups on the carbon black surface seem to lower the polymer-filler interaction for unsaturated rubbers[18,19], in the case of saturated or nearly saturated rubbers, such as butyl rubber, the oxygen functionality plays an important role in bound rubber formation. The mechanism of this oxygen group functionality still remains unclear, even though Gessler[20] postulated a cationic interaction between polymer and carbon black. He believed that the active hydrogen on the surface of oxygenated black, especially in carboxyl groups, is donated as a proton to the polymeric double bond in butyl rubber, and the carbonium ion formed on the polymer chain is proposed either to react directly with the anion left on the carbon black surface or to serve as an agent effecting electrophilic substitution at the crystal edges of the graphitic crystallites of the carbon black surface. However, this mechanism is doubtful for the simple reason that carbon blacks with a wide range of carboxyl group functionalities, resulting from ozonization for varying periods of time and at different temperatures, do not show any influence of the concentration of carboxyl groups on their reinforcing effect in butyl rubber. Similarly, no relationship was observed between the reinforcement in butyl rubber and the concentration of phenolic or quinonic groups[21].

The existence of physical adsorption resulting in bound rubber formation has been demonstrated by Ban, Hess, and coworkers[22,23]. In a study of electron micrographs of microtomed samples of N330-filled SBR compounds, in which the sol was extracted with benzene in a vapor extractor, they found that bound rubber is nearly totally absent on the greater part of the carbon black surface (especially in the convex region of the aggregates). The very small amount of rubber remaining in the concave region of the aggregates would be due to the retraction of the polymer during sample preparation, instead of a chemical reaction between polymer and carbon black surface, since it is hard to believe that there should be a difference in chemical reactivity between concave and convex regions. Wolff, Wang, and Tan[24] reported that at an extraction temperature above 70°C in xylene, the bound rubber of SBR compounds filled with 50 phr N330 decreases very rapidly with increasing temperature and only about 3% of bound rubber is left on the carbon black surface at 100°C. Even this low level of bound rubber cannot be attributed to chemisorption from primary valence bonding because the measurement of bound rubber was carried out at room temperature after extraction. In this case, some molecules of rubber adsorbed physically on the carbon black surface from the solution at room temperature could not be fully removed by the solvent[4,25]. Wolff et al. thus concluded that the polymer-filler interaction, involved in the formation of bound rubber in SBR-carbon black compounds, is essentially a physical phenomenon.

Regardless of the nature of the interaction between filler surface and polymer, bound rubber should be a measure of surface activity. At fixed and practical loadings, the bound rubber content increases with increasing surface area of carbon black[10,24]. This is, obviously, related to the difference in interfacial area in the compound among different grades of carbon blacks. With regard to the surface activity of carbon blacks, meaningful information can only be obtained by a comparison of the bound rubber per unit surface. By normalizing the bound rubber content with the interfacial area in the compound, Dannenberg[10] showed a decrease in bound rubber per unit surface with decreasing particle size of the carbon blacks for SBR compounds filled with 50 phr of carbon black. Wolff et al.[24] reached the same conclusion with different loadings. This is in contrast to observations made in surface energy measurements where both the dispersive and specific components were higher for small-particle carbon blacks[26]. Since bound rubber is essentially a physical phenomenon for SBR-carbon black compounds and since the physical adsorption of a substance is related to the surface energy of the solid, these contradictory results were interpreted in terms of interaggregate multiattachment of rubber molecules and the agglomeration of carbon blacks[24]. It is understood that a rubber molecule may repeatedly attach onto the carbon black surface at different sites. This would reduce the effectiveness of the surface since only a single attachment would render the whole molecule unextractable in solvent. The multiattachment of single chains may occur on different aggregates close to each other, which, of course, depends on the distance between the aggregates. This distance decreases as the surface area of carbon blacks increases at constant loading (see Eqs. 3.8 and 3.9), which in turn reduces the efficiency of the surface for the formation of bound rubber.

On the other hand, the agglomeration of aggregates in the polymer matrix enhances interaggregate multiattachment, and/or decreases the interfacial area by direct contact between aggregates. This agglomeration also shows a strong dependence on inter-aggregate distance[27]. The greater tendency of agglomeration of the smaller carbon blacks would also result in a reduction of the effectiveness of the filler surface. It is thus understandable that the bound rubber per unit surface is underestimated when normalization is based on the total area of interface calculated from filler loading and specific surface area of the carbon black. This has been confirmed by the loading dependence of bound rubber content per unit surface, showing lower values for the higher loadings which is obviously related to greater interaggregate multiattachment and more developed agglomeration rather than to lower surface activity[24]. The bound rubber per unit surface of carbon black-filled SBR compounds at the critical loading, i.e., the lowest loading where a coherent gel can be formed, is approximately proportional to the surface energy of the carbon blacks (Figure 3.2). It can be assumed that, at the critical loading, interaggregate multiattachments are similar and no significant agglomeration of the carbon blacks exists.

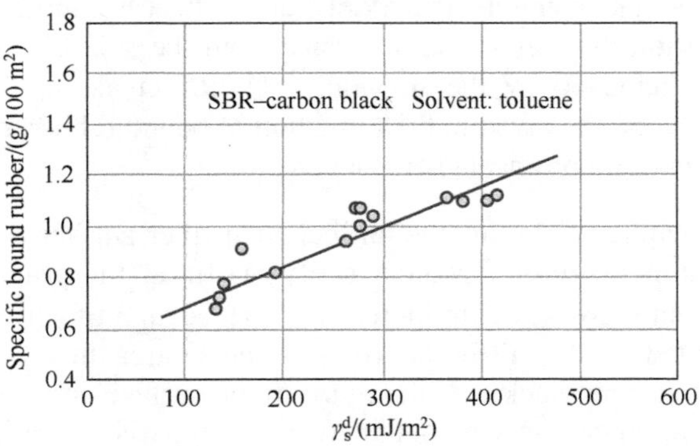


**Figure 3.2** Bound rubber contents on unit surface area of carbon blacks at critical loadings as a function of dispersive components of surface energies[24]

Therefore, when the specific surface activity is investigated with regard to bound rubber, the compounding parameters and conditions of measurement should be well defined. Nevertheless, the total bound rubber content in the compound may be an important parameter influencing rubber properties, especially compound processability.

Among carbon blacks with similar surface area, high-structure products generally have higher bound rubber content. Several explanations have been proposed in this regard, including:

- a less ordered graphite layer structure, leading to higher surface activity; and
- easier breakdown of the aggregates during mixing[10,13] in the case of high-structure carbon blacks[28].

The latter explanation seems to be more reasonable because the aggregate breakdown could result in two things: an increase in filler-polymer interface and higher surface activity of freshly created surface[24,29].

### 3.2.2 Rubber Shell

Another result of filler-polymer interaction is the formation of a more or less immobilized layer or rubber shell surrounding the filler, in compounds as well as in vulcanizates, resulting from the restriction of the molecular motion of the rubber in the vicinity of the filler surface. This immobilization of rubber may also be related to increased entanglements due to filler-polymer attachments.

The existence of a rubber shell on the filler surface has well been demonstrated experimentally. Westlinning and coworkers[30,31], in their investigation of the freezing point depression of benzene in swollen networks, concluded that crosslink density is higher in the vicinity of carbon black surface. Westlinning[32] suggested that, at 20°C, molecular movement was hindered by the filler surface up to a depth of 35 nm. At 100°C, this layer would disappear. Schoon and Adler[33] reported that the thickness of this hindered layer is 35–50 nm for SBR and 25–30 nm for NR.

Following the investigations of rheological behavior of rubber compounds and the dynamic properties of carbon-black-filled SBR and *cis*-BR vulcanizates, Smit[34–37] proposed a rubber shell of 2–5 nm on the carbon black surface to explain his results. This model was supported by the investigation of filled rubber by NMR[19,38–43], DSC[40], and DMTA[40]. Kaufman et al.[38], using the nuclear spin relaxation time measurement (T2) to investigate carbon-black-filled *cis*-BR and EPDM, demonstrated the presence of three distinct regions, characterized by different mobility of the rubber chains, i.e., a unbonded, mobile rubber region like gum, a bonded rubber in an outer shell around the carbon black, suffering some loss in mobility, and an inner shell of tightly bonded rubber with very limited mobility on the T2 time scale. Similar findings were made by another group using *cis*-BR[39]. Other NMR studies on polyisoprene[44], SBR[45], and PDMS-silica systems[46] have also found greatly-reduced T2 for the tightly bound polymer layer at the interface.

Haidar[47,48], based on a study of the effect of static strain on dynamic modulus, suggested that the segmental mobility of the polymer on the filler surface was diminished and that the polymer at this point had glassy characteristics.

Although there is sufficient evidence to show the existence of a rubber shell on filler surface, considerable disagreement exists with regard to the thickness or volume of this shell. Kraus and coworkers[49,50], for instance, concluded from their investigation of $T_g$ and the coefficient of expansion of the filled rubber that the segmental motion of the rubber was not severely restricted over a substantial distance from the filler surface.

More recently, rubber shell has been displayed visually and the thickness of shell measured easily with AFM images by Zheng et al., as shown in Figure 3.3[51], where carbon black aggregates are shown in brighter area. Comparing the height and modulus images of the vulcanizate, it reveals that the sizes of filler aggregates are increased due to the existence of rubber shell and modulus of the polymer layers in rubber shell gradually decreased with the distance from the filler surface.

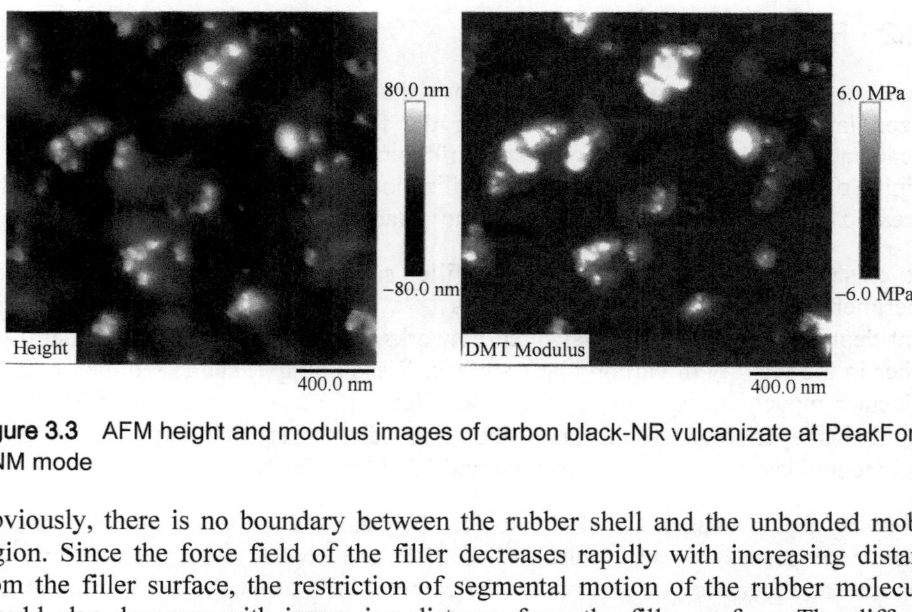


**Figure 3.3** AFM height and modulus images of carbon black-NR vulcanizate at PeakForce QNM mode

Obviously, there is no boundary between the rubber shell and the unbonded mobile region. Since the force field of the filler decreases rapidly with increasing distance from the filler surface, the restriction of segmental motion of the rubber molecules should also decrease with increasing distance from the filler surface. The different values for the thickness of the rubber shell given in the literature are, therefore, related to the criterion applied for the mobility of the rubber segments, the precision of the test method, the physical chemistry of the filler surface and the polymer concerned.

Although bound rubber and rubber shell are both related to polymer-filler interaction, they represent different concepts. Bound rubber refers to the whole molecule in which one or several segments are in contact with the filler surface, and to the whole molecule entangled with other molecules which are attached to the filler. The precondition for the rubber molecule being bound rubber is that the adsorptive attachments cannot be separated and that entanglements cannot be solvated during solvation of the rubber compound. In the case of the rubber shell, the definition is related to the change of segmental mobility of the rubber molecules. The segments, rather than the molecules, whose motion is affected by the force field of the filler, constitute the rubber shell, whether they satisfy the necessary conditions for bound rubber formation (contact or entanglement) or not.

Since the rubber shell has its origin in the restriction of the segmental motion of rubber molecules near the filler surface, it should also show a strong dependence on temperature and the rate of strain to which the rubber is subjected. At high frequencies of dynamic strain, the rubber shell may be "harder" and decrease rapidly with increasing temperature. Therefore, when estimating the thickness of the rubber shell, the measurement conditions should be taken into consideration.

## ■ 3.3 Occlusion of Rubber

The effect of filler structure on the rubber properties of filled rubber has been explained by the occlusion of rubber by filler aggregates. In 1970, Medalia[52,53] proposed that, when structured carbon blacks are dispersed in rubber, the polymer portion filling the internal void of the carbon black aggregates, or the polymer portion located within the irregular contours of the aggregates, is unable to participate fully in the macrodeformation (Figure 3.4 and Figure 3.5). The partial immobilization in the form of occluded rubber causes this portion of rubber to behave like the filler rather than like the polymer matrix. Due to this phenomenon, the effective volume of the filler, with regard to the stress-strain behavior of the filled rubber, is increased considerably.

The occlusion of rubber is a geometrical concept. As pointed out by Medalia[52,53], its quantitative representation is quite difficult since most aggregates do not contain any significant volume of three-dimensional "ink bottle"-like concavities. The definition of occluded rubber therefore depends on what is meant by "within" the aggregates. Conceptually, the occluded rubber could be defined as the rubber contained within the convex hull surrounding the aggregate, but it is difficult to estimate in three dimensions from electron micrographs. For practical purposes, Medalia defined the occluded rubber as the rubber within a sphere of the same projected area as the aggregate, i.e., the volume of occluded rubber within an aggregate, $V_{occ}$, was defined geometrically as the difference between the volume of this "equivalent" sphere, $V_{es}$, and the solid volume of the aggregate, $V_a$. $V_{es}$ can be measured directly on electron micrographs. $V_a$ can also be estimated from electron micrographs by means of floc simulation of the aggregate by primary particles. Based on the quantitative treatment of the so-called "soft ball" DBP absorption, in which the void volume consists of void volume within equivalent spheres, related to the projected area of the aggregate as measured in electron microscope, and void volume between spheres in a closely-packed state, Medalia proposed the following equation to calculate the occluded rubber:

$$\phi' = \phi\left(\frac{1+0.02139\text{DBP}}{1.46}\right) \tag{3.2}$$

where $\phi'$ is the volume fraction of filler plus occluded rubber, and $\phi$ is the volume fraction of the filler. The volume ratio between occluded rubber and carbon black is given by:

$$\frac{V_{\text{occ}}}{V_{\text{a}}} = \frac{\phi' - \phi}{\phi} = \frac{\text{DBP} - 21.5}{68.26} \tag{3.3}$$

Using the occluded rubber concept, Kraus[54,55] investigated the effect of carbon black structure on modulus of SBR vulcanizates and proposed a so-called carbon black "Structure Concentration Equivalence" principle. He applied it to the strength properties of vulcanizates filled with carbon blacks having the same surface areas, but different structure. Assuming that the effective filler concentration includes the internal void volume of the primary aggregate structure, and that this internal void volume in turn is proportional to the excess of the observed crushed DBP (24M4DBP) value over the corresponding value of a "structureless" carbon black, Kraus proposed another equation to calculate the volume of occluded rubber:

$$\frac{V_{\text{occ}}}{V_{\text{a}}} = \frac{\text{24M4DBP} - 31}{55} \tag{3.4}$$

Simply by relating the endpoint of crushed DBP absorption to the void space within and between equivalent spheres of aggregates and assuming the spheres to be packed at random, Wang, Wolff, and Tan[27] obtained a similar equation but with different numerical constants:

$$\frac{V_{\text{occ}}}{V_{\text{a}}} = \frac{\text{24M4DBP} - 33}{87.8} \tag{3.5}$$

The crushed DBP values were used because they are more representative of the aggregate state in rubber.

Studies carried out by Medalia[56] and Sambrook[57] showed that the relative modulus, calculated by means of the Guth-Gold equation, using $\phi'$ instead of $\phi$, was higher than experimental values. Medalia believed that the occluded rubber was only partially immobilized and that its contribution was therefore only partially effective. In order to fit his experimental data, he introduced a so-called *F* factor, i.e., the occluded volume effectiveness factor. This factor is, of course, dependent on the basis used for the calculation of occluded rubber and on the rubber properties concerned which are related to different levels of strain and temperature.

Moreover, Medalia proposed that most of the polymer sub-chains which were bound to the filler surface, led to crosslinks within the occluded rubber, and the stretching of polymer chains linked the occluded rubber within one aggregate to that occluded within another aggregate[53] (Figure 3.4). The occluded rubber, as shown in Figure 3.5, could be deformed after stretching to 300% elongation, where the non-localized or mobile bonds might slip and short chains might detach from one of the aggregates and then rupture on undergoing 480% elongation.

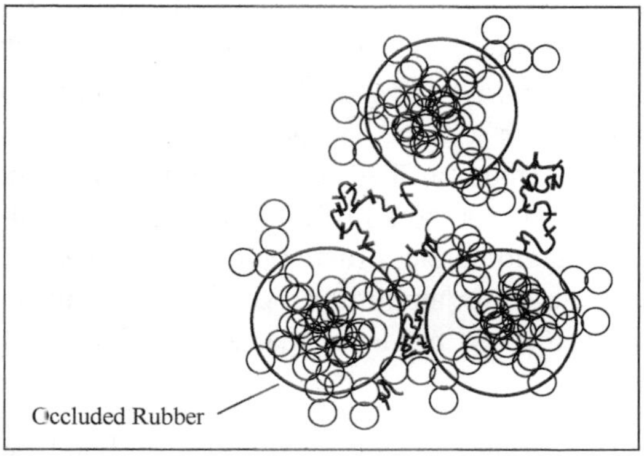


**Figure 3.4** Three 50-particle aggregates with occluded rubber (the circles are of the same area as the projected area of the aggregate and thus are projections of spheres representing the aggregate plus occluded rubber)[53]

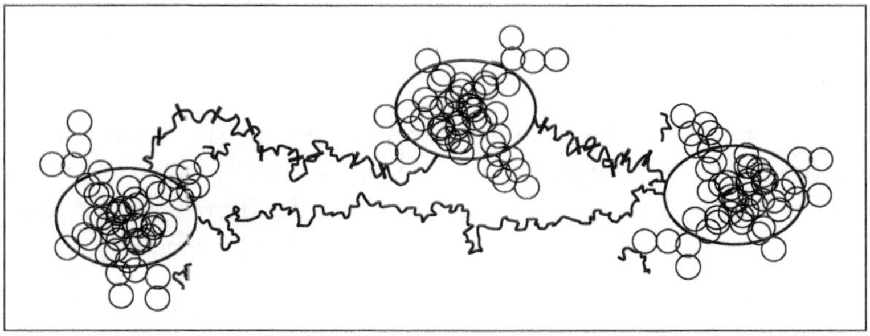

**Figure 3.5** Aggregates of Figure 3.4, after stretching to 300% elongation (the ellipses are projections of prolate ellipsoids of the same volume as the spheres of Figure 3.4 with an arbitrary elongation of 50%, compared to the spheres)[53]

## 3.4 Filler Agglomeration

### 3.4.1 Observations of Filler Agglomeration

Filler aggregates in a polymer matrix have a tendency to associate to agglomerates, especially at high loadings, leading to chain-like filler structures or clusters. These are generally termed secondary structure or, in some cases, filler network, even though the latter is not comparable to the continuous polymer network structure. Although this phenomenon has been known for a long time, it was not until investigations of the dynamic properties of filled rubber were carried out that its role in rubber reinforcement was better understood. In 1950, Warring[58] observed that the dynamic modulus $E'$ decreases with increasing strain amplitude from a high plateau $E'_0$ to a low plateau $E'_\infty$. (Note that in the literature of rubber science and technology, the elastic and viscous modulus are generally denoted with $E'$ and $E''$ for extension and compression strain and $G'$ and $G''$ for shear strain. The discussion of $E'$ and $E''$ in this chapter is generally valid for $G'$ and $G''$ and vice versa.) This phenomenon, later termed "Payne effect", is unique to filled rubber, and the difference ($E'_0 - E'_\infty$), or $\Delta E'$, increases exponentially with increases in filler loading. A simple explanation for this phenomenon is that the agglomerate structure formed by carbon black aggregates is destroyed at high dynamic strain amplitude. Although a number of objections to this explanation were raised, there is no doubt that a secondary agglomerate structure of carbon blacks exists.

In the studies of the dynamic mechanical properties on carbon-black-filled vulcanizates, Gerspacher and coworkers[59,60] used a shift factor for the Cole-Cole plot to characterize the agglomeration of carbon blacks. They also used the same method as that Fitzgerald used[61] to measure agglomerate cohesion energy. The investigation of various carbon blacks showed that agglomerate cohesion energy is related to the specific surface area of the carbon black, i.e., the higher the surface area of the carbon black, the higher is the cohesion energy.

Measuring the dynamic properties of filled rubber under static strain, Voet and Morawski[62] reported a decrease in elastic modulus for HAF loaded SBR, followed by a pronounced increase with increasing static strain. They attributed the decrease in $E'$ with increasing strain at low deformations to the disintegration of the filler agglomerate.

Wolff and Donnet[63] studied the stress-strain behavior of vulcanizates filled with carbon blacks and silicas, respectively, and observed a decrease in the effective factor, which converts the actual volume fraction into the effective volume fraction, at low strain. This effect was also interpreted as being caused by a breakdown of filler agglomerates.

Based on their evaluation of the electrical properties of dynamically[64] and statically[65] strained carbon-black-filled vulcanizates, Voet and coworkers confirmed the formation of carbon black agglomerates in unstrained vulcanizates. These results were in agreement with mechanical dynamic and static properties, showing that carbon blacks with smaller particles and higher structure had a greater tendency to undergo filler agglomeration.

### 3.4.2 Modes of Filler Agglomeration

There is still no experimental evidence which shows how filler agglomerates are constructed, i.e., whether they are formed by direct contact between filler aggregates or via a layer of immobilized rubber on the filler surface.

With different fillers, agglomerates may be formed in different ways which would respond differently to dynamic hysteresis. As cited by Wolff and Wang[66,67], in the case of a highly polar filler such as silica, which is incompatible with hydrocarbon rubber, filler agglomeration may take place primarily by direct contact between aggregates, in this particular case perhaps even via hydrogen bonding[66] to give a rigid construction. This type of agglomerate would be rapidly destroyed above a certain level of strain. In the case of a filler which has a greater affinity toward hydrocarbon rubber, agglomerates could be formed by a joint shell mechanism[68] or junction rubber mechanism[69].

It is known that as a result of polymer-filler interaction, the adsorption of polymer molecular chains on the filler surface may reduce the mobility of polymer segments. As discussed in Section 3.2.2, this would result in a rubber shell on the filler surface in which the polymer viscosity is increased, the spectrum of relaxation time would be broadened, and the modulus would be increased. The very high modulus of rubber near the filler surface in rubber shell would gradually decrease with increasing distance from the surface and finally reach the same level as that of polymer matrix at a certain distance[68] (Figure 3.6). This is related to the change in polymer chain mobility, transition zone, and $T_g$, which are determined by the mobility of the polymer segments, which would shift to a higher temperature as shown in Figure 3.7. When two or more filler particles or aggregates are close enough, they would form an agglomerate via a joint rubber shell in which the modulus of the polymer is higher than that of the polymer matrix (Figure 3.8 and Figure 3.9). The filler agglomerates formed via this joint shell construction would be much less rigid than those formed by the direct contact of aggregates. This type of filler agglomerate may begin to break down at a relatively lower level of strain but would proceed less rapidly. More recently, Clément[70,71] confirmed that the model of joint shell is the only reasonable interpretation to explain the Payne effect in silica filled silicon elastomers, by the existence of a gradient in the mobility of polymeric chains from the polydimethylsiloxane/silica interface to the bulk.

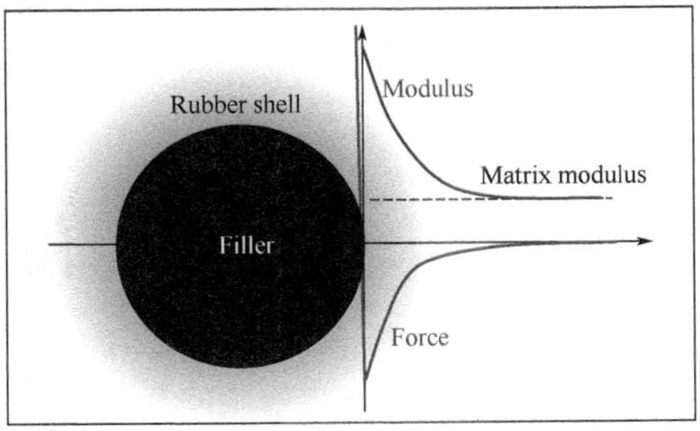


**Figure 3.6** Model of rubber shell[68]

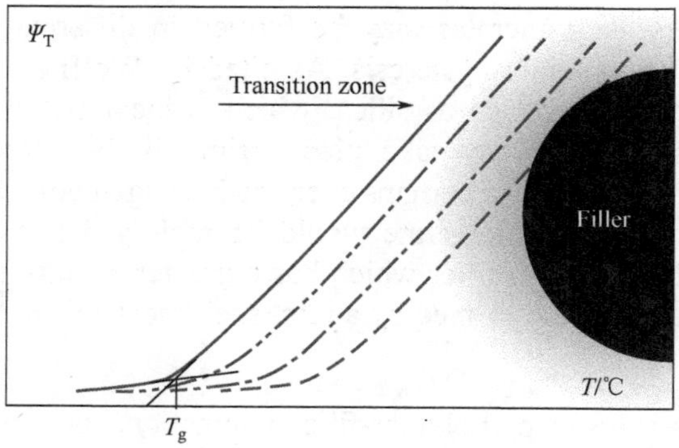


**Figure 3.7** Segmental mobility of rubber shell

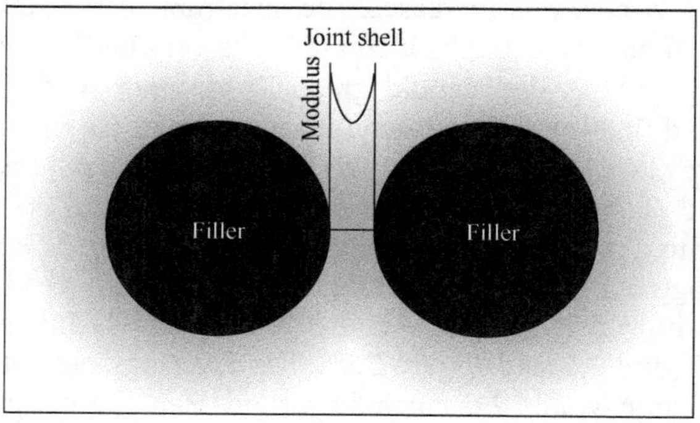


**Figure 3.8** Model of joint shell

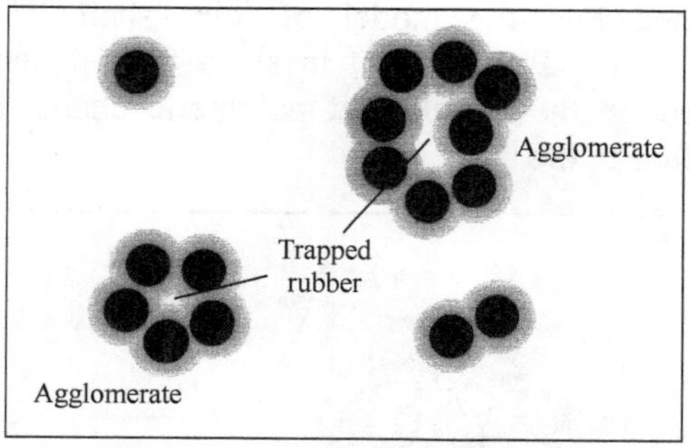


**Figure 3.9** Agglomeration of filler aggregates by joint shell mechanism

It should be pointed out that for a given polymer/filler system, both mechanisms, i.e., direct contact mode and joint rubber shell, may play a part in filler agglomeration. However, depending on the nature of the polymer and the surface characteristics of the filler which determine the polymer/filler and filler/filler interaction, one mechanism may be more important than the other.

One of the consequences of filler agglomeration is that the rubber trapped in the filler agglomerates, or secondary structure, is largely immobilized in the sense that this rubber portion is shielded from deformation under stress. Of course, the effect of agglomeration is highly dependent on strain and temperature. At moderate and high strain, most of the agglomerates are broken down and the rubber trapped in the agglomerates acts as polymer matrix. On the other hand, rising temperature weakens interaggregate interaction and reduces the modulus of the rubber shell. Consequently, filler agglomeration would be expected to be lower.

### 3.4.3 Thermodynamics of Filler Agglomeration

The formation of filler agglomerates is, of course, dependent on the intensity of interaggregate attractive potential and the distance between aggregates. Assuming that only van der Waals forces exists between the particles and that the shape of a carbon black particle lies statistically between a cube and a sphere, Van den Tempel[72] proposed that the average interaction energy between particles of various shapes and orientations, $\Delta F$, is given by:

$$\Delta F = \frac{Ad^{1.5}}{12\delta_{\mathrm{aa}}^{1.5}} \tag{3.6}$$

and that the attractive force between consecutive particles in a chain is given by:

$$\text{Force} = -\frac{Ad^{1.5}}{8\delta_{\mathrm{aa}}^{2.5}} \tag{3.7}$$

where $d$ is the average particle diameter, $\delta_{\mathrm{aa}}$ the average distance between two particles, and $A$ a constant which depends on the polarizabilities of the atoms present. The equations show that the smaller the distance between the particles and the greater the particle diameter, the higher is the attractive force.

Based on the arrangement of equivalent spheres and the concept of occluded rubber, Wang, Wolff, and Tan[27] obtained an equation for the calculation of the interaggregate distance, $\delta_{\mathrm{aa}}$, which is another important factor governing filler agglomeration:

$$\delta_{\mathrm{aa}} = \left[k\phi^{-1/3}\beta^{-1/3} - 1\right]d_{\mathrm{a}} \tag{3.8}$$

or

$$\delta_{\mathrm{aa}} = \frac{6000}{\rho S}\left[k\phi^{-1/3}\beta^{-1/3} - 1\right]\beta^{1.43} \tag{3.9}$$

where $\phi$ is the volume fraction of carbon black, $d_{\mathrm{a}}$ the projected equivalent of the aggregate diameter, $\rho$ the density of the filler, $S$ the specific surface area, and $k$ a constant which is dependent on the arrangement of aggregates in the polymer matrix. For a cubic packing, $k$ is 0.806, and for face-centered cubic packing (i.e., the most dense packing), $k$ is 0.905. In the case of random packing, $k$ is 0.85, which is generally used. In this equation, $\beta$ is an expansion factor which is defined as the effective volume of the filler divided by the solid volume of the aggregates. It can be calculated from Eq. 3.3, 3.4, or 3.5. Equation 3.9 shows that the distance between aggregates, at given loadings, decreases with decreasing aggregate size or increasing surface area. On the other hand, the filler structure influences the interaggregate distance through the factor $\beta$.

It should be noted that the formation of filler agglomerates depends not only on interparticle interaction, but to a considerable extent also on polymer-filler interaction. In other words, if filler-filler interaction remains constant, filler agglomeration would be impeded as filler-polymer interaction increases. Therefore, taking both filler-filler and filler-polymer interaction into consideration, Wang, Wolff, and Donnet[26] proposed a factor $S_f$ to represent the tendency of the filler to undergo agglomeration in hydrocarbon rubbers. This factor is related to the ratio between filler-filler interaction and filler-polymer interaction, and can be determined by means of inverse gas chromatography at infinite dilution (see Section 2.4.2.3.2). If the distance between aggregates is constant, a high value of $S_f$ would be equivalent to easy filler agglomeration in hydrocarbon rubbers. Wang et al.[26] found that the values of $S_f$ for the adsorption of acetonitrile on carbon blacks were mostly in the same range, with slightly higher values for large-particle carbon blacks. Based on the surface energies of polymer and filler, Wang[73] estimated the driving force for the filler agglomeration. In a given polymer system, according to the detailed description of interaction between molecules and surface energy of solids in Section 2.4.2, the driving force for filler agglomeration originates from polymer-filler interaction and filler-filler interaction, which are determined by the filler surface energy and chemical nature.

For the filled compounds, the filler agglomeration process may be simulated as shown in Figure 3.10 and Figure 3.11.

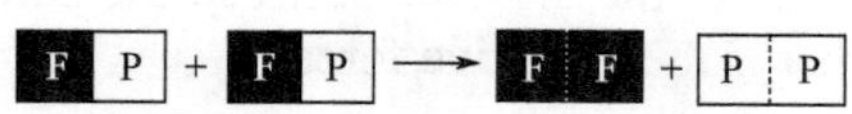


**Figure 3.10** Scheme of the filler agglomeration process

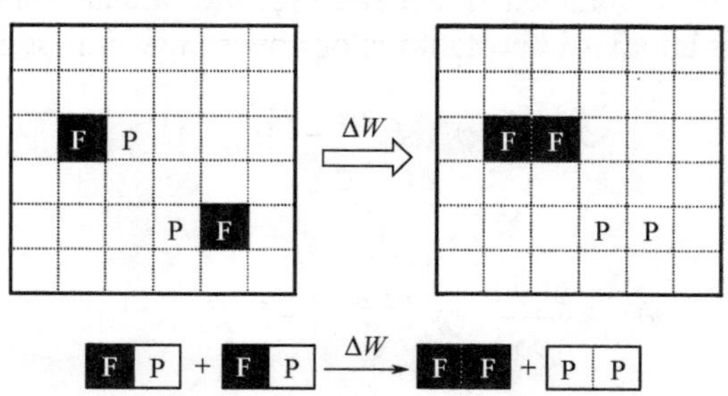


**Figure 3.11** Change in energy associated with filler agglomeration process

In this system, each filler particle is treated as a unit and surrounded by polymer which may also be considered to consist of polymer units. When two filler particles agglomerate, a pair of filler particles and a pair of polymer units will be formed. This process can be dealt with as a kinetic process in which the change in potential energy, $\Delta U$, is:

$$\Delta U = U_{\mathrm{ff}} + U_{\mathrm{pp}} - 2U_{\mathrm{fp}} \tag{3.10}$$

where $U_{\mathrm{ff}}$, $U_{\mathrm{pp}}$, and $U_{\mathrm{fp}}$ are the attractive potential energies between filler particles, polymer units, and polymer-filler respectively. Each term of the attractive potential energy contains an attractive constant (or Hamaker constant). If all other conditions of the attractive potentials for each pair of units, such as distance and size, are the same, the difference and the attractive potential of each unit will be estimated from the change in their adhesive and cohesive work. According to Eq. 2.134, the total change in adhesive energy, $\Delta W$, in the agglomeration process is given by:

$$\begin{aligned}\Delta W = {} & 2\left(\gamma_{\mathrm{f}}^{\mathrm{d}}\gamma_{\mathrm{f}}^{\mathrm{d}}\right)^{1/2} + 2\left(\gamma_{\mathrm{f}}^{\mathrm{p}}\gamma_{\mathrm{f}}^{\mathrm{p}}\right)^{1/2} + W_{\mathrm{f}}^{\mathrm{h}} + W_{\mathrm{f}}^{\mathrm{ab}} + 2\left(\gamma_{\mathrm{p}}^{\mathrm{d}}\gamma_{\mathrm{p}}^{\mathrm{d}}\right)^{1/2} + 2\left(\gamma_{\mathrm{p}}^{\mathrm{p}}\gamma_{\mathrm{p}}^{\mathrm{p}}\right)^{1/2} \\ & + W_{\mathrm{p}}^{\mathrm{h}} + W_{\mathrm{p}}^{\mathrm{ab}} - 2\left[2\left(\gamma_{\mathrm{f}}^{\mathrm{d}}\gamma_{\mathrm{p}}^{\mathrm{d}}\right)^{1/2} + 2\left(\gamma_{\mathrm{f}}^{\mathrm{p}}\gamma_{\mathrm{p}}^{\mathrm{p}}\right)^{1/2} + W_{\mathrm{fp}}^{\mathrm{h}} + W_{\mathrm{fp}}^{\mathrm{ab}}\right]\end{aligned} \tag{3.11}$$

or

$$\begin{aligned}\Delta W = {} & 2\gamma_{\mathrm{f}}^{\mathrm{d}} + 2\gamma_{\mathrm{f}}^{\mathrm{p}} + W_{\mathrm{f}}^{\mathrm{h}} + W_{\mathrm{f}}^{\mathrm{ab}} + 2\gamma_{\mathrm{p}}^{\mathrm{d}} + 2\gamma_{\mathrm{p}}^{\mathrm{p}} + W_{\mathrm{p}}^{\mathrm{h}} + W_{\mathrm{p}}^{\mathrm{ab}} \\ & - 4\left(\gamma_{\mathrm{f}}^{\mathrm{d}}\gamma_{\mathrm{p}}^{\mathrm{d}}\right)^{1/2} - 4\left(\gamma_{\mathrm{f}}^{\mathrm{p}}\gamma_{\mathrm{p}}^{\mathrm{p}}\right)^{1/2} - 2W_{\mathrm{fp}}^{\mathrm{h}} - 2W_{\mathrm{fp}}^{\mathrm{ab}}\end{aligned} \tag{3.12}$$

where $\gamma_{\mathrm{f}}^{\mathrm{d}}$ and $\gamma_{\mathrm{f}}^{\mathrm{p}}$ are the dispersive and polar components of the surface energy of the filler; $\gamma_{\mathrm{p}}^{\mathrm{d}}$ and $\gamma_{\mathrm{p}}^{\mathrm{p}}$ the dispersive and polar components of the surface energy of the polymer; and $W_{\mathrm{f}}^{\mathrm{h}}$, $W_{\mathrm{p}}^{\mathrm{h}}$, and $W_{\mathrm{fp}}^{\mathrm{h}}$ the hydrogen bonding work, and $W_{\mathrm{f}}^{\mathrm{ab}}$, $W_{\mathrm{p}}^{\mathrm{ab}}$, and $W_{\mathrm{fp}}^{\mathrm{ab}}$ the work from acid-base interactions between filler surface, polymer surface, and between filler surface and polymer surfaces, respectively.

By rearranging Eq. 3.12, one has:

$$\begin{aligned}\Delta W = {} & 2\left[\gamma_{\mathrm{f}}^{\mathrm{d}} + \gamma_{\mathrm{p}}^{\mathrm{d}} - 2\left(\gamma_{\mathrm{f}}^{\mathrm{d}}\gamma_{\mathrm{p}}^{\mathrm{d}}\right)^{1/2}\right] + 2\left[\gamma_{\mathrm{f}}^{\mathrm{p}} + \gamma_{\mathrm{p}}^{\mathrm{p}} - 2\left(\gamma_{\mathrm{f}}^{\mathrm{p}}\gamma_{\mathrm{p}}^{\mathrm{p}}\right)^{1/2}\right] \\ & + \left(W_{\mathrm{f}}^{\mathrm{h}} + W_{\mathrm{p}}^{\mathrm{h}} - 2W_{\mathrm{fp}}^{\mathrm{h}}\right) + \left(W_{\mathrm{f}}^{\mathrm{ab}} + W_{\mathrm{p}}^{\mathrm{ab}} - 2W_{\mathrm{fp}}^{\mathrm{ab}}\right)\end{aligned} \tag{3.13}$$

Therefore, the following equation may be given:

$$\begin{aligned}\Delta W = {} & 2\left[\left(\gamma_{\mathrm{f}}^{\mathrm{d}}\right)^{1/2} - \left(\gamma_{\mathrm{p}}^{\mathrm{d}}\right)^{1/2}\right]^2 + 2\left[\left(\gamma_{\mathrm{f}}^{\mathrm{p}}\right)^{1/2} - \left(\gamma_{\mathrm{p}}^{\mathrm{p}}\right)^{1/2}\right]^2 \\ & + \left(W_{\mathrm{f}}^{\mathrm{h}} + W_{\mathrm{p}}^{\mathrm{h}} - 2W_{\mathrm{fp}}^{\mathrm{h}}\right) + \left(W_{\mathrm{f}}^{\mathrm{ab}} + W_{\mathrm{p}}^{\mathrm{ab}} - 2W_{\mathrm{fp}}^{\mathrm{ab}}\right)\end{aligned} \tag{3.14}$$

If $\gamma_f^d = \gamma_p^d$, $\gamma_f^p = \gamma_p^p$, $W_f^h = W_p^h = W_{fp}^h$, and $W_f^{ab} = W_p^{ab} = W_{fp}^{ab}$, then $\Delta W = 0$; the attractive potential between filler particles would disappear. This suggests that the driving force of filler agglomeration is the difference in surface energies – both in the intensity and nature – between filler and polymer. From the thermodynamic point of view, the dispersed filler in a polymer matrix is only stable provided the energy characteristics of the filler and polymer surface are identical, or the adhesive energies between polymer and filler surface due to hydrogen bonding, acid-base interaction, and/or other specific interactions are so high that they are able to offset the effect of the difference in surface energies between polymer and filler and cohesive energies of filler and polymer themselves. The greater the difference in their surface energies and the lower the specific interactions in terms of hydrogen bonding and acid-base interaction between filler and polymer, the higher is the tendency for the filler agglomeration in the polymer.

### 3.4.4 Kinetics of Filler Agglomeration

For a filled compound, there is always a difference in surface energy between filler and polymer, so that even for a system in which the filler is well or uniformly dispersed in the polymer matrix, the filler aggregates will inexorably tend to flocculate[74] during storage and vulcanization of the compound forming a filler agglomerate. This effect, termed as flocculation in a general sense in colloid systems and as filler agglomeration in rubber compounds in particular, has been well demonstrated by Böhm et al.[75,76] in a study of compound annealing on Payne effect, which was generally taken as a measure of filler flocculation. They found that poorer initial dispersion of carbon black, lower molecular weight of the polymer, and higher annealing temperature were associated with higher flocculation rate. At the same annealing condition, longer annealing time gives a higher Payne effect. Besides the attractive potential between aggregates of the filler, the flocculation process is determined by diffusion of aggregates due to Brownian motion to form thermodynamically stable agglomerates. For a given polymer-filler system, the diffusion characteristics of filler aggregates in the polymer matrix play an important role in flocculation kinetics. As well established in colloid chemistry, the diffusion constant, $\Delta$, of a colloid system, which is the main factor to control flocculation, is related to the temperature, $T$, and resistance coefficient, $f$, by:

$$\Delta = kT / f \tag{3.15}$$

where $k$ is the Boltzmann constant. The resistance coefficient is determined by the medium viscosity, $\eta$, and the size and shape of the particle. For a spherical particle with a radius $r$, according to Stokes law, the resistance coefficient $f$ is:

$$f = 6\pi\eta r \tag{3.16}$$

so that

$$\Delta = kT \frac{1}{6\pi\eta r} \tag{3.17}$$

This equation suggests that at a given temperature, the flocculation rate of fillers would basically be controlled by the polymer viscosity and the size of the aggregates. For asymmetric particles, such as carbon black aggregates, the diffusion constant can be estimated from the Stokes equivalent radius as:

$$\frac{f}{f_0} = \frac{\Delta_0}{\Delta} \tag{3.18}$$

where $f$ and $\Delta$ are, respectively, the resistance coefficient and diffusion constant of the asymmetric particle, and $f_0$ and $\Delta_0$ those of an equivalent sphere with the same weight and same volume. For asymmetric particles, $f/f_0$ is always larger than 1, leading to a lower diffusion constant. This suggests that the filler agglomeration rate is lower for highly asymmetric fillers. In the case of high structure carbon black, the high occluded rubber, due to a large internal void volume within the aggregates, leads to a larger equivalent radius in comparison with its low structure counterpart.

It can be concluded from a diffusion point of view that the following properties of polymer and filler would be favorable for a reduced rate of flocculation or rate of filler agglomeration:

- higher viscosity of polymer,
- large aggregates size (or effective aggregate size) of filler, and
- higher filler structure.

Another factor which controls the filler agglomeration kinetic process is the mean distance between filler aggregates, $\delta_{aa}$, which can be estimated with Eq. 3.9, since an aggregate needs to diffuse to make contact or to engage with another aggregate. Besides, the attractive force between aggregates is also determined by the distance between them (see Eq. 3.7). As previously described in Section 3.4.3, $\phi$, $S$, $\rho$, and $k$ are accessible values or constant for a given filled rubber in Eq. 3.9. And $\beta$ is an expansion factor which is the ratio between the effective volume fraction of the filler and the actual volume fraction. This factor can be calculated from Eq. 3.3, 3.4, or 3.5. In most cases, the equation derived by Wang, Wolff, and Tan[27], which is based on the value for crushed DBP and the random packing of equivalent spheres, is better suited for this purpose:

$$\beta = \frac{\phi_{\text{eff}}}{\phi} = \frac{0.0181\text{DBP} + 1}{1.59} \tag{3.19}$$

It is obvious from Eq. 3.9 that with regard to the interaggregate distance, the filler loading and surface area are the primary control factors. As small-particle-size fillers have comparatively large surface areas, the interaggregate distance of such blacks at equal loading will be shorter. Equation 3.19 includes the effect of DBP, i.e., the higher the structure, the smaller will be the distance and the easier would be filler agglomeration. However, in comparison to surface area, the effect of DBP on the $\delta_{aa}$ is much smaller.

If the above-mentioned consideration is reasonable, there are a few approaches which can be applied to depress filler agglomeration and hence, compound hysteresis behavior:

- Thermodynamic approaches:
  - reducing the difference in surface characteristics, especially in surface energy, between polymer and filler by filler surface modification and/or polymer modification;
  - increasing filler-polymer interaction and compatibility by filler surface modification, polymer modification, or/and using chemical or physical coupling agents;
  - blending fillers with different surface characteristics.

- Kinetic approaches:
  - improving the initial filler dispersion in the compound;
  - increasing the mean surface-to-surface distance between aggregates by changing filler morphology;
  - increasing bound rubber to increase the effective aggregate size and the viscosity of polymer matrix;
  - introducing a small quantity of crosslinks between polymer molecules to increase the effective molecular weight and hence the viscosity of polymer matrix;
  - reducing unnecessary scorch time for compound processing and increasing cure rate to lock filler aggregates in place before a developed filler aggregation is formed.

It is known that the aggregate nature of the reinforcing fillers plays a critical role in determining the processing performance, mechanical and dynamic properties, etc., of the uncured or vulcanized rubber. Furthermore, there is no doubt that the filler agglomeration is formed in rubber at a practically normal loading. Therefore, it is essential and significant to adequately understand the probable mechanisms of filler agglomeration for rubber-filler systems.

Practically, rubber compounders have consciously or unconsciously employed these principles for enhancing the filler micro-dispersion, reducing the compound viscosity, and improving the mechanical properties of vulcanizates such as hardness, strain-stress properties, dynamic properties, and fatigue and abrasion resistance.

## References


[1] Eirich F R. *Interaction Entre les Elastomeres et les Surfaces Solides Ayant une Action Renforcante.* Paris: Colloques Internationaux du CNRS, 1975: 15.

[2] Eirich F R, Smith T. Fracture. Vol. VII, New York: Academic Press, 1972.

[3] Oberth A E. Principle of Strength Reinforcement in Filled Rubbers. *Rubber Chem. Technol.,* 1967, 40: 1337.

[4] Donnet J B, Voet A. Carbon Black, Physics, Chemistry, and Elastomer Reinforcement. New York: Marcel Dekker, 1976.

[5] Kraus G. Reinforcement of Elastomers. New York: Wiley Interscience, 1965.

[6] Twiss D F. Technologie der Kautschukwaren. By K. Gottlob. Second edition. Pp. xi+340. Brunswick: F. Vieweg and Sohn. *J. Soc. Chem. Ind.,* 1925, 44: 587.

[7] Stickney P B, Falb R D. Carbon Black-Rubber Interactions and Bound Rubber. *Rubber Chem. Technol.,* 1964, 37: 1299.

[8] Kraus G. Reinforcement of Elastomers by Carbon Black. *Adv. Polym. Sci.,* 1971, 8: 155.

[9] Blow C M. Polymer/particulate Filler Interaction-The Bound Rubber Phenomena. *Polymer,* 1973, 14: 309.

[10] Dannenberg E M. Bound Rubber and Carbon Black Reinforcement. *Rubber Chem. Technol.,* 1986, 59: 512.

[11] Collins R L, Bell M D, Kraus G. Unpaired Electrons in Carbon Blacks. *J. Appl. Phys.* 1959, 30: 56.

[12] Riess G, Donnet J B. Physico-Chimie du Noir Carbone. Paris: Edition du CNRS, 1963: 61.

[13] Gessler A M. *Proc. Fifth Rubber Technol. Conf.*, London: Maclaren and Sons, 1968. pp249.

[14] Gessler A M. Evidence for Chemical Interaction in Carbon and Polymer Associations. Extension of Original Work on Effect of Carbon Black Structure. *Rubber Chem. Technol.,* 1969, 42: 858.

[15] Donnet J B, Papirer E. Effect on Surface Reactivity of Carbon Surface by Oxidation with Ozone. *Rev. Gen. Caoutch Plast.,* 1965, 42: 729.

[16] Donnet J B, Furstenberger R. Mécanisme et Cinétique de la Thermolyse de L'azodiisobutyronitrile en Présence D'oxygène - I. Formation D'acide Cyanhydrique. & II. Fixation D'acide Cyanhydrique sur les Noirs de Carbone. *J. Chim. Phys.,* 1971, 68: 1630.

[17] Donnet J B, Rigaut M, Furstenberger R. Etude par Resonance Paramagnetique Electronique de Noirs de Carbone Traites par L'azodiisobutyronitrile en Absence D'oxygene. *Carbon,* 1973, 11: 153.

[18] Hess W M, Lyon F, Burgess K A. Einfluss der Adhäsion zwischen Ruß und Kautschuk auf die Eigenschaften der Vulkanisate. *Kautsch. Gummi Kunstst.,* 1967, 20(3): 135.

[19] Serizawa H, Nakamura T, Ito M, Tanaka K, Nomura A. Effects of Oxidation of Carbon Black Surface on the Properties of Carbon Black-Natural Rubber Systems. *Polym. J.*, 1983, 15: 201.

[20] Gessler A M. Evidence for Chemical Interaction in Carbon and Polymer Associations. Butyl and Acidic Oxy Blacks. The Possible Role of Carboxylic Acid Groups on the Black. *Rubber Chem. Technol.,* 1969, 42: 850.

[21] Voet A. Reinforcement of Butyl Rubber by Ozonized Carbon-Black. *Kautsch. Gummi Kunstst.,* 1973, 26: 254.

[22] Ban L L, Hess W M, Papazian L A. New Studies of Carbon-Rubber Gel. *Rubber Chem. Technol.,* 1974, 47: 858.

[23] Ban L L, Hess W M. *Interaction Entre les Elastomeres et les Surfaces Solides Ayant une Action Renforcante.* Paris: Colloques Internationaux du CNRS, 1975: 81.

[24] Wolff S, Wang M-J, Tan E H. Filler-Elastomer Interactions. Part Ⅶ. Study on Bound Rubber. *Rubber Chem. Technol.,* 1993, 66: 163.

[25] Kraus G, Dugone J. Adsorption of Elastomers on Carbon Black. *Ind. Eng. Chem.,* 1955, 47: 1809.

[26] Wang M-J, Wolff S, Donnet J B. Filler-Elastomer Interactions. Part Ⅲ. Carbon-Black-Surface Energies and Interactions with Elastomer Analogs. *Rubber Chem. Technol.,* 1991, 64: 714.

[27] Wang M-J, Wolff S, Tan E H. Filler-Elastomer Interactions. Part Ⅷ. The Role of the Distance between Filler Aggregates in the Dynamic Properties of Filled Vulcanizates. *Rubber Chem. Technol.,* 1993, 66: 178.

[28] Hess W M, Ban L L, Eckert F J, Chirico V E. Microstructural Variations in Commercial Carbon Blacks. *Rubber Chem. Technol.,* 1968, 41: 356.

[29] Wang W D. Ph.D. dissertation, France: University of Haute Alsace, 1992.

[30] Westlinning H, Butenuth G. Quellung und netzmaschengröße rußgefüllter naturkautschuk-vulkanisate. *Makromol. Chem.,* 1961, 47: 215.

[31] Westlinning H, Butenuth G, Leineweber G. Kristallisationserscheinungen an gefüllten, nicht gedehnten naturkautschukproben, kurzmitteilung. *Makromol. Chem.,* 1961, 50: 253.

[32] Westlinning H. *Kautsch. Gummi Kunstst.,* 1962, 15: WT475.

[33] Schoon T G F, Adler K. *Kautsch. Gummi Kunstst.,* 1966, 19: 414.

[34] Smit P P A. The Glass Transition in Carbon Black Reinforced Rubber. *Rheol. Acta,* 1966, 5: 277.

[35] Smit P P A. The Effect of Filler Size and Geometry on the Flow of Carbon Black Filled Rubber. *Rheol. Acta,* 1969, 8: 277.

[36] Smit P P A, Van der Vegt A K. Interfacial Phenomena in Rubber Carbon Black Compounds. *Kautsch. Gummi Kunstst.,* 1970, 23: 4.

[37] Smit P P A. *Interaction Entre les Elastomeres et les Surfaces Solides Ayant une Action Renforcante.* Paris: Colloques Internationaux du CNRS, 1975: 213.

[38] Kaufman S, Slichter W P, Davis D D. Nuclear Magnetic Resonance Study of Rubber–Carbon Black Interactions. *J. Polym. Sci.,* 1971, A-2, 9: 829.

[39] O'Brien J, Cashell E, Wardell G E, McBrierty V J. An NMR Investigation of the Interaction between Carbon Black and *cis*-Polybutadiene. *Rubber Chem. Technol.,* 1977, 50: 747.

[40] Kenny J C, McBrierty V J, Rigbi Z, Douglass D C. Carbon Black Filled Natural Rubber. 1. Structural Investigations. *Macromolecules,* 1991, 24: 436.

[41] Fujimoto K, et al. Studies on Heterogeneity in Filled Rubber Systems (Part Ⅰ) Molecular Motion in Filler-Loader Unvulcanized Natural Rubber. *Nippon Gomu Kyokaishi.,* 1970, 43: 54.

[42] Fujimoto K, et al. Studies on Young's Modulus and Poisson's Ratio of Particles Filled Vulcanizates. *Nippon Gomu Kyokaishi.,* 1984, 57: 23.

[43] Fujiwara S, Fujimoto K. NMR Study of Vulcanized Rubber. *Macromol. Sci. - Chem.*, 1970, A4, 5: 1119.

[44] Kida N, et al. Studies on the Structure and Formation Mechanism of Carbon Gel in the Carbon Black Filled Polyisoprene Rubber Composite. *J. Appl. Polym. Sci.*, 1996, 61: 1345.

[45] Lüchow H, Breier E, Gronski W. Characterization of Polymer Adsorption on Disordered Filler Surfaces by Transversal $^1$H NMR Relaxation. *Rubber Chem. Technol.*, 1997, 70: 747.

[46] Litvinov V M, Barthel H, Weis J. Structure of a PDMS Layer Grafted onto a Silica Surface Studied by Means of DSC and Solid-State NMR. *Macromolecules*, 2002, 35: 4356.

[47] Haidar B. *Meeting of the Rubber Division, ACS*, Las Vegas, May 29 - June 1, 1990.

[48] Haidar B. *IRC'90*, Paris, June 12-14, 1990.

[49] Kraus G, Gruver J T. Thermal expansion, free volume, and molecular mobility in a carbon black-filled elastomer. *J. Polym. Sci.*, A2, 1970, 8: 571.

[50] Waldrop M A, Kraus G. Nuclear Magnetic Resonance Study of the Interaction of SBR with Carbon Black. *Rubber Chem. Technol.*, 1969, 42: 1155.

[51] EVE Rubber Institute, Qingdao, China, 2020.

[52] Medalia A I. Morphology of Aggregates Ⅵ. Effective Volume of Aggregates of Carbon Black from Electron Microscopy- Application to Vehicle Absorption and to Die Swell of Filled Rubber. *J. Colloid Interface Sci.*, 1970, 32: 115.

[53] Medalia A I. *Interaction Entre les Elastomeres et les Surfaces Solides Ayant une Action Renforcante.* Paris: Colloques Internationaux du CNRS, 1975: 63.

[54] Kraus G. A Structure-Concentration Equivalence Principle in Carbon Black Reinforcement of Elastomers. *Polym. Lett.*, 1970, 8: 601.

[55] Kraus G. Carbon Black Structure-Concentration Equivalence Principle. Application to Stress-Strain Relationships of Filled Rubbers. *Rubber Chem. Technol.*, 1971, 44: 199.

[56] Medalia A I. Effective Degree of Immobilization of Rubber Occluded within Carbon Black Aggregates. *Rubber Chem. Technol.*, 1972, 45: 1171.

[57] Sambrook R W. Influence of Temperature on the Tensile Strength of Carbon Filled Vulcanizates. *J. Inst. Rubber Ind.*, 1970, 4: 210.

[58] Warring J R S. Dynamic Testing in Compression: Comparison of the ICI Electrical Compression Vibrator and the IG Mechanical Vibrator in Dynamic Testing of Rubber. *Trans. Inst. Rubber Ind.*, 1950, 26: 4.

[59] Gerspacher M, Yang H H, Starita J M. *IRC'90*, Paris, June 12-14, 1990.

[60] Gerspacher M, Yang H H, O'Farrell C P. *Meeting of the Rubber Division, ACS*, Washington D.C., Oct 9-12, 1990.

[61] Fitzgerald E R. Response of Carbon Black-in-Oil to Low-Amplitude Dynamic Stress at Audiofrequencies. *Rubber Chem. Technol.*, 1982, 55: 1547.

[62] Voet A, Morawski J C. Dynamic Mechanical and Electrical Properties of Vulcanizates at Elongations up to Sample Rupture. *Rubber Chem. Technol.*, 1974, 47: 765.

[63] Wolff S, Donnet J B. Characterization of Fillers in Vulcanizates According to the Einstein-Guth-Gold Equation. *Rubber Chem. Technol.*, 1990, 63: 32.

[64] Voet A, Cook F R. Investigation of Carbon Chains in Rubber Vulcanizates by Means of Dynamic Electrical Conductivity. *Rubber Chem. Technol.*, 1968, 41: 1207

[65] Voet A, Sircar A K, Mullens T J. Electrical Properties of Stretched Carbon Black Loaded Vulcanizates. *Rubber Chem. Technol.*, 1969, 42: 874.

[66] Wolff S. *International Exhibition "Tires' 91"*, Moscow, USSR, March, 14-12, 1991.

[67] Wolff S, Wang M-J, Tan E H. Filler-Elastomer Interactions. X: The Effect of Filler-Elastomer and Filler-Filler Interaction on Rubber Reinforcement. *Kautsch. Gummi Kunstst.,* 1994, 47: 102.

[68] Wolff S, Wang M-J. Filler-Elastomer Interactions. Part IV. The Effect of the Surface Energies of Fillers on Elastomer Reinforcement. *Rubber Chem. Technol.*, 1992, 65: 329.

[69] Ouyang G B, Tokita N, M-J Wang. Hysteresis Mechanisms for Carbon Black Filled Vulcanizates-A Network Junction Theory for Carbon Black Reinforcement. *Rubber Division of ACS*, Cleveland, Oct 17-20, 1995. pp108. Abstract in *Rubber Chem. Technol.,* 1996, 69, 166.

[70] Clément F, Bokobza L, Monnerie L. Investigation of the Payne Effect and Its Temperature Dependence on Silica-Filled Polydimethylsiloxane Networks. Part Ⅰ: Experimental Results. *Rubber Chem. Technol.*, 2005, 78: 211.

[71] Clément F, Bokobza L, Monnerie L. Investigation of the Payne Effect and Its Temperature Dependence on Silica-Filled Polydimethylsiloxane Networks. Part Ⅲ: Test of Quantitative Models. *Rubber Chem. Technol.*, 2005, 78: 232.

[72] Van den Tempel M. Mechanical properties of plastic-disperse systems at very small deformations. *J. Colloid Sci.,* 1961, 16: 284.

[73] Wang M-J. Effect of Polymer-Filler and Filler-Filler Interactions on Dynamic Properties of Filled Vulcanizates. *Rubber Chem. Technol.*, 1998, 71: 520.

[74] D. Bulgin, Electrically Conductive Rubber. *Trans. Inst. Rubber Ind.,* 1945, 21: 188.

[75] Böhm G G A, Nguyen M N. Flocculation of Carbon Black in Filled Rubber Compounds. I. Flocculation Occurring in Unvulcanized Compounds During Annealing at Elevated Temperatures. *J. Appl. Polym. Sci.,* 1995, 55, 1041.

[76] Böhm G G A, Nguyen M N, Cole W M. *Proc. Int. Rubber Conf.*, Kobe, Japan, Oct 23-27, 1995. Paper 27A-8.

# 4 Filler Dispersion

The ability to disperse filler in different media is an important issue in almost all applications. In some cases, the costs of the dispersion processes can be comparable to the purchase price of the filler. On the other hand, the inability to achieve optimal dispersion impairs the ability to realize the full performance of the filler. Sub-optimal dispersion also creates other undesirable results.

The term "dispersion" can be described by two processes: one is the incorporation, distribution and dispersion of free filler (either in pellets or agglomerates) into the media. The other is the agglomeration of the dispersed filler during downstream processing or storage. This chapter will discuss the former process and the factors influencing the quality of so-called macro-dispersion of filler. New developments of dispersion techniques will be reviewed, especially related to filler application in rubber reinforcement[1]. Finally, the new developments of filler dispersion with liquid phase mixing will also be described.

## ■ 4.1 Basic Concept of Filler Dispersion

Most fillers are used in the form of pellets or agglomerates, especially in the rubber industry. During mixing of the filler with polymer, after incorporation, the distributive and especially dispersive processes will control the dispersion quality (Figure 4.1). There are various factors that influence the dispersion process, such as mixing equipment, compound formulation, mixing sequence, and mixing conditions. For a given formulation, which determines the filler compatibility (wettability) and interaction with polymers, after the early stage of pellet breakdown and incorporation, the de-agglomeration of carbon black is especially critical. In this regard, we apply the model of breakup of agglomerates due to an applied hydrodynamic stress in a simple shear flow, developed by Bagster et al.[2] and Horwatt et al.[3] to demonstrate the effect of filler properties on their dispersibility. In this case, agglomerate refers to the undispersed filler, which consists of a great number of aggregates. The aggregate, which is a discrete and rigid colloidal entity, is the primary dispersable and functional unit in the well-dispersed systems, although certain aggregates may be broken down during mixing[4].

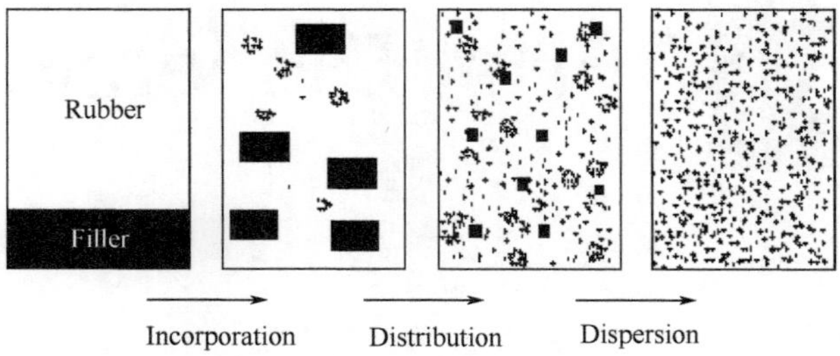


**Figure 4.1** Incorporative, distributive, and dispersive processes during carbon black mixing

It is assumed that an agglomerate and an aggregate are two centroids of equivalent spheres with radii $R$ and $r$ centered at $O$ and $o$, respectively, as shown in Figure 4.2 in a two-dimensional model[5]. The aggregate centroid is centered on the fracture vector at a distance $(R-r)$, and $\Psi$ is the angle which is describes the size of the aggregate relative to the agglomerate. Since real agglomerates are not perfectly spherical, there should be a tendency for the agglomerates to rotate through all possible orientations, and occasionally aggregates within the agglomerate will experience the maximal hydrodynamic force, $F_H$, during the course of rotation, which is given by:

$$F_H = \frac{5}{2}\pi R^2 \eta \dot{\gamma} \sin^2 \Psi \tag{4.1}$$

where $\eta$ and $\dot{\gamma}$ are the medium viscosity and the shear rate, respectively. The hydrodynamic force $F_H$ acting on the aggregate is a function of the portion of the agglomerate surface subtended by the aggregates. It can be used to determine the critical rupture stress, i.e., the minimum stress required to separate an aggregate from an agglomerate, initiating dispersion (Eq. 4.1).

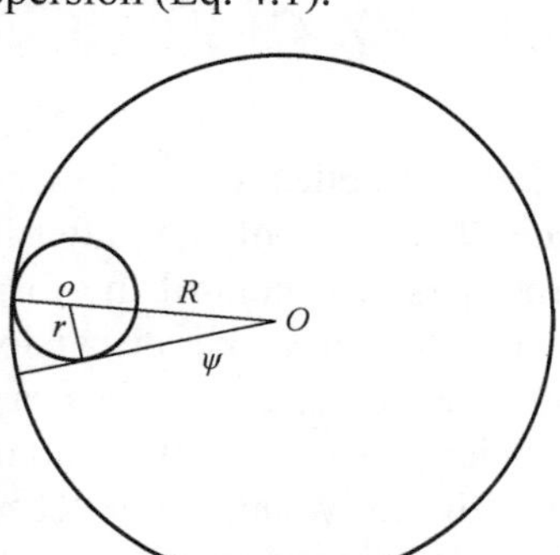


**Figure 4.2** Schematic representation of equivalent agglomerate and aggregates

Assuming that there is an average connection number $\upsilon$ with which an aggregate binds to the agglomerate, and the mean force of a connection is $H$, the cohesive force, $F_C$, for an aggregate resisting separation from the agglomerate will be:

$$F_C = H\upsilon \tag{4.2}$$

Based on Eqs. 4.1 and 4.2, at the critical rupture stress $(\eta\dot{\gamma})_{crit}$, for which $F_H > F_C$, rupture takes place for the values of the variables according to:

$$(\eta\dot{\gamma})_{crit} \geqslant \frac{2H\upsilon}{5\pi R^2 \sin^2 \Psi} \tag{4.3}$$

With increasing viscosity, the agglomerate size decreases to a critical value, $R_{crit}$, below which the aggregates in the agglomerates cannot be further dispersed. This critical size of agglomerates is a measure of the size of undispersed agglomerates, representing the macrodispersion of the filler. By rearranging Eq. 4.3, the smallest radius of undispersed agglomerates, which can be obtained during mixing, can be expressed by Eq. 4.4:

$$(R)_{crit} \geqslant \frac{1}{\sin\Psi}\sqrt{\frac{2H\upsilon}{5\pi\eta\dot{\gamma}}} \tag{4.4}$$

## ■ 4.2 Parameters Influencing Filler Dispersion

Based on Eq. 4.4, the filler dispersion, i.e., $R_{crit}$, is determined by $\psi$, $H$, $\upsilon$, $\eta$, and $\gamma$. In these parameters, $\gamma$ is determined by the mixing equipment and the mixing conditions, $H$ is related to the filler surface characteristics, surface energy in particular, and others are dominated by the filler morphology.

### About $\psi$: The Angle Characterizing the Size of the Aggregate Relative to the Agglomerate

For the same size of undispersed agglomerate, $R_{crit}$ is inversely proportional to sin $\psi$. The critical stress for an aggregate to be dispersed is higher for small particles, hence high specific surface area fillers, due to the smaller $\psi$. This means that smaller aggregate size gives higher critical rupture stress, resulting in a larger $R_{crit}$, hence poorer filler dispersion.

Morphologically, the filler can be considered as a set of properties related to the average magnitude and frequency distribution of the primary particle diameter, aggregate diameter and the way particle are connected in the aggregates. Generally, the size of aggregates is related to the size of the primary particles, which determines the surface area. For furnace carbon blacks and thermal blacks, which possess no or low surface porosity, higher surface area blacks have smaller aggregate size[6] and thus, according to Eq. 4.4, are more difficult to disperse.

On the other hand, for filler having the same surface area, high structure filler has larger aggregate size which is also favorable for filler dispersion.

### About $\eta$: The Viscosity of the Medium

Upon dispersion of a particle in the polymer matrix, the viscosity of the medium increases along with reduction of agglomerate size $R$. For a spherical filler, over a certain range of concentration of dispersed filler, the viscosity of the filled matrix can be estimated by the Guth-Gold equation (Eq. 4.5)[7,8]:

$$\eta = \eta_0 (1 + 2.5\phi + 14.1\phi^2) \qquad (4.5)$$

where $\eta_0$ and $\eta$ are the viscosity of unfilled and filled polymers, respectively, and $\phi$ is the volume fraction of the dispersed filler in the medium.

In fact, the aggregates of most rubber-grade fillers are not spherical particles but are composed of spheres that are fused together. Due to the irregular or branched shapes of the aggregates, there is some internal void volume within the aggregates. For aggregates formed with the same size primary particles, i.e., the same surface area, high structure carbon blacks have more void volume that can be characterized by higher DBP absorption, and larger aggregate size. An increase in void volume will certainly reduce the number of connections in the agglomerates, as will be discussed later, and the connections between individual aggregates. This, on the one hand, will make the agglomerate or pellets more easily broken down during incorporation. On the other hand, the higher void volume of the aggregates also causes the density of the high structure black agglomerates to be lower. When the formulation of a carbon black-filled system is based on the same mass, the number of similar-sized pellets or agglomerates in the medium after incorporation will be larger than that for its low structure counterpart. This will result in higher viscosity of the system, $\eta$, hence high stress, reducing the $R_{crit}$.

In addition, the effect of structure on the dispersibility of carbon black is also related to the occlusion of polymer by the black aggregates[9]. When structured carbon blacks are dispersed in rubber, the rubber portion filling the internal void of the carbon black aggregates, or the rubber portion located within the irregular contours of the aggregates, is unable to participate fully in the macro-deformation of the filled system. When occluded rubber is partially immobilized, it behaves more like a filler than like a polymer matrix. Due to this phenomenon, the effective volume of the filler, with regard to the stress-strain behavior of the filled system, is increased considerably.

Assuming the same mass of filler has been dispersed in the polymer, the effective volume of dispersed filler in the medium, $\phi_{eff}$, is substantially higher than the volume fraction calculated based on the dispersed filler mass and its density. Substituting $\phi$ by $\phi_{eff}$, Eq. 4.5 can be written as Eq. 4.6:

$$\eta = \eta_0 \left(1 + 2.5\phi_{eff} + 14.1\phi_{eff}^2\right) \qquad (4.6)$$

Several equations have been derived for calculating the effective volume in the medium[8,10]. Simply by relating the endpoint of crushed DBP absorption to the void space within and between equivalent spheres of aggregates and assuming the spheres to be packed at random, Wang et al.[11] obtained the following equation (Eq. 4.7) for the effective volume of carbon black:

$$\phi_{\text{eff}} = \phi \frac{0.0181\text{CDBP} + 1}{1.59} \tag{4.7}$$

where CDBP is the compressed DBP absorption number. Again, it is clear that high structure black with high CDBP yields lower $R_{\text{crit}}$ relative to low structure materials.

### About *H*: The Adhesion Energy between Aggregates

From Eq. 4.4, the mean force of aggregate-aggregate connection, or *H*, determines the size of undispersed agglomerates, which is related to the cohesive energy between filler aggregates.

It is known that the possible interaction between two materials 1 and 2 is determined by their surface energies, which consist of two components, dispersive, $\gamma^{\text{d}}$ and specific (polar), $\gamma^{\text{p}}$[12–14]. $W_{\text{c}}$ can, therefore, be estimated by Eq. 4.8:

$$W_{\text{c}} = 2\gamma_{\text{f}}^{\text{d}} + 2\gamma_{\text{f}}^{\text{p}} + W_{\text{c}}^{\text{h}} + W_{\text{c}}^{\text{ab}} \tag{4.8}$$

where $\gamma_{\text{f}}^{\text{d}}$ and $\gamma_{\text{f}}^{\text{p}}$ are, respectively, the dispersive and polar components of the filler surface energy, and $W_{\text{c}}^{\text{h}}$ and $W_{\text{c}}^{\text{ab}}$ are the cohesive energies from hydrogen bonding and acid-base interactions. Carbon black surface energies are measured by inverse gas chromatography (IGC)[15].

Figure 2.57 shows the dispersive component of the surface energy of a series of carbon blacks as a function of their specific surface areas, and the specific component of adsorption energy, $I^{\text{sp}}$, of acetonitrile results are given in Figure 2.64. The $I^{\text{sp}}$ of acetonitrile is a measure of the polar component of the filler surface energy. Hydrogen bonding may contribute to $I^{\text{sp}}$ of acetonitrile if it exists[15]. As can be seen, the surface energies of higher surface area blacks, both the dispersive and polar components, are generally higher than those of their lower surface area counterparts. This implies that with regard to surface energies, i.e., the mean force, *H*, between aggregate connections, the dispersibility of the high surface area blacks will be lower.

The statement about the effect of surface energy on dispersibility for different surface area blacks seems to be farfetched at first glance, as it is cannot easily be separated from the effect of morphology. However, an approach to the problem can be made by the investigation of simple heat-treatment of blacks on dispersion, from which the effect of a change in morphology has been omitted. When carbon black was heated in an inert atmosphere (nitrogen in this case), the $\gamma_{\text{s}}^{\text{d}}$ of the carbon black increased with increasing temperature up to 900°C, which is far below the graphitization temperature (about 1500°C) (Figure 4.3). When these treated blacks (N234) are compounded into emulsion SBR at 50 phr loading, the dispersion of carbon black deteriorates with increasing treatment temperature. At the range of temperatures used, the morphology of the carbon black cannot be changed, so the variation of black dispersion upon heat treatment can only be interpreted in terms of surface characteristics. According to Eq. 4.8, the negative impact of temperature increase appears related to the poorer dispersion of the filler due to stronger filler-filler interaction originating from the very high $\gamma_s^d$.

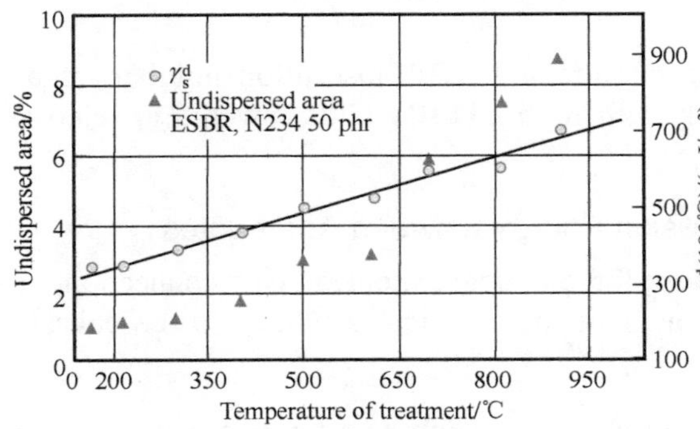


**Figure 4.3** Dispersive components of carbon black surface energies and their dispersion in ESBR as a function of heat-treatment under $N_2$ for N234

### About $\upsilon$: The Connection Number $\upsilon$ with Which an Aggregate Binds to the Agglomerate

Beyond morphology and surface characteristics, the dispersibility of filler also depends on the connection number $\upsilon$ with which an aggregate binds to the agglomerate. According to Rumpf[16] and Meissner et al.[17,18], the connection number $\upsilon$ can be calculated by Eq. 4.9:

$$\upsilon = 2\exp\left[2.4\left(1 - Vv\right)\right] \tag{4.9}$$

where $Vv$ is the void volume of the filler agglomerates, which can be measured by oil absorption (for example DBP) and mercury porosimetry. In the case of carbon black, furnace products for example, fluffy black separated from the filter and pneumatic system still contains considerable amounts of air or gas and has a very low bulk density. Therefore, it has to be further densified before it can be used commercially[19].

Higher densification of the carbon black is achieved by pelletization. With few exceptions, rubber-grade blacks are always pelletized, since only in this form can they be handled and incorporated into the various polymers using standard industrial equipment. Other advantages of pelletized blacks are smaller transport volumes, more favorable free-flow properties, and reduced dusting.

Because of their different characteristics, it is necessary to differentiate between the two pelletizing methods commonly used, namely dry pelletization and wet pelletization. The process of dry pelletization makes use of the fact that densified carbon black aggregates can be formed into small round pellets in a rotating drum (dry mill). This process is continuous, but limited in its application. The higher the carbon black structure, the more difficult is dry pelletization. In order to achieve the desired throughput, a certain proportion of pelletized carbon black needs to be recycled to act as seeds for new pellets.

Effective, high densification of filler can be obtained with a wet pelletization process. In this process, the carbon black is mixed with water in special pin mixers, namely pelletizers. The water/carbon black ratio used is approximately 1:1 for most blacks. The mechanical action of the pins forms the carbon black into wet pellets with a diameter of 0.5–2.0 mm. If necessary, pelletizing agents or binders, such as molasses, lignin sulfonates, or sugar are added to influence pellet hardness. Further densification is achieved during drying. In this case, the densification of filler is driven by the attractive force between particles which, as schematically illustrated in Figure 4.4, originates from the capillary phenomenon according to Eq. 4.10:

$$\Delta p = p_o - p_i = \gamma \cos\theta / r \tag{4.10}$$

where $p_o$ is the atmosphere pressure, $p_i$ is the liquid pressure, $\gamma$ the surface energy of the liquid, and $r$ the radius of the capillary.

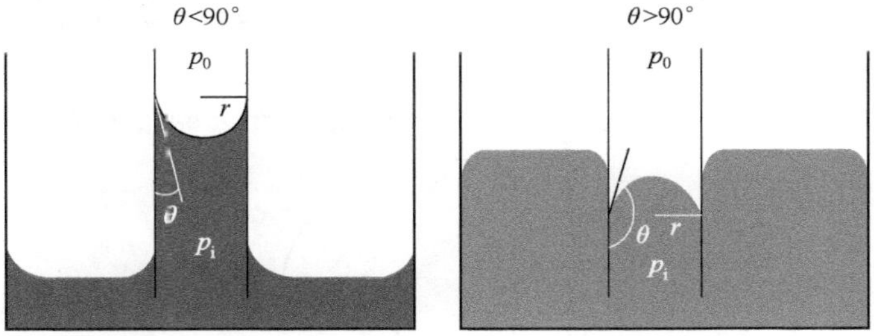


**Figure 4.4** Capillary phenomenon

Accordingly, when a regular particle and flat surface are bridged by a liquid (Figure 4.5), the attractive force, $F$, can be estimated by the attractive force between a solid sphere and smooth surface as[20]:

$$F = \frac{4\pi R \gamma \cos\theta}{1 + D/d} \tag{4.11}$$

where $R$ is the radius of the particle, and $D$ the smallest distance between the particle and surface (see Figure 4.5); $d$ is related to the area on which the attractive force applies (Eq. 4.11).

In the case where a particle is in contact with a flat surface ($D = 0$) (Figure 4.6), the maximum attractive force can be obtained as:

$$F = F_{max} = 4\pi R \gamma \cos\theta \tag{4.12}$$

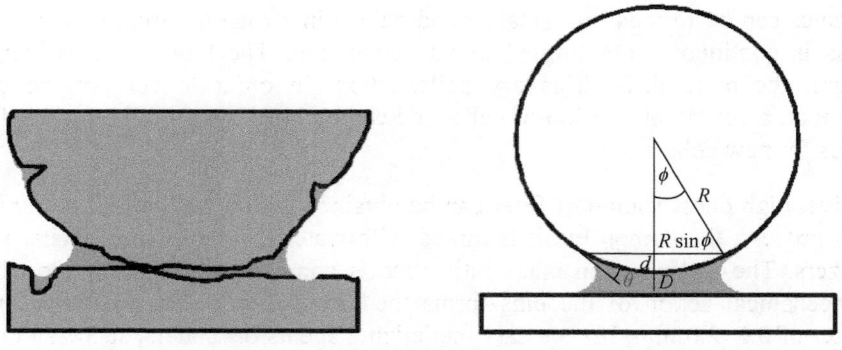


**Figure 4.5** Attractive force between a particle and flat surface bridged with water

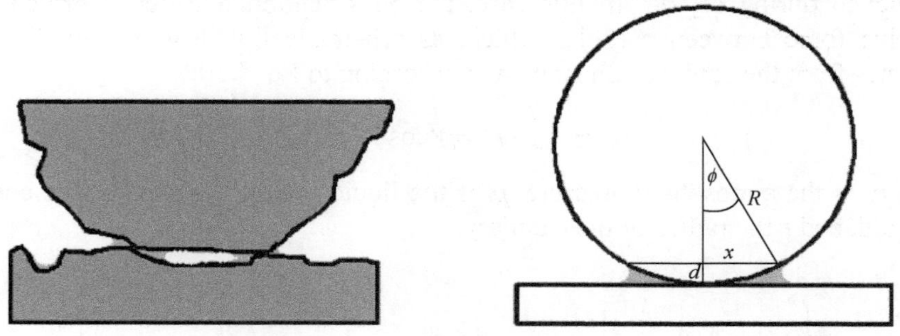


**Figure 4.6** Attractive force between a particle and flat surface bridged with water

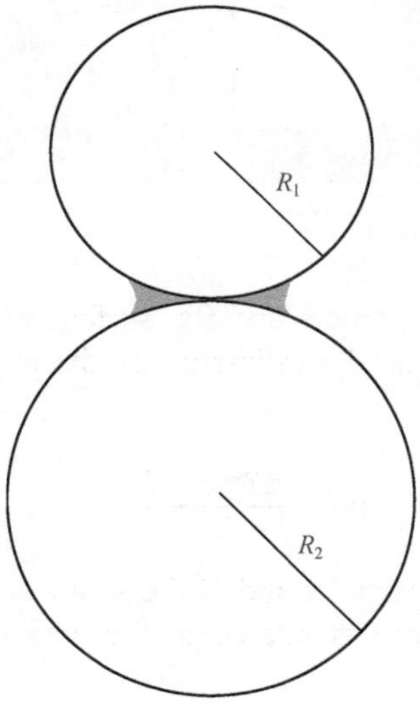


**Figure 4.7** Attractive force between a two spherical particles in contact

If two spherical particles are in contact (Figure 4.7), the attractive force generated by the residual liquid will be:

$$F = F_{max} = 4\pi \left( \frac{1}{R_1} + \frac{1}{R_2} \right)^{-1} \gamma \cos\theta \qquad (4.13)$$

where $R_1$ and $R_2$ are the radiuses of particle 1 and 2, respectively.

In fact, in a wet pelletization process, all of these may occur as the water surface energy is high enough and the contact angle is small. In some cases, the branches of neighboring aggregates can be pulled tightly together by the attractive force from the residual liquid, as illustrated schematically in Figure 4.8. This is the case with carbon black. For example, since the contact angle of water on a compressed disk of carbon black N220 is about 50° and the surface energy of water is about 0.068 J/m$^2$ at a temperature of 50°C, the attractive force between two particles may reach as high as $5.5 \times 10^{-9}$ N during wet pelletization. As a result, the densification of carbon black can be increased, which is good for handling of the filler, but on the other hand, the void volume of the filler will be reduced, which is not favorable for carbon black dispersion.

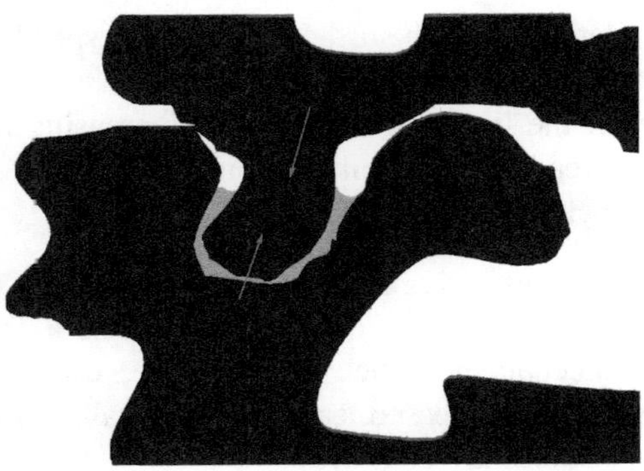

**Figure 4.8** Attractive force between a two aggregated particles in contact

With regard to precipitated silica, the situation is quite different. In the production process, when the growth of silica aggregates or precipitation is complete, the obtained suspension is filtered and the filter cake is washed. It is then resuspended and dried. This means that the filler dispersability should be highly determined by the conditions used for post treatment, drying in particular. Depending on the drying technology, the dried product can be then optionally milled and/or granulated into a low-dust form.

Compared with carbon black, the filler-filler interaction of silica is much higher, due to its high polar component of surface energy, especially its high hydrogen bonding between aggregates. This will increase adhesion energy of the connections. The high polarity of silica surface also leads to a very small contact angle with water. In fact, adding a small drop of water to a compressed silica disk reduces the equilibrium contact angle to almost 0°. This suggests that during drying, the attractive forces are considerably higher than for carbon black.

The conventional drying process of filtrated silica from aqueous suspension is to remove the residual water in long-retention dryers (e.g., tunnel driers or rotary dryers)

for several hours. In this case, the high attractive force generated by the residual water leads to strong agglomeration of the silica aggregates, resulting in very low void volume of the silica product, hence poorer dispersibility in rubber even though it may be milled and granulated.

Flash drying, operated under lower pressure, high temperature, and short drying time, is able to form more void volume in the filler agglomerates. This benefits silica dispersion by significantly reducing the number of connections between silica aggregates.

The most favorable drying technology for silica dispersability is spray-drying. The features of this process are as follows:

- The suspension is injected into the tank through a nozzle at very high speed, which generates a very low pressure in the stream, leading to atomization of the suspension as the hot air diffuses into the stream;
- The very small droplets have high surface area which speeds up the evaporation of water and moisture;
- The surface energy of the liquid decreases with increasing temperature. When the liquid's temperature reaches the boiling point at a given pressure, the liquid's surface energy approaches zero.

Under above conditions, the attractive force between aggregates is extremely small and the drying process of silica suspension is completed in a few seconds. With the combination of these mechanisms, the void volume, especially the void volume between aggregates, is greatly increased, causing the $\upsilon$ value to fall considerably. As a result, highly dispersible silicas are obtained.

### About $\dot{\gamma}$: The Shear Rate of the Medium

The shear rate, $\dot{\gamma}$, which is determined by the mixing equipment, has a significant effect on filler dispersion. The mixing equipment generally used in the rubber industry comprises two roll-mills, Banbury mixers (tangential and intermesh internal mixers), kneaders, continuous mixers, twin-screw extruders, and multi-screw extruders. The $\dot{\gamma}$ is determined by the roller gap between the rollers, between rotors, between rotor and wall, between screws, and between screw and walls. Speed also affects $\dot{\gamma}$. Careful selection of the equipment and operation conditions is very critical for filler dispersion.

Although affected by other properties, the dispersion of carbon black is essentially decided by its morphology. Traditionally, as carbon blacks with surface areas higher than 160 $m^2/g$ and CDBP lower than 60 mL/100 g cannot be dispersed by dry mixing with the existing equipment, they are not considered rubber grades. Shown in Figure 4.9 is the ASTM carbon black spectrum used in the rubber industry, expressed by compressed DBP *vs.* surface area. From left top to the right bottom, the dispersibility of the black is getting poorer. Below a reference line, some non-rubber blacks are shown.

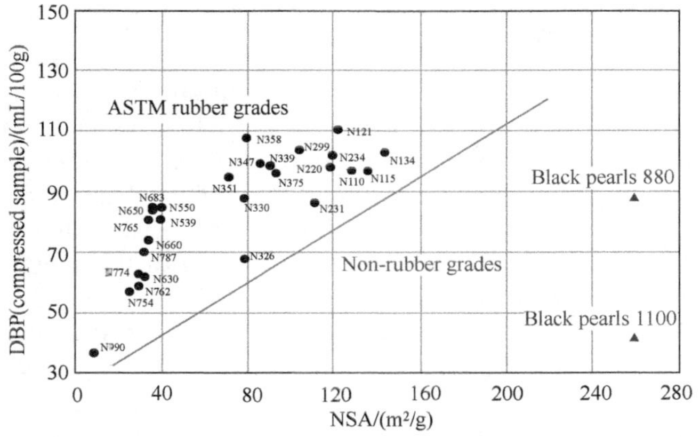


**Figure 4.9** Rubber grades of carbon black spectrum with some non-rubber grades

## 4.3 Liquid Phase Mixing

Historically, the improvement of filler dispersion has been achieved with liquid phase mixing[21–23]. However, the long mixing and coagulation times reduces productivity. In addition, for some polymer latexes, such as natural rubber and emulsion-SBR, there are some non-rubber substances, protein and surfactants in particular, which can be adsorbed on the carbon black surface, interfering with polymer-filler interaction. This is not favorable for the properties of vulcanizates.

Among others, two liquid phase mixing processes developed in the last few decades are great step forward in improving filler dispersion. One is the continuous liquid phase mixing/coagulation process developed by Cabot Corporation to produce a natural rubber-carbon black masterbatch[24,25] that has been named Elastomer Composite, known as CEC. Another is the continuous liquid phase mixing invented by EVE Research Institute to produce synthetic rubber-silica masterbatches[26–29] that have been named as Eco-Visco-Elastomer Composite, abbreviated as EVEC®. The two processes and the properties of masterbatches will be given in detail in the Sections 9.3 and 9.4.

As will be discussed later in Section 9.3, the features of the CEC process are the fast mixing and coagulation, and a short drying time at high temperature. This achieves excellent performance for the material as polymer-filler interaction can be better preserved, polymer degradation can be minimized, and, especially, carbon black dispersion can be substantially improved. Shown in Figure 4.10 are the comparison of macro dispersion measured by means of optical microscopy for two carbon-black-N134-filled vulcanizates, one prepared with CEC using two-stage mixing only for adding small chemicals and curatives, and another with dry mixing using four-stage mixing. It is clear that the dispersion of the filler in CEC is superior to the dry mixed compound, even though energy input for CEC mixing is much lower. In fact, the carbon black distribution and dispersion are fully completed at the very early stages of CEC process. This can be demonstrated by an investigation of the carbon black dispersion of CEC by transmission electron microscopy. Figure 4.11 shows N234-filled CEC, where the carbon black dispersion is already well dispersed and uniformly distributed throughout the polymer matrix on a very small scale after dewatering of the coagulum, with only minor mechanical energy input. This suggests that high quality of filler dispersion can be achieved with CEC technology without significant mechanical breakdown of the polymer molecules. Consequently, compared with dry-mixed compounds, the tear and tensile strengths of the vulcanizates are higher, the hysteresis and heat build-up are lower, and the flex fatigue life is substantially longer.

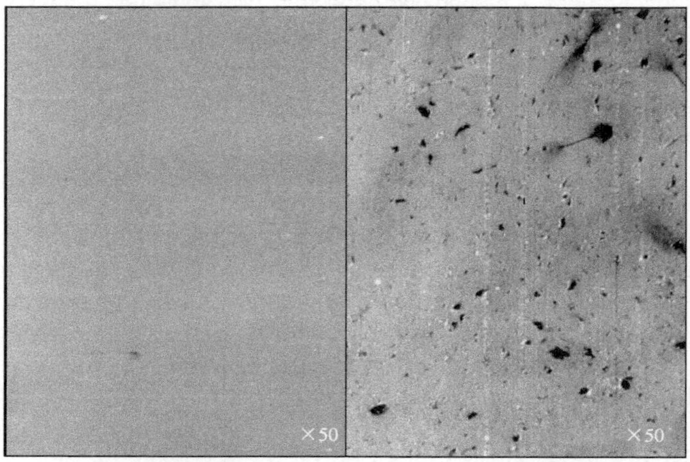


CEC-N134 50 phr
2 stage mixing

Dry mixing-N134 50 phr
4 stage mixing

**Figure 4.10** Dispersion of carbon black N134 in CEC and dry-mixed compounds

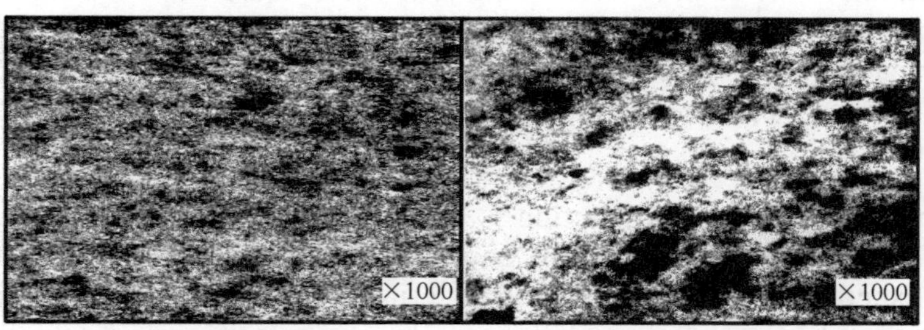


CEC-N234 50 phr / Oil 5 phr

Dry mixing-N234 50 phr / Oil 5 phr

**Figure 4.11** TEM images of dewatered CEC and dry-mixed compounds

Most importantly, with CEC technology the dispersion of carbon black in the compound is independent of its morphology for all the carbon blacks investigated. Technically, the grades of carbon black used for rubber reinforcement can be greatly expanded via CEC processing to the high surface area and low structure blacks[30]. This is unique for rubber reinforcement as these carbon blacks may impart some unusual properties to the filled compounds. For example, over the practical range of loading, a product Black Pearls®1100 (BP1100)❶ with surface area of 260 $m^2/g$ and CDBP of 42 mL/100 g (see Figure 4.9) can be dispersed in the polymer matrix with excellent quality. With this material, the vulcanizates possess not only significantly higher ultimate properties, but also higher tear strength than any that has ever been seen with conventional compounds. More interestingly, the tradeoff among tensile strength, elongation at break, and hardness can be improved, which will be discussed in Section 9.3.4.

For silica compounds, the challenge of improving silica dispersion is of utmost importance in rubber applications. The excellent dispersion of filler achieved by the EVE process is one of the greatest advantages for EVEC. Figure 4.12 shows the macro-dispersion of silica measured by the Dispergrader (from Alpha) for two vulcanizates: dry mixed silica filled compounds and EVEC-L, both using the same formulation[26]. The white areas in the figure are undispersed fillers. The distribution of undispersed fillers is shown in Figure 4.13. The filler dispersion in the dry mixed vulcanizate filled with highly dispersible silica is still much poorer than that in EVEC. In the dry mixed form, not only are there many undispersed fillers, but the size of these agglomerates is larger compared to EVEC. The maximum diameter of undispersed agglomerates in EVEC is about 25 μm versus 43 μm for dry mixed compounds. The superior silica dispersion of EVEC is also demonstrated with TEM analysis (Figure 4.14).

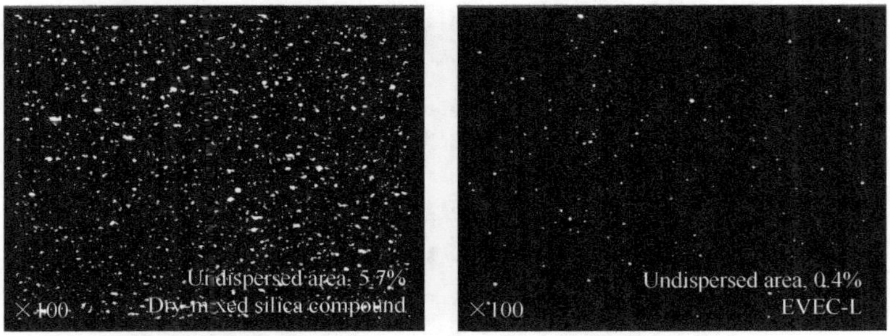


**Figure 4.12** Dispersion of silica in dry-mixed compound and EVEC-L

❶ Black Pearls is a registered trademark of Cabot Corporation.

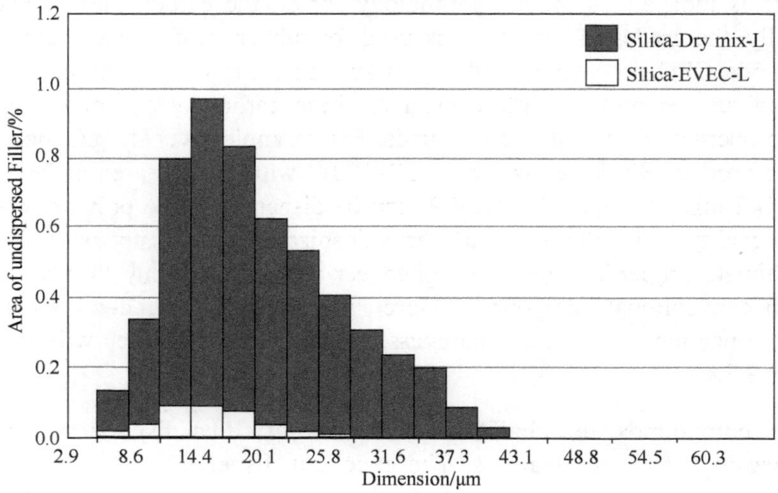


**Figure 4.13** Undispersed area as a function of size of agglomerate for dry-mixed compound and EVEC-L

The excellent dispersion of the silica in EVEC plays an important role in the strength and abrasion vulcanizates, which will discussed later in Chapters 8 and 9 (Figure 4.14).

**Figure 4.14** TEM images of SSBR compounds filled with different silicas and loadings. Oil: 37.5 phr, TESPT: 6.85 phr for silica 90 phr and 8.4 phr for silica 110 phr

## References


[1] Wang M-J. New developments in carbon black dispersion. *Kautsch. Gummi Kunstst.*, 2005, 58: 626.

[2] Bagster D F, Tomi D. The Stress Within a Sphere in Simple Flow Fields. *Chem. Eng. Sci.*, 1974, 29: 1773.

[3] Horwatt S W, Feke D L, Manas-Zloczower I. The Influence of Structural Heterogeneities on the Cohesively and Breakup of Agglomerates in Simple Shear-Flow. *Powder Technol.*, 1992, 72: 113.

[4] Hess W M, Chirico V E, Burgess K A. Carbon-Black Morphology in Rubber. *Kautsch. Gummi Kunstst.*, 1973, 26: 344.

[5] Tatek Y, Pefferkon E. Cluster-cluster Aggregation Controlled by the Number of Interculster Connections: Kinetics of Aggregation and Cluster Mass Frequency. *J. Colloid Interface Sci.*, 2004, 278: 361.

[6] Wang M-J, Gray C A, Reznek S A, et al. "Carbon Black", 5th ed., New York: John Wiley & Sons, 2003.

[7] Guth E, Simha R, Gold O. Untersuchungen über die Viskosität von Suspensionen und Lösungen. 3. Über die Viskosität von Kugelsuspensionen. *Kolloid Z.*, 1936, 74: 266.

[8] Guth E, Gold O. Viscosity and Electroviscous Effect of the AgI Sol. Ⅱ. Influence of the Concentration of AgI and of Electrolyte on the Viscosity. *Phys. Rev.* 1938, 53: 322.

[9] Medalia A I. Morphology of Aggregates: Ⅵ. Effective Volume of Aggregates of Carbon Black from Electron Microscopy; Application to Vehicle Absorption and to Die Swell of Filled Rubber. *J. Colloid Interface Sci.*, 1970, 32: 115.

[10] Kraus G, Naylor F E, Rollmann K W. Steady Flow and Dynamic Viscosity of Branched Butadiene-Styrene Block Copolymers. *Rubber Chem. Technol.*, 1972, 45: 1005.

[11] Wolff S, Wang M-J, Tan E-H. Filler-Elastomer Interactions. Part Ⅶ. Study on Bound Rubber. *Rubber Chem. Technol.*, 1993, 66: 163.

[12] Fowkes F M. Ideal Two-Dimensional Solutions. Ⅱ. A New Isotherm for Soluble and "Gaseous" Monolayers. *J. Phys. Chem.*, 1962, 66: 382.

[13] Owens D K, Wendt R C. Estimation of The Surface Free Energy of Polymers. *J. Appl. Polymer Sci.*, 1969, 13: 1741.

[14] Kaelble D H, Uy K C. A Reinterpretation of Organic Liquid-Polytetrafluoroethylene Surface Interactions. *J. Adhesion*, 1970, 2: 50.

[15] Wang M-J, Wolff S, Donnet J -B. Filler-Elastomer Interactions. Part Ⅲ. Carbon-Black-Surface Energies and Interactions with Elastomer Analogs. *Rubber Chem. Technol.*, 1991, 64: 714.

[16] Rumpf H. Grundlagen und methoden des granulierens. *Chemie Ingenieur Technik. Chem. Ing. Tech.*, 1958, 30: 144.

[17] Meissner H P, Michaels A S, Kaiser R. Crushing strength of zinc oxide agglomerates. *I. & EC. Process Design and Development*, 1964, 3: 202.

[18] Georgalli G A, Reuter M A. Modelling the co-ordination number of a packed bed of spheres with distributed sizes using a CT scanner. *Miner. Eng.*, 2006, 19: 246.

[19] Kühner G, Voll M. In: Donnet J-B, Bansal R C, Wang M-J (Editors). Carbon Black, Science and Technology. New York: Marcel Dekker, Inc., 1993.

[20] Israelachvill J. Intermolecular and Surface force, Second edition. London: Academic Press, 1991.

[21] Cohnn E S A. Coagulating Latex. British Patent, 214210, 1923.

[22] Jose L, Joseph R, Joseph M S. Studies on Latex Stage Carbon Black Masterbatching of NR and Its Blend with SBR. *Iran. Polym. J.*, 1997, 6: 127.

[23] Sone K. Materbatch of ESBR and Carbon black (wet mixing). *Nippon Gomu Kyokaishi*, 1998, 6: 308.

[24] Mabry M A, Rumpf F H, Podobnik I Z, et al. Elastomer Composites Method and Apparatus. US Patent 6048923, 2000.

[25] Wang M J, Wang T, Wong Y L, et al. NR/Carbon Black Masterbatch Produced with Continuous Liquid Phase Mixing. *Kautsch. Gummi Kunstst.*, 2002, 55: 388.

[26] Wang M-J. Report Presented at Tire Technology Conference, Hannover, Germany, Feb. 2016.

[27] Wang M-J, Song J, Dai D. Method for Preparing Rubber Masterbatch Continuously, the Prepared Rubber Master Batch and Rubber Products. China Patent, CN 103205001 A, CN 103203810 A, CN 103113597 A, CN 103600435 A, 2013.

[28] Wang M-J, Song J, Dai D. Continuous Production of Rubber Masterbatch. WO 2015/192437 A1, WO 2015/109789 A1, WO 2015/109792 A1, WO 2015/018278 A1, WO 2015/018282 A1, 2015.

[29] Wang M-J, Song J, Dai D. Continuous Production of Rubber Masterbatch. US 20160168341 A1, EP 3031591 A1, JP 2016527369 A , 2016.

[30] Wang T, Wang M-J, McConnell G A, et al. Carbon Black Elastomer Composites, Elastomer Blends and Methods. WO 2003050182, 2003.

# 5 Effect of Fillers on the Properties of Uncured Compounds

## 5.1 Bound Rubber

The term "bound rubber" has been used extensively in the literature since it was first introduced by Fielding in 1937[1]; however, a clear, universally accepted definition for the term does not exist. It is generally understood to mean the portion of rubber in an uncured, filled rubber compound that could not be extracted by a good solvent at room temperature. A good solvent is defined as one that is known to dissolve the polymer in question when it is unfilled. The phenomenon of bound rubber in carbon black-filled rubber compounds appears to have been known since the early days of rubber reinforcement. As early as 1925, Twiss observed improved mechanical properties when mixes of (natural) rubber and carbon black resisted solvents[2]. The term "carbon gel", like bound rubber, has been widely used in the literature without an agreed upon definition. It was first used by Sweitzer[3], and is considered a particular case of bound rubber in which the carbon black particles in a rubber compound do not disperse when subjected to a large excess of a good solvent for the polymer. Instead, they remain bound in a polymer gel. Since this is typically the situation in carbon black-filled compounds at the loadings used in many rubber products, bound rubber has also come to mean that portion of rubber contained in a swollen gel when an uncured compound is immersed in solvent.

The concept of such a simple test to measure something related to mechanical reinforcement in rubber compounds is attractive, and has led to a great number of studies on the subject over the years. Although the term "bound rubber" does not exclusively refer to carbon black, most of the studies have been focused on black-filled compounds, reflecting the dominant position of carbon black as the reinforcing filler for rubber during the 20th century. Silica is the only other reinforcing filler that has gained significant usage, and the applicability of bound rubber measurements to silica-filled compounds has consequently been the subject of numerous investigations since the 1970s.

A number of excellent reviews on bound rubber have appeared in the literature previously[4–8]. The objective of this section is to discuss bound rubber concepts as a means to better understand particulate reinforcement of rubbers.

### 5.1.1 Significance of Bound Rubber

Bound rubber is important for rubber reinforcement. It is often considered to be the classical measure of surface activity of fillers, which determines the magnitude of filler-polymer interaction in filled elastomer systems. Other methods, such as heat of immersion, heat of adsorption, and inverse gas chromatography (IGC), have been used to probe filler surface activity, but bound rubber measurement has the advantage of being simple, requiring no special skills or equipment. On the other hand, interpretation of bound rubber results can be complicated because factors other than surface activity affect the amount of bound rubber measured in a particular system. Many of these complications will be discussed in this section.

It has been well demonstrated in a general sense that bound rubber is a good indicator of the reinforcing effect of a filler, particularly carbon black. However, there has been some dispute over whether it plays a direct role in the physical properties of cured, filled elastomers, or whether it is just a good indicator of certain factors that are important to particulate reinforcement. On one side, it can be argued that bound rubber is a direct measure of the extent of "bonding" or adhesion between filler and polymer in a particular system. The amount of bound rubber in the uncured compound is dependent on the type of polymer and filler and on the specific mixing history of that compound, and is therefore related to the degree of reinforcement that can be expected in that particular compound after curing. The contrary argument is that the conditions used to determine bound rubber are quite different from those normally experienced by rubber vulcanizates. Since the solvents that are used to determine bound rubber content must dissolve the polymer, they therefore enable freer movement of polymer chains in the filled compounds than would be possible in the absence of solvent. Also, solvents can dissolve some non-rubber components in the compound but not others. Accordingly, it is argued that bound rubber content is just an indicator of the reinforcing potential of a particular filler in a given polymer system, and not a direct measure. Since bound rubber content is affected by the same factors that influence reinforcement, namely surface area, structure and surface activity of a filler, it can be a very useful indicator of reinforcing ability if it used appropriately.

It is important to bear in mind that by most methods of bound rubber measurement, any insoluble polymer will be counted as bound rubber, whether or not is it attached to filler. This is relevant not only to natural rubber, which often has significant gel content in its natural state, but also to other polymers which may form gel after compounding and heating. For example, when a gum compound of SBR, zinc oxide, stearic acid, and antioxidant was heated in a press for 20 minutes at 160°C, a significant amount of gel was measured[9]. The influence of polymer gel, unrelated to the filler, is one of the factors that can lead to misinterpretation of bound rubber results when they are used as an indicator of polymer-filler interaction.

It should be pointed out, however, that no matter what the composition is, the total amount of bound rubber in the uncured compound plays a significant role in the processability of compound, such as viscosity and die swell, as well as the vulcanizate properties, such as dynamic properties.

### 5.1.2 Measurement of Bound Rubber

The detailed measurement of bound rubber is described in Section 2.4.2.4. Some effects of test conditions on bound rubber content are discussed as following:

**Solvents:** For general hydrocarbon rubbers such as NR, SBR, BR, and butyl rubber, the solvents typically used include toluene, xylene, benzene, and *n*-heptane. A study on the effect of different solvents on the bound rubber in butyl compounds showed that benzene, cyclohexane, cyclohexene, ethyl benzene, xylene, chloroform, and carbon tetrachloride gave results with differences similar to the precision of the test with a single solvent[10]. The conclusion is that the choice of solvent is not critical for bound rubber measurement in hydrocarbon rubbers, provided it is a good solvent for the polymer in question. The choice of solvent for bound rubber determination in silicone rubber compounds, however, seems to be more important. Southwart[11] compared acetone, methyl isobutyl ketone, THF, toluene, and hexane, and found that the amount of rubber extracted increased as the solubility parameter of the solvent decreased. Thus, hexane, the least polar of the solvents tested, caused significantly more polymer to be extracted than the other solvents over extraction times from 1 day to 8 weeks.

**Extraction time:** The time of extraction in bound rubber determination is generally chosen to ensure that full extraction, or equilibrium conditions are achieved. Leblanc studied the kinetics of the extraction process and found that the results fit quite well to a model that assumed the rate of extraction was approximately proportional to the amount of extractable rubber remaining[12]. In a typical BR system with 50 phr of carbon black, approximately 50 hours was required to reach equilibrium at 23°C. Other, less rigorous, studies have found that several days are required for complete extraction of hydrocarbon rubbers. Since the dissolved polymer is in equilibrium with adsorbed polymer on the filler surface, it is usual to replace the rubber solution with fresh solvent one or more times during the extraction process to ensure that the maximum amount of polymer is extracted. With the large volume of solvent used relative to the test sample in most methods, it appears that changing to fresh solvent during the extraction process has minimal, if any, effect on the bound rubber result. This can be seen in Figure 5.1.

Bound rubber determination in the silicone rubber-silica system seems to be more complicated. Extraction of a silicone-silica compound for periods up to 4 years showed that unextracted polymer was linearly related to the square root of extraction time in toluene, and that by extrapolation, all of the rubber could be expected to be extracted after 20 years[11]. The facts that bound rubber in the silicone-silica combination is dependent on the solvent, and never really reaches equilibrium, are clear indicators that

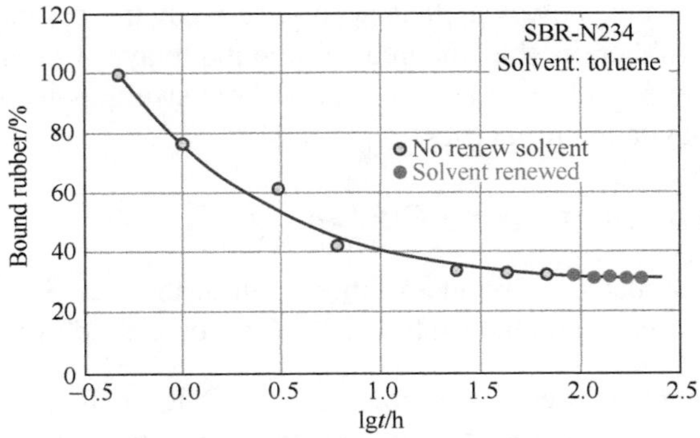


**Figure 5.1** Bound rubber content as a function of extraction time at room temperature

the type and strength of the interactions between polymer and filler are quite different from the carbon black–hydrocarbon rubber system.

**Storage effect of the compounds:** Many researchers have observed that the amount of bound rubber increases with time of storage of the carbon black-filled compounds after mixing. For example, carbon black compounds of NR, BR, and EPDM all showed significant increases in bound rubber over a period of 50–60 days of ambient storage after compounding (see Figure 5.8 and Figure 5.22). The observation of this phenomenon in several different systems supports the proposed mechanism of physisorption of polymer onto carbon black surface. Any chemical reaction seems unlikely under such mild conditions in these systems. At higher temperatures, the increase in bound rubber occurs more rapidly[13]. This observation was initially attributed to radical formation; however, the similarity in activation energy for this process between conventionally milled compounds and compounds prepared from solution[6] again favors an adsorption mechanism. In all cases, bound rubber eventually reaches an equilibrium or plateau. The extent of the increase in bound rubber with time depends on the particular combination of polymer and filler, and on the mixing conditions. In silicone rubber filled with silica, the increase in bound rubber content has been found to reach saturation only after an extended period of time, up to several years at room temperature[14]. The kinetics of this process has been thoroughly studied, and found to fit well with a model based on diffusion and random polymer adsorption[15].

**Critical filler loading of coherent gel:** It has been found that for some carbon blacks, especially those with low surface areas, a coherent gel cannot be obtained below a certain level of loading. This is obviously due to the inability of the bound rubber to develop a network capable of holding carbon black and rubber together. The lowest possible filler loading at which a coherent mass can be formed and where no carbon black is dispersed in the solvent can be defined as the critical coherent loading[16]. This critical level of loading can easily be derived experimentally from a plot of bound rubber *vs.* loading as shown in Figure 5.2. As illustrated in Figure 5.3, the critical coherent loading, $C_{crit}$, decreases with increases in surface area. The reasons for the high critical loading of low-surface-area carbon blacks are probably the greater distance between particles (discussed later) and the low surface energy, neither of which favor the formation of a three-dimensional gel network.

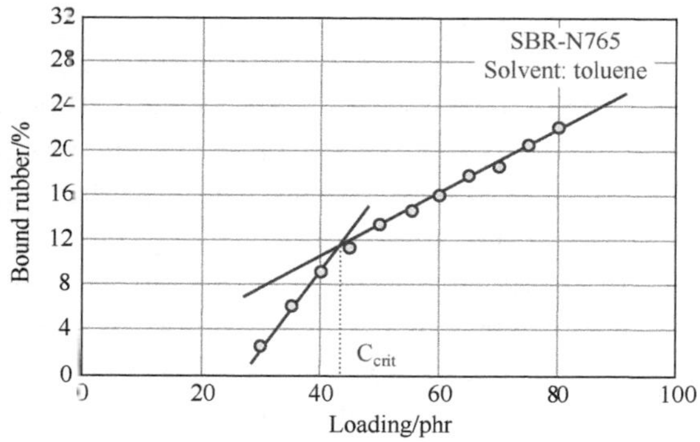


**Figure 5.2** Determination of the critical coherent loading, $C_{crit}$

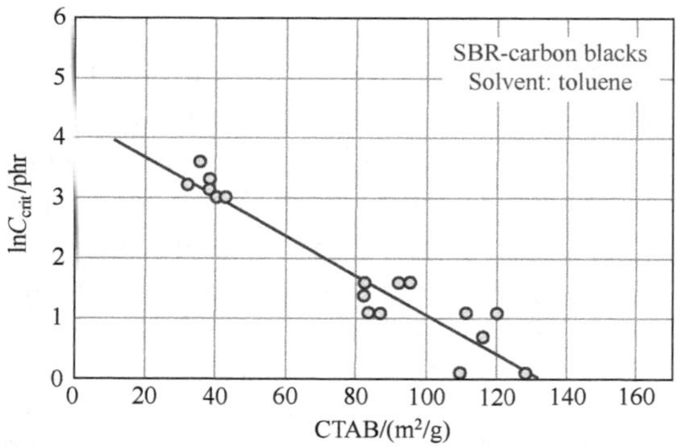


**Figure 5.3** Critical coherent loading as a function of CTAB surface area for various carbon blacks in SBR

### 5.1.3 Nature of Bound Rubber Attachment

Initially, some researchers attributed the inability of solvents to displace polymer from the surface of particles to a chemical attachment or bonding between the polymer and particle[13–15]. Watson, for example, proposed that free radicals formed on the polymer were responsible for the chemical attachment and showed that free-radical inhibitors could reduce bound rubber formation[14]. It was later found, however, that addition of other small molecules, such as cetyl trimethyl ammonium bromide (CTAB) and Aerosol OT to carbon black before or during mixing could also reduce bound rubber by a similar extent[17]. The effect was attributed to blocking of high-energy adsorption sites on the filler surface by the organic surfactant molecules. Results from a number of other studies lead to the conclusion that while chemisorption may play a role in bound rubber formation in some systems, physical adsorption of polymer chains on the filler surface, plays a major role, and is probably the dominant mechanism in most systems.

Support for this conclusion comes from the fact that the amount of bound rubber measured is dependent on the extraction temperature[18–20]. The quantity of specific bound rubber increases linearly with the reciprocal of absolute temperature up to 80°C [20]. Other studies indicated that the bound rubber in SBR and NR systems was much lower and did not form a coherent mass when extracting at temperatures greater than 70°C [18]. In these experiments, nitrogen was used in order to eliminate the effects of oxidation, which can cause chain scission and lead to lower molecular weight polymer and consequently lower bound rubber. The temperatures of these extractions are too mild to cause any significant division of covalent bonds of the type proposed by the chemical attachment hypothesis. Physical adsorption mechanisms of attachment in carbon black–hydrocarbon rubber systems best explains the substantial decrease in bound rubber content at moderate temperatures where oxidation is absent. Although these bonds are weak, the tenacity of bound rubber on carbon black under normal extraction conditions is easily explained by a multi-contact mechanism, as presented by Frisch[21]. Whilst the strength of a single van der Waals bond is not as strong as a typical covalent bond, multiple points of contact between polymer and filler cause the polymer chain to be effectively permanently attached. If one point of contact is temporarily broken due to thermal motion, there are enough other contact points to keep the polymer chain bound to the particle. The probability of all contacts being broken at the same time is remote under normal conditions. However, at higher temperature, the higher thermal energy of the polymer chains leads to an increased probability that all contact points will be simultaneously broken, and the polymer solubilized by solvent.

To further confirm the effect of temperature on bound rubber, tests were carried out using solution SBR and 50 phr N234 and 4-aminophenyl disulfide (APDS)-modified N234. SBR was chosen because it is gel-free and is relatively resistant against oxidation. To further limit the oxidation process, the experiment was carried out under a nitrogen atmosphere, and an antioxidant was added to the solvent. The results are shown in Figure 5.4. Below 70°C, only a slight reduction in bound rubber with increases in temperature is observed. Above 80°C, no coherent mass is obtained, causing a dramatic decrease in bound-rubber content.

To determine whether any polymer was present on the carbon-black surface in the noncoherent mass, the solvent containing the carbon-black dispersion was collected and centrifuged at room temperature. The analysis for polymer content remaining on the carbon black was conducted using TGA. The results are marked in Figure 5.4 by the plus signs. As can be seen, even in the noncoherent mass, some polymer still remains on the carbon black surface. The amount of polymer, however, decreases rapidly with temperature. Since oxidation has been minimized, the residual bound-rubber content at elevated temperatures can either be due to chemisorption or a readsorption of polymer on the carbon black during the separation and washing process conducted at room temperature. Kraus and Dugone[22], in their study of carbon-black adsorption in an SBR solution, found that the adsorption/desorption process at room temperature is characterized by considerable hysteresis (Figure 5.5). Under these conditions, it is assumed that chemisorption cannot take place. It can therefore be concluded that the residual polymer determined by TGA cannot be due to chemisorption alone.

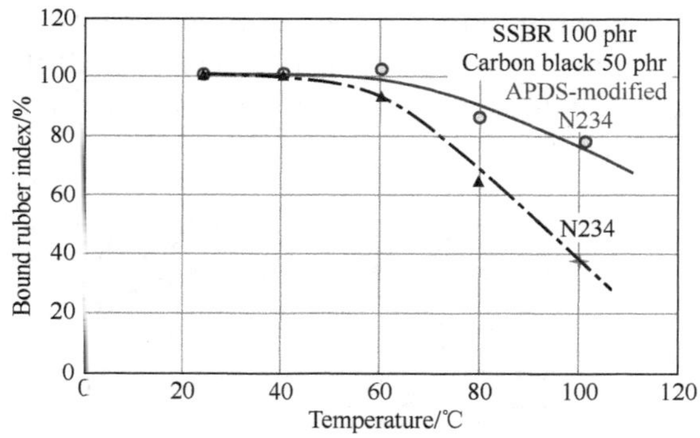


**Figure 5.4** Bound rubber as a function of the extraction temperature for N234 and APDS-modified carbon black- filled SSBR compounds

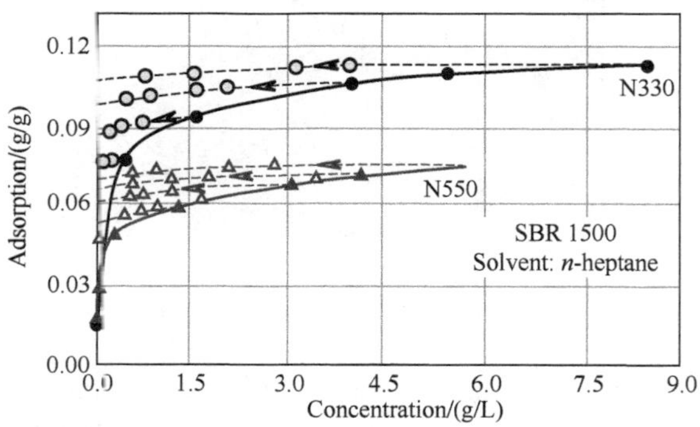


**Figure 5.5** Adsorption and desorption of SSBR on carbon black N330 and N550[22]

In the case of APDS-modified carbon black that is prepared by grafting APDS on carbon black surface with diazonium chemistry, disulfide groups react with polymer chains during mixing, thus forming bound rubber. The difference in bound rubber at high extraction temperature between N234 and APDS-N234 should suggest that the main portion of bound rubber for APDS-N234 is chemisorption in nature.

Based on the above observations, it is postulated that bound-rubber formation of conventional carbon blacks with polymers, at least with natural rubber and SBR, is mainly, if not entirely, a physical process. This corresponds with studies conducted by Ban and Hess et al.[23,24], who found direct evidence for multiple site adsorption of polymers from transmission electron micrographs. In their studies, the very thin sections, 100–200 nm in thickness, were supported on 300 mesh copper electron microscope grids which were covered with preformed carbon films of 10–20 nm thickness. The thin sections were extracted with benzene in a vapor extractor for 8 hours. The polymer which was not tied up with the carbon black in forming the composite was removed with benzene. The TEM image of the SBR 1500 filled with low structure N330-LM carbon black (DBP-71 mL/100 g) is shown in Figure 5.6. As can be seen, on all the convex area, the polymer is missing. It concluded that carbon black-polymer interactions are essentially physical in nature.

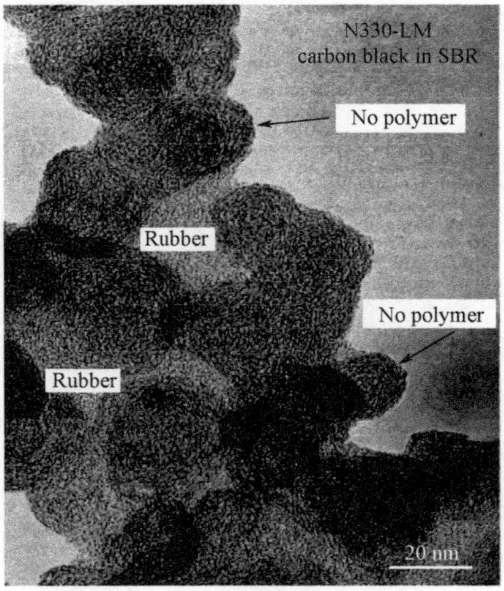


**Figure 5.6** TEM image of carbon black N330-LM-filled compound extracted with benzene

For clarifying the nature of bound rubber of silica compounds, Wolff, Wang, and Tan[25] carried out a study at room temperature in an ammonia atmosphere using precipitated silica Z1165/SBR and carbon black N110/SBR compounds. According to Scott's concept, the physical adsorption and chemical attachment of the polymer on the filler surface is differentiated using ammonia during bound rubber determination. Since ammonia is a strongly hydrogen-bonding chemical, it can preferentially adsorb and displace polymer from the silica surface[26]. The results are listed in Table 5.1. Ammonia treatment reduces bound rubber of silica-filled compound from 21.1% to 1.9%. This indicates that bound rubber is mainly caused by physical adsorption, since ammonia treatment is known to cleave physical linkages between silica and polymer.

**Table 5.1** Bound rubber content of SBR compound filled with a variety of fillers (solvent: toluene, room temperature)[25]

| Filler | Treatment | Bound rubber/% |
|---|---|---|
| Carbon black N110 | – | 32.4±0.2 |
| Carbon black N110 | Ammonia | 30.5±0.2 |
| Silica VN2 | – | 21.1±0.3 |
| Silica VN2 | Ammonia | 1.9±0.1 |
| Silica VN2-TESPT | – | 41.5±0.7 |
| Silica VN2-TESPT | Ammonia | 40.0±0.8 |
| Silica VN2-HDTMS | – | 9.7±0.2 |
| Silica VN2-HDTMS | Ammonia | 1.4±0.1 |

In the case of carbon blacks, the picture is quite different. The bound rubber content is only slightly affected by ammonia treatment. One possible interpretation of the above results could be that ammonia does not weaken the stronger interaction between carbon black and polymer, whereas polymer-silica interaction is much weaker than the interaction between silica and ammonia. This is in line with observations made in swelling measurements of carbon black and silica vulcanizates in which swelling increased tremendously for the silica compounds after ammonia treatment, but had little effect on carbon-black vulcanizates[27].

For bis(3-triethoxysilylpropyl)-tetrasulfide (TESPT) modified silicas, due to the covalent bonds formed between polymer chains and silica surface, the bound rubber content is not affected by ammonia treatment. On the contrary, with regard to hexadecyltrimethoxy silane (HDTMS) modified silica, there is no coupling reaction taking place between silica and polymer even though the silane molecules may attached chemically on silica surface. As a result, while bound rubber formation is substantially reduced, the chemisorption of polymer on silica surface does not occur.

It is believed that the fumed silica-silicone rubber system, the hydrogen bonding between silanol groups on the silica surface and the siloxane bonds of the polymer cause the strong adsorption of this polymer. Ammonia treatment has demonstrated that bound silicone rubber on silica is normally physisorbed through hydrogen bonding[18,28]. However, if the mixture is heated to temperatures between 130°C and 280°C, it becomes progressively chemically attached through a condensation reaction[28].

In summary, there is convincing evidence that in carbon black–hydrocarbon rubber systems, physisorption is the predominant process by which bound rubber is formed. Some contribution from chemisorption cannot be excluded, however, especially for polymers like NR, which are highly sensitive to mechano-chemical reactions. This is true for silicas as well. When the filler surfaces are modified with grafts having functional groups that are able to react chemically with polymers, such as TESPT modified silica, the bound rubber increases drastically and the ammonia treatment does not reduce the bound rubber significantly. In that case, the attachment of the polymer molecules on the filler surface is chemical in nature.

### 5.1.4 Polymer Mobility in Bound Rubber

A number of bound rubber studies using the spin-spin relaxation time, T2, of NMR signals have shown that proximity to a filler surface significantly affects mobility of polymer. One of the first studies on this subject identified reduced polymer mobility in filled BR and EPDM compounds[19,29–31]. As mentioned in Section 3.2.2, in bound rubber there are also three levels of polymer mobilities in bound rubber; the highest mobility polymer is comparable to the gum polymer; the lowest mobility polymer is in close contact with the filler surface; and there is material with intermediate mobility around the filler particles. Kraus interpreted the original NMR results as showing a continuous increase in polymer mobility outward from the filler surface[32], and found it to be consistent with his results from dilatometry[33]. The change in glass transition temperature measured by dilatometry, however, seems to have under-estimated the change in mobility of polymer on the surface. Other results obtained using black-filled NR have also indicated a tightly bound polymer layer with greatly reduced T2[34,35]. From the relaxation point of view, this layer behaved similarly to polymer in the glassy state, up to 150°C. The results also provided evidence of motional cooperation between the tightly bound, immobilized layer, and the loosely bound rubber, thus indicating the immobilized layer comprises segments of many different polymer chains, yet other segments of the same chains are not immobilized in the same way.

Investigations on the silicone rubber (PDMS)-silica system have also found a large reduction in T2 for the polymer at the interface. Based on the fraction of polymer that it represented and the total interfacial area, it was estimated that the thickness of this interfacial layer of polymer was about 1nm[36]. Dimensions of the tightly bound rubber estimated by others have been of the same order.

The general picture for the dimensions of the bound rubber structure that emerges from these studies is the presence of a tightly bound layer on the particle surface, only about 1 nm thick, in which the polymer mobility is severely limited on the T2 time scale. This layer, which contains segments from many polymer chains but probably no complete chains, behaves as though it were below, or close to its glass transition temperature. This layer has also been referred to as a “rubber shell”. The rubber shell constitutes only a small part of the bound rubber in a typical system. Beyond that, there is an outer layer with a radius approximately 5 times larger, in which the polymer chains have reduced mobility compared to the bulk polymer. This outer layer comprises mainly those parts of the bound polymer molecules outside of the rubber shell, and includes loops as well as chain ends. The remainder of the total bound rubber is outside the range over which the polymer mobility is affected by the filler surface. It comprises mainly chain ends of high molecular weight polymer molecules and entangled polymers.

### 5.1.5 Polymer Effects on Bound Rubber

There are two properties of a polymer that are primarily responsible for the bound rubber content found in a particular system. The first is its molecular weight and gel content, and the second is the chemical type or composition. It was demonstrated early on that for any given carbon black, the amount of bound rubber depends significantly on the polymer properties[37].

#### 5.1.5.1 Molecular Weight Effects

It can easily be understood that higher molecular weight and/or higher gel content in a polymer will lead to higher bound rubber content, if chemical effects are equal. This is thought to be one of the major reasons that in practice natural rubber compounds have the highest bound rubber contents. Kraus and Gruver showed that the amount of bound rubber was proportional to the square root of the weight-average molecular weight for narrow MW distribution polybutadienes[37]. The same rule applied reasonably well to SBR. In broad or bimodal polymer molecular weight distributions, there appears to be preferential adsorption of higher molecular weight polymers. This has been demonstrated by the change in MW distribution of the soluble rubber[17,37–41]. Many theories exist relating the fraction of adsorbed polymer to the molecular weight of the starting and adsorbed polymer and to the volume fraction and surface area of the filler[42]. Most provide reasonable agreement with experimental results[43,44]. All of the models are based on essentially a process of random adsorption of polymer segments. Preferential binding of higher molecular weight molecules at equilibrium occurs due to the greater probability of a large molecule with sufficient adsorbed segments becoming bound.

#### 5.1.5.2 Polymer Chemistry Effects

Chemical composition is the other important polymer property. Although it is difficult to make comparisons at the same molecular weight distributions, there is evidence that unsaturated polymers give higher bound rubber content than saturated polymers like butyl rubber, and that SBR gives higher bound rubber with carbon black than BR of the same molecular weight. This is consistent with the higher adsorption energy of aromatic rings that has been observed with the carbon black surface compared to alkanes or alkenes[45].

Polar polymers like NBR generally give lower bound rubber than pure hydrocarbon polymers with carbon black, which provides support for the model of non-specific physical interactions as the dominant mechanism in bound rubber formation. However, on silica, nitrile rubber gives higher bound rubber contents than hydrocarbons, presumably due to some polar interactions with this filler[46]. Silicone rubber (PDMS) gives very low levels of bound rubber on carbon black[47], but with silica, bound rubber contents are much higher. This can be explained by the different types of adsorption in the two systems. On carbon black, where adsorption is mainly through non-specific or dispersive interactions, silicone rubber does not interact strongly. However on a silica surface, hydrogen bonding is possible with the silanol groups on the filler, and the chemical structure of PDMS is well-suited to accepting this type of interaction. Strong interactions are therefore formed between polymer and filler, leading to high bound rubber contents for a given polymer molecular weight.

## 5.1.6 Effect of Filler on Bound Rubber

### 5.1.6.1 Surface Area and Structure

Figure 5.7 shows bound rubber as a function of carbon-black surface area at 50 phr loading in SBR. Carbon blacks with larger surface areas have a greater interfacial area with the polymer per unit volume of compound at the same loading; therefore, unsurprisingly, total bound rubber content increases with increasing carbon black surface area. This phenomenon is well known[8,16,30,48] and has been reconfirmed by Dannenberg[7] in a study using different types of rubber.

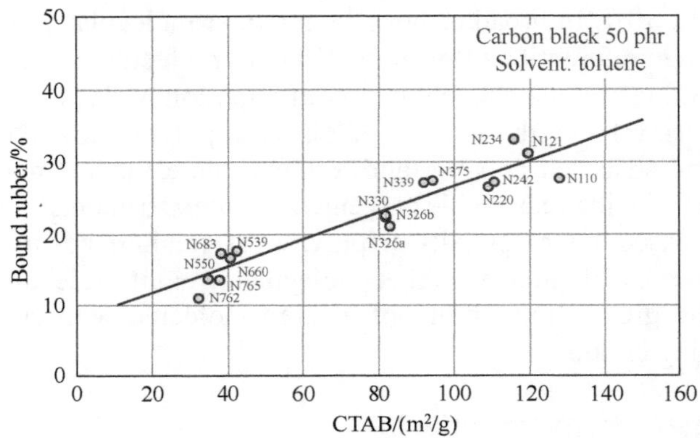


**Figure 5.7** Bound rubber as a function of CTAB surface area for various black at 50 phr loading in SBR

The carbon blacks in Figure 5.7 differ not only in surface area, but also with regard to structure, which covers the full range of rubber-grade furnace blacks. It can be observed that the carbon blacks with high structure, N121, N234, N242, N375, N339, N539, N550, N683, and N765, show higher values than their normal and low-structure counterparts such as N110, N220, N330, N326, N660, and N762. This can also be seen more clearly from Figure 5.8. This suggests that the carbon-black structure, next to surface area, is also an important parameter affecting bound rubber. For carbon blacks with similar surface areas, high-structure carbon blacks generally have a higher bound-rubber content. This confirms the observations made in the past[7,23,49]. Several explanations follow:

- the probability of multiple molecular-segment adsorption is greater for high-structure carbon blacks[7];

- low-structure carbon blacks have a more ordered graphite layer structure leading to lower surface activity[23,50];
- high-structure carbon blacks are more easily broken down during mixing, thus creating new active surfaces which are considered to be "facile new free radical sources"[49] for polymer-filler bonding[7].

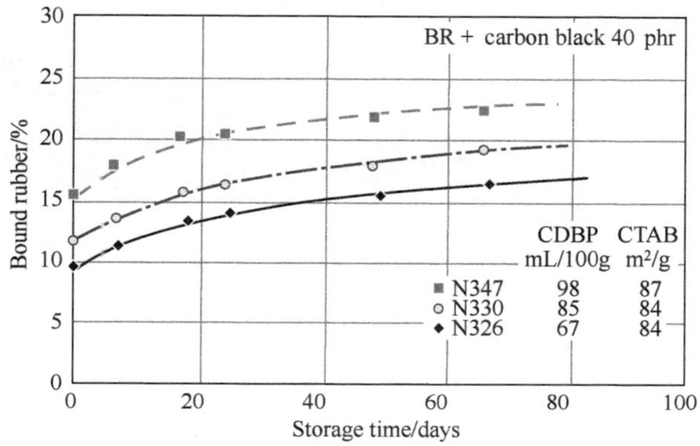


**Figure 5.8** Effect of carbon black structure on bound rubber formation

Multiple molecular-segment adsorption seems to be supported by an electron-microscope investigation of bound rubber. Ban and coworkers[23], in their study of TEM micrographs of microtome samples, extracted sol using benzene in a vapor extractor, and found that rubber remained in the concave regions after extraction and that it was nearly totally absent in the convex regions of the surface.

If the high bound-rubber values of high-structure carbon blacks are related to multiple molecular-segment adsorption, this would lead to lower interface effectiveness and therefore less bound rubber per unit surface. This cannot be confirmed by the investigation of bound rubber per unit surface of carbon blacks with different structures, as will be discussed later. The rubber in the concave region could be due to the retraction of the polymer during sample preparation. It is difficult to imagine that the rubber molecules are not spread over the whole carbon black surface in the swollen rubber and that these molecules should all be concentrated in the internal voids of the carbon blacks.

It seems reasonable to explain the low surface activity of low-structure carbon blacks by a more ordered graphite structure, but on the other hand, measurements of surface energy have shown that neither the dispersive nor the specific components correlate with structure[45].

It is therefore highly probable that the breakdown of the high-structure carbon blacks may be largely responsible for their higher bound-rubber content.

Several authors[51–53] have provided evidence that the primary structure of carbon blacks undergoes a significant breakdown during mixing. Ban and Hess[24] demonstrated the significant breakdown in aggregates in their EM investigation of carbon black recovered by pyrolysis of the vulcanizate and the carbon gel, as shown in Table 5.2. Moreover, this phenomenon was even more pronounced in high-structure carbon blacks.

**Table 5.2** Effect of mixing on aggregate size of carbon black in different polymers[24]

| Rubber | Carbon black loading/phr | HAF-HS(DBP = 155mL/100g) | | HAF-LM(DBP = 71mL/100g) | |
|---|---|---|---|---|---|
| | | Volume value/$nm^3 \times 10^{-3}$ | Retention volume/% | Volume value/$nm^3 \times 10^{-3}$ | Retention volume/% |
| Dry black | Control | 736 | 100 | 332 | 100 |
| BR/OEP | 70 | 560 | 76 | 298 | 90 |
| SBR 1712 | 70 | 485 | 66 | 306 | 92 |
| SBR 1500 | 50 | 416 | 56 | 284 | 86 |
| SBR 1500 | 70 | 378 | 51 | 224 | 67 |

There are two possible consequences for the structural breakdown of carbon black:

- an increase in the filler-polymer interface leading to higher bound rubber content, as mentioned earlier;
- greater surface activity of the freshly created surfaces and, hence, greater adsorption ability.

There is considerable evidence to support the assumption of high surface activity of the fresh surfaces. Serizawa et al.[54] recently investigated the carbon gel of natural rubber by pulse NMR with the spin-spin relaxation time, T2, and found that the surface activity was reduced after introduction of oxygen groups and that the mobility of the molecular segments in the loosely and tightly bonded rubber phase increased. This means that surfaces freshly formed during mixing have a greater ability to adsorb rubber molecules.

### 5.1.6.2 Specific Surface Activity of Carbon Blacks

The total bound-rubber content is dependent not only on the filler surface activity, but also on the interfacial area between filler and polymer in the compound. With regard to the surface activity of the filler, meaningful information can only be obtained by normalizing the bound rubber per unit filler surface. Figure 5.9 shows a plot of bound rubber per 100 $m^2$ as a function of carbon-black surface area at 50 phr loading. Although the experimental points are slightly scattered, it is apparent that the surface activity, expressed as bound rubber per unit surface, decreases with increasing surface area of the carbon blacks. The bound rubber per unit interface of semi-reinforcing carbon blacks is nearly twice as high as that of reinforcing types. This corroborates Dannenberg's observations[7]. Likewise O'Brien et al.[30] arrived at similar conclusions.

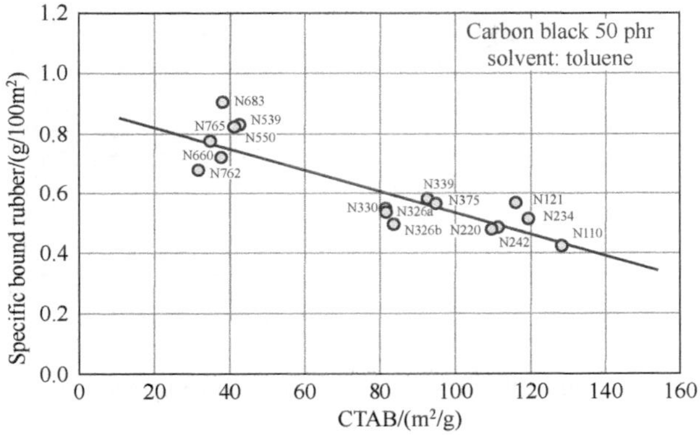


**Figure 5.9** Specific bound rubber as a function of CTAB surface area for various carbon blacks at 50 phr loading in SBR

Dannenberg attributed this phenomenon to the easier accessibility of large aggregates for the adsorption of molecules. He believed that low-surface-area carbon blacks possess a higher surface activity because of the lower temperature of carbon-black formation in the furnace process. Furthermore, he pointed out that high structure, and hence large particles, are more easily broken down during mixing, thus exposing new active surfaces. As mentioned above, the breakdown of aggregate structure could explain the high bound-rubber content of high-structured carbon blacks at equal surface area. It is, however, difficult to explain such great differences between the different grades. However, the high surface activity derived from bound rubber for low-surface-area carbon blacks contradicts the results from surface-energy measurements. The results of inverse gas chromatography at infinite dilution[45] show that the dispersive component of surface energy, $\gamma_s^d$, of low-surface-area carbon blacks is much lower than that of carbon blacks with high surface areas. In addition, Wang and Wolff's investigation of the adsorption-energy distribution[55] indicates that the concentration of low-energy sites may not vary significantly between the carbon blacks, but that the number of high-energy sites is much higher for small-particle carbon blacks. While the easier accessibility of large-particle carbon blacks to adsorb rubber molecules may initially seem to explain the differences observed in Figure 5.9, upon further inquiry this explanation does not hold true in the light of the considerable dependence of surface activity on filler loading. Figure 5.10 illustrates the dependence of bound rubber per unit surface area on CTAB surface area at different loadings. The basic trend is similar for all loadings, but it can be observed that the decrease in bound rubber per unit surface with increasing surface area is more pronounced at higher degrees of filler loading. Therefore, if bound rubber per unit surface is taken as a measure of surface activity, then its high loading dependency would be very difficult to understand.

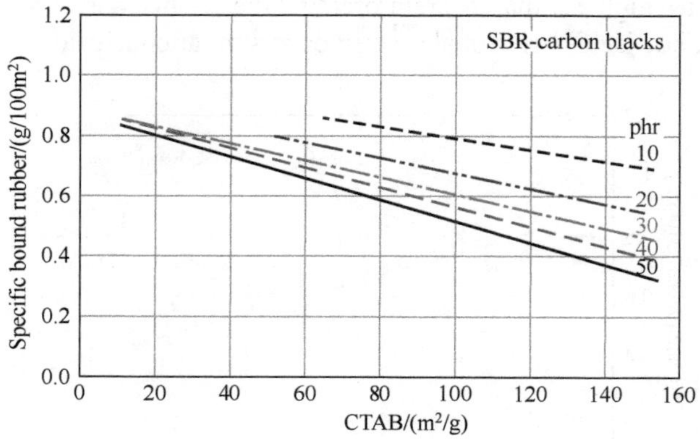


**Figure 5.10** Specific bound rubber as a function of CTAB surface area for various carbon blacks and loading

One possible interpretation for the lower bound-rubber content per unit surface of large-surface-area blacks could be due to the difference in modes of adsorption for high molecular weight rubber.

It is generally believed that molecular weight of rubber may decrease during mixing, resulting from the scission of rubber chains at mechanical action, especially at high temperature. This effect could be enhanced by the incorporation of carbon black. Another reason may be multiple-segment adsorption which leads to a reduced effectiveness of the filler surface to form bound rubber. It is possible to imagine that, once a segment is attached, the whole molecule becomes part of the bound rubber. In the case of multiple-segment adsorption, two or more active sites are occupied by the same molecule without increasing the bound rubber. Multiple segment adsorption can also take place between neighboring aggregates (interaggregate multiple-segment adsorption), as shown schematically in Figure 5.11. The multiple-segment adsorption would increase with decrease of interaggregate distance. In addition, it has repeatedly been demonstrated that carbon-black aggregates interact, which results in the formation of aggregate agglomeration[9,56–58]. It is believed that in a system where filler-filler interaction is strong and filler-polymer interaction is weak, for example, silica in hydrocarbon systems, filler agglomeration is built via direct contacts (hydrogen bonding)[59]. In the case of systems with strong polymer-filler interaction, this may lead to an immobilized rubber shell on the filler surface[29,60]. In such cases, agglomeration is probably formed by a joint shell mechanism. Nevertheless, by whichever mechanism agglomeration is formed, the bound-rubber content per unit surface, calculated on the assumption that no contacts exist, would be underestimated.

Moreover, multiple attachments and the formation of agglomerates are highly dependent on surface characteristics and on the distance between aggregates. With regard to the surface characteristics which influence the formation of agglomeration, Wang, Wolff, and Donnet[45] introduced a factor, $S_f$, which indicates the ability of fillers

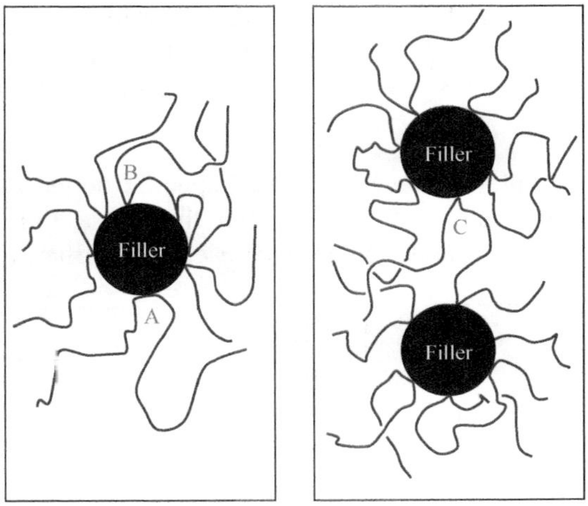


**Figure 5.11** Schematic diagram of the concept of multiple segmental adsorption; A: single attachment, B: multiple attachment, C: inter-aggregate attachment

to agglomerate in hydrocarbon rubbers. This factor is related to the specific interaction of polar molecules on the filler surface. A high value of $S_f$ reflects a greater ability of the filler to agglomerate. For most carbon blacks and most of the polar probes, however, $S_f$ is almost a constant. It is therefore clear that the interaggregate distance in carbon black-filled rubbers has a predominant influence on agglomeration. The distance between aggregates is also an important factor with regard to multiple-segment attachment, i.e., the shorter the distance, the higher would be the number of interaggregate multi-attachments. The distance between filler aggregates in a compound depends on the degree of loading and on aggregate size. Based on the occlusion of rubber and according to Medalia's claim[61] that each aggregate consists of uniform, randomly connected nodules, Wang, Wolff, and Tan[62] estimate the distance between aggregates, $\delta_{aa}$, (see Eq. 3.9), which shows the distance decreases with filler loading and surface area. As small particle-size carbon blacks have comparatively large surface areas, the interaggregate distance of such blacks at equal loading will be shorter. As discussed above, the bound-rubber content per unit surface of such blacks will, therefore, be underestimated. In most cases, Eq. 3.19 is better suited for the evaluation of the distance because it includes the effect of DBP, i.e., the higher the structure, the shorter will be the distance. Compared to CTAB, the DBP effect on $\delta_{aa}$ is much smaller.

If the distance between aggregates has an effect on bound rubber per unit surface, this effect should be reduced as loading decreases. As can be seen in Figure 5.12(a) for a sample which formed a coherent mass, the bound rubber content per unit area shows a linear correlation with the logarithm of weight loading (phr). The value of bound rubber content per unit surface can be extrapolated to the critical coherent loading (see Figure 5.3), at which multiple attachments per unit surface can be assumed to be similar for all blacks. At this loading, the filler aggregates can be assumed to be free from the effects of agglomeration, and the difference in molecular weight reduction of rubber for different compounds during mixing can be assumed to be similar. This, then, is the most useful specific bound rubber value for each carbon black.

Figure 5.12(b) shows a plot of the extrapolated specific bound rubber against CTAB surface area of carbon blacks. It can be seen that, to a first approximation, the extrapolated specific bound rubber content increases with surface area.

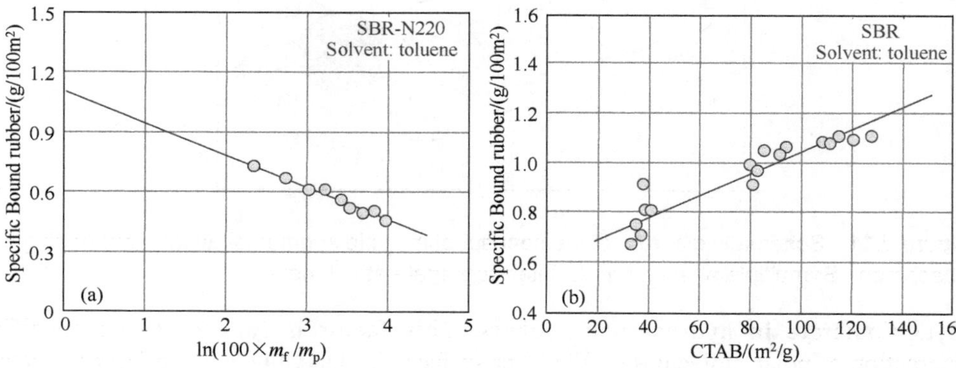


**Figure 5.12** Specific bound rubber as a function of the logarithm of loading (a) and the dependence on surface area (CTAB) at critical loading of carbon blacks (b)

### 5.1.6.3 Effect of Surface Characteristics on Bound Rubber

The type and degree of interaction between the polymer and filler surface is an important factor in determining bound rubber. It is, in fact, the subject of many research studies. It seems to be reasonable to expect that for fillers with the same morphology and loading, particles with a surface that has stronger interaction with the polymer would be expected to give higher bound rubber content, because, on average, more molecules could be irreversibly attached per unit surface area. The term "surface activity" describes how well a filler surface "bonds" with polymer, without defining the bonding as physical or chemical in nature. As mentioned in the last section about specific bound rubber content, there have been conflicting claims based on bound rubber measurements where, on the one hand, surface activity of carbon blacks generally increases with increasing surface area, while on the other hand, it generally decreases with increasing surface area. The inconsistencies arise from experimental protocols and the manner in which results are presented. If comparisons are made at constant, high loadings, it is usually found that specific bound rubber decreases with increasing filler surface area. However, as discussed in the previous section, the filler particle size itself plays a role in determining the amount of bound rubber. When compared at the critical loading, carbon blacks with higher surface area generally give higher bound rubber per unit area, suggesting that these blacks have a more active surface than lower area blacks.

This is also confirmed by inverse gas chromatography (IGC)[63]. It has been demonstrated that for hydrocarbon rubbers and conventional fillers such as carbon black and silica, the polymer-filler interaction is essentially physical in nature (see Section 5.1.3). For given hydrocarbon rubbers such as natural rubber (NR), polybutadiene (BR), and styrene-butadiene-copolymer (SBR) that are non- or low-polarity polymers, the terms of polar, hydrogen bonding, and acid-base interactions can be eliminated; therefore, Eq. 2.135 can be written as Eq. 5.1:

$$W_a^{pf} = 2\sqrt{\gamma_p^d \gamma_f^d} \tag{5.1}$$

where $W_a^{pf}$ is the adhesion energy between polymer and filler which is a measure of polymer-filler interaction, $\gamma_p^d$ is the dispersive component of the polymer surface energy, and $\gamma_f^d$ is the dispersive component of the filler surface energy. For given polymer, the higher the $\gamma_f^d$, the stronger is the polymer-filler interaction.

On the other hand, as Figures 2.56 and 2.57 show, the $\gamma_f^d$ increases with surface area in a linear fashion, over the whole range of carbon blacks. Likewise, bound rubber follows the same function, as shown in Figure 3.2.

The linear relationship between bound rubber and the $\gamma_f^d$ clearly demonstrates that the root cause of the bound rubber formation for hydrocarbon polymers is the polymer-filler interaction, which is determined by the dispersive component of the fillers' surface energy. This is in agreement with the direct observation of aggregate separation stress from the polymer matrix in carbon black-filled SBR vulcanizates by means of electron microscopy. The adhesion force between polymer-carbon black surface at which a given percentage of carbon black separated from the vulcanizates increases in the order of carbon black MT, GPF, FEF, and HAF. This order is the same as surface area (Figure 5.13)[64]. This also suggests that while bound rubber is an important parameter in the mechanism of rubber reinforcement, the specific bound rubber has little meaning per se as a criterion for the specific activity, except in cases when differences in filler morphology are absent or small.

#### 5.1.6.4 Carbon Black Surface Modification

It is possible to study the effect of surface activity and surface chemistry independent of morphology, by deliberately modifying the surface of particular fillers. For example, heat treatment of N772 carbon black at temperatures up to 900°C was found to increase the bound rubber of this filler in SBR by up to 6%[65]. No change in surface area or particle morphology is observed under these conditions. The increase in bound rubber is attributed to decomposition of oxygen-containing groups and volatilization of adsorbed substances on the filler surface, exposing more of the highly active carbon surface, and therefore providing stronger polymer adsorption. The explanation is supported by

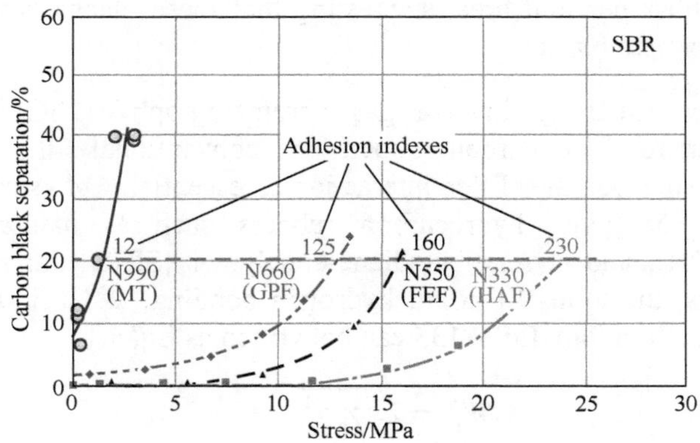


**Figure 5.13** Carbon black separation as a function of stress in SBR

IGC measurements on N234 carbon black heated at temperatures up to 900°C in nitrogen, which showed an increase in surface energy with increasing temperature of heating above 200°C [66]. Heating carbon black to graphitization temperatures, however, results in a large drop in surface activity, measured by IGC[27] and also by a large decrease in bound rubber[67]. For example, the effect of graphitization of two carbon blacks, N330 and N220, on bound rubber in two different polymers is shown in Figure 5.14.

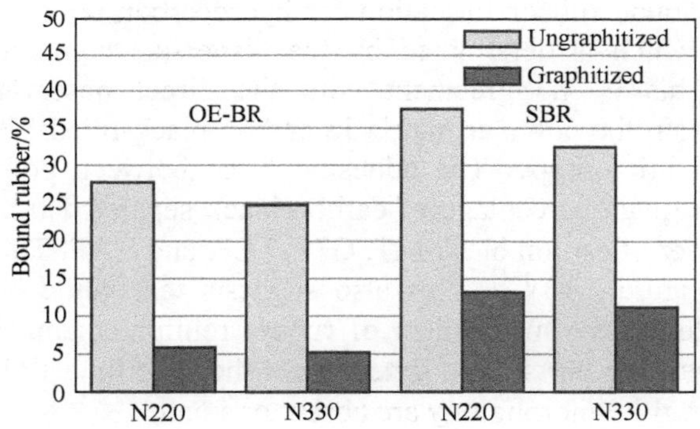


**Figure 5.14** Effect of graphitization on bound rubber content

Again, the process of graphitization has been shown to have little effect on the particle morphology or surface area, so the large effect on bound rubber must be due to surface deactivation. Graphitization severely reduces the number of edges of the graphitic crystallites and defects in carbon black[68]. These are thought to be the high energy sites on the carbon black surface[55], so it should not be surprising that their removal leads to substantially lower surface energy and lower bound rubber contents.

Whereas graphitization has been shown to reduce the surface activity of carbon black, plasma treatment has been shown to increase surface activity. For example, treatment of N772 carbon black with an argon plasma increased the surface activity by 35%, as measured by IGC[69]. Similar effects can be achieved by $\gamma$-irradiation and neutron irradiation[70]. A 40% increase in bound rubber content of N660 and N375 carbon blacks was observed after $\gamma$-irradiation. Both plasma treatment and radiation treatments have been shown to increase the concentration of defects and the degree of disorder on carbon black surface. In fact, even larger increases in surface activity are observed when these techniques are applied to graphite or graphitized surfaces[55]. The fact that an increase in surface defects corresponds to an increase in activity, coupled with a higher bound rubber content, provide strong evidence that surface defects on carbon black are primarily responsible for its strong interaction with hydrocarbon polymers.

The development of carbon-silica dual phase fillers (CSDPF) has led to some interesting observations with respect to bound rubber. It was found that CSDPF compounds with the same overall composition as compounds with blends of carbon black and silica had higher bound rubber contents (Figure 5.15)[71]. For CSDPF, certain parts of the surface covered by silica have a lower polymer-filler interaction, similar to pure silica filler[72]. However, the higher bound rubber content is due to higher surface activity of the carbon domain, as measured by IGC. Compared with carbon black N110, the $\gamma_s^d$ of CSDPF 2115 (CRX 2006) is 16% higher which should be related to the change in the surface microstructure. Upon silica doping, more surface defects and/or smaller crystallite dimensions in the carbon domain can be formed, leading to the observed increase in activity. This was confirmed by scanning tunneling microscopy (see Figure 5.16).

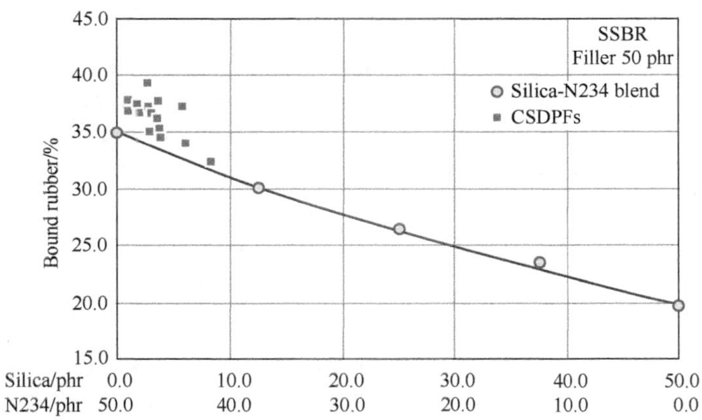


**Figure 5.15** Comparison of bound rubber contents between compounds filled with CSDPFs and silica-carbon black blends

Most other attempts at modifying carbon black surfaces through chemical treatment have resulted in reduced surface activity and reduced bound rubber. Oxidation of N330 by ozone, for example, led to reduced bound rubber content in NR compounds, and this was attributed to a reduction in the dispersive component of the carbon black surface energy[73].

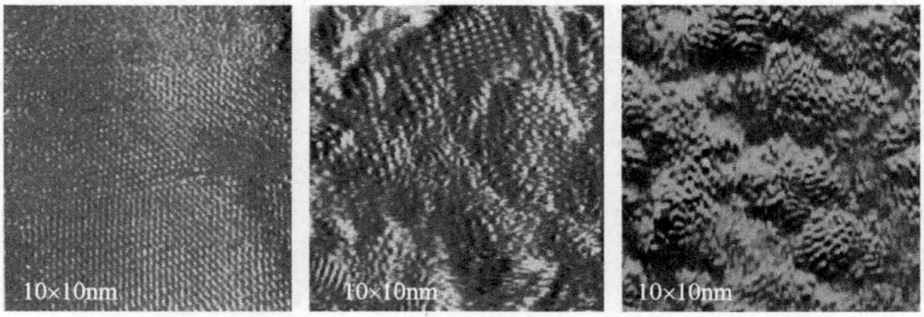


**Figure 5.16** Top-view STM images of graphitized N234, carbon black N234, CSDPF 2214, and their bound rubber contents in SBR

An alternative hypothesis that has been put forward is that hydrogen, either directly bonded to carbon, or as part of other functional groups on the surface, is mainly responsible for the high activity of carbon black[74]. A number of facts, however, cast doubt on whether hydrogen is a direct cause of the high surface activity. Firstly, hydrogen content is reduced by heat treatment at 900°C[74], yet surface activity remains high, or is even increased[65]. Secondly, hydrogenation of carbon black did not increase surface activity of carbon black[75]. Nevertheless, active hydrogen concentration is still used by many as an indicator of carbon black surface activity. It seems likely that the reason this is useful is that, for a series of carbon blacks made in a similar way, the hydrogen content is an indicator of the crystallite edges or defects which are primarily responsible for the surface activity.

All of the above results taken together lead to the conclusion that the high surface activity of carbon black which determines the bound rubber formation is probably due to defects in the graphitic structure and exposed edges of graphitic crystallites. The following facts, in particular, support this conclusion:

- for hydrocarbon rubbers, the polymer-carbon black interaction is predominantly dispersive or non-specific in nature;
- the surface activity of carbon black increases with increasing surface area (or decreases with primary particle size);
- removal of other functional groups, such as oxygen-containing groups by heating carbon black at temperatures up to 900°C tends to increase surface activity and bound rubber;
- graphitization leads to a large reduction in surface activity and bound rubber;
- plasma treatment and $\gamma$-irradiation lead to increases in activity;
- introduction of other chemical groups on the surface tends to decrease activity;
- doping some foreign atoms on the carbon black surface results in increased surface activity and bound rubber due to the reduced graphitic crystallite size and increased defects in the carbon domain.

#### 5.1.6.5 Silica Surface Modification

Modification of silica surfaces through chemical reactions with various chemicals is well known, and is used commercially in silica-filled SBR compounds. IGC analysis of silicas modified with some chemicals showed that the surface energy, particularly the specific component, was drastically reduced by the chemical treatment[76]. The effect of silica surface chemical modification has generally been to reduce bound rubber content. Bound rubber in NR is nearly eliminated by esterification treatment of silica with hexadecanol[77]. In other studies in NR[78] and PDMS[79], bound rubber was reduced to a lesser extent by silane reaction on the silica surface. In the case where the silane is a polymer-reactive one, like TESPT, bound rubber content of uncured compounds should not be considered as an indicator of polymer-filler interaction in the vulcanizate because chemical reaction between filler surface and polymer is known to occur mainly during vulcanization. However, some reaction of the coupling agent with the polymer can also occur during mixing with a bifunctional coupling agent, especially if the temperature is allowed to rise. This can lead either to grafting of polymer or some slight crosslinking of the polymer, either of which will lead to an increase in measured bound rubber content.

### 5.1.7 Effect of Mixing Conditions on Bound Rubber

Bound rubber formation begins almost immediately when rubber and filler are mixed in a typical internal mixer. The initial rate of formation is high, and as mixing continues, bound rubber increases at a slower rate, and may eventually plateau[80]. Figure 5.17 shows that if bound rubber reaches the plateau, it occurs well after the second power peak. When mixing is finished, the bound rubber formation process is usually not complete because storage of the compound after mixing leads to some further increase. However, it seems that a major part of the total bound rubber is usually formed during the mixing process.

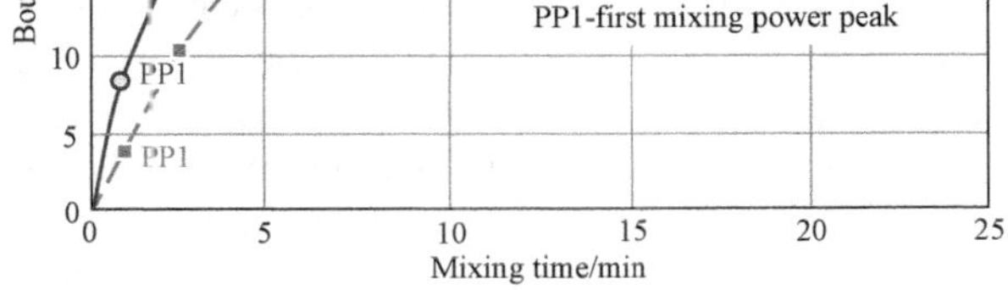


**Figure 5.17** Plot of bound rubber *vs.* mixing time

### 5.1.7.1 Temperature and Time of Mixing

There is no single, simple way to determine if a mix is in its optimized state. The appropriate measure and level of mixing are generally dependent upon the product application[16,81–84].

During mixing, while the bound rubber is forming upon filler dispersion, other physical and chemical processes, such as polymer gelation, chain scission, breakdown of the carbon black aggregates, and interaction of filler with other ingredients simultaneously occur. The formation of bound rubber arises mainly from the physical adsorption of polymer molecules on the filler aggregates[18,24] although various forms of chemisorption cannot be excluded, especially for some polymers like natural rubber which show a highly mechanochemical sensitivity.

Bound rubber formation takes place very rapidly during mixing, and as much as 80% of the bound rubber may form during the first few minutes[85]. However, bound rubber continuously increases with mixing time and remains incomplete. Its development continues throughout storage of the compounds[16,18,51,52,86,87] (also see Figure 5.8).

Heat treatment and prolongation of mixing time would speed up all these processes. By increasing temperature, the mobility of the polymer segment would increase which could raise the rate of diffusion of the polymer chains so that they can migrate to more favorable positions for physical adsorption and chemisorption. In addition, when the filler and polymer are mixed, the polymer can penetrate the interior of an aggregate which would lead to a physical entrapment of free molecules of polymer. This may also be involved in bound rubber formation where intensive mixing is favorable, and the viscosity of the rubber decreases with increasing temperature.

There is evidence for polymer gelation during mixing, especially at high temperature[9,88]. This is related to crosslinking of polymer molecules even in the absence of vulcanizing agents. Since bound rubber measured by the standard method is in a general sense related to its insolubility in normal rubber solvents, it consists of the polymer portion truly bonded on the filler surface plus rubber gel. In a study of air-aging of pure BR film prepared with polymer solution on a glass substrate, it was found that the gelation of the polymer starts in a few minutes and it is delayed by antioxidants[89]. When a gum compound containing only a gel-free solution SBR, zinc oxide, stearic acid, and antioxidant prepared at a low temperature was heated in a mold at 160°C, a detectable amount of gel, assessed with a 100-mesh stainless steel cage as used in bound rubber measurement, can be obtained after 20 minutes, the shortest heating time used. When the compound was heated for 40 minutes, the gel content was higher than 50%. Therefore, it may be reasonable to believe that similar chemical processes occur in the mixer at high temperature, especially when the mixing time is extended. In a filled compound, crosslinking may also take place so that polymer chains as well as polymer gel may crosslink with the filler-bound rubber complex (filler-gel), i.e., the filler aggregate with polymer molecules directly attached on it. As mentioned above, bound rubber consists of the polymer portion truly adsorbed on the filler surface plus the rubber gel, whether it is crosslinked by filler-gel or not. Here again, the intensive mixing would raise the bound rubber portion originating from polymer crosslinking[90]. The effect of mixing time and temperature was demonstrated by Stickney et al. as shown in Figure 5.18[91].

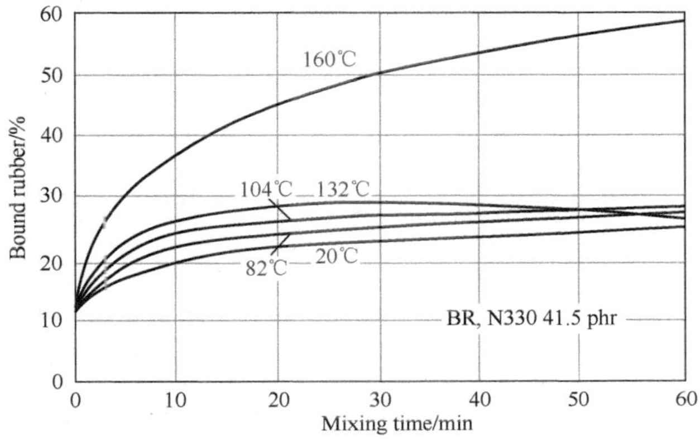


**Figure 5.18** The effect of mixing time and temperature on bound rubber content

On the other hand, molecular chain scission may take place during high shear mixing, and can be affected by mixing intensity. This is caused by mechanochemical degradation and results in a reduction in molecular weight of the polymer. This phenomenon has been demonstrated with natural rubber and in fact has been adopted as a routine approach for mastication[92]. Reduction in polymer molecular weight significantly lowers the bound rubber content[37,93]. However, while the chain scission process occurs very rapidly for polyisoprene containing polymers, this is not the case for SBR and BR due to their high mechno-oxidative stability. In a series of systematic studies, Cotten[94,95] was able to show that SBR molecular weight and molecular weight distribution did not change significantly upon mastication in a Brabender mixer. Consequently, taking the filler-gel and rubber-gel as the total bound rubber, the bound rubber content is considerably increased with mixing time and this effect is much more pronounced when mixing temperature is higher.

However, for natural rubber, bound rubber increases with increasing mixing time (or passes on the mill), and then decreases[14,96]. This can be seen in Figure 5.19(a), in which the bound rubber content is plotted as function of milling for a solution-mixed masterbatch of deproteinized NR with 50 phr N330 (Philblack O). Also, in contrast with BR and SBR, the drop of bound rubber with NR is much more pronounced at higher milling temperatures [Figure 5.19(b)][14].

The chain scission process begins immediately when the rubber is loaded in the mixer and runs through the mixing process. This can be seen from Figure 5.20(b), where the molecular weight in sol is plotted against mixing time. At the beginning of mixing, the rate of bound rubber formation increases very rapidly, and its content increases. This is due to the wetting of filler surface with polymer and adsorption of polymer molecules

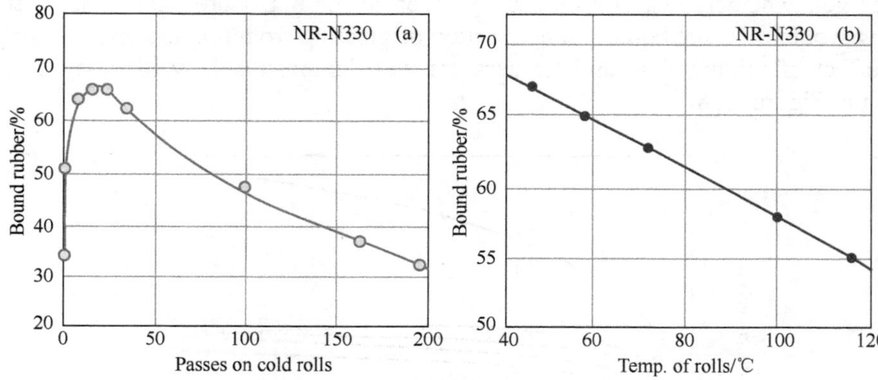


**Figure 5.19** The effect of mixing time and temperature on bound rubber contents of carbon black-filled NR

on the filler. This rate would slow down as the availability of filler surface for bound rubber formation is continuously reduced. Once the effect of chain scission on the bound rubber formation has exceeded the effect of adsorption, the bound rubber content will decrease with prolonging the mixing time as shown in Figure 5.20(a)[14].

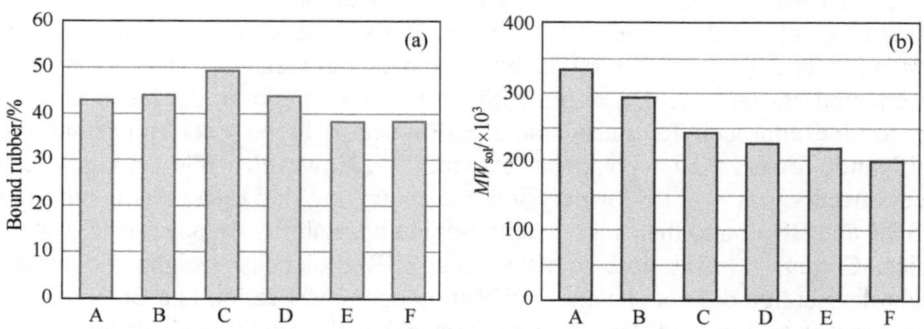


**Figure 5.20** Effect of mixing intensity on bound rubber formation. NR 100, carbon black N134 50, ZnO 3, stearic acid 2, antioxidant 1. A-compound from Banbury. From A to F: mastication time of compound on mill increases

### 5.1.7.2 Mixing Sequence Effect of Rubber Ingredients

In addition to mixing temperature and time, the sequence in which ingredients are added is another critical parameter influencing rubber properties. Designing the mixing sequence remains an art, and is based on the convenience of the operation and the processability and properties of the final cured materials. However, the science of the mixing sequence requires some basic understanding of the interactions between different components of the materials not only during mixing but also in the vulcanized state. The same can be said for the dynamic properties, since filler agglomerate formation is eventually controlled by these interactions.

#### 5.1.7.2.1 Mixing Sequence of Oil and Other Additives

It has been reported that bound rubber formation varies significantly upon changing the mixing sequence of ingredients[65,97]. This is related to the interactions between filler and other components of the compound such as polymer, activators, antioxidants, and curatives. Shown in Figure 5.21 are the adsorption energies of a variety of chemicals on the carbon black and silica surface as a function of their cross-section areas, $\sigma_m$.

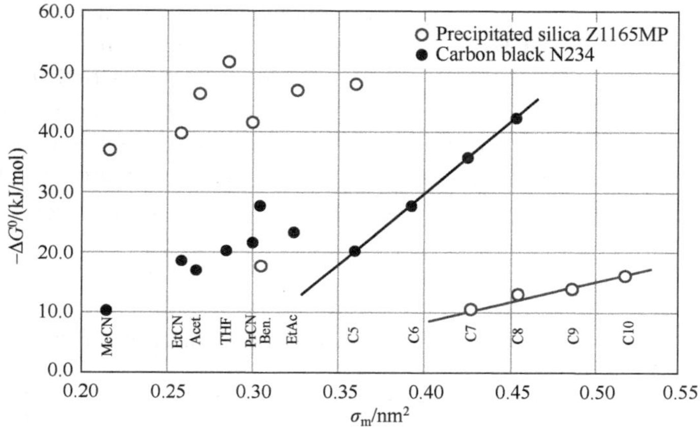


**Figure 5.21** Adsorption free energies of a variety of chemicals on carbon black and silica as function of their cross-sectional area, $\sigma_m$

As can be seen, for the chemicals having the same cross-sectional area, the polar chemicals give higher adsorption energies than alkanes that can be taken as model compounds of hydrocarbon polymers and oils. This suggests that when filler interacts with polar ingredients, such as antioxidants and stearic acid, the amount of available surface area and the available number of filler active centers for polymer absorption on filler surface are substantially reduced by the adsorption of these small molecules. Once these ingredient molecules are adsorbed on the filler surface, it is very difficult to replace them with polymer chains. Oil has good compatibility with polymer and low polarity, and so the competition between it and the polymer molecules for the higher energetic sites on the filler surface would make the mixing sequence equally important. This is especially true for carbon black, which has a higher interaction with hydrocarbon materials. Therefore, from the point of view of polymer-filler interaction or bound rubber formation, the mixing procedures of carbon black- and silica-filled compounds should be optimized so that oil and other ingredients are added after the filler is incorporated into the polymer.

Figure 5.22[97] presents the results obtained with NR compounds containing different ingredients but prepared with the same procedure, i.e., single pass Banbury mixing adding all ingredients and then immediately the polymer, keeping energy input constant. The highest bound rubber was found for the compound containing only 40 phr carbon black (N330), and the lowest bound rubber is for the one having an additional 5 phr oil, 5 phr zinc oxide, and 2 phr stearic acid. Even 2 phr antioxidant (*N*-isopropyl-*N'*-phenyl-*p*-phenylenediamine) mixed with carbon black can considerably depress bound rubber formation. Carbon black mixed in the presence of other ingredients gives lower bound rubber than with polymer alone because the antioxidants, especially amine-types, can serve as a free-radical scavenger to inhibit the gelation of polymer during mixing. As mentioned before, when carbon black is mixed with the other ingredients including antioxidant, the amount of available surface area and the available number of filler active centers for polymer adsorption can be substantially reduced by the adsorption of these small chemicals that are not easy to be replaced with polymer chains. This process would also be true even for a mixing procedure where the polymer is first masticated in the mixer and then the filler and other ingredients are added. During mixing, the higher-polar ingredients will be driven to the filler surface where polar component of the surface energy of the filler is relatively higher. If the oil and stearic acid are added to the mixer with carbon black, the high $\gamma_f^d$ of the filler will lead to strong interaction with oil and stearic acid, resulting in the lowest bound rubber.

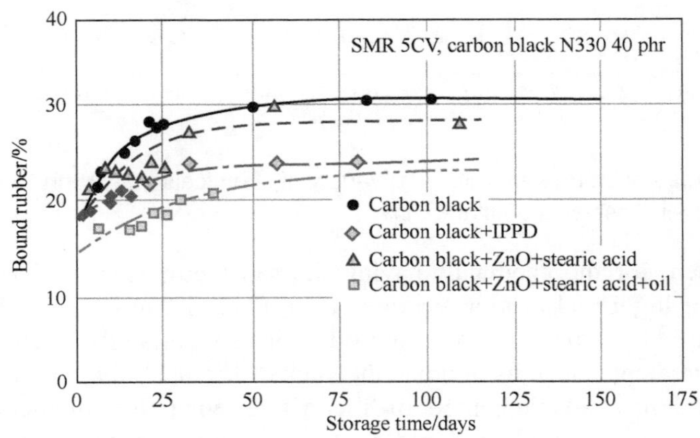


**Figure 5.22** The effect of mixing sequence of chemicals and storage time on the bound rubber formation

Donnet et al. compared a thermally treated carbon black to a mixing procedure where the oil is added with carbon black to the premasticated SBR and found a 40% increase in bound rubber when adding carbon black to the premasticated rubber first and mixing until incorporated before adding the oil[63,65].

Similarly, for a typical passenger tire tread compound with a blend of SSBR/BR (75/25), 75 phr carbon black N234, and 25 phr oil, the bound rubber content was 40% when the oil was added with carbon black into the masticated polymers followed by the addition of zinc oxide, stearic acid, and antioxidants. However, when the carbon black was added to the polymer first and oil was added after carbon black had been incorporated, the bound rubber content increased to 47%[98].

#### 5.1.7.2.2 Mixing Sequence of Sulfur, Sulfur Donor, and Other Crosslinkers

Not only does the mixing order of oil and other ingredients that are generally added in the earlier mixing stage(s) affect bound rubber formation, but also the mixing sequence of those ingredients used as curatives and generally added in the last stage is found to have a significant influence, especially at higher mixing temperatures.

In a research disclosure, MRPRA[99] described that a variety of polymer-carbon black masterbatches can be modified with some sulfur donors such as bis-4-[1,1-dimethyl(propyl)] phenoldisulfide (BAPD), dithiodicaprolactam (DTDC) and dithiodimorpholine (DTDM) in a Brabender Plasticorder during mixing at an elevated temperature in excess of 150°C. They found that after the reactive mixing, a certain amount of sulfur donor is bonded to the polymer chains and the polymer-filler interaction is enhanced, indicated by a considerable increase in bound rubber. The high bound rubber may originate from the polymer modification, some functional groups being grafted on the polymer chains. Before the chemical reactivities of sulfur donors with filler surface functional groups to create covalent linkages between filler and polymer are clearly identified, it can be safely assumed that these groups, generally having higher polarity than hydrocarbon polymer, may be strongly adsorbed on the filler surface. This is due to the fact that beyond dispersive interaction, additional specific (polar) interactions should be involved which would lead to higher bound rubber. However, the high bound rubber is certainly related to the polymer gelation (precrosslinking), since the additives are active as sulfur donors in a normal way, and are effective as a source of crosslinks during mixing. This is especially true at temperatures higher than 150°C, which is the cure temperature of the compound containing sulfur donors.

If this is reasonable, then any chemical which can generate crosslinks between polymer molecules during mixing stages would give the same effect. Wang et al. also compared the effect of the addition of sulfur and sulfur donors in different stages[100,101]. By simply moving the sulfur from the last stage to the first stage the bound rubber is significantly increased. The bound rubber content of the compound with sulfur added in the first stage is 43.1%, compared with 38.4% of its counterparts with sulfur added in the last (third) stage. When some sulfur donors such as DTDM and TMTD (tetramethylthiuram disulfide), which are used as curatives in the recipe, are moved from the last stage to the early mixing stage, higher bound rubber contents are always observed. This effect may be related to the promotion of polymer-gelation or pre-crosslinking which apparently increases bound rubber contents determined by solvent extraction, although the efficiency of sulfur for generating crosslinks in the polymer matrix should be also enhanced.

Another example of the application of crosslinker to increase bound rubber was given by Terakawa and Muraoka[102]. When epoxidized natural rubber (ENR) and 50 phr N330 carbon black was mixed with diamines in a Banbury mixer, compared with a non-amine counterpart, the bound rubber was considerably increased. Although the filler-polymer interaction may be enhanced by addition of amines, the crosslinking of the epoxy groups with amines during mixing is, with a high degree of certainty, involved. It is well known that the crosslinking of epoxy materials with amines can take place at room temperature and the rate of this reaction can rise very rapidly with elevating temperature[103]. Even though the temperature of mixing in Terakawa's study is not specified, the dump temperature can be anticipated to be higher than 120°C so that certain gelation of the polymer would be expected.

In an extreme case, Gerspacher and his co-workers[104,105] used dynamic vulcanization to get 100% bound rubber. In this case, they added all curatives, including sulfur and accelerators, into the mixer to allow the torque on the mixer blades to increase to a given level, indicating that a certain degree of crosslinking was generated. However, as the compound is relatively highly crosslinked, it cannot be processed in the downstream processing such as milling and extrusion with conventional equipment.

#### 5.1.7.2.3 Bound Rubber of Silica Compounds

In Figure 5.21 the free energies of adsorption of a series of normal alkanes and a variety of polar chemicals are presented as a function of the cross-sectional areas of their molecules, $\sigma_m$, for carbon black N234 and precipitated silica ZeoSil 1165. Both fillers represent the most popular fillers used in tread compounds of passenger tire. Their $\gamma_f^d$ and $I^{sp}$ for the polar chemicals are given in Table 2.14. The specific interaction factor, $S_f$, of acetonitrile on a series of rubber blacks and precipitated silicas are also presented in Figure 2.66 as a function of their surface areas. As can be seen, regardless of the difference in their surface areas, the silica is characterized by higher free energies of adsorption, higher $I^{sp}$ for the polar chemicals, but the $\gamma_f^d$ and the adsorption free energies of alkanes are substantially lower, relative to carbon black. Therefore, with respect to carbon black, silica is characterized by a much lower polymer-filler interaction with hydrocarbon polymers, which is not desirable for rubber reinforcement. However, when the silica surface is chemically modified with silanes, the specific component of surface energy is reduced drastically whereas $\gamma_f^d$ is only slightly decreased[76]. The effect of silica modification with bis (3-triethoxysilylpropyl)-tetrasulfide (TESPT), a bi-functional silane, on its surface energy is also listed in Table 2.16, taking benzene as the polar chemical for estimation of surface polarity. With this modification, the tendency for filler agglomeration is significantly depressed due to the reduction of surface polarity, and the poor polymer-filler interaction of the modified silica, indicated by the low level of $\gamma_f^d$, is offset by creating covalent bonds between the filler surface and elastomer matrix through sulfur linkages.

When using coupling agents, premodification of the filler surface with the coupling agent proves most efficient[106]. However, when the modification of the filler surface has to be executed in situ, the specific mixing sequence of the coupling agent and other ingredients should be considered[107,108].

When the addition of other ingredients is considered, it would be better to introduce the coupling agent before them since their molecules may occupy the filler surface and interrupt the filler-coupling agent reaction. This may not be very critical for the oil as higher surface polarity of the filler which may allow silanes, a polar material, to be driven to and react with silanols. In any case a better mixing procedure would be that the coupling agent goes with filler into the polymer first, followed by the oil and other ingredients. In Table 5.1, the bound rubber contents of carbon black, silica, and TESPT modified silica added *in situ* during mixing are presented.

#### 5.1.7.3 Bound Rubber in Wet Masterbatches

In any dry mixing process of carbon black, the process of simply incorporating the black into the polymer generates some heat and applies some shear to the material. Therefore, even at minimal mixing times, some bound rubber is formed. In a wet masterbatch process, however, where the polymer is mixed with a carbon black slurry at low temperature, it should be possible to initially form a mix with no bound rubber. This was in fact demonstrated in a polybutadiene masterbatch. In order to observe no bound rubber, it was necessary to dry the masterbatch at room temperature in a vacuum[91]. Drying at a temperature even as low as 70°C resulted in significant bound rubber formation. It was found that static heating of dried masterbatch at temperatures between 70 and 160°C caused increases in bound rubber, but that roll-milling at the same temperatures resulted in faster bound rubber formation, and a higher final level.

After adequate mixing, the bound rubber contents of wet masterbatch compounds are higher than dry-mixed compounds of the same composition, as demonstrated by CEC (Cabot Elastomer Composite) for continuously mixed NR/carbon black masterbatch[109]. Figure 5.23 shows the differences in bound rubber content between CEC and dry-mixed compounds with different levels of oil. The results are taken from a statistical analysis of about 200 pairs, each pair having one CEC and one dry-mixed compound with the same formulation. For all compounds with and without oil, CEC gives higher bound rubber than its dry-mixed counterpart. The difference increases as the oil content increases. This is explained by the fact that in the wet masterbatch material, bound rubber is formed before oil is added, so there is no competition for the filler surface. In the dry mixing process, as discussed before, the small oil molecules can effectively compete with the polymer for the active filler surface, even if the mixing sequence is optimized. In contrast, the polymer-filler interaction in CEC is much less affected by the addition of oil during mixing, as the adsorption of polymer chains on carbon black surface has already been completed.

The effect of continuous liquid phase mixing of silica with solution polymers on bound rubber formation is quite different. The masterbatch produced with this process is termed as Eco-Visco-Elastomer Composite (EVEC). Shown in Figure 5.24 are the bound rubber contents of carbon black N234 and silica compounds[110]. The formulation for carbon black is conventional for passenger tire tread compound, and a typical green tire formulation is used for silica compounds, except EVEC-H has 6 phr more silica. Silane coupling agent TESPT was used for all silica compounds with the same dosages. For all compounds, oil-extended SSBR and BR (70/30) basic polymer system are employed.

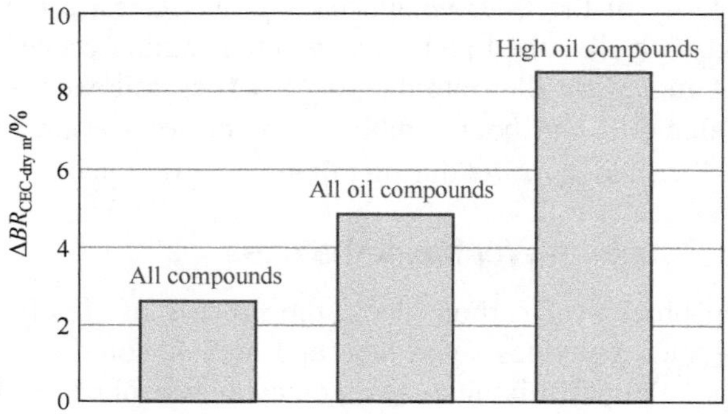


**Figure 5.23** Effect of oil addition on bound rubber contents

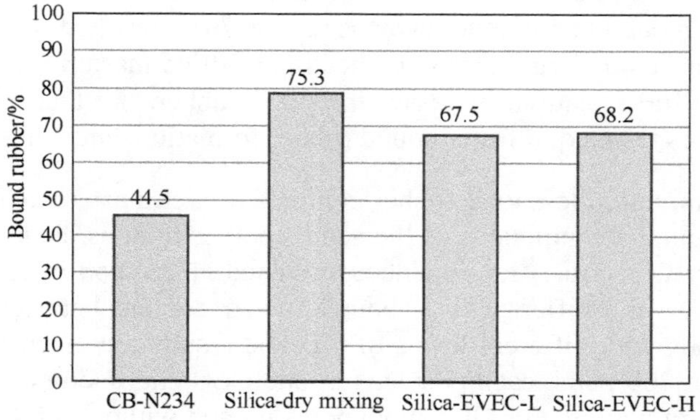


**Figure 5.24** Bound rubber contents of compounds filled with carbon black and silica

As can be seen, the dry mixed compounds filled with silica possess the highest bound rubber, followed by EVEC-H and EVEC-L, and the dry mixed compounds filled with carbon black has the lowest value. Table 5.1 shows a lower bound rubber content for the carbon black compound than silica compounds with a coupling agent. Evidently, the increase of the silica compound's bound rubber is related to increasing the polymer-filler interaction between silica and rubber caused by the coupling reaction. In addition, the polysulfide of the coupling agents cause the rubber molecules to crosslink and generate some gel during the heat treatment of the dry mixing process, thus resulting in an increase in bound rubber.

Compared with the dry mixed silica compound, the lower bound rubber for EVEC may not be caused by a lower extent of coupling reaction; it should be related to less gelation. Most of the silane coupling agent is adsorbed on silica surfaces in EVEC, and the concentration of free coupling agent in the polymer matrix is much lower, resulting in much less polymer gelation. This means that the efficiency of the coupling agent for silica modification is higher in EVEC. This has been further confirmed by the lower viscosity, better processability, and excellent cure characteristics of EVEC, which will be discussed later.

#### 5.1.7.4 Bound Rubber of Fumed Silica-Filled Silicone Rubber

Wang, Morris, and Kutsovsky studied the effects of fumed silica on bound rubber formation in silicone rubber (HCR) with the silicas' surface area from 90 $m^2/g$ to 420 $m^2/g$ at a fixed loading[111]. The bound rubber contents as a function of the surface area of fumed silica are presented in Figure 5.25. There is a critical surface area of fumed silica around 170 $m^2/g$, below which no coherent mass could be obtained during solvent extraction. Bound rubber could not therefore be determined by the conventional test procedure. Beyond the critical surface area, the bound rubber content seems to decrease with increasing silica surface area. Besides polymer-filler interaction, one of the necessary conditions for coherent mass formation in the swollen state of the compounds is that the filler aggregates must be held together by polymer, through polymer bridging or by entanglement of polymer molecules that are attached on the neighboring aggregates which cannot be de-entangled by swelling. For a given polymer system, the ability to form a coherent mass will be determined by the inter-aggregate distance. Equation 3.9 shows that with decreasing surface area of the filler at a given loading, the average interaggregate distance increases. As the surface area decreases, inter-aggregate distance increases to where polymer bridging among most of the aggregates cannot be achieved, and the aggregates and agglomerates are dispersed in the solvent along with the attached polymers. Consequently, no coherent mass would be expected. This explanation is based purely on geometric considerations, and assumes no difference in polymer-filler interactions between the different systems. However, these observations could also be accounted for by the lower polymer-filler interactions arising from the less active surface of the low-surface-area silica[111] and the lower interfacial area in the compounds.

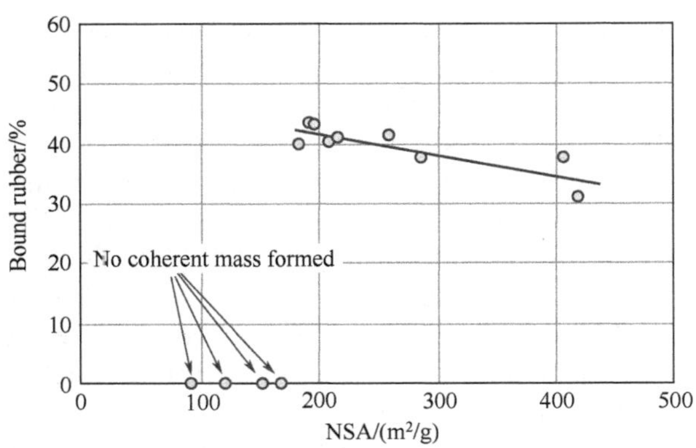


**Figure 5.25** Bound rubber content as a function of surface area of silicas

An alternative explanation for the bound rubber results is based on the fact that the processing aid, which is a low molecular weight, hydroxy-terminated OH-PDMS, is known to reduce polymer-filler interactions as well as filler-filler interactions by silica surface modification. Therefore, at constant dosage of the processing aid, the surface coverage of the silica with low molecular weight PDMS will increase with decreasing surface area, leading to a significant drop in polymer-filler interaction. If this interpretation is reasonable, the critical surface area for coherent mass formation will be reduced either by increasing filler loading, increasing molecular weight of the polymer, or by enhancement of polymer-filler interaction. The former has been demonstrated by a filler loading study, showing for the low-surface-area silica in the same polymer with the same dosage of OH-terminated PDMS used in this study, a coherent mass was obtained at higher loading. But for the high-surface-area silica, the coherent mass may disappear when additional OH-PDMS is added. The effects of polymer-filler interaction and length of polymer chains can be further verified from carbon black–hydrocarbon rubber systems. The critical loading for mass formation increases with decreasing surface area of carbon blacks; and for carbon black–SBR compounds and those with comparable surface area, it is considerably smaller due to their stronger polymer-filler interaction and longer polymer chain length[18]. A typical chain length of SBR 1500 is about three times that of silicone rubber used in this study.

Beyond the critical surface area, the bound rubber content decreases with increasing surface area of the fillers. This is contradictory to what might be expected from the interfacial area between polymer and filler, which is proportional to the surface area of the filler. It is also unexpected from the surface activity of the filler as the dispersive and polar components of surface energies increase with surface area. The lower bound rubber contents for the higher surface area silica may be related to the poorer dispersion, both in macro and micro scales. The poorer macro dispersion, as shown later, will certainly reduce the effective surface area of the filler for bound rubber formation. The effect of the micro scale dispersion, i.e., agglomeration of filler on bound rubber formation can be explained in terms of multiple adsorption of polymer chains. Once an appropriately sized segment of a polymer chain becomes adsorbed to the filler, the whole polymer becomes part of the bound rubber (provided there is enough bridging to form a coherent mass). Further adsorption of the other segments from the same polymer chain takes up space on the filler surface, which could otherwise be used to adsorb other chains, reducing the effectiveness of filler surface area for bound rubber formation. As surface area of the filler increases, filler agglomeration, due to the higher surface energies and shorter inter-aggregate distances (Eq. 3.9), would give a greater probability for multiple adsorption. The development of agglomeration can be demonstrated by the viscosity of the uncured compounds and the strain dependence of dynamic moduli, which will be discussed later. In addition, even in a well-dispersed system, the average inter-aggregate distance decreases with increasing surface area, so the same effect which led to the sudden onset of bound rubber formation at a given surface area, causes the specific bound rubber to decrease as surface area continues to increase. Thus, the maximum bound rubber content on a surface area basis is achieved at just the point where there is enough bridging to form a coherent mass. Similar observations have been made in carbon black-filled rubber systems[18]. The results presented here indicate that, in the fumed silica/PDMS system, the effects of multiple adsorption outweigh the effects of increased interfacial area and surface activity of higher surface area silicas on bound rubber formation.

## 5.2 Viscosity of Filled Compounds

While unfilled compounds generally undergo Newtonian flow at low shear rates, fillers are known to alter the rheological behavior of compounds, resulting not only in highly non-Newtonian flow, but also in comparatively high viscosity (Figure 5.26)[6].

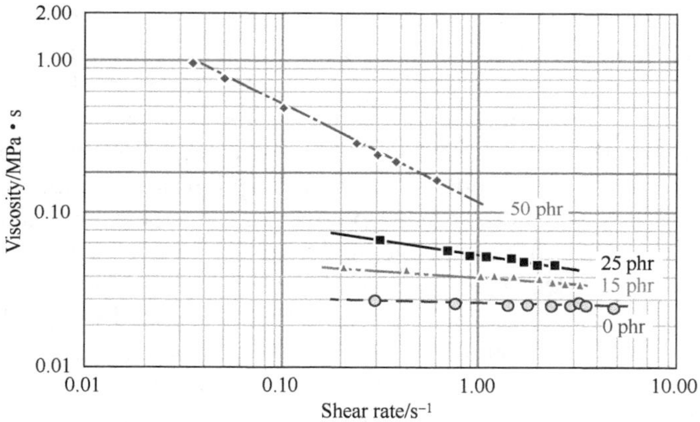


**Figure 5.26** Viscosity *vs.* shear rate for SBR compounds filled with different loading of N220

### 5.2.1 Factors Influencing Viscosity of the Carbon Black-Filled Compounds

When fillers are added to polymers, the rheological properties of compounds can be changed by several mechanisms:

*Hydrodynamic effect:* This is related to the filler loading that reduces the volume fraction of the flow medium and causes a shear strain amplification of the polymer matrix, thus increasing the viscosity of the filled compounds. The viscosity of rubber compounds is known in principle and in practice to depend on the volume fraction of hard particles within the compound. For a spherical filler with low loading where there is no interaction between filler particles as well as between polymer and filler, the filled compounds follow the original Einstein equation to describe the dependence of viscosity on filler volume fraction. As the filler loading is increased, the Guth-Gold equation is applied in which a square term of filler loading is introduced to fit the high viscosity.

*Geometric effect – structure effect:* This is associated with filler structure, i.e., the anisometry and bulkiness of the filler aggregates and the occluded flow medium[61]. It plays a very important role in compound viscosity. In contrast to spherical particles, the anisometric aggregates increase the flow resistance[112]. The occlusion of rubber within the aggregates, in turn, greatly increases the effective hydrodynamic volume of the carbon blacks. Consequently, high-structure carbon blacks always lead to higher compound viscosity. Many have realized that it was more accurate to use an effective volume fraction for the filler than the actual volume fraction in these equations, and attempts have been made to quantify the effective volume fraction.

In fact, Kraus[6], using SBR filled with 50 phr of different carbon blacks with specific surface areas ranging from 14 $m^2/g$ to 164 $m^2/g$, found an empirical correlation between Mooney viscosity and structure as measured by compressed DBP absorption.

*Adsorption of rubber molecules – filler surface effect:* Irrespective of the details of molecular adsorption, this effect seems sensible since once a segment is anchored on the filler surface, the movement of the whole molecule in the flow field should be restricted. Although this reflection is rendered somewhat more complicated by the reduction in molecular weight of the free rubber phase in the matrix, as large molecules are adsorbed preferentially[37], this adsorption process may help to explain the increase in viscosity with decreasing particle size. At the same loading, the larger interfacial area per unit volume of compound and the strong polymer-filler interaction caused by the higher surface energy of small-particle carbon blacks should result in a greater amount of adsorbed rubber[113]. This can be confirmed by the deactivation of carbon blacks; both the graphitization of blacks by heating[32,67], and their surface modification by alkylation[114] lead to a considerable reduction in their capacity for rubber adsorption and hence to a drastic decrease in compound viscosity. On the other hand, as indicated in the above discussion of bound rubber per unit surface, the smaller distance between aggregates for the finer carbon blacks should lead to multi-attachment between aggregates. Such a three-dimensional structure could contribute to the higher viscosity of compounds filled with small-particle carbon blacks. When the polymer molecules are adsorbed on the filler surface, they form bound rubber and immobilized rubber or increase polymer gel in the flow medium[37,115]. This effect is quite complex. The adsorbed rubber surrounding the carbon black surface, which can either be treated as a shell of immobilized rubber[115–117], or as bound rubber, was considered as additional filler volume in order to calculate compound viscosity.

In fact, since bound rubber formation can be considered as a mechanism for increasing the effective size of filler particles, attempts have been made to explain compound viscosity effects based on the size or effective volume increase due to bound rubber[4,38,67]. The effective volume fraction was considered to be the total of the filler and the bound rubber volumes. The fit between experiment and theory for such approaches has been limited. Pliskin and Tokita[38] attempted to improve it by proposing the following equation (Eq. 5.2):

$$\frac{MV}{MV_0} = (1 - \phi_{eff})^{-N} \tag{5.2}$$

where $MV$ is the Mooney torque, $MV_0$ is the Mooney torque of the unfilled polymer, $\phi_{eff}$ is the effective volume fraction (including filler and bound rubber), and $N$ is a parameter that depends on the type of elastomer, the ratio of $\phi_e$ to $\phi$, and the shear rate. $N$ was found to decrease with increasing shear rate. Others have used the peak in Mooney instead of the regular Mooney torque in order to obtain better correlations[4].

Other phenomena are involved, which is why this approach of using an effective volume fraction of filler based on the bound rubber layer does not fully explain viscosity effects in uncured compounds. In fact, in highly loaded compounds, there is a strong tendency for filler flocculation.

*Polymer gelation:* This is the same as that for pure NR hardening during storage[118]. In this mechanism, the condensation of the biochemically formed aldehyde groups along the polymer chains, probably via association with non-rubbers, gives rise to increased viscosity of the rubber. Similar effects may also take place in other polymers during mixing and/or storage when sulfur or a sulfur donor is mixed at higher temperature. In this case, some crosslinking may be formed which gives rise to significantly high viscosity, as described in Section 5.1.7.2.2.

*Agglomeration of the aggregates:* This is another phenomenon involved in augmenting of the viscosity of filled compounds. It has been shown that filler flocculation can have a major effect on viscosity and on the hardening effect in filled rubber compounds. In fact, in highly loaded compounds in which there is strong attraction between filler particles, filler flocculation is probably the main mechanism by which viscosity is increased. Filler flocculation leads to an increase in trapped rubber and therefore an increase in effective filler volume fraction at low deformations. The phenomenon of "crepe hardening", which is well-known is silicone rubber-silica compounds, has been directly related to bound rubber formation[119]. However, the fact that both bound rubber content and compound hardness increase over a similar period of time, does not establish cause and effect. It seems likely that filler agglomeration may be more important than polymer "bridging" in these systems[120]. Polymer adsorption onto the filler surface and filler agglomeration are the processes that occur simultaneously, but polymer adsorption itself cannot account for the large increase in hardness or viscosity that is often observed in silica-filled silicone rubber.

The poor relationship between bound rubber content and uncured compound properties is clearly illustrated by the fact that re-milling of crepe-hardened silicone compounds leads to a rapid decrease in hardness and viscosity, but has little effect on bound rubber content. Similar effects are observed in carbon black-filled hydrocarbon rubbers, although usually to a lesser extent. In a natural rubber masterbatch, for example, it has been shown that both bound rubber content and Mooney viscosity increase substantially upon storage, but after mixing or re-milling, the viscosity is greatly reduced, but bound rubber content remains high[109].

### 5.2.2 Master Curve of Viscosity *vs.* Effective Volume of Carbon Blacks

Wolff and Wang[121] studied the effects of carbon black loading on the properties of uncured compounds and vulcanizates. Two rubbers were selected for the investigation: SBR 1500 (an amorphous polymer) and NR (strain-induced crystallization). For SBR, the carbon black loading varied between 0 and 50 phr for reinforcing (hard) blacks and between 0 and 70 phr for semi-reinforcing (soft) grades, at intervals of 5 phr. In the case of NR, the carbon black content was increased from 0 to 70 phr in steps of 5 phr. For convenience, a relative Mooney viscosity, $ML^{R}$, was used for their work. It is defined as:

$$ML^{R} = \frac{ML_{f}}{ML_{0}} - 1 \tag{5.3}$$

In order to construct a master curve, $ML^{R}$-*vs.*-loading curves of the carbon blacks were shifted horizontally until their curves coincided with that of the reference carbon black (for example, N330). The shift factor, $f$, is defined as the factor by which the real filler volume fraction, $\phi$, has to be multiplied to obtain a curve identical with that of the reference black. This was done visually in a property-*vs.*-$f\phi$ diagram by successively adjusting $f$ on a computer until obtaining optimum superposition of the experimental curve onto the reference one. It is self-evident that the shift factor, $f$, depends on the choice of the reference carbon black. In addition, the measurement error in determining the property-loading curve for the reference black is reflected in the shift factors of all carbon blacks. In order to eliminate this effect, the average shift factor of all carbon blacks after the first shift, i.e., the average property-loading curve, served as the reference point to reshift the data. After the conversion, the data in Figure 5.27 are represented by a Mooney viscosity master curve referenced to the average shift factor, as in Figure 5.28. It should be pointed out that, by this operation, the factor, $f$, serves to relate the effective volume fraction of the carbon black in filled rubber.

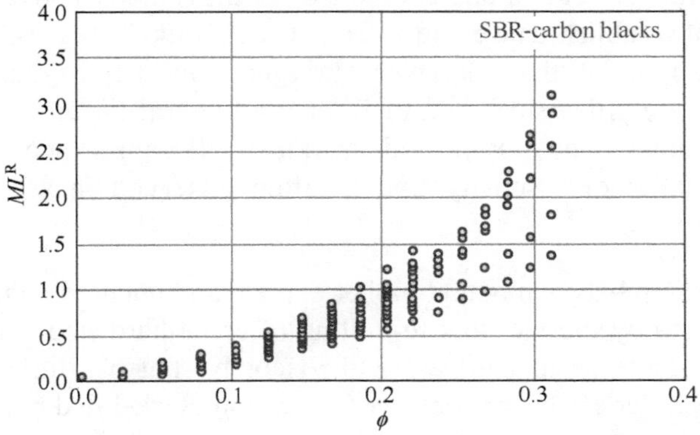


**Figure 5.27** Mooney viscosity *vs.* loading of a variety of carbon blacks for SBR compounds

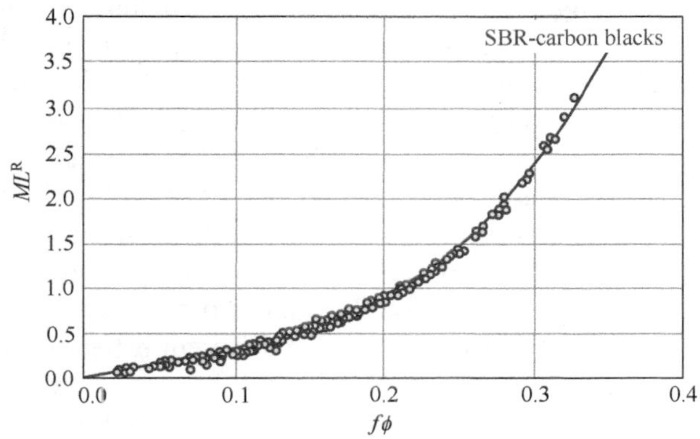


**Figure 5.28** Master curve of Mooney viscosity for SBR compounds

As can be seen from Figure 5.28, the Mooney viscosity increases monotonically with the normalized filler loading. This suggests that by either increasing the real filler loading, or increasing the factor $f$, the Mooney viscosity would increase. In other words, the factor $f\phi$ is the effective filler volume in the compounds for Mooney viscosity and any parameters influencing apparent filler volume are included in factor $f$. This also leads to a conclusion that all furnace carbon blacks evaluated, both reinforcing and semi-reinforcing grades, are governed by the same mechanism for rubber reinforcement. Since the carbon blacks are classified on the basis of their particle size, structure and surface activity, these parameters (and perhaps others) seem to play a role through their influence on the effective filler volume.

Medalia proposed that the structure of carbon blacks, as characterized by DBP absorption, is related to "occluded rubber" in compounds which is shielded against deformation. The effective strainability of a rubber portion, acting as part of the filler rather than of the polymer matrix[61,122], increases the effective volume of the filler beyond the value calculated from the mass and density of the filler. Another structure-related factor is anisometry of filler aggregates. Applying the Einstein-Smallwood equation to Young's modulus, Guth-Gold[123–125] introduced a shape factor for fillers to account for the effect of the anisometry of colloidal particles. In compounds containing fillers with identical surface area and chemical nature but different shape, the modulus increases with increasing anisometry[126]. It was also argued that, depending on the strength of polymer-filler interaction, physical adsorption and/or chemisorption of rubber molecules may take place on the filler surface. This interaction leads to an effective immobilization of elastomer segments[30,34,35,54,115–117,127–132]. Therefore, the filler "surface activity", or surface energy, may be considered as factors influencing the effective volume of the filler.

Moreover, there is a tendency for filler aggregates to associate, forming agglomerates or clusters, especially at high filler loading[27,45,62].

From the above discussion, it is perhaps not too surprising that all effect of filler parameters on rubber properties can be related to changes in the effective filler volume. However, the mechanisms that filler parameters influence the effective filler volume are different, as shown in Table 5.3, which in turn are reflected by different shift factors. The effects of individual filler parameters on the effective filler volume are associated with mechanisms that differ in their strain and temperature dependence; therefore, the shift factors vary from property to property. An investigation of these effects on rubber reinforcement thus becomes a study of the dependence of the shift factors for the rubber properties concerned. Table 5.4 shows the shift factors of all carbon blacks for different rubber properties evaluated.

**Table 5.3** Effect of filler parameter and filler-polymer interaction on effective filler volume

| Filler | Effect | Mechanism | Related filler parameters | Temp.&strain dependence | Effective filler volume |
|---|---|---|---|---|---|
|  | Volume | Hydrodynamic | | | $\phi$ |
|  | Shape | Orientation | Structure | Strain | $\phi_{eff}'$, $\phi_{eff}' > \phi$ |
|  | Branched aggregation | Orientation Occluded rubber | Structure | Strain Temperature | $\phi_{eff}''$, $\phi_{eff}'' > \phi_{eff}'$ |
|  | Polymer-filler interaction | Immobilized rubber Rubber shell | Surface area Surface activity | Temperature (strong) | $\phi_{eff}'''$, $\phi_{eff}''' > \phi_{eff}''$ |
| 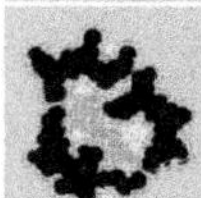 | Interaggregate interaction | Filler networking Trapped rubber | Aggregate size Surface energy | Temperature (strong) Strain (strong) | $\phi_{eff}''''$, $\phi_{eff}'''' > \phi_{eff}'''$ |

**Table 5.4** Shift factors of compound properties

| Black | ML (1+4) | $D_{min}$ | T100 | T200 | T300 | Hard-ness | TS | Tear | Abra-sion | Re-bound | HBU | Swel-ling | tan $\delta$ |
|---|---|---|---|---|---|---|---|---|---|---|---|---|---|
| N110 | 1.1 | 1.05 | 1.02 | 1 | 1.03 | 1.15 | 1.12 | 0.94 | 1.2 | 1.27 | 1.16 | 1.38 | 1.29 |
| N121 | 1.08 | 1.29 | 1.11 | 1.11 | 1.13 | 1.15 | 1.11 | 1.04 | 1.28 | 1.27 | 1.2 | 1.25 | 1.22 |
| N220 | 1.08 | 1.14 | 1.01 | 0.96 | 1.01 | 1.17 | 1.09 | 1.11 | 1.16 | 1.22 | 1.09 | 0.91 | 1.18 |
| N234 | 1.15 | 1.04 | 1.06 | 1.01 | 1.04 | 1.13 | 1.12 | 1.06 | 1.22 | 1.28 | 1.22 | 0.92 | 1.27 |
| N242 | 1.07 | 1.18 | 1.05 | 1.05 | 1.09 | 1.18 | 1.12 | 1.06 | 1.32 | 1.21 | 1 | 1.01 | 1.2 |
| N326n | 0.9 | 0.94 | 0.89 | 0.87 | 0.85 | 0.97 | 0.97 | 0.94 | 0.85 | 1.06 | 0.9 | 0.81 | 1.08 |
| N326t | 0.93 | 0.9 | 0.89 | 0.87 | 0.88 | 0.9 | 0.94 | 0.91 | 0.91 | 1.03 | 0.96 | 0.76 | 1.04 |

Continued

| Black | ML (1+4) | $D_{min}$ | T100 | T200 | T300 | Hard-ness | TS | Tear | Abra-sion | Re-bound | HBU | Swel-ling | tan $\delta$ |
|---|---|---|---|---|---|---|---|---|---|---|---|---|---|
| N330 | 0.83 | 1 | 1 | 1 | 1 | 1 | 1.06 | 1.02 | 1.1 | 1.06 | 1.04 | 0.96 | 1.17 |
| N339 | 1.06 | 1.11 | 1.11 | 1.11 | 1.1 | 1.16 | 1.1 | 0.99 | 1.32 | 1.1 | 1.03 | 1.04 | 1.04 |
| N347 | 0.89 | 1.02 | 1.09 | 1.06 | 1.09 | 1.06 | 1.09 | 1.02 | 1.23 | 1.08 | 1.16 | 1.13 | 1.23 |
| N375 | 1.03 | 1.16 | 1.03 | 1.03 | 1.07 | 1.09 | 1.11 | 1.01 | 1.2 | 1.12 | 1.16 | 0.92 | 1.21 |
| N539 | 0.99 | 0.97 | 1.09 | 1.06 | 1.04 | 0.96 | 0.86 | 0.94 | 0.8 | 0.78 | 0.87 | 1.02 | 0.74 |
| N550 | 1.03 | 1.01 | 1.04 | 0.98 | 0.98 | 0.86 | 0.84 | 1 | 0.76 | 0.81 | 0.84 | 0.79 | 0.89 |
| N660 | 0.88 | 0.75 | 0.95 | 0.94 | 0.92 | 0.88 | 0.89 | 0.93 | 0.75 | 0.76 | - | 0.92 | 0.7 |
| N683 | 1.05 | 0.96 | 1.06 | 1.06 | 1.04 | 1 | 1.01 | 0.97 | 0.86 | 0.81 | 0.85 | 1.07 | 0.76 |
| N762 | 0.79 | 0.6 | 0.85 | 0.84 | 0.78 | 0.79 | 0.76 | 0.82 | 0.63 | 0.65 | 0.74 | 0.79 | 0.65 |
| N765 | 0.99 | 0.83 | 1.08 | 1.04 | 1 | 0.98 | 0.86 | 0.97 | 0.65 | 0.77 | 0.91 | 1.12 | 0.74 |

Shown in Figure 5.29 is the shift factor as a function of surface area and CDBP. For the carbon blacks with low surface area, the $f$ values increase rapidly with CDBP. But with high-surface-area blacks, the $f$ value is higher for low structure blacks. With increasing surface area, $f$ increases at low structure and decreases somewhat at high structure. The highest factor $f$ is obtained for the blacks with either lowest surface area/highest CDBP or highest surface area/lowest structure. The lowest CDBP and surface area blacks yield the lowest shift factor, and there may be several reasons for this. Firstly, according to Eqs. 3.3–3.5, the lower CDBP gives lower occluded rubber, leading to lower effective loading. Secondly, the low surface areas have lower surface energies, which brings about lower bound rubber and less filler flocculation. All these factors lead to low shift factors.

The high extrapolated $f$ factor for the blacks having the high CDBP and low surface area seems to suggest that the effect of occlusion of polymer on viscosity exceeds the effect of high surface area that includes the surface energies, bound rubber, and flocculation of the aggregates.

In contrast, the very high Mooney viscosity of low-CDBP and high-surface-area carbon blacks implies that the higher surface energies, along with more bound rubber of high-surface-area blacks and more developed filler flocculation, have a much greater impact on compound viscosity than the occlusion effect.

### 5.2.3 Viscosity of Silica Compounds

The discussion so far has mainly focused on carbon black-filled compounds. In recent years the application of silica in the rubber industry has increased very rapidly, particularly

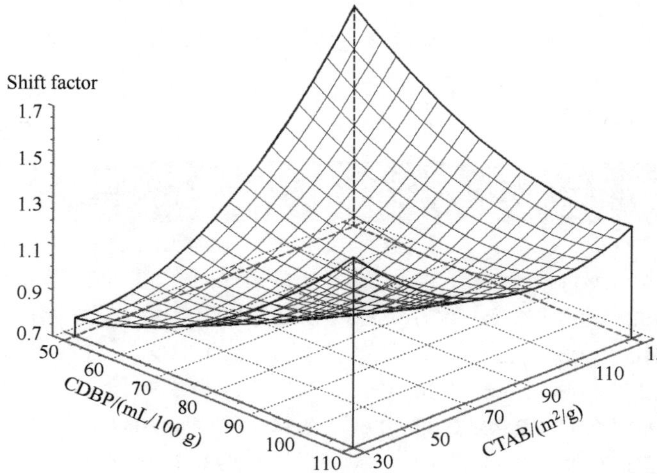


**Figure 5.29** Shift factor *f* for Mooney viscosity as a function of CTAB surface area and CDBP for SBR compounds

for tires. Compared to carbon black, silica is characterized by lower polymer-filler interaction and higher filler-filler interaction in the compound, due to the high-polar component and lower nonpolar component of its surface energy (see Table 2.14). This gives a big difference in processability, including viscosity of the uncured compound, compared to carbon black.

Wolff and Wang studied compound viscosity with precipitated silica P1 (VN2 from Degussa), using carbon black N110 as a reference, as both fillers have a similar surface area.

Figure 5.30 shows the dependence of Mooney viscosity measured at 100°C on filler loading for N110 and P1 filled NR. In order to illustrate the effect of filler loading, the relative viscosity, $ML^R$, was used for the plot. At low concentration, there is no significant difference in $ML^R$ between the two fillers. However, considerable increases in relative viscosity are observed for silica at high loadings. Since the hydrodynamic effect and filler structure effect (occlusion of rubber) of these two fillers are similar, the great difference in viscosity is mainly, if not entirely, due to the agglomeration of silica aggregates, caused by the high $S_f$ value. The abrupt rise of $ML^R$ indicates that agglomeration is so strong that it cannot be dispersed during shearing, leading to great increase of effective volume of filler. With similar concepts, Lee[133] and Sircar[134] defined a critical loading at which an index $L$, i.e., the difference between relative viscosity and relative modulus (modulus of the filled compound divided by that of the gum), increases sharply. They assumed that the concentration had then reached a point where there was not enough rubber to fill all available voids in the filler. This may be true for carbon blacks which, due to their high surface energies, possess a high affinity for hydrocarbon rubbers, resulting in good wetting of the surfaces. This does not seem to be the case for silicas since the strong aggregate-aggregate interaction may cause agglomeration to occur at lower concentrations[59] where there is still more than enough rubber to fill all the voids of the filler.

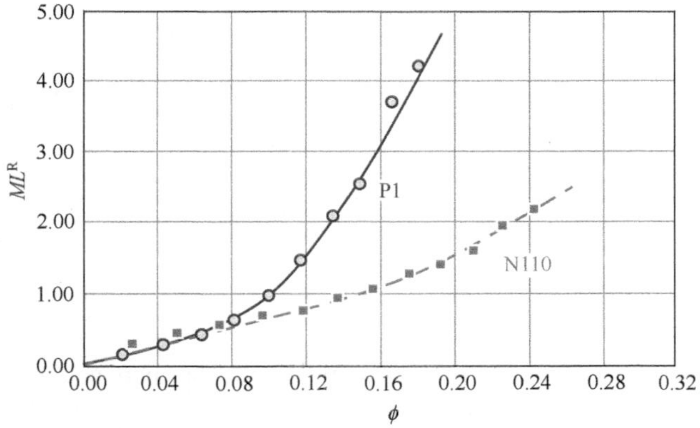


**Figure 5.30** Relative Mooney viscosity as a function of volume fraction for carbon black N110 and silica P1

The minimum torque obtained with a Monsanto Oscillating-Disc Rheometer is also a measure of compound viscosity. Figure 5.31 gives the relative minimum torque, which is defined similarly to relative Mooney viscosity, as a function of loading. The pattern of the change in relative minimum torque is similar to that of relative Mooney viscosity, although the difference in viscosity is greater than that found in Mooney measurements. This is undoubtedly related not only to the different test temperatures, but also to the different patterns of strain. Whereas a biconical rotor oscillates in the Monsanto rheometer, a shearing disc rotates continually in one direction in Mooney viscometer. Therefore, the shear is an infinite strain for the Mooney measurement, and filler agglomerates can be broken down to a greater extent. This reduces the difference in viscosity between the compounds filled with carbon blacks and silicas. Moreover, the effect of the interaction between polymer and filler on compound viscosity should be taken into account. The high adsorption capacity of N110 results in a slightly larger immobilized shell of polymer on the filler surface, thus leading to a somewhat greater hydraulic volume of the filler. Consequently, the strong interaction of N110 with the polymer would partially compensate the difference caused by the less developed agglomeration of the carbon black aggregates. This effect could be more pronounced for high-strain viscosity.

The effect of silanization of silica on the viscosity is also shown in Figure 5.31. Obviously, the decrease in filler-filler interaction, hence filler agglomeration, due to silane modification is responsible for the lower compound viscosity of the modified silicas P1-HDTMS and P1-TESPT in comparison to the silica P1. The compound viscosity of P1-HDTMS is greatly reduced and is in the same range as N110. The difference in viscosity between compounds filled with P1-HDTMS and P1-TESPT may be caused by different surface modification and bound rubber contents.

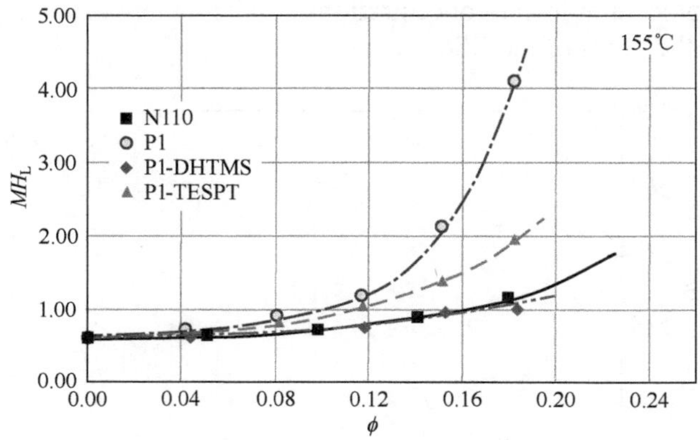


**Figure 5.31** Relative minimum torque of curometer measurement as a function of volume fraction for carbon black N110 and silica P1

The observation above can be further confirmed with a typical passenger tire tread compounds of SSBR/BR blends. Table 5.5 gives the Mooney viscosities measured at 100°C for the compounds filled with the same loading of carbon black N234, silica, and TESPT-modified silica, along with the bound rubber contents determined by toluene extraction.

**Table 5.5** Processability of filled compounds[a]

| Property | Carbon black N234 | Silica | Silica/TESPT |
|---|---|---|---|
| Bound rubber/% | 42 | 27 | 58 |
| Mooney viscosity [ML(1+4),100°C] | 82 | 150 | 59 |
| Die swell/% | 19 | 16 | 28 |
| Green strength/MPa | 0.41 | 0.62 | 0.34 |
| Extrudate appearance, (Garvey die,110°C) | Excellent | Good | Very poor |

a. Basic formulation: SSBR 75, BR 25, filler 80, oil 32.5, TESPT (only for silica) 6.4.

On the other hand, if agglomeration is a main cause of the extremely high viscosity of silica-filled compound at high loading, it would be verified with different polymer due to its high $S_f$. In the definition of $S_f$, the adsorption energy of alkane was taken as representative of a nonpolar or low-polar hydrocarbon rubber (see Eq. 2.198). This would mean that in polar elastomers, silicas will exhibit less filler agglomeration and higher filler-elastomer interaction in comparison to nonpolar or low-polar elastomers[46].

Figure 5.32 depicts the comparison of the Mooney viscosity of carbon black- and silica-filled NR and NBR. In order to normalize the effect of both loading and the differences between polymers and fillers, $\Delta ML$ was used. $\Delta ML$ is defined as the difference between the normalized Mooney viscosity for silica, $ML_{silica}$, and carbon black, $ML_{CB}$, relative to the Mooney viscosity for the unfilled gum, $ML_0$ (Eq. 5.4):

$$\Delta ML = \frac{ML_{silica} - ML_{CB}}{ML_0} \tag{5.4}$$

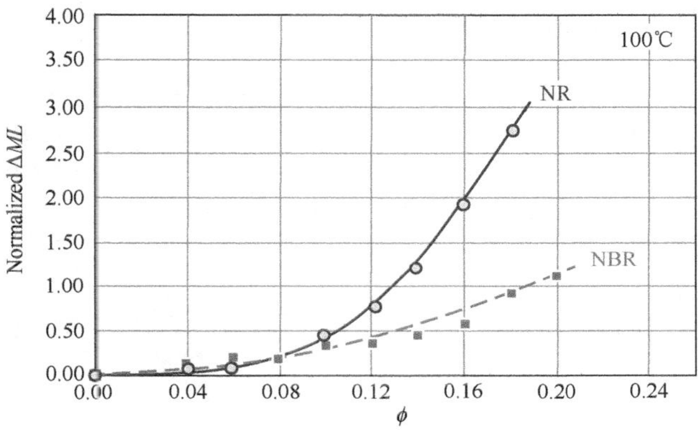


**Figure 5.32** Normalized minimum torque as a function of volume fraction for silica-filled NR and NBR

It is clear that in the case of NR, there is a sharp increase at high loading indicating stronger filler-filler interactions of silica, as mentioned before. In the case of NBR, the increase in $\Delta ML$ is less dramatic with loading, indicating reduced filler-filler interactions as a result of greater polymer-filler interactions. In other words, polymer-filler interaction via the specific interaction between CN-groups of NBR and the silica surface would increase at the expense of filler-filler interactions.

As indicated from the free energy of adsorption[45,76, 135], the interaction between filler surfaces and polar probes increased with the content of polar surface functional groups (also see Figures 2.77 and 2.78). It would therefore be expected that in the case of filled NBR, an increase in acrylonitrile-content (and therefore an increase in polarity) would lead to increased filler-polymer interactions and reduced filler-filler interactions for both silicas and carbon blacks. A reduction of the filler-filler interactions, this being more pronounced for silica, will ultimately lead to a lowering of the viscosity, particularly at low shear stresses. This is demonstrated in Figure 5.33 where the dependence of the minimum torque measured at 160°C for carbon black N110 and silica P1 at 50 phr loading is plotted as a function of the acrylonitrile content. For comparative purposes, compounds with polybutadiene rubber were also used and denoted as 0 acrylonitrile content in the figure. In order to eliminate the varying minimum torque of the gum compound due to the varying ACN content, the relative minimum torque, $MH_L$, defined as the difference in minimum torque between filled and unfilled compounds was applied. Although the viscosity of the silica-filled BR compound is much higher than the corresponding carbon black compound, the rate of decrease of the viscosity with the ACN content is much higher for silica compounds. The point of interception of these two lines is approximately at 50% ACN content. The effects of ACN content can be again interpreted in terms of polymer-filler and filler-filler interaction. As mentioned earlier, the increase in ACN content leads to increased polymer-filler interactions, which in turn results in reduced aggregate-aggregate interactions. As the polymer-filler interactions increase more rapidly for silica with increasing ACN content, the filler agglomeration will be decreased, thus leading to a significantly lower viscosity.

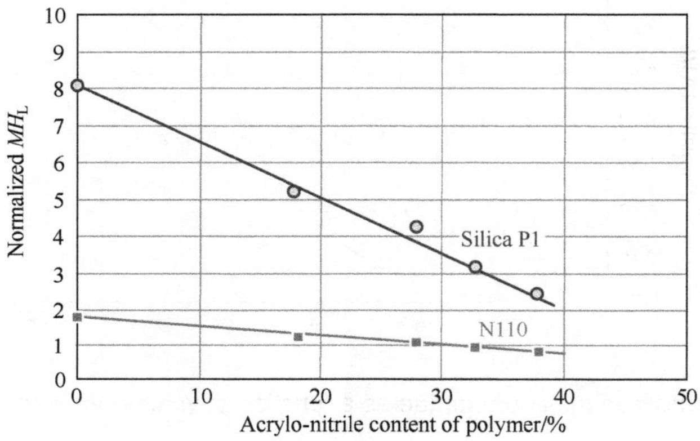


**Figure 5.33** Normalized minimum torque as a function of acrylonitrile content for N110 and silica P1

## 5.2.4 Viscosity Growth – Storage Hardening

The Mooney viscosity for NR/carbon black masterbatch (CEC) produced with continuous liquid phase mixing increases drastically with storage time, as shown in Figure 5.34. The peak Mooney values grow even more rapidly.

As mentioned before, besides the hydrodynamic effect of the filler, three mechanisms are responsible for the hardening effect of CEC: polymer gelation, bound rubber formation and carbon black flocculation.

*Polymer gelation:* At first glance, this phenomenon is similar to what has been observed for pure NR. It is well known that during storage of NR called technically specified rubber (TSR) or standard rubber (SXR, X is the first letter of the production country), the viscosity increases significantly during storage. Also referred to as storage hardening, it is thought to be caused by the condensation of biochemically formed aldehyde groups along the polymer chains, probably via association with non-rubbers, thus giving rise to increased viscosity of the rubber[118]. Chemicals such as hydroxylamine, which reacts with aldehyde groups on the polymer chains via a condensation mechanism, can efficiently inhibit this effect[136]. However, such chemicals'

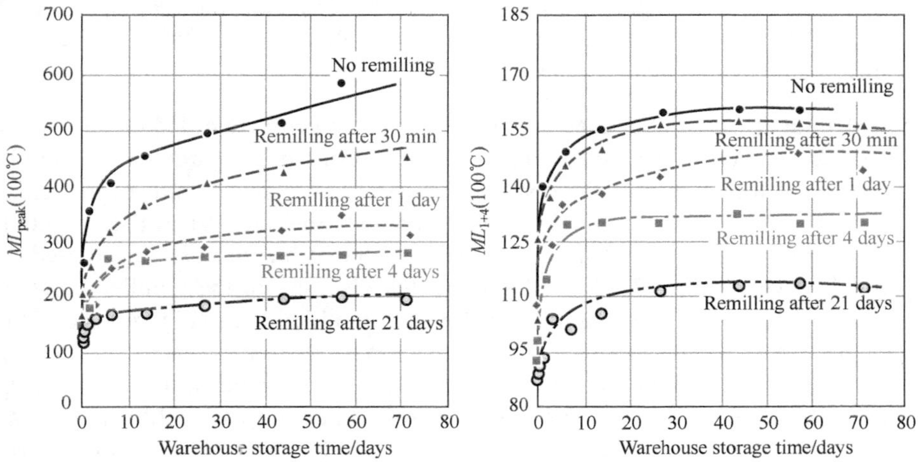


**Figure 5.34** Mooney viscosity of CEC with N220/54 phr as a function of storage time. The *ML* values are calculated from MS values

usefulness in preventing storage hardening of CEC is very limited due to the filler effect.

*Bound rubber formation:* This mechanism is related to the addition of carbon black in the polymer resulting in adsorption of polymer chains on the filler surface.

This process may continue during storage as in the case of dry mixing where the bound rubber increases rapidly at the beginning of storage and reaches equilibrium in about 1 month[137]. The formation of bound rubber will significantly raise the compound viscosity as discussed earlier[115,138] due to the immobilization of the polymer segments on the filler surface. Indeed, for the sample without remilling, while the molecular weight of sol component does not change significantly, the bound rubber increases about 18%. The increase in bound rubber would be mainly due to the continuous adsorption of polymer on filler surface and partly due to the association between rubber molecules and bound rubber, probably through non-rubber substances and/or functional groups on polymer chains formed by polymer oxidation. It is also related to the entanglement between the adsorbed polymer chains and polymer molecules in the matrix (sol). All these mechanisms likely play a role in the hardening effect, but it is unlikely that they could cause the viscosity of the masterbatch can reach such high levels in such a short storage time in the warehouse at ambient temperatures.

*Carbon black flocculation:* Filler flocculation or agglomeration takes place during storage through filler-filler interaction. This is especially true at the very early stage of storage when the bound rubber has not been fully developed. As a consequence, the rubber trapped in agglomerates will increase effective volume fraction, hence the viscosity of the compound, as far as the agglomerates cannot be broken down under an applied stress[9]. This is more obvious for the masterbatch without remilling (see

Figure 5.34). In this case, due to less developed bound rubber, flocculation of the filler aggregates takes place easily from the kinetic point of view. This can be further confirmed by the remilling of the masterbatch.

The hardening effect can be reduced when the masterbatch is remilled on an open mill after storing for a certain period of time. The extent of storage hardening for remilled CEC correlates to the storage time before remilling. For a short storage time, e.g., 30 minutes, there was no obvious reduction in storage hardening. When the masterbatch stored in the warehouse for 21 days was remilled, the at-line viscosity decreased from 160 to 87; however, after 72 days of storage, the viscosity of the remilled compound increased to 113, which was much lower than non-remilled CEC (Figure 5.34). This is understandable as during storage, while the agglomeration has gradually developed, the bound rubber also increases. For CEC rested in the warehouse for 21 days, upon remilling, while the filler agglomerates were broken down, resulting in drastic reduction in Mooney viscosity, the bound rubber content did not decrease significantly. This suggests that the reduced agglomeration rate of the remilled CEC is due to the increased bound rubber formed during storage, which increases the apparent or effective size of the carbon black aggregates. The longer the storage time, the higher is the bound rubber, hence the lower is the hardening rate of the masterbatch (see Section 3.4).

The same hardening effect is observed with EVEC specified to the solution rubber/silica composites produced with continuous liquid phase mixing. The basic formulation of EVEC-H masterbatch comprised SSBR/BR 70/30 phr, silica Newsil 165MP 84 phr, coupling agent TESPT 6.4 phr, and oil 28 phr. As Figure 5.35 shows, the hardening effect is similar to that of CEC, but the viscosity increments are much lower, even after 4 months, and not reaching equilibrium. The lower hardening effect is certainly related to the surface modification of silica with coupling agent. As in the case of CEC, remilling leads to lower viscosity, even somewhat lower than the value of unstored sample.

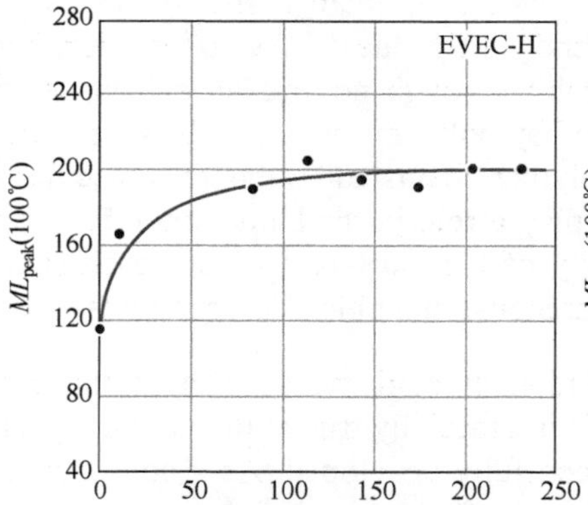


**Figure 5.35** Mooney viscosity of EVEC-H with SSBR/BR/silica/TESPT/oil 70/30/84/6.4/28 phr as a function of storage time[139]

In fact, storage hardening is a common phenomenon which takes place not only for liquid phase mixing compounds but also for the conventional dry-mixed compounds.

## 5.3 Die Swell and Surface Appearance of the Extrudate

In practice, the elastic response of rubber compounds is reflected in their processing behavior in terms of die swell (extrusion shrinkage) and surface roughness of the extrudate.

### 5.3.1 Die Swell of Carbon Black Compounds

Die swell is defined as the ratio between the area of the cross-section of the extrudate and that of the die and is generally greater than unity. This phenomenon, which is associated with elastic recovery, is caused by the incomplete release of long-chain molecules orientated by shear in the die (or capillary) and occurs in the rubber phase alone[140]. It is therefore obvious that, besides the test conditions such as temperature, extrusion rate, and geometrical features of the die, the primary factor influencing the die swell of unfilled elastomers is the entanglement of elastomer molecules, which, in turn, is determined by their molecular weight and molecular weight distribution. The die swell is improved by the addition of carbon blacks, due to the reduction of the elastic component of the compound and the decrease in effective relaxation time[141]. Figure 5.36 illustrates the effect of HAF loading and naphthenic oil on the die swell of SBR compounds[142].

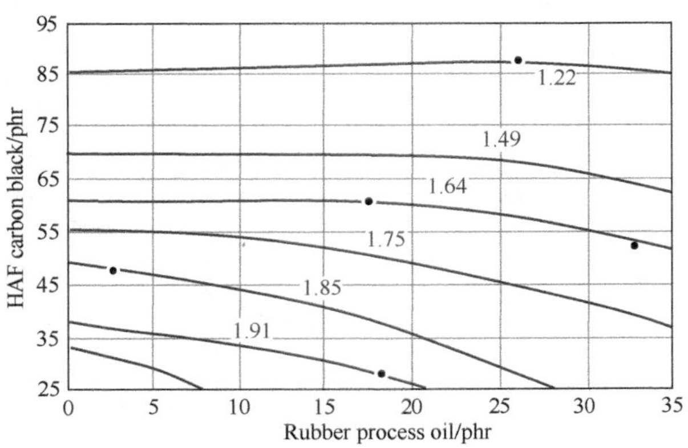


**Figure 5.36** Effects of oil and black on extrudate swelling at 100°C, 286.7 sec$^{-1}$ and *L*/*D* of 22.41

It shows that filler concentration is much more important than processing oil, particularly for highly loaded compounds. Simply by normalizing the die swell with the rubber volume fraction of SBR filled with MT (N990) black, Cotten[143] was able to superimpose all data at different shear stresses onto a single master curve (Figure 5.37). The die swell shows a closed relationship with DBP absorption, indicating that higher DBP gives lower die swell. Indeed, the die swell of filled compounds has been adopted as a measure of carbon black structure[61]. This implies that the rubber occluded within the aggregates is "dead" in terms of elastic memory. This correction is valid for carbon blacks with low, normal, and high structure, ranging from soft to hard grades[61]. Since the correction was only implemented with DBP absorption, it is apparent that, among the fundamental properties of carbon blacks, their structure is the main parameter contributing to die swell, and the effect of surface area seems to be less significant. Cotten[143] showed that the die swell of a given rubber compound extruded through a long capillary depends only on the shear stress at the entrance to the capillary and is independent of other factors such as temperature or viscosity. Changes in molecular weight distribution had little effect on die swell measured in this way; thus, the results should be relatively independent of polymer breakdown during mixing. All these observations suggest that the role played by the polymer properties is determined by the strained molecular chains in the capillary having sufficient relaxation time to reach equilibrium. Comparison of compounds prepared with different carbon blacks showed that the die swell depends on structure and loading but not on surface activity. Dannenberg and Stokes[144] suggested that the effect of loading could be accounted for by basing the calculations on the rubber phase alone, which accords with the aforementioned observation by Cotten (Figure 5.37). Cotten's superimposition plots the shear stresses against the data of experimental die swell divided by the volume fraction of rubber $(1 - \phi)$. In this case, the thermal blacks have a very low DBP, and almost all of the DBP absorption originates from the void volume between aggregates. The shape of thermal black aggregates is basically spherical.

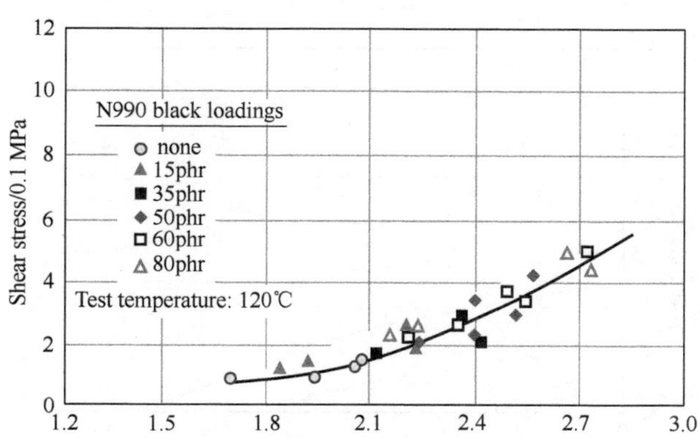


**Figure 5.37** Die swell of SBR filled with MT black at 120°C, normalized with the rubber volume fraction

Assuming the DBP absorption is the main filler property influencing the die-well of rubber compounds, then it is related to the part of the void volume within aggregates rather than the void volume between the aggregates. Based on an analysis of TEM images, Medalia[61] calculated the effective filler volume fraction that is the sum of volume fraction of filler and occluded rubber. In his treatment, an aggregate of carbon black in rubber behaves as an equivalent sphere within which rubber is occluded, where the amount of occluded rubber is equal to the amount of occluded DBP at a "soft ball" end point. He then treated the aggregate with its occluded rubber as an inert filler particle, and based the die swell calculation on the volume fraction of rubber remaining outside the aggregates.

This postulate calculates the volume of occluded rubber directly from the DBP absorption, as follows. For a bulk sample of carbon black, the void ratio, *e*, can be expressed by Eq. 5.5:

$$e = \frac{\begin{array}{c}(\text{Volume of Equivalent Spheres} - \text{Volume of Carbon})\\ +(\text{Volume of Voids between Equivalent Spheres})\end{array}}{\text{Volume of Carbon}} \tag{5.5}$$

so that

$$1+e = \frac{\begin{array}{c}\text{Volume of Equivalent Spheres}\\ +\text{Volume of Voids between Equivalent Spheres}\end{array}}{\text{Volume of Carbon}}. \tag{5.6}$$

From the definition of $\epsilon$ (see Eqs. 2.98–2.101),

$$\epsilon = \frac{\text{Volume of Voids between Equivalent Spheres}}{\text{Volume of Equivalent Spheres}}. \tag{5.7}$$

Thus,

$$\frac{\text{Volume of Equivalent Spheres}}{\text{Volume of Carbon}} = \frac{1+e}{1+\epsilon} \tag{5.8}$$

where the volume of equivalent spheres includes the carbon and the occluded rubber.

The volume fraction of equivalent spheres in the rubber compound is:

$$\phi' = \frac{\text{Volume of Equivalent Spheres}}{\text{Volume of Carbon} + \text{Total Volume of Rubber}} \tag{5.9}$$

and since the volume fraction of carbon black is:

$$\phi = \frac{\text{Volume of Carbon}}{\text{Volume of Carbon} + \text{Total Volume of Rubber}} \tag{5.10}$$

and it is found that:

$$\phi' = \phi\left\{\frac{1+e}{1+\epsilon}\right\}, \tag{5.11}$$

taking $e = 0.02139 \times$ DBP and $\epsilon = 0.46$ [61], one obtains:

$$\phi' = \phi\left\{\frac{1+0.02139DBP}{1.46}\right\} \tag{5.12}$$

Figure 5.38 presents a plot of shear stress as a function of die swell corrected for volume of carbon, for SBR 1500 compounds filled with N220 black. Shown in Figure 5.39 is the data plotted with the die swell normalized with the effective rubber volume fraction from effective volume fraction $\phi'$ calculated with Eq. 5.12. When the filler loading is lower than 35 phr black, Medalia was able to superimpose the data at different shear stresses to form a master curve. The deviations found at 50 phr and 65 phr (Figure 5.39) are not too surprising, in view of the high values of the corrected volume fraction at these loadings (see Table 5.6). As can be seen, the die swell is lower than expected from the correction for occluded rubber above loadings of 35 phr. Medalia[61] attributed the low die swell of the higher loaded compound to interaggregate interaction, in which the anisometry of the aggregates would be expected to play an important part, and where the simple "equivalent sphere" model, from which the occluded rubber is calculated, would not be applicable. The occluded rubber not being totally "dead", and hence not acting as a non-elastic filler, may be another reason for the small reduction in die swell, particularly at high shear stress. Probably, the better and simple interpretation is filler flocculation. This reinforcing black N220 with its higher surface energy, smaller aggregate size, and shorter distance between aggregates should produce stronger and more developed agglomerates, which would significantly increase the effective filler volume fraction, resulting in lower die swell, insofar as the agglomerates cannot be broken down under the experimental stress. This argument will be verified later with silica compounds.

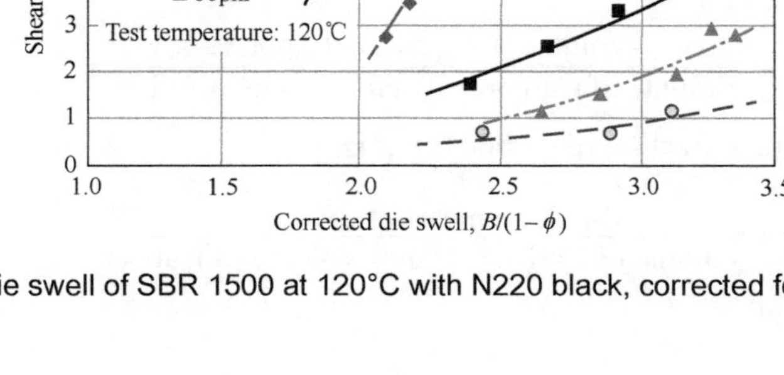


**Figure 5.38** Die swell of SBR 1500 at 120°C with N220 black, corrected for volume of carbon only

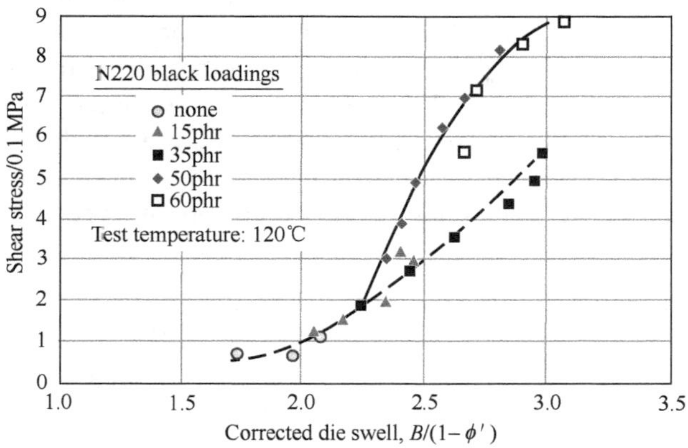


**Figure 5.39** Die swell of SBR 1500 with N220 black at 120°C, corrected for volume of equivalent spheres

Table 5.6 Data for carbon blacks used in die swell measurements

| Carbon black | ASTM classification | CTAB /($m^2$/g) | DBP /(mL/100 g) | Loading /phr | $\phi$ | $\phi'$ |
|---|---|---|---|---|---|---|
| Regal 600 | N219 | | 85 | 35 | 0.15 | 0.29 |
| Sterling 105 | N683 | 32 | 135.5 | 35 | 0.15 | 0.4 |
| Vulcan 6 | N220 | 110 | 115.2 | 15 | 0.07 | 0.166 |
| | | | | 35 | 0.15 | 0.356 |
| | | | | 50 | 0.202 | 0.479 |
| | | | | 65 | 0.247 | 0.585 |
| Vulcan XC-72 | N472 | 145 | 187 | 35 | 0.15 | 0.514 |

Figure 5.40 showes that the effects of carbon black morphology at a loading of 35 phr can be accounted for (over a certain range) by the simple correction factor for occluded rubber. For most carbon blacks (apart from N472), this superposition is quite satisfactory considering the simplicity of the model developed about the effective volume fraction of polymer, $1 - \phi'$, compared with the complexity of the physical processes underlying both die swell and DBP absorption. According to Medalia, the discrepancy between the data for the ultra-high structure N472 black and those for the other three blacks (Figure 5.40) is due to a small overcorrection that could be accounted for by assuming that the true DBP absorption is 10% less than the measured value[61]. The DBP titration does not give a sharp end point with

ultra-high structure blacks, so that this may be within experimental error for this black. Another source of this overcorrection may be found in the breakdown of carbon black aggregates on mastication in rubber. The DBP absorption values in Table 5.6 were determined on the raw black and do not allow for any breakdown of carbon black structure which may have taken place when incorporating the black in rubber and the subsequent remilling just prior to the die swell measurements. Ban and Hess[24] confirmed this by demonstrating that the breakdown of aggregates is significant and that this phenomenon is even more pronounced in the case of high-structure carbon blacks.

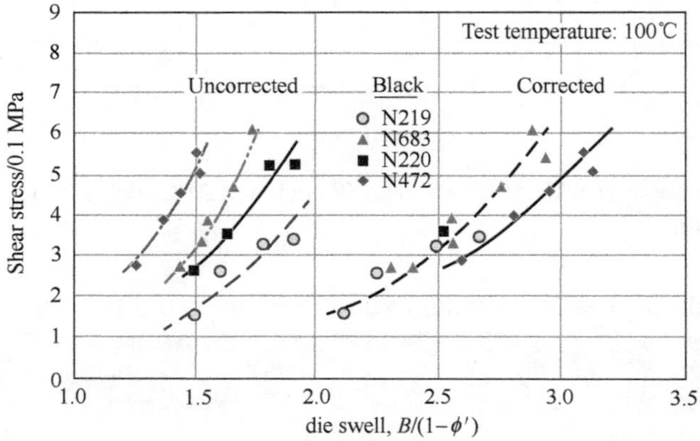


**Figure 5.40** Die swell of SBR 1500°C at 100°C with various blacks (35 phr). Uncorrected curves and curves corrected for volume of equivalent spheres are shown[61]

## 5.3.2 Die Swell of Silica Compounds

The data in Table 5.5 show the die swell and extrudate appearances for passenger tire tread compounds containing silica, TESPT-modified silica, and carbon black[66]. The die swell was measured with a Plasti-Corder extruder operating at extrusion output rate of 87 mL/min and die temperature of 70°C (die diameter of 4.75 mm). The die swell is lower for the unmodified-silica-filled compounds and highest for modified silica. Apparently, the high die swell of modified-silica compounds cannot be attributed to the silica structure. It may be due to its high bound rubber content related to chemical reaction via the coupling agent. In this case, the filler particles behave like multiple crosslinks, increasing the elastic memory of the polymer matrix. It may also be related to filler agglomeration, as the agglomerates are treated as highly structured fillers and the trapped rubber in the agglomerates act as occluded rubber so long as they do not break down under the shear rate and temperature used for extrusion. Therefore, the lower die swell of the silica compound is expected from its highly developed filler agglomeration. The more developed filler agglomeration with less bound rubber content of the unmodified silica compound leads to a much lower die swell compared to TESPT-silica compounds.

### 5.3.3 Extrudate Appearance

The surface appearance of an extrudate is associated with elastic recovery due to the incomplete stress release caused when long chain molecules are orientated by shear in the die. The distortion of the extrudate increases as the unreleased stress increases, and in some cases may even lead to the complete fracture of the extrudate (melt fracture).

For a given compound, high temperature, which effectively reduces the relaxation time of the polymer, and lower extrusion rate, which allows the polymer enough time to relax in the die, both work to reduce the elastic recovery of the extrudate, giving a smoother surface. In terms of the polymer, the primary factor influencing the surface roughness of unfilled elastomers is the entanglement of elastomer molecules, which, in turn, is determined by their molecular weight and its distribution. Since the elastic memory occurs in the rubber phase alone[140], the surface roughness of the extrudate for any given polymer is generally improved by the addition of filler, due to the reduction of the elastic component of the compound and the decrease in effective relaxation time[141]. Fillers with higher structure will substantially increase the effective volume of the fillers[61,143], leading to shorter relaxation times. In addition, filler agglomeration will further reduce elastic memory as long as the agglomerates are not broken up under the shear stress during extrusion[66]. On the other hand, filler aggregates may serve as multi-functional crosslinks due to the adsorption of polymer molecules, causing an increase in elasticity of the compounds.

The effect of N330 carbon black and processing (naphthenic) oil on the melt fracture of SBR compounds is illustrated in Figure 5.41 by a contour diagram[142]. The compound with the highest carbon black loading and the least amount of oil yields the best appearance of the extrudate surface. As the shear rate increases, the boundary of the fracture region shifts toward the upper left-hand corner.

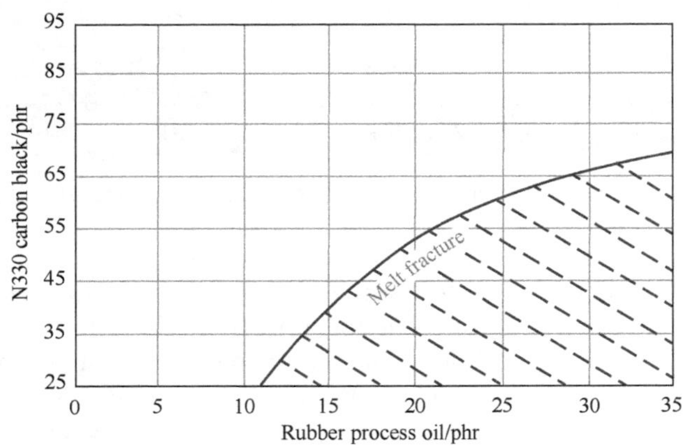


**Figure 5.41** Effect of carbon black and oil on extrudate surface roughness of SBR 1500 compound at 100°C and 286.7 $s^{-1}$[142]

The surface modification of silica with TESPT leads to less filler agglomeration and higher bound rubber, which, as will be discussed later, is most desirable for tire tread compounds in terms of tire performance. In addition, during mixing of the silica/TESPT compounds, as mentioned before, polymer gelation may take place due to pre-crosslinking of polymer. Consequently, the effective filler loading would be significantly reduced and elasticity of the polymer matrix would be increased. This yields very poor appearance of the extrudate surface (see Table 5.5). The lower die swell and lower surface roughness of the extrudate for the unmodified silica compound is related to its more developed filler agglomeration and lower bound rubber content. Therefore, one of the deficiencies of silica tread compounds is their poorer processability related to extrusion as the silane coupling agent has to be used to improve polymer-filler interaction and to depress filler-filler interaction for improving abrasion resistance, rolling resistance, and wet skid resistance as well.

Compared with the conventional silica compounds prepared by dry mixing, the die swell of EVEC-L in the extrusion process is much lower and its surface is very smooth. Figure 5.42 shows the extruded treads of passenger tires. The formulations of the compounds are exactly the same: SSBR/BR-70/30; silica 78, TESPT 6.4. The better processability of the EVEC-L could be related to the low bound rubber, low viscosity, and lack of pre-crosslinking of rubber molecules in the compounds, even though the flocculation of the filler aggregates in the compound is extremely low[145].

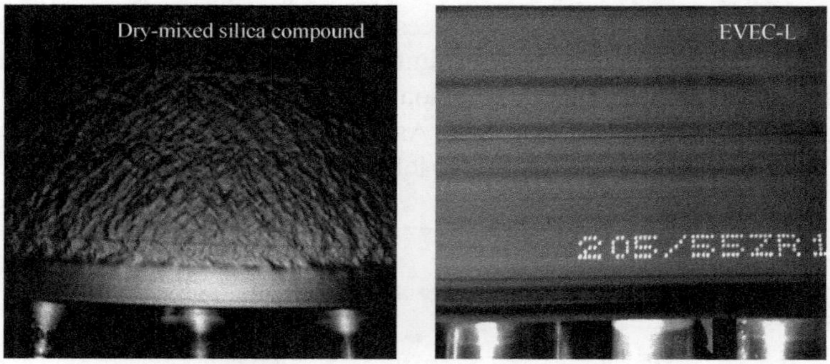


**Figure 5.42** Surface roughness of extrudates of dry-mixed compound and EVEC-L with the same formulation

Since the agglomeration of silica in EVEC-L is much less than that in the dry-mixed compound, in this case, it seems that the effect of filler flocculation on surface roughness and die swell is much less than that of polymer elasticity originating from bound rubber, especially polymer gel. As discussed before, for EVEC, there is almost no free coupling agent in the polymer matrix, therefore, no polymer gelation may take place. This, with its low bound rubber content, would result in higher plasticity of the polymer phase in EVEC-L which allows the raw tread to have lower die swell and smooth surface.

# 5.4 Green Strength

Green strength is the resistance of the uncured rubber to deformation and fracture. It is a very important property during the building of rubber products. For example, in the construction of radial-ply tires, the uncured compounds between the cords in the carcass may be subject to extensions of up to three times their original dimension during the building process[146]. Moreover, a high green strength is necessary to prevent the uncured tires from creep and distortion prior to molding.

## 5.4.1 Effect of Polymers

NR and IR: Generally, NR has high green strength which has been attributed to the rapid crystallization upon strain. Compared with synthetic polyisoprene (IR), the crystallization rate of NR is much higher[147,148]. One of the reasons for the different crystallization is its micro-molecular structure. While IR contains about 93%–98% *cis*-polymer, NR contains 100%[149,150] *cis*-polymer even though some researchers reported that it has 1%–2% 3,4-units[151,152]. It has been proposed that the rapid cold crystallization of NR is closely related to the presence of a long sequence of *cis*-1,4 isoprene units[153] and free fatty acids[154]. However, when the NR is coagulated from fresh latex, its green strength is as low as that of IR[155], which breaks on continuous straining without further increase in stress.

Furthermore, give these factors, with NR sample SMR and IR 80, Cameron and McGill reported that the yield stress of the unfilled NR is even much lower than that of IR (Figure 5.43)[150]. It seems that, beyond the molecular structure, other factors may also increase green strength of rubber either through changing the strain crystallization, increasing apparent molecular weight of polymer, hence entanglements, or through introducing temporary or permanent crosslinks in the polymer matrix. In these regards,

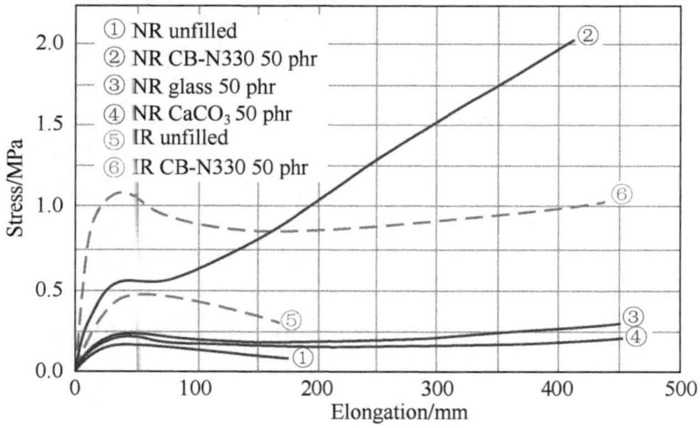


**Figure 5.43** Stress-strain curves measured at room temperature and a strain rate of 500mm/min for unfilled and filled NR and IR[150]

the high portion of non-rubber components, i.e., proteins and phospholipids, and some polar groups on the polymer chains may play an important role in improving green strength[156,157].

As shown in Figure 5.43, for unfilled NR and IR the stress increases rapidly with strain at low strain and then necks down and breaks at relatively lower elongation, yielding at elongations around 35%. In the case of filled NR and IR with carbon black N330, after the yield point, the stress continuously increases with strain, and breaks at elongation about 400%. At very low elongation, the modulus of filled IR is much higher than that of NR. The yield stress is also much higher even though the yield strain is somewhat lower. While the higher yield stress and modulus at low strain is certainly related to the higher yield stress of the unfilled IR, more mechanisms are likely to be involved. Among others, filler agglomeration may, in certain degree, be responsible for the difference in their yield properties between NR and IR.

It is known that NR contains some non-rubber substances that originate from latex and the ammonia preservation. These substances, such as proteins, phospholipids and ester-like impurities serve as stabilizers for the latex. When the filler is incorporated in NR, these substances may be adsorbed on the filler surface, reducing the surface energy of the filler, hence less agglomeration. In the case of IR which is produced with solution polymerization, there are no surfactants in the polymer, resulting in somewhat more developed filler agglomeration. Therefore, at low strain, the modulus is higher due to the higher amount of trapped rubber in the agglomerates, leading to higher apparent filler loading.

Beyond yield points, negative slopes are observed upon increasing elongation. This is caused by breakdown of the agglomerates as well as some entanglements or associates. At the same time, strain-induced crystallization increases in the polymer matrix, leading to increased modulus. When these two effects on the stress are equal, a minimum stress appears after which the strain crystallization effect dominates. As a result, the stress increases with increasing elongation. As discussed above, compared with IR, the NR has higher strain-crystallization rate and therefore the strain hardening effect is more pronounced. Indeed, the stresses at about 400% elongations for NR and IR are 1.95 MPa and 0.98 MPa, respectively, about a factor of two.

The effect of carbon black on the stress-strain performance of the uncured compounds is related to the basic effects of the filler in polymer, as described in Chapter 3. This can be demonstrated by non-reinforcing fillers. When $CaCO_3$ or glass powder is used as the filler in NR stocks, while the elongations at break are as high as that of the carbon black-filled compound, the strain moduli and yield stresses are considerably low. This is certainly due to the very weak polymer-filler interaction and much lower surface area which leads to much less bound rubber and filler agglomeration.

NR and SBR: Figure 5.44 shows the stress-strain curves for typical NR and emulsion polymerized SBR (ESBR) uncured compounds that are filled with 50 phr of carbon black N330. Similar to the data presented in Figure 5.43, after the yield point, the stress of NR stock increases progressively with elongation, although the test methods are different[158]. With regard to the uncured compound of ESBR, the stress increased very sharply at low strain, giving very high modulus. After the yield point, however, it necks down rapidly and breaks at low elongation. In this case the green properties of the compounds can be represented by yield stress and yield strain.

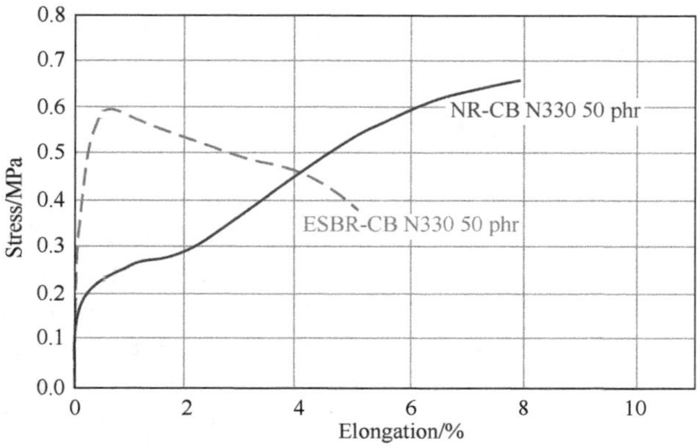


**Figure 5.44** Stress-strain curves measured at room temperature and a strain rate of 500 mm/min for unfilled and filled NR and ESBR, strain rate-50/min[158]

Multiple mechanisms may be responsible for the observed difference between the stress-strain curves of NR and ESBR compounds. As mentioned before, besides strain-induced crystallization of the polymers, especially at high elongation, the green properties are generally affected by the polymer coherent force, filler agglomeration, polymer chain entanglements and association. They are also closely related to bound rubber formation and gelation.

The formation of bound rubber in carbon black-filled polymer has been discussed in this chapter and Chapter 3 in detail. Due to the polymer-filler interaction, a small amount of rubber is truly immobilized in the vicinity of the filler surface, forming a rubber shell. The rest of the bound rubber chains are constrained but not immobilized. Thus, according to Cameron and McGill[150], the polymer chains of the bound rubber no longer act independently, but they have a degree of coherence via their association with the carbon black aggregates. This coherence assures that the influence of the bound rubber–carbon black entity is much greater than that of isolated molecules, though each bound rubber–carbon black entity is still independent. Few carbon black particles are linked by single polymer molecules which form part of the bound rubber on two or more particles. As such, a three-dimensional network structure will result if the bound rubber–carbon black entities can be linked together. This is especially true for NR as in its pure polymer there is always 15%–25% gel which would be joined with the bound rubber, forming a three-dimensional network, even though the junction points are far fewer than in crosslinked rubber. On the contrary, SBR is a gel-free polymer, so forming a three-dimensional polymer network structure is very difficult.

Compared with NR, ESBR possesses higher intermolecular interaction caused by its aromatic segments. It is also a gel-free polymer which would lead to more developed filler agglomerates. Both of these factors are favorable to augment the higher modulus at low strain. On the other hand, ESBR is an amorphous rubber that is incapable of strain crystallization. This, along with its relatively lower molecular weight, short molecular chains and less entanglements explains the negative slope after the yield point.

## 5.4.2 Effect of Filler Properties

Filler Morphology: As shown in the last section, fillers have an important effect on the stress-strain performance of uncured compounds. Besides filler loading, filler morphology also significantly influences the green strength of the stocks. Cho and Hamed[159] compared filled SBR 1502 rubber to the pure gum compound. Compounds filled with 60 phr N990, a low surface area and low structure black, produced a higher green strength but lower yield strain. When the high surface area and high structure black N110 was used, both green strength and yield strain increased drastically, giving a yield strain much higher than that of the gum compounds[152].

Recently, with a series of carbon blacks having different surface areas ranging from 8 $m^2/g$ to 143 $m^2/g$ and different structure with DBP ranging from 43 mL/100 g to 127 mL/100 g, Zhang et al.[160] carried out a systematic study about the effects of filler morphology on the green strength of black-filled SSBR compounds. The stress-strain curves of stocks measured at strain rates from 10 to 200 mm/min are shown in Figure 5.45 for four compounds filled with four typical carbon blacks. For all compounds, the stress increases with strain, passes a maximum at the yield points and then decreases monotonically until the specimen breaks. The fact that there is no upturn in stress-strain response after the yield point is certainly due to the absence of strain-induced crystallization as solution-polymerized SBR is also an amorphous rubber. The initial modulus, the yield stress, and yield strain all increase with increasing strain rates.

The effects of carbon black surface area and structure are presented in Figure 5.46 and Figure 5.47, respectively for different strain rates. With an increasing carbon black surface area, both the yield stress (i.e., green strength) and the yield strain increase. However, as the structure increases and the green strength improves, the yield strain gradually declines. Several reasons may be responsible for the improvement of green strength when the stocks are filled with high-surface-area blacks. The higher surface energies[45], larger interfacial area, higher bound rubber[18], as well as shorter interaggregate distance[62] would help the formation of polymer network through chain entanglements and agglomeration of filler particles. With increasing filler structure, the effective filler volume would increase which, on one hand, increases the yield stress, and on the other hand, reduces the effective volume fraction of polymer, resulting in lower yield strain.

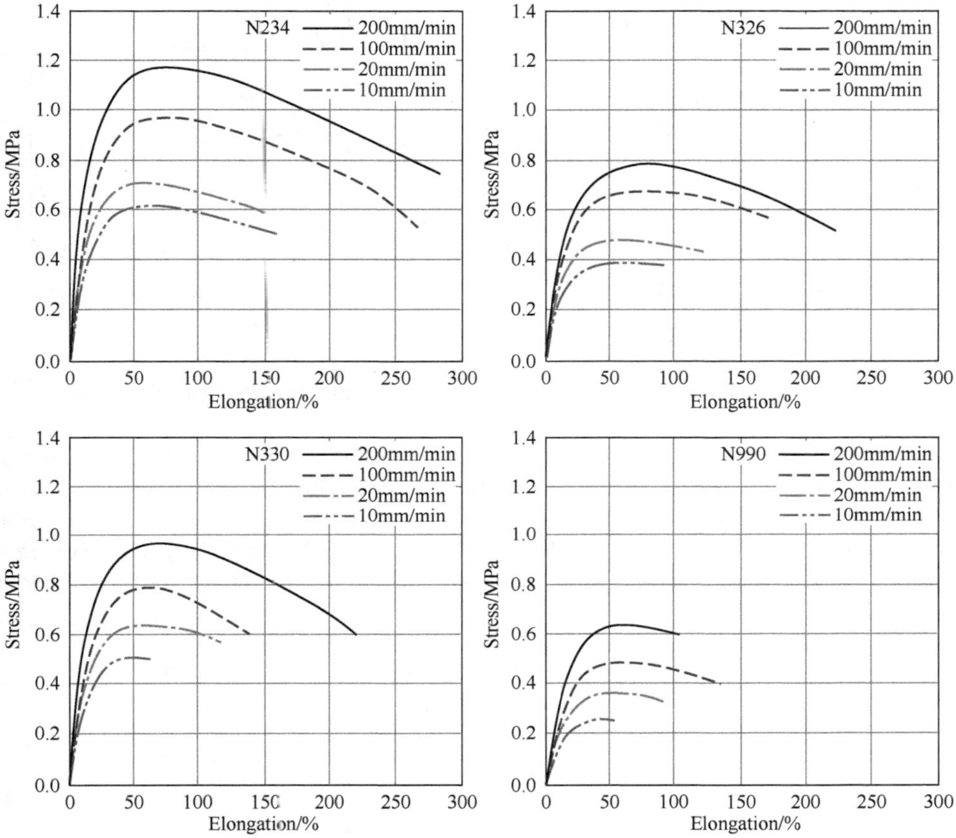


**Figure 5.45** Stress-strain curves for SSBR filled with a variety of carbon blacks

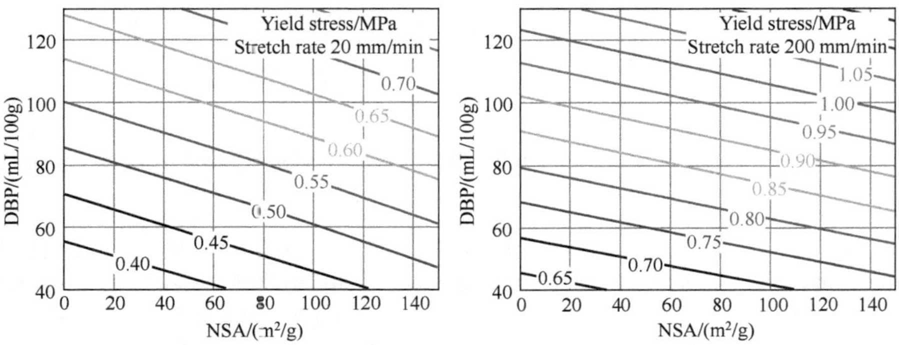


**Figure 5.46** Effect of surface area and structure on yield stress for carbon black-filled SSBR

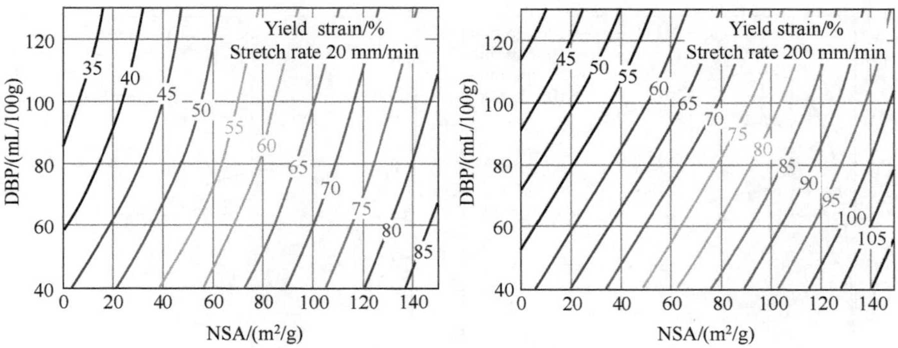


**Figure 5.47** Effect of surface area and structure on yield strain for carbon black-filled SSBR

Filler Surface Energies: It was found that, besides filler morphology, the filler surface energy is another important factor influencing green strength[25]. The green strengths of passenger tire tread compounds based on a blend of SBR and BR are shown in Table 5.5[25]. Similar results were obtained with SBR filled with 50 phr fillers. Presented in Figure 5.48 and Figure 5.49 are the green strength and yield strain of filled SBR as functions of the strain rates[66].

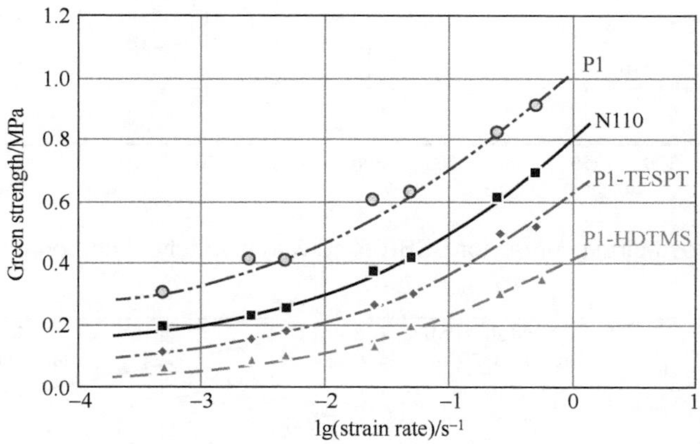


**Figure 5.48** Green strength as a function of the strain rate for SBR filled with a variety of fillers (50 phr, room temperature)

In both cases, compared with surface modified silica compounds, the unmodified silica and carbon black gives higher green strength (yield stress), but the yield strain at which the yield stress occurs is totally different. Whereas the highest yield strain is for carbon black, the lowest is for silica. It seems that high green strength of carbon black-filled rubber is related to higher polymer-filler interaction, while the high green strength of the silica-filled compound originates from strong filler agglomeration that is rigid, brittle, and characterized by very low yield strain. The very low green strength and yield strain of the compound filled with the HDTMS-modified silica are, of course, associated with low polymer-filler and filler-filler interactions, i.e., the low specific and

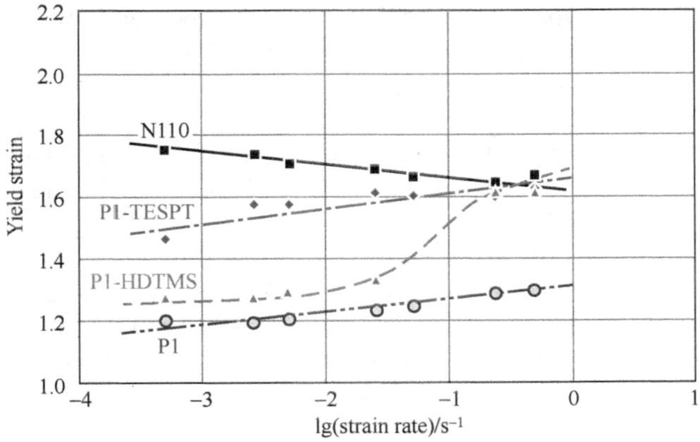


**Figure 5.49** Yield strain as a function of the strain rate for SBR filled with a variety of fillers (50 phr, room temperature)

dispersive components of surface energy of this silica. Compared with HDTMS modified silica compound, the higher green strength and yield strain of the compound containing silica modified with the bifunctional silane TESPT may be explained by a high bound rubber content and pre-crosslinking between the silica surface and polymer during mixing.

## References


[1] Fielding J H. Impact Resilience in Testing Channel Black. *Ind. Eng. Chem.*, 1937, 29: 880.

[2] Twiss D F. *J. Soc. Chem. Ind,*, 1925, 44: 106.

[3] Sweitzer C W, Goodrich W C, Burgess, K A. *Rubber Age*, 1949, 65: 651.

[4] Leblanc J L. Rubber-filler Interactions And Rheological Properties in Filled Compounds. *Prog. Polym. Sci.*, 2002, 27: 627.

[5] Karsek L. *Inter. Poly. Sci. Technol.*, 1994, 21: T/35.

[6] Kraus G. Interactions of Elastomers and Reinforcing Fillers. *Rubber Chem. Technol.*, 1965, 38: 1070.

[7] Dannenberg E M. Bound Rubber and Carbon Black Reinforcement. *Rubber Chem. Technol.*, 1986, 59: 512.

[8] Stickney P B, Falb R D. Carbon Black-Rubber Interactions and Bound Rubber. *Rubber Chem. Technol.*, 1964, 37: 1299.

[9] Wang M-J. Effect of Polymer-Filler and Filler-Filler Interactions on Dynamic Properties of Filled Vulcanizates. *Rubber Chem. Technol.*, 1998, 71: 520.

[10] Gessler A M, Hess W M, Medalia, A I. *Plast. Rubb. Process.*, 1978, 3: 141.

[11] Southwart D W. Comparison of Bound Rubber and Swelling in Silicone Rubber/Silica Mixes And in Silicone Rubber Vulcanizates. *Polymer*, 1976, 17: 147.

[12] Leblanc J L, Stragliati B. An Extraction Kinetics Method to Study the Morphology of Carbon Black Filled Rubber Compounds. *J. Appl. Poly. Sci.*, 1997, 63: 959.

[13] Papirer E, Voet A, Given P H. Transfer of Labeled Hydrogen Between Elastomers And Carbon Black. *Rubber Chem. Technol.*, 1969, 42: 1200.

[14] Watson W F. Combination of Rubber and Carbon Black on Cold Milling. *Ind. Eng. Chem.*, 1955, 47: 1281.

[15] Roychoudhury A, De P P. Elastomer-Carbon Black Interaction: Influence of Elastomer Chemical Structure and Carbon Black Surface Chemistry on Bound Rubber Formation. *J. Appl. Poly. Sci.*, 1995, 55: 9.

[16] Dessewffy O. *Magyar Kemiai Folyoiral*, 1961, 67: 259.

[17] Cotton G R. Influence of Carbon Black on Processability of Rubber Stocks. I. Bound Rubber Formation. *Rubber. Chem. Technol.*, 1975, 48: 548.

[18] Wolff S, Wang M-J, Tan E H, Filler-Elastomer Interactions. Part VII. Study on Bound Rubber. *Rubber. Chem. Technol.*, 1993, 66: 163.

[19] Kida N, Ito M, Yatsuyanagi F, et al. Studies on the Structure and Formation Mechanism of Carbon Gel in the Carbon Black Filled Polyisoprene Rubber Composite. *J. Appl. Poly. Sci.*, 1996, 61: 1345.

[20] Dessewffy O. Dependence of Bound Rubber on Concentration of Filler and on Temperature I. *Rubber. Chem. Technol.*, 1962, 35: 599.

[21] Frisch H L, Simha R. Statistical Mechanics of Flexible High Polymers at Surfaces. *J. Chem. Phys.*, 1957, 27: 702.

[22] Kraus G, Dugone J. Adsorption of Elastomers on Carbon Black. *Ind. Eng. Chem.*, 1955, 47: 1809.

[23] Ban L L, Hess W M, Papazian L A. New Studies of Carbon-Rubber Gel. *Rubber Chem. Technol.*, 1974, 47: 858.

[24] Ban L L, Hess W M. *Les Interactions Entre les Élastomères et les Surfaces Solides Ayant Une Action Renforçante, Colloques Internationaux du CNRS, No. 231*, Paris, 1975.

[25] Wolff S, Wang M-J, Tan, E H. Surface Energy of Fillers and Its Effect on Rubber Reinforcement Part 2. *Kautsch. Gummi. Kunstst.*, 1994, 47: 873.

[26] Polmanteer K E, Lentz, C W. Reinforcement Studies-Effect of Silica Structure on Properties and Crosslink Density. *Rubber Chem. Technol.*, 1975, 48: 795.

[27] Wolff S, Wang M-J. Filler-Elastomer Interactions. Part IV. The Effect of the Surface Energies of Fillers on Elastomer Reinforcement. *Rubber Chem. Technol.*, 1992, 65: 329.

[28] Li Y F, Xia Y X, Xu D P, et al. Surface Reaction of Particulate Silica with Polydimethylsiloxanes. *J. Polym. Sci.: Polym. Chem. Ed.*, 1981, 19: 3069.

[29] Kaufman S, Slichter W P, Davis, D D. Nuclear Magnetic Resonance Study of Rubber-Carbon Black Interactions. *J. Polym. Sci. Part A-2*, 1971, 9: 829.

[30] O'Brien J, Caschell E, Wardell G E, et al. An NMR Investigation of the Interaction Between Carbon Black and *cis*-Polyutadiene. *Rubber Chem. Technol.*, 1977, 50: 747.

[31] Luchow H, Breier E, Gronski W, meeting of the *Rubber Division, ACS*, Anaheim, 1997.

[32] Kraus G. Reinforcement of Elastomers of Carbon Black. *Rubber. Chem. Technol.*, 1978, 51: 297.

[33] Kraus G, Gruver, J T. Thermal Expansion, Free Volume, and Molecular Mobility in a Carbon Black-Filled Elastomer. *J. Polym. Sci. Part A-2*, 1970, 8: 571.

[34] Kenny J C, McBrierty V J, Rigbi Z, et al. Carbon Black Filled Natural Rubber. 1. Structural Investigations. *Macromolecules*, 1991, 24: 436.

[35] O'Brien J, Caschell E, Wardell G E et al. An NMR Investigation of the Interaction between Carbon Black and cis-Polybutadiene. *Macromolecules*, 1976, 9: 653.

[36] Litvinov V M, Barthel H, Weis J. Structure of a PDMS Layer Grafted onto a Silica Surface Studied by Means of DSC and Solid-State NMR. *Macromolecules*, 2002, 35: 4356.

[37] Kraus G, Gruver J T. Molecular Weight Effects in Adsorption of Rubbers on Carbon Black. *Rubber Chem. Technol.*, 1968, 41: 1256.

[38] Pliskin. I, Tokita N. Bound Rubber in Elastomers: Analysis of Elastomer-Filler Interaction and Its Effect on Viscosity and Modulus of Composite Systems. *J. Appl. Polym. Sci.*, 1972, 16: 473.

[39] Leblanc J L. A Molecular Explanation for the Origin of Bound Rubber in Carbon Black Filled Rubber Compounds. *J. Appl. Polym. Sci.*, 1997, 66: 2257.

[40] Meissner B. Theory of Bound Rubber. *J. Appl. Polym. Sci.*, 1974, 18: 2483.

[41] Cohen-Addad J P, Frébourg P. Gel-like Behaviour of Polybutadiene/Carbon Black Mixtures: NMR and Swelling Properties. *Polymer*, 1996, 37: 4235.

[42] Meissner B. Bound Rubber Theory and Experiment. *J. Appl. Polym. Sci.*, 1993, 50: 285.

[43] Cohen-Addad J P. Silica-Siloxane Mixtures. Structure of the Adsorbed Layer: Chain Length Dependence. *Polymer*, 1989, 30: 1820.

[44] Cohen-Addad J P. Sol or Gel-Like Behaviour of Ideal Silica-Siloxane Mixtures: Percolation Approach. *Polymer*, 1992, 33: 2762.

[45] Wang M-J, Wolff S, Donnet J-B. Filler-Elastomer Interactions. Part III. Carbon-Black-Surface Energies and Interactions with Elastomer Analogs. *Rubber. Chem. Technol.*, 1991, 64: 714.

[46] Tan E H, Wolff S, Haddeman M, et al. Filler-Elastomer Interactions. Part IX. Performance of Silicas in Polar Elastomers. *Rubber. Chem. Technol.*, 1993, 66: 594.

[47] Pouchelon A, Vond-acek P. Semiempirical Relationships Between Properties and Loading in Filled Elastomers. *Rubber. Chem. Technol.*, 1989, 62: 788.

[48] Kraus G. Reinforcement of Elastomers. New York: John Wiley and Sons, 1965.

[49] Gessler, A M. *Proc. Int. Rubber Conf., 5$^{th}$*, Brighton, England, 1967, p. 249.

[50] Hess W M, Ban L L, Eckert F J, et al. Microstructural Variations in Commercial Carbon Blacks. *Rubber Chem. Technol.*, 1968, 41: 356.

[51] Hess W M, Chirico V E, Burgess K A. *Kautsch. Gummi Kunstst.*, 1973, 26: 344.

[52] Heckman F A, Medalia A I. *J. Inst. Rubber Ind.*, 1969, 3: 66.

[53] Gessler A M. Effect of Mechanical Shear on the Structure of Carbon Black in Reinforced Elastomers. *Rubber Chem. Technol.*, 1970, 43: 943.

[54] Serizawa H, Nakamura T, Ito M, et al. Effects of Oxidation of Carbon Black Surface on the Properties of Carbon Black-Nature Rubber Systems. *Polym. J.*, 1983, 15: 201.

[55] Wang M-J, Wolff S. Filler-Elastomer Interacitons. Part VI. Characterization of Carbon Blacks by Inverse Gas Chromatography at Finite Concentration. *Rubber Chem. Technol.*, 1992, 65: 890.

[56] Payne A R. The Dynamic Properties of Carbon Black-Loaded Natural Rubber Vulcanizates. Part I. *J. Polym. Sci.*, 1962, 6: 57.

[57] Voet A, Cook F R. Investigation of Carbon Chains in Rubber Vulcanizates by Means of Dynamic Electrical Conductivity. *Rubber Chem. Technol.*, 1968, 41: 1207.

[58] Kraus G. *J. Appl. Polym. Sci., Applied Polymer Symposia*, 1984, 39: 75.

[59] Freund B, Wolff S, *meeting of the Rubber Division, ACS,* Mexico City, Mexico, May 9-12, 1989.

[60] Smit P P A. Glass Transition in Carbon Black Reinforced Rubber. *Rubber Chem. Technol.*, 1968, 41: 1194.

[61] Medalia A I. Morphology of Aggregates VI. Effective Volume of Aggregates of Carbon Black from Electron Microscopy; Application to Vehicle Absorption and to Die Swell of Filled Rubber. *J. Colloid Interf. Sci.*, 1970, 32: 115.

[62] Wang M-J, Wolff S, Tan E H. Filler-Elastomer Interactions. Part VIII. The Role of the Distance Between Filler Aggregates in the Dynamic Properties of Filled Vulcanizates. *Rubber Chem. Technol.*, 1993, 66: 178.

[63] Wolff S, Wang M-J, Tan E H. Surface Energy of Fillers and Its Effect on Rubber Reinforcement. Part I. *Kautsch. Gummi Kunstst.*, 1994, 47: 780.

[64] Hess W M, Lyon F, Burgess K A. Einfluß der Adhäsion zwischen Ruß und Kautschuk auf die Eigenschaften der Vulkanisate. *Kautsch. Gummi Kunstst.*, 20, 135 (1967).

[65] Donnet J-B, Wang W, Vidal A, et al. Study of Surface Activity of Carbon Black by Inverse Gas Chromatography. Part II. Effect of Carbon Black Thermal Treatment on its Surface Characteristics and Rubber Reinforcement. *Kautsch. Gummi Kunstst.*, 1993, 46: 866.

[66] Wang M-J, Application of Inverse Gas Chromatography to the Study of Rubber Reinforcement. In: Nardin M, Papirer E (Editors). Powder and Fibers: Interfacial Science and applications. Boca Raton, FL: CRC Press-Taylor and Francis Group, 2007.

[67] Brennan J J, Jermyn T E, Boonstra B B. Carbon Black-Polymer Interaction: A Measure of Reinforcement. *J. Appl. Polym. Sci.*, 1964, 8: 2687.

[68] Wang M-J, Wolff S, Freund B. Filler-Elastomer Interactions. Part XI. Investigation of the Carbon-Black Surface by Scanning Tunneling Microscopy. *Rubber Chem. Technol.*, 1994, 67: 27.

[69] Wang W, Vidal A, Donnet J-B, et al. Study of Surface Activity of Carbon Black by Inverse Gas Chromatography, Part III: Superficial Plasma Treatment of Carbon Black and Its Surface Activity. *Kautsch. Gummi. Kunstst.*, 1993, 46: 933.

[70] Cataldo F. Effects of Radiation Pretreatments on the Rubber Adsorption Power and Reinforcing Properties of Fillers in Rubber Compounds. *Polym. Inter.*, 2001, 50: 828.

[71] Wang M-J, Tu H, Murphy, L J, et al. Carbon-Silica Dual Phase Filler, A New Generation Reinforcing Agent for Rubber: Part VIII. Surface Characterization by IGC. *Rubber. Chem. Technol.*, 2000, 73: 666.

[72] Mahmud K, Wang M-J, Francis R A. Elastomeric compounds incorporating silicon-treated carbon blacks. U.S. Patent, 5830930, 1998.

[73] Sheng E, Sutherland I, Bradley R H, et al. Effects of A Multifunctional Additive on Bound Rubber in Carbon Black and Silica Filled Natural Rubbers. *Eur. Polym. J.*, 1996, 32: 35.

[74] Ayala J A, Hess W M, Dotson A O, et al., New Studies on the Surface Properties of Carbon Blacks. *Rubber. Chem. Technol.*, 1990, 63: 747.

[75] Lezhnev N N, Lyalina N M, Zelenev Y V, et al. *Colloid J.*, 1966, 28: 342.

[76] Wang M-J, Wolff S. Filler-Elastomer Interactions. Part V. Investigation of the Surface Energies of Silane-Modified Silicas. *Rubber. Chem. Technol.*, 1992, 65: 715.

[77] Wang M-J, Doctor's Thesis, France: Universite de Haute Alsace, 1984.

[78] Wolff S, Wang M-J, Tan E H. Filler-Elastomer Interactions. Part X. *Kautsch. Gummi. Kunstst.*, 1994, 47: 102.

[79] Aranguren M I, Mora E, Macosko C W. Compounding Fumed Silicas into Polydimethylsiloxane: Bound Rubber and Final Aggregate Size. *J. Colloid Interf. Sci.*, 1997, 195: 329.

[80] Leblanc J L, Evo C, Lionnet R. Composite Design Experiments to Study the Relationships between the Mixing Behavior and Rheological Properties of SBR Compounds. *Kautsch. Gummi. Kunstst.*, 1994, 47: 401.

[81] Funt J M, *meeting of the Rubber Division, ACS*, Cleveland, Ohio Oct.1-4, 1985.

[82] Gerke R H, Ganzhorn G H, Howland L H, et al. Manufacture of rubber. U.S. Patent, 2118601, 1938.

[83] Dannenberg E M. Carbon Black Dispersion and Reinforcement. *Ind. Eng. Chem.*, 1952, 44: 813.

[84] Barton B C, Smallwood H M, Ganzhorn G H. Chemistry in Carbon Black Dispersion. *J. Polym. Sci.*, 1954, 13: 487.

[85] Cotton G R. Mixing of Carbon Black with Rubber. II. Mechanism of Carbon Black Incorporation. *Rubber. Chem. Technol.*, 1985, 58: 774.

[86] Gessler A M. *Rubber Age*, 1969, 101: 54.

[87] Berry J P, Cayré P J. The Interaction Between Styrene-Butadiene Rubber and Carbon Black on Heating. *J. Appl. Polym. Sci.*, 1960, 3: 213.

[88] Wang M-J. The Role of Filler Networking in Dynamic Properties of Filled Rubber. *Rubber Chem. Technol.*, 1999, 72: 430.

[89] Wang M-J. Studies on Basic Properties of Several Antioxidants. Technical Bulletin, BRDIRI, May 1975.

[90] Wang M-J. *Third International Conference on Carbon Black*, Mulhouse, France, Oct. 25-26, 2000.

[91] Stickney P B, McSweeney E E, Muller W J, et al. Bound-Rubber Formation in Diene Polymer Stocks. *Rubber Chem. Technol.*, 1958, 31: 369.

[92] Crowther B G, Edmondson H M. Rubber Technology and Manufacture. Cleveland, Ohio: CRC Press, 1971.

[93] Villars D S. Studies on Carbon Black. III. Theory of Bound Rubber. *J. Polym. Sci.*, 1956, 21: 257.

[94] Influence of Carbon Black Activity on Processability of Rubber Stocks Part I, Cabot, Alpharetta, Georgia, 1974.

[95] Cotton G R. Mixing of Carbon black with Rubber, IV. Effect of Carbon Black Characteristics. *RUBBEREX 86 Proceeding*, ARPMA, April, 29-May, 1, 1986.

[96] Ashida M, Abe K, Watanabe T. Studies on Bound Rubber in Elastomer Blends(III) Bound Rubbers in IR, BR and SBR Mixed with Silica. *Nippon Gomu Kyokaishi*, 1976, 49: 821.

[97] Leblanc J L, Hardy P. *Kautsch. Gummi Kunstst.*, 1991, 44: 1119.

[98] Wang M-J. Effect of Filler-Elastomer Interaction on Tire Tread Performance Part III. *Kautsch. Gummi Kunstst.*, 2008, 61:159.

[99] Functionalization of Elastomers by Reactive Mixing, MRPRA, Malaysia, 1994.

[100] Wang M-J, Brown T A, Dickinson R E. Elastomer Composition and Method. U.S. Patent, 5916956, 1999.

[101] Brown T A, Wang M-J. Elastomers Compositions with Dual Phase Aggregates and Pre-Vulcanization Modifier. U.S. Patent, 6172154, 2001.

[102] Terakawa K, Muraoka K. *IRC'95*, Kobe, Japan, Oct. 23-27, 1995. pp 24.
[103] Hamerton I. Recent Developments in Epoxy Resins. *RAPRA Review Reports (Report 91)*, 8, No. 7, 1996.
[104] Wampler W A, Gerspacher M, Yang H H, et al. *Rubber & Plastics News*, pp 45, April 24, 1995.
[105] Gerspacher M. *IRC'96*, Manchester, June 17-20, 1996.
[106] Degussa AG, *Tech. Inf. Bull.*, No. 6030.1, Sempt. 1995.
[107] Wolff S. Optimization of Silane-Silica OTR Compounds. Part 1: Variations of Mixing Temperature and Time During the Modification of Silica with Bis-(3-triethoxisilylpropyl)-tetrasulfide. *Rubber Chem. Technol.*, 1982, 55: 967.
[108] Patkar S D, Evans L R, Waddel W H. *International Tire Exhibition and Conference*, Akron, Ohio, Sept. 10-12, 1996.
[109] Wang M-J, Wang T, Wong Y L, et al. NR/Carbon Black Masterbatch Produced with Continuous Liquid Phase Mixing. *Kautsch. Gummi Kunstst.*, 2002, 55: 388.
[110] Wang M-J. Application of EVEC to Passenger Tire Treads. *Tire Technology Conference*, Hannover, Germany, Feb. 16~18, 2016.
[111] Wang M-J, Morris M, Kutsovsky Y. Effect of Fumed Silica Surface Area on Silicone Rubber Reinforcement. *Kautsch. Gummi Kunstst.*, 2008, 61: 107.
[112] Donnet J-B, Voet A. Carbon Black Physics, Chemistry, and Elastomer Reinforcement. New York: Marcel Dekker, 1976.
[113] Wolff S, Wang M-J, Tan E H, *meeting of the Rubber Division, ACS*, Detroit, 1991.
[114] Shi Z, Doctor's Thesis, France: University of Haute Alsace, 1989.
[115] Smit P P A. *Les Interactions Entre les Élastomères et les Surfaces Solides Ayant Une Action Renforçante, Colloques Internationaux du CNRS, No. 231,* Paris, 1975.
[116] Smit P P A. The Effect of Filler Size and Geometry on the Flow of Carbon Black Filled Rubber. *Rheol. Acta*, 1969, 8: 277.
[117] Smit P P A, Van der Vegt A K. *Kautsch. Gummi Kunstst.*, 1970, 23: 4.
[118] Gregiry M J, Tan A S, *Proc. Int. Rubber Conf.*, Kuala Lumpur, Vol. IV, p28.
[119] Vondráček P, Schätz M. Bound Rubber and "Crepe Hardening" in Silicone Rubber. *J. Appl. Polym. Sci.*, 1977, 21: 3211.
[120] DeGroot Jr. J V, Macosko C W. Aging Phenomena in Silica-Filled Polydimethylsiloxane. *J. Colloid Interf. Sci.*, 1999, 217: 86.
[121] Wolff S, Wang M-J. Physical Properties of Vulcanizates and Shift Factors. *Kautsch. Gummi Kunstst.*, 1994, 47: 17.
[122] Medalia A I. Morphology of Aggregates I. Calculation of Shape and Bulkiness Factors; Application to Computer-Simulated Random Flocs. *J. Colloid Interf. Sci.*, 1967, 24: 393.
[123] Guth E, Simha R. Untersuchungen über die Viskosität von Suspensionen und Lösungen. 3. Über die Viskosität von Kugelsuspensionen. *Kolloid Z.*, 1936, 74: 266.
[124] Guth E, Gold O. Viscosity and Electroviscous Effect of the AgI Sol. II. Influence of the Concentration of AgI and of Electrolyte on the Viscosity. *Phys. Rev.*, 1938, 53: 322.
[125] Guth E. Theory of Filler Reinforcement. *J. Appl. Phys.*, 1945, 16: 20.
[126] Boonstra B B. Chapter 7. In: Blow C M, Hepburn C (Editors). Rubber Technology and Manufacture, 2$^{nd}$ ed., London:Buttenvorths, 1982.
[127] Fujimoto K, Nishi T. Studies on Heterogeneity in Filled Rubber Systems. (Part I). Molecular Motion in Filler-Loaded Unvulcanized Naural Rubber. *Nippon Gomu Kyokaishi*, 1970, 43: 54.

[128] Fujimoto K, Ueki T, Mifune N. Studies on Young's Modulus, and Poisson's Ratio of Particles Filled Vulcanizates. *Nippon Gomu Kyokaishi*, 1984, 57: 23.

[129] P. P. A. Smit, *Rhoel. Acta*, 1966, 5: 277.

[130] AFM Observation of Bound Rubber. EVE Rubber Institute, Qingdao, China, 2017.

[131] Voet A, Sircar A K, Mullens T J. Electrical Properties of Stretched Carbon Black Loaded Vulcanizates. *Rubber Chem. Technol.*, 1969, 42: 874.

[132] Voet A, Morawski J C. Dynamic Mechanical and Electrical Properties of Vulcanizates at Elongations up to Sample Rupture. *Rubber Chem. Technol.*, 1974, 47: 765.

[133] Lee B L. Reinforcement of Uncured and Cured Rubber Composites and Its Relationship to Dispersive Mixing-An Interpretation of Cure Meter Rheographs of Carbon Black Loaded SBR and *cis*-Polybutadiene Compounds. *Rubber Chem. Technol.*, 1979, 52: 1019.

[134] Sircar A K. *Rubber World*, 1987, 196, Nov., 30

[135] Wang M-J, Wolff S, Donnet J-B. Filler-Elastomer Interactions. Part I: Slilica Surface Energies and Interactions with Model Compounds. *Rubber Chem. Technol.*, 1991, 64: 559.

[136] Chin P S. *J. Rubber Res. Inst. Malaysia*, 1960, 22 (1): 56.

[137] Leblenc J L. *IRC'98*, Paris, France, 1998, pp 12-14.

[138] Westlinning H. Verstärkerfüllstoffe für Kautschuk. *Paper presented at D.K.G.*, Freiburg, Germany, Oct. 3-6, 1962.

[139] Song J, He F, Zhang H, et al. Storage Hardening Effect of EVEC. EVE Rubber Institute, Unpublished.

[140] Spencer R S, Dillon R E. The Viscous Flow of Molten Polystyrene. *J. Colloid Sci.*, 1948, 3: 163.

[141] Minagawa N, White J L. The Influence of Titanium Dioxide on the Rheological and Extrusion Properties of Polymer Melts. *J. Appl. Polym. Sci.*, 1976, 20: 501.

[142] Hopper J R. Effect of Oil and Black on SBR Rheological Properties. *Rubber Chem. Technol.*, 1967, 40: 463.

[143] Cotton G R. *Rubber Age*, 1968, 100 (11): 51.

[144] Dannenberg E M, Stokes C A. Characteristics of Reinforcing Furnace Blacks. *Ind. Eng. Chem.*, 1949, 41: 812.

[145] Wang M-J. Application of EVE Compounds to Passenger Tire Treads. *The International Tire Exhibition & Conference*, Akron, Ohio, USA, Sep. 13-15, 2016.

[146] Buckler E J, Briggs G J, Dunn J R, et al. Green Strength of Emulsion SBR. *Rubber Chem. Technol.*, 1978, 51: 872.

[147] Andrews E H, Owen P J, Singh A. Microkinetics of Lamellar Crystallization in a Long Chain Polymer. *Proc. R. Soc. Lond. A*, 1971, 324: 79.

[148] Gent A N. Crystallization in Natural Rubber II. The Influence of Impurities. *Rubber Chem. Technol.*, 1955, 28: 457.

[149] Chen H Y. Determination of *cis*-1,4 and *trans*-1,4 Contents of Polyisoprenes by High Resolution Nuclear Magnetic Resonance. *Rubber Chem. Technol.*, 1965, 38: 90.

[150] Cameron A, McGill W J. A Theory of Green Strength in Polyisoprene. *J. Polym Sci: Part A, Polymer Chemistry*, 1989, 27: 1071.

[151] Bruzzone M, Corradini G, de Chirico A, et al. *4th Synthetic Rubber, Symposium,* London, 1969, pp 83.

[152] Hackathome M, Brock M J. *Rubber Age*, 1972, 104: 60.

[153] Burfield D R, Tanaka Y. Cold Crystallization of Natural Rubber and Its Synthetic Analogues: The Influence of Chain Microstructure. *Polymer*, 1987, 28: 907.

[154] Kawahara S, Nishiyama N, Matsuura A, et al. Crystallization Behavior of Natural Rubber, Part I. Effect of Mixed Fatty Acid on the Crystallization of *cis*-1,4 Polyisoprene. *Nippon Gomu Kyokaishi*, 1999, 72: 288.

[155] Ichikawa N, Eng A H, Tanaka Y. Natural Rubber-Current Developments in Product Manufacture and Application. Kuala Lumpur: Rubber Research Institute of Malaysia, 1993.

[156] Amnuaypornsri S, Sakdapipanich J, Tanaka Y. Green Strength of Natural Rubber: The Origin of the Stress-Strain Behavior of Natural Rubber. *J. Appl. Polym. Sci.*, 2009, 111: 2127.

[157] Kawahara S, Isono Y, Sakdapipanich J T, et al. Effect of Gel on the Green Strength of Natural Rubber. *Rubber Chem. Technol.*, 2002, 75: 739.

[158] Hamed G R. Tack and Green Strength of NR, SBR, and NR/SBR Blends. Rubber Chem. Technol. 1981, 54: 403.

[159] Cho P L, Hamed G R. Green Strength of Carbon-Black-Filled Styrene-Butadiene Rubber. *Rubber Chem. Technol.*, 1992, 65: 475.

[160] Green Strength of Carbon Black-Filled SBR Compound. EVE Rubber Institute, Qingdao, China, 2018.

# 6 Effect of Fillers on the Properties of Vulcanizates

## 6.1 Swelling

While uncrosslinked rubber dissolves in a good solvent, vulcanized rubber swells in solvents to an extent determined by crosslink density and the nature of the solvent. For a given solvent, the higher the crosslink density of the rubber, the lower the swelling; conversely, for a given degree of crosslink density, a more powerful solvent yields higher degree of swelling. Flory and Rehner[1–3] provide the fundamental equation relating the equilibrium degree of swelling defined by $V_r$, volume fraction of rubber network in the swollen gel, to crosslink concentration, $\nu$, as:

$$\nu = -\frac{1}{\rho V_s}\left\{[\ln(1-V_r)+V_r+\chi V_r^2]/\left(V_r^{1/3}-\frac{1}{2}V_r\right)\right\} \tag{6.1}$$

where $V_s$ is the molar volume of the swelling liquid, $\rho$ is the density of rubber, and $\chi$ is a constant which is characteristic of the interaction between the rubber and the swelling liquid, commonly termed the "rubber-solvent interaction parameter" (Eq. 6.1). The relation, which is derived from statistical theory and the liquid mixing thermodynamics with network chains, is strictly applicable only to a hypothetical network that is free from entanglements and chain ends. The success of these techniques with gum vulcanizates makes their extension to the technologically important filler-reinforced rubbers highly desirable. Unfortunately, the introduction of rigid filler particles into the network leads to some serious theoretical and practical complications.

With a carbon black (HAF; N330) filled NR compounds cured with accelerated sulfur, Lorenz and Parks[4] observed that the relationship of swelling ratios, $Q_{CB}$ and $Q_{gum}$, can be expressed by:

$$Q_{CB}/Q_{gum} = a e^{-z} + b \tag{6.2}$$

where $z$ is the mass loading of the reinforcing filler, and $a$ and $b$ are constants. Besides the effect of crosslink density, the ratio $Q_{CB}/Q_{gum}$ is also determined by polymer-filler interaction (Eq. 6.2).

Assuming that the filler does not swell in the solvent, the volume fraction of the filled rubber in the swollen gel, $V_{rf}$ (corrected for the filler volume), should differ from that of the pure gum, $V_{ro}$. In the case of reinforcing fillers, the strong interaction caused by strong physical and/or chemical adsorption may have the same effect as crosslinks. It is therefore understood that for filled vulcanizates in which the crosslink density in the polymer matrix is not affected by the filler, the ratio $V_{ro}/V_{rf}$ should change with filler loading as well as polymer-filler interaction. Based on these considerations, Kraus[5] derived different expressions relating the ratio $V_{ro}/V_{rf}$ to filler loading for non-adhering (no filler-polymer interaction) and adhering (with polymer-filler interaction) filled vulcanizates.

**Non-adhering filler:**

If the linear swelling coefficient of the rubber is $q_0$ and the volume fraction of filler is $\phi$, the final volume of swollen rubber is $(1-\phi)q_0^3$. In addition, as the rubber swells, it forms a vacuole around each particle, which will fill with solvent. The solvent volume is clearly $\phi(q_0^3-1)$ and has to be added to the volume of the swollen rubber. The volume swelling ratio $Q$ is then (Eq. 6.3):

$$Q = V_{rf}^{-1} = \left(q_0^3 - \phi\right)/\left(1-\phi\right) = \left(V_{ro}^{-1} - \phi\right)/\left(1-\phi\right) \tag{6.3}$$

or

$$V_{ro}/V_{rf} = (1 - V_{ro}\phi)/(1-\phi) \tag{6.4}$$

Equation 6.4 shows that a total lack of filler-polymer adhesion produces a considerable increase in the apparent swelling of the rubber.

**Adhering filler:**

For reinforcing fillers with a strong polymer interaction, $V_{rf}$, should be higher than $V_{ro}$. When a filler particle of radius $R$ embedded in a matrix of rubber, if the filled rubber is swollen, but the bond between the particle and the rubber remains intact, swelling is obviously restricted completely at the surface. At a distance $r$ (>$R$) away from the center of the particle, swelling is still partially restricted, and as $r$ approaches infinity, swelling becomes normal.

Consider an element of volume in the unswollen rubber: d$r$, $r$d$\theta$, and $r\sin\theta d\psi$. After swelling, the dimensions of this element are $q_r$d$r$, $q_t r$d$\theta$, and $q_t r\sin\theta d\psi$, where the $q$'s are functions of $r$. Conservation of solid angle requires that $q_r$ (radial) differs from $q_t$ (tangential). To find the relation between them one first finds the distance ($r'$) by which the element is removed from the center of the particle after swelling. This is:

$$r' = \int_R^r q_r \mathrm{d}r + R \tag{6.5}$$

The requirement that:

$$r'\mathrm{d}\theta = q_t r \mathrm{d}\theta \tag{6.6}$$

leads to

$$q_t r - R = \int_R^r q_r \mathrm{d}r \tag{6.7}$$

It follows from Eq. 6.7 that:

$$q_r = q_t + r(\mathrm{d}q_t / \mathrm{d}r) \tag{6.8}$$

The swelling deficiency ($\Delta V$) can now be calculated due to the particle for a shell of rubber bounded by $R$ and any arbitrary $r$ by considering the difference between the swelling of this shell and the extent to which it would have swelled in the absence of the constraint:

$$\Delta V = \int\int\int \left(q_r q_t^2 - q_0^3\right) r^2 \sin\theta \mathrm{d}r \mathrm{d}\theta \mathrm{d}\psi \tag{6.9}$$

Here $q_0$ is the "normal" linear swelling coefficient. Substituting for $q_r$ from Eq. 6.7 and integrating:

$$\begin{aligned} \Delta V &= 4\pi \int_R^r \left[ r^2 q_t^2 \left( q_t + r\frac{\mathrm{d}q_t}{\mathrm{d}r} \right) - r^2 q_0^3 \right] \mathrm{d}r \\ &= (4\pi/3)\left[ r^3 q_t^3(r) - R^3 - q_0^3\left(r^3 - R^3\right)\right] \end{aligned} \tag{6.10}$$

In evaluating the integral, use was made of the identity:

$$r^2 q_t^3 + r^3 q_t^2 (\mathrm{d}q_t / \mathrm{d}r) = 1/3\left[\mathrm{d}\left(r^3 q_t^3\right)/\mathrm{d}r\right] \tag{6.11}$$

and the fact that $q_t$ must be unity at the lower limit (no swelling at the surface).

It is convenient at this point to replace the function $q_t$ by the following expression:

$$q_t = q_0 - (q_0 - 1) f(r) \tag{6.12}$$

where $f(\infty) = 0$ and $f(R) = 1$. Factoring out $R^3 q_0^3$ from Eq. 6.10 and rearranging, it is found that:

$$\Delta V = (4\pi/3) R^3 q_0^3 \left\{1 - q_0^{-3} + r^3 R^{-3}\left[q_0^{-3} q_t^3(r) - 1\right]\right\} \tag{6.13}$$

and passing to the limit:

$$\Delta V = (4\pi/3) R^3 q_0^3 \left[1 - q_0^{-3} + R^{-3} \lim_{r\to\infty} r^3 \left(q_t^3 q_0^{-3} - 1\right)\right] \tag{6.14}$$

Using Eq. 6.12, it is found that:

$$\lim_{r\to\infty} r^3[q_t^3 q_0^{-3} - 1] = \lim_{r\to\infty}\left\{r^3\left[-3\frac{q_0 - 1}{q_0}f(r) + 3\left(\frac{q_0 - 1}{q_0}\right)^2 f^2(r) - \left(\frac{q_0 - 1}{q_0}\right)^3 f^3(r)\right]\right\} \quad (6.15)$$

It is easily shown that if the first term converges, the second and third terms will converge to zero; hence,

$$\lim_{r\to\infty} r^3\left(q_t^3 q_0^{-3} - 1\right) = -3(1 - q_0^{-1})\lim_{r\to\infty} r^3 f(r) \quad (6.16)$$

Substituting Eq. 6.14 leads to:

$$\Delta V = (4\pi/3)R^3 q_0^3\left[1 - q_0^{-3} - 3c\left(1 - q_0^{-1}\right)\right] \quad (6.17)$$

where

$$c = R^{-3}\lim_{r\to\infty} r^3 f(r) \quad (6.18)$$

It is now possible to write the swelling deficiency for $N$ particles (in unit volume of rubber) if the particles are far enough apart so as not to interact. If $\phi$ is the volume fraction of filler,

$$N = 3\phi / 4\pi R^3(1 - \phi) \quad (6.19)$$

and

$$\Delta V(\text{all particles}) = q_0^3[1 - q_0^{-3} - 3c(1 - q_0^{-1})]\phi/(1 - \phi) \quad (6.20)$$

The ratio of the volume swelling ratios ($Q$) of the filled rubber to the unfilled rubber is obviously:

$$Q/Q_0 = \left(q_0^3 + \Delta V\right)/q_0^3 = 1 + \left[1 - q_0^{-3} - 3c\left(1 - q_0^{-1}\right)\right]\phi/(1 - \phi) \quad (6.21)$$

It is standard to denote the reciprocal swelling ratios as $V_{\text{rf}}$, which are equal to the volume fraction of rubber in the swollen gel. Noting that:

$$q_0^{-3} = V_{\text{ro}} \quad (6.22)$$

it is found that:

$$V_{\text{ro}}/V_{\text{rf}} = 1 - \left[3c\left(1 - V_{\text{ro}}^{1/3}\right) + V_{\text{ro}} - 1\right]\phi/(1 - \phi) = 1 - m\phi/(1 - \phi) \quad (6.23)$$

The theory thus predicts that $V_{\text{ro}}/V_{\text{rf}}$ should vary linearly with $\phi/(1-\phi)$, the slope, $m$, being a function of $V_{\text{ro}}$ as:

$$m = 3c\left(1 - V_{\text{ro}}^{1/3}\right) + V_{\text{ro}} - 1 \quad (6.24)$$

It is important to note that the addition of a filler may additionally change the crosslink density in the polymer matrix, which in turn affects the vulcanization reaction. For cases where the crosslink density of the matrix is independent of the presence of a filler, e.g., peroxide-cured natural rubber[6,7], $c$ is a characteristic parameter of the filler and related to filler-polymer interaction and other factors. For fillers having similar structure, higher values of $c$ suggest stronger interaction between filler and rubber matrix.

Besides the crosslink density of the polymer matrix, the morphology and the surface characteristics of filler, the solvent power of the swelling medium has also a significant effect on swelling, which is related to $\chi$, the rubber-solvent interaction parameter in Eq. 6.1, and the solvent-filler interaction.

Figure 6.1 shows the volume fraction $\phi$ of filler in the vulcanizate increasing as the restriction of swelling increases for the reinforcing carbon black N220 (ISAF). This restriction also changes with solvents. Non-adhering fillers, such as glass beads, show an increase in $V_{ro}/V_{rf}$ because of pockets of solvent forming around the particles. Graphitized carbon blacks constitute a completely unique class of fillers in that they neither restrict nor enhance swelling but leave it unaffected at loading in any solvent[8].

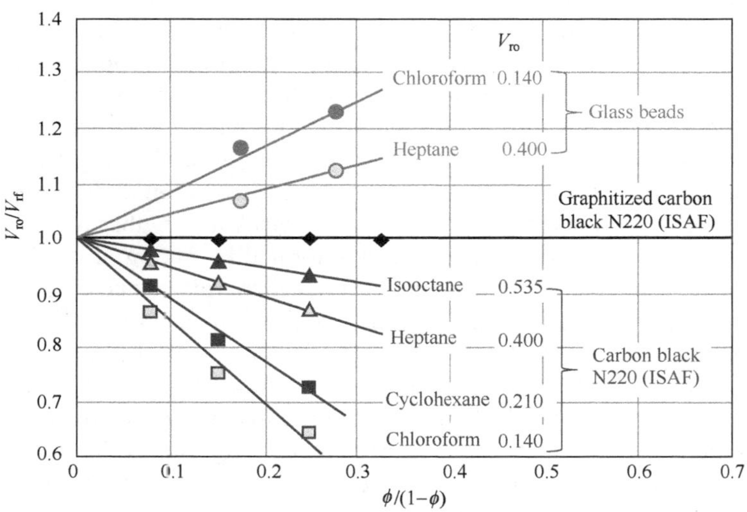


**Figure 6.1** The relation between the volume fraction $\phi$ of filler in a vulcanizate and the swelling, for glass beads and carbon black N220 (ISAF) in SBR

Figure 6.2 shows Kraus' plots, i.e., the change in $V_{ro}/V_{rf}$ with filler loading, for different vulcanizates using toluene as a solvent[9]. The fillers used were carbon black N110, silica P1 (a precipitated silica with NSA 140 $m^2/g$ and DBP 92 mL/100 g), P1-TESPT [silica P1 modified with TESPT, i.e., bis(3-triethoxysilylpropyl)tetrasulfide], and P1-HDTMS (silica P1 modified with HDTMS, i.e., hexadecyltrimethoxy silane). These vulcanizates were cured with dicumyl peroxide (DCP). For the fillers investigated, $V_{ro}/V_{rf}$ decreases with increased filler loading, suggesting a restriction of rubber matrix swelling due to the presence of filler.

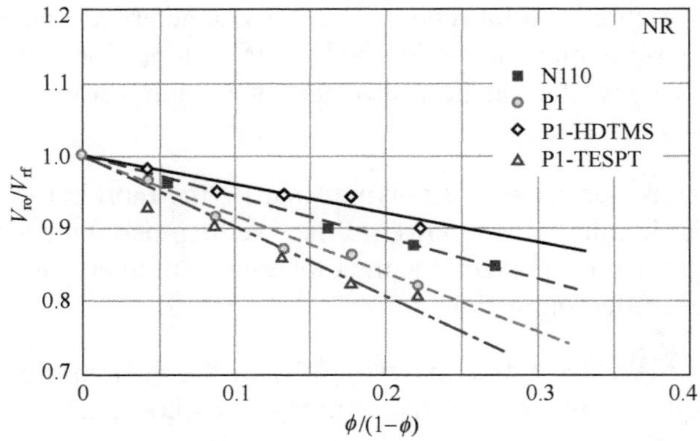


**Figure 6.2** Kraus' plots of swelling in toluene for DCP-cured NR filled with a variety of fillers

In addition to the surface characteristics of filler, the swelling of filled vulcanizates is also dependent on filler morphology, i.e., surface area and structure. In other words, $c$ is also a morphology-related parameter. While the impact of surface area is obvious, Rigbi and Boonstra[10] have shown that a high filler structure leads to a considerable restriction of swelling. In addition, the agglomeration (or secondary structure) of filler also exerts a significant restraint on swelling. This has been demonstrated by a comparison of swelling of vulcanizates filled with carbon black and silica, respectively, both having comparable surface areas and structures. It was found that, although filler-polymer interaction is much higher for the carbon black, $c$ is higher for the silica-filled vulcanizates. This suggests that agglomerates formed by strong silica interaggregate interaction may still exist, at least partially, in swollen vulcanizates, and may play an important role in reducing swelling[11]. Moreover, the chemical modification of filler, especially silica, also affects the swelling of vulcanizates. Wolff and Wang[9,12] have shown that although the potential for filler agglomeration is reduced in the case of the silica P1-TESPT vulcanizates, its higher value for $c$ in comparison to the non-modified silica P1 is indicative of a thermally induced coupling reaction which must have taken place between the filler surface and the polymer and which, indirectly, leads to an increase in the apparent crosslink density. This assumption seems to be confirmed by the swelling behavior of vulcanizates filled with silica P1-HDTMS. In this case, the swelling restriction is even lower than for N110 and is attributed to greatly reduced filler agglomeration and to much lower filler-polymer interaction as expressed by $\gamma_s^d$ and $\gamma_s^{sp}$ [9].

Additional information concerning the effect of polymer-filler interaction was obtained by investigating equilibrium swelling in ammonia atmosphere. Figure 6.3 shows Kraus' plots of the toluene swelling data for NR vulcanizates containing different fillers treated with ammonia at room temperature. After ammonia treatment, the swelling of the silica-filled vulcanizate increases considerably, and $V_{ro}/V_{rf}$ even reaches unity, which is usually characteristic of a non-adhering filler. Therefore, the silica surface does not seem to show any restriction of swelling after ammonia treatment. This suggests once more that polymer-filler and filler-filler interactions are primarily physical phenomena. In the case of the carbon black N110, however, swelling in ammonia atmosphere does not differ significantly from swelling under normal atmosphere.

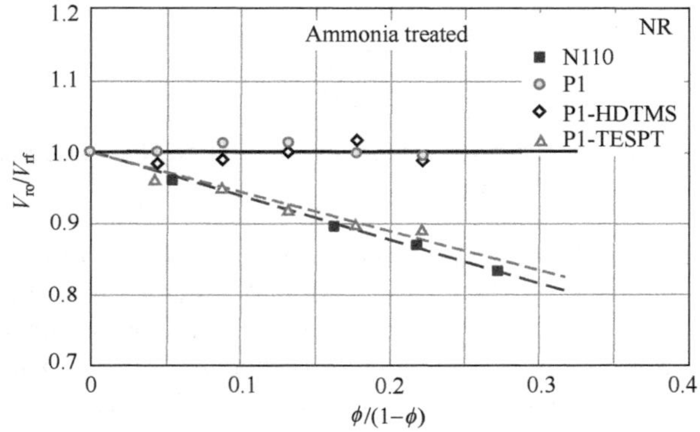


**Figure 6.3** Kraus' plots of swelling in toluene under ammonia atmosphere for DCP-cured NR filled with a variety of fillers

Similar to N110, swelling of the vulcanizate containing silica P1-TESPT is only slightly affected by the ammonia treatment, but the mechanisms are different. In the case of N110, polymer-filler interaction is stronger than the interaction between ammonia and carbon black. Hence, there is no change in apparent swelling. In the case of the silica P1-TESPT vulcanizate, only the physical polymer-filler interaction is eliminated[13], whereas the chemical linkages cannot be separated by ammonia. The fact that the swelling restriction largely remains after ammonia treatment must, therefore, be due to the formation of covalent bonds between filler and polymer chains. In the case of the silica P1-HDTMS vulcanizate, the reduction of the swelling restriction is undoubtedly caused by the partition of physical polymer-filler bonds by ammonia.

The above model regarding the effect of filler-polymer and filler-filler interactions on swelling was also tested using a polar polymer[9]. Figure 6.4 contains Kraus' plots of nitrile rubber vulcanizates filled with N110 and the silica P1, using toluene as a solvent. It is apparent that the silica P1 swells more than the corresponding carbon black. This is contrary to what is shown in Figure 6.2. The high polarity of NBR causes increased polymer-filler interaction at the expense of filler-filler agglomeration, leading to increased swelling. Furthermore, the effect of filler-filler interaction on swelling is not sufficiently compensated for by the relatively weak polymer-filler interaction of the silica in comparison to carbon black, so that, overall, the net swelling restriction in the presence of silica is lower.

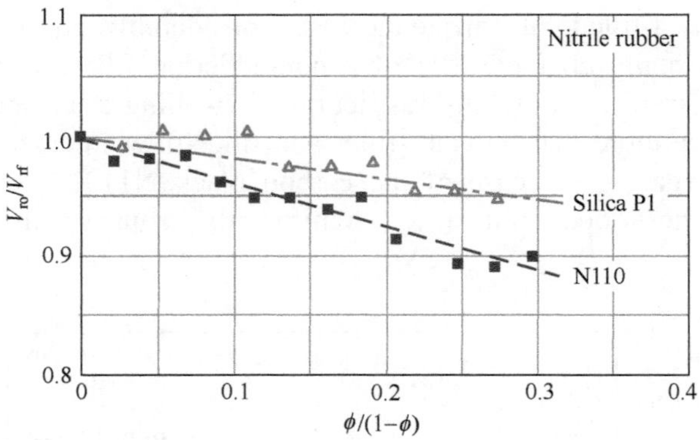


**Figure 6.4** Kraus' plots of swelling in toluene for DCP-cured NBR vulcanizates filled with N110 carbon black and the silica P1

Whereas the crosslink density in the polymer matrix is not affected by fillers in the case of peroxide-cured natural rubber[6,7], there is no proof that it is constant in the presence of fillers for peroxide-cured nitrile rubber. Hence, the reversal in swelling behavior could perhaps be attributed to a varying degree of crosslinking. Although this argument seems valid, the moduli at high strain for silica vulcanizates in NBR are higher than in natural rubber[9], and this contradicts the concept that crosslink density is diminished by the addition of silica in NBR. Crosslink density could play a certain role in this case; however, the dominant factor in this reversal of swell behavior is the reduction in filler agglomeration.

Wolff and Wang, using 18 carbon blacks with one carbon black-filled polymer as a reference[14], found that the filler loading-rubber property curves of SBR vulcanizates cured with accelerated sulfur can be superposed to a single master curve by introduction of a shift factor. The master curve for swelling was established by plotting $V_{ro}/V_{rf}$ as a function of $f\phi/(1+f\phi)$, shown in Figure 6.5[15]. $V_{ro}/V_{rf}$ *vs.* $\phi/(1-\phi)$ had been expected to be a linear function according to Kraus' theory, but significant deviations

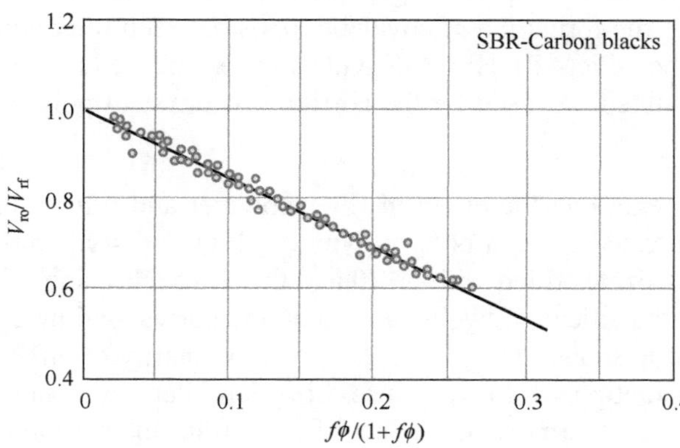


**Figure 6.5** Master curve for swelling of SBR vulcanizates filled with a series of carbon blacks in benzene

were observed at higher filler loading. However, by using $f\phi/(1+f\phi)$ instead of $\phi/(1-\phi)$ for the construction of the master curve, an excellent correlation with the swelling data of all carbon-black-filled vulcanizates was obtained. The mechanism of the deviation of the master curve from the Kraus' plots is not understood. It is not known if this is caused by effect of filler on the crosslinking reaction or the filler-polymer and/or filler-filler interactions.

## 6.2 Stress-Strain Behavior

### 6.2.1 Low Strain

The main change in the stress-strain performance of rubber due to the incorporation of fillers is observed in its stiffness, which scales with modulus, i.e., the ratio between stress and strain. (In rubber technology, for quasi-static measurements, it is customary to refer to the stress at a specified strain as "modulus". This stress is always calculated based on the original cross-sectional area.) The modulus of filled rubber, whether measured in shear, compression, or extension, under static or dynamic conditions, is considerably higher than that of the "pure gum". Smallwood[16] and Guth[17] attempted a theoretical description of the increase in modulus caused by filler. Using Einstein's viscosity equation as modified by Guth and Gold[18,19], which took into account the mutual disturbance caused by spheres at high concentration, they proposed the following equation (Eq. 6.25) for Young's modulus:

$$E = E_0\left(1+2.5\phi+14.1\phi^2\right) \tag{6.25}$$

where $E$ and $E_0$ are the respective Young's moduli of the filled rubber and gum. Several authors[20] proposed different values for the coefficient of the second term. Other analytical equations were also proposed[21–24]. Kraus compared these equations for modulus and found that they deviated considerably from each other at high filler loadings. However, all of equations are inadequate for reinforcing fillers, especially at filler loadings commonly used in practice, since they are based on the same assumptions as Einstein's equations, i.e., spherical particles, well-wetted particles, low strain, perfect dispersion, negligible particle-particle interaction, and no effect on the characteristics of the polymer.

In practice, it is extremely difficult to satisfy these conditions because (a) most filler aggregates depart considerably from the spherical shape, and (b) the mobility of the rubber matrix is not uniform due to different filler-polymer interactions. Furthermore, particle-particle interaction enables the particles to agglomerate, which affects the uniformity and continuity of the polymer matrix. So far, the applicability of Eq. 6.25 has been limited to almost structure-less, very-large-particle carbon blacks (MT/FT)[25,26] and filler concentrations of $\phi$ smaller than 0.2.

Considering the asymmetry of fillers for the purpose of practical applications, Guth[27] later proposed another equation (Eq. 6.26) for rod-shaped particles:

$$E = E_0\left(1+6.7f_s\phi+16.2f_s^2\phi^2\right) \tag{6.26}$$

where $f_s$ is the shape factor. Mullins and coworkers[26] applied this equation to calculate the strain amplification of an NR matrix for an HAF-filled compound and discovered the optimal fit with the experimental data occurred when $f_s$ had a value of 6.7. Meinecke[28] found a value for $f_s$ of 4.7 for HAF in SBR. Medalia[29], on the other hand, used electron microscopy in his investigation of HAF blacks and obtained values between 1.7 and 1.9. This was confirmed by Hess[30]. Ravey et al.[31], in their investigations using light scattering, obtained values between 1.8 and 2.2 for the range of ultra-low-structure to high-structure HAF carbon blacks. The applicability of Eq. 6.23 and the significance of the shape factor for reinforcing fillers are therefore doubtful.

Considering the effect of filler structure on modulus, Medalia[29] replaced $\phi$ in Eq. 6.25 with the effective volume fraction of the filler, $\phi_{eff}$ (Eq. 6.27):

$$E = E_0\left(1+2.5\phi_{eff}+14.1\phi_{eff}^2\right) \tag{6.27}$$

He believed that $\phi_{eff}$ related to the occlusion of rubber[32]. Wolff and Donnet[33,34] investigated the extension modulus of SBR compounds filled with a series of carbon blacks chosen at random. They also used the effective volume by introducing a factor $f_v$ which converts the filler volume fraction into an effective volume fraction, i.e.,

$$f_v\phi = \phi_{eff}' \tag{6.28}$$

They found that the experimental data could be fitted to the Guth-Gold equation for a wide range of quasi-static tensile strains and filler concentrations, even though this was not justified by the original theory. The original theory is limited to very low levels of strain where the stress-strain behavior is linear.

Since the factor $f_v$ is related to the effective volume of the filler in the rubber matrix, all factors influencing the strain of the polymer under stress will have a determining effect on the value of $f_v$. As mentioned in Section 3.3, Medalia[29] related the effective volume to the occlusion of rubber and believed that only part of the occluded rubber, calculated from Eq. 3.3, could be effectively shielded from bulk deformation, and is defined as an effectiveness factor $F$. For tetramethylthiuramdisulfide (TMTD)-cured SBR filled with different grades of carbon black, an effectiveness factor of 0.5 was introduced empirically to fit the experimental data of low-strain equilibrium modulus.

In Figure 6.6, the effective factor $f_v$ is plotted as a function of compressed dibutylphthalate absorption (CDBP), and the straight line represents the values calculated from Eq. 3.5 with an effectiveness factor $F$ of 0.70. Although the effectiveness factors are, to a large extent, determined by the occlusion of rubber, significant deviations are observed. In contrast to the large-particle carbon blacks, the systematically higher $f_v$ values of small-particle carbon blacks may be associated with other factors influencing the effective volume.

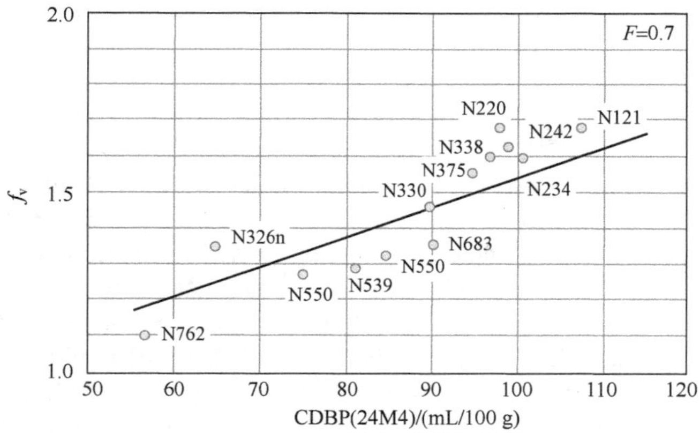


**Figure 6.6** Effective factor $f_v$ *vs.* CDBP (24M4) for carbon black filled SBR. The straight line is calculated from Eq. 3.5 with an effectiveness factor of 0.7[35]

First, it was argued that depending on the strength of polymer-filler interaction, physical adsorption and/or chemisorption of rubber molecules may take place on the filler surface. This interaction leads to an effective immobilization of the elastomer segments. As discussed in Chapter 3, depending on the intensity of the filler-polymer interaction and the distance from the filler surface, the mobility of the polymer segments near the interface is lower than in the matrix. This rubber portion was identified as a quasi-glass state by means of NMR[36–40]. Kaufman et al.[41] and Smit[42–44] demonstrated the presence of different regions within the polymer characterized by different degrees of molecular mobility. The modulus of the inner shell is very high and decreases gradually with increasing distance from the filler surface. The amount of rubber in this quasi-glass state, or volume of the shell, obviously depends on the strength of the polymer-filler interaction and the surface area of the filler. Therefore, the filler "surface activity" or surface energy and its particle size may be considered as factors influencing the effective volume of the filler.

Second, considerable evidence supports the tendency of filler aggregates to associate to form agglomerates, especially at high filler loading. This may lead to a chain-like filler structures or clusters, which may affect the properties of vulcanizates[11,45–50]. A large difference in surface energy between filler and polymer and short interaggregate distance could result in a high degree of agglomeration. This agglomeration is, of course, highly strain- and temperature-dependent. At moderate and high strains, the agglomerates are largely destroyed, and the rubber trapped within the secondary filler structure acts as the polymer matrix. On the other hand, raising the temperature would weaken interaggregate interaction and diminish the modulus of the rubber shell by reversing immobilization. Consequently, reduced filler agglomeration is expected.

As will be discussed further in Chapter 7, the agglomeration of filler aggregates and the high strain- and temperature-dependence of filler agglomeration play a critical role in determining the dynamic properties of filled rubber. The effects of individual filler parameters on the effective filler volume are associated with different mechanisms that differ in their strain and temperature dependence. These mechanisms are summarized in Table 5.3[51].

## 6.2.2 Hardness

For vulcanizates, the strain involved in the hardness test includes shear, extension, and compression, though the strain amplitudes are very small. Generally, hardness is related to Young's modulus measured at very low deformation. Figure 6.7 shows stress-strain curves for typical tread compounds. All vulcanizates are prepared with the basic formulation of 100 phr SSBR/BR blends (70/30), 28 phr oil, and 78 phr fillers including carbon black (N234) and silica (Zeosil 1165MP) with and without bi-functional silane coupling agent bis(3-triethoxysilylpropyl)-tetrasulfide (TESPT). The silane dosage is 6.4 phr. All compounds are mixed with conventional procedures except EVEC, which is silica-filled compound produced with liquid phase mixing. (Detailed information about these compounds can be found in Section 9.4.2.) The slope at the origin of the tensile stress-strain curve is Young's modulus. At such low deformation, the filler agglomerates cannot be destroyed. The trapped rubber in agglomerates does not participate in the deformation, which has the effect of increasing the effective filler volume. Therefore, both Young's modulus and hardness increase. This is confirmed by the strain dependence of elastic modulus, $G'$, i.e., the decrease of $G'$ with increasing strain amplitude (i.e., the Payne effect, see Section 7.2.1), which has been taken as a measure of filler agglomeration. As can be seen in Figure 9.25, the Payne effect of carbon black compounds is the highest, followed by the silane-coupled silica compound prepared with dry mixing. The silica compound with the same formulation and mixed with liquid phase (EVEC) has the lowest Payne effect. It can be concluded that among others, such as crosslink density of the polymer matrix, and polymer-filler interaction, the filler agglomeration plays an important role in the hardness of filled vulcanizates.

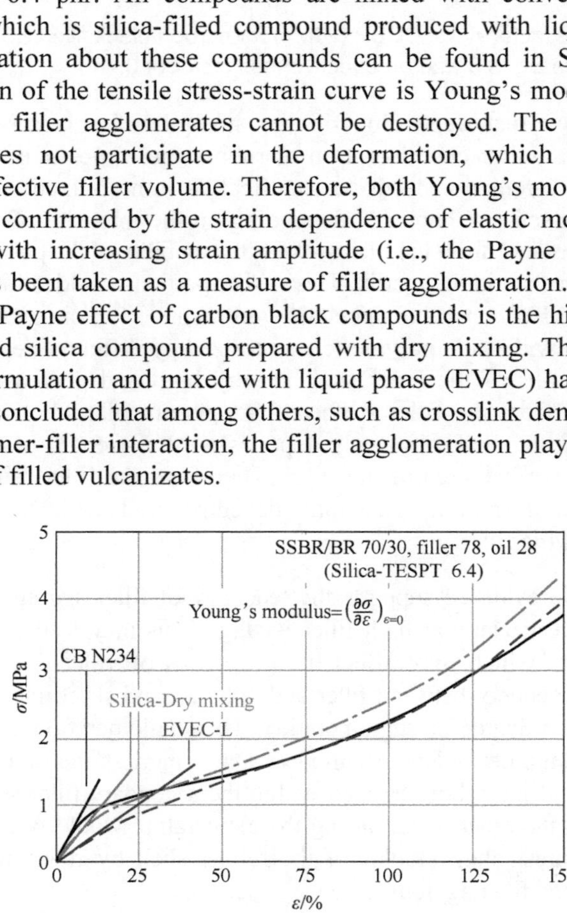


**Figure 6.7** The stress-strain curves and Young's modulus of a variety of tread vulcanizates for passenger tires

### 6.2.3 Medium and High Strains—The Strain Dependence of Modulus

Figure 6.8(a) shows the stress-strain curve of NR vulcanizate with 50 phr carbon black N110, typical for almost all filled rubbers. Instead of a linear stress-strain behavior achieved under certain limitations of strain for unfilled vulcanizates, addition of fillers leads to non-linearity at very low strain. A decrease in modulus occurs when the strain is increased from as little as 1% to an intermediate level. This can be seen from Figure 6.8(b) in which the moduli, $d\sigma/d\lambda$, are presented as a function of strain $\lambda$.

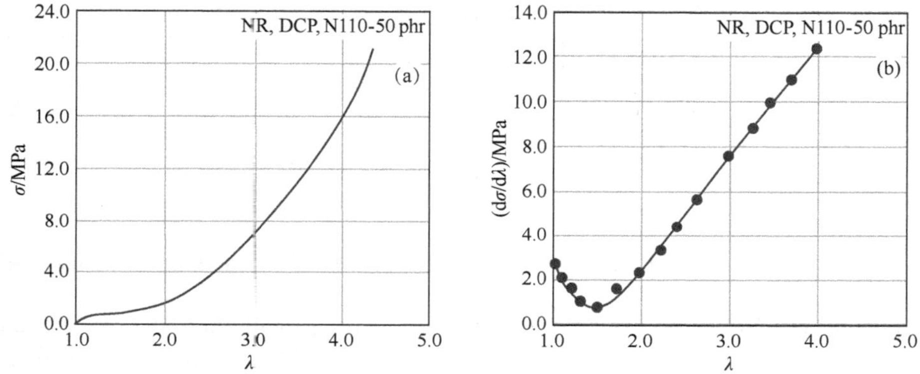


**Figure 6.8** Stress and modulus as a function of strain for NR vulcanizates filled with carbon black N110

As mentioned before, Wolff and Donnet[33] in their systematic study of the effect of carbon blacks on stress-strain behavior investigated the applicability of the Guth-Gold equation to the non-equilibrium tensile modulus for the first stretch of the stress-strain curve. For all carbon-black-loaded NR compounds, the $f_v$ factors, as defined by Eq. 6.28, first decrease with increasing deformation and then increase at high strain, with a minimum $f_{v\text{-min}}$ typically occurring at $\lambda \approx 1.75$. This is similar to the modulus-strain plots. The low strain modulus is mainly related to filler agglomeration that is generally controlled by the interaggregate interaction. The breakdown of agglomerates is responsible for reduced modulus upon strain in this region as the trapped rubber is released, reducing the apparent filler volume.

While the low-strain behavior can be explained mainly by the destruction of filler agglomerates, non-affine deformation (non-Gaussian behavior) of polymer chains and crystallization are involved at high strain. This leads to an increase in stress at a given strain, which is equivalent to an apparent rise in filler loading. This effect is reduced by slippage and/or detachment of the rubber molecules on the filler surface and, hence, is related to the polymer-filler interaction. Therefore, the slope of $d\sigma/d\lambda$ versus strain, $\lambda$, is closely related to the surface activity of the fillers, in particular the dispersive components of the filler surface energy, as far as hydrocarbon elastomers are concerned.

Custodero[52] studied the effects of surface treatments on the stress-strain characteristics of SBR vulcanizates filled with carbon blacks N234 and its derivatives with different surface modifications, such as oxidation, esterification, and graphitization (Figure 6.9).

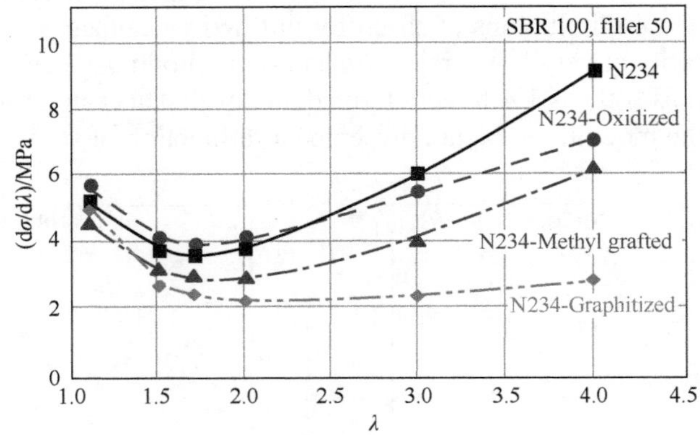


**Figure 6.9** The effect of surface modifications of carbon black on moduli of vulcanizates[52]

With regard to surface activity, in terms of polymer-filler interaction, the active centers of carbon black surface have been assigned to the defects and edges of surface crystallites[53]. Carbon black oxidation occurs at the active centers and introduces a variety of oxygen-containing groups on the filler surface. Upon oxidation, while the polar component of the surface energy increases, the dispersive component decreases significantly. This explains why the modulus is higher at low strain as more agglomerates are formed, and the modulus is lower at high strain as the polymer-filler interaction is lower, compared to unmodified carbon black.

With alkylation, the methyl group on the carbon black may not only reduce the surface polarity, but also block the active center, resulting in lower $\gamma_s^d$. This, on the one hand, depresses the filler agglomeration, and on the other hand, reduces the polymer-filler interaction. As a result, over the whole range of strain, the modulus of the vulcanizate is lower than that of the carbon black filled rubber.

During graphitization at very high temperature such as 2700°C, all functional groups are decomposed[54,55], and hence, the dimensions of the graphitic crystallites are drastically increased[56–58]. In this process, the highly energetic sites and polar oxygen-containing groups are basically eliminated[59]. Therefore, the surface energy and polarity are all decreased. This should depress the filler agglomeration and lower the modulus at higher elongation. In fact, at very low strain the modulus is comparable with its ungraphitized counterpart but it decreases more rapidly with increasing strain. It seems that the agglomerates of the graphitized carbon black in the polymer matrix are more developed than the agglomerates of the unmodified carbon black, but weaker, which is likely due to the lower filler-filler interaction. As discussed by Wang[60], compared to polymer, the surface energy of graphitized carbon black is still high[61], suggesting that the graphitized carbon black will have a high agglomeration tendency even though it is not comparable to ungraphitized black. However, due to the reduction of high energy sites on the black surface, which are mainly responsible for the adsorption of the polymer chains on the filler surface, the bound rubber content of the filled rubber is drastically reduced[60,62]. In a kinetic view of flocculation of filler, as the bound rubber can substantially increase the effective dimension of the filler aggregate and may be entangled with the polymer matrix which serves as an anchor to further reduce the agglomeration of the carbon black, the reduction of bound rubber for graphitized carbon black-filled compound would more favorable for agglomeration. Thus, weaker but more developed filler agglomerates are expected for graphitized carbon black. Of course, modulus should decrease more rapidly with strain at low elongation. At higher strain, compared with all other blacks, the modulus of the vulcanizates filled with graphitized black increases more slowly, which is certainly attributed to its low polymer-filler interaction.

Shown in Figure 6.10 are the data of stress-strain curves. All vulcanizates are the same as those used for the hardness study presented in Section 6.2.2 (Figure 6.7), which are the SSBR/BR filled with carbon black, and silica with and without coupling agent and mixed with different techniques.

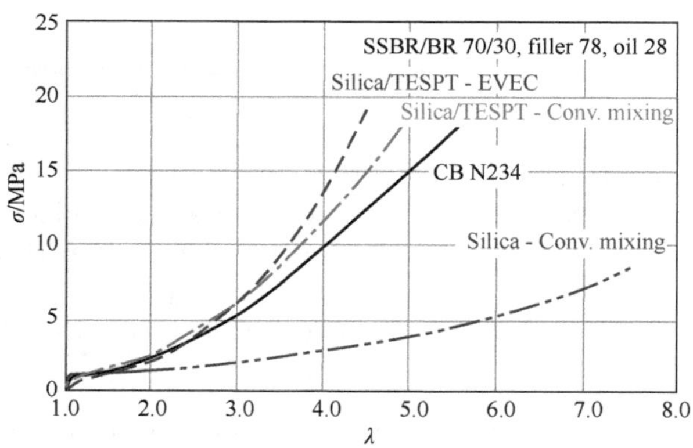


**Figure 6.10** The stress-strain curves of a variety of tread vulcanizates of passenger tires

Figure 6.11 presents the strain dependence of modulus. For all vulcanizates filled with different fillers, the modulus decreases very rapidly with increasing strain, passing a minimum at $\lambda$ about 1.5, and then increases. Silica compounds without the TESPT coupling agent and mixed by a conventional procedure yield the highest modulus at lowest strain, followed by carbon black vulcanizates. The EVEC vulcanizate shows the lowest modulus. The same sequence is found for differences between modulus measured at lowest strains and the minima. While several factors, such as crosslink density and filler dispersion, may influence the modulus of the filled rubber, at low strain the difference between the modulus at lowest strains and the minima would be mainly attributed to agglomeration of the fillers. With increasing strain, the number of agglomerates broken-down would increase and so the modulus decreases. The Payne effect, which will be discussed in detail in Chapter 7, is the difference between the modulus at lowest strains and the minimum modulus, and would be taken as a measure of filler agglomeration. This means that the silica-filled vulcanizate without a coupling agent has the most developed filler agglomeration since the filler possesses the highest polarity and lowest non-polar component of surface energy. With a lower polar component and a higher dispersive component of the surface energy, carbon black produces lower differences in modulus for the low strain region.

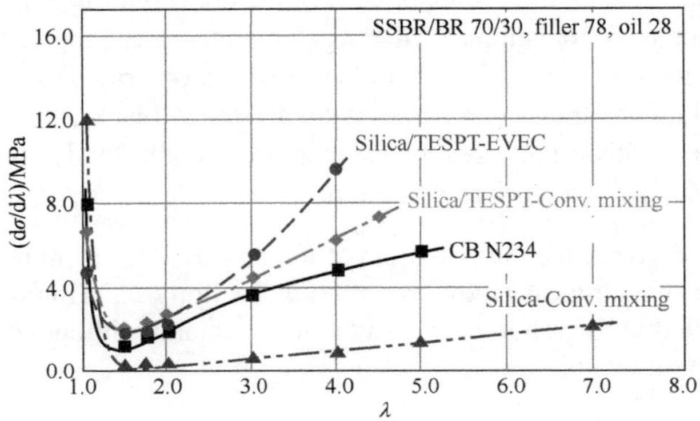


**Figure 6.11** The moduli as a function of strain for a variety of tread vulcanizates of passenger tires

The silica with coupling agent produces lower values because surface modifications that improve the affinity and compatibility between the filler and hydrocarbon polymers result in considerable reduction of filler agglomeration. Therefore, the higher efficiency of the surface modification and best filler dispersion of EVEC make silica-filled vulcanizate the lowest difference between the modulus at lowest strains and the minimum.

Beyond the strain where the minimum occurs, the modulus of the vulcanizate increases with increasing strain. As mentioned before, in this region the breakdown of the filler agglomerates has been complete, and the rate of the modulus increase is determined by polymer-filler interaction. Accordingly, the silica compound gives the lowest polymer-filler interaction and this interaction is considerably improved with the bifunctional coupling agent TESPT, which introduces chemical bonding between the rubber molecular chains and the filler surface. It is thus not surprising that EVEC shows the highest rate of modulus increase due to its improved efficiency of the coupling reaction and excellent dispersion that reduces the slippage of the polymer chains along the filler surface and vacuole formation. This is in agreement with the direct observation of aggregate separation stress from the polymer matrix in the filled vulcanizates by means of scanning electron microscopy (see Section 6.3.2.2).

From Figure 6.9 and Figure 6.11, it can be observed that beyond $\lambda = 2$, i.e., 100% elongation, the modulus increases almost linearly with increasing strain. The slopes of the linear section are determined by the polymer-filler interaction. Since there is a direct relationship between the modulus and the stress at any given elongation, the slope of the linear section can be estimated from the tensile stresses (T) at two different elongations over the range of testing. Practically, the ratio of tensile stress at 300% elongation (T300) and at 100% elongation (T100), i.e., T300/T100 is a reasonable measure of the polymer-filler interaction.

## 6.3 Strain-Energy Loss—Stress-Softening Effect

It is generally accepted that rubber reinforcement by fillers is closely related to the processes of energy dissipation during deformation. The importance of energy dissipation in the fracturing of vulcanizates, such as ultimate strength, tearing, cracking, abrasion, and fatigue, is also widely recognized. In practice, stress-softening effects are used as a quantitative measure of strain energy loss.

It has been known for many years that for vulcanizates either with or without fillers, deformation results in the softening of rubber and that the initial stress-strain curve determined during the first deformation is unique and cannot be retraced. This effect is generally called stress softening. Further repeated deformations will cause the vulcanizates gradually to approach a steady state with a constant or equilibrium stress-strain curve. Softening in this way occurs in both unfilled and filled vulcanizates, although the effect appears to be much more pronounced in vulcanizates containing high proportions of reinforcing fillers.

The stress-softening phenomenon was first observed by Bouasse and Carrière in 1903[63]. Holt[64] studied the effect of repeated stretching and the speed of stretching on the stress-strain properties of typical rubber compounds and their recovery after straining. He showed that the initial strain history has a remarkable effect upon the stress-strain performance of vulcanizates, and that rubber will not recover completely after straining. However, he did not offer a theoretical explanation for this phenomenon. Subsequently, this phenomenon was studied in more detail by Mullins[65,66], and has been termed the "Mullins effect". He showed that while the softening of filler reinforced vulcanizates mainly occurred during the first cyclic stretch, after a few stretching cycles a steady state will be reached. The performance of stress-strain is relatively unaffected by previous stretching as long as the stretch is higher than the maximum stretch applied.

Typical stress softening can be seen in Figure 6.12 for NR gum vulcanizate and vulcanizates filled with different loadings of carbon black HAF (N330). The three vulcanizates were stressed initially to the same load on a tensile tester at the same strain rate and at room temperature. Also shown on the figure are the first retraction (curves 2) and the second complete stressing cycles (curves 3 and 4) for extensions up to the same strain as the initial strain.

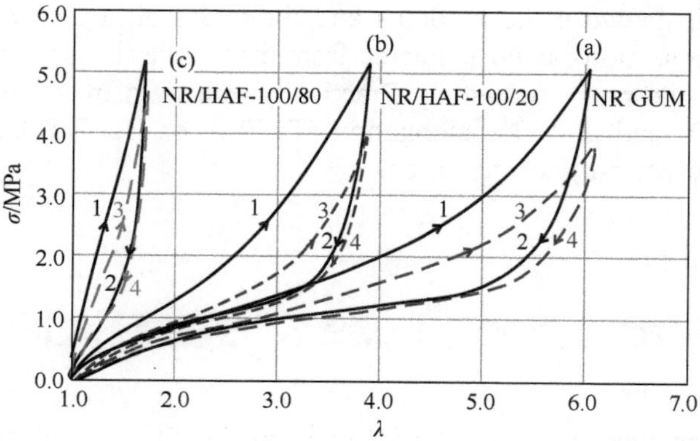


**Figure 6.12** Stress (referred to original cross-section) as a function of strain for (a) gum vulcanizate, (b) vulcanizate containing 20 phr N330, and (c) vulcanizate containing 80 phr N330; (1) initial stressing curve; (2) first retraction curve; (3) second stressing curve; (4) second retraction curve[67]

The stress softening measured with the second mode of tensile tests is shown in Figure 6.13 for vulcanizates of NR gum and NR filled with 60 phr carbon black N330. When the stress-strain curves of the vulcanizates are increased to the same stress level, the shapes of stress-softening curves between the first, second, and third extension cycles are quite similar.

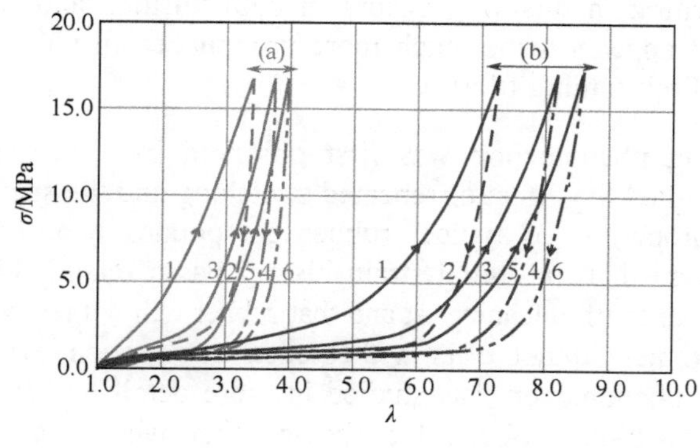


**Figure 6.13** Stress-strain curves for (a) NR vulcanizate filled with 60 phr N330 carbon black and (b) NR gum vulcanizate additional extensions[68]

Another stress-softening tensile test is performed that is similar to how new samples of gum or filled vulcanizates are stretched to a given elongation and then retracted. After the first stress-strain cycle, the subsequent extension cycle is performed in a way that each succeeding stress-strain cycle is stressed to a higher level in incremental steps. The results are shown in Figure 6.14, in which the stress-strain curves of fresh samples extended to break on initial stressing are also plotted. It can be seen that the correlation of the maximum stress-strain loops with the original curve is excellent for SBR vulcanizate filled with 60 phr carbon black N220. In the case of NR vulcanizate containing 60 phr carbon black N220, the performance of the NR vulcanizate is very similar to the SBR vulcanizate at low strains. At high extensions, the stress at the maximum of the stress-strain loops is significantly lower than the original curve. In other words, no coincidence of the previously unstretched and the cycled data occurs.

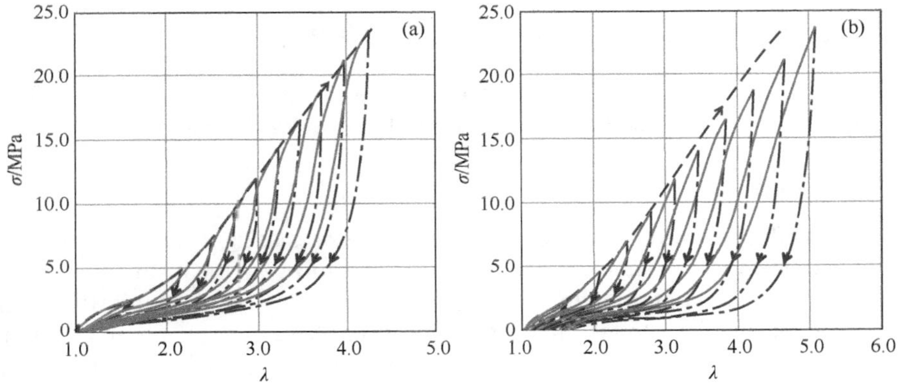


**Figure 6.14** Stress-strain curves for (a) SBR vulcanizate filled with 60 phr N220 carbon black and (b) NR vulcanizate filled with 60 phr N220 carbon black. Dash lines – the stress-strain curves for a new sample[69]

Although there are some other strain states employed in the literature, such as equibiaxial tension[70,71], pure shear[72], uniaxial compression[73,74], etc., uniaxial extension is predominately used in Mullins effect research.

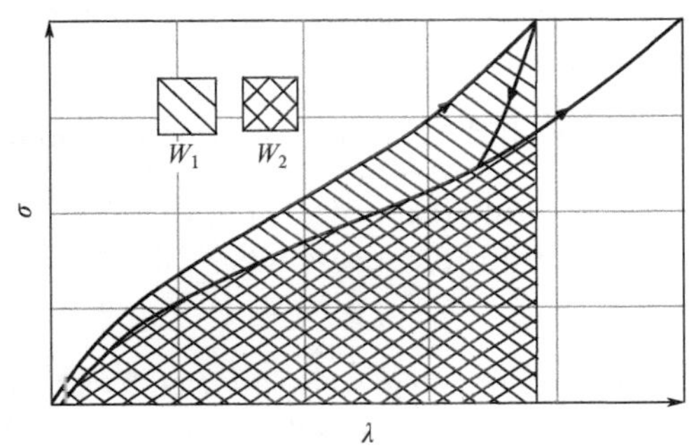


**Figure 6.15** Schematic representation of stress-softening effect

Generally, stress softening is expressed by the reduction, $\Delta W$, in energy input between the first and second stretch to a given elongation in cyclic uniaxial tension test[75]. After retracting to zero stress, the specimens are immediately re-stressed to the setting elongation (shown in Figure 6.15):

$$\Delta W = \frac{W_1 - W_2}{W_1} \times 100 \tag{6.29}$$

where $W_1$ is the energy required in the first extension to a given elongation; $W_2$ is the energy required in the second extension to the same elongation measured shortly after the first extension (Eq. 6.29).

### 6.3.1 Mechanisms of Stress-Softening Effect

No single mechanism is sufficient to account for all of the observed stress softening. This applies not only for the filled vulcanizates, but also for the cured gum.

#### 6.3.1.1 Gum

The Mullins effect has been observed with gum vulcanizates using uniaxial extensions[67,76]. Phenomenologically, in non-crystalline gum vulcanizates most of the softening appears to be due to conformational changes of the rubber fine structure, which permit subsequent deformations to take place more readily. These changes are related to non-affine rearrangement of the molecular networks, which involves displacement of network junctions and entanglements from their initial random positions due to localized non-affine deformation as short network chains became fully extended. This also explains why the softening or set in gum vulcanizates may recover almost completely by heat treatment or swelling in solvents. There is evidence of network junction breakdown in some vulcanizates with weak crosslinks[77]. In this case, more permanent softening or set may take place. This is especially true in the regions where the rubber is highly strained, due to the finite extensibility of highly strained molecular chains. This statement was confirmed by Harwood and Payne[76,78], who showed that in gum vulcanizates, the chemical nature of the network crosslinks had an effect on the extent of stress softening in the following sequence:

Polysulfidic crosslink > monosulfidic crosslink > carbon-carbon crosslink.

This sequence is consistent with the order of the bond energies as polysulfidic crosslink <268 kJ/mol, monosulfidic crosslink −285 kJ/mol, and carbon-carbon crosslink −352 kJ/mol[79].

For crystalline gum vulcanizates, softening is also affected by strain-induced crystallization, especially at high deformation. Figure 6.16 shows the stress softening of unfilled NR vulcanizate in cyclic uniaxial tension. The $\Delta W$ remains essentially constant at the extension ratio $\lambda < 4$, which may be related to the quasi-irreversible rearrangement of the molecular network. However, the degree of softening increases rapidly with an extension increase when $\lambda > 4$. For higher elongations, other processes may take place, such as strain-induced crystallization. The strain-induced crystallites are not restored to their original state without delay in the next stretch, resulting in increased softening with more strain-induced crystallites. Harwood and Payne[69] reported that the maximum of the loops in cyclic extension corresponds well with the original curve from a new specimen stressed initially for gum NR vulcanizates at low deformation, though not at high elongation (> 400%). Moreover, the deviation between the maximum loops and the original curve increases with increasing extension, which is similar to the softening effects of carbon black N220 filled NR vulcanizates shown in Figure 6.14. They attributed this deviation to strain-induced crystallization of NR. Based on the observation of volume reduction upon stress relaxation in stretched vulcanizates, Gent[80] relates stress reduction to strain-induced crystallization. Rault et al.[81] measured the crystallinity of pure and carbon black filled natural rubber during cyclic uniaxial tension by WAXS and $^{2}H$ NMR. They found that the crystallization increases linearly with the extension ratio, and fillers play a role in the nucleation centers of the strain-induced crystallization of vulcanizates. More recently, Toki and colleagues[82,83] employed in situ synchrotron wide-angle X-ray diffraction to study the strain-induced crystallization and molecular orientation in natural rubber and synthetic rubbers, such as polyisoprene, polybutadiene, and butyl rubber, during uniaxial deformation. It was found that the molecules in vulcanizates under high strains are composed of three states: unoriented amorphous state (50%–75% in mass), oriented amorphous state (5%–25%), and strain-induced crystallites. The microfibrillar crystalline structure results from the heterogeneity of the network topology. In addition, the molecular orientation and strain-induced crystallization depend on the intrinsic characteristics of chain crystallizability and crosslinking topology.

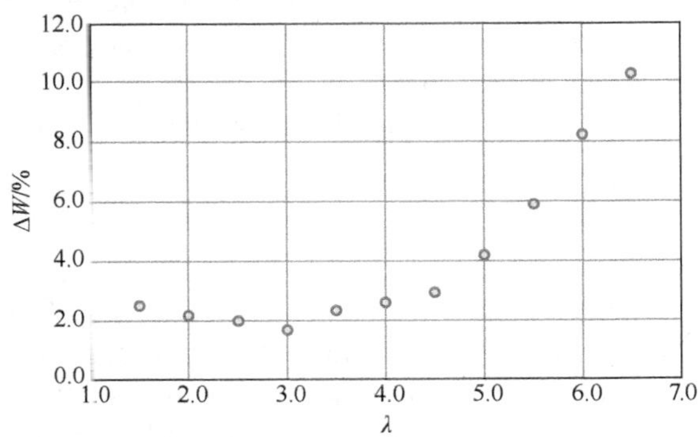


**Figure 6.16** Degree of stress softening of unfilled NR vulcanizate[84]

#### 6.3.1.2 Filled Vulcanizates

For filled vulcanizates, the mechanisms involved in stress softening are more complicated. At small deformations, pronounced softening occurs due to a breakdown of filler agglomerates, though softening due to this cause is relatively small at larger deformations.

In large range of deformations, Mullins and Tobin proposed a phenomenological approach to describe the stress-strain performance of filled vulcanizates[85]. Based on the observation that after previous extension, the stress-strain curve shows the characteristic shape normally associated with gum vulcanizates except that the upturn is more rapid than that of a gum vulcanizate and occurs at lower extensions, they devised a simple model which enabled the whole course of the stress-strain curves determined both before and after previous deformation to be quantitatively described. In their model, a filled vulcanizate consists of two regions, hard and soft rubber respectively. On application of a stress most deformation takes place in the soft regions; the hard regions undergo little deformation but could break down to form additional soft regions if stress application exceeds the maximum. During the initial extension, the fraction in the soft regions increases continuously with increasing extension. Thus, if a fraction $\phi$ of the vulcanizate is in the hard region, the fraction $(1 - \phi)$ is in the soft region. Thus, the measured strain $\varepsilon$ is related to the actual strain by the relation:

$$\varepsilon = (1-\phi)\varepsilon_s + \phi\varepsilon_H \tag{6.30}$$

and if the contribution of the hard regions to deformation, $\varepsilon_H$, is neglected, one has:

$$\varepsilon = (1-\phi)\varepsilon_s \tag{6.31}$$

This suggests that at the same strain of the vulcanizate, the presence of filler leads to an increased strain $\varepsilon_s$ in the soft regions. This is known as strain amplification.

Mullins and Tobin[26] assumed that the average strain present in the rubber phase increases by a certain factor due to the presence of filler. They recommended using the volume concentration factor of the Guth-Einstein expression to the strain amplification factor $X$ that describes the ratio of the effective average strain in the rubber phase, $\lambda'$ to the measured overall strain, $\lambda$:

$$X = \lambda' / \lambda = 1 + 2.5\phi + 14.1\phi^2 \tag{6.32}$$

where $\phi$ is the volume fraction of the filler in the vulcanizate. This factor only corresponds with large-particle-size thermal blacks at moderate extensions. For reinforcing blacks, one must assume abnormally large shape factors in the modified form of the Guth equation. In this case, the following relation (Eq. 6.33) has been proposed:

$$X = \lambda' / \lambda = 1 + 0.7 f_s\phi + 1.62 f_s\phi^2 \tag{6.33}$$

where $f_s$ is a factor describing the shape of the asymmetric particles as expressed by the ratio of their length to their diameter.

The use of this strain amplification factor involves the assumption that the effective strain in the rubber phase of a filled vulcanizate at any stress is equal to the strain in the base vulcanizate without filler at the same stress. With this approach, when all the vulcanizates filled with a series of carbon blacks are subjected to the same initial maximum stress and therefore to the same effective maximum strain, it is possible to normalize the strain data from both gum and filled vulcanizates by using factor $\varepsilon/\varepsilon_{max}$, where $\varepsilon$ is the measured extension and $\varepsilon_{max}$ is the maximum extension corresponding to the same initial maximum stress. The normalized stress-strain curves obtained during second extension cycles are plotted for both gum and filler-loaded natural rubber vulcanizates (Figure 6.17). As can be seen, the curves are very similar for all the vulcanizates studied. Thus, Mullins and colleagues[67,77] concluded that the stress-softening effect of filled vulcanizates originates from the same source as gum vulcanizates, except that the strain in the rubber network is amplified by the presence of filler in the filled ones. In other words, the softening process is primarily due to the rubber phase itself, and that the contribution of the filler to this softening is relatively small.

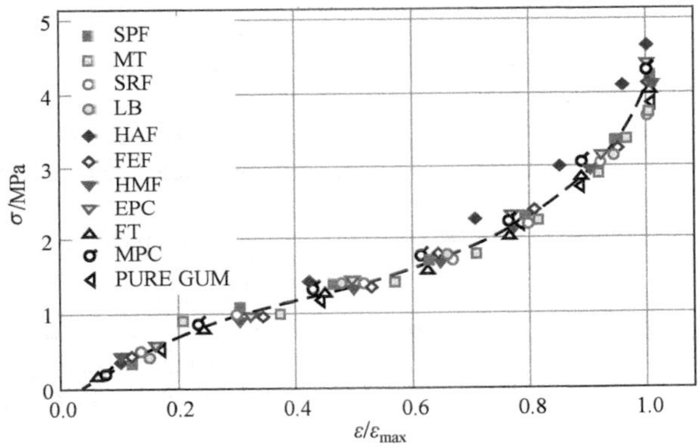


**Figure 6.17** Stress-normalized strain curves obtained during second extension on gum and carbon black filled vulcanizates[67]

With different molecular approaches, Blanchard and Parkinson[86,87] attributed stress softening primarily to breakage of weak linkages related to the filler. The linkages formed through fillers are of two types. Stressing can break the relatively weak linkage due to physical attachments.

Bueche[88,89] proposed a mechanism of stress softening based on the concept of network chain breakage, or their attachment to adjacent filler particles, during extension. These interparticle chains have a distribution of lengths, and the shorter chains will rupture first, at small elongations. Chains broken on the first stretch will not be able to affect stiffness on the second stretch and softening will occur. In addition to chain breakage, Bueche offers another possible process of stress relaxation involving the movement of the aggregate, rotationally and otherwise through the surrounding matrix, especially at high temperatures and in slow tests. Small motions of filler aggregates are sufficient for the relief of excessive stress in localized areas. This increases the strength as the filler aggregates distribute the load more equitably. This load-sharing mechanism is very important in determining tensile strength.

Following the works of Alexandrov and Lazurkin[90], Houwink[91], and Peremsky[92,93], who described stress softening in terms of molecular slippage on the filler surface, Dannenberg and colleagues[94–96] further developed the so-called "molecular slippage mechanism" to explain the complex mechanical properties of reinforced vulcanizates. This mechanism assumes that, under stress, the surface-adsorbed network segments move relative to the filler surface, accommodating the imposed stress and preventing molecular rupture. A schematic diagram illustrating this reinforcement mechanism is shown in Figure 6.18.

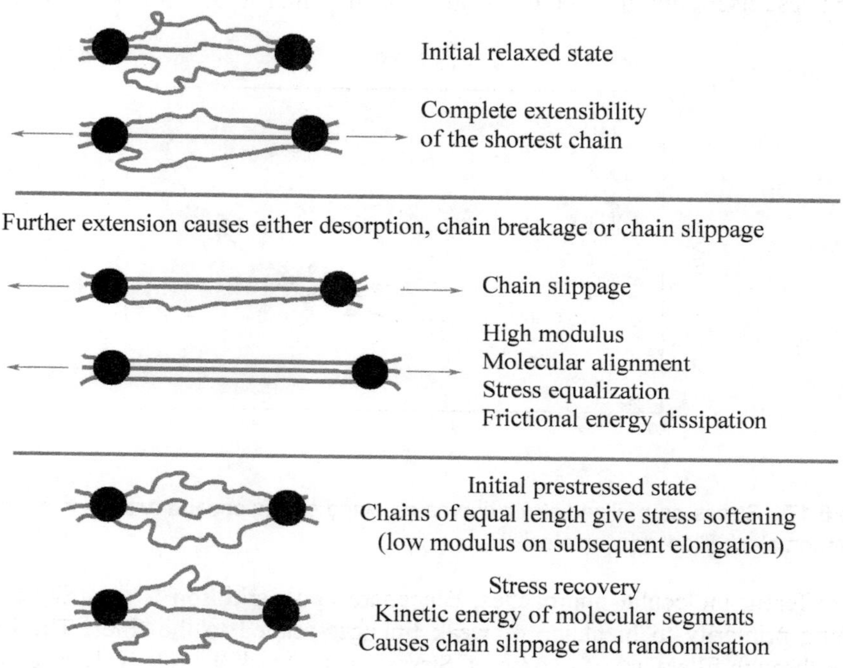


**Figure 6.18** Schematic diagram showing slippage of chains attached to two filler particles[94]

When the polymer-filler interaction is too low, besides slippage of the molecular segments, dewetting or detachment of the polymer from filler surface may take place, resulting in vacuole formation during stretching to softening. This was confirmed by electron microscope studies[97,98]. More recently, Zheng, Zhang, and Wang[99] have studied the vacuole formation during extension of the filled vulcanizate with scanning electron microscopy, and found that the cavity strain of the filled rubber correlates with the polymer-filler interaction. As can be seen in Figure 6.19, at 300% elongation, the cavitation is more developed for the lower surface area blacks that have lower polymer-filler interaction (see Table 2.14). Of course, the effect of aggregate geometry on the vacuole formation cannot be excluded. Vacuole formation at large deformations was also demonstrated by volume expansion due to pulling away of polymer from the filler surface[100].

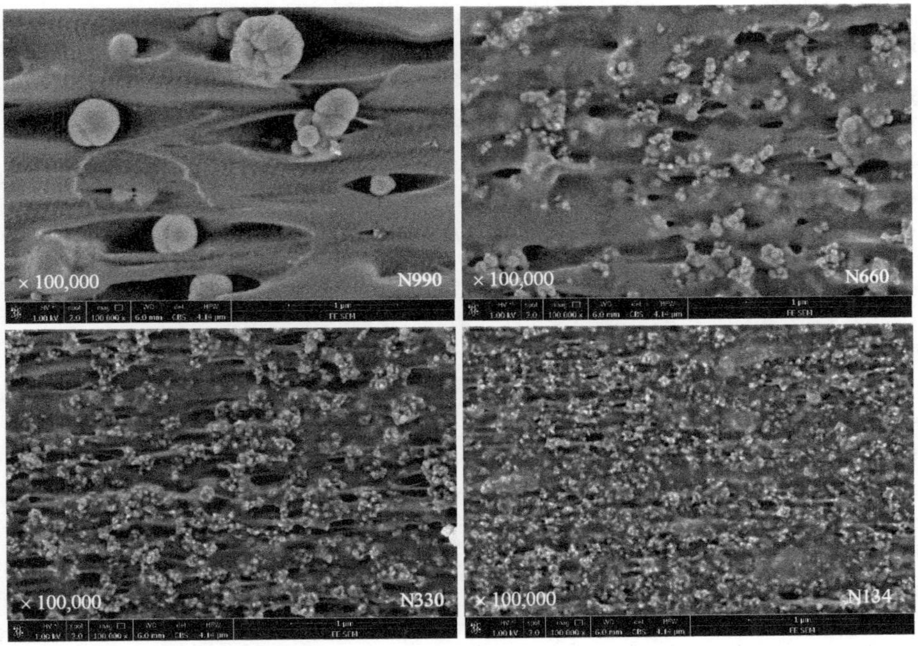


**Figure 6.19** SEM images of NR vulcanizate filled with 50 phr carbon black N990, N660, N330, and N134 at 300% elongation

#### 6.3.1.3 Recovery of Stress Softening

The softening resulting from deformation recovers slowly on standing, but may be accelerated and made more complete by an increase in temperature or by swelling in suitably chosen solvents. Although recovery of stress softening is often nearly complete in gum vulcanizates, in corresponding vulcanizates containing fillers the recovery is only partial and much slower[77]. Therefore, it is understandable that the mechanisms responsible for stress softening differ for the reversible and irreversible portions.

Obviously, the reversible softening in gum vulcanizates suggests that no structural breakdown has occurred during deformation. In this case, the softening is due solely to a change in the conformation of the molecular network, which involves a rearrangement or displacement of the network junctions. The incomplete recovery of softening in gum vulcanizates reflects a breakdown of some network junctions, and some reformation in the deformed state, notably polymers with weak crosslinks, which are readily broken and reformed.

For the irreversible or permanent portion of stress softening of filled vulcanizates, other processes should be involved, such as breakage of network chains, of filler-to-rubber bonds, and of filler agglomerates and filler structure[89,101], slippage of rubber-filler linkages on the surface of filler particles[94], and detachment of polymer chains from the filler surface (vacuole formation).

## 6.3.2 Effect of Fillers on Stress Softening

According to the mechanisms of energy loss during strain discussed above, the stress-softening effects of filled vulcanizates are determined by the types, loading, surface area, structure, and the surface characteristics of the fillers.

### 6.3.2.1 Carbon Blacks

#### 6.3.2.1.1 Effect of Loading

Zhong and his colleagues[102] investigated the effects of carbon black loading on stress softening of NR vulcanizates filled with carbon black N234. Figure 6.20 shows the volume fraction of the filler varying from 0 to 0.253. For the unfilled compound, the $\Delta W$ is very low, but is much higher for the filled vulcanizates, and it increases with increasing filler loading. On the other hand, when the strain increases, the stress softening increases with strain until $\lambda = 3.5$. Above this strain, while the $\Delta W$ of less loaded vulcanizates ($\phi$ = 0.046–0.127) continuously increases with extension, the softening effect decreases with further extension for the higher loaded compounds ($\phi$ = 0.195–0.253). It seems that several mechanisms may be involved in the loading effects. As discussed in Chapter 3, with increasing loading, the distance between aggregates decreases, which leads to a developed agglomeration; the interfacial area between the polymer and fillers increases, which gives the polymer chains more opportunity for slippage along filler surface; strain amplification augments as the polymer fraction reduces; and the dispersion of the filler in the polymer deteriorates which enhances vacuole formation.

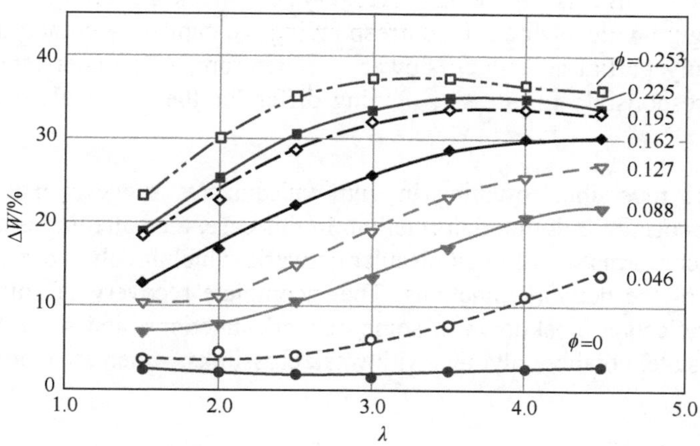


**Figure 6.20** Stress softening of carbon black N234 filled NR vulcanizates as a function of strain

At low strain, the main energy loss may be due to a breakdown of the filler agglomerates. This effect is more pronounced at higher loadings as more agglomerates are formed and the strain amplification effect also increases. In this case, besides the breakdown of agglomerates occurring at small strain, the molecular chain slippage over the filler surface may start at lower strain. Therefore, for high loaded vulcanizates, the stress softening at small extension is higher and rises fast with strain. At higher strain, the filler agglomerates seem to break down completely and polymer slippage increases, resulting in more energy loss. For vulcanizates with volume fractions higher than 0.16, the softening effect decreases with strain increase at high strain, and the maximum of strains decrease with loading. This may be related to different mechanisms. For the higher loaded compounds, the slippage and detachment of polymer chains over the filler surface should take place at lower strain due to the increased stress-amplification. In addition, the strain-induced crystallization starts at relatively lower extension when the filler loading increases and the crystallinity of NR is higher at the same elongation when more filler is added, leading to less softening.

#### 6.3.2.1.2 Effect of Surface Area

The effect of surface area on stress softening appears more complicated. The strain dependence of $\Delta W$ for low surface-area black N990 is similar to that for vulcanizates filled with less-loaded black N234 (see Figure 6.21). For the high surface area blacks such as N330 and N134, the strain dependences of energy loss is almost identical except at low strain where $\Delta W$ is higher. The higher energy loss of black N134 is caused by more agglomeration as its larger surface area and higher surface energy are more favorable for filler agglomeration.

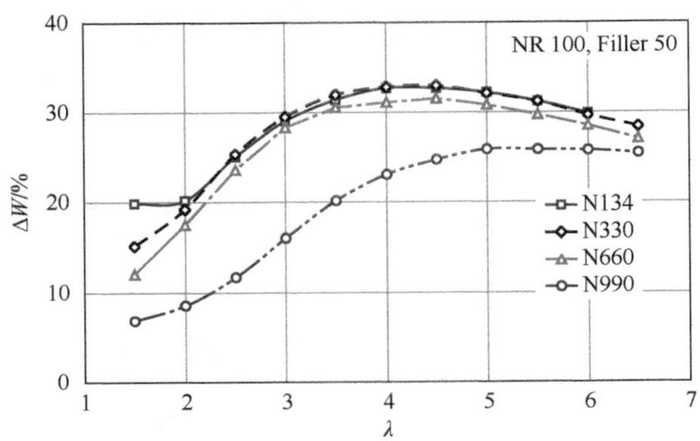


**Figure 6.21** Stress softening as a function of strain for NR vulcanizates filled with carbon blacks having different surface area

At medium and higher deformation, it is surprising to see that the surface area effect appears insignificant for the two reinforcing blacks even though the difference in their surface area is quite large. Several parameters may be responsible for these results.

Compared with lower surface area blacks, it possesses a higher $\gamma_s^d$ hence a stronger polymer filler interaction, while the finer particle filler N134 has a higher polymer interfacial area. More interface leads to a higher likelihood of polymer slippage, though the strong polymer-filler interaction would limit the slippage. In addition, the large particle black may be favorable for vacuole formation and crystallization of polymer, while the larger number of small particles of the higher surface black may interfere with orientation of polymer chains.

#### 6.3.2.1.3 Effect of Structure

The effect of carbon black structure on stress softening is shown in Figure 6.22. The three carbon blacks, N347, N330, and N326, have similar surface area, but their structures are quite different. The oil absorption values of compressed samples of the three blacks are 99 mL/100 g, 88 mL/100 g and 68 mL/100 g, respectively.

Surface energies of the three carbon blacks are similar (see Table 2.14), and the higher energy loss of higher structure blacks seems to be related to the higher effective volume as a result of more polymer occluded in the aggregates. This leads to higher stress-amplification, hence higher energy loss, although for high structure blacks the effective surface area of the fillers for slippage of polymer chains is significantly reduced because the stress of polymer in the recessed part of the aggregates is shielded.

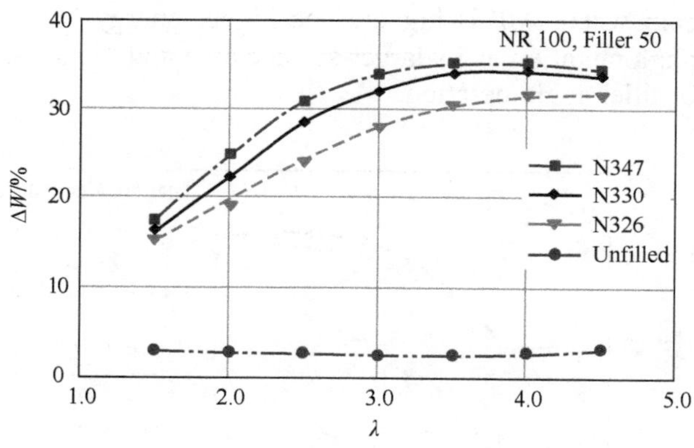


**Figure 6.22** Stress softening as a function of strain for NR vulcanizates filled with carbon blacks having different structure

### 6.3.2.2 Precipitated Silica

The main differences between silica and carbon black are their surface characteristics. Compared to carbon black, silica's low $\gamma_s^d$ and high polarity yield lower filler-polymer interactions and higher filler-filler interactions, and hence more developed agglomeration of filler aggregates in hydrocarbon rubbers. These, of course, are reflected in the stress softening.

Figure 6.23 presents the stress-softening data obtained using the same test methods as shown for Figure 6.14. All vulcanizates are the same as those used for the stress-strain study presented in Section 6.2 (Figure 6.7 and Figure 6.10), which are the SSBR/BR filled with different fillers and mixed with different techniques.

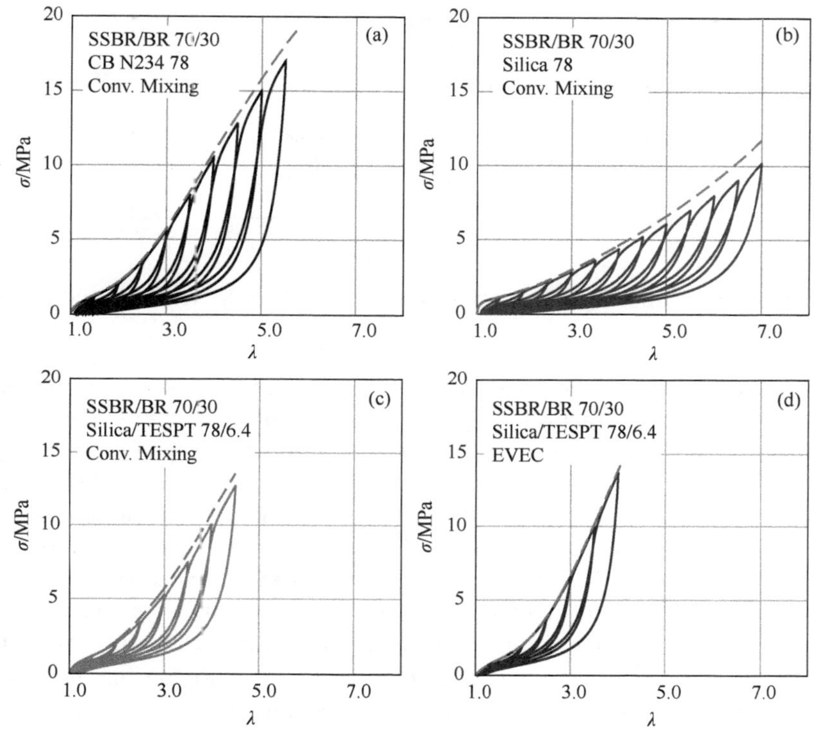


**Figure 6.23** Stress-strain curves for SSBR/BR vulcanizate filled with carbon black N234 (a) and silica (b)–(d). Dash lines – the stress-strain curves for a new sample

The $\Delta W$ as a function of strain $\lambda$ plots are given in Figure 6.24. In the case of the N234 compound, $\Delta W$ decreases with increasing strain, undergoes a minimum at $\lambda$ about 2.2, and then increases with strain. As mentioned above, the softening effect at low strain is mainly related to the breakdown of filler agglomerates. With increasing strain, the number of broken-down agglomerates is reduced, and so the $\Delta W$ decreases. At the same time, other processes may take place such as the breakage of polymer chains, or their attachments to adjacent filler particles, and the movement of the aggregate through the surrounding matrix. All these processes make the first stretch unable to affect the stiffness on the second stretch, leading to softening. Therefore, $\Delta W$ increases with strain. If this exceeds the effect of the agglomerates breakdown, a minimum will occur.

Due to the more developed agglomeration of silica in the polymer matrix, the energy loss of silica vulcanizate is significantly higher than its carbon black counterpart. $\Delta W$ decreases rapidly from first stretch to the $\lambda$ around 2.7, and then it increases very slightly.

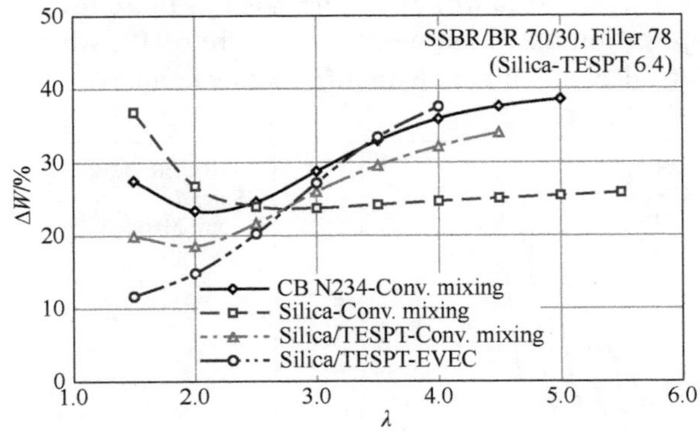


**Figure 6.24** Stress softening as a function of strain for SBR/BR blend vulcanizates filled with a variety of fillers

Both the lower value and the slight rise of $\Delta W$ at higher strains are related to the weak polymer-filler interaction. Upon strain, the polymer chains can easily slip over and are detached from the filler surface, which is accompanied by energy loss. As seen in Figure 6.23, the stress is much lower for silica at the same strain amplitude, such as $\lambda = 3$, compared with carbon black, even though the vacuole formation is similar (Figure 6.25). Upon further stretching, the growth of vacuoles due to low polymer-filler interaction easily relieves excessive stress in the localized area, causing the load to be more equally distributed. This allows the polymer molecules to orient along the direction of stress, forming micro-fibrillar crystalline structures as shown in Figure 6.26. This load-sharing mechanism is very important in determining elongation at the vulcanizate breakpoint. It seems that in order to improve elongation at break, it is beneficial to reduce polymer-filler interaction to a certain extent.

When coupling agent TESPT is added in the compound by conventional dry-mixing, its ethoxy groups react with the silanol groups on the filler surface causing the polar component of its surface energy to be significantly reduced. Therefore, upon silanization of the silica surface, filler agglomeration is substantially depressed, resulting in lower energy loss and a slow dip of $\Delta W$ at low strain, compared with an unmodified silica vulcanizate, even with a carbon black one. On the other hand, the tetrasulfide groups in TESPT will split and react with the rubber chains, forming mono-, di-, and poly-sulfidic covalent bonds during mixing and vulcanization of the compound. The introduction of the chemical bonds between silica and rubber limits the slippage and detachment of the polymer chains over the filler surface. This increases the stress during further stretching and dissipates more energy due to the breakdown of polymer chains and stronger polymer-filler linkages. Therefore, the $\Delta W$ rises more rapidly with strain compared with unmodified-silica-filled vulcanizates.

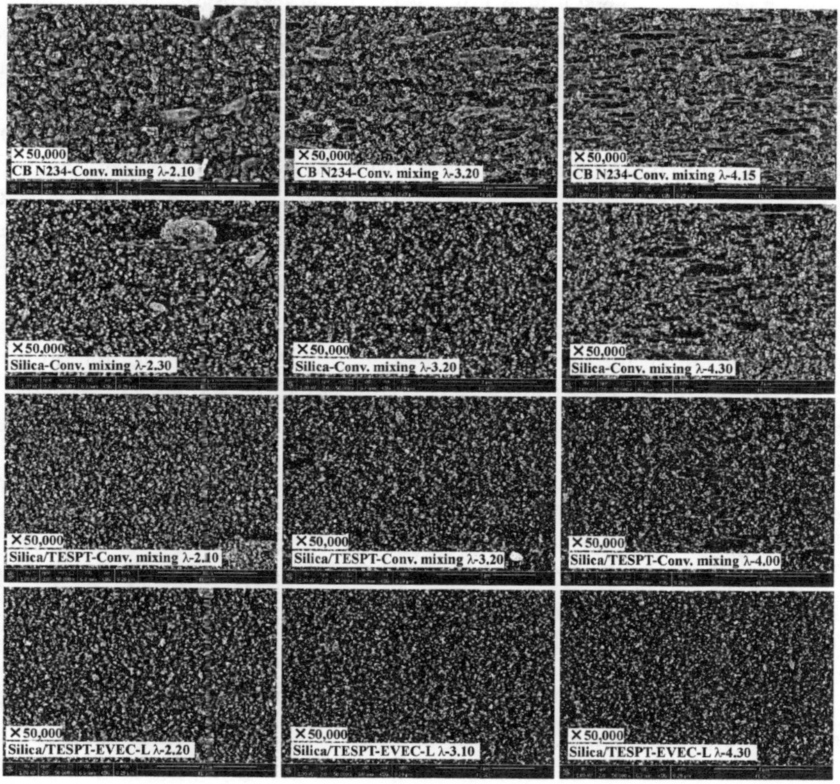


**Figure 6.25** SEM images of SSBR/BR (70/30) vulcanizate filled with 78 phr fillers and 28 phr oil at different strains[103]

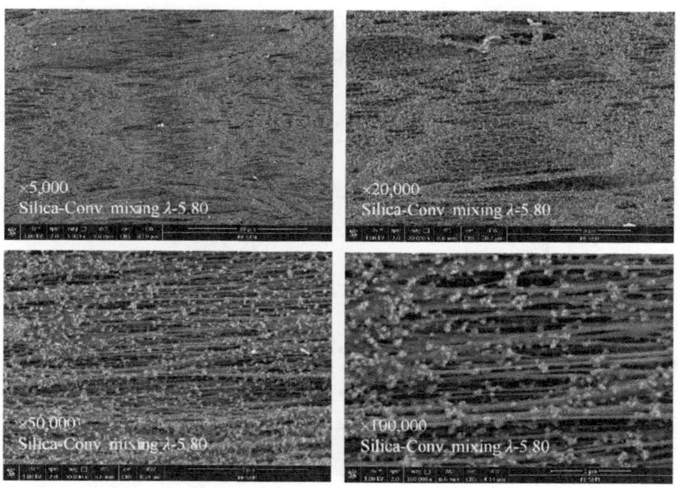


**Figure 6.26** SEM images of SSBR/BR (70/30) vulcanizate filled with 78 phr silica and 28 phr oil at $\lambda = 5.8$[103]

For EVEC, while $\Delta W$ is very low at low strain, it increases monotonically and rapidly with strain. This is likely related to the high efficiency of surface modification and excellent dispersion with liquid phase mixing. With this technique, silica agglomeration is almost eliminated, which can be demonstrated by the Payne effect in a strain sweep (see Section 9.4.3). The higher silanization also means that more covalent bonds are formed between polymer chains and silica surfaces. This suggests that, on one hand, the energy dissipation at low strain due to the breakdown of filler agglomerates should be drastically reduced. On the other hand, at high strain, due to less slippage of polymer chains along filler surface, the excessive stress in the localized area can not be easily relieved. The rupture of polymer crosslinks and polymer chains in the matrix rises rapidly with increasing extension, which is accompanied by more orientation of rubber molecules along the direction of stress. No or less vacuole formation occurs in the stretched vulcanizate (see Figure 6.25). Consequently, the energy required in the first extension to a given strain is much higher than that required in the second extension to the same strain.

Instead of using silica modified with a bi-functional coupling agent, Donnet and Wang[104] investigated the effects of filler-polymer and filler-filler interactions on the stress softening with silica modified by esterification. Upon reaction of precipitated silica P (NSA 137 $m^2/g$) with hexadecanol, alkyl chains cover the silica surface, so that the surface energy of the silica $PC_{16}$, both dispersive and polar components, is substantially reduced[50], similar to the silica modified with mono-functional coupling agent octadecyltrimethoxy silane (ODTMS)[105]. When fillers are added in emulsion-polymerized SBR, the silica $PC_{16}$ results in much less filler agglomeration and weaker polymer-filler interaction without creating chemical bonding between polymer and filler surface. This can be seen in Figure 6.27. While $\Delta W$ of $PC_{16}$ silica-filled vulcanizate does not decrease with increasing extension at low strains, the energy loss increases much more slowly over the whole range of strain amplitudes applied. The latter is certainly related to the weak polymer-filler interaction because without strong chemical linkages, the polymer chains are much easier to slip on the filler surface, leading to more developed detachment and vacuolation.

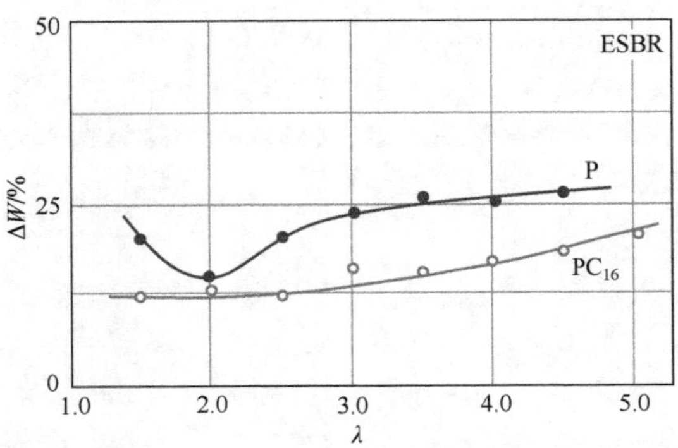


**Figure 6.27** Stress softening as a function of strain for ESBR vulcanizates filled with 50 phr silica[50]

## ■ 6.4 Fracture Properties

In theory, fracture is a process that creates a new free surface area of the material. The energy required for the creation of a new surface area is usually considerably greater than the intrinsic fracture energy necessary to merely rupture the bonds in this area. This suggests that other processes of energy dissipation are involved in rubber fracture. The addition of fillers enhances energy dissipation by different mechanisms, as discussed previously, and improves the fracture properties of rubber materials. In practice, this effect has a considerable influence on the service performance and service life of rubber products, which are dependent on different failure modes, such as cracking, fatigue, tensile failure, tearing, and abrasion.

### 6.4.1 Crack Initiation

It is generally believed that cracks frequently start at an inherent flaw, e.g., inclusions, microvoids, shortcomings of network, contamination, and other local inhomogeneities. When a vulcanizate undergoes stress, the local stress at the tip of the flaw is greatly magnified. Once it reaches a critical level, which is dependent on the radius of the tip and the rupture energy per unit volume of rubber, cracks will form, creating new surfaces. It is, therefore, understandable that an increase in the inherent strength by changing the network structure and a reduction of the stress concentration at the tip by dissipating input energy as heat effectively delay crack expansion. Both of these mechanisms help distribute the stress uniformly to the molecular chains at the front of the flaw tip.

When vulcanizates containing rigid-particle fillers are subjected to tensile stress, this not only results in strain amplification in the polymer matrix, but also causes non-uniform stress distribution with the stress tensor including effective bi- and triaxial strain as well as tensile and shear strains. Oberth and Bruenner[106] indicated that the maximum triaxial stress is generated at a short distance into the rubber matrix, above and below the filler particle. These regions may act as sites that favor the initiation and growth of internal voids and cracks (matrix cavitation mechanism)[107]. According to this mechanism, the rubber does not immediately detach itself from the filler particle, but undergoes an internal rupture near the filler surface, nucleated by a precursor microvoid present within the rubber matrix. This microvoid is expanded by the triaxial stresses generated, and when the expansion of its walls reaches the breaking extension of the matrix, the original cavity will burst, and a crack is initiated at an applied critical stress $\sigma_c$. This critical stress is, of course, dependent on the aggregate size of the filler and the modulus of the vulcanizate. The higher the modulus, the higher is $\sigma_c$[106]. Moreover, as pointed out by Gent[108], if the radius of the precursor microvoids is less than 0.1 μm, their surface energy will exert a significant additional restraint on expansion. Thus, when the rubber contains no microvoids with a radius greater than 10 nm, it will be more resistant against initiation of cracks by matrix cavitation.

On the other hand, the matrix cavitation mechanism can only take effect if the polymer-filler interaction is strong. When the dewetting stress at the direct interface between filler and polymer is lower than the nucleated failure stress of the polymer matrix or rubber shell, in other words, vulcanizates with poor polymer-filler interactions, cavitation will occur at the filler surface instead of matrix cavitation near the filler surface (interfacial debonding or dewetting mechanisms)[107,109,110]. Based on the analysis of an energy balance, Gent[109] proposed the following equation (Eq. 6.34) for the critical stress of dewetting, $\sigma_c$:

$$\sigma_c = \left( \frac{2\pi G_{if} E}{3d_p} \right)^{1/2} \tag{6.34}$$

where $G_{if}$ is the interfacial fracture energy, $E$ is Young's modulus of the vulcanizate, and $d_p$ is the diameter of the spherical filler particle. This suggests that $\sigma_c$ increases not only with increasing polymer-filler interaction, characterized by $G_{if}$, but also with decreasing particle size of the filler. In this regard, high-surface-area fillers have a tendency to delay crack initiation via dewetting processes, due to the smaller radius of their aggregates and strong filler-polymer interactions[111]. This is consistent with electron microscope observations by Hess and colleagues[112] on strained supported microtome sections of SBR vulcanizates filled with different grades of carbon black. While extensive cavitation was observed at the particle edges for non-reinforcing MT blacks, the number of vacuoles between filler surface and polymer matrix is greatly reduced in the case of small-particle blacks.

Crack propagation, or crack growth is similar to the tearing process when static stress is applied, and to fatigue failure under dynamic deformation, both of which are affected considerably by fillers.

### 6.4.2 Tearing

Once cracks have been initiated in a vulcanizate, they will propagate under stress. The basic process underlying crack growth is tearing. As mentioned previously, the tearing energy, i.e., the rate at which strain energy is released, is much greater than that required to create a new surface. It is thus obvious that energy is dissipated irreversibly in a highly localized region around the tip of a growing tear. On the other hand, it is known that tearing energy is a fundamental measure of the strength of macromolecular materials, irrespective of the shape of the test piece and of how force is applied to it[113,114]. It is thus an inherent strength property of rubber, in addition to other failure properties, such as tensile strength and fatigue, which are initiated from small, naturally occurring flaws[115,116].

#### 6.4.2.1 State of Tearing

Two types of tear phenomena are observed in tearing tests (in "trouser" tear, for instance), namely, stable tearing and unstable tearing. In the case of stable tearing, the tear force fluctuates only slightly and the rate of tear propagation is basically continuous. The torn surface in this type of tearing is rather smooth. Unstable tearing is often referred to as "stick-slip" tearing. Instead of the tear propagating at a steady rate, tear growth is arrested and reinitiated at fairly regular intervals. Correspondingly, the force necessary to propagate the tear varies widely, from a minimum at tear arrest to a maximum at tear initiation. This type of tearing leads to surface irregularities with periodic “knots”. This effect is viscoelastic since it depends on changes in both tearing rate and temperature.

Using trouser specimens, the effect of temperature and tearing rate on the states of tearing were investigated by Wang and Kelley[117,118]. The specimens, about 2 cm wide and 15 cm long, were cut from molded sheets of vulcanizates backed with cloth, to minimize deformation of the legs. These specimens with thickness of about 1 mm were scored with a razor blade along the center line to a depth about one-third of the original thickness from both sides, so that only about 0.3 to 0.4 mm thickness remained to be torn through (Figure 6.28). Practically, the scoring was effective to guide the tear path. The specimens were torn at different temperatures and tear rates for the observation of tear states. The stick-slip phenomenon occurs in the region of low tear rates and high temperatures (see Figure 6.29).

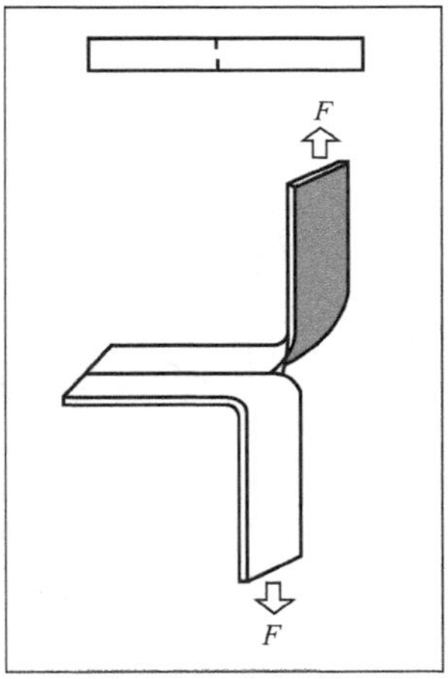


**Figure 6.28** Test piece backed with cloth and method of measuring tear

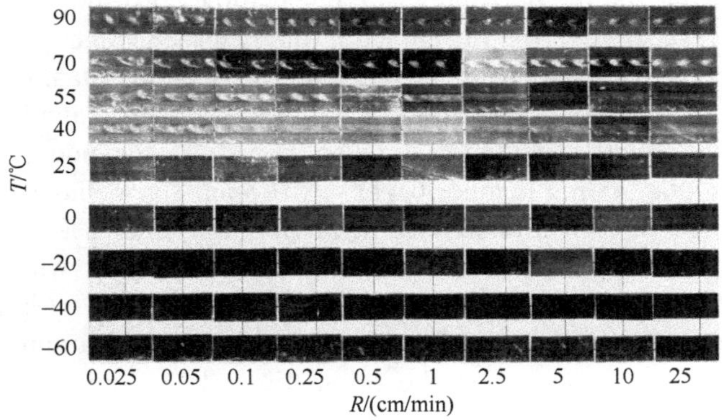


**Figure 6.29** Microscopic images of the torn surface of SBR vulcanizates showing two distinct types of tear surfaces, depending on the temperatures and tear rates

The commonly accepted mechanism of unstable tear is based on the assumption of the existence of an anisotropically reinforced structure at the tear tip, involving an energy dissipation process[113,116]. As stress is applied, an orientated structure (crystallization for strain-crystallizing rubber like NR, or alignment of molecules for amorphous rubber such as SBR) is created in the region of the tear tip associated with large local deformations. This oriented structure in the tear tip web is parallel to the applied stress and perpendicular to the direction of tear growth. Since much more energy is required to propagate a tear perpendicular to the orientation direction[119], the onset of tear rupture is delayed. As stress continues to increase, this force is exceeded at some point and catastrophic rupture occurs. After this point of initiation, the crack advances rapidly (slip) around the reinforced region until the highly strained web is removed and the propagation rate of the tear drops to zero. After this process, a knot remains on the torn surface. When stress continues to be applied, this process is repeated (see Figure 6.30), leaving periodic knots on the torn surface of vulcanizates.

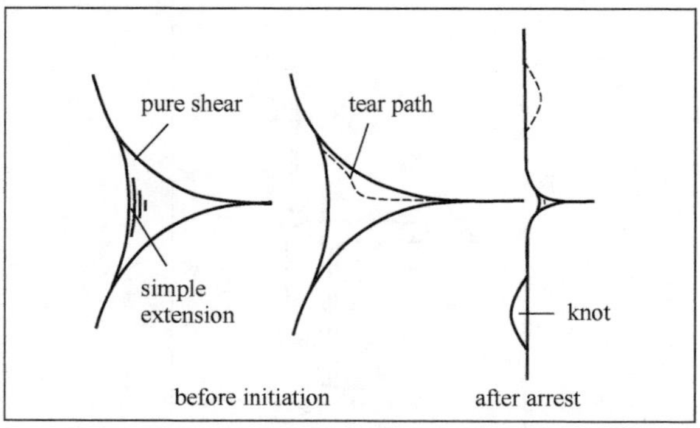


**Figure 6.30** Schematic diagram showing tear propagation and arrest for unstable tearing

Since the minimum and the maximum of the tear force correspond to tear arrest and tear initiation, respectively, the difference in the recorded forces between peak and minimum are considered a measure of the level of stick-slip. Quantitatively, several pairs of peaks and minimums of the tear force are used to calculate the standard deviation, and the relative deviation is taken as a stick-slip index, SSI, which is a measure of unstable tearing (Figure 6.31)[117]. If stick-slip behavior becomes more predominant, the extent of a knotty surface topography increases and so does SSI. The SSI also relates to the distance between knot formations, which subsequently gives information about the relative diameter of the tear tip (Figure 6.32). This parameter, of course, is determined by the formation of reinforcing structure near the tear tip. The more developed the orientation of the polymer molecules, the higher the SSI.

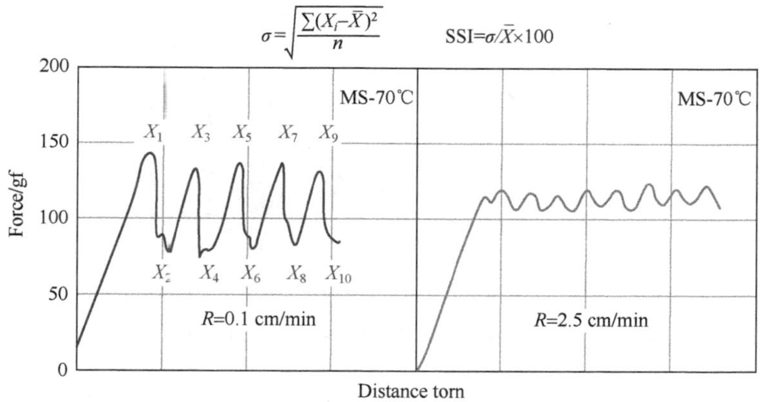


**Figure 6.31** Schematic illustration of the stick-slip index definition for SBR vulcanizates with monosulfidic (MS) crosslinks

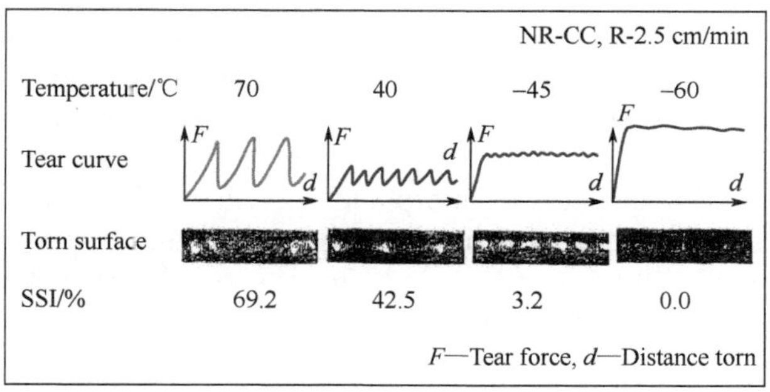


**Figure 6.32** Temperature dependence of arrest tear energy, initiation energy, and SSI at tear rate of 2.5 cm/min for NR vulcanizates with carbon-carbon (CC) crosslink

Figure 6.33 represents the best fit SSI surface as functions of both temperature and tear rate for SBR vulcanizates with monosulfidic crosslinks used in Figure 6.29. As can be seen, the SSI increases with an increase of temperature, undergoes a maximum on the SSI surface diagram, and then decreases with the temperature. The SSI also shows a tear rate-dependence. At very low temperatures and high tear rates, the molecular reorganization into anisotropic crystallites cannot take place in the short time during which the rubber is stretched as the tear tip advances, thereby producing a smooth tear surface. At moderate to high temperatures the SSI decreases with increasing tear rate.

The effect of tear rate on the tear state may be better understood when SBR-MS vulcanizate tearing is carried out at 55°C and 0.5 cm/min (Figure 6.34). Under these conditions, tearing is in the stable state, showing stable tearing force and smooth torn surface. Then the separation of the clamps stops. Within the few minutes of the stopping period, the tear force decreases due to stress relaxation. Then the separation of clamps restarts the tearing process, and the crack does not grow immediately until

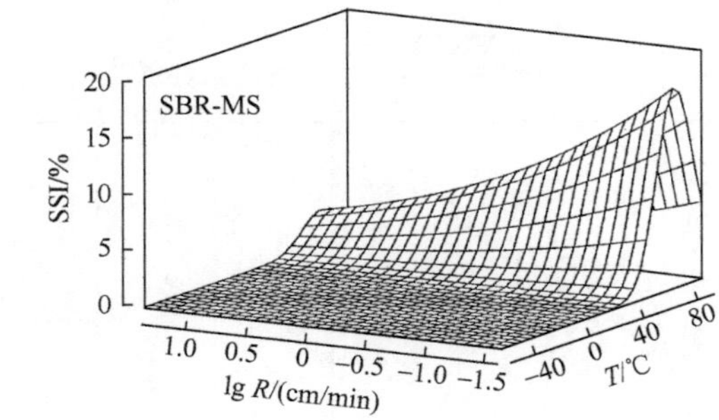


**Figure 6.33** Dependence of SSI for SBR-MS vulcanizate on tear rate and temperature

the tear force reaches a level that is higher than the force of the prior tear stops. Next the tear growth re-initiates and tearing propagation returns to a stable state. As a result, a small "knot" appears on the torn surface. This suggests that during the stress relaxation, some orientation of the polymer chains takes places, perhaps due to slippage of the rubber segments along the filler surface and the orientation of filler aggregates in the stress field. This is demonstrated by anisotropy investigations of the amorphous peak of X-ray patterns for stretched carbon black loaded NR in which the molecular orientation factor increases with extension and carbon black content[120]. The fact that carbon black assists in the orientation helps to explain the effect of carbon blacks on the reinforcement of amorphous polymers. This would lead to the development of an anisotropic structure over a large volume, causing the stress to be uniformly distributed to the network chains. This is certainly true at the crack tip where the polymer matrix is subjected to high tensile strain.

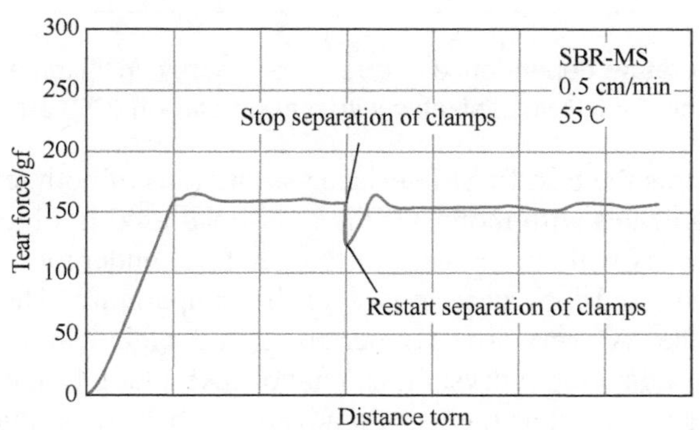


**Figure 6.34** Tear curve measured at tear rate of 0.5 cm/min and 55°C for SBR-MS vulcanizate, showing the effect of stress relaxation on knotty formation

Rivlin and Thomas[121] considered the effect of stress relaxation on the tearing of vulcanizates with respect to understanding unstable tearing. They supposed, as did Green and Tobolsky[122], that when an elastomer is held in the deformed state, some of the crosslinks break and new crosslinks may form in such a way that the resulting vulcanizate consists of two coexisting networks. For one of these, the original conformation of the polymer is its state of ease. For the other, the state of deformation in which the new crosslinks are formed is considered the state of ease. When the deforming forces are removed, the composite network adopts a shape that is different from the initial undeformed shape and is intermediate between the states of ease of the two component networks. From this two-network model, they calculated the characteristic initiation tear energy. According to their calculation, if stress relaxation occurs in the immediate neighborhood of the region of high stress, the incipient tearing will take place parallel to the direction of overall extension of the test piece rather than transversely to it. As the tearing proceeds, the high extension at the tip of the tear becomes transverse to the direction of the overall extension of the test piece, corresponding to a knotty tearing process. From this point of view, whatever processes are responsible for stress relaxation will enhance the anisotropic tearing or stick-slip behavior. It is probably for this reason that the stick-slip phenomenon is more enhanced for carbon black reinforced rubber than it is for the gum, and is more prominent for strain-induced crystallizing rubbers than for amorphous elastomers.

On the other hand, when a web is formed during tearing, the deformation state in the middle of the web is considered a simple extension, while the deformation state at the side is pure shear. Of course, the orientation of the molecules or crystallization at the tip near the middle of the web is much more prominent than at the side of the web. Thus, tearing is initiated at the side of the web and propagates through to the center line of the web and then continues on the path until arrest occurs (Figure 6.30). Microscopic observation of the actual tear path confirms this (Figure 6.29).

According to the discussion above, the stick-slip behavior depends on the formation of a reinforced structure and relaxation process in the strained web. Therefore, the filler, crystallizability, as well as network structure would play important roles in the tearing state.

#### 6.4.2.1.1 Effect of Filler

It has been reported that, for crystallizable elastomer such as NR, the presence of carbon black enhances rubber crystallization[120,123], and promotes stick-slip tearing. The enhancement of strain-induced crystallization of NR by carbon blacks is confirmed by X-ray studies[124,125] and differential scanning calorimetry. Using NR vulcanizates filled with different amounts of HAF black, Lee et al.[120] observed in X-ray analyses that the amount of crystallization of the rubber, and the size and energy density of the process zone near the tear tip, increase considerably with carbon black loading. This effect is primarily due to strain magnification by the filler as crystallization increases rapidly with increasing strain[126,127].

In the case of amorphous rubber, pronounced stick-slip tearing appears at certain levels of temperature and tear rates for compounds filled with carbon black, particularly with the reinforcing grades[117,128]. This can be seen in Figure 6.29, for SBR filled with 50 phr N330 carbon black, even though no unstable tearing is observed for non-crystallizable gum. For all filled vulcanizates the stick-slip phenomenon occurs in the region of low tear rates and high temperatures.

#### 6.4.2.1.2 Effect of Polymer Crystallizability and Network Structure

##### 6.4.2.1.2.1 Non-Crystallizable Rubber-SBR

The stick-slip regions of carbon black filled SBR vulcanizates, having polysulfidic (PS), monosulfidic (MS), and carbon-carbon (CC) crosslinks, are represented in Figure 6.35[117]. All vulcanizates have about the same crosslink density. The SBR-MS and SBR-CC vulcanizates show about the same boundary between the smooth tear region and the stick-slip one. This boundary shifts to lower temperatures by about 15°C for SBR-PS vulcanizates. This means that unstable tearing can occur at lower temperatures at the same tear rates, or at higher tear rates at the same temperatures as compared to SBR-MS and SBR-CC vulcanizates.

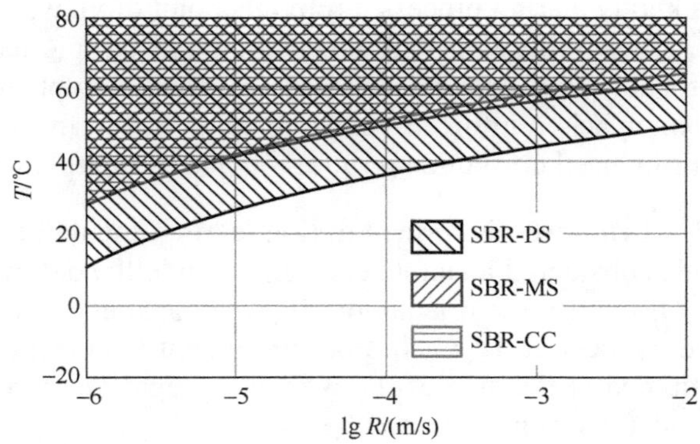


**Figure 6.35** Schematic illustration of stick-slip region for SBR-PS, SBR-MS, and SBR-CC vulcanizates. Ruled areas represent stick-slip tearing

Figure 6.36, Figure 6.37, and Figure 6.38 represent a best fitting surface for the raw data of the SSI of the three SBR vulcanizates, showing each SSI surface compared to the other two. The SBR-PS vulcanizate has a higher SSI than either SBR-MS or SBR-CC materials.

Additionally, the SBR-PS vulcanizate has a significant higher SSI over a larger area of rate and temperature than do the SBR-MS or SBR-CC vulcanizates. All the vulcanizates exhibit a maximum on the SSI surface. Above the temperature of the maximum, the SSI decreases with temperature.

Based on the discussion before, it is reasonable to believe that the orientation degree of polymer chains at the tear tip depends on network structure and experimental conditions.

The more pronounced stick-slip behavior for the SBR-PS vulcanizate should be due to the stability of polysulfidic crosslinks under strain of the vulcanizate as their bond energy is significantly lower than MS and CC[129]. From this point of view, the more labile polysulfidic crosslinks are more favorable to the formation of oriented structure, extending the stick-slip region to lower temperatures and higher tear rates and increasing the stick-slip index.

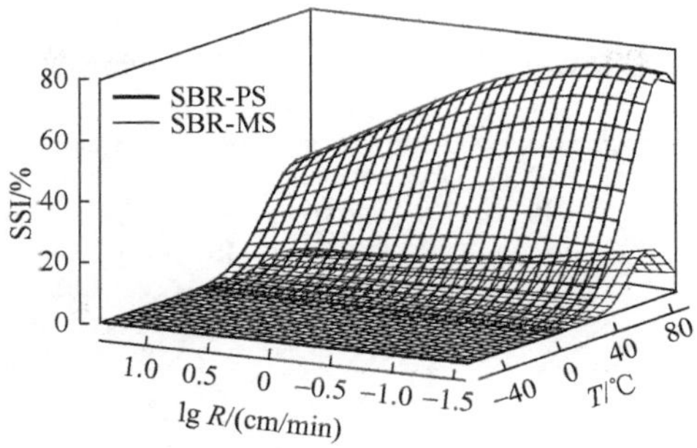


**Figure 6.36** Comparison between SSI of SBR-PS and SBR-MS vulcanizates

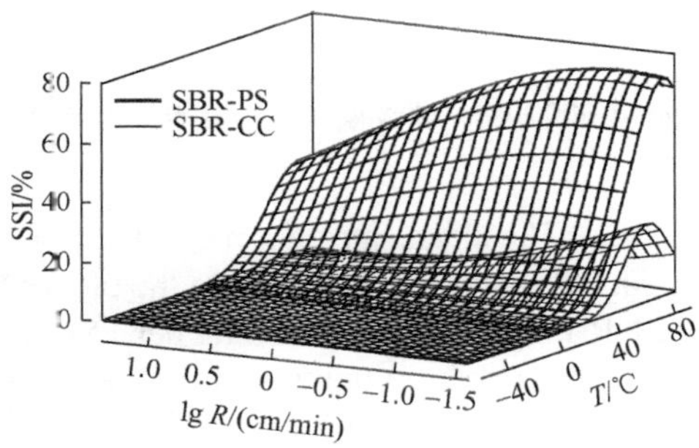


**Figure 6.37** Comparison between SSI of SBR-PS and SBR-CC vulcanizates

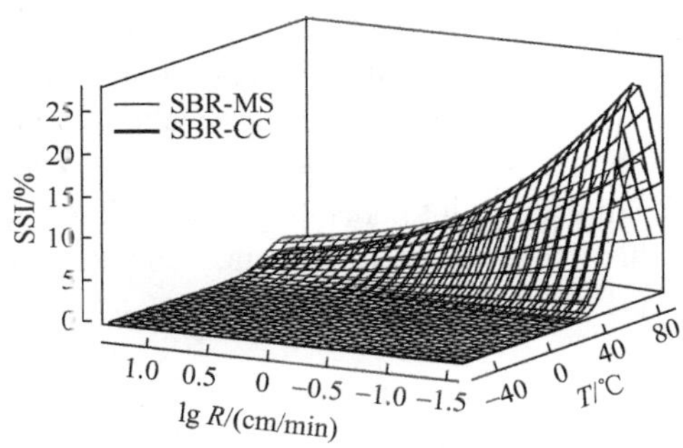


**Figure 6.38** Comparison between SSI of SBR-MS and SBR-CC vulcanizates

#### 6.4.2.1.2.2 Crystallizable Rubber-NR

Figure 6.39 shows the stick-slip region in a temperature-tear rate plot for filled NR vulcanizates with polysulfide (PS), monosulfide (MS), and carbon-carbon (CC) crosslinks. Similar to the SBR vulcanizates, the stick-slip phenomenon occurs in the region of lower tear rates and higher temperatures and the NR-MS and NR-CC vulcanizates have about the same boundary position between stable and unstable tearing behavior. However, this boundary position shifts to lower temperatures for the

NR-PS samples. Further, all NR vulcanizates extend the unstable region to much lower temperatures (ca. 80–90°C lower) than the SBR vulcanizates[118].

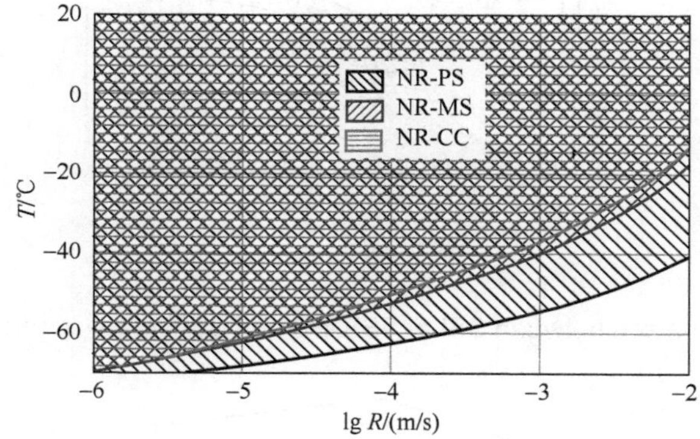


**Figure 6.39** Schematic illustration of stick-slip region ruled areas for NR-PS, NR-MS, and NR–CC vulcanizates

The SSI surfaces of the three NR vulcanizates are presented in Figure 6.40, Figure 6.41, and Figure 6.42, showing each SSI surface compared to the other two. Here again, as with the SBR vulcanizates, the NR-PS vulcanizate has a higher SSI than the NR-MS and NR-CC samples. All the vulcanizates exhibit a maximum on the SSI surface diagram. Above this maximum, the SSI decreases with the temperature. The SSI also shows a similar tear rate-dependence. At lower temperatures, the SSI increases with tear rate and at moderate to high temperatures, the SSI increases slightly with rate and then decreases, exhibiting a maximum at moderate tear rates, which is also similar to the case of SBR-PS, even though the SSI is higher and the range of unstable tearing extends greatly to low temperature for NR vulcanizates. This is certainly due to the fact that reinforced structure at the tear tip is highly developed due to the strained-induced crystallization of NR molecules, compared with SBR counterparts. If the temperature is too low or tear rate is too high, the reorganization of polymer molecules into anisotropic crystallites is unable to take place in the short time during which the rubber is stretched as the tear tip advances, thereby producing a smooth tear surface even for NR.

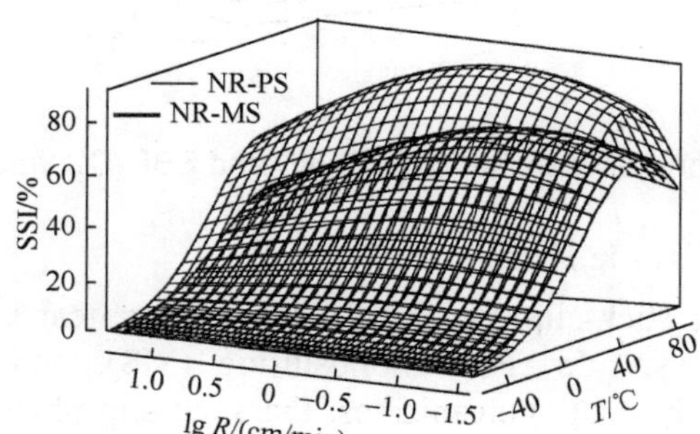


**Figure 6.40** Comparison between SSI of NR-PS and NR-MS vulcanizates

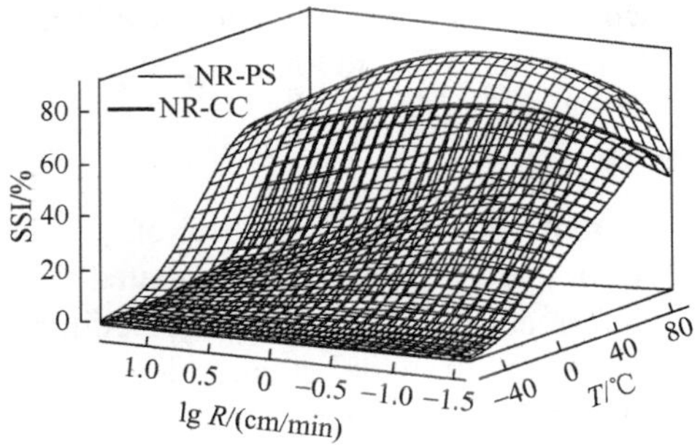


**Figure 6.41** Comparison between SSI of NR-PS and NR-CC vulcanizates

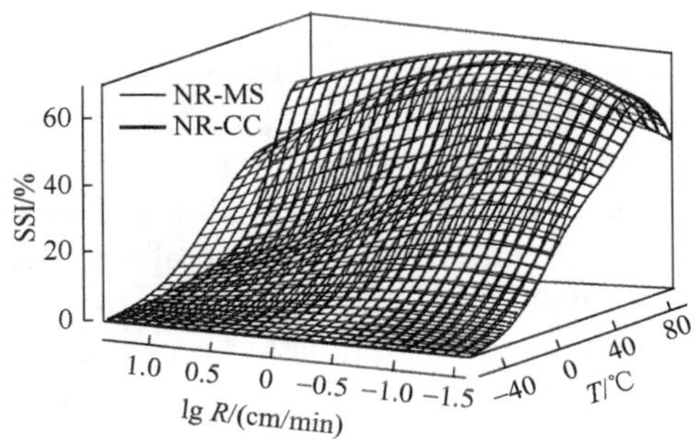


**Figure 6.42** Comparison between SSI of NR-MS and NR-CC vulcanizates

As found in the case of SBR, the weaker polysulfidic crosslinks are favorable to the formation of strain-induced crystallization. Therefore, the stick-slip region is extended to lower temperature and shorter time, enhancing the extent of stick-slip behavior. Moreover, the breakage and reformation of the low energy polysulfidic crosslinks during tear can release some of the stress concentrations in the network, making the crystallization process much easier in the highly strained region.

The slightly lower SSI for the low tear rate in the high SSI region is probably due to the effects of stress relaxation and crystallization. During slow tear, stress relaxation makes the force peak recorded lower than it should be (see Figure 6.43) and crystallization can take

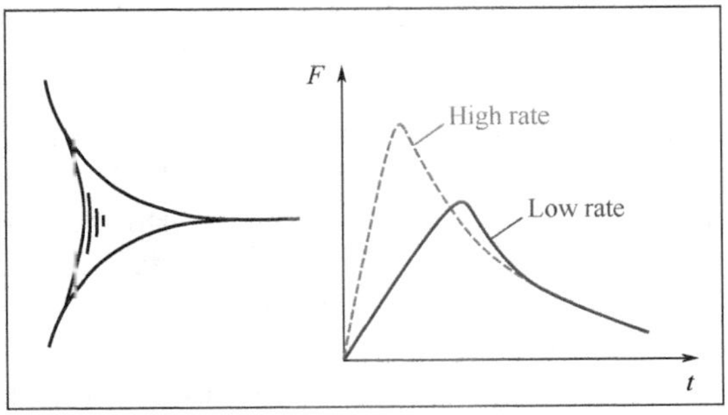


**Figure 6.43** Schematic illustration of the effect of stress relaxation on the recorded tear force

place at the tip when the tear arrests at lower tear rates (see Figure 6.30), which raises the minimum. Both of these are a parallel effect to lower the SSI.

### 6.4.2.2 Tearing Energy

Experimentally, the energy required for tearing is not only related to tear propagation, but also involves elastic strain energy stored and dissipated in the legs of the test piece (trouser tear). When the deformation of the legs can be eliminated, for instance in the case of a "trouser" test piece with wide enough legs or backed by undeformed materials as shown in Figure 6.28, the tearing energy, $G$, can be calculated from the force, $F_t$, required to propagate a tear in rubber[113]:

$$G = 2F_t / t \tag{6.35}$$

where $t$ is the thickness torn through. In the stable tearing region, the average force can be used for the calculation of tearing energy because the force varies only slightly with tear propagation. For unstable tearing, the force of tear initiation is much higher than that of arrest. In this case, as pointed out by Kelley and colleagues[116], the tearing energy at arrest is more representative of the inherent tear strength of the materials, even though the initiation value is an important feature of the rubber since cracks do not develop below this energy level. The initiation energy can be estimated from both the arrest tear energy and stick-slip index. Figure 6.32, as an example, illustrates the relationship among them.

#### 6.4.2.2.1 Effect of Filler

For all rubber-grade carbon blacks, the tear energy increases with filler concentration, due to increasing energy dissipation, and then decreases for amorphous rubbers at high loadings. This is also true for crystallizable rubber such as NR. However, at very low loadings (below 5–10 phr, depending on the type of carbon black), the tear resistance of NR is not improved by the addition of carbon blacks (the values are even lower than those of the gum).

Although the small-particle carbon blacks have a favorable effect on the tear energy, especially with regard to the initiation value, increases in structure are slightly detrimental because, as compared to low-structure carbon blacks, the high modulus will largely reduce the effective radius of the tear tip (the tip being sharpened), which is an important factor determining the tearing energy[130,131].

The effect of silica on tear properties is important for its application in tire. Before the commercialization of green tire in which the principal filler of tread compounds is precipitated silica, its use in tire in any quantities worth mentioning had long been limited to two types of compounds: one is textile and steel cord bonding compounds containing a certain amount of silica, in combination with resorcinol/formaldehyde systems, to improve its adhesive property; and another is carbon black filled off-the-road tread compounds containing 10–15 phr of silica in order to improve tear properties, hence chipping and chunking resistances[132]. For the latter application, the weaker polymer-filler interaction between silica and polymer can enhance the formation of molecule orientation at the tear tip in the direction of tear stress, as the polymer chains are easily slipped along the filler surface and/or detached from the filler surface, releasing the stress concentration in the polymer network. In this application, no or less coupling agents are needed.

#### 6.4.2.2.2 Effect of Polymer Crystallizability and Network Structure

##### 6.4.2.2.2.1 Non-Crystallizable Rubber-SBR

The presence of carbon black can significantly increase the tearing energies of arrest and initiation. This effect is even more pronounced in amorphous rubber, such as SBR. The increase in initiation tearing energy can be attributed to the easy formation of a reinforcing structure around the tear tip, but the enhancement of energy dissipation during tearing, by other molecular processes which are promoted by the carbon black (see Section 6.3), may also be considered as a major contribution to the high values of the tearing energies of both initiation and arrest.

The measured tear energies for carbon black filled SBR vulcanizates with different network structures, namely polysulfidic (PS), monosulfidic (MS), and carbon-carbon (CC) crosslinks, are presented in Figure 6.44, Figure 6.45, and Figure 6.46 for pairwise comparisons between them. These surfaces show the effects of rate and temperature on tear energies.

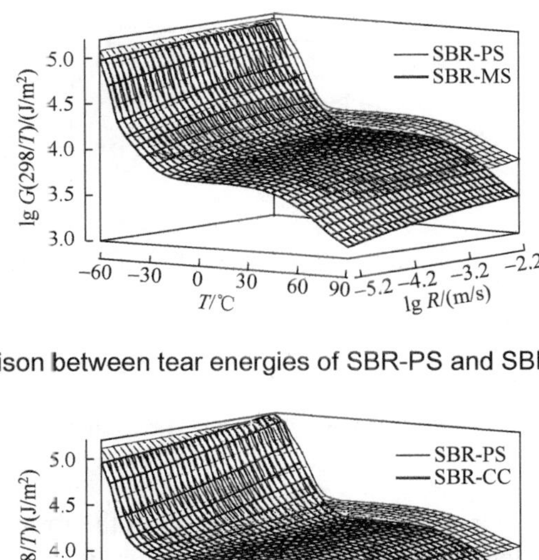


**Figure 6.44** Comparison between tear energies of SBR-PS and SBR-MS vulcanizates

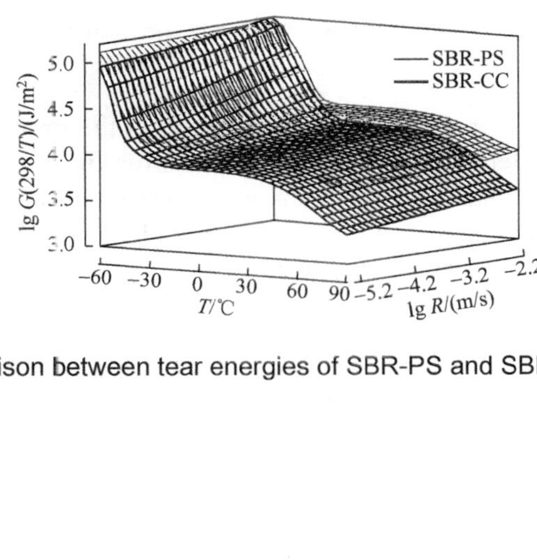


**Figure 6.45** Comparison between tear energies of SBR-PS and SBR-CC vulcanizates

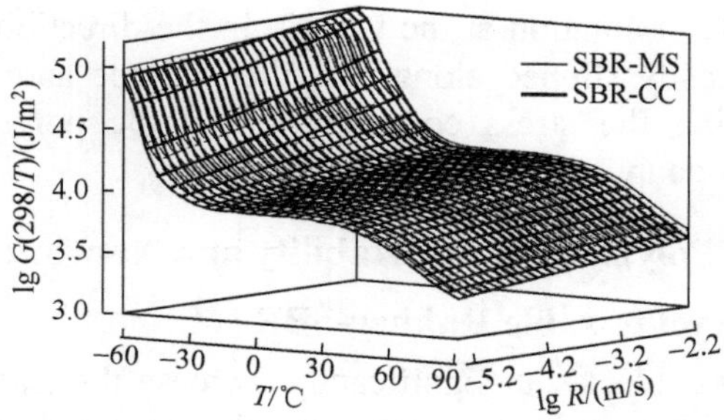


**Figure 6.46** Comparison between tear energies of SBR-MS and SBR-CC vulcanizates

Figure 6.47 presents experimental tear energies plotted against tear rate for the SBR-PS vulcanizate at 14 temperatures. The viscoelastic behavior of tearing energy has demonstrated that the equivalence between changes in the rate of tearing and in temperature follows the well-known Williams-Landel-Ferry (WLF) relationship[133] for unfilled amorphous vulcanizates[134,135]. With regard to amorphous rubber filled with carbon blacks, this can only be assumed to be true for the values of tearing energy at arrest in the whole tearing region.

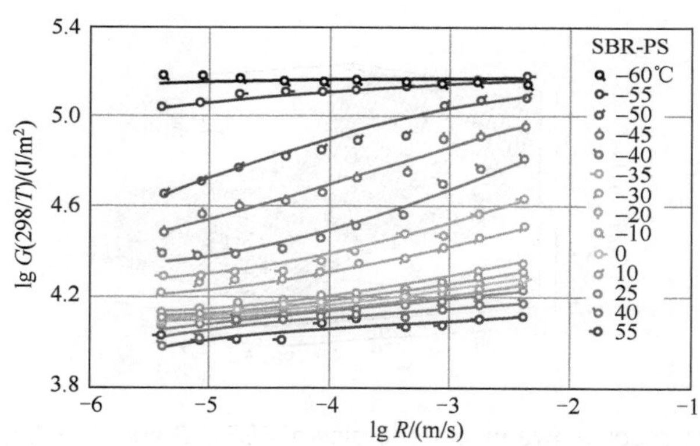


**Figure 6.47** Tear energy, *G*, for SBR-PS vulcanizate as a function of tear rate at 14 temperatures

The data in Figure 6.47 is represented in Figure 6.48 in the form of an arrested tearing energy master curve referenced to 25°C for SBR sulfur-cured vulcanizates filled with 50 phr of N330 carbon black[117]. The temperature-related horizontal shift factors $a_T$, i.e., the factors by which the tear rate has to be multiplied to yield a master curve, were calculated from the WLF equation (Eq. 6.36):

$$\lg a_T = \frac{-C_1 (T - T_S)}{C_2 + (T - T_S)} \tag{6.36}$$

where $C_1$ and $C_2$ are experimental constants for the reference temperature $T_S$. The constants were obtained by substituting the WLF equation with $a_T$ from the best superposition of the data. Experimental shift factors for the three SBR vulcanizates are shown in Figure 6.49. The constants $C_1$ and $C_2$ from each sample are listed in Table 6.1.

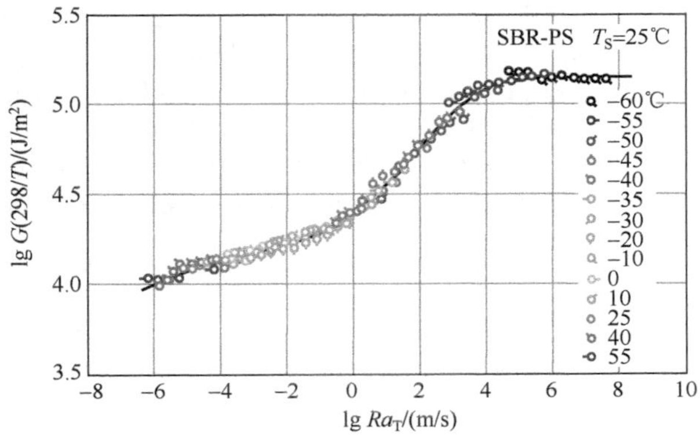


**Figure 6.48** Tear energy master curve for SBR-PS vulcanizate referenced to 25°C

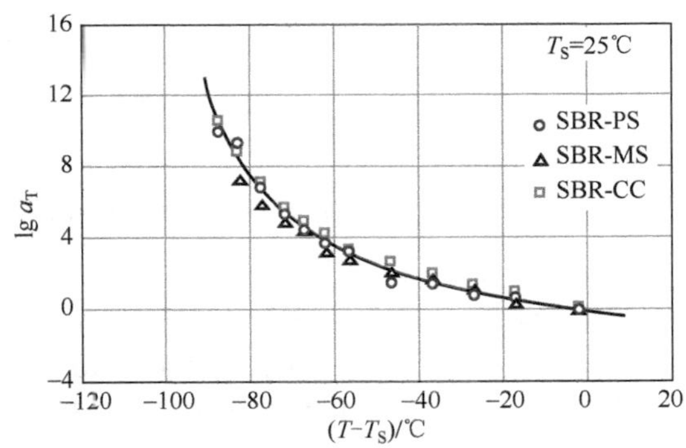


**Figure 6.49** Shift factors of WLF superposition for SBR-PS, SBR-MS, and SBR-CC vulcanizates as a function of temperature

**Table 6.1** WLF constants for SBR-PS, SBR-MS, and SBR-CC vulcanizates

| Vulcanizates | $C_1$ | $C_2$ |
|---|---|---|
| SBR-PS | 3.46 | 112.4 |
| SBR-MS | 3.21 | 115.3 |
| SBR-CC | 4.45 | 121.8 |
| Average① | 3.71 | 116.5 |

① The average values of $C_1$ and $C_2$ were used to shift the data of all samples.

The tear energy master curves referenced to 25°C for the three SBR vulcanizates are shown in Figure 6.50. There is a qualitative similarity among all the master curves. Similar to the unfilled gum[135], the curves of N330-filled SBR vulcanizates exhibit an "*S*" shape over a broad time scale. As can be seen, the high rate region of the tear energy curves display a glassy plateau in the vicinity of 120−130 kJ/m$^2$, and in the low rate region all the curves have a tendency to drop towards lower values of tear energy, predicted by the theory of Lake and Thomas[136], which is the lower limit of the mechanical strength of the rubber. This value is about the same as that of the gum, showing that at high temperature and low tear rate, the effect of carbon black on the arrested-tearing energy can be eliminated. Between these two extremes, each curve shows evidence of an intermediate plateau in the same region. Although the intermediate plateaus have not been completely understood, a comparison can be made among the different networks.

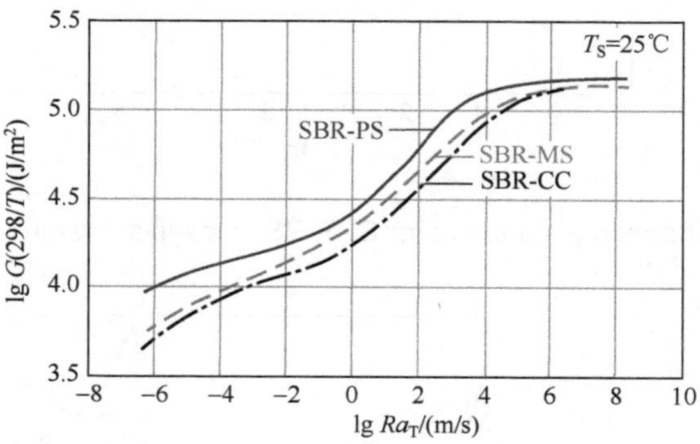


**Figure 6.50** Comparison of tear energy master curves referenced to 25°C for SBR-PS, SBR-MS, and SBR-CC vulcanizates

For SBR filled with carbon black, over the whole rate scale, except the very high rate range where the tear energies of MS are equal to those for CC, the tear energies decrease from PS to MS to CC in this order. This is the reverse of the order for crosslink bond energies.

Thomas and Greensmith[130,131] proposed that the relationship between the tear energy, $G$, tensile rupture energy per unit volume, $W_b$, and effective diameter of tear tip, $d$, can be expressed by:

$$G = W_b d \tag{6.37}$$

In this case, any change of the network structure that either increases the breaking energy or increases the effective tear tip diameter, $d$, will enhance tear resistance (Eq. 6.37).

For breaking energy, Taylor and Darin[137] postulated that stress in the specimen at rupture will be borne mainly by those chains that are oriented in the direction of extension and nearly at their ultimate elongation. Tensile rupture energy is assumed to be proportional to the number of chains per unit volume whose chain displacement vectors, at rupture, are oriented at a small angle to the direction of extension. The ability of weak polysulfide crosslinks to release localized high stress concentrations under test conditions and to encourage the chain orientation has been considered to contribute to high rupture energy. In fact, several investigations have compared the tensile strength of vulcanizates (mainly natural rubber) with a variety of network structures and have concluded that vulcanizates with polysulfidic linkages are superior to those with C–C or monosulfidic crosslinks[138].

In the high rate region, or at low temperature, the influence of the crosslink type on energy dissipation should decrease, especially in the glassy state. This may be the reason that vulcanizates with different crosslink types show about the same tear energy in this region.

In the discussion of tearing energy for the carbon black filled SBR vulcanizates, with polysulfidic, monosulfidic, and carbon-carbon crosslinks, Wang and Kelley[117] considered that the only difference between vulcanizates was crosslink type. However, it is known that different cure systems also result in chemical modification of the polymer chains. Compared with the conventional curing system, the highly accelerated sulfur compound produces less cyclic sulfide groups and pendant groups. In what way and to what extent these modifications influence tearing are still not clear for amorphous rubbers.

#### 6.4.2.2.2.2 Crystallizable Rubber-NR

The experimental relations for the tear energy, $G$, of carbon black filled NR vulcanizates with different network structures as a function of the tear rate and test temperature are presented as three-dimensional diagrams in Figure 6.51, Figure 6.52, and Figure 6.53. It can be seen that all the vulcanizates display the same general features. When the temperature is near the glassy region, the vulcanizates are found to approach a tear energy of approximately 120 kJ/m$^2$, which is about the same as that of filled SBR. At much higher temperatures, the tear energy drops rapidly, and between these two extremes there is a plateau. In this region the tear energy from the arrest values is approximately the same as SBR, but the initiation tear energy is much higher. On the other hand, over the entire range of tear rate and temperature, the tear energy decreases as NR-PS > NR-MS > NR-CC[118].

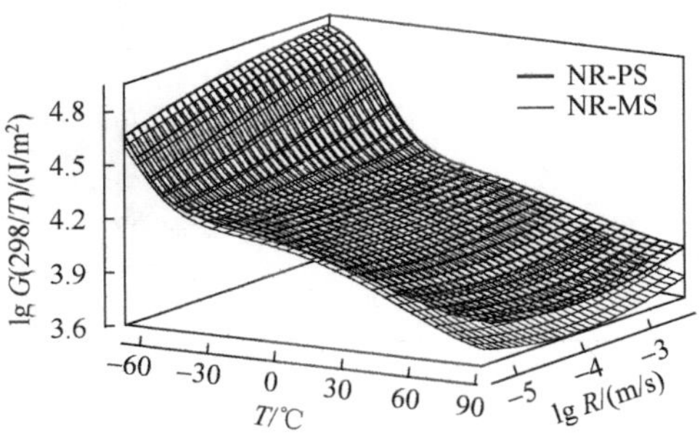


**Figure 6.51** Comparison between tear energy of NR-PS and NR-MS vulcanizates

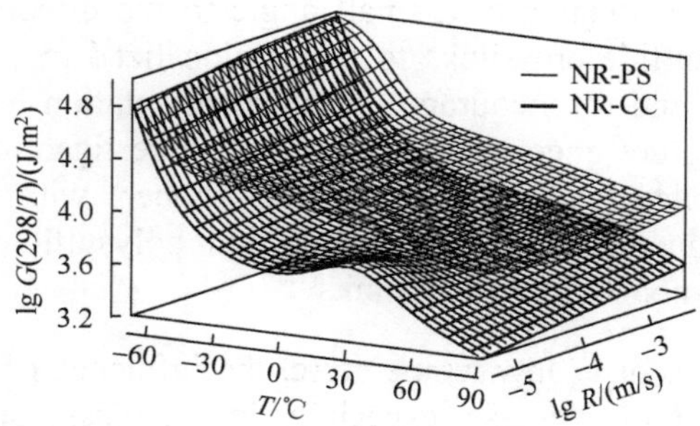


**Figure 6.52** Comparison between tear energy of NR-PS and NR-CC vulcanizates

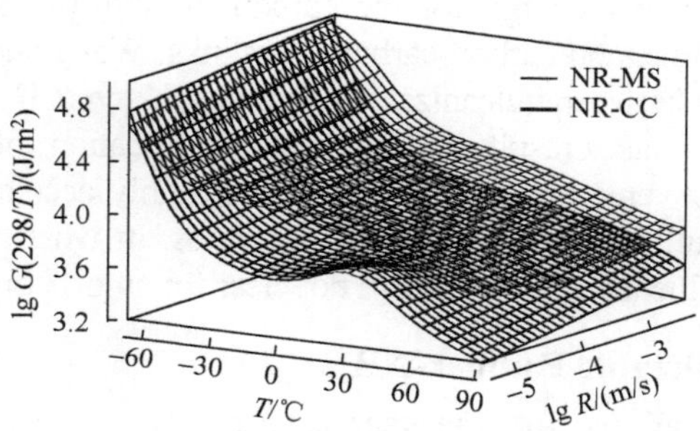


**Figure 6.53** Comparison between tear energy of NR-MS and NR-CC vulcanizates

Figure 6.54 presents the experimental tear energies plotted against tear rate for the NR-CC vulcanizates. At low temperatures where tearing is in a stable or a slightly unstable state, the tear energy relations at different temperatures can be superimposed to form a single master curve. Shift factors were obtained by fitting the Eq. 6.36 with $a_T$ from the

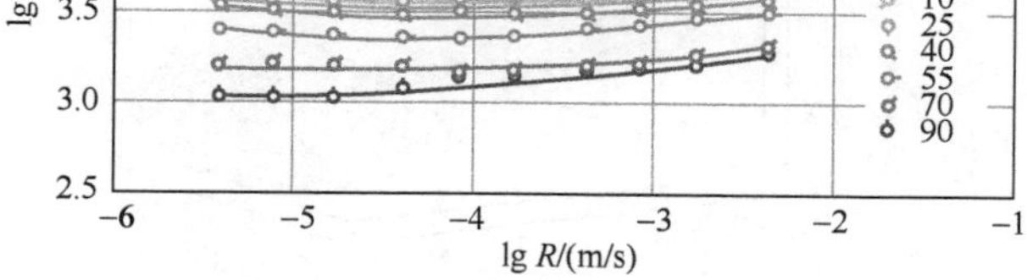


**Figure 6.54** Tear energy, *G*, for NR-CC vulcanizate as a function of tear rate at different temperatures

optimal superposition of the data in the stable and slightly unstable region, using −40°C as the reference temperature. The experimental shift factors for the three vulcanizates are shown in Figure 6.55. Table 6.2 presents constants $C_1$ and $C_2$ obtained by shifting the data from −40°C to −70°C.

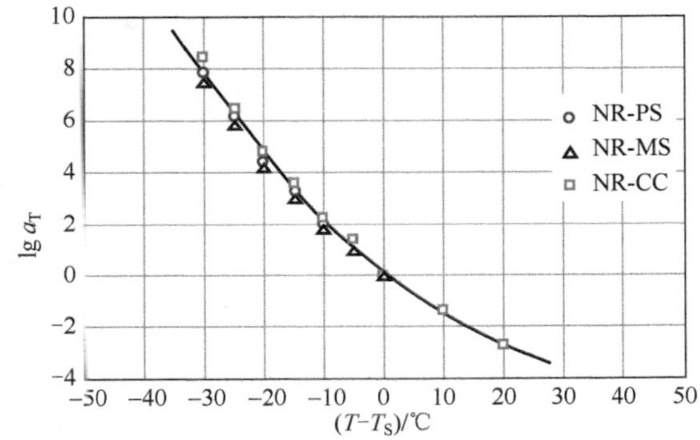


**Figure 6.55** Experimental WLF shift factors for NR-PS, NR-MS, and NR-CC vulcanizates as a function of temperature

Table 6.2 WLF constants for NR-PS, NR-MS, and NR-CC vulcanizates

| Vulcanizates | $C_1$ | $C_2$ |
|---|---|---|
| NR-PS | 15.0 | 88.0 |
| NR-MS | 14.2 | 85.6 |
| NR-CC | 16.8 | 90.2 |
| Average① | 15.3 | 88.6 |

① The average values of $C_1$ and $C_2$ were used to shift the data of all samples.

As shown in Figure 6.56, in the high rate range or at low temperature, the superposition forms a single master curve. At the low rate range where stick-slip tear takes place, the shifted data do not superimpose well. These data are unshiftable, since the tear energy at lower tear rates slowly increases. Two reasons may be considered for this kind of data scatter. One is that the stick-slip propagation of the tear makes the tear rate meaningless. In the unstable tearing region, the tear propagation arrests and reinitiates repeatedly so that it is not clear what the actual tear propagation rate is (see Figure 6.57). Another reason may be that there is a parallel effect of strain-induced crystallization on the viscoelastic tearing. At a lower separation rate of the clamps, there is still some strain-induced crystallization at the tear tip when the tear slip arrests (see Figure 6.30), which makes the arrest value higher. The superimposed curves of the three vulcanizates with different network structures, referenced to −40°C, are presented in Figure 6.58, in which the solid lines indicate the single master curves and the dashed lines are the averages of the scattered data at the low rates. Since the average shift factors of the three samples were used to shift the tear energy data, the shifted curves are comparable, because at the same reduced rate, the tear energies were obtained from the same experimental conditions.

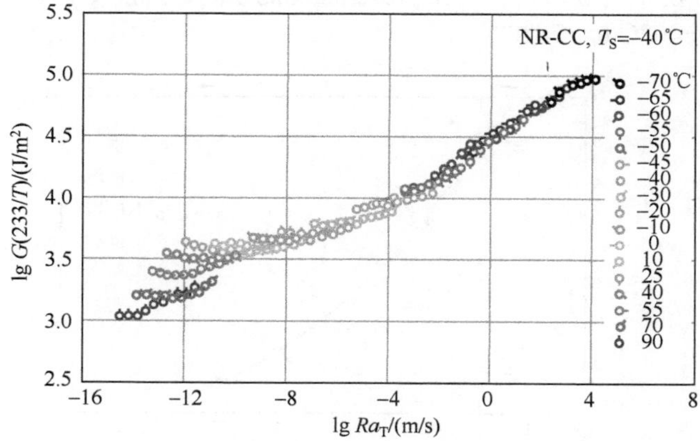


**Figure 6.56** Tear master curve for NR-CC vulcanizates referenced to –40°C

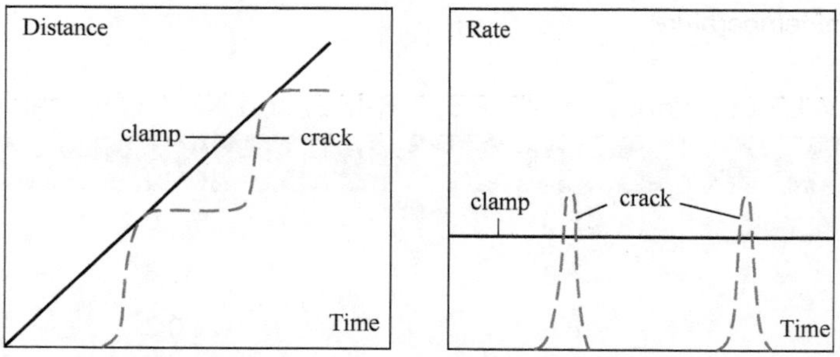


**Figure 6.57** Schematic diagram showing the relationship between the tear propagation rate and the separation rate of the clamps in the unstable tearing region

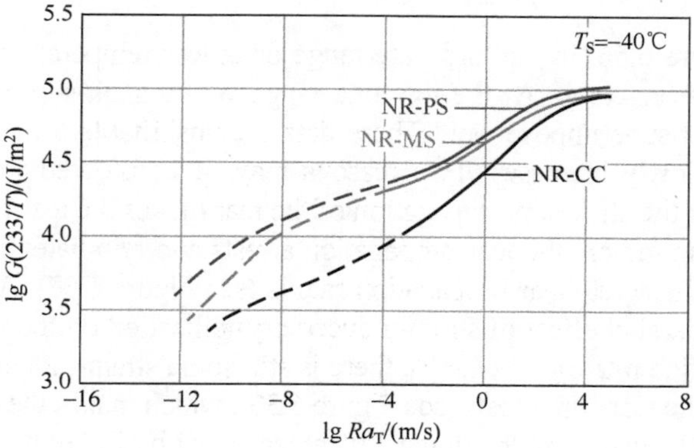


**Figure 6.58** Comparison of tear master curves referenced to –40°C for NR-PS, NR-MS, and NR-CC vulcanizates

As can be seen, all curves show a qualitative similarity, with a small glassy plateau in the high rate region and another plateau in the intermediate rate region. Over the entire time scale the tear energy decreases as NR-PS > NR-MS > NR-CC, as shown in the three-dimensional diagrams.

The high tear energy for the vulcanizates with weak crosslinks may be due to the easy breakage under stress, thus causing internal energy dissipation, and ease of orientation near the tear tip, thereby increasing the arrest forces.

It was found for all samples, when the temperature was −75°C and the rubber was in the glassy state, the tear resistance was high at the beginning of tear and then dropped sharply. At the same time, the crack propagated very rapidly to the end of the specimen, termed a running crack, resembling the fracture of typical polymeric glasses, even though the clamps were separated by only a small distance. This process also occurred at −70°C at higher tear rates.

### 6.4.3 Tensile Strength and Elongation at Break

Although the mechanism of tensile failure of elastomers has not yet been fully established, it can be regarded as catastrophic tearing by growth of cracks initiated from accidental flaws, microvoids, dewetting, or cavitation from the filler surface. If the elastomeric network is capable of dissipating input energy into heat through irreversible molecular processes in the bulk of the specimen (instead of near the crack tip in tearing), less elastic energy will be available to break the polymer network, thus increasing the fracture energy[139]. Strain-induced crystallization of elastomers is also one of the most effective processes of dissipation of stored energy. The formation of a filamentous structure prevents cracks from growing until catastrophic rupture occurs, thus leading to high tensile strength of the crystallizable rubbers. As mentioned earlier, this process can be enhanced by the addition of carbon blacks. With regard to amorphous rubbers, lack of crystallization under strain has been attributed to inherent weakness of these materials. The incorporation of reinforcing carbon blacks has been considered a major source of energy dissipation, raising the strength to a level equivalent to that of crystallizable rubber. The relationship between tensile strength and energy dissipation during stretching was represented by Harwood and Payne[140] as:

$$U_{\mathrm{b}} = B\left(H_{\mathrm{b}}\varepsilon_{\mathrm{b}}\right)^{1/2} \tag{6.38}$$

where $U_b$ is the energy density to break, $\varepsilon_b$ is the elongation at break, $H_b$ is the energy dissipated in the deformation to break, and $B$ is a constant which is independent of filler loading and type of filler (Eq. 6.38).

Several mechanisms have been proposed in order to interpret the function of carbon blacks in energy dissipation. Some are related to stress softening, i.e., the energy loss due to the breakdown of agglomeration, local highly viscous flow due to the filler-frozen entanglements[124], and slippage of the rubber molecular chains along the filler surface[75,94,95,141,142], which also results in uniform stress distribution to the network chains, etc. With regard to crack growth, the alignment of the network chains may have a similar function as crystallization in absorbing elastic energy as well as in strengthening and blunting the crack tip, thereby delaying the onset of catastrophic failure. On the other hand, as in other materials, the presence of rigid filler particles plays a role in retarding cracks and dissipating energy by increasing the length of the crack path. Since the crack cannot occur through filler particles, it has to move around them, thus requiring more energy[124]. This effect is even more pronounced for a filler, such as carbon black, whose surface is covered by a strong rubber shell.

As suggested by Gent and Pulford[143], the filler particles may also act as stress risers by forming potential nucleating sites for secondary cracks, as will be discussed below, which are believed to increase energy dissipation considerably.

In order to prevent crack growth and increase hysteresis, loading and carbon black morphology may play an important role through interaggregate distance. As pointed out by Eirich[124], the interaggregate distance must not be so large as to allow the development of cracks of a critical size within the matrix before another aggregate is met. If the cracks, initiated by filler aggregates via dewetting or matrix cavitation, are small enough to be harmless, but numerous enough to consume a large portion of the invested energy, the material will exhibit considerable strength. At the same loading, the small-particle carbon blacks, having both small aggregate size and high surface area, will exhibit short interaggregate distance (Eq. 3.9) and thus confer greater resistance against crack propagation.

Small-particle carbon blacks with large surface areas are conducive to all these processes, including greater rubber shell caused by higher polymer-filler interaction and large interfacial area; greater probability of molecular slippage, leading to developed molecular alignment; and short interaggregate distance. Therefore, it is not surprising that the tensile strength increases with increasing surface area of carbon black, especially for non-crystallizing elastomers, as their structure is less important.

The degree of carbon black loading has the same effect as surface area. In fact, tensile strength increases with increasing filler loading up to a certain level and then drops at higher filler concentrations. At high loading, the poor dispersion of carbon blacks results in flaws that can easily create cracks and then develop into a catastrophic failure. This was confirmed by the fractographic observation of N330-filled and unfilled SBR vulcanizates[144], showing that the flaw size initiating tearing increases with filler concentration. The concentration at which maximum tensile strength is obtained varies with the carbon black. Small-particle blacks reach the maximum earlier, which can be interpreted in terms of carbon black dispersion. For high-surface-area carbon blacks, the interaggregate distance is short and the surface energy is much high, so that it is more difficult to disperse in rubber. Products with higher structure have the same effect, although this effect is not as evident as that of surface area.

The effects of carbon black on the strength of strain-crystallizing rubber, such as natural rubber, are rather complicated since strain-induced crystallization itself is an important source of energy dissipation and thus increases the fracture energy. The crystallites developed during extension of the unfilled rubber in bulk increase tensile strength to such an extent that most carbon blacks are unable to raise it much further. In comparison to amorphous rubber, the maximum tensile strength occurs much earlier as loading increases. In some cases, no maximum is observed at all[145]. It seems that the addition of carbon blacks may also interfere with the occurrence of strain crystallization in bulk rubber.

Since the strength and extensibility of the elastomer network depends on the test conditions, such as temperature and rate of extension, Smith[146,147] used a failure envelope to characterize the ultimate tensile properties of vulcanizates. He found that the stress at break, $\sigma_b$, and the ultimate strain, $\lambda_b$, measured over a wide range of temperatures and strain rates, yielded a single curve, the so-called failure envelope, when $\lg(\sigma_b T_0/T)$ was plotted against $\lg(\lambda_b-1)$, where $T$ and $T_0$ are the test and reference temperature, respectively. The failure envelope provides a basic characterization of the ultimate tensile properties because a change in strain rate or temperature results in the shift of a point along the curve. This suggests that the envelope is independent of the test conditions, and depends only on the structural characteristics of the vulcanizates. On the failure envelope, the fracture point moves clockwise around the envelope, either with rising temperature or with decreasing rate of extension. The extreme occurs at $\lambda_{b\text{-max}}$, the largest extension ratio attainable. Figure 6.59 illustrates the effect of carbon blacks on the failure envelopes of SBR vulcanizates[148]. When the filler concentration increases, the failure envelope moves toward higher tensile strength and lower elongation at break, while $\lambda_{b\text{-max}}$ decreases. This is comparable to that of unfilled strain-crystallizing rubber (Figure 6.60)[149]. The function of strain-induced rubber crystallites in tensile fracture is therefore similar to that of fillers, and vice versa.

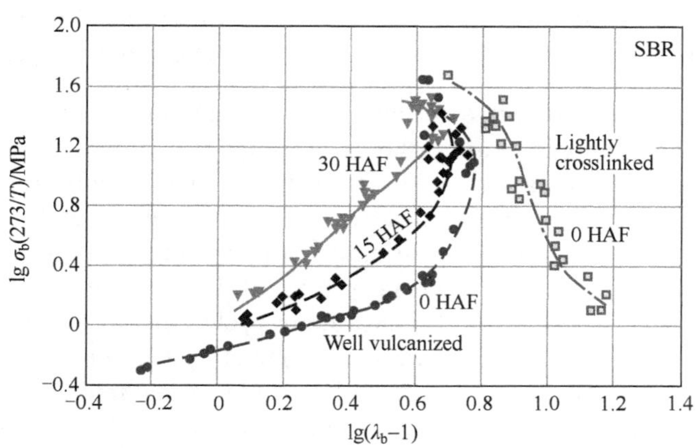


**Figure 6.59** Effect of carbon blacks on the failure envelopes of SBR[148]

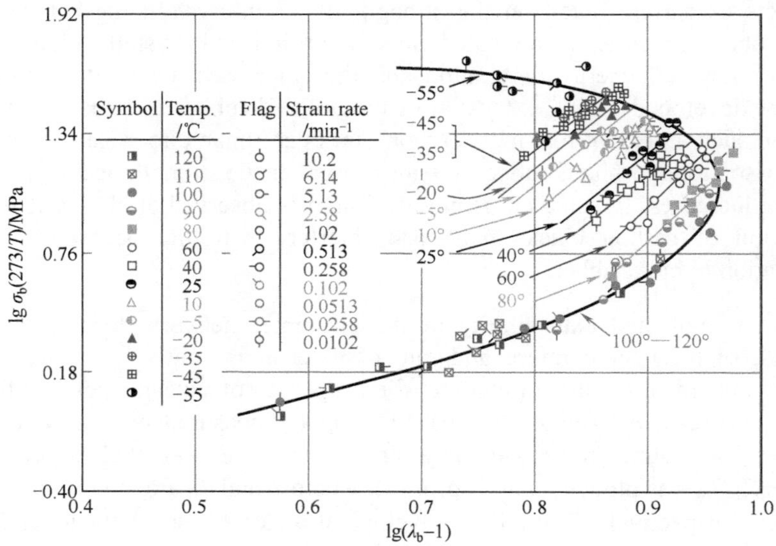


**Figure 6.60** Failure envelopes of unfilled NR[149]

## 6.4.4 Fatigue

As defined by Lake[150], "fatigue" refers, broadly speaking, to a progressive deterioration of one or several properties of the materials during service life or test. In rubber technology specifically, it indicates the failure resulting from crack growth initiated by one or several naturally occurring flaws under repeated deformation (or stress) which is much smaller than that of a single extension to break. When the flaw is a small cut introduced deliberately, this process is sometimes termed "cut growth". This vulcanizate property directly affects the service life of many rubber products, for instance tires, by affecting groove, carcass cracking, tread, and ply separation[151].

Once the force applied exceeds a certain level (critical stress or tear strength) in the case of static or monotonic tearing, the crack propagates continuously until rupture occurs (stable tearing) or travels rapidly for some distance before arresting (unstable tearing). In the case of fatigue, the crack (or cut) growth occurs under a repeated load which is less than the tear strength. During dynamic deformation, the crystallization or alignment of rubber molecules can occur in front of the crack tip in the loading cycle. This reorganization disappears in the unstrained cycle when the rubber is relaxed to zero strain. This causes the stress distribution around the crack tip to move forward and causes further growth during the next cycle[107]. If the strengthening structure does not totally disappear at zero strain, crack growth may be slowed down or prevented. It is, therefore, understandable that the rate of formation and disappearance of the alignment structure, as well as the test conditions, have a considerable influence on rubber fatigue.

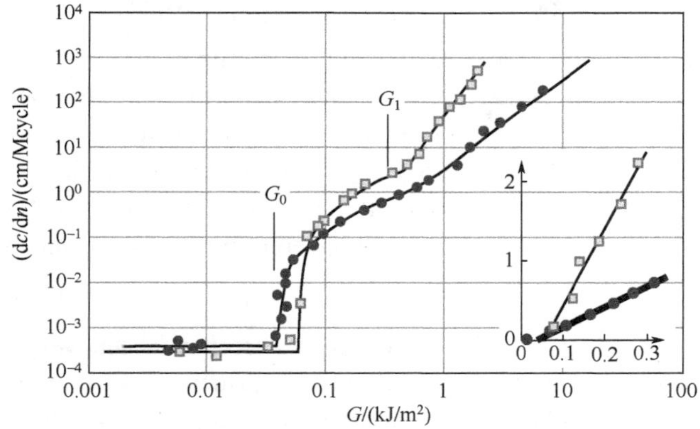


**Figure 6.61** Relationship between crack growth and maximum tearing energy attained in each cycle for NR and SBR[152]

Lake et al.[150,152–155] found experimentally that three distinct regions exist in the relationship between crack (or cut) growth and maximum tearing energy, $G$, attained in each cycle (Figure 6.61). Below $G_0$, cyclic crack growth is very low and the growth rate is largely independent of tearing energy. This growth was attributed entirely to ozone attack. Above $G_0$, but below $G_1$ which is a distinctive value of $G$ as defined in Figure 6.61, the growth rate of the crack is proportional to $(G - G_0)$. Between $G_1$ and $G_c$, the static tearing energy, the crack growth rate increase exponentially with tearing energy. Above the static tearing energy, catastrophic tearing occurs. When a long strip of rubber in the unstrained state is subjected to cyclic extension, the crack growth performance over the whole range of tearing energies can be expressed approximately as:

$$\frac{dc}{dn} = BG^{\beta} \tag{6.39}$$

where $c$ is the crack length, $n$ is the number of cycles, and $B$ and $\beta$ are constants. $\beta$ is mainly dependent on the rubber used and generally lies between 1 and 6. For natural rubber, $\beta$ is about 2, and for amorphous SBR, it is about 4[152,156,157], i.e., the effect of the flaw size and strain on fatigue properties is much more pronounced for SBR than for NR. The tearing energy, $G$, of a long strip of rubber with edge cracks of a length $c$ is given by[113]:

$$G = 2KcW \tag{6.40}$$

where $W$ is the strain energy density in the bulk of the test piece, and $K$ is a constant which changes only slightly with strain. The fatigue lifetime can thus be estimated by integration of Eq. 6.39 and by substituting $G$ from Eq. 6.40:

$$n = \frac{1}{(\beta - 1)B(2KW)^{\beta}} \times \left( \frac{1}{c_0^{\beta-1}} - \frac{1}{c^{\beta-1}} \right) \tag{6.41}$$

where $c_0$ is the effective initial flaw size. At fatigue failure, $c \gg c_0$, the number of cycles to failure is obtained from this equation by eliminating the $c$ term. Moderate and high strains yield satisfactory agreement with the experimental data for unfilled rubber. The flaw size required to fit the experimental data is a few μm, which is similar to the imperfections in the rubber caused by particulate impurities and test piece processing[150]. Another factor affecting fatigue is the atmosphere. Not only $G_0$ but also the constant $B$ is affected by the oxygen concentration through "mechano-oxidative" mechanisms[150,153,158].

With regard to the effect of filler on fatigue, there have been very few systematic investigations under well controlled conditions. In principle, as reviewed by Kraus[159], the presence of carbon black plays an important role through the effects of poor dispersion and agglomeration on $c_0$, the effect of rising strain energy, and its action as a mild thermal antioxidant in sulfur vulcanizates. Moreover, the value of $\beta$ in Eq. 6.41 for non-crystallizable SBR decreases upon incorporation of carbon black from 4 to about 2 in fully reinforced vulcanizates, which corresponds to the value for natural rubber. On the other hand, by increasing the energy dissipation, as discussed previously, the stress concentration at the tip of the growing crack and hence the rate of crack growth is reduced by the addition of carbon blacks[160].

Goldberg et al.[144], using SBR vulcanizates filled with different amounts of N330 carbon black, reported that the residual strength of sample, subjected to 100 cycles of deformation at different cycling stresses, increased with increasing filler concentration. The stress, with which the samples were fatigued to failure within 100 cycles of deformation, also increased with filler loading.

Lake and Lindley[161] showed that 50 phr HAF in NR is able to reduce the rate of cut growth considerably at a tear energy above $G_0$. However, in fatigue, this improvement is cancelled out by the large size of the flaws created by this filler, which is in a good agreement with Wolff's observations[162]. Using the DeMattia test, Wolff showed that the number of cycles $n$, enabling the cut to grow from 8 mm to 12 mm, first increases with carbon black loading, passes through a maximum, and then decreases (Figure 6.62). As surface area increases, the maximum shifts towards lower filler concentration. While the improvement of fatigue at low concentration is due to increase in tearing energy, the lower fatigue life at higher loading is related to the high value of $c_0$ resulting from the agglomeration of carbon black. It is thus apparent that the lower concentration of the maximum for small-particle blacks is due to their difficulty to disperse. In addition, when the comparison is carried out at constant stain, common in laboratory testing, and at practical loadings, the high fatigue resistance of rubber filled with large-particle and low-structure carbon black may also be caused by their lower stiffness which requires less energy input[159,163].

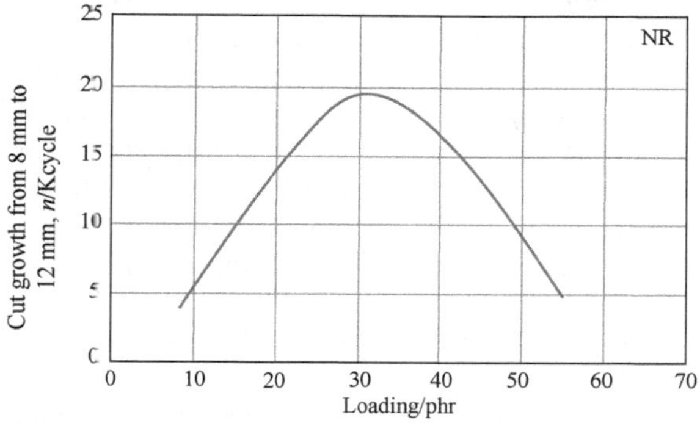


**Figure 6.62** Effect of carbon black loading on DeMattia cut growth for NR filled with N330 carbon black[162]

## References


[1] Flory P J, Rehner J. Statistical Mechanics of Cross-Linked Polymer Networks I. Rubberlike Elasticity. *J. Chem. Phys.*, 1943, 11: 512.

[2] Flory P J, Rehner J. Statistical Mechanics of Cross-Linked Polymer Networks II. Swelling. *J. Chem. Phys.*, 1943, 11: 521.

[3] Flory P J. Statistical Mechanics of Swelling of Network Structures. *J. Chem. Phys.*, 1950, 18: 108.

[4] Lorenz O, Parks C R. The Crosslinking Efficiency of Some Vulcanizing Agents in Natural Rubber. *J. Polym. Sci.*, 1961, 50: 299.

[5] Kraus G. Swelling of Filler-Reinforced Vulcanizates. *J. Appl. Polym. Sci.*, 1963, 7: 861.

[6] Hummel K, *Kautsch. Gummi Kunstst.*, 1962, 15: 1.

[7] Wolff S. *Kautsch. Gummi Kunstst.*, 1970, 23: 7.

[8] Boonstra B B. Rubber Technology and Manufacture. Cleveland, Ohio: CRC press, 1971:252, Chapter 7.

[9] Wolff S, Wang M-J, TAN E-H. Surface Energy of Fillers and its Effect on Rubber Reinforcement. II, *Kautsch. Gummi Kunstst.* 1994, 47: 873.

[10] Rigbi Z, Boonstra B B. *Meeting of the rubber division, ACS*, Chicago, Ill, Sept. 13-15, 1967.

[11] Wolff S, Wang M-J. Filler-Elastomer Interactions. Part IV. The Effect of the Surface Energies of Fillers on Elastomer Reinforcement. *Rubber Chem. Technol.*, 1992, 65: 329.

[12] Wolff S, Wang M-J, *Meeting of the Rubber Division, ACS*, Toronto, Canada, May 21-24 1991.

[13] Polmanteer K E, Lentz C W. Reinforcement Studies-Effect of Silica Structure on Properties and Crosslink Density. *Rubber Chem. Technol.*, 1975, 48: 795.

[14] Wolff S, Wang M-J. Physical Properties of Vulcanizates and Shift Factors. *Kautsch. Gummi Kunstst.*, 1994, 47: 17.

[15] Wolff S, Panenka R. *IRC' 85*, Kyoto, Japan, Oct. 15-18, 1985.

[16] Smallwood H M. Limiting Law of the Reinforcement of Rubber. *J. Appl. Phys.*, 1944, 15: 758.

[17] Guth E. Theory of Filler Reinforcement. *Rubber Chem. Technol.*, 1945, 18: 596.
[18] Guth E, Simha R, Gold O. Untersuchungen über die Viskosität von Suspensionen und Lösungen. 3. Über die Viskosität von Kugelsuspensionen. *Kolloid Z.*, 1936, 74: 266.
[19] Guth E, Gold O. Viscosity and Electroviscous effect of the AgI sol. II. Influence of the Concentration of AgI and of Electrolyte on the Viscosity. *Phys. Rev.*, 1938, 53: 322.
[20] Bachelor G K, Green J T. The Determination of the Bulk Stress in a Suspension of Spherical Particles to Order c2, *J. Fluid Mech.,* 1972, 56: 401.
[21] Kerner E H. The Electrical Conductivity of Composite Media. *Proc. Phys. Soc. B*, 1956, 69: 802.
[22] Brinkman H C. The Viscosity of Concentrated Suspensions and Solutions. *J. Chem. Phys.*, 1952, 20: 571.
[23] Eilers H. Die Viskosität von Emulsionen hochviskoser Stoffe als Funktion der Konzentration. *Kolloid Z.*, 1941, 97: 313.
[24] Van der Poel C. On the Rheology of Concentrated Dispersions. *Rheol. Acta*, 1958, 1: 198.
[25] Cohan L H. The Mechanism of Reinforcement of Elastomers by Pigments. *Indian Rubber World*, 1947, 117: 343.
[26] Mullins L, Tobin N R. Stress Softening in Rubber Vulcanizates. Part I. Use of a Strain Amplification Factor to Describe the Elastic Behavior of Filler-reinforced Vulcanized Rubber. *J. Appl. Polym. Sci.*, 1965, 9: 2993.
[27] Guth E. Theory of Filler Reinforcement. *J. Appl. Phys.*, 1945, 16: 20.
[28] Meinecke E A, Taftaf M I. Effect of Carbon Black on the Mechanical Properties of Elastomers. *Rubber Chem. Technol.*, 1988, 61: 534.
[29] Medalia A I. Effective Degree of Immobilization of Rubber Occluded within Carbon Black Aggregates. *Rubber Chem. Technol.*, 1972, 45: 1171.
[30] Hess W M, McDonald G C, Urban E. Specific Shape Characterization of Carbon Black Primary Units. *Rubber Chem. Technol.*, 1973, 46: 204.
[31] Ravey J C, Premilat S, Horn P. Light Scattering by Suspensions of Carbon Black. *Eur. Polym. J.*, 1970, 6: 1527.
[32] Medalia A I. Morphology of Aggregates. *J. Colloid Interface Sci.*, 1967, 24: 393.
[33] Wolff S, Donnet J B. Characterization of Fillers in Vulcanizates According to the Einstein-Guth-Gold Equation. *Rubber Chem. Technol.*, 1990: 63: 32.
[34] Wolff S. Renforcement des élastomères et facteurs de Structure des Charges: noir de Carbone et Silice. Ph. D. dissertation, France: Univeristé de Haute-Alsace, 1987.
[35] Wang M-J, Wolff S, Tan E-H. *Meeting of the Rubber Division, ACS*, Louisville, Apr. 19-21, 1992.
[36] Serizawa H, Nakamura T, Ito M, et al. Effects of Oxidation of Carbon Black Surface on the Properties of Carbon Black-Natural Rubber Systems. *Polym. J.,* 1983, 15: 201.
[37] O'Brien J, Cashell E, Wardell G E, et al. An NMR Investigation of the Interaction Between Carbon Black and Cis-Polybutadiene., *Rubber Chem. Technol.*, 1977, 50: 747.
[38] O'Brien J, Cashell E, Wardell G E, et al. An NMR Investigation of the Interaction Between Carbon Black and Cis-polybutadiene. *Macromolecules*, 1976, 9: 653.
[39] Kenny J C, McBrierty V J, Rigbi Z, et al. Carbon Black Filled Natural Rubber. 1. Structural Investigations. *Macromolecules*, 1991, 24: 436.
[40] Fujimoto K, Nishi T. *Nippon Gomu Kyokaishi*, 1970, 43: 54.

[41] Kaufman S, Slichter W P, Davis D D. Nuclear Magnetic Resonance Study of Rubber–Carbon Black Interactions. *J. Polym. Sci. Part A-2*, 1971, 9: 829.

[42] Smit P P A. The Glass Transition in Carbon Black Reinforced Rubber. *Rheol. Acta*, 1966, 5: 277.

[43] Smit P P A, Van der Vegt A K. Interfacial Phenomena in Rubber Carbon Black Compounds. *Kautsch. Gummi Kunstst.*, 1970, 23: 147.

[44] Smit P P A. Les Interactions Entre les Elastomères et les Surfaces Solides Ayant Une Action Renforçant. Paris: Colloques Internationaux du CNRS, 1975.

[45] Gerspacher M, Yang H H, O'Farrell C P. *Meeting of the Rubber Division, ACS*, Washington D.C., 1990.

[46] Voet A, Cook F R. Investigation of Carbon Chains in Rubber Vulcanizates by Means of Dynamic Electrical Conductivity. *Rubber Chem. Technol.,* 1968, 41: 1207.

[47] Voet A, Sircar A K, Mullens T J. Electrical Properties of Stretched Carbon Black Loaded Vulcanizates. *Rubber Chem. Technol.,* 1969, 42: 874.

[48] Boonstra B B, Medalia A I. *Rubber Age,* 1963, 46: 892.

[49] Donnet J B, Wang M-J, Papirer E, et al. Influence of Surface Treatment on the Reinforcement of Elastomers, *Kautsch. Gummi Kunstst.,* 1986, 39: 510.

[50] Wang M-J. Etude du renforcement des élastomères par les charges: Effet Exercé par l'Emploi de Silices Modifiées par Greffage de Chaines Hydrocarbonées. Ph. D. dissertation, Mulhouse, France: Univeristé de Haute-Alsace, 1984.

[51] Wang M-J. presented at a workshop *"Praxis und Theorie der Verstärkung von Elastomeren"*, Hanover, 1996.

[52] Custodero E. Caractérisation de la Surface des noirs de Carbone: Nouveau Modèle de Surface et Implications pour le Renforcement. Ph. D. dissertation, France: Université de Haute-Alsace, 1992.

[53] Wang M-J. Powder and Fibers: Interfacial Science and Applications. Boca Raton: CRC Press-Taylor and Francis Group, 2006.

[54] Rivin D. Use of Lithium Aluminum Hydride in the Study of Surface Chemistry of Carbon Black. *Rubber Chem. Technol.*, 1963, 36: 729.

[55] Rivin D. Surface Properties of Carbon. *Rubber Chem. Technol.,* 1971, 44: 307.

[56] Dannenberg E M. Primary Structure and Surface Properties of Carbon Black. *Rubber Age,* 1966, 98: 82.

[57] Wolff S, Wang M-J, Tan E-H. Filler-Elastomer Interactions. X: The Effect of Filler-elastomer and Filler-filler Interaction on Rubber Reinforcement. *Kautsch. Gummi Kunstst.,* 1994, 47: 102.

[58] Wang M-J, Wolff S, Freund B. Filler-Elastomer Interactions. Part XI. Investigation of the Carbon-Black Surface by Scanning Tunneling Microscopy. *Rubber Chem. Technol.*, 1994, 67: 27.

[59] Wang M-J, Wolff S. Filler-Elastomer Interactions. Part VI. Characterization of Carbon Blacks by Inverse Gas Chromatography at Finite Concentration. *Rubber Chem. Technol.*, 1992, 65: 890.

[60] Wang M-J. Effect of Polymer-Filler and Filler-Filler Interactions on Dynamic Properties of Filled Vulcanizates. *Rubber Chem. Technol.,* 1998, 71: 520.

[61] Wang M-J, Wolff S. Donnet J-B, Filler-Elastomer Interactions. Part III. Carbon-Black-Surface Energies and Interactions with Elastomer Analogs. *Rubber Chem. Technol.*, 1991, 64: 714.

[62] Kraus G. Reinforcement of Elastomers. New York: Wiley Interscience, 1965: 140.

[63] Bouasse H, Carriére Z. Courbes de Traction du Caoutchouc Vulcanisé. *Ann Fac Sci Toulouse,* 1903, 5: 257.

[64] Holt W L. Behavior of Rubber under Repeated Stresses. *Ind. Eng. Chem.,* 1931, 23: 1471.

[65] Mullins L. Effect of Stretching on the Properties of Rubber. *J. Rubber Res.,* 1947, 16: 275.

[66] Mullins L. The Thixotropic Behavior of Carbon Black in Rubber. *J. Phys. Colloid Chem.*, 1950, 54: 239.

[67] Harwood J A C, Mullins L, Payne A R. Stress Softening in Natural Rubber Vulcanizates. Part II. Stress Softening Effects in Pure gum and Filler Loaded Rubbers. *J. Appl. Polym. Sci.*, 1965, 9: 3011.

[68] Harwood J A C, Payne A R. Whittaker R E. Stress-Softening and Reinforcement of Rubber. *J. Macromol. Sci., Part B: Phys.*, 1971, 5: 473.

[69] Harwood J A C, Payne A R. Stress Softening in Natural Rubber Vulcanizates. Part V. The Anomalous Tensile Behavior of Natural Rubber. *J. Appl. Polym. Sci.*, 1967, 11: 1825.

[70] Johnson M A, Beatty M F. The Mullins Effect in Equibiaxial Extension and its Influence on the Inflation of a Balloon. *Int. J. Eng. Sci.*, 1995, 33: 223.

[71] Palmieri G, Sasso M, Chiappini G, et al. Mullins Effect Characterization of Elastomers by Multi-axial Cyclic Tests and Optical Experimental Methods. *Mech. Mater.*, 2009, 41: 1059.

[72] Mars W V, Fatemi A. Observations of the Constitutive Response and Characterization of Filled Natural Rubber Under Monotonic and Cyclic Multiaxial Stress States. *J. Eng. Mater. Tech.*, 2004, 126: 19.

[73] Chai A B, Verron E, Andriyana A, et al. Mullins Effect in Swollen Rubber: Experimental Investigation and Constitutive Modelling. *Polym. Test.*, 2013, 32: 748.

[74] Rickaby S R, Scott N H. Cyclic Stress-softening Model for the Mullins Effect in Compression. *Int. J. Non-Lin. Mech.*, 2013, 49: 152.

[75] Dannenberg E M, Brennan J J. *Meeting of the Rubber Division, ACS*, Philadelphia, 1965.

[76] Harwood J A C, Payne A R. Stress Softening in Natural Rubber Vulcanizates. Part IV. Unfilled Vulcanizates. *J. Appl. Polym. Sci.*, 1966, 10: 1203.

[77] Mullins L. Softening of Rubber by Deformation. *Rubber Chem. Technol.*, 1969, 42: 339.

[78] Harwood J A C, Payne A R. Stress Softening in Natural Rubber Vulcanizates. Part III. Carbon Black-filled Vulcanizates. *J. Appl. Polym. Sci.*, 1966, 10: 315.

[79] Hofmann W. Vulcanization and Vulcanizing Agents. London: Maclaren and Sons Ltd., 1967.

[80] Gent A N. Crystallization and the Relaxation of Stress in Stretched Natural Rubber Vulcanizates. *Trans. Faraday Soc.*, 1954, 50: 521.

[81] Rault J, Marchal J, Judeinstein P, et al. Stress-induced Crystallization and Reinforcement in Filled Natural Rubbers: $^{2}H$ NMR Study. *Macromolecules*, 2006, 39: 8356.

[82] Toki S, Sics I, Hsiao B S, et al. Structural Developments in Synthetic Rubbers During Uniaxial Deformation by in situ Synchrotron X-ray Diffraction. *J. Polym. Sci., Part B: Polym. Phys.*, 2004, 42: 956.

[83] Toki S, Sics I, Ran S, et al. New Insights into Structural Development in Natural Rubber During Uniaxial Deformation by in Situ Synchrotron X-ray Diffraction. *Macromolecules*, 2002, 35: 6578.

[84] EVE Rubber Institute. Qingdao, CN. 2015.

[85] L. Mullins and N. R. Tobin, Theoretical Model for the Elastic Behavior of Filler-Reinforced Vulcanizated Rubbers. *Rubber Chem. Technol.*, 1957, 30: 555.

[86] Blanchard A F, Parkinson D. Structures in Rubber Reinforced by Carbon Black. *Rubber Chem. Technol.*, 1950, 23: 615.

[87] Blanchard A F, Parkinson D. Breakage of Carbon-Rubber Networks by Applied Stress. *Ind. Eng. Chem.*, 1952, 44: 799.

[88] Bueche F. Molecular Basis for the Mullins Effect. *J. Appl. Polym. Sci.*, 1960, 4: 107.

[89] Bueche F. Reinforcement of Elastomers. New York: Interscience, 1965, Chapter 1.

[90] Alexandrov A P, Lazurkin J S. *Dokl. Akad. Nauk SSSR,* 1944, 45: 291.

[91] Houwink R. Slipping of Molecules during the Deformation of Reinforced Rubber. *Rubber Chem. Technol.*, 1956, 29: 288.

[92] Peremsky R. *Kaucuk a Plastiche Hmoty*, 1963, 12: 499.

[93] Peremsky R. *Kaucuk a Plastiche Hmoty*, 1963, 2: 37.

[94] Dannenberg E M, Molecular Slippage Mechanism of Reinforcement. *Trans. Inst. Rubber Ind.*, 1966, 42: 26.

[95] Dannenberg E M, The Effects of Surface Chemical Interactions on the Properties of Filler-reinforced Rubbers. *Trans. Inst. Rubber Ind.*, 1975, 48: 410.

[96] Dannenberg E M, Brennan J J. Strain Energy as a Criterion for Stress Softening in Carbon-Black-Filled Vulcanizates. *Rubber Chem. Technol.*, 1966, 39: 597.

[97] Hess W M, Ford F P. Microscopy of Pigment-Elastomer Systems. *Rubber Chem. Technol.*, 1963, 36: 1175.

[98] Smith R W. Vacuole Formation and the Mullins Effect in SBR Blends with Polybutadiene. *Rubber Chem. Technol.*, 1967, 40: 350.

[99] EVE Rubber Institute. Qingdao, CN. 2017.

[100] Bryant K C, Bisset D C. *Proceedings of the Third Rubber Technology Conference*, London, 1954: 655.

[101] Kraus G, Childers C W, Rollmann K W. Stress Softening in Carbon Black-Reinforced Vulcanizates. Strain Rate and Temperature Effects. *J. Appl. Polym. Sci.*, 1966, 10: 229.

[102] EVE Rubber Institute. Qingdao, CN. 2018.

[103] EVE Rubber Institute. Qingdao, CN. 2019.

[104] Donnet J B, Wang M-J. *The international Rubber Conference*, Stuttgart, Germany, 1985.

[105] Wang M-J, Wolff S. Filler-Elastomer Interactions. Part V. Investigation of the Surface Energies of Silane-Modified Silicas. *Rubber Chem. Technol.*, 1992, 65: 715.

[106] Oberth A E, Bruenner R S. Tear Phenomena Around Solid Inclusions in Castable Elastomers. *Trans. Soc. Rheol.*, 1965, 9: 165.

[107] Kinloch A J, Young R J. Fracture Behavior of Polymers, London: Elsevier Science Publishers Ltd., 1983, Chapter 10.

[108] Gent A N. Science and Technology of Rubber. New York: Academic Press, 1978: Chapter 10.

[109] Gent A N. Detachment of an Elastic Matrix from a Rigid Spherical Inclusion. *J. Mater. Sci.*, 1980, 15: 2884.

[110] Nicholson D W. On the Detachment of a Rigid Inclusion from an Elastic Matrix. *J. Adhes.*, 1979, 10, 255.

[111] Kraus G. Interactions of Elastomers and Reinforcing Fillers. *Rubber Chem. Technol.*, 1965, 38: 1070.

[112] Hess W M, Lyon F, Burgess K A. *Kautsch. Gummi Kunstst.*, 1967, 20: 135.

[113] Rivlin R S, Thomas A G. Rupture of Rubber. I. Characteristic Energy for Tearing. *J. Polym. Sci.*, 1953, 10: 291.

[114] Kainradl P, Handler F. The Tear Strength of Vulcanizates. *Rubber Chem. Technol.*, 1960, 33: 1438.

[115] Griffith A A. The Phenomena of Rupture and Flow in Solids. *Phil. Trans. R. Soc. Lond. A*, 1921, 221: 163.

[116] Stacer R G, Yanyo L C, Kelley F N. Observations on the Tearing of Elastomers. *Rubber Chem. Technol.*, 1985, 58: 421.

[117] Wang M-J, Kelley F N. Effect of Crosslink Type on the Time-Dependent Tearing. Part I. Carbon Black Filled SBR Vulcanizates. *Kautsch. Gummi Kunstst.*, 2016, 69(1-2): 38.

[118] Wang M-J, Kelley F N. Effect of Crosslink Type on the Time-Dependent Tearing. Part II Carbon Black Filled Natural Rubber Vulcanizates. *Kautsch. Gummi Kunstst.*, 2016, 69(3): 46.

[119] Gent A N, Kim H J. Tear Strength of Stretched Rubber. *Rubber Chem. Technol.*, 1978, 51: 35.

[120] Lee D J, Donovan J A. Microstructural Changes in the Crack Tip Region of Carbon-Black-Filled Natural Rubber. *Rubber Chem. Technol.*, 1987, 60: 910.

[121] Rivlin R S, Thomas A G. The Effect of Stress Relaxation on the Tearing of Vulcanized Rubber. *Eng. Fracture Mech.*, 1983, 18: 389.

[122] Green M S, Tobolsky A V. A New Approach to the Theory of Relaxing Polymeric Media. *J. Chem. Phys.*, 1946, 14: 80.

[123] Gehman S D. Reinforcement of Elasiomers. New York: Wiley Interscience, 1965.

[124] Eirich F R. Les Interactions Entre les Elastomères et les Surfaces Solides Ayant Une Action Renforçant. Paris: Colloques Internationaux du CNRS, 1975: 15.

[125] Liu H, Lee R F, Donovan J A. Effect of Carbon Black on the-Integral and Strain Energy in the Crack Tip Region in a Vulcanized Natural Rubber. *Rubber Chem. Technol.*, 1987, 60: 893.

[126] Andrews E H, Gent A N. The Chemistry and Physics of Rubber-like Substances. , London: Maclaren and Sons Ltd., 1963, Chapter 9.

[127] Mark J E, Erman B. Rubberlike Elasticity. New York: John Wiley & Sons, 1988.

[128] Henry A W. Tear Behavior of Filled and Unfilled Elastomers. Ph. D. dissertation, Ohio, USA: University of Akron, 1967.

[129] Hofmann W. Vulcanization and Vulcanization Agents. London: Maclaren and Sons Ltd., 1967: Chapter 1, pp. 21.

[130] Greensmith H W. Rupture of rubber. IV. Tear Properties of Vulcanizates Containing Carbon Black. *J. Polym. Sci.*, 1956, 21: 175.

[131] Thomas A G. Rupture of Rubber. II. The Strain Concentration at an Incision. *J. Polym. Sci.*, 1955, 18: 177.

[132] Wolff S, Görl U, Wang M-J, et al. Silica-Based Tread Compounds: Background and Performancec, *the TyreTech'93 Conference*, Basel, Switzerland, 1993.

[133] Williams M L, Landel R F, Ferry J D. The Temperature Dependence of Relaxation Mechanisms in Amorphous Polymers and Other Glass-forming Liquids. *J. Am. Chem. Soc.,* 1955, 77: 3701.

[134] Mullins L. *Trans. Inst. Rubber Ind.*, 1959, 35: 213.

[135] Stacer R G, von Meerwall E D, Kelley F N. Time-Dependent Tearing of Carbon Black-Filled and Strain Crystallizing Vulcanizates. *Rubber Chem. Technol.*, 1985, 58: 913.

[136] Lake G J, Thomas A G. The Strength of Highly Elastic Materials. *Proc. R. Soc. Lond. A: Math. Phys. Eng.*, 1967, 300: 108.

[137] Taylor G R, Darin S R. The Tensile Strength of Elastomers. *J. Polym. Sci.*, 1955, 17: 511.

[138] Bateman L, Cunneen J I, Moore C G, et al. The Chemistry and Physics of Rubber-like Substances. London: Maclaren and Sons Ltd., 1963.

[139] Hamed G R. Energy Dissipation and the Fracture of Rubber Vulcanizates. *Rubber Chem. Technol.*, 1991, 64: 493.

[140] Harwood J A C, Payne A R. Hysteresis and Strength of Rubbers. *J. Appl. Polym. Sci.*, 1968, 12: 889.

[141] Brennan J J, Jermyn T E, Perdagio M F. *Meeting of the Rubber Division, ACS*, Detroit, 1964. paper no. 36.

[142] Ambacher H, Strauss M, Kilian H-G, et al. Reinforcement in Filler-Loaded Rubbers. *Kautsch. Gummi Kunstst.*, 1991, 44: 1111.

[143] Gent A N, Pulford C T R. Micromechanics of Fracture in Elastomers. *J. Mater. Sci.*, 1984, 19: 3612.

[144] Goldberg A, Lesuer D R, Patt J. Fracture Morphologies of Carbon-Black-Loaded SBR Subjected to Low-Cycle, High-Stress Fatigue. *Rubber Chem. Technol.*, 1989, 62: 272.

[145] Studebaker M L. Reinforcement of Elastomers. New York: Intersience Publishers, 1965, Chapter 12.

[146] Smith T L. Ultimate Tensile Properties of Elastomers. II. Comparison of Failure Envelopes for Unfilled Vulcanizates. *J. Appl. Phys.*, 1964, 35: 27.

[147] Smith T L. Strength of Elastomers: a Perspective. *Polym. Eng. Sci.*, 1977, 17: 129.

[148] Halpin J C. Molecular View of Fracture in Amorphous Elastomers. *Rubber Chem. Technol.*, 1965, 38: 1007.

[149] Smith T L. Rheology. New York: Academic Press, 1969.

[150] Lake G J. Progress of Rubber Technology, England: Applied Science, 1983: 89.

[151] Beatty J R. Fatigue of Rubber. *Rubber Chem. Technol.*, 1964, 37: 1341.

[152] Lake G J, Lindley P B. Cut Growth and Fatigue of Rubbers. II. Experiments on a Noncrystallizing Rubber. *J. Appl. Polym. Sci.*, 1964, 8: 707.

[153] Lake G J, Lindley P B. The Mechanical Fatigue Limit for Rubber. *J. Appl. Polym. Sci.*, 1965, 9: 1233.

[154] Lake G J, Lindley P B. Role of Ozone in Dynamic Cut Growth of Rubber. *J. Appl. Polym. Sci.*, 1965, 9: 2031.

[155] Lake G J, Lindley P B. Fatigue of Rubber at Low Strains. *J. Appl. Polym. Sci.*, 1966, 10: 343.

[156] Thomas A G. Rupture of Rubber. V. Cut Growth in Natural Rubber Vulcanizates. *J. Polym. Sci.*, 1958, 31: 467.

[157] Gent A N, Lindley P B, Thomas A G. Cut Growth and Fatigue of Rubbers. I. The Relationship Between Cut Growth and Fatigue. *J. Appl. Polym. Sci.*, 1964, 8: 455.
[158] Lake G J, Thomas A G. *Kautsch. Gummi Kunstst.*, 1967, 20: 211.
[159] Kraus G. Advances in Polymer Science, Berlin: Springer-Verlag, 1975:155, Chapter 8.
[160] James A G. *Dtsch Kautsch. Ges. Conference*, Wiesbaden, 1971.
[161] Lake G J, Lindley P B. Mechanical Fatigue Limit for Rubber. *Rubber Chem. Technol.*, 1966, 39: 348.
[162] S. Wolff, "Kautschukchemikalien und Füllstoffe inüder Modernen Kautschuktechnologie", Weiterbildungsstudium Kautschuktechnologie, Universität Hannover, 1989/1990.
[163] Mullins L. The Chemistry and Physics of Rubber-like Substances, London: Maclaren and Sons Ltd., 1963, Chapter 11.

# 7 Effect of Fillers on the Dynamic Properties of Vulcanizates

"Dynamic properties" refers generally to periodically varying strains and stress. These properties are most simply defined for the small, sinusoidally varying strain or stress, for which the response is a small sinusoidally varying stress or strain, respectively, with the same frequency but generally out of phase. The dynamic properties are strongly dependent on temperature, frequency, strain amplitude, and the presence of filler, and, of course, the viscoelastic properties of the polymers. Fillers, when added to polymer systems, are known to cause a considerable change in dynamic properties, not only the dynamic modulus, both viscous and elastic, but also their ratio, i.e., loss factor, which is related to the portion of the energy dissipated during dynamic deformation. In practice, for rubber products used under dynamic stress or strain, the dynamic properties are of great importance, as, for example, in automotive tires, where they affect the service performance of the products with regard to rolling resistance, traction, skid resistance, handling, and heat generation.

## 7.1 Dynamic Properties of Vulcanizates

When a shear stress is imposed periodically with a sinusoidal alternation at a frequency $\omega$, on a polymer which possesses viscoelastic behavior, the strain will also alternate sinusoidally but will be out of phase, the strain lagging the stress[1] (Figure 7.1).

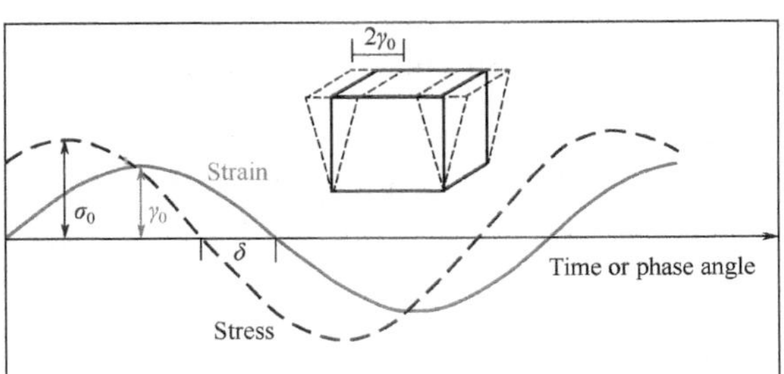


**Figure 7.1** Sinusoidal stress and strain correspondence

Thus, the strain, $\gamma$, and stress, $\sigma$, can be written as:

$$\gamma = \gamma_0 \sin(\omega t) \tag{7.1}$$

and

$$\sigma = \sigma_0 \sin(\omega t + \delta) \tag{7.2}$$

where $t$ is time, $\delta$ is the phase angle between stress and strain, and $\gamma_0$ and $\sigma_0$ are the maximum amplitude of strain and stress, respectively. In an alternative form, the stress can be subdivided further into two components, i.e., in phase and out phase with strain, namely,

$$\sigma = \sigma_0 \sin(\omega t)\cos\delta + \sigma_0 \cos(\omega t)\sin\delta \tag{7.3}$$

Accordingly, the dynamic stress-strain behavior of an elastomer material can be expressed by a modulus $G'$, which is in phase with strain (elastic modulus or storage modulus), and a modulus $G''$, which is 90° out of phase (viscous modulus or loss modulus):

$$\sigma = \gamma_0 G' \sin(\omega t) + \gamma_0 G'' \cos(\omega t) \tag{7.4}$$

with

$$G' = (\sigma_0 / \gamma_0)\cos\delta \tag{7.5}$$

and

$$G'' = (\sigma_0 / \gamma_0)\sin\delta \tag{7.6}$$

Thus,

$$\tan\delta = \frac{G''}{G'} \tag{7.7}$$

Alternatively, the modulus can be expressed by the complex, $G^*$, as:

$$G^* = (\sigma_0 / \gamma_0) = G' + iG'' \tag{7.8}$$

The energy loss during one cycle of strain, $\Delta E$, is given by:

$$\Delta E = \int \sigma \, \mathrm{d}\gamma = \int_0^{2\pi\omega} \frac{\sigma \, \mathrm{d}\gamma}{\mathrm{d}t} \mathrm{d}t \tag{7.9}$$

From Eqs. 7.1 and 7.4, one has:

$$\Delta E = \omega \gamma_0^2 \int_0^{2\pi\omega} \left[ G' \sin(\omega t)\cos(\omega t) + G'' \cos^2(\omega t) \right] \mathrm{d}t = \pi \gamma_0^2 G'' \tag{7.10}$$

By the definitions of $G''$ and $G^*$, $\Delta E$ can be also written as:

$$\Delta E = \pi \sigma_0 \gamma_0 \sin \delta \approx \pi \sigma_0 \gamma_0 \tan \delta \tag{7.11}$$

or

$$\Delta E = \omega \sigma_0^2 G'' / G^{*2} = \omega \sigma_0^2 J'' \tag{7.12}$$

where $J''$ is loss compliance, which is defined as $G''/G^{*2}$ or $G''/(G''^2 + G'^2)$. Therefore, depending on whether $\gamma_0$, $\sigma_0$, or $\gamma_0\sigma_0$ is kept constant during dynamic deformation (corresponding to constant strain, constant stress, or constant energy input), the energy loss or dynamic hysteresis is proportional to $G''$, $J''$, or tan $\delta$, respectively.

It has generally been recognized that depending upon molecular motion of the elastomers, the dynamic hysteresis of unfilled rubber shows a strong temperature-time (frequency) dependence. At a given frequency of the dynamic strain, typical plots of modulus and dynamic hysteresis, in terms of $G'$, $G''$, tan $\delta$, and $J''$, against temperatures are shown in Figure 7.2 along with the temperature dependence of the molecular motion, $\psi_T$. At sufficiently low temperature, the tan $\delta$ value is very low because the viscosity of the rubber is so high and the free volume in the polymer is so small that the movement of the polymer segments and adjustment of their relative position can hardly take place in the time scale involved in the normal dynamic experiment (frequency). This results in low energy dissipation, hence low hysteresis. Under this condition, the polymer falls in the glassy state with a very high elastic modulus.

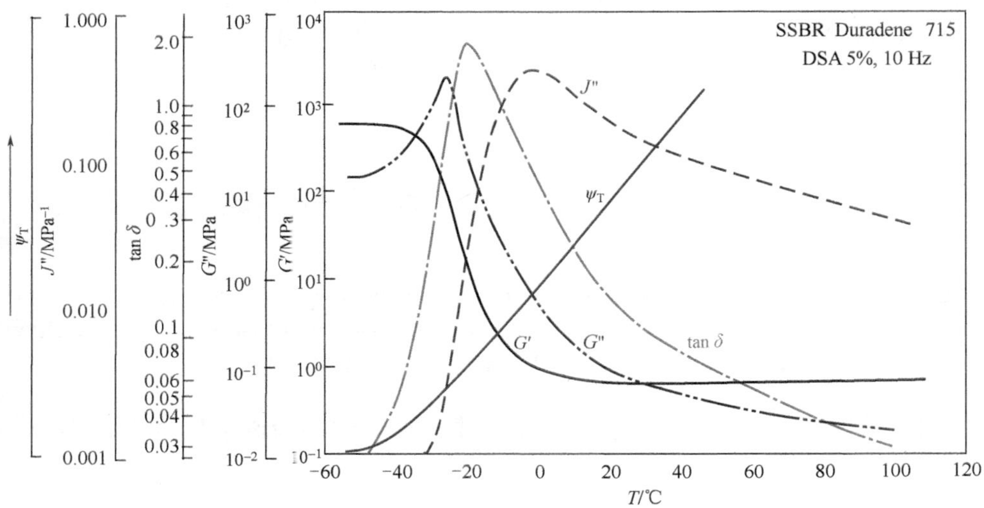


**Figure 7.2** Polymer segment mobility $\psi_T$ and dynamic parameters as a function of temperature for a gum compound of solution SBR Duradene 715

With increasing temperature, the movement of the polymer segments increases. When the temperature reaches a certain level, the free volume of the polymer increases more rapidly than the volume expansion of the molecules, facilitating segment motion. From this point, which is known as the glass transition temperature, $T_g$, the viscosity of the polymer decreases very rapidly and the molecular adjustments take place more easily, so that the elastic modulus decreases and the energy dissipation among polymer molecules will increase with temperature, resulting in increasing hysteresis.

However, at temperatures high enough such that the Brownian motion is so rapid and the viscosity is so low in the polymeric solid that the thermal energy is comparable to the potential energy barriers to segment rotation, the molecular adjustment is quick enough to be able to follow the dynamic strain. In this case, long-range contour changes of polymer molecules may take place, which associates with high entropic elasticity and low resistance to strain. The material falls into a so-called rubbery region with low modulus and low energy dissipation during dynamic deformation.

Between the glassy and rubbery regions is the transition zone where the elastic modulus drops monotonically by several orders and $G''$ and tan $\delta$ go through pronounced maxima. For most polymers, when the temperature is near the maximum of the tan $\delta$-temperature curve, more than half of the energy input during dynamic strain would be converted to heat as tan $\delta$ is generally higher than unity.

For a given polymer, it is inherent in the above discussion that the temperature function of dynamic properties must depend on the time scale of the experiment. In fact, the dynamic modulus and hysteresis at different temperatures can be related to that at different frequencies in a manner described by the temperature-time superposition principle, expressed by the WLF Equation[2,3].

## ■ 7.2 Dynamic Properties of Filled Vulcanizates

The main change in the dynamic stress-strain behavior of rubber due to the incorporation of fillers is observed with regard to elastic and viscous modulus. Other dynamic parameters can be derived from these two moduli.

### 7.2.1 Strain Amplitude Dependence of Elastic Modulus of Filled Rubber

#### Carbon Black

Shown in Figure 7.3 is a plot of elastic modulus (measured at 70°C and 0°C and 10 Hz) *vs.* the logarithm of the double strain amplitude (DSA), for a solution SBR (Duradene 715, 23.5% styrene and 46% vinyl-butadiene) filled with N234 carbon black over a loading range from 0 phr to 70 phr[4]. The dynamic properties were measured using a

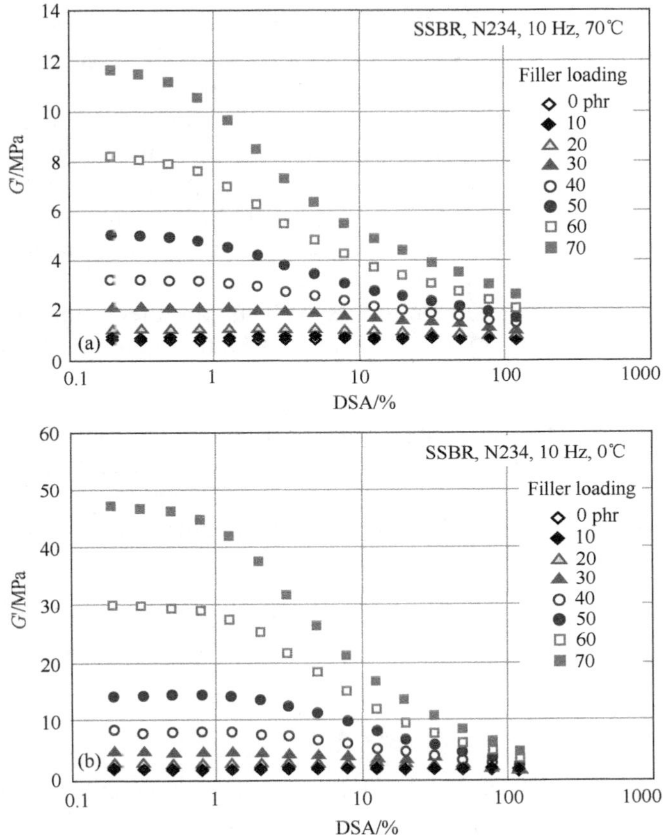


**Figure 7.3** Strain dependence of $G'$ at 70°C and 0°C and 10 Hz for SSBR compounds with different loadings of carbon black N234[4]

Rheometrics Dynamic Spectrometer operated in torsion mode (shear). As can be seen, while the modulus of the unfilled compound does not change significantly upon increasing strain amplitude over the range of DSA tested, it decreases for the filled rubber, showing a typical non-linear behavior. This phenomenon was observed in 1950 by Warring[5] and later it was studied extensively by Payne, after whom the effect is often named[6,7]. This effect is exponentially increased by increasing filler loading. On the other hand, since the moduli of all vulcanizates at high strain amplitude are in a narrow range, an exponential augmentation of the modulus at low strain amplitudes with increasing filler concentration is also observed. The decrease in elastic modulus upon increasing strain amplitude was attributed by Payne to "the structure of the carbon black, and may be visualized as filler-filler linkages of physical nature which are broken down by straining"[8]. This structure was further clarified by Medalia as that "interaggregate association by physical forces, not the 'structure' or aggregate bulkiness" as generally termed in the rubber industry[9]. This suggests that the Payne effect is mainly, if not only, related to filler agglomerate formed in the polymer matrix. In addition, it is understandable that the rubber trapped or caged in the filler agglomerates would be at least partially "dead", losing its identity as an elastomer and behaving as filler in terms of stress-strain properties. Therefore, the effective volume of the polymer bearing the stresses imposed upon the sample is reduced by filler agglomeration, resulting in increased modulus which is governed primarily by the filler concentration. The breakdown of the filler agglomerate by increasing strain amplitude would release the trapped rubber so that the effective filler volume fraction and hence, the modulus would decrease. This mechanism suggests that the Payne effect can serve as a measure of filler agglomeration which originates from filler-filler interaction as well as polymer-filler interaction. This observation can be verified by the changes in effective volume of filler upon filler loading and their strain dependence.

In fact, the approaches for describing the dependence of quasi-static modulus on filler loading and filler parameters (see Table 5.3) can also be used to describe dynamic-elastic modulus. Among others, Wolff and Donnet[10] introduced a shift factor, $f_v$, to convert the filler volume fraction, $\phi$, to an effective volume fraction, $\phi_{eff}$, for fitting to the Guth-Gold equation (Eqs. 6.25 and 6.27), i.e.,

$$G'_f = G'_{gum}\left(1+2.5 f_v \phi + 14.1 f_v^2 \phi^2\right) \tag{7.13}$$

with

$$f_v \phi = \phi_{eff} \tag{7.14}$$

The factor $f_v$ reflects all effects of filler properties on modulus through effective loading, with the structure being most important. When this concept was applied to the dynamic elastic modulus, especially that measured at low strain amplitude, it was found that the experimental data of most fillers only fit Eq. 7.13 up to certain levels of loading, which is defined as the critical loading, $\phi_{crit}$. Beyond this concentration, deviations occur with the experimental data being higher than the values calculated from the modified equation. This can be seen in Figure 7.4(a) where the ratio of $G'$ of filled rubber to that of gum, $G'_f / G'_{gum}$, measured at 70°C and 10 Hz with different strain amplitudes, is plotted as a function of effective volume. The effective volume is obtained from fitting Eq. 7.13 by adjusting the shift factor $f_v$. In this particular case $f_v$ is 1.38 which was derived from the data at 2% DSA. Obviously, the modified Guth-Gold equation is only applicable to the two lowest loadings, namely, 10 phr and 20 phr. It was also found that the shift factor is about the same for all strain amplitudes until 2% DSA, beyond which it is slightly reduced with increased strain amplitude. For example, the shift factor became 1.2 at 12.5% DSA. This may be explained by the reduction of the shielding effectiveness of rubber occlusion at high strain.

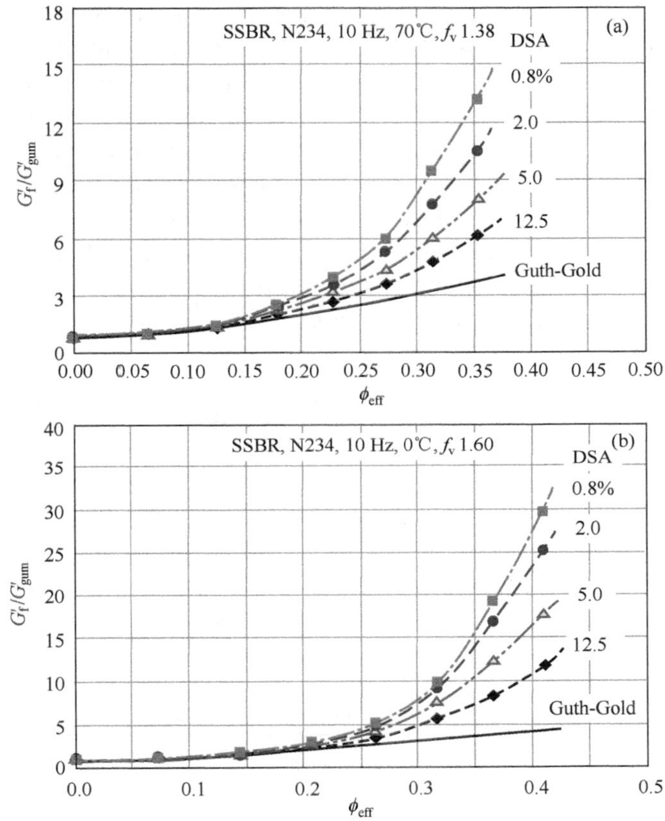


**Figure 7.4** $G'_f / G'_{gum}$ measured at different strain amplitudes at 70°C and 0°C and 10 Hz for N234 filled SSBR compounds as a function of filler concentration. The effective filler volume used for plotting is derived from the data at 2% DSA[4]

It is clear that the equations discussed above hold only for low concentration of filler. For a composite filled with rigid spheres at high concentration (up to a filler volume fraction of 0.7), Van der Poel[11, 12] derived theoretical values for the ratio of moduli of the composite to the medium modulus, $G'_f / G'_{gum}$, which is shown in Figure 7.5. Van der Poel's theory was based on the assumption that the fillers should be completely dispersed with complete wetting of the filler by the medium, and the rigidity of the filler is much higher than that of the medium. The theoretical values shown in Figure 7.5, labeled as Van der Poel, are based on the assumption that the ratio of the rigidity of filler to the medium is 100,000. Shown in Figure 7.5 is also the ratio $G'_f / G'_{gum}$ of carbon black filled rubber measured at 70°C at DSA 0.8% and plotted as a function of $\phi_{eff}$ instead of $\phi$. The $\phi_{eff}$ values were obtained from Eq. 7.13, which includes the effect of hydrodynamic, rubber occlusion and immobilized rubber shell on the filler surface as schematically shown in Figure 7.6(a). As can be seen, all experimental data, except that at very low loading, lie far above the theoretical line. For the vulcanizate filled with 50 phr carbon black having $\phi = 0.197$ based on filler density and $\phi_{eff} = 0.276$ derived from Eq. 7.13, the measured $G'_f / G'_{gum}$ is equivalent to that of the vulcanizate having a filler volume fraction of 0.462 obtained from the Van der Poel theory. This equivalent filler volume is termed as $\phi_{VdP}$, as schematically represented in Figure 7.6(b). This suggests that the amount of rubber trapped in the filler agglomerate is about 94% of the carbon black volume or equivalent to 67% of the effective volume, $\phi_{eff}$, of the filler, even though part of the trapped rubber which overlaps with that immobilized on the filler surface and decreases substantially upon reduction of filler loading and increase in strain amplitude, as shown in Table 7.1.

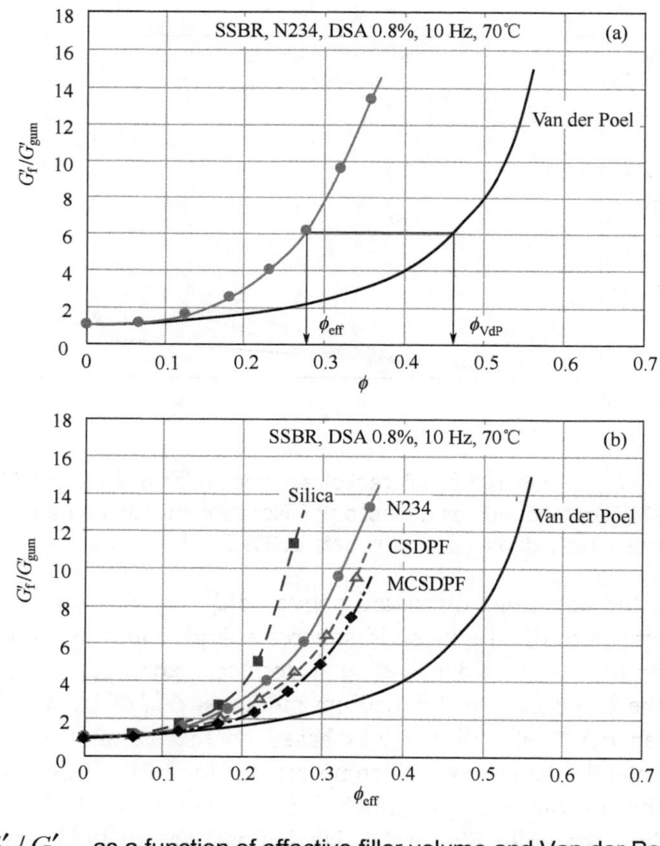


**Figure 7.5** $G'_f / G'_{gum}$ as a function of effective filler volume and Van der Poel filler volume-$G'_f / G'_{gum}$ plot[4]

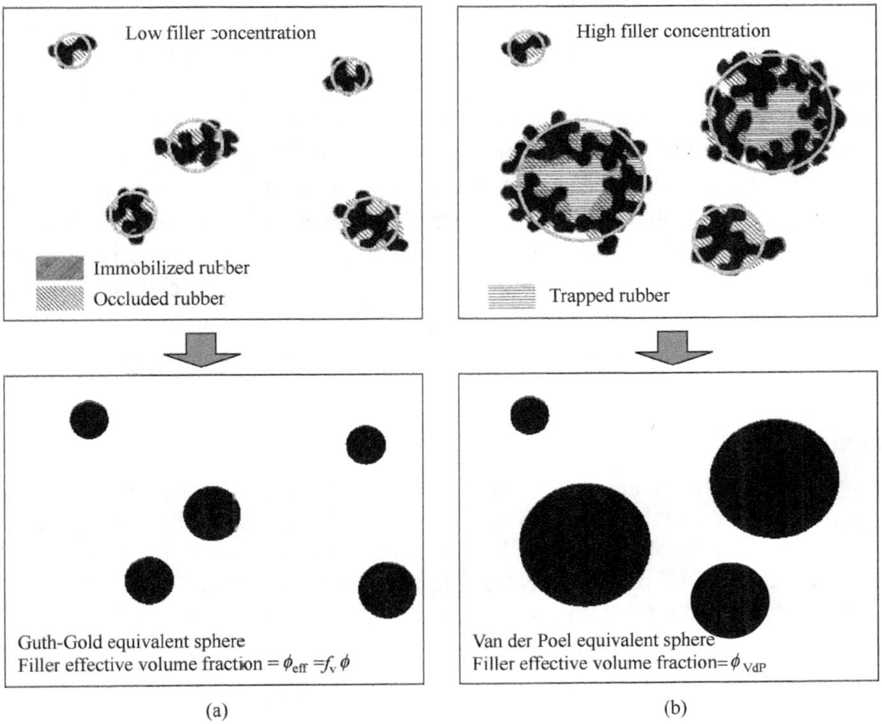


**Figure 7.6** Schematic presentation of effective volume of filler $\phi_{eff}$ ($f_v\phi$, Eq. 7.13) and Van der Poel effective volume $\phi_{VdP}$[4]

**Table 7.1** Van der Poel equivalent filler volume $\phi_{VdP}$[4]

| Filler loading | DSA/% | Temperature/°C | N234 | Silica | CSDPF | MCSDPF |
|---|---|---|---|---|---|---|
| 70 | 0.8 | 70 | 0.552 | – | 0.520 | 0.483 |
| 50 | 0.8 | 70 | 0.462 | 0.538 | 0.415 | 0.371 |
| 30 | 0.8 | 70 | 0.315 | 0.330 | 0.240 | 0.200 |
| 70 | 5.0 | 70 | 0.500 | – | 0.478 | 0.458 |
| 50 | 5.0 | 70 | 0.410 | 0.520 | 0.382 | 0.345 |
| 30 | 5.0 | 70 | 0.278 | 0.290 | 0.235 | 0.192 |
| 70 | 0.8 | 0 | 0.620 | – | 0.578 | 0.552 |
| 50 | 0.8 | 0 | 0.523 | 0.540 | 0.455 | 0.430 |
| 30 | 0.8 | 0 | 0.342 | 0.365 | 0.290 | 0.280 |
| 70 | 5.0 | 0 | 0.576 | – | 0.530 | 0.518 |
| 50 | 5.0 | 0 | 0.495 | 0.523 | 0.430 | 0.408 |
| 30 | 5.0 | 0 | 0.310 | 0.360 | 0.275 | 0.260 |

All observations concluded from the results obtained at 70°C are also evident at 0°C [Figure 7.3(b) and Figure 7.4(b)], but the difference in moduli between low strain and high strain is much greater, due to the temperature dependence of filler-filler and filler-polymer interaction. It should be pointed out that even though at 0°C the gum is in its transition zone so that the modulus is higher than that obtained at 70°C, from the point of view of kinetic theory, the modulus of the gum or the crosslinked polymer matrix should increase with increasing temperature in the rubber state. However, what is really observed is that when the $G'$ of filled rubber is normalized by that of gum, the $G'_f / G'_{gum}$ rises much more rapidly at lower temperature. The equivalent volume, $\phi_{VdP}$, of carbon black in the vulcanizate filled with 50 phr filler is now about 0.523, which is 13% higher than that obtained at 70°C. This may suggest that filler agglomerates are more developed or stronger at low temperature.

## Silica and CSDPF

If the dynamic modulus and its strain dependence of filled rubber is mainly related to filler agglomeration, its formation would be essentially governed by filler types which possess various filler-polymer and filler-filler interactions. Shown in Figure 7.7 are the strain dependence of $G'$ of vulcanizates filled with 50 phr carbon black, silica, carbon-silica dual phase filler (CSDPF, see Section 9.2), and TESPT-modified dual phase filler (MCSDPF) at both 70°C and 0°C.

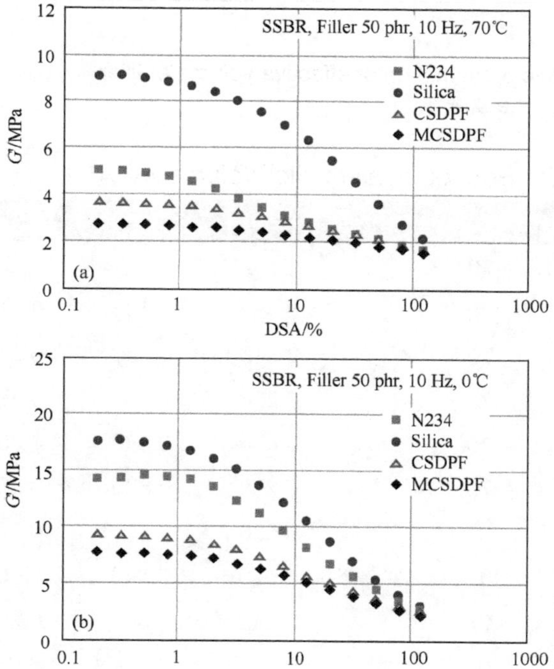


**Figure 7.7** Strain dependence of $G'$ at 70°C and 0°C and 10 Hz for SSBR compounds with different fillers[4]

Compared to carbon black, silica gives high modulus at low strain amplitude and large Payne effect at both temperatures. This would be an indication that more developed and stronger filler agglomerates form in the silica-filled vulcanizates. This observation can be verified by the equivalent filler volume fraction, $\phi_{VdP}$, using the shift factor deduced from the modified Guth-Gold equation (Table 7.1). For example, at 50 phr, while the real volume fraction of the filler calculated from density and loading is about 0.176 for silica and 0.197 for carbon black, their $\phi_{VdP}$ are 0.538 and 0.462 respectively at 70°C and DSA of 0.8%.

The high level of agglomeration in silica is first attributed to its surface characteristics. As discussed in Chapter 2, silica possesses a low dispersive component of surface energy while carbon black is high. By contrast, the specific or polar component of surface energy is much higher for silica than for carbon black[13,14]. According to the thermodynamics of filler agglomeration (see Chapter 3), this suggests that in a hydrocarbon polymer, the polymer-silica interaction which is determined by the dispersive component of filler surface energy is lower, and the interaction between silica aggregates is higher[15]. Consequently, less compatibility with polymer and strong filler-filler interaction due to high surface polarity and strong hydrogen bonding between silanol groups would lead to a more developed filler agglomerate. On the other hand, the higher surface area of silica may also be responsible for the higher agglomeration at the same loading, the surface area is the primary factor to control the mean distance between aggregates[16]. The higher the surface area, the shorter the interaggregate distance would be. However, at the same loading, this effect will be, to a certain extent, offset by the difference in density between silica and carbon black.

Also shown in Figure 7.7 are the results for the vulcanizates filled with CSDPF and its TESPT modified counterpart. Although the dual phase filler is between carbon black and silica from the point of view of chemical composition, what is actually observed here is that the CSDPF gives a lower Payne effect than any of these two conventional fillers. This suggests that filler agglomeration is lower for CSDPF, which would be mainly due to its lower interaggregate interaction. This can also be seen from the plot of $G'_f / G'_{gum}$ *vs.* $\phi_{eff}$ and $\phi_{VdP}$, which is related to the volume fraction of the rubber trapped in agglomerates [Figure 7.5(b) and Table 7.1].

The lower filler-filler interaction of CSDPF may involve multiple mechanisms, but it is most readily interpreted on the basis of lower interaction between filler surfaces having different surface characteristics, surface energies in particular. Based on surface energies and interaction between two filler aggregates, Wang[17] derived an equation to show that the same categories of surface would preferentially interact with each other (see Section 8.1.2.3.3 about blends of carbon black and silica). Accordingly, when dual phase aggregates are dispersed in a polymer matrix, the interaction between the silica domains and the carbon phase of the neighboring aggregates should be lower than that between carbon black aggregates as well as lower than that between silica aggregates. In addition, the portion and the probability of the aggregate surface being able to face directly the same category of surface of the neighboring aggregate is substantially reduced by doping silica in the carbon phase. Consequently, there would be less hydrogen bonding, the main cause of the high filler-filler interaction between silica aggregates, between silica domains on neighboring aggregates since their average interaggregate distance would be greater. In addition, from the point of view of flocculation kinetics, the higher surface activity of the carbon phase of CSDPF, which leads to higher bound rubber[18], may contribute to less developed filler agglomeration.

The filler agglomeration of CSDPF can be further depressed by surface modification with coupling agent TESPT as shown by the Payne effect and equivalent filler volume. This is primarily due to the reduction of surface energies, both dispersive and polar components. The higher bound rubber content of the MCSDPF *vs.* its unmodified counterpart (45% *vs.* 32%) may also be a reason as the flocculation rate of the filler aggregates may be significantly reduced during vulcanization.

### 7.2.2 Strain Amplitude Dependence of Viscous Modulus of Filled Rubber

Fillers are known to cause a considerable increase not only in $G'$, as discussed above, but also in $G''$. Since $G''$ is representative of the viscous component of the modulus, all processes of energy dissipation will affect $G''$. It can therefore be argued that interaggregate interaction must be involved to a considerable extent in energy loss, which would be highly dependent on strain amplitude, filler concentration, and filler properties.

#### Carbon Black

The strain dependence of viscous modulus for vulcanizates having different loadings of N234 are shown in Figure 7.8 for the results obtained at both 70°C and 0°C [4]. Evidently, the incorporation of carbon black in rubber will substantially increase the viscous modulus of the material regardless of the strain amplitude. As observed with $G'$, this effect is also augmented exponentially with filler loading and is partially attributed to the hydrodynamic effect, as the addition of unstrained (or solid) particles in the polymer matrix would result in a high viscosity compound. It is also evident that in contrast to $G'$ of filled rubber which decreases monotonically with increasing strain, $G''$ shows maximum values at moderate strain amplitudes (2%–5% and 3%–6% DSA for 70°C and 0°C, respectively, for these particular compounds and test conditions). After passing through a maximum, the $G''$ decreases rapidly with further increase in strain amplitude. Such behavior of the strain dependence of filled rubber cannot be explained only by the hydrodynamic effect since the $G''$ values of the gum compound, while being lower, do not show a significant strain dependence over a large practical range of strain amplitudes.

Payne[19], in his investigation of different polymers and carbon blacks at different loadings, found that $G''_{max}$ is linearly related to $(G'_0 - G'_\infty)$, i.e., the maximum change in elastic modulus with increasing strain amplitude, and can be expressed as:

$$G''_{max} = 0.17(G'_0 - G'_\infty) \tag{7.15}$$

where $G'_0$ and $G'_\infty$ are the leveling values of $G'$ at low and high amplitudes, respectively.

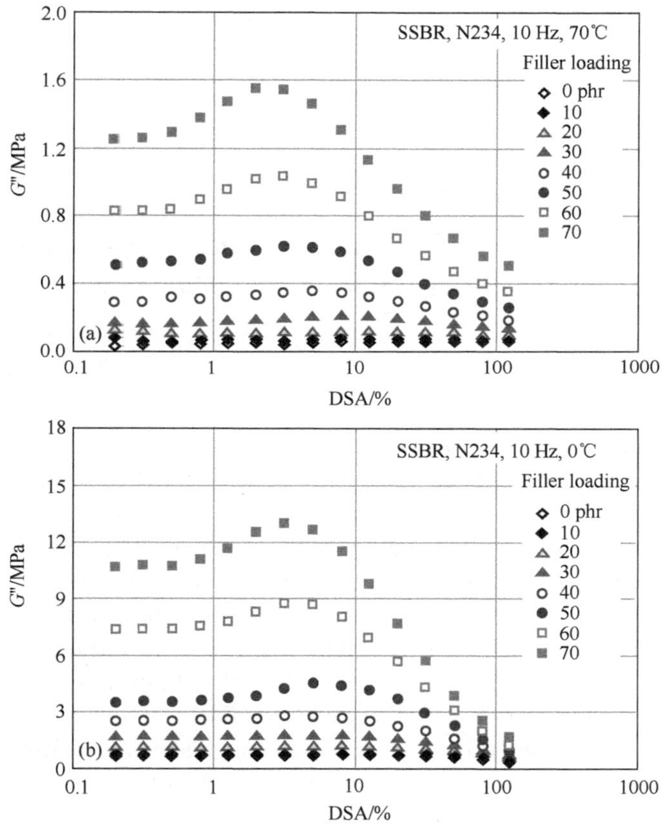


**Figure 7.8** Strain dependence of $G''$ at 70 and 0°C and 10 Hz for SSBR compounds with different loadings of carbon black N234[4]

Several authors[20–24] later investigated different polymers and polymer blends filled with various carbon blacks and arrived at similar equations with different slopes and intercepts. Payne[19] believed that this type of energy loss was related to the breakdown and reformation of the filler agglomerate. At small amplitude, the disruption of the agglomerate is insignificant. $G''$ is therefore low, but increases rapidly at higher amplitudes where breakdown and reformation of the agglomerates is significant. High amplitudes will largely destroy the agglomerates, so that less energy is required for dynamic oscillation, thus leading to a reduction in $G''$. This suggests that internal friction is the dominant mechanism of energy dissipation during dynamic deformation. In a systematic study dealing with the effect of carbon blacks on rubber hysteresis,

Ulmer et al.[25] concluded that the formation of three-dimensional carbon black-rubber agglomerates is one of the factors affecting $E''$.

Based on the assumption of the breakdown and reformation of secondary structure, Kraus[26] derived the following equations for viscous modulus $E''$:

$$E'' = E''_\infty + \frac{C\varepsilon_0^m \left(E'_0 - E'_\infty\right)}{1 + \left(\varepsilon_0 / \varepsilon_c\right)^{2m}} \tag{7.16}$$

and

$$E'' = E''_\infty + C'\varepsilon_0^m \left(E'_0 - E'_\infty\right) \tag{7.17}$$

where $m$, $C$, and $C'$ are constants; $E''_\infty$ is the $E''$ after total destruction of the filler agglomerates; $\varepsilon_0$ is the strain amplitude; and $\varepsilon_c$ is the characteristic strain given by:

$$\varepsilon_c = \left(\frac{k_m}{k_b}\right)^{0.5m} \tag{7.18}$$

in which $k_b$ is the rate constant of breakage of interaggregate contacts, and $k_m$ is the rate constant of reagglomeration. It is obvious that the loss modulus is not only dependent on $\left(E'_0 - E'_\infty\right)$ and $E''_\infty$, but also on the rates of agglomerate breakdown and reformation, which are related to the strain amplitudes. The breakdown of filler agglomerates would increase with increasing strain amplitude, and the reformation of this structure would diminish more rapidly than its disruption. Once the strain amplitude is high enough that the filler agglomerate is destroyed to such an extent that it cannot be reconstructed in the time scale of dynamic strain (frequency), the effect of filler agglomerate on the $G''$ will disappear. Similarly, if the filler agglomerate is strong enough and the strain (or stress) is small enough so that the filler agglomerate is unable to be broken, the $G''$ would be determined mainly by the hydrodynamic effect of the filler so that the strain dependence would be eliminated (Figure 7.8). In this case, however, the effective volume fraction of the filler hence the absolute $G''$ would be higher due to the hydrodynamic effect.

### Silica and CSDPF

At the same mass loading, silica gives higher $G''$ than carbon black, whereas lower values were obtained with CSDPF at 70°C. The TESPT-modification of CSDPF is able to further reduce the hysteresis of the vulcanizate (Figure 7.9). This is true at both high and low temperatures. Lower $G''$ suggests that for CSDPF, less filler agglomerates would be broken down and reformed during dynamic strain.

At 0°C, silica gives much lower viscous modulus than carbon black does, which is the opposite of what happens at higher temperature. On the other hand, over the whole range of strain applied, the $G''$ of silica compound is almost the same as the value of CSDPF compound. This may be caused by two reasons. One is related to the energy

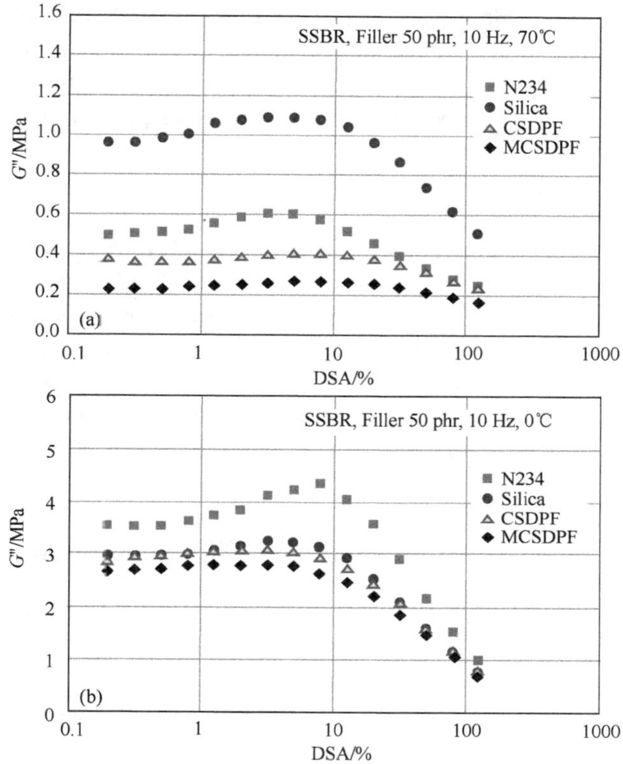


**Figure 7.9** Strain dependence of $G''$ at 70°C and 0°C and 10 Hz for SSBR compounds with different fillers[4]

dissipation of the polymer under dynamic strain, and the other depends on the agglomeration of the filler. At such low temperature, the rubber used in this study is in the transition state, which absorbs more energy under dynamic strain. Moreover, the agglomerates become so strong that they cannot be broken down under the applied stress, resulting in a decrease in the overall energy dissipation as the polymer trapped in the agglomerates is unable to participate in energy dissipation. This will be further discussed later.

### 7.2.3 Strain Amplitude Dependence of Loss Tangent of Filled Rubber

By its definition, the loss tangent is determined by both viscous and elastic moduli. While elastic modulus is representative of dynamic stress that is in phase with strain and a measure of the energy returned to the system, the viscous modulus is 90° out of phase and related to the component of lost energy. Therefore tan $\delta$ is a ratio of the work converted into heat (or the work absorbed by the compound) to that recovered, for a given work input. Besides the hydrodynamic effect, the influence of filler on $G'$ and $G''$ involves different mechanisms and different strain dependencies, with both mechanisms influencing tan $\delta$. While $G'$ is mainly related to filler agglomeration which is reduced during dynamic strain, $G''$ is related to the breakdown and reformation of these structures. Consequently, the factors predominantly governing tan $\delta$ would be the state of filler-related structures, or more precisely, the ratio between the portion capable of being broken down and reconstituted and those remaining unchanged during dynamic strain[16]. The change in tan $\delta$ from changing strain amplitude would reflect the ratio of these two processes.

### Carbon Black

Presented in Figure 7.10(a) is the strain dependence of tan $\delta$ at 70°C for vulcanizates filled with different loadings of carbon black N234[4]. Similar to the effect of strain amplitude on $G''$ of filled rubber, tan $\delta$ is shown to increase with increasing DSA at lower strain amplitude up to a high level after which there is a gradual reduction. Also like the observations of $G'$ and $G''$, strain dependence drastically increases with increasing filler concentration, with basically no strain dependence being observed for gum rubber. In addition, compared with $G''$, the maximum appears at higher DSA (8%–12% *vs.* 2%–5% for tan $\delta$ *vs.* $G''$).

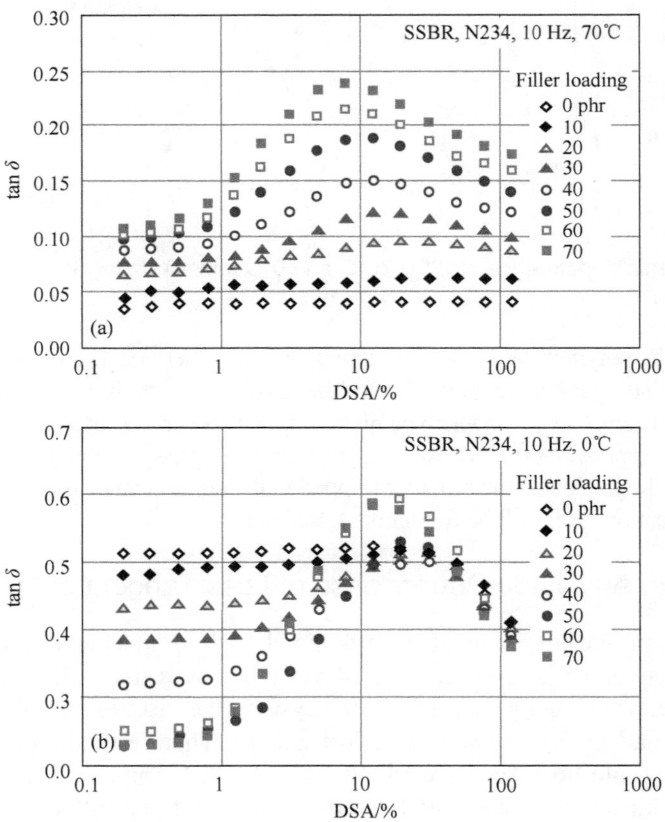


**Figure 7.10** Strain dependence of tan $\delta$ at 70°C and 0°C and 10 Hz for SSBR compounds with different loadings of carbon black N234[4]

Since for the unfilled compound there is no strain dependence for either $G'$ or $G''$, for the filled compound, the great increase in tan $\delta$ upon increasing strain amplitude at low DSA reflects an increase in $G''$ and a decrease in $G'$, both contributing to higher loss tangent. With further increase in DSA, $G'$ continuously drops as the filler agglomerates are further disrupted, the $G''$ passes its maximum and then decreases as the rate of filler reformation drops more rapidly than its disruption. When the rate of decrease of $G''$ is more rapid than that of $G'$, the tan $\delta$-DSA curve drops with increasing DSA. Consequently, the tan $\delta$ maximum appears at higher strain amplitude than that of $G''$.

The discussion so far concerns results obtained at relatively high temperatures where the main cause of energy dissipation is related to the change of filler agglomerate structure during cyclic strain as the polymer is in its rubbery state with very high entropic elasticity and low hysteresis. When dynamic stress is applied to the filled rubber at temperatures which fall in the transition zone of the polymer, the hysteresis of the polymer needs to be taken into consideration.

As clearly shown in Figure 7.10(b), at 0°C – a temperature in the transition zone of the polymer used – the picture is quite different from that observed at 70°C. Over a certain range of strain amplitudes, tan $\delta$ decreases with increasing filler loading. The highest value is obtained with the gum compound. However, with increasing DSA the difference between the filled vulcanizates with increased loading and the gum compound is minimized. On the one hand, gum does not show a strong strain dependence, and on the other hand, the tan $\delta$ of loaded rubber increases rapidly with DSA up to a strain amplitude of about 20%–30% DSA in this particular case. For highly loaded vulcanizates, tan $\delta$ is even higher than that of its unfilled counterpart at high DSA. Such an observation may also be related to the filler agglomeration. At 0°C, solution SBR Duradene 715 is in its transition zone where the polymer energy loss of polymer per se is high due to its high viscosity, hence long relaxation time, as discussed previously. Under a strain amplitude which cannot break the filler agglomerate, the polymer fraction is reduced so that tan $\delta$ would be lower. When the strain amplitude increases to such an extent that a portion of the filler agglomerate can be broken down, the polymer trapped in this agglomerate would be released to take part in the energy dissipation process, thus resulting in higher tan $\delta$. Moreover, under this condition, cyclic disruption and reformation of the filler agglomerates would also impart additional energy dissipation to the filled composite. As these two processes increase with strain amplitude, tan $\delta$ would increase steeply with DSA. As a consequence, for the highly loaded vulcanizates, tan $\delta$ is even higher than that of gum rubber at high strain amplitudes.

It is also noted that tan $\delta$ of the vulcanizates at high strain amplitude drops drastically with DSA. While this would be attributed to the reduction of filler agglomerate reformation at high DSA as discussed before, the longer relaxation time at low temperature would also be responsible for the lower hysteresis as the polymer chains are unable to adjust their conformations in a time scale related to both the dynamic frequency and the strain amplitude. This also applies to the tan $\delta$ at high temperature but its effect may be minor in comparison with the filler agglomerate contribution to hysteresis.

## Silica

As mentioned previously, silica can form stronger and more developed agglomerates which would be reflected in loss tangent. This is clearly shown in Figure 7.11. At 70°C, while at low loading the tan $\delta$ does not change drastically with strain amplitude, at loadings higher than 20 phr, the tan $\delta$ increases continuously without reaching a maximum over the range of DSA investigated. It is also observed that at low strain amplitude, the highly loaded rubbers, relative to vulcanizate having 30 phr silica, give lower hysteresis but at high strain amplitude tan $\delta$ is higher, with a crossover at DSA around 10%. Such an observation can also be explained based on the strength and development of filler agglomeration. As silica can form stronger filler agglomerates, at low strain amplitude, they are unable to be disrupted and therefore lower hysteresis would be expected. When high strain amplitude is applied, more agglomerates will break down and reform, resulting in higher hysteresis. The agglomerates are so strong and so developed that, at the strain amplitude that the carbon black agglomerate is mostly destroyed, a great portion of agglomerates is still involved in an energy dissipation process. The higher hysteresis at low strain amplitude of 30 phr silica compound, and its low increase rate with increasing DSA, indicates that the concentration of weak agglomerates is higher and that of stronger ones is lower, compared to its highly loaded counterparts.

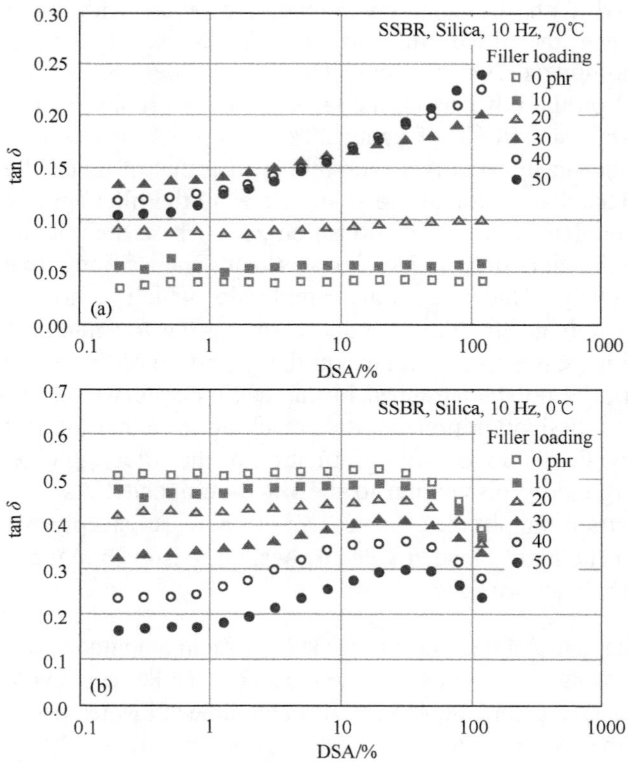


**Figure 7.11** Strain dependence of tan $\delta$ at 70°C and 0°C and 10 Hz for SSBR compounds with different loadings of silica[4]

At 0°C, it is observed that, over the DSA range investigated, tan $\delta$ decreases when the filler loading increases. Compared with carbon black-filled rubber, the maximum tan $\delta$ appears at higher strain amplitudes and the overall tan $\delta$ is substantially lower, especially for the highly loaded vulcanizates. All these features of silica compounds are mainly and undoubtedly due to their strong filler agglomerates, resulting in reduction of the polymer fraction and less filler agglomerates being able to be broken down and reformed under the strain amplitude applied.

## CSDPF

The arguments about the role of filler agglomeration on dynamic hysteresis can be further strengthened by the results of CSDPF, which are shown in Figure 7.12. Relative to their carbon black counterparts, the lower tan $\delta$ of CSDPF-filled vulcanizates over the entire range of DSA at high temperature, their higher tan $\delta$ at low strain amplitude, and lower values at larger strain and low temperature are primarily due to the fewer and weaker agglomeration of the filler aggregates. This phenomenon is significantly enhanced by TESPT modification of CSDPF (Figure 7.13).

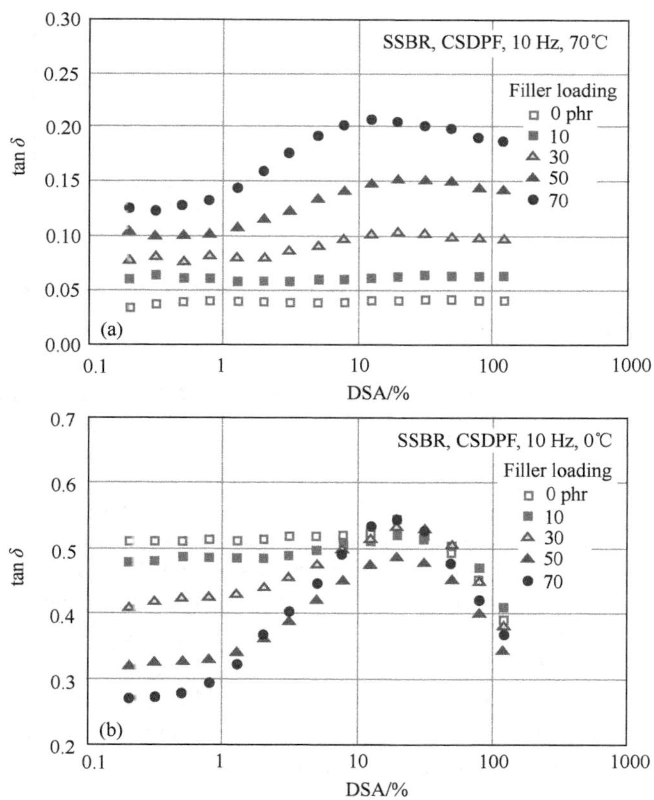


**Figure 7.12** Strain dependence of tan $\delta$ at 70°C and 0°C and 10 Hz for SSBR compounds with different loadings of CSDPF[4]

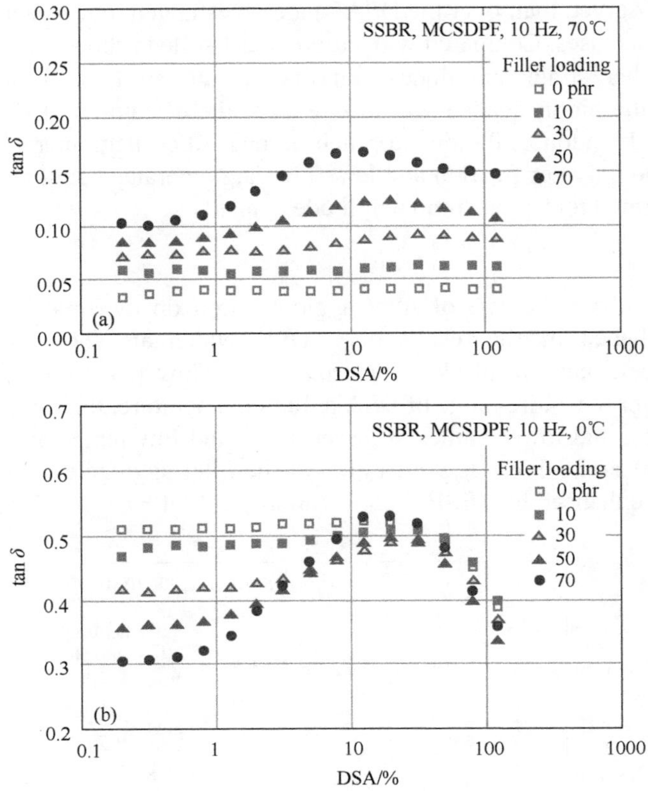


**Figure 7.13** Strain dependence of tan $\delta$ at 70°C and 0°C and 10 Hz for SSBR compounds with different loadings of TESPT modified CSDPF[4]

### 7.2.4 Hysteresis Mechanisms of Filled Rubber Concerning Different Modes of Filler Agglomeration

From the foregoing discussion of strain dependence of dynamic properties of filled vulcanizates it should be recognized that filler agglomeration plays an important role in the modulus and energy dissipation or hysteresis during dynamic strain. With regard to the structure of agglomerates, two modes of filler agglomeration have been introduced (see Section 3.4.2): direct contact of filler aggregates and the joint shell mechanism. The formation of rubber shell on the filler surface would result in a modulus gradient with increasing distance from the filler surface, which is equivalent to a shift of the transition zone and $T_g$ to a higher temperature for the polymer in the rubber shell. When two or more filler particles or aggregates are close enough, they would form an agglomerate via a joint rubber shell in which the modulus of the polymer is higher than that of the polymer matrix.

Accordingly, the mechanism of dynamic hysteresis may also differ from one type of agglomerate to another. For the filler agglomerate constructed by the direct contact mode, it is believed that the higher energy dissipation of the filled rubber originates from the breakdown and reformation of the filler agglomerate[19,27]. This suggests that at higher temperatures, internal friction between aggregates is the dominant mechanism. With decreasing temperature, while the polymer may fall in the transition zone which gives high energy dissipation from the polymer, the filler-filler interaction increases to such a degree that the filler agglomerates cannot be broken down under the applied strain amplitude. Consequently, due to trapped polymer, the hysteresis of the filled vulcanizate will be substantially lower than that expected from the volume fraction of the filler calculated from its weight.

The mechanism of filler agglomerate formed via a joint rubber shell mechanism is different from that described above. When the filled rubber undergoes dynamic strain, the rubber in the joint shell may influence the temperature dependence of the filled rubber in a different way. At a temperature where the polymer matrix is in the rubbery state, but the polymer in the rubber shell is in its transition zone caused by the adsorption of the polymer molecules on the filler surface or the interaction between polymer chains and filler which will lead to reduction of segment mobility of the polymer, the joint shell rubber would absorb more energy resulting in higher hysteresis. The more developed the filler agglomerate the more rubber in the joint shell, and the higher the dynamic hysteresis of the filled rubber would be expected as shown in Figure 7.14. With increasing temperature, the thickness of the rubber shell would decrease and the molecular mobility of the rubber in the shell would increase, the hysteresis would decrease, and the rate at which tan $\delta$ decreases would be much higher in comparison to gum rubber.

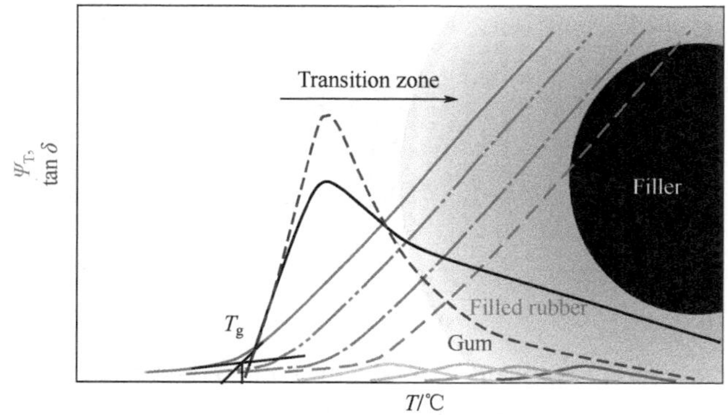


**Figure 7.14** Effect of filler-polymer interaction on the mobility of polymer segments and dynamic hysteresis of filled rubber

On the other hand, with decreasing temperature, while the filler-filler and filler-polymer interactions increase, the amount of rubber involved in the rubber shell will increase and the mobility of the rubber segments will attenuate so that the hysteresis would increase due to an increase in energy dissipation in the rubber shell as well as in polymer matrix. When the temperature is low enough that the rubber shell falls into the glassy state but the polymer matrix is in the transition zone, the effective filler volume fraction would substantially increase as the rubber shell in the glassy state and the rubber trapped in the agglomerate "cages" are totally "dead", behaving as hard filler. In this case, the hysteresis of the filled rubber would be lower than that expected from the system with filler less agglomerated.

### 7.2.5 Temperature Dependence of Dynamic Properties of Filled Vulcanizates

The temperature dependence of $G'$, $G''$, and tan $\delta$ for vulcanizates having different loadings of N234 are shown in Figure 7.15, Figure 7.16, and Figure 7.17, respectively[28]. The dynamic properties were measured at 10 Hz with DSA of 5%±0.1%. This strain amplitude remained constant until the temperature decreased lower than the peak temperature of the tan $\delta$ curve. At further decreasing temperature (below −25°C for this particular polymer), the strain amplitude decreased with decreasing temperature as the materials became very hard. It can be seen that while the moduli, both elastic and viscous, increase with filler loading over the range of temperature investigated, the effect of filler concentration on tan $\delta$ can be classified by the temperature. At low temperature, tan $\delta$ is reduced by increasing filler loading. However, the reverse is true at high temperature with a cross-over point of around 0°C in this particular system and under these particular test conditions. The temperature where the maximum tan $\delta$ appears does not change with filler concentration.

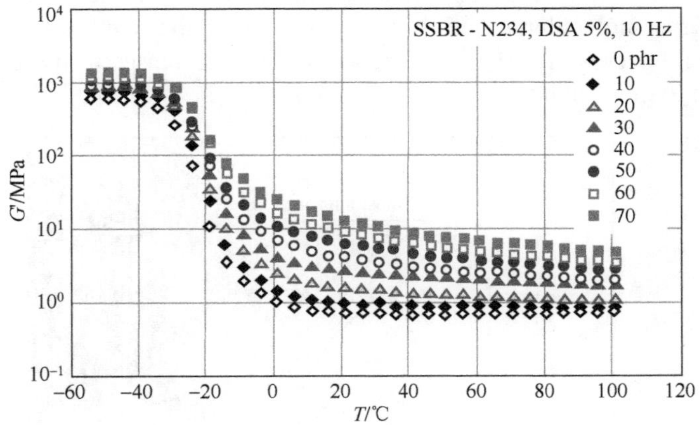


**Figure 7.15** Temperature dependence of $G'$ at DSA of 5% and 10 Hz for SSBR compounds with different loadings of carbon black N234[28]

The results in Figure 7.17 suggested that the loading effects at different temperature regions are governed by different mechanisms. At temperatures near the tan $\delta$ peak in the transition zone, the presence of filler gives a lower hysteresis for a given energy input. This may be interpreted in terms of a reduction in polymer fraction in the composite as the polymer per se would be responsible for a high portion of energy dissipation, and, individual solid filler particles in the polymer matrix may not absorb energy significantly. However, at high temperatures the hysteresis is increased by the introduction of filler. The main cause of energy dissipation in this area is related to the change of filler agglomerate structure during cyclic strain as the polymer is in its rubbery state with very high entropic elasticity and low hysteresis.

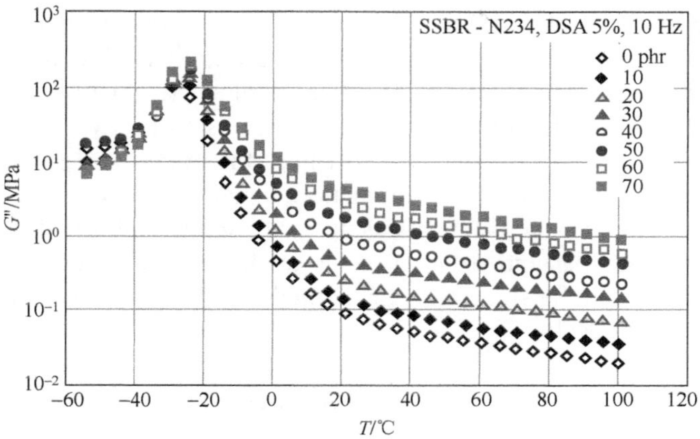


**Figure 7.16** Temperature dependence of $G''$ at DSA of 5% and 10 Hz for SSBR compounds with different loadings of carbon black N234[28]

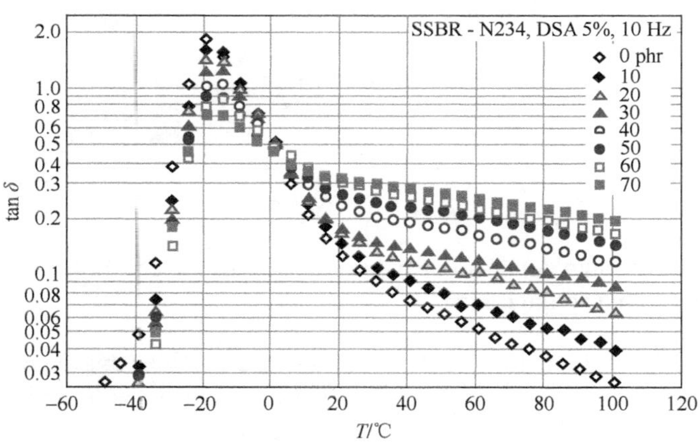


**Figure 7.17** Temperature dependence of tan $\delta$ at DSA of 5% and 10 Hz for SSBR compounds with different loadings of carbon black N234[28]

The evidence of the impact of filler agglomeration on the dynamic properties, hysteresis in particular, can also be reinforced by temperature dependence of tan $\delta$ obtained from temperature sweeps for the vulcanizates filled with different fillers. As shown in Figure 7.18, the silica compounds are quite distinctive in their temperature dependence of loss factors. Relative to the carbon black vulcanizates, the very low tan $\delta$ of the silica compounds over the temperature range from −30°C to 40°C is certainly associated with the feature of more and stronger filler agglomerates with a large volume of trapped rubber unable to participate in energy dissipation, even though the real volume fraction of polymer is higher, due to the higher density of silica. On the other hand, stronger filler agglomerates would also result in depressing another energy dissipation process involving disruption and reformation of filler agglomerates. Obviously, the increase of tan $\delta$ with temperature represents the increase in the agglomerate portion involving internal friction. This portion augments continuously reaching a maximum in the range of the temperature around 75°C to 85°C, indicating that the agglomerate is more developed and/or stronger than that of the carbon blacks. Consequently, for the silica-filled rubber, there seems to be two peaks in the tan $\delta$-temperature curve: one is related to polymer, the other corresponds to a filler agglomerate in which polymer-filler interaction, hence mobility of polymer segments, is also involved[17]. The same mechanisms probably apply to all filled vulcanizates, but depending on the number and strength of the agglomerates, these two peaks may be distinctly separate as in the silica case, or merged together, leading to a broad peak, as in the case of carbon black.

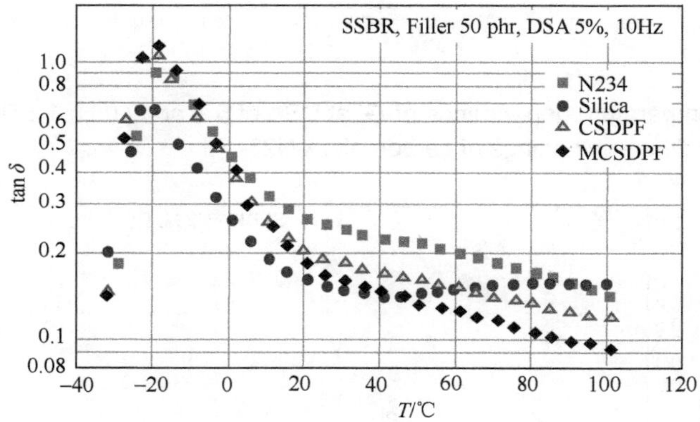


**Figure 7.18** Temperature dependence of tan $\delta$ at DSA of 5% and 10 Hz for SSBR compounds filled with different fillers[4]

The low filler-filler interaction, hence less and weaker filler agglomerate of CSDPF, manifests itself primarily in a tan $\delta$-temperature curve featuring high hysteresis at low temperature, and lower hysteresis at high temperature, with a narrower peak in the polymer transition zone. As observed in the strain sweep, this phenomenon can be further developed by TESPT modification.

The role of filler agglomeration in the temperature dependence of dynamic properties of filled vulcanizates will be further demonstrated in Chapter 8 with different filler-polymer systems and processing techniques of the compounds, which is related to the tire performance, rolling resistance and skid resistance in particular.

To sum up, for a given polymer system, filler agglomeration is the dominant factor in the determination of the hysteresis of filled vulcanizates. Its role can be summarized as follows:

- to the extent that the filler agglomerates cannot be broken down under the applied deformation, the agglomeration of filler would substantially increase the effective volume fraction of the filler due to the rubber trapped in the agglomerates;
- the breakdown and reformation of the filler agglomerates would cause an additional energy dissipation hence higher hysteresis during cyclic strain in the rubbery state;
- alternatively, in the transition zone at low temperature, where the main portion of the composite for energy dissipation is polymer matrix and the filler agglomerates may not be easily broken down, the hysteresis may be significantly attenuated by the filler agglomeration due to the reduction of the effective volume of the polymer;
- also in the transition zone, once the filler agglomerate can be broken down and reformed under a cyclic strain, the hysteresis can be substantially augmented through release of polymer to participate in energy dissipation and change in the agglomerate structure.

Accordingly, besides the hydrodynamic effect, the necessary conditions for high hysteresis of filled rubber, among others, are:

- presence of filler agglomerates; and
- breakdown and reformation of the agglomerates under dynamic strain.

Practically, a good balance of tan $\delta$ at different temperatures with regard to tire tread performance, namely higher hysteresis at low temperature and low hysteresis at high temperature, is determined for a given polymer system by filler agglomeration. In order to obtain a crossover behavior at a specified test temperature and frequency for a given filler relative to a conventional filler, the following conditions have to be satisfied:

- in the vulcanizate filled with the given filler, the filler agglomerate should be less developed;
- the specified low temperature or temperature range should fall in the range of the transition zone of the polymer and the high temperature is in the region of the rubbery state of the polymer;
- the strain amplitude of the test should remain in a limited range. The critical strain amplitude beyond which no crossover behavior can be obtained is determined by the strength of the filler agglomerates.

In other words, when the temperature sweep results of tan $\delta$ for two filled vulcanizates at a given frequency are compared, the crossover point would be changed by the $T_g$ of polymer, strain amplitude, and of course, the difference in their filler agglomeration. The higher the $T_g$ of polymer, the smaller the strain amplitude, and the larger the difference in the degree of filler agglomeration, the higher the crossover temperature that would be expected.

Briefly, the discussion above suggests that for a given polymer system, filler agglomeration plays a critical role in determining the dynamic properties of the filled rubber, hysteresis in particular. Now the questions may arise as to why and how filler agglomeration can be formed in the filled rubber. The answer should be related to the thermodynamics, i.e., the driving force of filler agglomeration and its kinetics which were described in detail in Chapter 3.

## ■ 7.3 Dynamic Stress Softening Effect

Generally, the strain-softening effect has been investigated with quasi-static and usually large deformations, as introduced in Section 6.3. Studies at high deformation may provide important information about the changes in vulcanizate structure and energy dissipation, which would be the most essential factors to govern the failure properties of vulcanizates, such as tensile, tearing, as well as wear. With regard to tire performance, while in certain conditions such as curbing, indentations, as well as cornering and braking of vehicles, the local stresses of the vulcanizates can be quite high, in most cases the actual deformation of the rubber is relatively small and the strain is under a dynamic state with relatively high frequencies. Under these conditions, the changes in dynamic modulus and hysteresis in terms of tan $\delta$ and $J''$ can, in a broad sense, be a measure of stress-softening. The decreases in elastic modulus which occurs when the strain is increased over a strain range from as small as < 0.1% to an intermediate level is another mode to show the stress-softening. This phenomenon, well known as the Payne effect, along with the hysteresis, has been studied extensively[6,7], and drawn great attention recently in relation to the development of tire tread compounds. Experimentally, most studies of dynamic stress-softening involve subjecting the vulcanizates to sinusoidal strains of varying amplitudes, sweeping from small strain continually to higher strain. It has been reported by Engelhardt et al.[29] and Moneypenny et al.[30] that, when the subsequent strain sweeps are performed immediately after the initial sweep, large changes were found for dynamic parameters, shown either by change in modulus, or/and by increase in loss tangent. Such a phenomenon may be considered as another form of stress-softening which would have great influence on the dynamic performance of rubber products during service.

Wang et al.[31] have further investigated dynamic stress-softening and revealed the factors to govern this phenomenon, especially related to the filler characteristics, with two types of fillers, namely silica and carbon black, and a variety of test modes. Dynamic stress-strain behaviors of the filled rubbers were measured by means of a Rheometrics rheometer operated in torsion (shear) mode. The tests were programmed to do a series of measurements of dynamic parameters with different strain amplitude. Three test modes were performed. With mode 1, the strains were swept from 0.95% to 79% DSA at constant frequencies (1 Hz or 10 Hz) and temperatures (0°C, 30°C, or 60°C). For each strain, three cyclic strains were applied and the data were measured at the third cycle. In mode 2, the same strain sweep as mode 1 was immediately followed by two repeated strain sweeps. After these three strain sweeps, the specimens were allowed to rest at room temperature for 16 h for recovery (relaxation), and then an additional strain sweep (fourth sweep) was carried out with the same condition as in the first three (Figure 7.19). In mode 3, the strain sweeps were conducted repeatedly three times as shown previously for mode 2 (but with different maximum strain and without recovery), followed by another three strain sweeps with maximum strain higher than the previous. These procedures were continued until the maximum strain reached 79% DSA. In each sweep, the strains were preprogrammed from the lowest strain as used for the first sweep to the maximum.

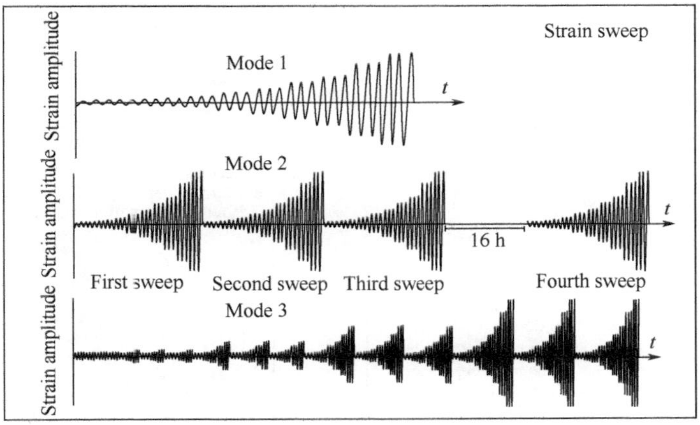


**Figure 7.19** Three modes of strain sweeps used in reference[31]

## 7.3.1 Stress-Softening Effect of Filled Rubbers Measured with Mode 2

### Elastic Modulus

A typical set of data showing stress-softening effects with elastic modulus $G'$, viscous modulus $G''$, loss factor tan $\delta$, and loss compliance $J''$ are presented in Figure 7.20, using SBR (Duradene 715) vulcanizates filled with 75 phr carbon black N234 and 25 phr oil as an example. The data were collected at 0°C with a frequency of 1 Hz.

As is well documented, the elastic modulus decreases considerably with increasing strain amplitudes, showing a typical non-linear behavior of the filled rubber. This phenomenon (Payne effect), as mentioned before, is a special case of stress-softening under dynamic strain. It is generally believed to be related to filler agglomeration which is mainly determined by filler-filler interactions. Such a consideration has been verified by the very weak strain dependence of the elastic modulus for gum rubber. The breakdown of the filler agglomerate upon strain has been considered to be responsible for the lower modulus at higher strain amplitudes.

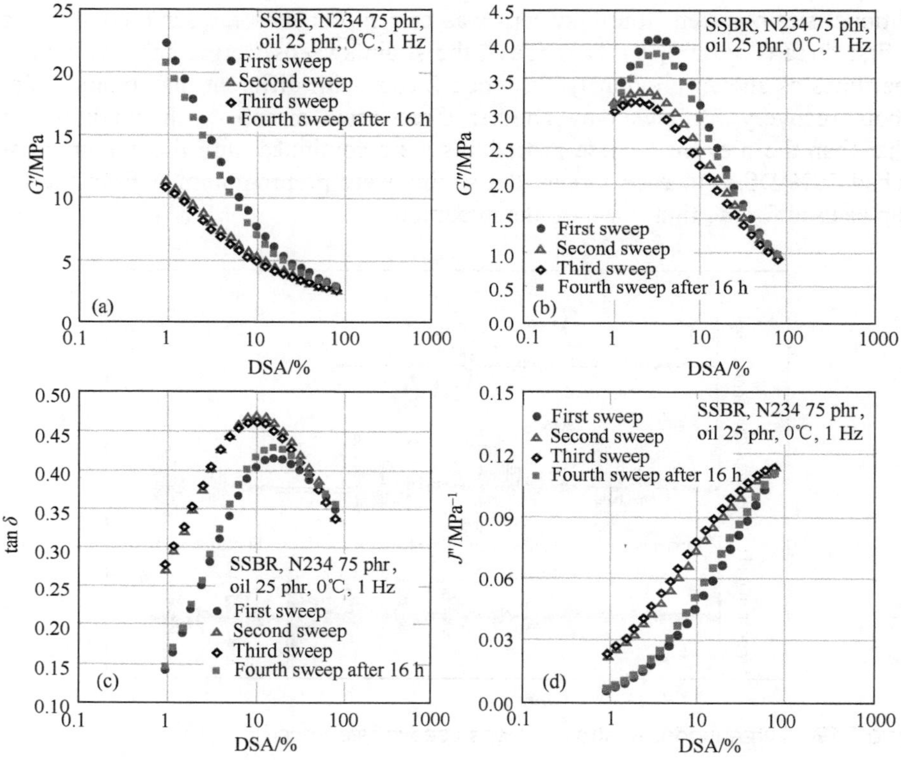


**Figure 7.20** Effect of number of strain sweeps and storage of the prestrained samples on dynamic strain softening effect for N234 filled rubber at 0°C

The immediate successive (second) sweep shows a drastic reduction in modulus at the same strain amplitudes compared with the first strain [Figure 7.20(a)]. The difference between these two sweeps is reduced as the strain amplitude increases, and finally at the highest strain amplitude, the modulus of the second sweep is almost identical to that obtained from the first sweep. Consequently, the Payne effect is greatly diminished by the previous strain. This suggests that after the first shear strain, the material softens until the maximum strain applied in the first sweep.

If the Payne effect is related to filler agglomeration, the reduction of this effect may be representative of less developed filler association which would result from the destruction of the original filler agglomerate. In addition, changes in some structures, for example caused by polymer-filler interaction near the interface between them, could be involved which have not clearly been understood. This kind of structure may be related to polymer molecule adsorption on the filler surface, and the changes in conformation of the adsorbed molecules and displacement of segments in the rubber shell could also be responsible for the stress softening process. Even at high strain, the "molecular slippage mechanism" proposed by Dannenberg[32,33] and others[34], could also be responsible for low modulus at higher strain. All these processes are, of course, related to the presence of the filler in the compound, since in the case of unfilled rubber, no significant Payne effect can be detected in a strain range applied to the filled rubber, even when the strain amplification caused by the filler is taken into account. Nevertheless, after the previous strain sweep, the remaining Payne effect would suggest that during the dynamic strain, even at the highest strains applied in the sweep, a certain amount of filler agglomerate can be reformed which can withstand lower strain. On the other hand, almost all of the softening is achieved in the first sweep. Only very little additional softening takes place on the third and subsequent strain sweeps. This may indicate that most of the structures reformed after the initial sweep can be reformed again for the subsequent strain sweeps at strains which do not exceed the initial sweep maximum.

The effect of storage of the tested sample on the softening effect is also shown in Figure 7.20(a). The strain sweep (the fourth sweep) was carried out with the sample which had been rested for 16 h after the third sweep (see Figure 7.19). As can be seen, for carbon black filled rubber, most of reduction in modulus is recovered during rest. Accordingly, the change of structures, at least most of the changes which cannot be reformed during the first few sweeps, is also reversible or reformable. It is unlikely that the breakage of polymer network, both polymer chains and crosslinks, plays any significant role in the stress softening, since for one reason the Payne effect does not show in a gum compound even at higher strain amplitudes, and for another reason, it has been reported that the stress of the softened vulcanizates, even measured at high quasi-static deformation, can be recovered by heat treatment[34,35] or swelling in solvent[36].The stress-softening is therefore mainly, if not entirely, a physical phenomenon.

### Viscous Modulus

While elastic modulus is representative of dynamic stress that is in phase with strain and a measure of the energy returned to the system, the viscous modulus is 90° out of phase and related to the component of lost energy. It has generally been recognized that the energy loss during dynamic strain is predominantly related to the breakdown and reformation of the filler agglomerate, and probably other superstructures introduced by the filler as discussed previously. The viscous modulus of gum compounds on the one hand shows a very low value and does not give a significant strain dependence over a large practical range of strain amplitudes[19].

The $G''$ values of the carbon black filled rubber in the first sweep are also highly dependent on strain amplitude, showing maximum values at about 4% DSA [Figure 7.20(b)]. After passing through the maximum, $G''$ decreases rapidly with further increase in strain amplitudes. If the cyclic change of the filler-related structures is responsible for the change in energy loss or dissipation at low DSA, this process would increase with increasing strain amplitude, and then the reformation of structure would diminish more rapidly than its disruption beyond certain DSA where the maximum appears.

It is also evident that the energy dissipation process is significantly less pronounced during the following strain sweeps and the maxima are shifted to lower strain amplitudes. Apparently, structures involved in energy dissipation are considerably diminished and these structures are relatively weakened by the initial stress-softening. As found with the elastic modulus, most of these changes take place in the first sweep, and the subsequent sweeps give only a minor effect.

### Loss Tangent

The loss tangent is the ratio of viscous and elastic modulus. Besides the hydrodynamic effect, the role played by filler in these two parameters involves different mechanisms, as discussed above, so both mechanisms influence tan $\delta$. While $G'$ is related to the filler-related structure (mainly filler agglomerate) which is reduced during dynamic strain, $G''$ is related to the breakdown and reformation of these structures. Consequently, the factors governing tan $\delta$ would be the state of filler-related structures, or more precisely, the ratio between the portion capable of being broken down and reconstituted, and those remaining unchanged during dynamic strain. The changes in tan $\delta$ in the multi-strain-sweeps would reflect the ratio of these two processes, another way to express the stress-softening effect.

From Figure 7.20(c), some features of the strain-sweep-dependence of tan $\delta$ may be abstracted as follows:

- the values of tan $\delta$ for repeated sweeps are dependent upon first strain sweeps;
- the peaks of tan $\delta$-strain curves for the subsequent sweeps appear at lower DSA in relation to the first sweep;
- the increase in tan $\delta$ upon the first sweep is not further enhanced by the following strain sweeps;
- during storage of the prestrained sample, the tan $\delta$ values can essentially be restored to their original level.

As both $G'$ and $G''$ decrease when resweeps are performed, the increase in tan $\delta$ in the second and third sweeps obviously suggests that the reduction rate of $G'$ is more pronounced in the initial sweep than that of $G''$. On the other hand, due to the stress-softening, at the same strain the energy input during one cyclic deformation of the specimen is much lower so that the absolute energy loss may not be higher in the subsequent sweeps. In fact, under this condition, i.e., constant strain, the energy loss is proportional to $G''$, which is lower for the pre-swept vulcanizate.

### Loss Compliance

At constant stress, the energy loss during dynamic strain is directly proportional to loss compliance. Although it depends on both $G'$ and $G''$, the effect of $G'$ is more significant in the determination of $J''$ in comparison with $G''$. The higher the $G'$, or the stiffness, the less the energy loss would be when the stress is kept constant. Since the reduction rate of $G'$ is higher than that of $G''$ upon pre-sweep, it is not surprising that $J''$ is enhanced after the initial sweep as shown in Figure 7.20(d).

In addition, the effect of prestrain on $J''$ is almost completed in the first strain sweep as further sweeps do not give any substantial increase in $J''$. Such observation is in line with the $G'$ and $G''$ from which loss compliance is derived. Here again, the recovery of the $J''$ during rest of the pre-strain-swept sample is evident although it is not fully completed after 16 hours.

### 7.3.2 Effect of Temperature on Dynamic Stress-Softening

As the stress-softening has been attributed to changes in the filler-related structures, which are related to filler-filler and filler-polymer interactions, and these interactions are, in turn, strongly dependent on temperature, the stress-softening effect would also show a strong temperature dependence. The results discussed so far were obtained at 0°C, which is in the transition zone of the polymer used, and the hysteresis at this temperature is correlated well with tire traction, as generally accepted[37,38]. However, it is also worthwhile to investigate the stress-softening effect at relatively higher temperatures, for example at 60°C where the dynamic hysteresis has been demonstrated to be a good measure for tire rolling resistance[37,38].

Shown in Figure 7.21 are the changes in dynamic parameters as a function of strain amplitude and the effect of the prestrain-sweeps at 60°C. Generally, all observations drawn from the results obtained at 0°C are valid for the results at 60°C. However, some features of the higher temperature results, in relation to low temperature ones, may be emphasized:

- the softening effect with regards to all dynamic parameters is less pronounced;
- upon rest, the fourth strain sweep gives a further reduction in $G''$.

The less pronounced stress-softening at high temperature may be easily understood from the temperature dependence of all filler related-structures, the filler agglomerate in particular. At high temperature the filler-filler as well as polymer-filler interactions which are considered to be the cause of the softening effect would substantially decrease. However, it is not understood yet why $G''$ can be further reduced after the preswept sample is rested for 16 h at room temperature. This has been observed with many (but not all) samples. It seems that this phenomenon is dependent on compounding and filler characteristics, such as curatives, coupling agents, and filler types as well. If the breakdown and reformation of the filler-related structures are the cause of $G''$, a question may arise as whether a new structure has been reconstructed. If so, this structure may be more stable than that in the original compound as it results in less energy dissipation. This is, however, unlikely to be true as such structure has to be formed at room temperature. The reasons for this unexpected result may be multiple. Nevertheless, more work has to be done to understand this phenomenon.

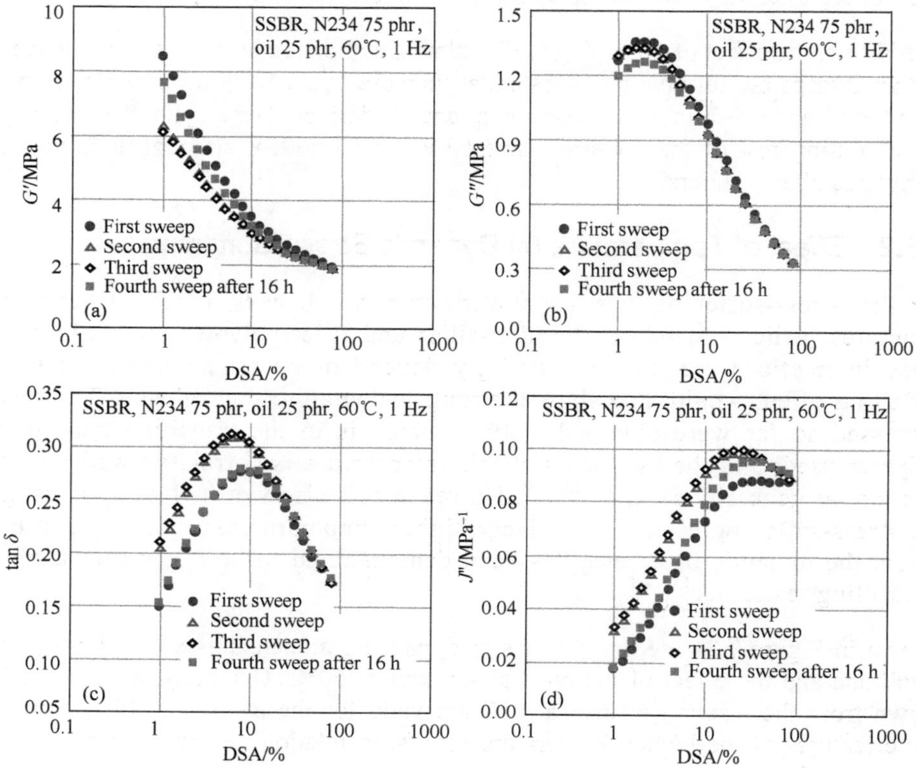


**Figure 7.21** Effect of number of strain sweeps and storage of the prestrained samples on dynamic strain softening effect for N234 filled rubber at 60°C

### 7.3.3 Effect of Frequency on Dynamic Stress-Softening

The viscoelastic properties of the filled rubbers generally follow the rule of the time (frequency)-temperature superposition principle, so that a strong frequency-dependence of the dynamic parameters would be expected from its temperature dependence. The stress-softening effect was also investigated experimentally at different frequencies because, on the one hand, different laboratories run rheometric measurement at different frequencies, and on the other hand, different applications of the filled rubber may involve a different rate of strain in time scale. The effect of the frequency of the cyclic strain on $G'$, $G''$, tan $\delta$, and $J''$ of Duradene 715 filled with 50 phr N234 is shown in Figure 7.22 and Figure 7.23 for 0°C and 60°C respectively. Two frequencies, 1 Hz and 10 Hz, were used. As can be seen, without exception, the general view of the stress-softening effect is similar to that observed for 75 phr carbon black filled rubber, showing that the second strain sweep gives lower $G'$ and $G''$, and higher tan $\delta$ and $J''$, no matter what the temperature and frequency used. However, as expected from their temperature dependence, there is a great difference in most of the dynamic parameters obtained with these two frequencies. The higher frequency, which corresponds to the low temperature, gives higher modulus, both $G'$ and $G''$ and high tan $\delta$. This effect seems to be more pronounced when the comparison is made at low temperature. This phenomenon can be also interpreted in terms of temperature dependence. At 60°C, the polymer is at rubbery state (low plateau at dynamic $G'$- and $G''$-temperature curves) and filler agglomeration is less developed, the dynamic properties would be less sensitive to temperature, hence to the test frequency. At 0°C, however, Duradene 715 is in its transition zone where the dynamic properties depend strongly on temperature. Small changes in temperature will lead to substantial change in modulus as well as hysteresis, and so would the change in frequency. This observation may also suggest the importance of properly setting-up the test frequency in a low temperature study.

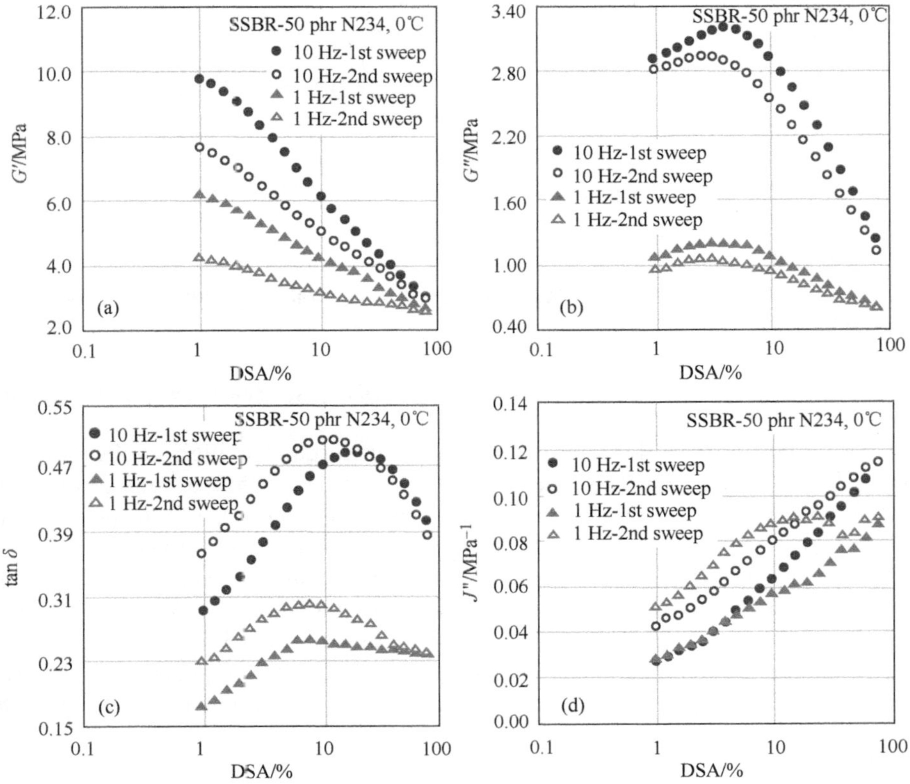


**Figure 7.22** Effect of dynamic strain frequency on strain softening effect for N234 filled rubber at 0°C

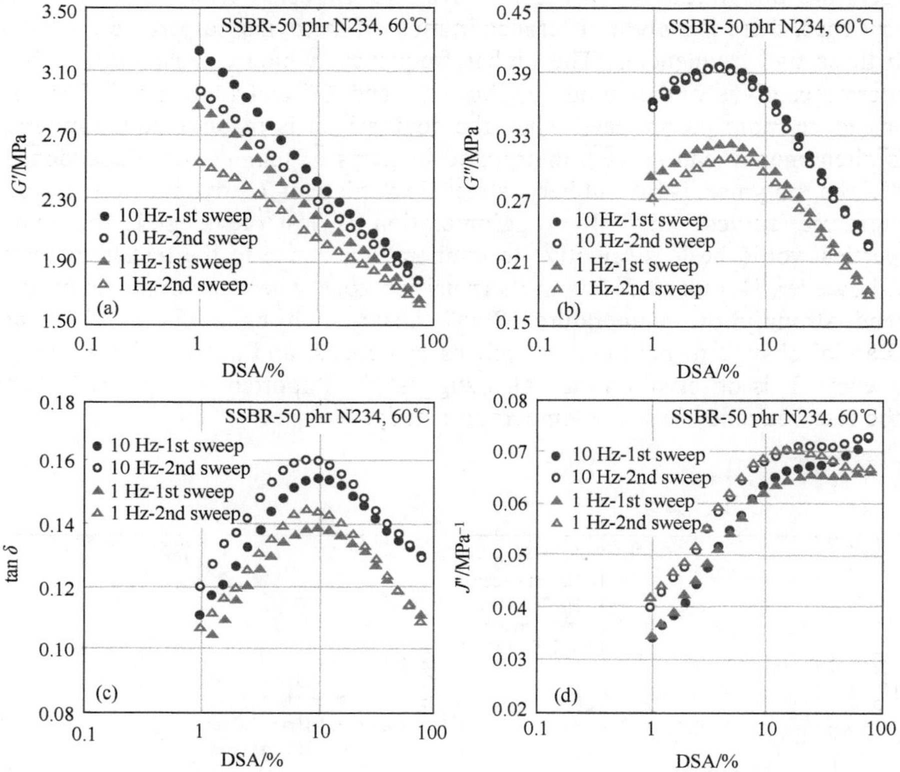


Figure 7.23 Effect of dynamic strain frequency on strain softening effect for N234 filled rubber at 60°C

## 7.3.4 Stress-Softening Effect of Filled Rubbers Measured with Mode 3

As shown above, once the vulcanizates have undergone a high cyclic strain, the changes in dynamic parameters cannot be fully recovered during a limited rest time in 16 h. However, the maximum strain of DSA 79% may be too high to be realistic in rubber product service, especially for tire application. It is therefore necessary to know the stress-softening effect during strain sweeps using lower maximum strain amplitudes. For this purpose, a strain sweep program as mode 3 was designed. In this program, after three strain sweeps, another three sweeps are immediately performed with the maximum strain amplitude higher than that used in the previous three sweeps (Figure 7.19). The maximum double strain amplitude DSA varies from 1.2% to 78.8%.

### Elastic Modulus

In Figure 7.24 the elastic moduli of the vulcanizate (SSBR/BR 75/25) filled with 75 phr silica and 25 phr oil measured at different strain sweeps are shown as a function of DSA. The reason for using this vulcanizate is only because of its great stress softening, which may help to express the results graphically and making it easier to observe the multiple-sweep effect. The observations drawn from this material also hold for other vulcanizates but the degree of the softening effect may differ from one to the other.

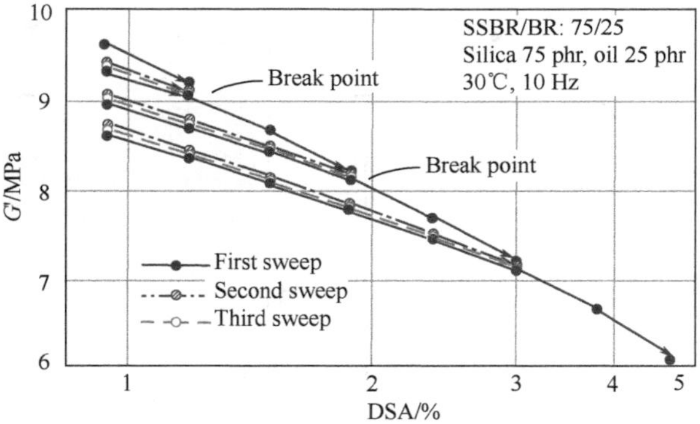


**Figure 7.24** *G'* as a function of number of strain sweeps and DSA ranges of sweeps (mode 3)

As found with mode 2, when the same maximum strain amplitude is used for the sweep, the moduli at the second sweep are much lower than those in the first sweeps and only a slight reduction is observed in subsequent sweeps (third sweep). As the maximum strain amplitude increases, however, the moduli almost follow the same curve until it reaches the maximum strain of the last sweep and then the modulus drops more rapidly. Consequently, the *G'*-DSA curves show a break point at the maximum strain of the previous sweep. This observation may be seen clearly in Figure 7.25(a), where only the first strain sweeps in each strain range are presented. After the first sweep, a large reduction in modulus (or Payne effect) can be seen with the succeeding sweep. This implies that the increase in strain amplitude is more important for the stress softening than a repeat of strain. If disruption of filler-related structures (mostly agglomerate) in the vulcanizate is responsible for the stress-softening, the increase in the number of sweeps does not seem to enhance this process significantly, but the increase in strain amplitude does. This may suggest that the modulus of the compound would not be reduced drastically during regular service, at least after a short period when the temperature reaches equilibrium, but an occasional increase in strain amplitude could cause a large change in the rubber properties which cannot be recovered in a short time, as shown previously.

It is also observed that the higher strain amplitude part of the modulus-DSA curves of the first sweep, i.e., the curve beyond the break point, yields a single curve which can be treated as an envelope [Figure 7.25(a)]. This envelope is found to coincide, within experimental error, with the *G'*-DSA curve of the first strain sweep of mode 2 [Figure 7.26(a)]. Similarly, the end points at the third strain sweeps for all strain ranges also form an envelope [Figure 7.27(a) and Figure 7.28(a)], and this envelope is also shown to be very close to the *G'*-DSA curve of the first sweep of mode 2 [Figure 7.28(a)], hence to the envelope formed by end of the first sweeps obtained with mode 3. This is naturally a consequence of the first strain sweep curve of the *G'* following the last curve when the strain does not go over the previous largest strain.

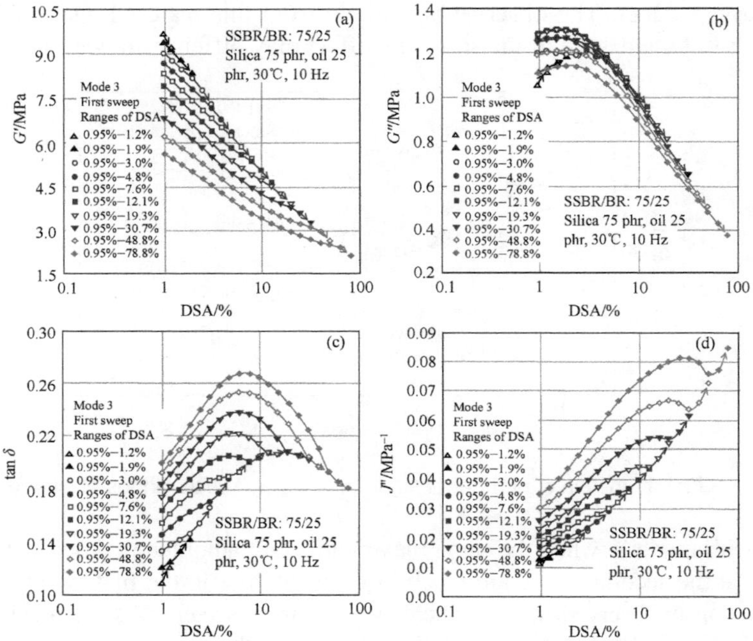


**Figure 7.25** Effect of increase in maximum strain amplitude on dynamic properties-DSA curves of the first sweeps (mode 3)

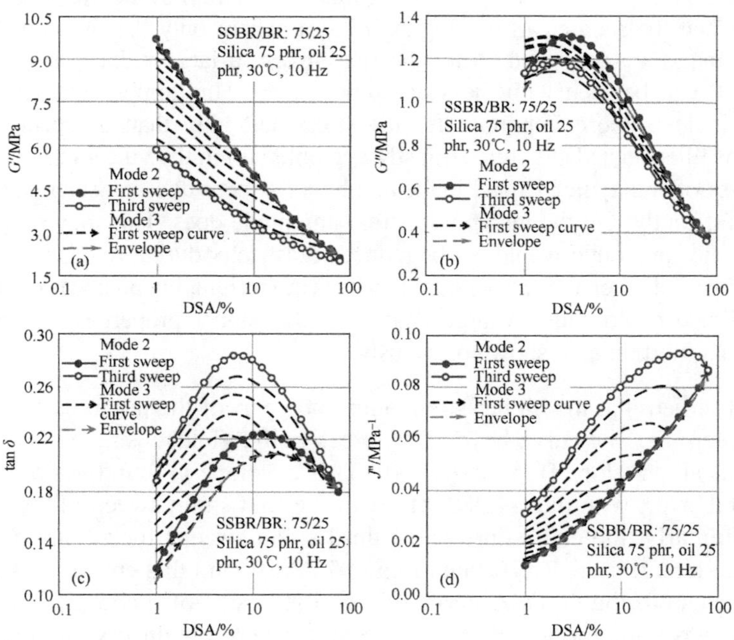


**Figure 7.26** Comparison of first sweep and third sweep curves of mode 2 with first strain sweep curves and their envelope of mode 3

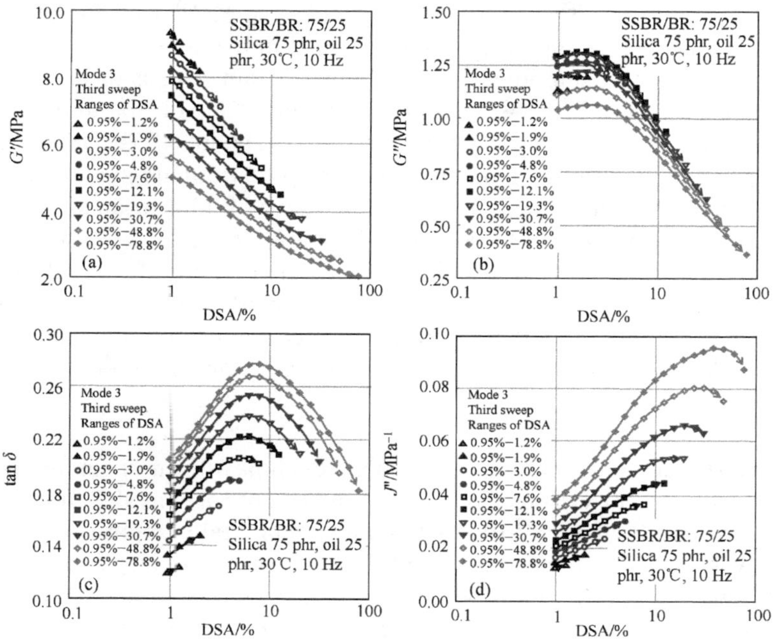


**Figure 7.27** Effect of increase in maximum strain amplitude on dynamic properties-DSA curves of the third sweeps (mode 3)

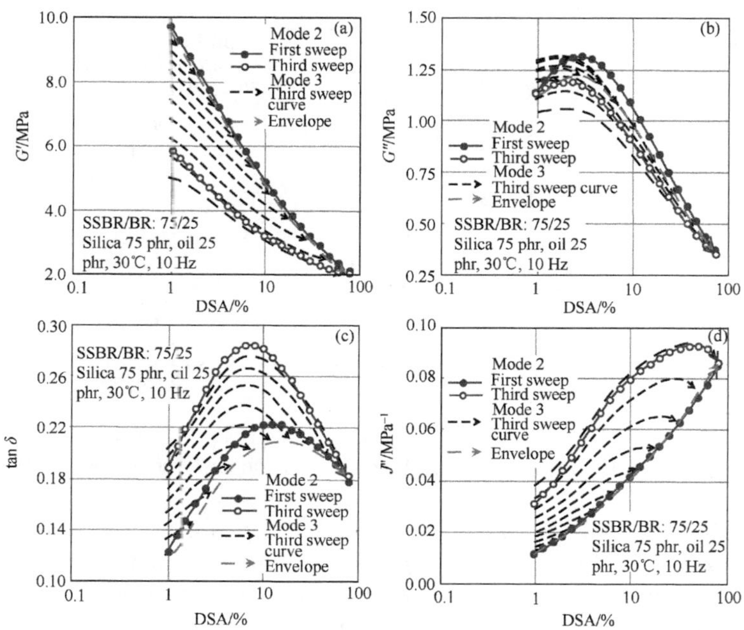


**Figure 7.28** Comparison of first sweep and third sweep curves of mode 2 with third strain sweep curves and their envelope of mode 3

In Figure 7.26(a), the lower part of the envelope, expressed with the arrowed thick dashed line, is the first strain sweep curve obtained in the sweeps of mode 3 with the highest maximum strain amplitude (called "highest strain sweeps of mode 3" in short or simply abbreviated as "HSS-3"). Shown in Figure 7.28(a) is the third strain sweep curve of HSS-3. For each sweep, the starting point was about 0.95% DSA. With regard to the comparison between the third strain sweep curve of mode 2 and the third strain sweep curve of HSS-3, it is found that while the curve of third strain sweep of mode 2 is somewhat higher than any of those of HSS-3 at low strain, at higher strain amplitude it is close to the third HSS-3 curve and lower than the first one. This evidence strongly suggests that the effect of sweep number is less important than the maximum strain amplitude. It also implies that the structure which can be broken down at higher strain would be stable during dynamic deformation at lower strain amplitude.

### Viscous Modulus

While the elastic modulus shows a monotonic reduction with strain amplitude in strain sweeps, the situation for viscous modulus is quite complicated as shown in Figure 7.29, Figure 7.25(b) and Figure 7.27(b). At low maximum strain amplitudes, the $G''$-DSA curves rise with increase in the maximum DSA, as shown in Figure 7.25(b) for the first sweeps and in Figure 7.27(b) for the third sweeps. After it reaches the maximum, the curves drop as the maximum strain increases. It appears that the structures being broken down and reformed (which is attributed to the energy dissipation processes) are increased with strain amplitude and that at high DSA such structures diminish, probably caused by lack of time for reformation or relaxation.

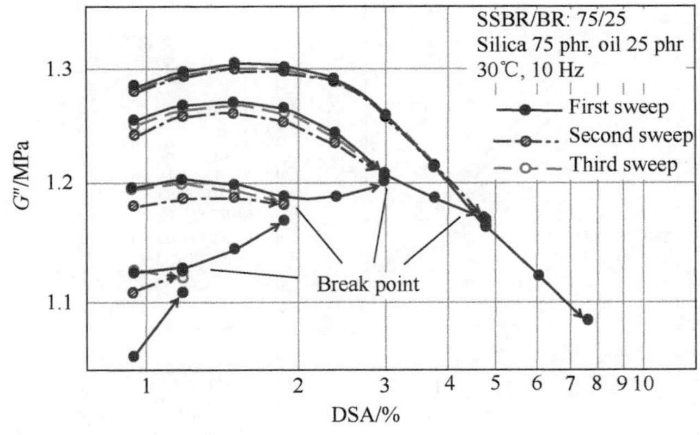


**Figure 7.29** $G''$ as a function of number of strain sweeps and DSA ranges of sweeps (mode 3)

Similar to what is found with the $G'$ of the first sweeps for each maximum strain, the $G''$-DSA curves seem to be broken at a strain amplitude which is identical with the maximum strain in the previous sweeps. This can be seen clearly in Figure 7.29 in which only the low strain results are presented. Also the higher strain parts of the broken curves form an envelope. This envelope lies below the first sweep curve of mode 2, but above the third sweep curve [Figure 7.26(b)]. It seems that compared with

$G'$, the $G''$ is more dependent on the strain history that may have a strong dependence of the structure reformation during test. The envelope composed by the end points of the third sweeps is given in Figure 7.28(b), also sits between the first and third sweep curves of mode 2. However, the third strain sweep curve of HSS-3 is substantially lower than its counterpart obtained with mode 2. As found with mode 2, the most part of the change in $G''$ is achieved in the first strain sweep and the contribution of the following sweeps is shown to be minor (Figure 7.29).

### Loss Tangent

As derived from $G'$ and $G''$, tan $\delta$ also shows strong dependence on the prestrain-sweeps and maximum strain amplitudes. The first strain sweeps of tan $\delta$ for the groups with the same maximum strain are shown in Figure 7.25(c) and the third sweeps in Figure 7.27(c). Clearly, for either the first strain sweeps or the third sweeps, tan $\delta$ is drastically increased when the maximum strain amplitude increases. For the sweeps with the same sweep range, the increase in tan $\delta$ is almost completed in the first sweep, as found for $G'$ and $G''$ (Figure 7.30). This again leads to a conclusion that the maximum strain amplitude of the sweeps plays a more important role in the determination of the hysteresis than the number of sweeps.

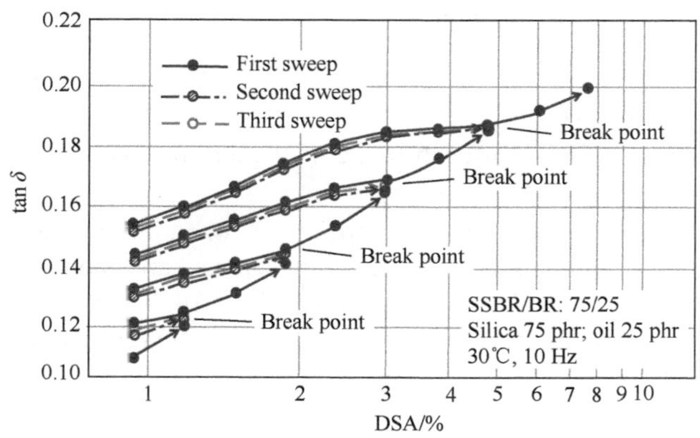


**Figure 7.30** tan $\delta$ as a function of number of strain sweeps and DSA ranges of sweeps (mode 3)

Similarly, the tan $\delta$-DSA curves are also broken at the point at which the strain amplitude is about the same as the maximum strain in the previous sweeps, as shown in Figure 7.30. An envelope is formed with the higher strain parts of the broken curves, and this envelope shows a maximum at DSA around 15%. This indicates that beyond the previous maximum strain, more energy dissipation processes than those expected from previous strain sweeps are involved. This is also true for strain sweeps with maximum strains higher than the maximum of the envelope for which tan $\delta$ decreases continuously with increasing strain amplitudes.

Compared with tan $\delta$-DSA function of the first strain sweep measured with mode 2, the envelope composed of the higher strain parts of the broken curves of the first sweeps [Figure 7.26(c)], hence that composed of the end points of the third sweeps [Figure 7.28(c)], are significantly lower, and the strains of the maxima of this envelope are somewhat higher. With regard to the comparison between the third sweep curve of mode 2 and the third strain sweep curve of HSS-3, while the tan $\delta$ measured with mode 3 seems to be higher at very low strain (< 2% of DSA) amplitude, and to be lower at intermediate strain, at high strain amplitudes (> 20% of DSA) these two curves almost follow the same function.

It is very interesting to notice that when DSA is beyond certain limitation, 3% in this special case, there is a maximum on the tan $\delta$-DSA curves below the break points, even though in some cases the curve will go up beyond the break points. This is observed for both first sweeps [Figure 7.26(c)] and the subsequent sweeps [Figure 7.28(c) for third sweeps]. The strain amplitude of the maximum increases as the maximum strain amplitude increases. In the case of the third strain sweep curve of HSS-3, the strain amplitude of the maximum is identical with that of the third sweep of mode 2.

### Loss Compliance

The observations obtained from $J''$ are similar to what is observed for the other three parameters discussed earlier; these are:

- with increasing maximum strain amplitude, $J''$ increases considerably [Figure 7.25(d) and Figure 7.27(d)];
- most of the changes in $J''$ for each range of the strain sweep are completed during initial sweep (see Figure 7.31); and
- the $J''$-DSA curves of the first sweeps are broken at the point which is the maximum strain amplitude of previous strain sweeps, and the higher strain parts of the $J''$-DSA yield a $J''$-DSA envelope [Figure 7.25(d)]. These envelopes almost fall on that composed of the end point of third sweep curves [see Figure 7.26(d) *vs.* Figure 7.28(d)].

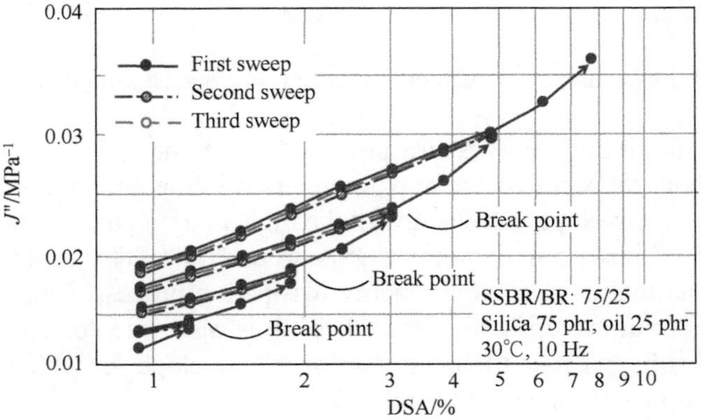


**Figure 7.31** $J''$as a function of number of strain sweeps and DSA ranges of sweeps (mode 3)

In addition, it is found that the $J''$-DSA envelope yielded by the higher strain parts of $J''$-DSA curves of the first sweeps and that formed by the end points of the third sweeps are quite close to the $J''$-DSA curve of the first sweep while the third strain sweep curve of HSS-3 is similar to the third sweep curve of mode 2. Such observation has a resemblance to that observed with $G'$.

### 7.3.5 Effect of Filler Characteristics on Dynamic Stress-Softening and Hysteresis

For investigation of the effect of filler characteristics on the stress-softening and hysteresis, mode 2 testing was performed with vulcanizates filled with silica and carbon black.

Presented in Figure 7.32 are the results of first strain sweep at 0°C and 60°C for vulcanizates prepared with SBR/BR 75/25 and 75 phr fillers. The fillers used for this comparative study are silica with 12 phr coupling agent X50S (carbon black N330/TESPT 50/50) or carbon black N234. The results are expressed by the change of

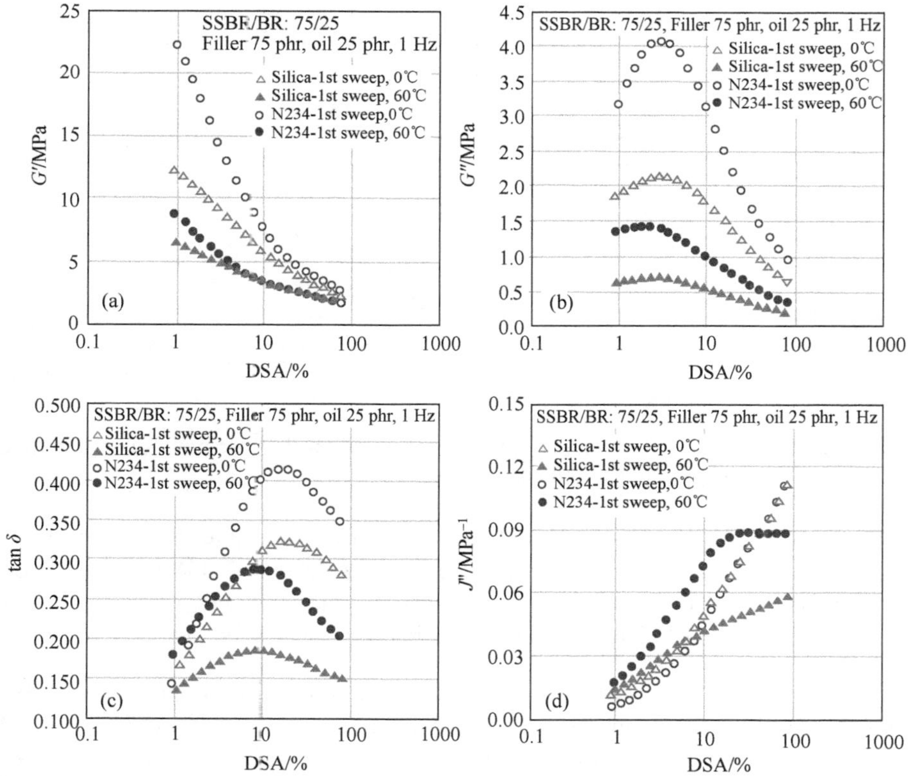


**Figure 7.32** Comparison of dynamic properties obtained from first strain sweeps at 0°C and 60°C for N234 black- and silica-filled vulcanizates

given dynamic parameters in percentage, in reference to the first strain sweep, as a function of the strain amplitudes at 0°C and 60°C, respectively (Figure 7.33 and Figure 7.34). The absolute data of the first strain sweeps for these two compounds are collected in Figure 7.32 for reference. Besides the results of the fourth sweep, which shows the recovery of the stress-softening after the samples have rested at room temperature for 16 h, only the second sweep results are included.

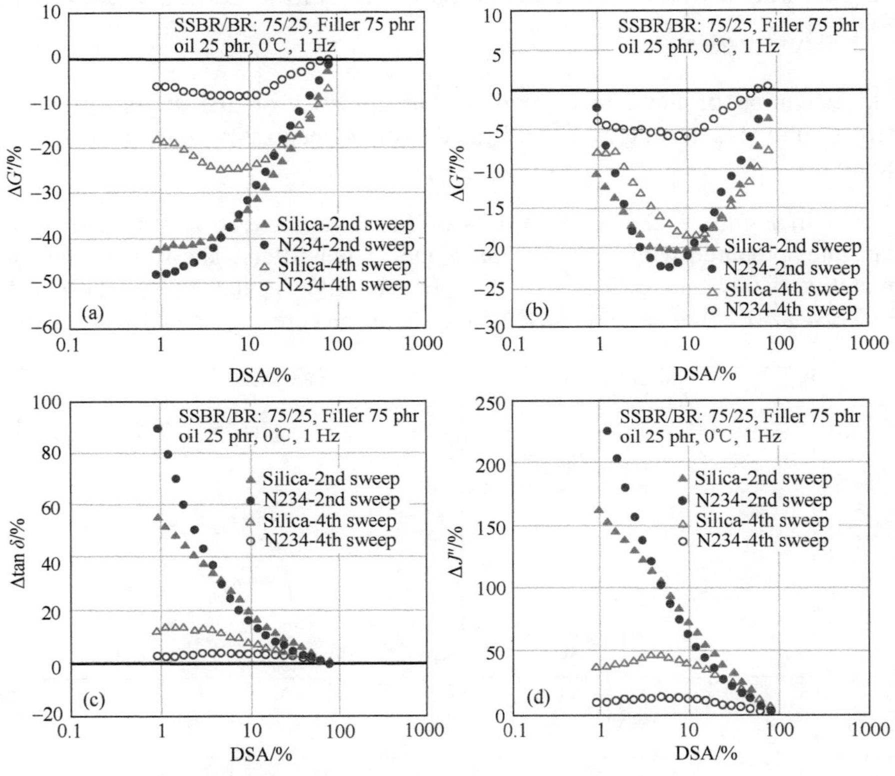


**Figure 7.33** Comparison of strain softening effect at 0°C and its recovery during storage between silica- and carbon black-filled vulcanizates

At 0°C, upon prestrain-sweep, carbon black seems to give greater reduction in $G'$ and more increase in tan $\delta$ and $J''$ than silica at lower DSA (Figure 7.33). During rest of the specimens at room temperature, the recovery of the dynamic properties is more efficient for carbon black filled rubber, with silica showing the lower recovery.

At 60°C, however, the silica compound has generally a more pronounced stress-softening effect, showing a greater drop in $G'$ and increase in tan $\delta$ and $J''$ after the first strain sweep (Figure 7.34). Apparently, the proportionally greater increase in tan $\delta$ and $J''$ for silica-filled rubber originates from high stress-softening in $G'$, since the $G''$ does not show less reduction than carbon blacks do, at least at lower strain amplitudes where the tan $\delta$ and $J''$ are much higher for the silica compound. In addition, the silica compound also gives a slower recovery versus carbon black.

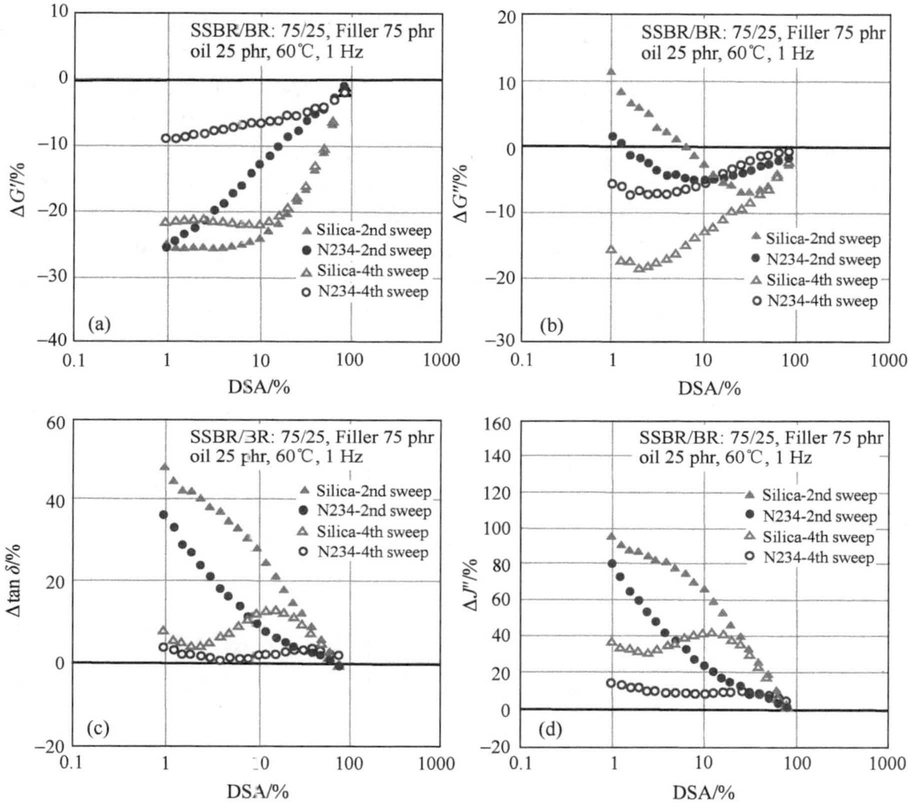


**Figure 7.34** Comparison of strain softening effect at 60°C and its recovery during storage between silica- and carbon black N234-filled vulcanizates

### 7.3.6 Dynamic Stress-Softening of Silica Compounds Produced by Liquid Phase Mixing

For silica filled rubbers, it has been well recognized that the dynamic properties of the vulcanizates are mainly determined by the application of silane coupling agent and the silanization reaction efficiency with the silica surface[39]. By using an innovative liquid phase mixing technology[40], a series of silica-filled synthetic rubber compounds named Eco-Visco-Elastomeric Composite (EVEC) have been developed. In this process, the efficiency of the silane coupling agents is greatly improved.

The main features of EVEC are excellent filler dispersion, lower filler-filler interaction and stronger polymer-filler interaction, compared to conventional dry-mixed silica compounds and carbon black compounds[41,42]. As stress-softening has been attributed to changes in the filler-related structures, which is related to filler-filler and filler-polymer interactions, the stress-softening effect of silica-filled rubbers would be affected by the extent of filler agglomeration. Recently, dynamic stress-softening of EVEC-L (composed of SBR/BR 70/30, 78 phr silica, 6.4 phr silane coupling agent TESPT, 28 phr oil and other ingredients) was investigated by Xie et al[43], compared with that of the compounds prepared by dry-mixing process (DM). Compounds without coupling agent were also introduced for reference. Figure 7.35 shows the results of stress-softening effect measured by mode 2 at 0°C and 60°C for EVEC-L and the dry mixed counterparts. Taking the results of first sweeps shown in Figure 7.35 as references, the changes of each dynamic parameter in percentage as a function of strain amplitude are given in Figure 7.36 and Figure 7.37. The data of second sweeps represent the softening effects and the data of the fourth sweeps represent the degree of recovery of the stress-softening after the preswept samples rested for 24 h at 30°C.

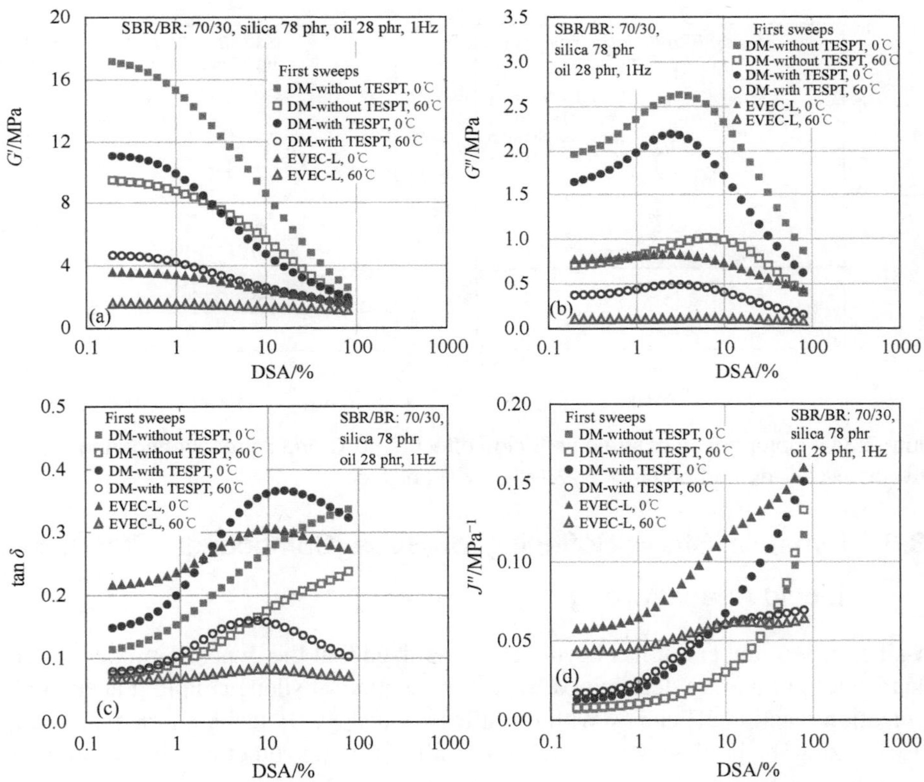


**Figure 7.35** Comparison of dynamic properties obtained from first strain sweeps at 0°C and 60°C between EVEC-L and dry mixed (DM) silica vulcanizates with and without coupling agent TESPT

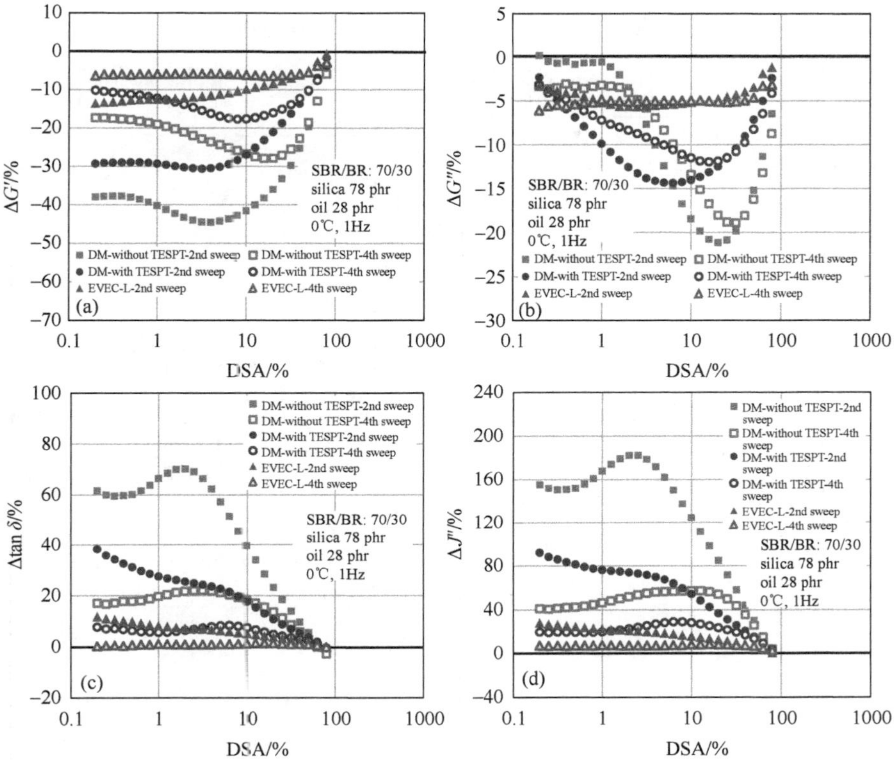


**Figure 7.36** Comparison of strain softening effect at 0°C and its recovery during storage between EVEC-L and dry mixed (DM) silica vulcanizates with and without coupling agent TESPT

It can be seen from Figure 7.35(a) that while the Payne effect is the highest for the silica compounds without coupling agent at both temperatures, for the compounds with coupling agent, the EVEC-L gives lower strain dependence. As mentioned before, $G'$ of filled vulcanizates is closely related to the effective volume fraction of the filler. In the investigation, the filler loading is the same for the three silica compounds, hence the effect of hydrodynamic effect and rubber occlusion can be assumed to be identical, but the immobilized rubber shell on the filler surface should be somewhat different due to the surface modification. With regard to the Payne effect, the main parameter to influence the effective volume fraction of the filler, $\phi_{\text{eff}}$, is filler agglomeration. According to the Van der Poel theory[11,12], Wang[4] estimated the amount of rubber trapped in the filler agglomerates is about 94% of the carbon black volume at a filler loading of 50 phr which is equivalent to 64% of the effective volume $\phi_{\text{eff}}$ of the carbon black. For silica without coupling agent, the effective filler volume should be much higher due to its high filler-filler interaction. Therefore, the filler agglomerate plays a critical role in the enhancement of $G'$ at high filler loadings. The very low $G'$ of EVEC-L and the much less strain dependence of $G'$ compared with the dry mixed silica compounds prove that formation of silica agglomerate is greatly inhibited. This

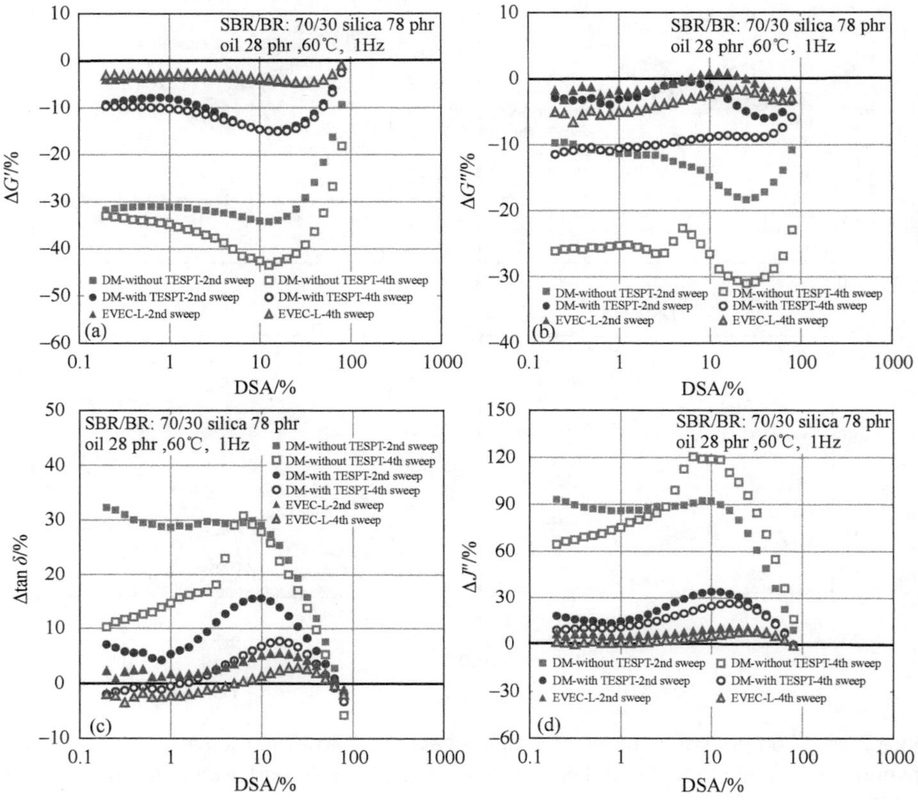


**Figure 7.37** Comparison of strain softening effect at 60°C and its recovery during storage between EVEC-L and dry mixed (DM) silica vulcanizates with and without coupling agent TESPT

also results in considerably lower viscous modulus of EVEC-L and again less strain dependence of $G''$ [Figure 7.35(b)], compared to the dry mixed counterparts. As shown in Figure 7.35(c), EVEC-L shows lower tan $\delta$ and less strain dependence than the dry mixed compound with TESPT at 60°C. Since the loss compliance is predominantly determined by $G'$, the low $G'$ of EVEC-L gives higher $J''$ from small to intermediate strain amplitudes at 60°C.

At 0°C, the stress-softening effect, including the reduction of $G'$ and $G''$ and the increase in tan $\delta$ and $J''$ after the first sweep, is obviously most pronounced for the dry mixed compound without TESPT, followed by the dry mixed compound with TESPT, and EVEC-L the least (Figure 7.36). These results are consistent with the order of extent of filler agglomeration in these three vulcanizates, i.e., the more filler agglomeration, the more softening effect. During rest of the specimens, the reduced $G'$ of the two dry mixed silica vulcanizates were both partially recovered as in the case of EVEC-L, but the recovery of $G''$ seems to be a little more efficient for the two dry mixed compounds than that of EVEC-L. The recovery of the changes in tan $\delta$ and $J''$ are evidently dependent on the changes of $G'$ and $G''$ during rest. For all these dynamic parameters, the softening effects are all small for the EVEC-L, and the recovery of the softening of tan $\delta$ and $J''$ is also efficient.

At 60°C, the softening behaviors of dynamic parameters for EVEC-L (Figure 7.37) are generally similar to those observed at 0°C, i.e., the softening is very small and the recovery of tan $\delta$ and $J''$ during rest is more efficient. The highest softening of dynamic properties is found for the compound filled with silica without coupling agent and the modified silica gives less softening compared with the unmodified counterpart, but much higher than EVEC-L. Upon rest at 30°C for 24 h, for EVEC-L and the modified silica compound, the $G'$ of the pre-swept specimens show little recovery and the $G''$ is further reduced in the fourth sweep, as shown in Figure 7.37. But the $G'$ of the fourth sweep for the silica compound without coupling agent seems to be lower than that of the second sweep. Actually, due to the developed filler agglomeration of the unmodified compound and the higher stress softening temperature, $G'$ of the third sweep drops a little more relative to the second one. This suggests that more filler agglomerates can be broken down during the third sweep. If the comparison is made between the $G'$ of the fourth sweep and the third sweep, the softening of $G'$ also show little recovery but not further decrease, which is in line with the other two compounds.

The recovery of $G'$ and $G''$ may be concerned with whether new structures can be rebuilt during rest and the level of energy dissipation caused by the disruption of these reformed structures on the fourth sweep. It is speculated that the structures disrupted at 60°C is not easy to be reformed during rest at 30°C taking the little recovery of $G'$ into consideration. The reason for the further reduction of $G''$ in the fourth sweep is currently unclear. If no new structures formed, the energy loss of the original structures may reduce upon rest at 30°C. Probably, after disruption of some filler agglomerates at 60°C, part of the rubber molecules within the agglomerates become more flexible; these molecules may change their network conformations in an eased state during rest, thus another strain sweep would cause less energy dissipation. This is especially true for the unmodified silica that has very low interaction with polymer.

Upon addition of coupling agent, while the filler agglomeration can be significantly depressed and polymer-filler interaction can be strengthened, the softening effects of the compound are considerably reduced. With liquid phase mixing, however, the efficiencies of TESPT for weakening filler-filler interaction and enhancing polymer-filler interaction are greatly improved so that the softening effects of all dynamic properties for EVEC-L are substantially reduced or almost eliminated[43].

## 7.4 Time-Temperature Superposition of Dynamic Properties of Filled Vulcanizates

During service of rubber products, vulcanizates may be involved in dynamic strains at different frequencies and temperatures. For example, it is generally accepted that rolling resistance is related to the movement of the tire, which corresponds to deformation at a frequency of 10~100 Hz. In the case of skid or wet grip, dynamic strain is generated between the rubber and road surface. The frequency of this dynamic deformation is quite high, probably around $10^4$–$10^7$ Hz at room temperature[37,44]. To better understand wet skid performance, the dynamic properties at the appropriate high frequency need to be generated. In practice, the testing is often limited by the frequency range of the instruments. It is generally believed that dynamic properties at different frequencies can be derived from data obtained at different temperatures based on the time-temperature superposition principle. However, the validity of the principle needs to be verified in these complicated vulcanizates.

The time-temperature superposition was first proposed by Williams, Landel, and Ferry in 1955[3] and is generally referred to as the WLF principle. The principle states that for materials exhibiting linear viscoelastic behavior, such as amorphous polymers and elastomers, data obtained at different temperatures and frequencies can be superimposed into one smooth master curve by a simple horizontal shift along the log frequency axis. The shift factor $a_T$ can be expressed as follows:

$$\lg a_{\mathrm{T}} = \frac{-C_1(T-T_0)}{C_2+T-T_0} \tag{7.19}$$

where $C_1$ and $C_2$ are constants, $T_0$ is the reference temperature, and $T$ is the measurement temperature (Eq. 7.19). $C_1$ and $C_2$ vary with the choice of $T_0$. They were first thought to be universal constants for all polymers when $T_0$ was set at $T_g$ + 50°C as a reference ($T_g$ is the glass-transition temperature). However, later studies[45] showed that the values could change from one polymer system to another.

The application of the WLF equation is limited to the glass-transition region, typically ranging from $T_g$ to $T_g$ + 100°C. The applicability of the WLF equation depends on the exact matching of adjacent curves when the master curve is constructed. Also, the same shift factor should be applicable to all the viscoelastic functions. Alternatively, the WLF equation can be derived from Doolittle's[46] viscosity equation:

$$\ln \eta = \ln A + B\left(\frac{v}{v_{\mathrm{f}}}\right) \tag{7.20}$$

where $\eta$ is the viscosity, $A$ and $B$ are constants, $v$ is the total macroscopic polymer volume, and $v_f$ is the proportion of the holes or voids, that is, free volume, in the polymer. Equation 7.20 was based on experimental data for monomeric liquids. It implies that viscoelastic properties of the polymer system as well as molecular rearrangement and transport phenomenon are solely dependent on free volume. If one assumes that free volume is a linear function of temperature, that is,

$$f = f_0 + \alpha_f (T - T_0) \tag{7.21}$$

where $f$ is the fraction of free volume (i.e., $v_f/v$) at $T$, $f_0$ is the fraction of free volume (i.e., $v_f/v$) at $T_0$, and $\alpha_f$ is the thermal expansion coefficient of the polymer, then Eq. 7.21 can be substituted into Eq. 7.20 to derive the WLF equation. However, at high temperatures, the shift factor has been proposed to follow an Arrhenius-type temperature dependence[45]:

$$\lg a_T = A + \frac{E_a}{RT} \tag{7.22}$$

where $E_a$ is the activation energy, $A$ is a constant, and $R$ is a gas constant.

It was proposed by Gross that the elastic shear modulus and loss modulus for single polymers can be expressed in terms of polymer relaxation time[47]:

$$G'(\omega) = G_0 \left[ 1 + \int_{-\infty}^{+\infty} H(\lambda) \frac{\omega^2 \lambda^2}{1 + \omega^2 \lambda^2} d\lambda \right] \tag{7.23}$$

and

$$G''(\omega) = G_0 \int_{-\infty}^{+\infty} H(\lambda) \frac{\omega \lambda}{1 + \omega^2 \lambda^2} d\lambda \tag{7.24}$$

where $G_0$ is the relaxed modulus, $H(\lambda)$ is the distribution function of relaxation time $\lambda$, and $\omega$ is the frequency. These relations determine the time-temperature dependencies of the dynamic properties of the polymer as relaxation time is closely related to the temperature.

The WLF equation has been successfully applied to many single polymers[45]. With regard to polymer blends, the time-temperature superposition principle has also been applied by many researchers[48–51]. Practically, blends of different rubbers are used very often for balancing the product performances. The dynamic properties of an unfilled polymer blend of SSBR/BR (75/25) are presented in Figure 7.38, with the reference temperature at 25°C. A discontinuity is found in the master curve of tan $\delta$ at a high frequency, starting around the tan $\delta$ peak position. This is similar to what was observed in filled compounds. The exact reason for this discontinuity is currently unknown, but it may be caused by the two polymers being immiscible. Nevertheless, the WLF equation is applicable to the blend from low to medium frequencies. When a similar study was carried out in the polymer blend with the addition of 32.5 phr oil, the same observations were obtained. However, there have been no general conclusions on the validity of the application to filled rubber systems, and the results seem to vary from one system to another. Only a few studies on the time-temperature superposition of filled systems have been reported on highly loaded elastomers[12,52,53].

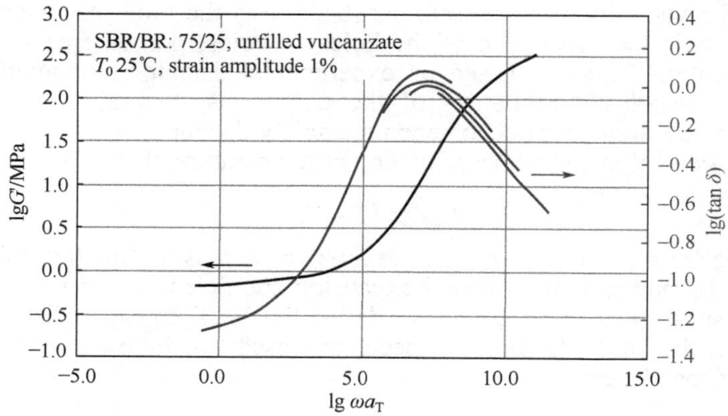


**Figure 7.38** Master curves of $G'$ and tan $\delta$ from an unfilled vulcanizate of the polymer blend

The temperature-frequency dependence of the dynamic properties of vulcanizates filled with conventional carbon black, silica filler, and carbon-silica dual-phase filler (CSDPF) was investigated by Wang et al.[54] using practical tire-tread-compound formulations. The filled vulcanizates were prepared with the following formulation: SSBR/BR 75/25, 80 phr filler, 12.8 phr X50S as silane coupling agent for the silica compound, 2.0 phr TESPT as coupling agent for the CSDPF compound, 32.5 phr oil and other ingredients. The changes of $G'$, $G''$, and tan $\delta$ were measured at a temperature range from −60°C to 60°C, and at different frequencies from 0.032 Hz to 32 Hz. A 1% strain amplitude was used in all the measurements. The dynamic property data were shifted along the frequency scale, and the reference temperature was 25°C. For $G'$, most of the data points from each individual curve were able to overlap with the adjacent curves, except for a few points at a high frequency. Therefore, each curve was consistently shifted on the basis of the low-frequency data. When $G''$ master curves were constructed, there seemed to be more data points at the high-frequency side departing from the master curves. In practice, the shift factors obtained from $G'$ were applied to $G''$. In this case, the master curves have the same features as the manually shifted curves. The same sets of shift factors were also used to construct tan $\delta$ master curves. In fact, some tan $\delta$ curves at high temperatures were severely concave and could not be manually shifted.

The master curves of $G'$, $G''$, and tan $\delta$ obtained from the carbon-black-filled polymer blends are presented in Figure 7.39(a)–(c), respectively. In Figure 7.39(a), $G'$ increases slowly from a low frequency to a high frequency, rises more than 2 decades when the temperature drops down into the transition zone, and then flattens out. The master curve looks quite smooth, and the shift seems to be valid. However, when the data are carefully examined, each individual curve, which represents data taken at one temperature, does not overlap completely with the adjacent curves. Overlapping data points at a low frequency tend to force the curve to tilt up at a high frequency, which results in a feather-like master curve. This is better seen in Figure 7.40(a), where the same data are presented with those of other filled vulcanizates, but with the actual data points left out. Strictly speaking, the curve can only be called a pseudomaster curve. However, for the sake of simplicity, it is referred to as a master curve in this section.

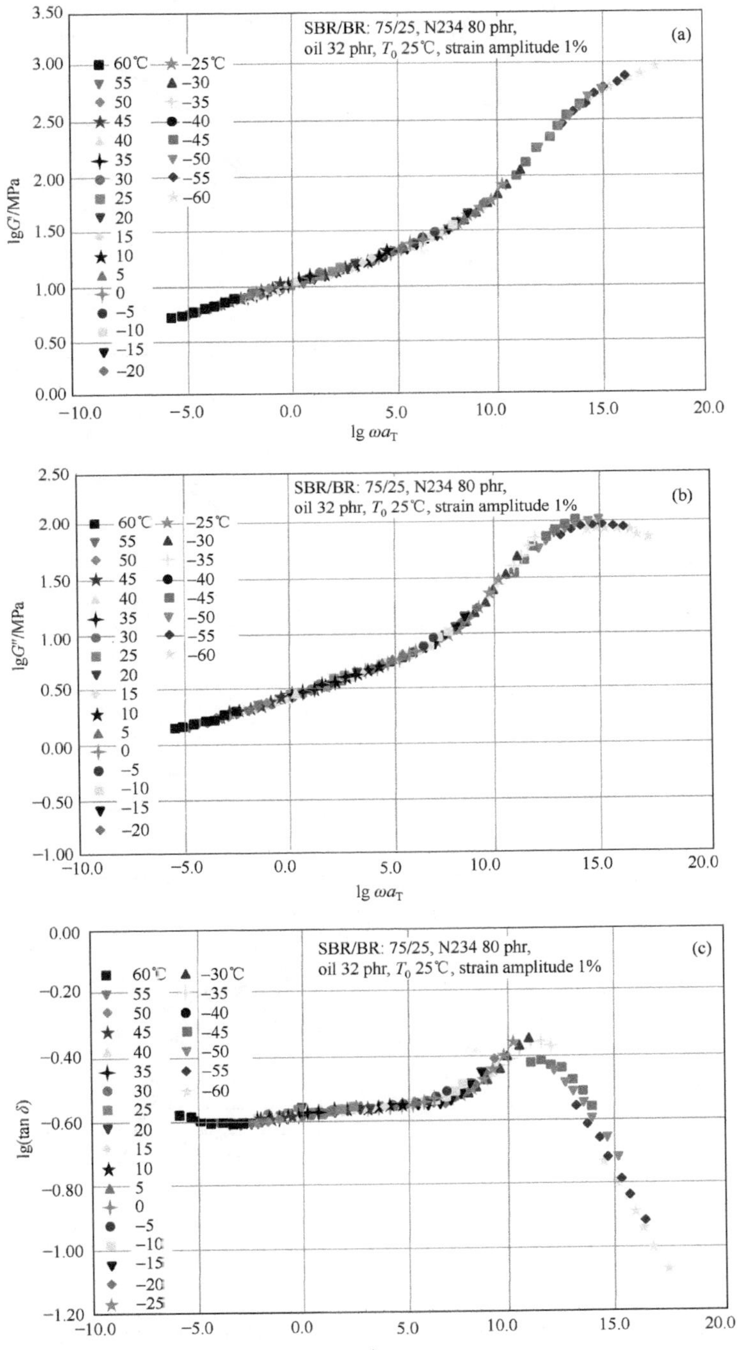


**Figure 7.39** Master curves of (a) $G'$, (b) $G''$, and (c) tan $\delta$ of carbon black N234-filled vulcanizate

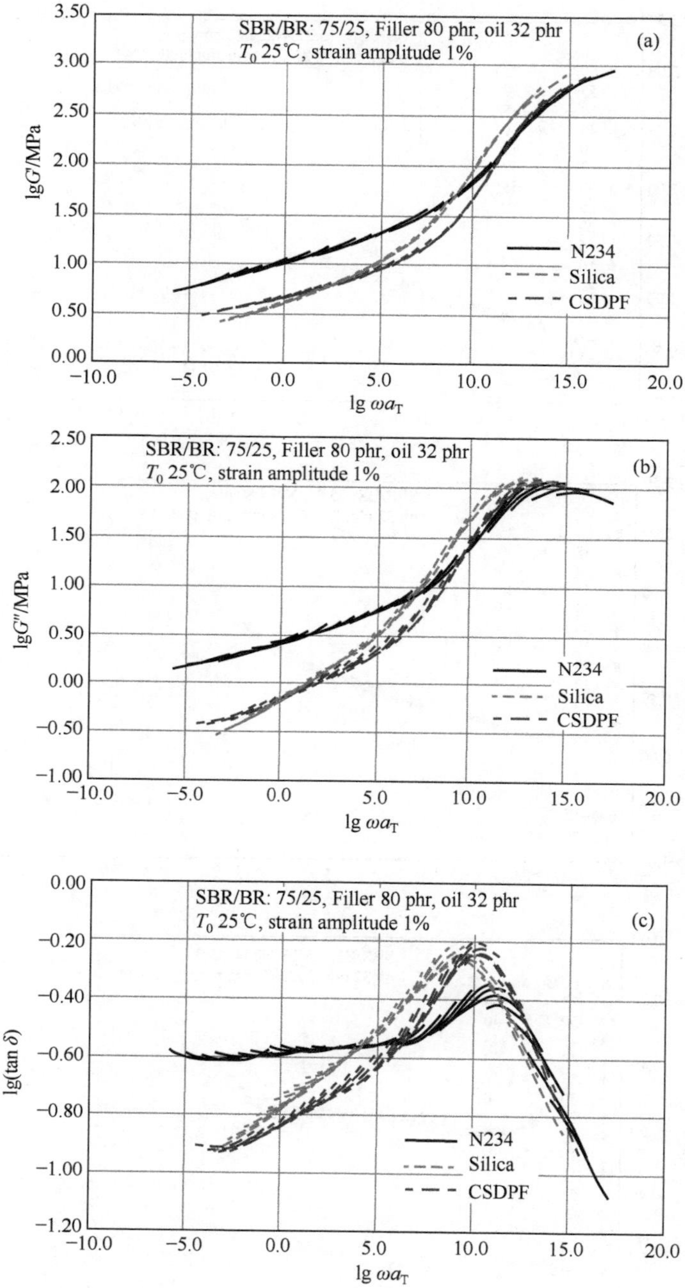


**Figure 7.40** Master curves of (a) $G'$, (b) $G''$, and (c) tan $\delta$ for vulcanizates filled with different fillers

In Figure 7.39(b), viscous modulus $G''$ first increases with increasing frequency, reaches a maximum, and then decreases slightly at a high frequency. Compared to $G'$, the feather-like feature shows more clearly in the $G''$ master curve, which again can be seen in Figure 7.40(b). In addition, the curve becomes discontinuous at the high-frequency end, as each individual curve seems to drop vertically. The attempt to shift these high-frequency data points both horizontally and vertically into one curve was not successful. The data at one temperature does not join smoothly with data obtained at the other temperature, and they often cross each other.

Shift factors from $G'$ were applied to tan $\delta$ in the construction of its master curve, and the result is shown in both Figure 7.39(c) and Figure 7.40(c). In Figure 7.39(c), the master curve is relatively flat at a low frequency and then increase, after reaching a peak, drops down quickly with a further increase in the frequency. At a low frequency, each individual tan $\delta$ curve is concave, first decreasing and then increasing as a function of frequency. The feather-like feature shows up in the medium-frequency regime. After tan $\delta$ passes its maximum, each curve decreases monotonically with frequency, accompanied by a vertical drop. Figure 7.40(c) shows again these observations, which suggest that in addition to the general time-temperature superposition principle, more complicated mechanisms may be involved in determining the dynamic properties of filled vulcanizates.

Master curves of $G'$, $G''$, and tan $\delta$ for silica-filled polymer blends are presented in Figure 7.40(a)–(c), respectively, together with dynamic properties from compounds with other types of fillers for direct comparison. Compared to carbon black, both $G'$ and $G''$ of the silica compounds have higher values at high frequency and lower values at low frequency. In tan $\delta$, the peak position shifts to a lower frequency, and the peak height is higher than a carbon-black compound. At a low frequency, silica-filled rubber has a lower hysteresis. Besides the differences in the overall shape of the master curves, the departure of each individual curve from the master curve seems to be less pronounced with silica compounds versus the carbon-black compounds.

Carbon-silica dual phase filler (CSDPF) consists of two phases, a carbon phase with a finely divided silica phase dispersed therein[18,55]. As will be discussed in Sections 8.1.2.3.6 and 9.2, from a compounding point of view, the dual-phase filler is characterized by higher polymer-filler interactions in relation to a physical blend of carbon black and silica at the same silica content and lower filler-filler interactions in comparison with either conventional carbon black or silica with comparable surface areas[18]. As has been observed by electron spectroscopy for chemical analysis (ESCA) and infrared spectroscopy, CSDPF aggregates contain carbon and silica phases as intrinsic components of the aggregates[56]. If the filler is viewed as a composite, each component would be expected to affect the dynamic properties in an additive function. However, the test results are quite different, as observed in Figure 7.40. In general, the overall shapes of the master curves are very close to that of the silica compound, even though this particular CSDPF contains only 10% silica by weight. The master curves from both $G''$ and tan $\delta$ bear a close resemblance to that of a silica compound with a slight shift along the frequency axis. $G'$ of CSDPF seems to be close to a carbon black at a high frequency but is more like a silica compound at a low frequency. At a more detailed level, however, CSDPF has a feather-like feature similar to the carbon-black compound.

The complexity of these constructed master curves may originate from different mechanisms that may be related to the polymer systems and filler characteristics. According to the dynamic properties of the unfilled polymer blend of SSBR/BR (75/25) in Figure 7.38, a discontinuity is found in the master curve of tan $\delta$ at a high frequency, which may be caused by the two polymers being immiscible. However, the WLF equation is applicable to the blend from low to medium frequencies, and the same observations were obtained with the addition of 32 phr oil. This implies that neither the blend nor the oil induces the feather-like feature in the master curves of the filled compounds. It seems reasonable to believe that it is the filler that disrupts the formation of the master curve.

With regard to filled rubber, the properties of the materials depend on the individual components and their interactions. In other words, the time-temperature dependence of filled vulcanizates should be determined not only by polymer property per se, but also by the effect of filler in the polymer, such as hydrodynamic effect[57,58], aggregate shape, occlusion of polymer inside the geometrical complex of the aggregate[59,60], rubber shell on the aggregate surface[15], and filler agglomeration, or filler flocculation[15,17], as described in Chapter 3 and Section 5.2.2. The rubber shell and filler agglomerate possess a strong temperature dependence.

Due to polymer-filler interaction, a layer of immobilized rubber shell, will form surrounding the filler aggregates[15], the chain mobility increases gradually as polymer chains extend away from the filler surface into the bulk region. This effect can be represented by Eq. 7.23 and Eq. 7.24, as $H(\lambda)$ may become a skewed relaxation spectrum due to the different segment mobility within the rubber shell. Therefore, temperature would affect rubber shell thickness, which may change the amount of polymer participating in the process. Westlinning[61] suggested that the thickness of the rubber shell on carbon black is 35 nm at 20 °C and becomes 0 nm at 100 °C. The types of polymer and filler also affect shell thickness[62].

In addition, as described before, filler aggregates can join together via joint rubber shell to form filler agglomerates (Section 3.4.2). This filler agglomerate may enclose some polymer, termed *trapped rubber*, and shield it from external stress during dynamic strain. As a consequence, the trapped rubber in the agglomerates would be at least partially "dead", so the effective volume of the filler increases[17] .

Accordingly, the time-temperature superposition principle can only be partially applied to the dynamic properties of filled rubber. The divergence may be explained in terms of the polymer-filler and filler-filler interactions. Firstly, three temperature regions are identified for the master curves: a high-temperature region, a medium-temperature region, and a low-temperature region. Schematic plots of filler agglomeration at these temperatures are presented in Figure 7.41.

At high temperature (approximately higher than 10°C), there is less interaction between the filler aggregates. Each has a thin layer of rubber shell. In the master curves, this is equivalent to the low-frequency region. Both $G'$ and $G''$ have lower values compared to other temperatures. The high tan $\delta$ values in the carbon-black compound may be the result of its stronger filler-filler interaction compared to other compounds.

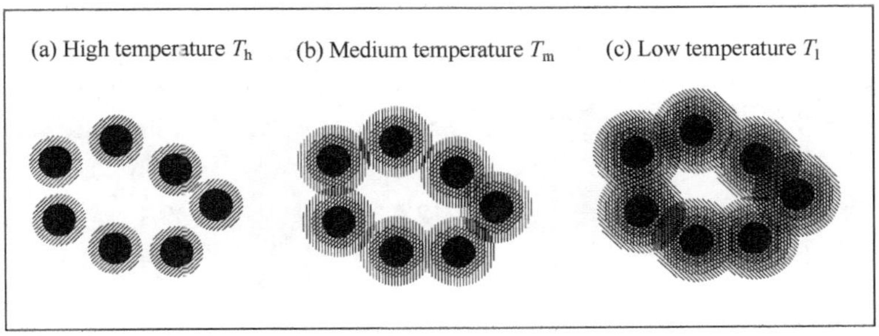


**Figure 7.41** Schematic plots of filler aggregates in the polymer matrix at (a) high temperature, (b) medium temperature, and (c) low temperature

At medium temperature (approximately between −30°C and +10°C), a layer of rubber shell, which has a higher modulus than the polymer matrix because of the polymer-filler interaction, develops around the filler aggregates. Some aggregates may agglomerate together via the joint shell, and some polymer may be trapped inside the filler agglomerates. As the frequency rises, the modulus of the rubber shell increases more rapidly than the modulus of the polymer matrix. At the same time, the thickness of the rubber shell also increases. This is because the polymer in the rubber shell is closer to or inside the transition zone, whereas the polymer matrix is still in its rubbery state. The same would also be true for hysteresis, as the rubber shell is close to or inside the transition zone. As a result, the high-frequency modulus and hysteresis values rise faster than what one can predict from low-frequency values, and, therefore, a feather-like feature is observed. Nevertheless, the effect of frequency on the rubber shell compared to temperature seems to be small. Therefore, satisfactory master curves can be obtained.

At low temperature (lower than −30°C), the thickness and the modulus of the rubber shell increase dramatically to the extent that the rubber shell is so hard, hence the deformation of the joint shell is so small that rubber trapped inside the filler agglomerate loses its identity as a polymer and behaves as a filler. This results in a significant increase in the effective volume of the filler. As a consequence, elastic modulus will increase, and loss modulus will decrease, both in a noncontinuous fashion. Together with the polymer effect discussed earlier, discontinuity in the master curves of both $G''$ and tan $\delta$ is observed.

It is thought that medium temperature is within the temperature regime where the dynamic properties of the polymers are governed by the WLF-type equation and that high temperature is inside the temperature regime where the shift factor follows an Arrhenius-type temperature dependence. In a detailed analysis, shift factors obtained from the three compounds were used to derive the coefficients $C_1$ and $C_2$ from Eq. 7.19 and the activation energy $E_a$ from Eq. 7.22. The results are tabulated in Table 7.2. $C_1$, $C_2$, and $E_a$ quantify the temperature dependence of the dynamic properties for each compound. In both types of temperature dependency, CSDPF stands in the middle between carbon-black and silica compounds.

**Table 7.2** Fitting parameters for the three compounds

| Filler type | WLF $C_1$ | WLF $C_2$ | Arrhenius $E_a$ /(kJ/mol) |
|---|---|---|---|
| N234 | 37.6 | 301.0 | 106.3 |
| Silica | 18.5 | 212.4 | 68.1 |
| CSDPF | 28.1 | 260.6 | 84.1 |

The overall dynamic properties of CSDPF show more similarity to silica than carbon black. At a low frequency, both have a low tan $\delta$, and at a high frequency, they have a high tan $\delta$. The results agree well with the studies on the temperature dependence of CSDPF[63]. Here, the filler-filler interaction is the key factor that affects the dynamic properties of the filled rubber. Compared to carbon black, the filler-filler interaction in CSDPF is low as a result of the dissimilar domains within the aggregates[18]. Fillers are less likely to flocculate, and filler-filler agglomeration is reduced. At low frequency, when the polymer is largely out of the transition zone, the major source for energy dissipation is the breakdown and reformation of the filler agglomerates. Therefore, a lower tan $\delta$ is expected for CSDPF because the filler agglomerate is less developed[4,17]. At medium to high frequency, however, the major source of energy dissipation is the polymer because it is inside the glass-transition zone. Developed filler agglomerate is likely to trap rubber, preventing it from absorbing energy. Therefore, CSDPF with a reduced filler agglomerate has a higher tan $\delta$ peak.

By carefully examining the deviations from the master curves, one can see that silica has the least effect on the degree of deviation, whereas carbon black has the most. This may be understood through the differences in the polymer-filler interface between the two compounds. In the silica-filled compound, the polymer-filler interaction is lower even though a certain amount of covalent bonding is generated through the coupling agent[56]. This has been demonstrated by the adsorption energies of polymer model compounds, measured by inverse gas chromatography, on these fillers[14,64,65]. Therefore, the thickness of the rubber shell is considered to be small. However, in carbon-black-filled compounds, the filler interacts with the polymer through strong physical adsorption, which may continue through a longer distance. As a result, the deviation is more pronounced. In the case of the CSDPF, the interaction between the polymer and the carbon-black domain of the filler is somewhat higher compared with that for the carbon-black-filled rubber, provided that the interaction between the polymer and the silica domain of the filler is identical to that in silica-filled rubber[18,65]. With only a small amount of the silica component, the development of the feather-like feature in the master curves of the CSDPF compound is closer to that of the carbon-black compound, as the carbon-black phase is the dominant component in CSDPF.

At the high-temperature region of the tan $\delta$ master curve, unique concave curves are observed in both the carbon-black- and CSDPF-filled compounds, but less so in the silica-filled compounds. The shape of the curves strongly suggests that two mechanisms may come into effect at this temperature and frequency regime. One mechanism may be the same as what causes the feather-like features at the medium temperature. When the rubber shell thickens as the frequency increases, the polymer moves into or close to the transition zone. The increase in $G''$ is more significant than the increase in $G'$, which results in an increase in tan $\delta$. However, when the frequency is low enough that it becomes comparable to the relaxation time of polymer chains, chain disentanglement is likely to occur. As a result, $G'$ decreases more rapidly than $G''$, which results in an increase in tan $\delta$. Because of the higher polymer-filler interaction for carbon black and CSDPF through their high surface energy[65], the thickness of the rubber shell and the polymer chain mobility show a strong temperature and frequency dependence. However, in the case of silica, even though a certain number of chemical bonds are formed between the silica surface and polymer chains via a coupling agent, the low polymer-filler interaction results in a low temperature and frequency dependency. Therefore, a less concave curve would be expected at a high temperature.

## 7.5 Heat Build-up

Heat build-up measured in the flexometer (e.g., a Goodrich Flexometer) is generally related to the energy loss under constant stress. The increase in temperature during the measurement, $\Delta T$, would thus be linearly related to tan $\delta$, as reported by Wolff and Panenka[66] for NR filled with a series of different carbon blacks. The factors influencing tan $\delta$ can, therefore, also be applied to heat generation. In a systematic study of the physical properties of SBR vulcanizates filled with 17 carbon blacks with surface areas ranging from 30 $m^2$/g to 139 $m^2$/g, compressed DBPA values from 57 mL/100 g to 107 mL/100 g, and loadings from 0 phr to 50 phr for hard blacks and 0 phr to 70 phr for soft blacks, Wolff and Wang[67] found that the filler loading-rubber property curves of SBR vulcanizates can be superposed to a single master curve by introduction of a shift factor, taking one carbon-black-filled polymer as a reference. The master curve for heat build-up of SBR vulcanizates is shown in Figure 7.42[67].

The shift factor *f* which is used to reduce the volume fraction of a given carbon black to that of a reference black is a relative measure of the effective filler volume in rubber, reflecting the influence of filler parameters on rubber reinforcement. For the heat build-up of the vulcanizates, the shift factor increases with surface area for low and medium surface area, whereas structure plays a role in the case of fine-particle carbon blacks, as presented in Figure 7.43[67].

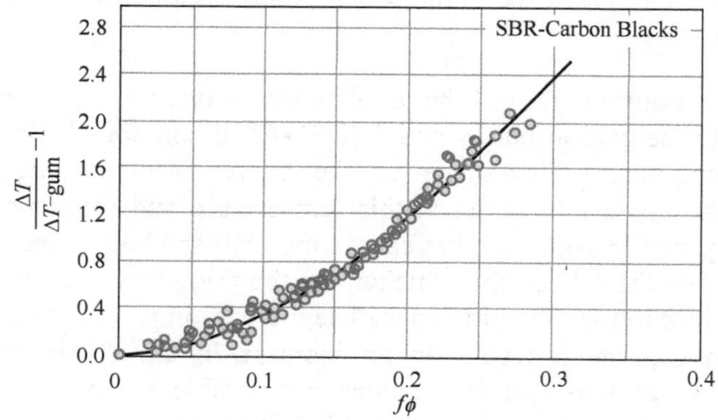


**Figure 7.42** Master curve for heat build-up for SBR vulcanizates[67]

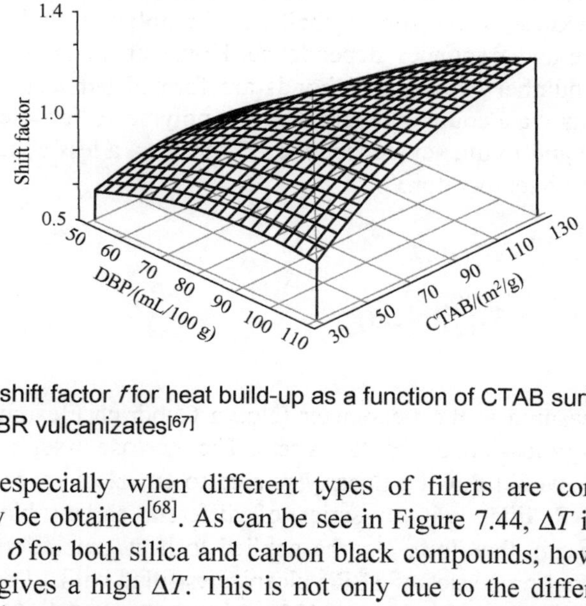


**Figure 7.43** The shift factor *f* for heat build-up as a function of CTAB surface area and DBP (24M4) for SBR vulcanizates[67]

In some cases, especially when different types of fillers are compared, different relationships may be obtained[68]. As can be see in Figure 7.44, $\Delta T$ increases with the increasing of tan $\delta$ for both silica and carbon black compounds; however, at constant tan $\delta$, the silica gives a high $\Delta T$. This is not only due to the different experimental conditions, but also reflects the structure of the vulcanizates. The stress or strain is usually lower in the standard tan $\delta$ measurement which is unable to destroy the strong filler agglomerate formed by silica, consequently leading to lower tan $\delta$. In the case of the heat build-up measurement, the high static load and large stroke applied should largely destroy the secondary filler structure, giving rise to considerable energy dissipation in developed silica agglomerate. On the other hand, the low modulus of silica vulcanizates after stress softening during the first few cycles of compression under static load should lead to high dynamic strain and, hence, to considerable heat generation. Moreover, while tan $\delta$ represents the hysteresis of one cycle of deformation at a constant temperature (even when measured after several cycles of deformation), $\Delta T$ constitutes a cumulative value for heat generation. It is determined not only by the hysteresis of the sample, but also by the heat exchange with the environment. The temperature of the sample at equilibrium could, therefore, be affected by the thermal conductivity of the filled vulcanizates.

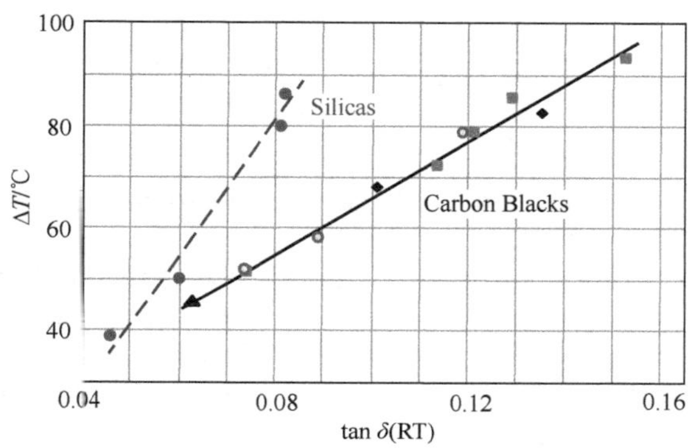


**Figure 7.44** Heat build-up *vs.* tan $\delta$ at room temperature for silica and carbon black filled NR[68]

# 7.6 Resilience

Resilience, or the rebound in single pendulum tests using a hemispherical striker or falling ball, also represents a measure of the hysteresis under conditions of constant energy input, although the deformation is complex, involving extension, compression, and shear. Gehman[69] used a pendulum test and showed that tan $\delta$ was directly proportional to the logarithm of the fractional rebound, $R$:

$$\tan\delta = -\ln R/\pi \tag{7.25}$$

Other authors[70–72] also observed a good correlation between resilience and tan $\delta$. It is therefore not surprising that the correlation between tan $\delta$ and the interaggregate distance, $\delta_{aa}$, can also be applied to resilience[16]. As illustrated in Figure 7.45, a single curve may be used to represent the results of Firestone Ball Rebound as a function of $\delta_{aa}$ for SBR vulcanizates filled with the whole range of rubber-grade carbon blacks at various degrees of loading[16]. The ball rebound increases rapidly in the low $\delta_{aa}$ region and reaches a maximum value which corresponds to that of the gum. The change in resilience with interaggregate distance follows the same mechanism which was discussed in the case of tan $\delta$.

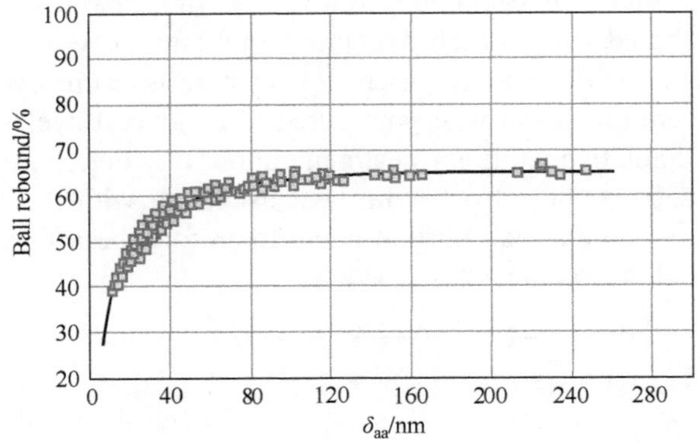


**Figure 7.45** Firestone ball rebound as a function of interaggregate distance[16]

Wolff and Wang[67] have found that the curves of ball rebound *vs.* filler loading of SBR vulcanizates filled with a series of different furnace carbon blacks can be superposed to one single master curve by introducing a shift factor, *f*, taking one-carbon-black filled polymer as a reference. The master curve for ball rebound of SBR vulcanizates is shown in Figure 7.46.

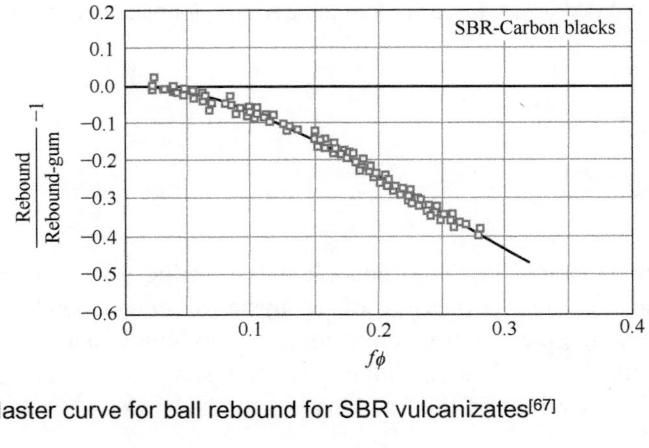


**Figure 7.46** Master curve for ball rebound for SBR vulcanizates[67]

The change in *f* with CTAB and compressed DBP for ball rebound is shown in Figure 7.47. This shift factor is mainly dependent on surface area; the effect of structure appears to be insignificant. Instead of a structure-loading equivalence, a surface area-loading equivalence seems to exist in this case. Furthermore, the shift factor for ball rebound is linearly related to the value of *A* (Figure 7.48), probably for the same reason. *A* is defined as[73]:

$$A = (R_0 - R) \times \frac{m_P}{m_F} \tag{7.26}$$

where $R$ and $R_0$ are the ball rebound values of filled and unfilled vulcanizates, and $m_F$ and $m_P$ are the mass of the filler and the polymer, respectively (Eq. 7.26). It was found that the constant $A$, i.e., the slope of the plot of $(R_0 - R)$ versus $m_F/m_P$, is independent of the crosslink density of the vulcanizates, of filler loading, and of filler structure, but that it depended on the type of elastomer and on carbon black surface area. This parameter is characteristic for each carbon black and was defined as a measure of the in-rubber surface area of furnace carbon blacks. The effective volume loading for ball rebound is thus associated with the interfacial area in filled vulcanizates.

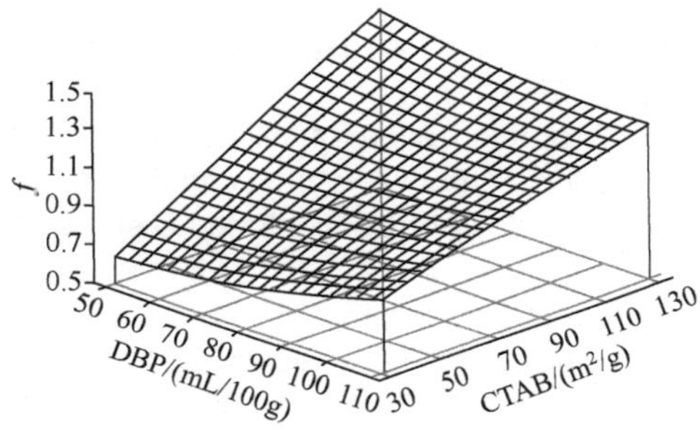


**Figure 7.47** *f* for ball rebound as a function of CTAB surface area and DBP (24M4) for SBR vulcanizates[67]

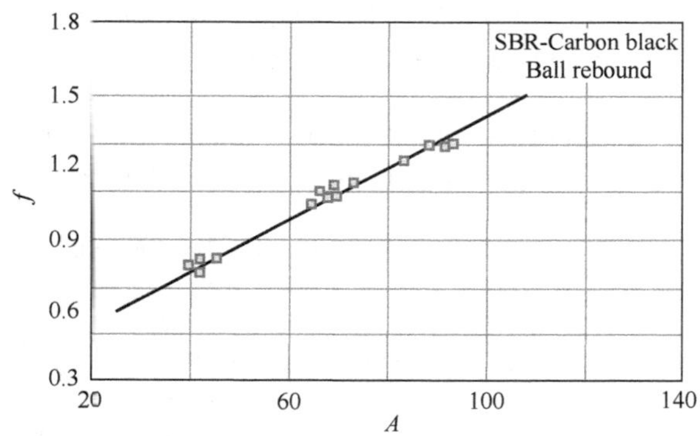


**Figure 7.48** Shift factor *f* for ball rebound as a function of *A* for SBR vulcanizates[67]

## References

[1] Ferry J D. Viscoelastic Properties of Polymers. 3rd Edition. Wiley, 1980, chapter 1.

[2] Williams M L. The Temperature Dependence of Mechanical and Electrical Relaxations in Polymers. *J. Phys. Chem.*, 1955, 59: 95.

[3] Williams M L, Landel R F, Ferry J D. The Temperature Dependence of Relaxation Mechanisms in Amorphous Polymers and Other Glass-forming Liquids. *J. Am. Chem. Soc.*, 1955, 77: 3701.

[4] Wang M-J. The Role of Filler Networking in Dynamic Properties of Filled Rubber. *Rubber Chem. Technol.*, 1999, 72: 430.

[5] Warring J R S. Dynamic Testing in Compression: Comparison of the ICI Electrical Compression Vibrator and the IG Mechanical Vibrator in Dynamic Testing of Rubber. *Trans., Inst. Rubber Ind.*, 1950, 26: 4.

[6] Payne A R. The Dynamic Properties of Carbon Black-Loaded Natural Rubber Vulcanizates. Part I. *J. Appl. Polym. Sci.*, 1962, 6: 57.

[7] Payne A R, Whittaker R E. Low Strain Dynamic Properties of Filled Rubbers. *Rubber Chem. Technol.*, 1971, 44: 440.

[8] Payne A R. *Rubber Plast. Age*, 1961, 42: 963.

[9] Medalia A I. Effect of Carbon Black on Dynamic Properties of Rubber Vulcanizates. *Rubber Chem. Technol.*, 1978, 51: 437.

[10] Wolff S, Donnet J-B. Characterization of Fillers in Vulcanizates According to the Einstein-Guth-Gold Equation. *Rubber Chem. Technol.*, 1990, 63: 32.

[11] Van der Poel C. On the Rheology of Concentrated Dispersions. *Rheol. Acta*, 1958, 1: 198.

[12] Payne A R. Reinforcement of Elastomers. New York: Interscience, 1965, chapter 3.

[13] Wang M-J, Wolff S, Donnet J-B. Filler-Elastomer Interations. Part I. Silica Surface Energies and Interactions with Model Compounds. *Rubber Chem. Technol.*, 1991, 64: 559.

[14] Wang M-J, Wolff S, Donnet J-B. Filler-Elastomer Interations. Part III. Carbon-Black-Surface Energies and Interactions with Elastomer Analogs. *Rubber Chem. Technol.*, 1991, 64: 714.

[15] Wolff S, Wang M-J. Filler-Elastomer Interations. Part IV. The Effect of the Surface Energies of Fillers on Elastomer Reinforcement. *Rubber Chem. Technol.*, 1992, 65: 329.

[16] Wang M-J, Wolff S, Tan E-H. Filler-Elastomer Interations. Part VIII. The Role of the Distance Between Filler Aggregates in the Dynamic Properties of Filled Vulcanizates. *Rubber Chem. Technol.*, 1993, 66: 178.

[17] Wang M-J. Effect of Polymer-Filler and Filler-Filler Interactions on Dynamic Properties of Filled Vulcanizates. *Rubber Chem. Technol.*, 1998, 71: 520.

[18] Wang M-J, Mahmud K, Murphy L J, et al. Carbon-Silica Dual Phase Filler, a New Generation Reinforcing Agent for Rubber. *Kautsch. Gummi Kunstst.*, 1998, 51: 348.

[19] Payne A R. The Role of Hysteresis in Polymers. *Rubber J.*, 1964, 146(1): 36.

[20] Payne A R, Swift P M, Wheelans M A. NR Vulcanisates with Improved Dynamic Properties. *J. Rubber Res. Inst. Malays.*, 1969, 22: 275.

[21] Medalia A I, Laube S G. Influence of Carbon Black Surface Properties and Morphology on Hysteresis of Rubber Vulcanizates. *Rubber Chem. Technol.*, 1978, 51: 89.

[22] Kraus G. *Proc. Int. Rubber Conf.*, Brighton, U. K., 1977.

[23] Sircar A K, Lamond T G. Strain-Dependent Dynamic Properties of Carbon-Black Reinforced Vulcanizates. I. Individual Elastomers. *Rubber Chem. Technol.*, 1975, 48: 79.

[24] Sircar A K, Lamond T G. Strain-Dependent Dynamic Properties of Carbon-Black Reinforced Vulcanizates. II. Elastomer Blends. *Rubber Chem. Technol.*, 1975, 48: 89.

[25] Ulmer J D, Hess W M, Chirico V E. The Effects of Carbon Black on Rubber Hysteresis. *Rubber Chem. Technol.*, 1974, 47: 729.

[26] Kraus G. Mechanical Losses in Carbon-Black-Filled Rubbers. *J. Appl. Polym. Sci.: Appl. Polym. Symp.*, 1984, 39: 75.

[27] Payne A R. Effects of Dispersion on Dynamic Properties of Filler-Loaded Rubbers. *Rubber Chem. Technol.*, 1966, 39: 365.

[28] Wang M-J. Presented at a workshop *"Praxis und Theorie der Verstärkung von Elastomeren"*, Hanover, Germany, Jun. 27-28, 1996.

[29] Engelhardt M L, Day G L, Samples R, et al. Study evaluates series of carbon blacks. *International Tire Exhibition and Conference*, Akron, OH, Sept. 1994. paper no. 17A.

[30] Moneypenny H G, Harris J, Laube S, et al. Reinforcing Fillers, Viscoelastic Behavior and Tyre Rolling Resistance Performance. *Rubbercon'95*, Gothenburg, Sweden, May 9-12, 1995.

[31] Wang M-J, Patterson W J, Ouyang G B. Dynamic Stress-Softening of Filled Vulcanizates. *Kautsch. Gummi Kunstst.*, 1998, 2: 106.

[32] Dannenberg E M. Molecular Slippage Mechanism of Reinforcement. *Trans., Inst. Rubber Ind.*, 1966, 42: 26.

[33] Dannenberg E M. The Effects of Surface Chemical Interactions on the Properties of Filler-Reinforced Rubbers. *Rubber Chem. Technol.*, 1975, 48: 410.

[34] Brennan J J, Jermyn T E, Perdigao M F. *Meeting of the Rubber Division, ACS*, Detroit, Michigan, Apr., 1964. paper no. 36.

[35] Bueche F J. Mullins Effect and Rubber-Filler Interaction. *Appl. Polym. Sci.*, 1961, 5: 271.

[36] Dannenberg E M, Brennan J J. Strain Energy as a Criterion for Stress Softening in Carbon-Black-Filled Vulcanizates. *Rubber Chem. Technol.*, 1966, 39: 597.

[37] Nordsiek K H. The "Integral Rubber" Concept—an Approach to an Ideal Tire Tread Rubber. *Kautsch. Gummi Kunstst.*, 1985, 38(3): 178.

[38] Saito Y. New Polymer Development for Low Rolling Resistance Tyres. *Kautsch. Gummi Kunstst.*, 1986, 39: 30.

[39] Wolff S, Goerl U, Wang M-J, et al. Silica-Based Tread Compounds: Background and Performance. *TYRETECH '93 Conference*, Basel, Switzerland, Oct. 28-29, 1993.

[40] Wang M-J, Song J J, Dai D Y. Continuous Manufacturing Process for Rubber Masterbatch and Rubber Masterbatch Prepared Therefrom, US Patent, 9758627 B2, 12 Sept. 2017.

[41] Wang M-J. Liquid Phase Mixing. *Tire Technology EXPO*, Hannover, Germany, Feb. 16-18, 2016.

[42] Wang M-J, Song J J, Wang Z, et al. Silica-Filled Masterbatches Produced with Liquid Phase Mixing. Part I. Characterization. *Kautsch. Gummi Kunstst.*, 2020, 73: Sept. 9.

[43] Xie M X, He F J, Yi L M, et al. Silica-Filled Masterbatches Produced with Liquid Phase Mixing. Part II. Dynamic Stress-Softening Effect. Kautsch Gummi Kunstst., 2020, 73: Dec. 10.

[44] Bulgin D, Hubbard D G, Walters M H. Road and Laboratory Studies of Fruction of Elastomers. *Proc. 4th Rubber Technology Conference*, London, 1962. p173.

[45] Ferry J D. Viscoelastic Properties of Polymers. 2nd Edition. Wiley, 1970, chapter 11.

[46] Doolittle A K, Doolittle D B. Studies in Newtonian Flow. V. Further Verification of the Free-Space Viscosity Equation. *J. Appl. Phys.*, 1957, 28: 901.

[47] Gross B. Mathematical Structure of the Theories of Viscoelasticity. Paris: Hermann, 1953.

[48] Roland C M, Ngai K L. Segmental Relaxation and the Correlation of Time and Temperature Dependencies in Poly(vinyl methyl ether)/Polystyrene Mixtures. *Macromolecules*, 1992, 25: 363.
[49] Bazuin C G, Eisenberg A. Dynamic Melt Properties of Ionic Blends of Polystyrene and Poly(ethyl acrylate). *J. Polym. Sci. Part B: Polym. Phys.*, 1986, 24: 1021.
[50] Kapnistos M, Hinrichs A, Vlassopoulos D, et al. Rheology of a Lower Critical Solution Temperature Binary Polymer Blend in the Homogeneous, Phase-Separated, and Transitional Regimes. *Macromolecules*, 1996, 29: 7155.
[51] Han C D, Kim J K. On the Use of Time-Temperature Superposition in Multicomponent/ Multiphase Polymer Systems. *Polymer*, 1993, 34: 2533.
[52] Duperray B, Leblanc J L. The Time-Temperature Superposition Principle as Applied to Filled Elastomers. *Kautsch. Gummi Kunstst.*, 1982, 35: 298.
[53] Kobayashi N, Furuta I. Comparison Between Silica and Carbon Black in Tire Tread Formulation. *J. Soc. Rubber Ind., Jpn.*, 1997, 70: 147.
[54] Wang M-J, Lu S X, Mahmud K. Carbon-Silica Dual-Phase Filler, a New Generation Reinforcing Agent for Rubber. Part VI. Time-Temperature Superposition of Dynamic Properties of Carbon-Silica-Dual-Phase-Filler-Filled Vulcanizates. *J. Polym. Sci. Part B: Polym. Phys.*, 2000, 38: 1240.
[55] Murphy L, Wang M-J, Mahmud K. Carbon-Silica Dual Phase Filler: Part V. Nano-Morphology. *Rubber Chem. Technol.*, 2000, 73: 25.
[56] Murphy L, Wang M-J, Mahmud K. Carbon-Silica Dual Phase Filler: Part III. ESCA and IR Characterization. *Rubber Chem. Technol.*, 1998, 71: 998.
[57] Guth E, Simha R, Gold O. The Viscosity of Suspensions and Solutions. III. The Viscosity of Sphere Suspensions. *Kolloid-Z.*, 1936, 74: 266.
[58] Guth E, Gold O. On the Hydrodynamical Theory of the Viscosity of Suspensions. *Phys. Rev.*, 1938, 53: 322.
[59] Medalia A I. Morphology of Aggregates. VI. Effective Volume of Aggregates of Carbon Black from Electron Microscopy; Application to Vehicle Absorption and to Die Swell of Filled Rubber. *J. Colloid Interface. Sci.*, 1970, 32: 115.
[60] Medalia A I. Effective Degree of Immobilization of Rubber Occluded Within Carbon Black Aggregates. *Rubber Chem. Technol.*, 1972, 45: 1171.
[61] Westlinning H. *Kautsch. Gummi Kunstst.*, 1962, 15: 475.
[62] Schoon T G F, Adler K. *Kautsch. Gummi Kunstst.*, 1966, 19: 414.
[63] Wang M-J, Patterson W J, Brown T A, et al. Carbon-Silica Dual Phase Filler, a New Generation Reinforcing Agent for Rubber. Part II. Examining Carbon-Silica Dual Phase Fillers. *Rubber & Plastics News*, 1998, Feb. 9: 12.
[64] Wang M-J, Wolff S. Filler-Elastomer Interactions. Part V. Investigation of the Surface Energies of Silane-Modified Silicas. *Rubber Chem. Technol.*, 1992, 65: 715.
[65] Wang M-J, Tu H, Murphy L, et al. Carbon-Silica Dual Phase Filler, a New Generation Reinforcing Agent for Rubber: Part VIII. Surface Characterization by IGC. *Rubber Chem. Technol.*, 2000, 73: 666.
[66] Wolff S, Panenka R. Present Possibilities to Reduce Heat Generation of Tire Compounds. *IRC' 85*, Kyoto, Japan, 1985.
[67] Wolff S, Wang M-J. Physical Properties of Vulcanizates and Shift Factors. *Kautsch. Gummi Kunstst.*, 1994, 47: 17.
[68] Wolff S. *IRC' 88*, Sydney, 1988. F-15.

[69] Gehman S D. Dynamic Properties of Elastomers. *Rubber Chem. Technol.*, 1957, 30: 1202.

[70] Medalia A I. Selecting Carbon Blacks for Dynamic Properties. *Rubber World*, 1973, 168 (5): 49.

[71] Barker L R, Payne A R, Smith J F. Dynamic Properties of Natural Rubber: Processing Variations. *J. Inst. Rubber Ind.*, 1967, 1(4): 206.

[72] Ulmer J D, Chirico V E, Scott C E. The Effect of Carbon Black Type on the Dynamic Properties of Natural Rubber. *Rubber Chem. Technol.*, 1973, 46: 897.

[73] Wolff S. Filler Development Today and Tomorrow. *Kautsch. Gummi Kunstst.*, 1979, 32(5): 312.

# 8 Rubber Reinforcement Related to Tire Performance

Continuous improvement of tire quality in the areas of better fuel economy, improved safety, and increased durability requires development of new tire compounds, particularly tread compounds. These requirements can only be met by improvements in rolling resistance, skid resistance, especially in wet conditions, and wear resistance, which are closely related to hysteresis, wet friction behavior, and abrasion of tread compounds. In these regards, it has been recognized that the filler takes an equal place alongside the polymer as the underlying determinant of the tire performance. In fact, the filler is no longer a "filler" in the sense of increasing the volume and reducing cost of the compounds. The filler is also not a "reinforcing agent" in the sense of increasing modulus and improving tensile strength of the compounds. The filler is actually a functional material or component in the tire compound to control tire functionality. In this chapter, the roles played by filler parameters, such as types, loading, morphology, and surface characteristics, in the rolling resistance, wet skid resistance and wear resistance of tires will be discussed.

## 8.1 Rolling Resistance

### 8.1.1 Mechanisms of Rolling Resistance−Relationship between Rolling Resistance and Hysteresis

The energy loss in rubber products during dynamic strain is of great importance for automotive tires, where it affects the service performance with regard to rolling resistance, traction, and skid resistance.

In fact, with regard to tire applications, it has been well established that repeated straining of the compound due to rotation and braking can be approximated as a process of constant energy input involving different temperatures and frequencies[1–3]. With respect to the tire tread, its deformation can be, as cited by Medalia[4], resolved "approximately into a constant strain (bending) and constant stress (compression) condition; and since the geometric mean of the hysteresis under these two conditions is approximately proportional to tan $\delta$, tire tread hysteresis is as a first approximation, proportional to "tan $\delta$".

Rolling resistance is related to the movement of the whole tire corresponding to deformation at a frequency of 10–100 Hz and a temperature of 50–80°C. In the case of skid or wet grip, the stress is generated by resistance from the road surface and movement of the rubber at the surface, or near the surface of the tire tread. The frequency of this movement depends on the roughness of the road surface but should be very high, probably around $10^4$ Hz to $10^7$ Hz at room temperature[2,3]. It is therefore obvious that any change in dynamic hysteresis of the compounds at different frequency and temperature will alter the performance of the tire. Since certain tire properties involve frequencies which are too high to be measured, these frequencies are reduced to a measurable level (e.g., 1 Hz or 10 Hz) at lower temperatures by applying the Time-Temperature Superposition Principle (or WLF Temperature-Frequency Conversion) even though in the case of filled vulcanizates the shift factors for building the elastic modulus master curve are not exactly the same as those for the master curve of viscous modulus[5], hence of tan $\delta$. However, the master curves for each property can be constructed experimentally according to the WLF Temperature-Frequency principle.

The reduced temperature for different tire properties at 10 Hz (Figure 8.1) has been used as the criterion for polymer[6] and filler[7] development for tire compounds. From the viscoelastic property point of view, an ideal material which is able to meet the requirement of a high-performance tire should give a low tan $\delta$ value at a temperature of 50–80°C in order to reduce rolling resistance and save energy. The ideal material should also demonstrate high hysteresis at lower temperature, for example, from −20°C to 0°C, in order to obtain high skid resistance and wet grip.

The topic of this section is rolling resistance that is determined by the tan $\delta$ at high temperature. Since the filler factors controlling the tan $\delta$ at high temperature are also related to the tan $\delta$ at low temperature, which is a key compound property for wet skid resistance, both loss tangents will be discussed in this section. However, the factors involved in skid resistance are recognized to be more complex than a single compound property and will be discussed in Section 8.2.

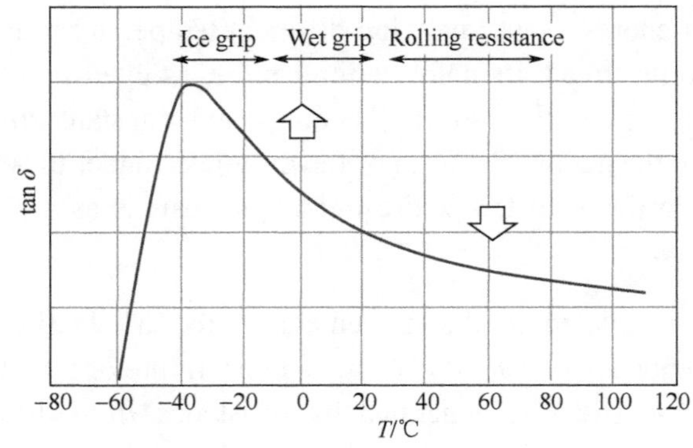


**Figure 8.1** Reduced temperatures at 10 Hz for different tire performances

## 8.1.2 Effect of Filler on Temperature Dependence of Dynamic Properties

### 8.1.2.1 Effect of Filler Loading

The effects of filler loading on the temperature dependence of dynamic properties are presented in Section 7.2.5 (see Figures 7.15−7.17) for carbon black N234 filled compounds[8]. It can be seen that with regard to the tire application, simply by reducing the filler loading, the dynamic properties can be easily satisfied, at least from the point of view of tan $\delta$. In fact, while the gum rubber gives lower hysteresis at relatively high temperature, which represents rolling resistance, its friction coefficient would be the best as tan $\delta$ is the highest at lower temperature[9]. However, in tire compounding, enough filler must be incorporated to meet the requirements for stiffness, wear resistance, handling, tearing resistance, and strength of the vulcanizate. These properties influence not only the service life of the tire but also driving safety, as hysteresis properties may not be the only factor to govern the skid resistance and cornering performance.

The results in Figure 7.17 suggested that the loading effects at different temperature regions should be governed by different mechanisms. As discussed in Section 7.2.5, among others, for a given polymer system, agglomeration of filler aggregates is the dominant factor in the determination of the hysteresis of filled vulcanizates. Due to filler agglomeration, the effective volume fraction of the filler will increase because of the rubber trapped in the agglomerates, which loses its identity as rubber, and behaves like filler. At high temperature, where rubber is in its rubbery state, the hysteresis is low under dynamic strain as the breakdown and reformation of the agglomerates would cause the energy dissipation, resulting in increased hysteresis. However, at low temperature where the main portion of the filled vulcanizate for energy dissipation is polymer matrix and the filler agglomerates may not be easily broken down, tan $\delta$ would be significantly attenuated by filler agglomeration due to the reduction of the effective volume of the polymer. Also in the transition zone, once the filler network can be broken down and reformed under a cyclic strain, the hysteresis can be substantially augmented through release of polymer to participate in energy dissipation and change in the agglomerate structure. Therefore, over a limited range of strain amplitudes, filler agglomeration plays a dominant role in improving the temperature dependence of tan $\delta$.

### 8.1.2.2 Effect of Filler Morphology

Besides loading and surface characteristics of filler, the main filler parameters influencing filler agglomeration, hence the hysteresis related to tire performance, are filler morphology, namely, surface area or particle size and structure. The role of these two parameters may involve different mechanisms and therefore will be discussed separately.

#### 8.1.2.2.1 Effect of Surface Area

It has been widely documented that with increasing surface area of carbon blacks at constant loading, Payne effect increases[10–12]. The increase in Payne effect suggests a strong tendency for agglomeration of small particle (aggregate) carbon black in the polymer matrix. It is also widely reported that the dynamic hysteresis, characterized by tan $\delta$, increases with filler surface area[1,10,13–15]. This is generally true at temperatures where the polymer is in its rubbery state as shown in Figure 8.2. Specifically, the DSA dependence of tan $\delta$ is compared using solution SBR (Duradene 715) vulcanizates filled with 50 phr of a series of carbon blacks having different surface areas (see Table 8.1). However, at low temperatures, where rubber falls in the transition zone, tan $\delta$ increases with decreasing carbon black surface area over a large range of strain amplitudes (see Figure 8.3). Such an observation can also be seen from temperature sweep data at a moderate-strain amplitude (5% DSA) and 10 Hz. As shown in Figure 8.4, there is a critical point around 0°C beyond which tan $\delta$ is higher for small-particle blacks and below which tan $\delta$ increases with decreasing surface area. This is even much clearer when the data are plotted at different temperatures on a linear scale as a function of CTAB surface area (Figure 8.5).

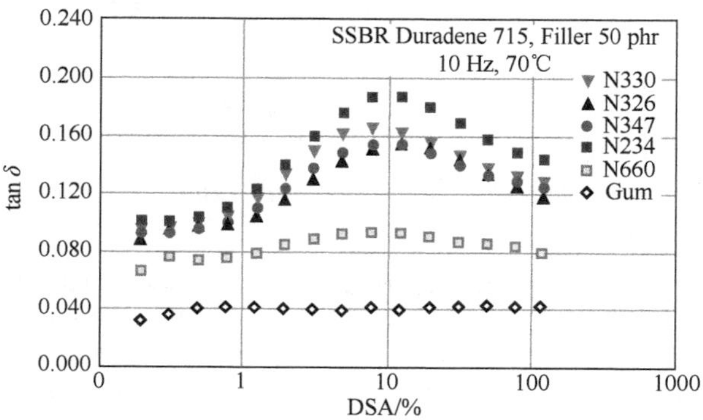


**Figure 8.2** Strain dependence of tan $\delta$ at 70°C and 10 Hz for vulcanizates filled with a variety of carbon blacks with different morphologies

**Table 8.1** Analytical properties of carbon blacks

| Carbon black | CTAB/($m^2$/g) | DBP/(mL/100 g) | CDBP/(mL/100 g) |
|---|---|---|---|
| N660 | 37 | 90 | 74 |
| N330 | 82 | 102 | 88 |
| N326 | 83 | 72 | 69 |
| N347 | 88 | 124 | 98 |
| N234 | 119 | 125 | 103 |

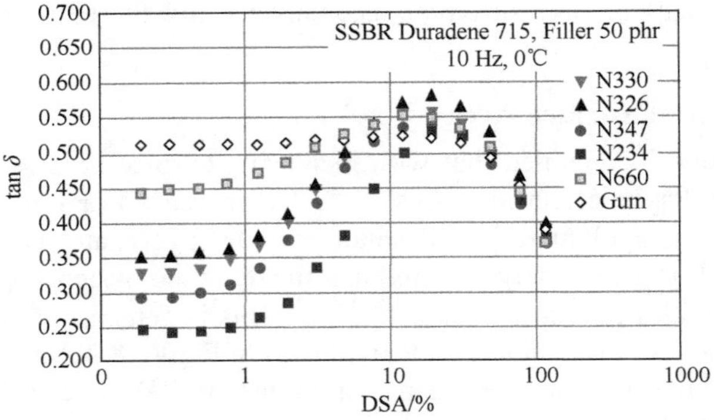


**Figure 8.3** Strain dependence of tan $\delta$ at 0°C and 10 Hz for vulcanizates filled with a variety of carbon blacjks with different morphologies. Same compounds as Figure 8.2

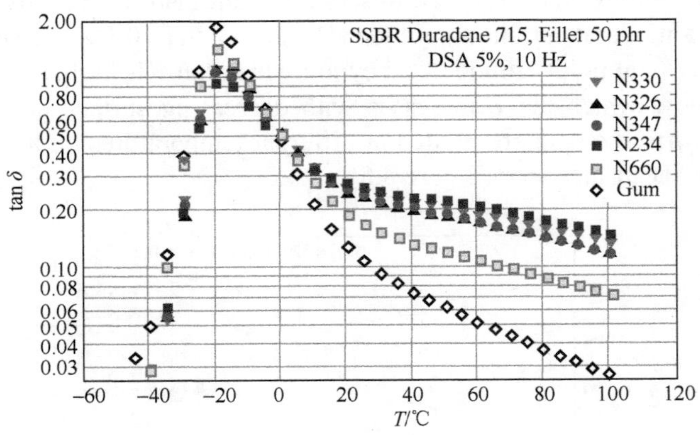


**Figure 8.4** Temperature dependence of tan $\delta$ at 5% DSA and 10 Hz for vulcanizates filled with a variety of carbon blacks with different morphologies. Same compounds as Figure 8.2

The effect of surface area of carbon black on the temperature dependence of dynamic hysteresis can be interpreted mainly by filler agglomeration in terms of its thermodynamics and kinetics. Thermodynamically, hydrocarbon polymers are low surface energy materials while the surface energy of carbon black is much higher. At room temperature, the dispersive components of surface energy measured by means of contact angle are 29.5 mJ/m$^2$ and 28.8 mJ/m$^2$ for NR and emulsion SBR 1500, respectively. The polar components of their surface energy are also quite low[16]. Although the surface energies of these two polymers cannot directly be transferred to the solution SBR used, it may be reasonable to assume that they are not much different. In contrast, the surface energy of carbon black and carbon products (even graphite) determined with different methods[17–22] is much higher than for the polymers. On the other hand, it is also reported that both the dispersive and polar components of carbon black increase with increasing surface area[22,23]. Therefore, according to Eq. 3.14, the attractive potential between filler aggregates would be higher and thus a more developed filler agglomerate would be expected for small-particle carbon black.

Accordingly, the filler-polymer interaction is also augmented when the surface area increases, as suggested by Eq. 2.134. Intuitive reasoning would lead one to expectations such as:

- Stronger polymer-filler interaction would result in a thicker rubber shell for small-particle carbon black, which, in combination with larger interfacial area in the unit volume compound at the same loading, would give a more immobilized rubber shell in comparison with large-particle black. The higher volume of immobilized rubber could first increase the effective volume of filler loading which, as discussed before, would cause higher hysteresis at higher temperature and lower hysteresis at lower temperature.
- Total bound rubber would also be expected to increase due to higher polymer-filler interactions as well as the large interfacial area between filler and polymer which will increase the viscosity of the polymer matrix. This in conjunction with higher effective volume of the aggregates could substantially reduce the flocculation rate of small-particle blacks.

The above-mentioned mechanisms, thermodynamic and kinetic, seem to be contrary to each other with respect to their effect on filler agglomeration. However, when the distance between aggregates is taken into account, filler agglomeration would be greatly affected by the surface area of carbon blacks as the interaggregate distance at constant filler loading is predominantly related to and inversely proportional to the surface area, see Eq. 3.9. The higher the surface area of the carbon black, the shorter the interaggregate distance; hence, more developed filler agglomerates would be expected. This may be the reason why tan $\delta$ at higher temperature (60°C) shows a good correlation with interaggregate distance for SBR 1500 filled with carbon black covering the full range of rubber grades[24]. It may also not be surprising that tan $\delta$ at 60°C (14.5% static compression, 20% DSA, and 0.25 Hz), for SBR 1500 vulcanizates filled with reinforcing carbon black from the N400 series to N200 series with different loadings, and tan $\delta$ at 24°C (14.5% static compression, 25% DSA and 1 Hz), for oil-extended SBR 1712 vulcanizates filled with carbon black over a range from the N900 to N100 series with different concentration, correlate well to a so-called loading-interfacial area parameter which, as cited by Caruthers, Cohen, and Medalia[13], has a dimension of length and might be related to the distance between aggregates.

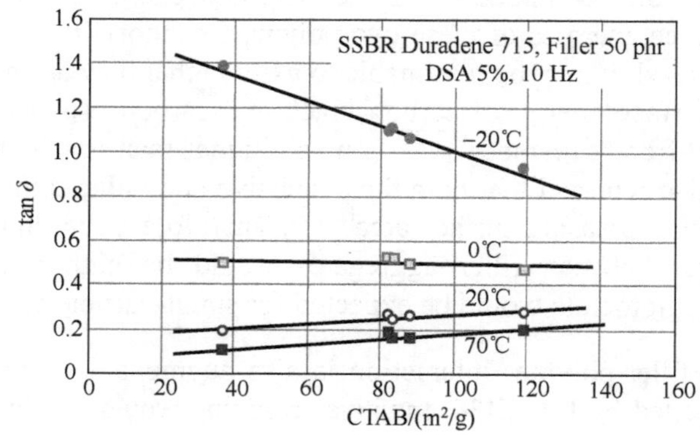


**Figure 8.5** tan $\delta$ as a function of surface area of carbon blacks at different temperatures

#### 8.1.2.2.2 Effect of Structure

Numerous researchers have investigated the effect of carbon black structure on dynamic hysteresis, especially loss tangent. It is generally observed that in comparison with conventional products, carbon blacks with a broad aggregate size distribution, obtained either from the reactor or by blending of products with different aggregate size, generally give lower tan $\delta$ values at temperatures where the rubber is in the rubbery state[24–31]. Kraus[25] hypothesized that this could be due to differences in structure of the secondary filler agglomerates of carbon black. Based on the fact that no significant effect of structure on the surface energy of the carbon black was found[22], Wang et al.[24] mainly attributed the effect of aggregate size distribution on hysteresis to the increase in interaggregate distance by broadening aggregate size distribution leading to a less developed filler agglomerate and, hence, lower tan $\delta$ value.

However, when tan $\delta$ at higher temperatures (beyond room temperature) is plotted as a function of DBP number, which is generally taken as a measure of structure, no significant correlation has been observed between loss tangent and carbon black structure[1,31]. This can be seen from Figure 8.2 as well as from Figure 8.4, in which the results of three N300 series of blacks (N326, N330, and N347) having similar surface area but differing in their DBP number (Table 8.1) are presented. Such a phenomenon may be interpreted in terms of the effect of structure on effective volume fraction and filler agglomeration. Hydrodynamically, the carbon black with higher structure, which gives higher effective volume fraction due to more rubber occluded in the aggregate, would give higher $G'$ as well as $G''$. Therefore, as tan $\delta$ is the ratio between $G''$ and $G'$, the effect of the hydrodynamic term on tan $\delta$ due to structure would be minimal.

On the other hand, from a filler agglomeration point of view, the effect of structure on the filler agglomerate formation is somewhat complex. Due to occlusion of rubber in an aggregate, two opposing effects would be expected: 1) a shorter distance between aggregates for higher structure black (Eq. 3.9 and Eq. 3.19 in Chapter 3) due to the higher effective volume would favor flocculation; 2) a larger effective size of aggregate would reduce the diffusion rate of carbon black in the polymer matrix, which could be an unfavorable factor for filler agglomeration.

It is also established that the bound rubber content of higher-structure carbon blacks is significantly higher than in their low-structure counterparts[31–34] (Figure 5.8). This would lead to a lower diffusion rate of the aggregate in the polymer matrix by increasing effective aggregate size and viscosity of the polymer medium. The higher bound rubber of high-structure blacks has been considered to be mainly due to the easier break down of aggregates during mixing[34]. While a few studies have demonstrated that the primary structure of carbon blacks or aggregate size is appreciably reduced after mixing[35–37], Ban and Hess[38], in their EM comprehensive investigation of carbon black recovered by pyrolysis of the vulcanizate and the carbon gel, indicated this phenomenon is more pronounced in the case of high-structure carbon black (Table 5.2). The same conclusion was also obtained in other studies[29–40]. Obviously, the structure breakdown of carbon black could have at least three consequences:

- increase in polymer-filler interface per unit volume of compound;
- creation of fresh surface;
- reduction in aggregate size.

While the higher bound rubber of higher structure carbon blacks is certainly related to the increased polymer-filler interfacial area, it may also be associated with a greater surface activity or greater adsorption ability of the freshly built surface during mixing. The high surface activity of a fresh surface has been demonstrated by compression of carbon black. Upon compression of carbon blacks at a fairly high pressure (165.5 MPa), surface energy and adsorption energies of chemicals are significantly raised[41,42] with increasing surface area of the carbon black[41]. Gessler considered the new surface to be a "facile new free-radical source resulting in increased carbon black-polymer interaction"[43]. Although the breakdown of filler aggregates during mixing leads to high bound rubber, hence depressing filler agglomeration, the reduction of aggregate size and increase in surface energy would have a positive effect on flocculation. It seems that the effect of structure on filler agglomeration is so complicated that correlation between structure and dynamic hysteresis cannot clearly be identified.

With regard to the effect of structure on low temperature hysteresis, the picture is somewhat different from that obtained at high temperature. As can be seen from Figure 8.3, hysteresis increases with reduction of structure over a large range of strain amplitudes at 0°C. This can also be seen from Figure 8.4, which shows that a higher structure black gives lower tan $\delta$ at a temperature between −20°C and 5°C. For the results obtained at lower temperatures, this effect seems not to be evident probably due to the uncertainty of strain amplitude, as mentioned before. Such an observation may be expected from the effect of different portions of the rubber on dynamic properties in a filled compound, namely, rubber shell, occluded rubber, and trapped rubber. It may be reasonable to believe that, in the filled vulcanizates, there are overlaps among these three rubber types, i.e., part of occluded rubber may also be trapped rubber or rubber shell and vice versa, as schematically shown in Figure 8.6. If the filler agglomeration is similar or if the trapped rubber is about the same for carbon blacks with the same surface area and surface activity; the higher structure black would have more occluded rubber which is not overlapped with trapped rubber and rubber shell, resulting in lower effective volume fraction of rubber matrix. While the non-overlapped occluded rubber may not impart a significant change to the tan $\delta$ at high temperature where the low hysteresis of the polymer matrix contributes only a minimal effect on the overall hysteresis of the filled rubber, it would significantly reduce the hysteresis at low temperature since in the transition zone the free polymer matrix plays a dominant role in energy dissipation.

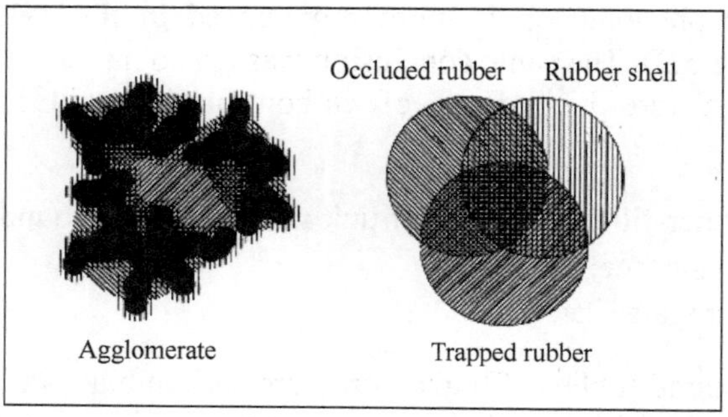


**Figure 8.6** Overlap of different types of rubber related to carbon black

### 8.1.2.3 Effect of Filler Surface Characteristics

It is also known that the dynamic properties of compounds, dynamic hysteresis in particular, as well as their temperature dependence can to a great extent be influenced by filler parameters. The effect of filler morphology, namely, its fineness related to surface area and/or particle size and distribution, and its structure related to its aggregate irregularity of shapes and their distribution, has been investigated and reviewed[1]. However, the importance of surface characteristics in the dynamic properties of the filled vulcanizates has not been emphasized as it should be; although, great numbers of data have been provided in the literature. For example, graphitization of carbon black, which drastically changes the surface physical chemistry and reduces the surface area in a certain degree while leaving its structure basically unchanged[44], leads to an increase in hysteresis[45,46]. A large change in dynamic properties, both modulus and loss factors, can also be achieved via chemical modification of filler surface[47] by which the filler morphology is unlikely to have been altered.

#### 8.1.2.3.1 Effect of Carbon Black Graphitization on Dynamic Properties

When carbon black is heated in an inert atmosphere at very high temperatures, for example, at around 2700°C, all functional groups are decomposed[45,48–53]; hence, the dimensions of the graphitic crystallites are drastically increased[54–57]. The treated product is generally called graphitized carbon black. When mixed into polymer, this carbon black gives a very low modulus at high strain, low tensile stress, and extremely poor abrasion resistance. Such behavior has always been attributed to the lower polymer-filler interaction. However, with regard to dynamic properties, it was found that the elastic modulus[46], Payne effect[12], as well as loss modulus are higher[58] for graphitized blacks than for their ungraphitized counterparts, as is the tan $\delta$ at relatively high temperatures[59]. The pronounced Payne effect and higher hysteresis at relatively high temperatures seem to suggest that a more developed but weaker filler agglomerate is formed with graphitized carbon black. If this were true, then it would be reflected in its effect on low-temperature dynamic hysteresis, i.e., tan $\delta$ should be lower for graphitized carbon black-filled rubber.

This observation has been confirmed with vulcanizates filled with 50 phr of N330 carbon black and its graphitized counterpart N330G. The samples were prepared with two polymer systems, solution SBR Duradene 715 and a blend of functionalized solution SBR NS 116 (70 phr)/NS 114 (30 phr). NS 116 is solution SBR that has 100% of its polymer chain ends terminated with 4,4′-bis(diethylamino)benzophenone and contains tin-coupling, 21% styrene, and 60% vinyl-butadiene. NS 114 is a solution SBR that has 80% of its polymer chain ends terminated with 4,4′-bis (diethylamino)benzophenone and contains tin-coupling, 18% styrene, and 50% vinyl-butadiene. Shown in Figure 8.7 are the temperature sweep results which provide additional supporting evidence for the higher flocculation tendency of graphitized carbon black. At constant strain amplitude (5% DSA) and frequency (10 Hz), the compound with N330G always gives a higher tan $\delta$ at high temperatures and the reverse effect at low temperatures with a crossover point at a certain temperature, depending on the $T_g$ of the polymer. This is a typical feature of a highly agglomerated filler system.

If the effect of graphitization of carbon black on the dynamic hysteresis is attributed to the weaker and more developed structure of agglomerates, the question arises as to why such a filler agglomerate can be formed. It has been reported[22] that both the dispersive and specific components of surface energy are considerably decreased upon graphitization of carbon black, which would lead to a lower attractive potential for filler agglomeration. However, compared with polymer, the surface energy of graphitized carbon black is still very high[22] suggesting that the graphitized carbon black would have a high agglomeration tendency even though it is not comparable to ungraphitized black. On the other hand, it has also been demonstrated that according to the adsorption energy distribution measured by means of inverse gas chromatography (IGC), the high energy sites on the carbon black surface, which are mainly responsible for the reinforcing capability of fillers in elastomers, almost totally disappear during graphitization due to the disappearance of the graphitic basal plain edges, and a decrease in crystal-structure defects[60]. As a result, the bound rubber content of the filled rubber would be drastically reduced[61]. In fact, by graphitization, the bound rubber content of the N330-filled Duradene 715 is reduced from 27.4% to 3.6%. In the case of the NS 116/NS 114 system, while there is 24.5% bound rubber in N330-filled compound, no bound rubber can be detected in the compound with graphitized black. As the bound rubber can substantially increase the effective dimension of the aggregate and raise the viscosity of the polymer matrix, the flocculation rate of the aggregate would be much lower for ungraphitized carbon black. In addition, the bound rubber may be entangled with the polymer matrix which serves as an anchor to further reduce the agglomeration of the carbon black. As a consequence, compared with graphitized black, even though from a thermodynamic point of view ungraphitized carbon black could more readily form a filler agglomerate, the kinetic effect may play a dominant role to govern the flocculation; thus, a more developed filler agglomerate would be expected for graphitized carbon black. On the other hand, as the filler-filler interaction is higher for ungraphitized carbon black, once the filler agglomerate is formed either through joint rubber shell or via direct contact between aggregates, it is stronger than that formed by graphitized aggregates. Therefore, taking the overall response of different factors influencing the filler flocculation into consideration, a weaker but more developed filler agglomerate of graphitized black would eventually be expected.

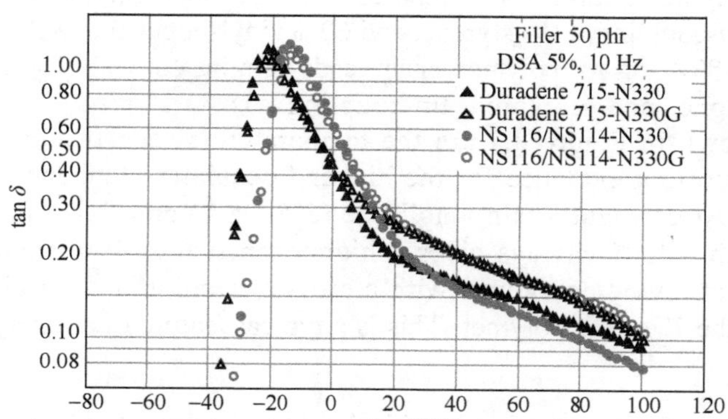


**Figure 8.7** Temperature dependence of tan $\delta$ at 5% DSA and 10 Hz for vulcanizates filled with carbon black N330 and graphitized N330 (N330G). Basic formulations: polymer, 100; filler, 50

It should be pointed out that a bound rubber/entanglement model was proposed to explain the effect of filler on the dynamic properties[62]. In this model, the carbon black affects the dynamic behavior of the filled rubber through its effect on the effective crosslink density of the rubber, which is controlled by the entanglement of bulk rubber with bound rubber. The higher the bound rubber content, the more adsorption sites, the higher the low strain modulus, and the more pronounced the Payne effect would be expected from the bound rubber/entanglement model. Obviously, this model cannot be justified by the higher Payne effect and higher elastic modulus at low strain amplitude for the graphitized carbon black-filled rubber. Compared with ungraphitized carbon black, the population of adsorption sites and bound rubber content are drastically reduced by graphitization of carbon black. Moreover, the concept of bound rubber is related to the uncured rubber. As cited by Medalia and Kraus[63], "in a filler-rubber mix, chains of bound rubber extend far into the polymer matrix where they freely intermix with unabsorbed rubber molecules. On vulcanization, they become part of the network, indistinguishable from (originally) free rubber." Thus, in cured rubber, the bound rubber loses its identity although the polymer segments on and near the filler surface can be immobilized.

#### 8.1.2.3.2 Comparison of Carbon Black and Silica

Precipitated silica, as a white reinforcing filler, has been used in the rubber industry for a long time. Whereas it was able to replace up to 100% of carbon black in shoe sole compounds, its use in tire compounds had been limited to two types of compounds: off-the-road tread compound to improve chipping-chunking resistance and textile and steel cord bonding compound for enhancing adhesive between the cord surface and rubber materials[64]. Even in these two compounds, the silica has to be blended at a low loading, normally 10 phr to 15 phr, with carbon black. The reason that it cannot completely replace carbon black as the main filler in tire compounds, tread compound in particular, is that besides the poor cure characteristics and poor processability, it imparts very low failure properties to the filled rubber. Such behaviors of silica in rubber have always been associated with its weak polymer-filler interaction and strong filler-filler interaction; both are related to the chemistry and physical chemistry of its surface.

Compared with carbon black, whose surface consists of a certain portion of unorganized carbon but mainly graphitic basal planes with some functional groups, mostly oxygen containing groups, located on the edges and crystal defects, the silica surface consists of siloxane and silanol groups. In the sense of surface physical chemistry, based on the adsorption investigation of different chemicals by means of IGC, it has been found that carbon black possesses a high dispersive component of surface energy, $\gamma_s^d$, while that of silica is low. By contrast, the specific or polar component of the surface energy, estimated from the specific interaction between filler surface and polar chemicals, is much higher for silica than for carbon black[22,65]. With regard to the polymer-filler and filler-filler interaction, Wang et al.[22] were able to differentiate these two fillers by means of adsorption energies of model compounds which are related to different components of their surface energies. In principle, as cited earlier, while fillers have polar surfaces, the hydrocarbon rubbers are generally non-polar materials (or very low-polar materials). The higher adsorption energy of non-polar chemicals such as heptane ($C_7$), resulting from higher $\gamma_s^d$ of the filler surface, would indicate a greater ability of the filler to interact with hydrocarbon rubbers. However, the higher adsorption energy of a more highly polar chemical such as acetonitile (MeCN), resulting from the higher $\gamma_s^d$, would be representative of filler aggregate-aggregate interaction in hydrocarbon rubber. Presented in Figure 8.8 are the energies of adsorption of acetonitrile, $\Delta G^{o}_{MeCN}$, plotted against $\Delta G^{o}_{C_7}$, the energies of adsorption of heptane. Evidently, the data for these two types of fillers are located in different regions of the plot, which permits a clear-cut distinction between their surface characteristics. It can be understood that, when $\Delta G^{o}_{C_7}$ is constant, the increase in $\Delta G^{o}_{MeCN}$ would promote incompatibility of the filler with the hydrocarbon polymer thus enhancing aggregate agglomeration. At the same level of $\Delta G^{o}_{MeCN}$, the higher $\Delta G^{o}_{C_7}$ would be representative of a stronger filler-polymer interaction. Moreover, a much higher population of silanol on the silica surface would lead to strong hydrogen bonding between silica aggregates and a stronger filler agglomerate in comparison with its carbon black counterpart. In addition, the lower filler-polymer interaction of silica with hydrocarbon rubber would also result in a lower level of bound rubber in the compound[66,67], which would facilitate filler flocculation. Consequently, the filler agglomerate of silica should be more developed and much stronger in hydrocarbon rubber. In fact, using a silica and carbon black having similar surface areas and structures, Wolff et al.[66,68] have demonstrated in NR that the Payne effect of a silica compound is much higher than for carbon black. This is also demonstrated in Chapter. 7.

It is inevitable that the characteristics of the silica agglomerates will give rise to a strong temperature dependence of tan $\delta$. Shown in Figure 8.9 are the temperature sweep curves obtained at 5% DSA and 10 Hz for solution SBR vulcanizates filled with carbon black and silica[69]. As expected, in the transition zone the silica compound gives considerably lower tan $\delta$ values than the carbon black vulcanizate. In the rubbery state at temperatures beyond 20°C, though the hysteresis is still higher for carbon black mainly due to the energy dissipation during repeat destruction and reconstruction of the filler agglomerates, tan $\delta$ decreases rapidly with temperature due primarily to a reduction of filler-filler interaction as well as filler-polymer interaction. Conversely, it is interesting to note that the hysteresis of the silica-filled rubber increases with increasing temperature, finally showing a crossover point with carbon black at about 90°C. Again, this may be anticipated from strongly and highly constructed filler clusters. As the temperature increases, weakening of the filler-filler interaction would result in an increase in the portion of filler agglomerate which can be broken down and reformed at 5% DSA during cyclic deformation, the main source of energy dissipation at high temperature. This process seems not to reach an equilibrium; in other words, a certain number of agglomerates still remain unchanged under this dynamic strain at a temperature as high as 100°C. This may be true as it is in line with the observation of viscosity measurements. It has been reported[66] that, with the same surface area and comparable structure, Mooney viscosity measured at 100°C, even the minimum torque obtained by means of a Monsanto rheometer at 155°C, is much higher for a silica-filled compound than for its carbon black counterpart, even though the bound rubber content is favorable for the latter. This phenomenon has generally been linked to filler flocculation in the polymer matrix[66].

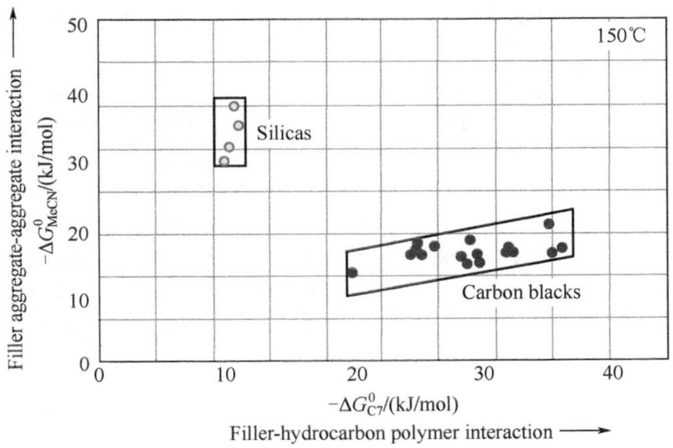


**Figure 8.8** Energies of adsorption of acetonitrile *vs.* those of heptane for carbon blacks and precipitated silicas[22]

In light of above discussion, certain strategies can be suggested for minimizing filler agglomeration of silica in order to obtain a better balance of dynamic properties. Among others, the following two approaches should be taken into consideration, at least from the thermodynamic standpoint:

- change the polymer system to increase the affinity between silica and polymer; and
- modify the surface of silica to increase the compatibility with the given polymer.

Before dealing with these two approaches, we will first discuss the effect of blending two fillers on the dynamic properties of filled rubber.

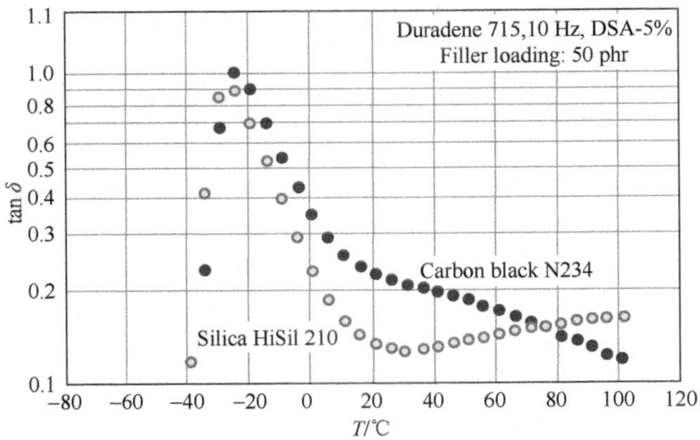


**Figure 8.9** Temperature dependence of tan $\delta$ at 5% DSA and 10 Hz for vulcanizates filled with carbon black N234 and precipitated silica HiSil 210[69].
Basic formulations: polymer, 100 phr; filler, 50 phr

#### 8.1.2.3.3 Effect of Filler Blends (Blend of Silica and Carbon Black, without Coupling Agent)

When two fillers having different surface characteristics are blended, the interactions between aggregates of different fillers would impart an influence on the filler agglomeration in the polymer matrix, hence the dynamic properties of the filled vulcanizates. When other factors such as filler morphology and loading are kept constant, the question is whether the filler agglomeration in a compound loaded with such a blended filler system is more or less developed than any of the single filler systems, or is it between that of the compounds loaded with its parent fillers. Taking $F_1$ and $F_2$ to represent filler 1 and filler 2, the change in adhesive energy, $\Delta W$, for reagglomeration of the two fillers can be estimated schematically as shown in Figure 8.10.

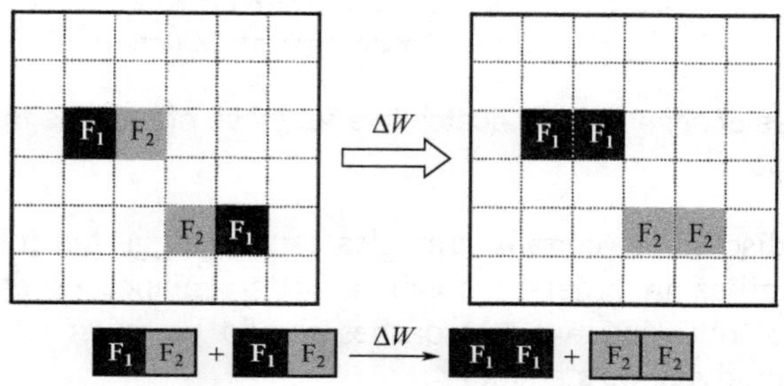


**Figure 8.10** Change in energy associated with reagglomeration process in a filler blend system

Similar to the derivation of adhesive energy in the agglomeration process of fillers (see Eqs. 3.10–3.14), one may immediately obtain:

$$\begin{aligned}\Delta W &= 2\left[(\gamma_{f_1}^{d})^{1/2} - (\gamma_{f_2}^{d})^{1/2}\right]^2 + 2\left[(\gamma_{f_1}^{p})^{1/2} - (\gamma_{f_2}^{p})^{1/2}\right]^2 \\ &+(W_{f_1}^{h} + W_{f_2}^{h} - 2W_{f_1f_2}^{h}) + (W_{f_1}^{ab} + W_{f_2}^{ab} - 2W_{f_1f_2}^{ab})\end{aligned} \quad (8.1)$$

where $\gamma^{d}$ is the dispersive component of the filler surface energy, $\gamma^{p}$ is the polar component from dipole-dipole and induced dipole interactions between molecules, $W^{h}$ is the adhesive energy due to hydrogen bonding, and $W^{ab}$ is the adhesive energy due to acid-base interaction. The terms $f_1$ and $f_2$ refer to fillers 1 and 2 and $f_1 f_2$ to the interaction between filler 1 and 2. This equation suggests that when an elastomer is charged with two different fillers, the two fillers could form a randomly-joint-filler agglomerate in the polymer matrix only if the two fillers have the exact same surface energy characteristics both in intensity and nature, i.e., $\gamma_{f_1}^{d} = \gamma_{f_2}^{d}$, $\gamma_{f_1}^{p} = \gamma_{f_2}^{p}$, $W_{f_1}^{h} = W_{f_2}^{h} = W_{f_1f_2}^{h}$, and $W_{f_1}^{ab} = W_{f_2}^{ab} = W_{f_1f_2}^{ab}$, hence, $\Delta W = 0$. The two fillers could preferentially form a joint filler agglomerate only if the adhesive energies between these two fillers due to hydrogen bonding, $W_{f_1f_2}^{h}$, acid-base interaction, $W_{f_1f_2}^{ab}$, and/or other specific interactions are so high that they are able to offset the adhesive energies between the same categories of fillers, resulting in $\Delta W < 0$. These conditions are unlikely to be satisfied, especially for the fillers generally used in rubbers such as carbon black and silica, so that intuitive reasoning leads one to the expectation that two types of filler agglomerates or a mixture of two different filler agglomerates would most probably be formed, at least from the thermodynamic point of view because $\Delta W$ is generally positive.

An example of the effect of filler blends on the dynamic properties of filled vulcanizates was reported with N234 carbon black (surface area 119 $m^2/g$) and HiSil 210 silica (surface area 150 $m^2/g$)[226]. The total filler loading of the fillers was 50 phr for all filled compounds. Figure 8.11 summarizes the effect of filler composition on $G'$ at a DSA of 0.2% and Payne effect represented by $\Delta G'_{0.2-120}$, the difference in $G'$ at DSA of 0.2% and 120%. With changing carbon black/silica ratios, the $G'$ at low strain amplitude and the Payne effect do not follow a linear function. This shows that with an increase of either carbon black or silica content in the blend systems, both $G'$ and the Payne effect decrease in comparison with those of their single filler system, and then increase. In other words, instead of following a simple additive function derived from the results of the two single fillers, the experimental points corresponding to the blend systems always lie well below the additive line.

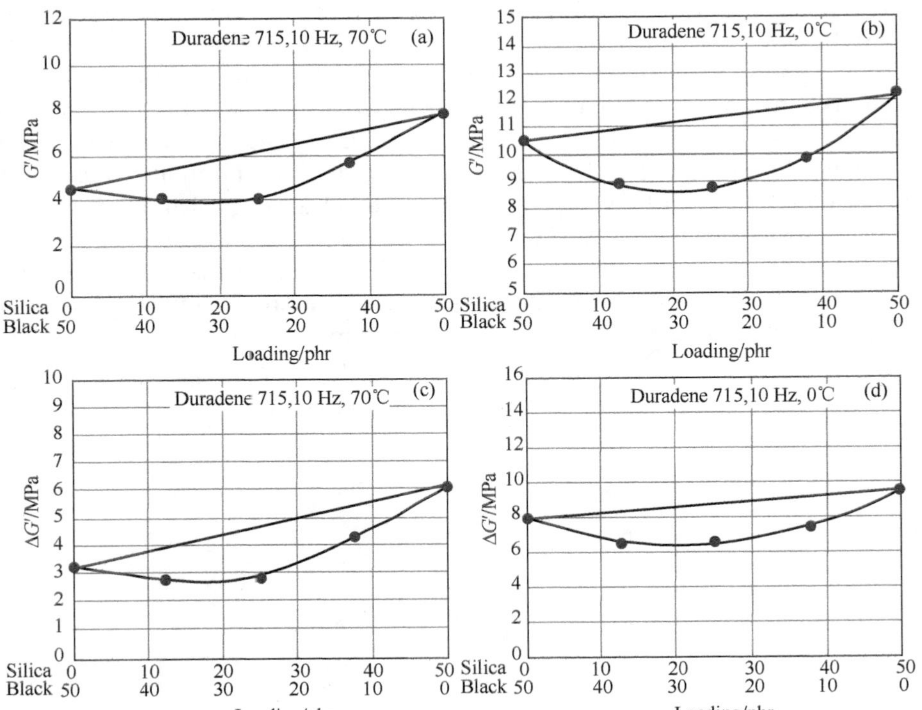


**Figure 8.11** $G'$ at 0.2% DSA and $\Delta G'_{0.2-120}$, the difference in $G'$ at DSA of 0.2 and 120%, as a function of blending ratio of carbon black and silica. Basic formulations: SSBR Duradene 715, 100; filler, 50

The deviation of the Payne effect from the additive function of the two single fillers suggests less developed filler agglomeration is formed in the filler blend compounds. These results are most easily interpreted on the basis of Eq. 8.1. On the one hand, the interaction between silica and carbon black aggregates is weaker than that between any of these two filler aggregates per se. On the other hand, although the surface area of silica is larger than carbon black, which may cause a monotonic decrease in inter-aggregate distance with increase in silica content, the inter-aggregate distance between the same types of fillers may also play a role in filler agglomeration, since the total filler loading is identical for both single and blend systems. In the latter systems, the mean distances between the same types of aggregates which have higher attractive potential are larger than that of any of their single filler counterparts. Accordingly, in view of the thermodynamics and kinetics of filler agglomerate formation, less flocculation of the aggregates would be expected in the blend, as would the $G'$ at low strain amplitudes and the Payne effect.

The effect of a blend of fillers on filler agglomerate formation would also be expected to have significant consequences on the dynamic hysteresis of the vulcanizates. Presented in Figure 8.12 are the strain dependencies of tan $\delta$ measured at 70°C. The vulcanizate filled with carbon black shows a single peak at DSA of about 8% and the tan $\delta$ of the silica compound does not reach a maximum as far as the maximum strain amplitude applied. In comparison, the vulcanizates filled with blended filler systems seem to have two peaks (or shoulders), one being coincident with that of carbon black and the other not appearing at the maximum strain used. The latter was also observed for the pure silica compound. If the tan $\delta$ at low strain is also taken into consideration, there seem to exist three filler agglomerates in the vulcanizates charged with blended filler systems, namely, carbon black filler agglomerates, silica agglomerates, and silica-carbon black filler agglomerates. The latter is so weak, due to the lower filler-filler interaction, that it would easily breakdown and reform at low strain amplitude, leading to a higher hysteresis than those of the compounds filled with the individual

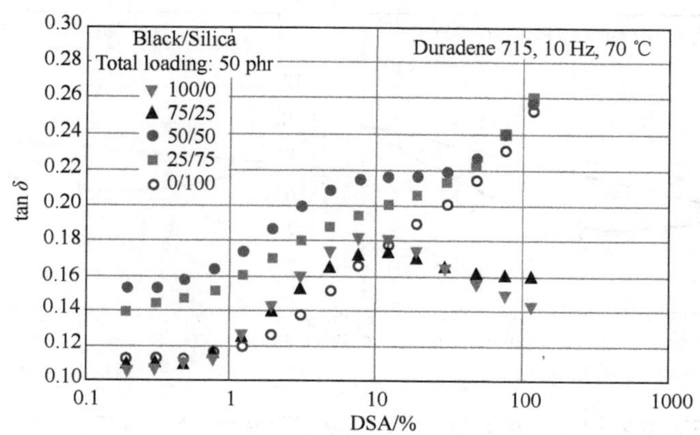


**Figure 8.12** Strain dependence of tan $\delta$ at 70°C and 10 Hz for vulcanizates filled with N234/silica HiSil 210 blends. Same compounds as Figure 8.11

fillers. The tan $\delta$-DSA curves of the blended filler systems seem to be a result of the superposition of these three filler agglomerates upon dynamic strain.

The above observation may be strengthened by the temperature dependencies of tan $\delta$ measured at 5% DSA and 10 Hz, shown in Figure 8.13. The experimental results can be summarized as follows:

- at lower temperatures, the carbon black rich blended system gives higher hysteresis compared with the pure black-filled rubber. The tan $\delta$ of the silica rich blend is between those of the individual fillers;
- at higher temperatures, the tan $\delta$ of the silica rich blend increases with increasing temperature, but this phenomenon is significantly less pronounced than for the silica compound. In the case of the carbon black-rich blended system, the hysteresis decreases with temperature, but at a lower rate relative to the carbon black-filled vulcanizate.

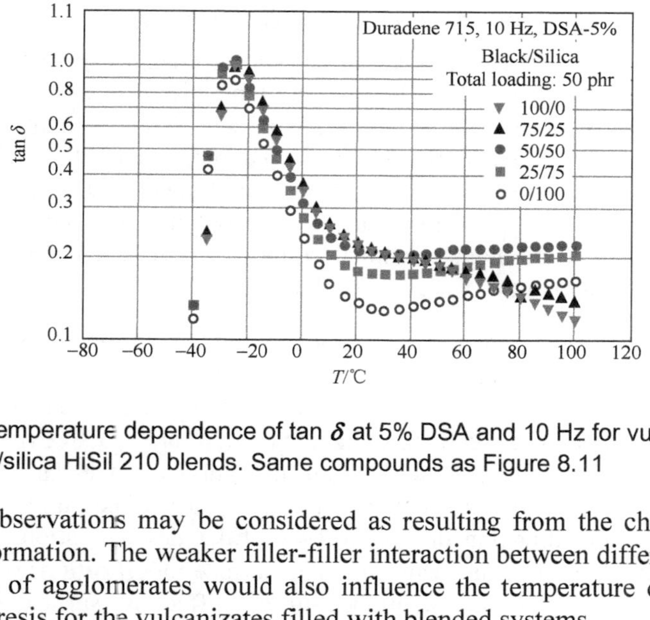


**Figure 8.13** Temperature dependence of tan $\delta$ at 5% DSA and 10 Hz for vulcanizates filled with N234/silica HiSil 210 blends. Same compounds as Figure 8.11

All of these observations may be considered as resulting from the changes in filler agglomerate formation. The weaker filler-filler interaction between different fillers and multiple types of agglomerates would also influence the temperature dependence of dynamic hysteresis for the vulcanizates filled with blended systems.

#### 8.1.2.3.4 Effect of Surface Modification of Silica

Surface modification of silica is one of the most effective approaches for changing surface characteristics to meet application requirements. Many approaches published in the literature can be used. Two approaches are frequently practiced in the rubber industry; namely, surface chemical modification and physical modification by adsorption of some chemicals on the filler surface.

### A. Effect of Surface Modification by Physical Adsorption

When certain chemicals are added to a silica-filled compound, they may be strongly adsorbed on the surface via dispersive interaction, polar interaction, hydrogen bonding, and acid-basic interaction. Examples include glycols, glycerol, triethanolamine, secondary amines, as well as diphenyl guanidine (DPG) or di-*o*-tolylguanidine (DOTG) and diethylene glycol (DEG)[70,71]. Generally, the polar or basic groups of these materials are directed towards the silica surface and the less polar or alkylene groups towards the polymer matrix, thereby increasing affinity with the hydrocarbon polymer. Consequently, the filler agglomeration of silica can be substantially depressed resulting in better dispersion in the polymer matrix, lower viscosity of the compound, and lower hardness, and better temperature balance of hysteresis. However, since the polymer-filler interaction can be drastically weakened, this approach cannot be applied in tire application, especially for tread compound, as the wear resistance will be greatly reduced.

### B. Effect of Chemical Modification on the Silica Surface

With chemical modification, the filler surface can be changed in such a way that it is tailored to its application. For the filler used in rubber products, two types of chemicals have been used for surface modification: grafts of chemical groups on the filler surface to change the surface characteristics and grafts that may react with polymer. The latter are frequently called coupling agents or bifunctional coupling agents as they provide chemical linkages between the filler surface and polymer molecules. The former is sometimes referred to as monofunctional coupling agents even though no chemical reaction with the polymer takes place with these grafts.

*1. Surface modification of silica with monofunctional chemicals:* Surface chemical modification to change the surface chemistry of silica[72] has been investigated, and grafted silica has been found to have many applications in different fields. According to the thermodynamic analysis of filler agglomeration, one would expect that grafting of oligomers or polymer chains would result in significant changes in the nature and intensity of the surface energy of silica, providing a surface similar to that of the polymer, thus eliminating the driving force of filler agglomeration. Although such consideration has not yet been realized, some surface modifications of silica provide evidence to demonstrate the importance of the effect of surface energy on the flocculation of filler particles in the polymer.

Much has been done in this area[73–80]. It has been found that although the modification of silica by monofunctional chemicals may greatly improve the microdispersion of the filler and provide low hysteresis materials, the lack of polymer-filler interactions would result in a lower static modulus and poorer failure properties, particularly abrasion or wear resistance.

*2. Surface modification of silica with bifunctional chemicals (coupling agents):* The bifunctional chemicals referred to here are a group of chemicals which are able to establish molecular bridges at the interface between the polymer matrix and filler surface. These chemicals are generally defined as coupling agents to enhance the degree of polymer-filler interaction and hence impart improved performance properties to the filled materials. The chemicals belonging to this group are numerous and include titanate-based coupling agents[81], zirconate-based coupling agents[82], and other metal complex coupling agents[81,83] which have found application in inorganic filler reinforced polymer composites. Perhaps the most important coupling agents for inorganic filler modification, silica in particular, is the group of bifunctional organosilanes with the general formula as $X_{3-m}R_mSi(CH_2)_nY$, where X is a hydrolyzable group, such as halogen, alkoxyl, or acetoxyl groups, and Y is a functional group which itself is able to chemically react with the polymer either directly or through other chemicals. It may also be a chemical group able to develop a strong physical interaction with polymer chains. For the Y groups, important silane coupling agents include amino, epoxy, acrylate, vinyl, and sulfur-containing groups, such as mercapto, thiocyanate, and polysulfide[84–88]. The bifunctional silane coupling agents most often contain three ($m$ = 0) X groups and the functional group Y is generally in the $\gamma$ position ($n$ = 3). Practically, when silica is incorporated in a hydrocarbon rubber, the most popular and effective coupling agents are $\gamma$-mercaptopropyltrimethoxy silane and bis(3-triethoxysilylpropyl)-tetrasulfide (TESPT), known as Si-69. In fact, as far as the commercialized coupling agents are concerned, TESPT is the most commonly used silane enabling silica to be applied to tire compounds, tread compounds in particular. The application of TESPT is also the key factor for silica being a successful replacement of carbon black in the tread compound of the "green tire"[89,90].

With TESPT modification of silica, either pre-modified[91] or modified in situ[92], i.e., the coupling agent is added to the compound in the mixer, the ethoxy groups of TESPT react with silanol groups on the silica surface, either directly or after a preceding hydrolysis[93,94]. During the second step, the tetrasulfide group is split due to heat treatment and/or the influence of the sulfur/accelerator system and reacts with the rubber chains by forming mono-, di-, and polysulfidic covalent bonds during mixing and vulcanization of the compound.

While the introduction of covalent bonds between silica and rubber imparts a stronger rubber-filler interaction which leads to considerable improvements in the system's failure properties, especially abrasion/wear resistance, the silanization of silica with TESPT is also able to drastically reduce the filler agglomeration tendency by the following mechanism:

- Reducing the filler surface energy, both dispersive and specific components, not only due to the reduction in number of silanol groups which are highly polar groups but also due to making the remaining silanols less accessible to the rubber chains by means of a TESPT layer. By nature, silane grafts are low in energy and polarity[73,80].
- Increasing the bound rubber content caused by the coupling reaction between the filler surface and polymer and perhaps in combination with very slight crosslinking of the polymer during mixing, especially at higher temperatures and longer mixing times. This would substantially raise the viscosity of the polymer matrix and provide a certain number of anchors (bound rubber) in the polymer matrix. In turn,

this would prevent the filler aggregates from flocculating even though the overall viscosity of the compound may be lower than that of an unsilanized compound due to the considerably reduced filler agglomeration[67,68].

The increase in bound rubber due to the coupling reaction has been confirmed. Basically, the bound rubber content is unchanged upon ammonia treatment which is able to eliminate the bound rubber caused by physical adsorption but unable to cleave chemical linkages between the silica and polymer[34,68,95]. The slight crosslinking of the polymer during mixing at relatively higher temperatures, such as 160°C, has been demonstrated by gel formation of a gum compound of SBR (a gel-free polymer). Upon heat treatment at 160°C, the gel content of a gum SBR compound containing only 3 phr TESPT increases very rapidly after 5 minutes, which is certainly related to slight crosslinking by TESPT, a sulfur donor. This reaction would take place in the mixer depending on the temperature and mixing time.

With coupling agent, the silica agglomeration is substantially depressed and the dynamic properties are improved. While the Payne effect is drastically reduced to a level that is even much smaller than the carbon black counterpart[47], the dynamic hysteresis is greatly improved. Shown in Figure 8.14 is a comparison in tan $\delta$ measured at 5% DSA and 10 Hz for the modified silica and carbon black compounds which are typically used for passenger tire treads. Both compounds were solution SBR (Duradene 715)/BR blends (75/25) with 25 phr oil and 75 phr fillers, either silica (Zeosil 1165, Rhône-Poulenc) or carbon black N234. In the case of the silica compound, 12 phr of coupling agent X50S (a mixture of 50% TESPT and 50% carbon black N330) was added as coupling agent. Obviously, there is a crossover point at about −3°C, which shows that at both high and low temperatures the hysteresis properties of the modified silica compounds are greatly improved. These results, together with lower rolling resistance, better wet skid resistance, and comparable wear resistance relative to carbon black compounds, are certainly the basis for the improved tire performance properties of silica-filled tire compounds.

#### 8.1.2.3.5 Effect of Surface Modification of Carbon Black on Dynamic Properties

##### A. Difference in Chemical Modification Efficiency between Silica and Carbon Black

If silica outperforms carbon black due to surface modification, the question naturally arises as to whether surface modification of carbon black can be used to enhance its performance. In fact, a lot of work has been done in this direction to improve the reinforcing ability of carbon blacks in hydrocarbon polymers. Most of the modifications focus on the development of new coupling agents. Among the most effective chemicals are benzofuroxan (BFO)[96], *N,N'*-bis(2-methyl-2-nitropropyl)-1,6-diaminohexane (Sumifine 1162)[97], and more recently, *p*-aminobenzenesulfonyl azide (amine-BSA)[98]. Probably more effort has been made to use silane coupling agents – especially TESPT–to increase filler-polymer interaction, reduce filler-filler interaction, and improve rubber reinforcement[99–101]. Although certain successes have been achieved it appears that the improvement of dynamic properties, hysteresis, in particular, of carbon black-filled rubber by coupling agents was not as appreciable as in the case of silica-filled vulcanizates. Several factors may be considered to be responsible for this difference. As far as the coupling reaction is concerned, besides the nature of coupling agents, this reaction is determined by the filler surface chemistry, namely,

- the types of chemically functional group;
- the reactivity of the functional groups with a given coupling agent;
- the concentration of the functional groups on the filler surface; and
- the distribution of the functional groups over the filler surface.

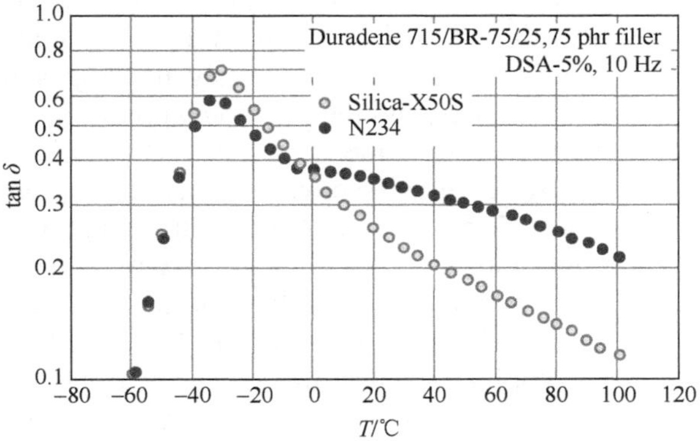


**Figure 8.14** Temperature dependence of tan $\delta$ at 5% DSA and 10 Hz for vulcanizates filled with silica/TESPT (X50S) and carbon black N234[69]. Basic formulations: SSBR Duradene 715, 75; BR, 25; filler, 75; oil, 25

Unlike the silica surface which is characterized by a uniform layer of siloxane and various types of silanol groups (isolated, geminal, and vicinal)[72], carbon black surfaces contain not only hydrogen (mostly in aromatic rings) but also a number of different oxygen-containing groups, namely, phenol, carboxyl, quinone, lactone, ketone, lactol, and pyrone[102]. For a given coupling agent the different groups would have different reactivity. For example, with TESPT all types of silanols on the silica surface have about the same condensation reactivity with the ethoxyl groups. However, in the case of carbon black, although the silanization chemistry of the carbon black surface has not been well established, the same coupling reactivity would not be expected for the various functional groups since their chemical nature differs greatly. Such consideration may be confirmed by the silane coupling reaction of oxidized carbon blacks. The greater effectiveness of TESPT with nitric acid-oxidized carbon black compared to its gas phase ozone-oxidized counterpart suggests a higher reactivity associated with the carboxyl and phenol groups as the concentration of such types of oxides is much higher on the nitric acid-treated carbon black relative to the ozone-oxidized product. With other coupling agents, the reactive groups are either carboxyl, phenyl[97,98], or lactone[98].

With regard to the concentration of functional groups, if the surface oxygen content of carbon black is normalized to a reactive oxide such as carboxyl or phenol for a typical furnace carbon black like N220, the surface concentration of the functional groups would be about 1~2-COOH/$nm^2$ or 2~4-OH/$nm^2$[2][51]. In fact, Rivin[52] reported about 3-OH/$nm^2$ and 0.05-COOH/$nm^2$ on N220 carbon black surface. This concentration of the reactive functional groups is much lower than that for silica. For most of the precipitated silicas used in the rubber industry, the surface concentration of silanol groups varies from 4–7/$nm^2$[2][103].

The difference in effectiveness of the coupling modification between carbon black and silica may also be associated with their microstructure. Silica is an amorphous material. The functional groups (i.e., the isolated, geminal, and vicinal silanols) are randomly located on the filler surface (Figure 8.15). Alternatively, the aggregates of carbon black consist of quasi-graphitic crystallites, and the functional groups are located only on the edges of the graphitic basal planes of the crystallites as shown in Figure 2.1. This suggests that besides the difference in reactivity and concentration of the functional groups, the functional group distribution on the two filler surfaces is also different. Obviously, this would result in different surface coverage of the coupling agent grafts. The silanols are randomly distributed on the silica surface so that the coupling grafts are spread uniformly over the surface which leads to better surface coverage. In the case of carbon black, the coupling agent grafts are located only on the edges of the graphitic basal planes which results in poorer surface coverage. The higher concentration of the reactive functional groups and their random distribution over the surface would be a key advantage of silica over carbon black for the coupling reaction.

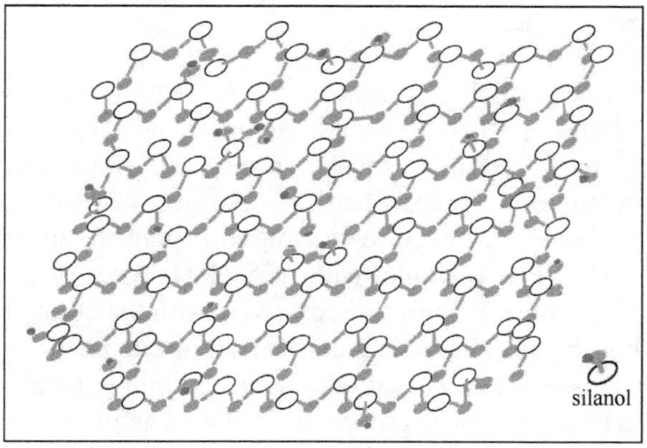


**Figure 8.15** Surface chemistry of silica

In light of the above discussion, certain strategies can be suggested in order to be able to obtain a better and more efficient coupling of carbon black; namely, developing a new coupling agent and/or modifying the carbon black surface for coupling reaction. Any new effective coupling agent for carbon black-filled rubber composites, besides having a high reactivity with the polymer molecules at the processing temperature of the rubber compound and an ease of reaction with the carbon black surface via the functional groups at the edges of the graphitic basal planes, must also easily react on the basal plane in order to obtain a random and uniform distribution of the coupling grafts over the surface to provide optimum coverage. The coupling efficiency for given coupling agents may also be improved if the surface chemistry as well as the microstructure of the surface can be modified.

### B. Surface Modification of Carbon Black for Coupling Reaction

For a coupling agent which can only react with certain functional groups, for example, TESPT and *N,N'*-bis(2-methyl-2-nitropropyl)-1,6-diaminohexane[97] with oxygen groups, there are two approaches from the filler point of view which may be used to increase coupling efficiency and improve coupling graft coverage:

- reduce the size of the graphitic crystallites, or
- increase the concentration of the active functional groups for a given coupling agent.

The reduction in size of the graphitic crystallites would increase the density of the crystallite edges and defects, and in parallel improve the functional group distribution. Several methods may be applied to change the surface microstructure to reduce the basal plane size, including:

- changes in reactor design and processing conditions of carbon black production,
- use of plasma treatment,
- oxidation of the carbon black, and
- surface treatment with silicon compounds.

*1. Carbon black production parameter changes:* Crystallite size increases with growth of the particles (or aggregates)[22,104–106]. The reduction of crystallite size may be achieved during the formation of carbon black in the reactor by proper reactor design and adjusting the processing parameters. This has been considered an effective approach to increase surface activity in spite of the fact that (a) it has limitations for tire-grade carbon black because of the need for ease of rubber processing of the compound, and (b) there is a trend that the amorphous or unorganized portion of the carbon increases with increasing surface area. With the latter, the role played by the high content of unorganized carbon of fine particle black in rubber reinforcement is not well understood. If the particles are very fine, they may have a high density of crystallite defects on the filler surface as the dimensions of the crystallite decreases with particle size[22,104]. In the downstream processing, after quenching, more oxygen-containing groups may be created on the surface. However, this results in very poor processability in terms of mixing and very poor dispersion which along with the increased cost hinders the practical application of this approach.

*2. Plasma treatment:* It has been reported that upon treatment with certain plasma (e.g., air, ammonia, and argon) the finely organized structure of a graphitized carbon black can be destroyed, resulting in a dramatic increase in the number of defects on the graphitic surface[107]. Such an observation can also be applied to ungraphitized carbon black. Since the plasma bombardment gives no preference to the different surface structures, neither the graphitic basal plane nor the disordered carbon portion, the corrosion may take place randomly over the carbon black surface. In addition, by properly selecting the reactive gas, some chemical groups may simultaneously be grafted on the surface providing an active surface for a coupling reaction[108]. With a low-temperature vacuum plasma process using oxygen and air as treating gases, Takeshita et al.[108] were able to introduce a large number of -OH groups on the carbon black surface. The compounds containing the treated carbon blacks exhibit substantially lower hysteresis at room temperature in comparison with their untreated counter-parts when compounded with the same amount of silane coupling agents.

Although this approach seems to be very attractive, its acceptability for large volume production is questionable due to its high cost and low efficiency.

*3. Oxidation of carbon black:* With regard to the oxidation approach either by gas phase oxidation or through liquid phase oxidation, the carbon black is degraded by oxygen attack resulting in different degradation products, mostly $CO_2$[109]. Consequently, the size of the graphitic basal plane would be reduced and the concentration of oxygen containing groups would considerably increase. When oxidized carbon black is used in combination with silane coupling agents, the heat generation measured by tan $\delta$ at 30°C can be considerably reduced[108]. This is certainly related to the higher efficiency of coupling with the surface oxygen-containing groups, which leads to good microdispersion of the filler, and hence to lower hysteresis of the compound.

However, it should be pointed out that as the well-organized crystallites are less susceptible to oxidative attack than the amorphous portion, a developed micropore surface can be formed before the crystallite dimension decreases significantly[109–111].

#### 8.1.2.3.6 Carbon/Silica Dual Phase Filler

From the carbon-black-surface modification point of view, introduction of some compounds or hetero-atoms on the carbon black surface, both physically and/or chemically, via co-fuming or a symbiotic process in the carbon black reactor would be another approach to reduce the crystallite dimension, and hence to increase the population of the crystal defects on the surface. In addition, if the dopant is properly selected and the doping reaction can be well controlled, some coupling reactive compounds may be doped on the surface. Both a high concentration of surface defects in the carbon structure and a high reactivity of the doped domain would greatly enhance the coupling reactions. In fact, the carbon/silica dual phase filler developed by Cabot[112] is a successful example of this approach.

The carbon/silica dual phase filler (CSDPF), as it is named, consists of two phases, the carbon phase with a finely divided silica phase (domains) dispersed therein[113]. The development and commercialization of this filler is one of the most important events that has occurred in the carbon black industry since the introduction of the oil furnace process in 1942. While the surface area, structure, and surface activity can be adjusted in a general sense by changing the carbon black production parameters, the introduction of the silica phase adds an additional dimension to conventional carbon black. Thus, the filler properties are altered to meet a variety of performance requirements for rubber products. The changes in silica content, domain size, distribution in aggregates, and surface chemistry would greatly affect filler-filler and polymer-filler interactions and reactivity with coupling agents, and hence would significantly impact the properties of filled rubber. In fact, when added to hydrocarbon rubber, this filler is characterized by higher filler-polymer interaction in relation to a physical blend of carbon black and silica, and lower filler-filler interaction in comparison with either conventional carbon black or silica having comparable surface areas[113].

Evidence of the higher filler-polymer interaction with the CSDPF was shown by the higher bound rubber content of a series of dual phase fillers compared with those of carbon black/silica blends having identical silica contents[113]. The higher surface activity of this filler was attributed to the change in surface microstructure of the carbon black phase. As mentioned before, upon introduction of a foreign substance, more surface defects may be formed in the carbon phase in the graphitic crystal lattice and/or smaller crystallite dimension which has been proposed as the active center for rubber adsorption. This consideration was based on qualitative evidence from a scanning tunneling microscopic investigation. It was found that the surface topographies of a dual phase black are considerably different from conventional carbon black. The CSDPF shows a statistically smaller area of well-organized carbon structure (quasi-graphitic) in comparison to its conventional carbon black counterpart[113].

The ease of mixing a CSDPF, indicated by lower torque and shorter filler incorporation times in comparison with its silica and carbon black counterparts, appears to suggest lower filler-filler interactions between doped aggregates than either between carbon black aggregates or between silica aggregates[113]. However, good evidence can be obtained from the strain dependence of the filled vulcanizates as shown in Figure 8.16, which includes the results of Duradene 715 with 50 phr CSDPF with 3.55% silicon content, its undoped counterpart with similar surface area and structure, as well as Hi-Sil 210 silica. Although from the chemical composition point of view doped black is between carbon black and silica, what is actually observed here is that the doped carbon black gives the lowest Payne effect, with still lower values observed when coupling agent TESPT is employed. The low Payne effect of uncoupled CSDPF, in relation to the other two fillers, is mainly related to the lower filler-filler interaction between aggregates[113]. This behavior may involve multiple mechanisms, but it is most readily interpreted on the basis of lower interaction between filler surfaces having different surface characteristics, surface energies in particular. According to Eq. 8.1, when CSDPF aggregates are dispersed in a polymer matrix, the interaction between the silica domains and the carbon phase of the neighboring aggregates should be lower than that between both carbon black and silica aggregates. In addition, the portion and probability of the aggregate surface being able to directly face the same category of surface from a neighboring aggregate is substantially reduced by doping. Consequently, there would be less hydrogen bonding, the main cause of the high filler-filler interaction between silica aggregates and between silica domains on neighboring aggregates since their average interaggregate distance would also be greater. Also, the higher surface activity of the carbon phase of CSDPF which results in higher bound rubber may, from the point of view of flocculation kinetics, contribute to a less developed filler agglomeration.

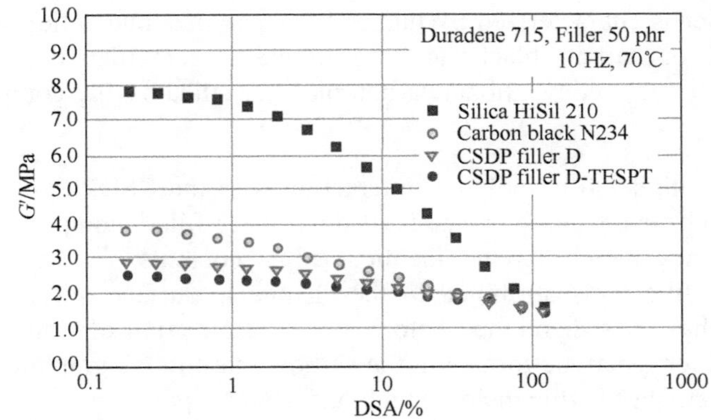


**Figure 8.16** Strain dependence of $G'$ at 70°C and 10 Hz for vulcanizates filled with silica HiSil 210, carbon black N234, CSDP filler, and CSDP filler/TESPT[113]. Basic formulations: SSBR Duradene 715, 100; filler, 50

In view of the depressed filler agglomeration in CSDPF-filled rubber, an improved temperature balance of dynamic hysteresis would be expected. It was reported that for all types of elastomers used in tire tread compounds, the tan $\delta$ at 70°C is dramatically reduced by silica doping even at a silicon level of 3.55%, and further suppression of tan $\delta$ is observed with TESPT-coupling[113]. On the other hand, the hysteresis at low temperatures is increased by use of a dual phase filler, especially in combination with coupling agent TESPT. This observation can be verified by the tan $\delta$-temperature curves of CSDPF filled vulcanizates versus their conventional black-filled counterparts for a variety of polymers (Figure 8.17). It is clearly shown that at the test condition indicated, i.e., DSA of 5% and frequency of 10 Hz, there exists a crossover point between the tan $\delta$-temperature curves of CSDPF and its unmodified counterpart for all polymer systems. While the temperature of the tan $\delta$ peak is not changed by the filler modification, the height of the maximum is increased and the hysteresis at high temperatures is depressed by doping and coupling. On the other hand, the crossover point varies from polymer to polymer, showing a strong dependence on polymer $T_g$. This can be seen clearly from Figure 8.18 in which the crossover temperatures are plotted against glass transition temperatures. For this particular formulation and processing conditions, there is a critical $T_g$ for the polymers below which higher hysteresis of the CSDPF filled vulcanizates at 0°C in comparison with the undoped black cannot be obtained. For this particular system and test conditions, this critical temperature is around −35°C.

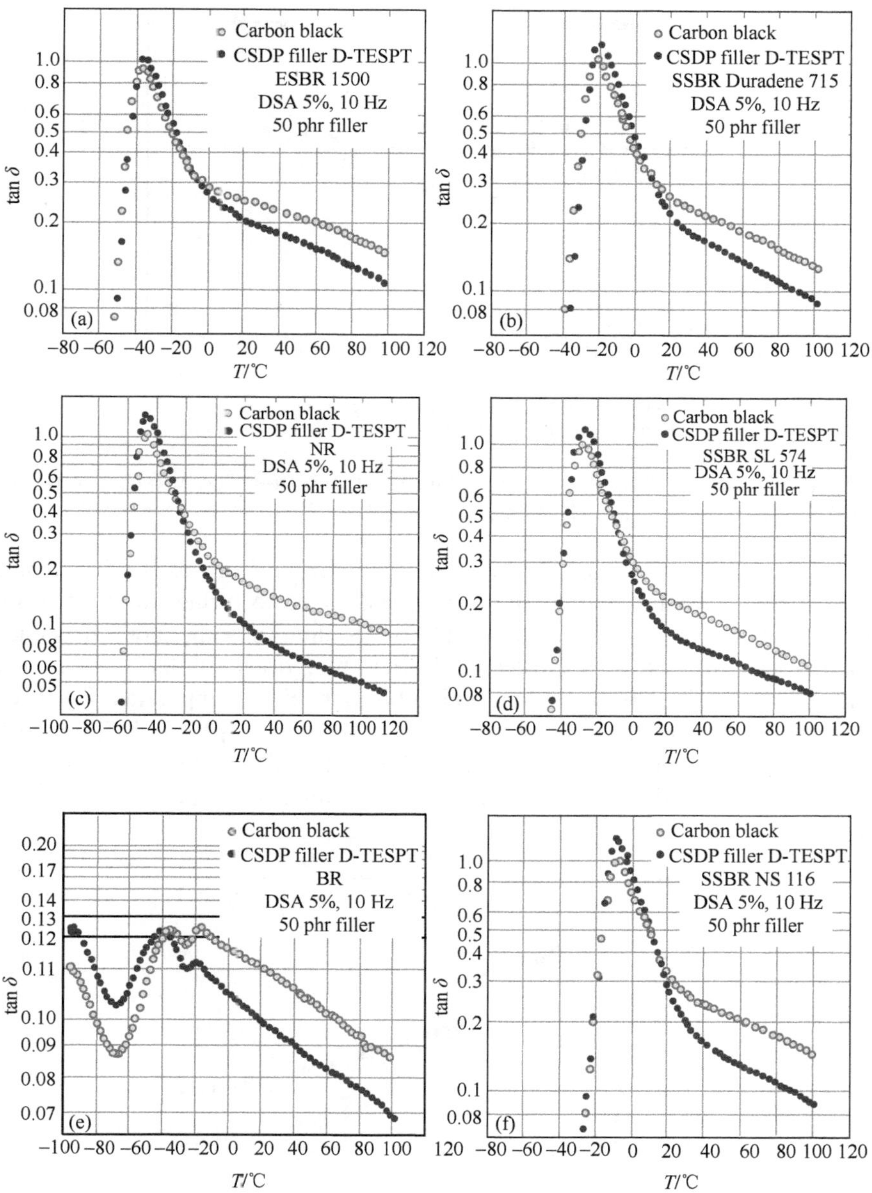


**Figure 8.17** Comparison of temperature dependence between carbon black and CSDPF/TESPT filled rubbers[113]. SL 574 is a tin coupled solution SBR with styrene 15%, vinyl 57%

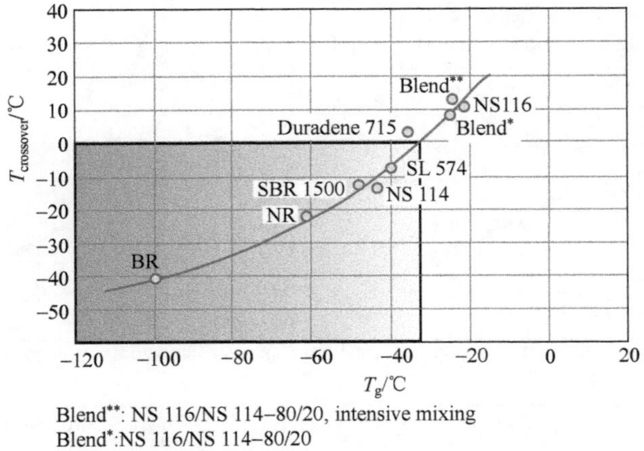


**Figure 8.18** Crossover temperature derived from vulcanizates filled with carbon black and CSDPF/TESPT as a function of $T_g$ for a variety of rubbers[113]

On the other hand, as the filler-filler interaction of CSDPF per se is lower than silica and carbon black, when the coupling agent is employed to further improve hysteresis of the compound, less coupling agent should be required to bring about the same level of hysteresis as for a silica compound. With formulations similar to those used in passenger tire treads (80 phr filler and 32.5 phr oil), compounds filled with a commercialized CSDPF A (silicon content 4.88%) with 2 phr TESPT, gave lower hysteresis results at high temperatures and comparable hysteresis results at low temperatures compared to silica compounds containing 6.4 phr TESPT[114] (Figure 8.19).

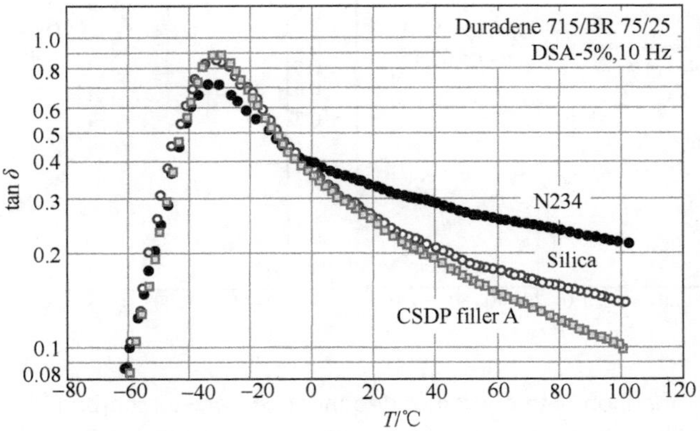


**Figure 8.19** Temperature dependence of tan $\delta$ at 5% DSA and 10 Hz for vulcanizates filled with silica/TESPT, carbon black N234, and CSDP filler/TESPT[114]. Basic formulations: SSBR Duradene 715, 75; BR, 25; filler, 80; oil, 32.5

#### 8.1.2.3.7 Polymeric Filler

Based on the Equation 3.4, the best way to eliminate the driving force for filler agglomeration is to make the filler surface having the same surface energy as the polymer. There are two approaches for this application: one is to modify the filler surface by grafting polymer chains on filler surface, and another is to make the filler particles with the same chemical composition as the polymer used in the compounding.

With regard to the first approach, much work has been done to introduce polymer chains on the filler surface – for carbon black in particular – either by a grafting reaction or by polymerization in situ. A few comprehensive reviews have been published not only for carbon black[115–117] but also for inorganic fillers[118]. Polymer-coated fillers show good dispersion in polymers of similar structure. However, most of the application studies are related to the effect of surface modification on the static mechanical and electrical properties of the composites. Very few have dealt with its effect on dynamic properties.

With the second approach, it is very difficult to obtain a filler having exactly the same composition as the rubber but with a significantly different modulus. However, it may be possible to prepare fine particulate materials which are close in chemical composition to the rubber matrix but are in either a glassy or highly crosslinked state at the application temperature.

An example of the former is particulate materials consisting either of polystyrene or styrene-butadiene (low concentration) copolymer which have been used as model fillers in SBR vulcanizates to study the mechanism of rubber reinforcement[119]. These fillers have demonstrated the role of strong (chemical) and weak (physical) interactions that influence the mechanical properties of filled rubber[119]. A good example of the latter consideration is the polymeric microgels prepared by emulsion copolymerization of common diene monomers followed by additional crosslinking[117]. By carefully controlling the experimental condition, Schuster[120] was able to obtain a polymeric microgel filler with a well-defined particle size and size distribution, surface chemistry, as well as $T_g$ of the microgels. When the polymeric fillers were incorporated in a polymer, it was found that the filler agglomeration (characterized by the Payne effect) is strongly dependent on characteristics of the polymer as well as the filler particle size. While there is no Payne effect when a polystyrene (PS) filler with a particle diameter of 60 nm is added to SBR with a chemical composition close to the filler, a significant Payne effect is observed in the case of NR (Figure 8.20). This is certainly due to the difference between the chemical composition of the filler surface and polymer matrix which results in a different degree of filler agglomeration. On the other hand, the addition of large-particle PS filler (diameter of 400 nm) to NR is shown to eliminate the Payne effect. This is expected from the kinetics of filler agglomerate formation since with increasing particle size at constant filler loading, filler agglomeration would be reduced due to both the reduction in the diffusion constant (Eq. 3.17) and the increase in the distance between particles (Eq. 3.9).

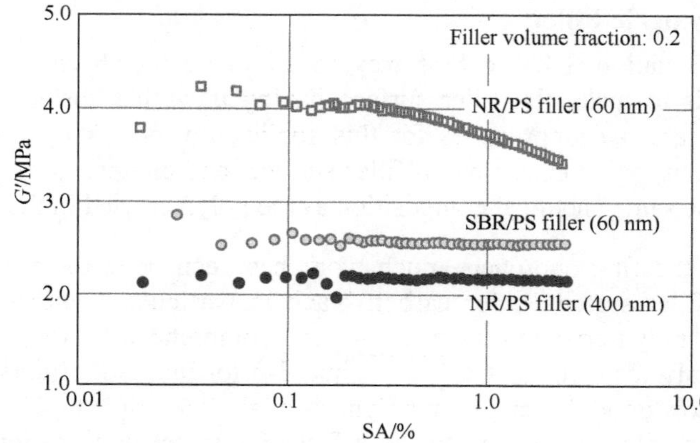


**Figure 8.20** Strain dependence of $G'$ for NR and SBR vulcanizates filled with PS-polymeric fillers[120]

This observation may be further verified by changing the surface characteristics of the polymeric filler. It was demonstrated[120] that while polybutadiene-microgel (BR-microgel)-filled NR demonstrates only a slight Payne effect, this effect is substantially enhanced by surface epoxidation of the BR-microgel, which would enlarge the difference in the surface characteristics between polymer matrix and filler surface (Figure 8.21).

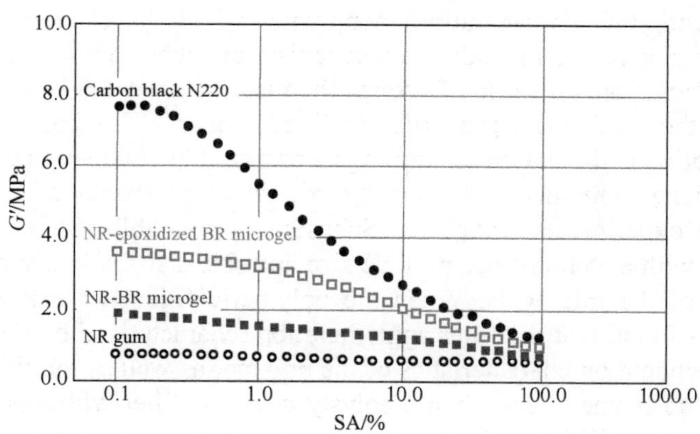


**Figure 8.21** Strain dependence of $G'$ for NR vulcanizates filled with carbon black and polymeric fillers[120]

A similar observation was made by Nuyken et al.[121] Using azomodified seed latex they were able to synthesize a polymeric filler with a highly crosslinked PMMA core and polystyrene shell. When 30 wt% of this filler is incorporated into SBR, the dicumylperoxide-cured vulcanizate, as expected, does not show a Payne effect or a significant strain dependence of tan $\delta$ (Figure 8.22).

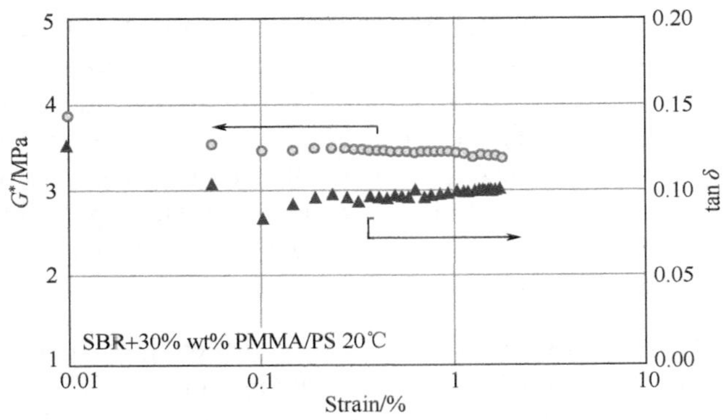


**Figure 8.22** Strain dependence of $G^*$ and tan $\delta$ at 20°C for SBR vulcanizates filled with polymeric filler PMMA/PS[121]

## 8.1.3 Mixing Effect

In practical rubber compounds, besides the polymer and filler there are various other ingredients each of which fulfills a specific function either in processing, vulcanization, or the end use of the product. When these ingredients are mixed together with the polymer and filler, they interact with each other to form a complex or composite which is essentially a multiphase material with each phase having a different morphology and composition. These parameters, which impart a significant effect on the composite properties, can be influenced by compound processing steps including mixing, milling, extrusion, calendering, molding, and injection. Among these processes, mixing is perhaps more crucial than any of the others. In this stage some basic physical changes of the materials, perhaps also some chemical reactions, are occurring and the incorporation and dispersion of filler and other ingredients is substantially completed. This is also true when the dynamic properties of the filled rubber are considered. In this chapter, the effect of processing on dynamic properties refers specifically to the effect of compound mixing.

Mixing is a complex process. Among the factors which control mixing productivity and quality are the variability of materials such as polymer, filler and processing aids, equipment design, and operating conditions. Discussion of these factors is beyond the scope of this book, but some excellent guidance to this important topic is given in other reviews[122–124]. In this chapter, only the above named factors as well as the basic physical processes and chemical reactions which occur during mixing and affect filler agglomerate formation, and hence the dynamic properties of the compound, will be discussed.

Generally, the quality of mixing is determined by mixing temperature, time and shear stress. For tire tread compounds, the conventional criterion is to obtain an acceptable level of filler dispersion, with acceptable compound physical properties, while maximizing the productivity. In conventional mixing for carbon black-filled compounds, the first stage mixing operation is set so that the dump time is just past the incorporation time of the filler[125].

After the first stage, the additional processing from downstream equipment such as mills and extruders as well as the internal mixer for incorporating curatives, all operated at relatively low temperature (<120°C), will provide sufficient dispersion and good physical properties. This is to say that during processing upstream of vulcanization, the basic physical processes of the materials and compounds are dominant as they occur at relatively lower temperature. However, when the mixing is done at high temperature, some beneficial effects on the physical properties of the vulcanizates with a variety of rubbers are obtainable[126–129].

With the series of high structure carbon blacks, Welsh et al.[130] investigated the effect of high temperature mixing with extended mixing times on the rubber properties related to tire tread applications. They found that the intensive mixing is characterized by lower Mooney viscosity and hardness, reduced hysteresis at high temperature and increased hysteresis at lower temperature as measured by both rebound and tan $\delta$. These are the properties of vulcanizates with a less developed filler agglomeration and a stronger polymer-filler interaction.

The improved hysteresis by intensive mixing has been confirmed in Cabot's laboratory with solution SBR filled with both conventional carbon black and CSDPF[114]. However, in the case of NR, the hysteresis results for the vulcanizates measured at high temperature (70°C) were slightly higher for the samples prepared with intensive mixing. The unexpected behavior of this natural rubber compound may be related to polymer properties which will be discussed later. When the superior performance of the vulcanizates obtained with intensive mixing is attributed to the depression of filler agglomeration, it may reflect the effect of mixing temperature and time on the physical and chemical processes occurring during mixing which govern the filler agglomerate formation. For polymer and filler only, while the mixing can be treated as a process by which the filler is incorporated, distributed, and dispersed in the polymer, the processes summarized in Table 8.2 would simultaneously occur. These may be the main factors governing filler agglomeration.

As discussed in Section 5.1.7, during mixing, bound rubber is forming while dispersion is occurring. The formation of bound rubber arises mainly from the physical adsorption of polymer molecules on the filler aggregates[34,38], although various forms of chemisorption cannot be excluded, especially for some polymers like natural rubber which show a highly mechano-chemical sensitivity. The formation of bound rubber occurs very rapidly during mixing [131,132]. The bound rubber continuously increases with mixing time, but formation rate is decreasing. This process does not seem to be completed during mixing since during storage of the compounds, bound rubber is still developing[35,36,127,133,134] (also see Figure 5.8). The heat treatment and prolongation of mixing time would speed up all these processes.

There is strong evidence for polymer gelation at high temperatures. This is related to crosslinking of polymer molecules even in the absence of vulcanizing agents. Similar chemical processes occur in the mixer at high temperatures, especially when the mixing time is extended. In a filled compound, crosslinking may also take place so that polymer chains as well as polymer gel may crosslink to the filler-bound rubber complex (filler-gel). Therefore, the intensive mixing would raise the bound rubber portion originating from polymer crosslinking.

**Table 8.2** Physical and chemical processes occurring during mixing

| Incorporation, distribution, and dispersion of filler in polymer | | |
|---|---|---|
| **Polymer** | **Mutual interactions of polymer-filler** | **Filler** |
| Gelation of polymer molecules<br>Scission of polymer chains | Formation of bound rubber<br>Physical adsorption and chemsorption of polymer chains on filler surface<br>Interaction with other ingredients | Break-down of filler aggregates |

Molecular chain scission may take place during high shear mixing and is a process that would be affected by mixing intensity. This part could be referenced the description of Figure 5.19 and Figure 5.20[135–143,192,193]. As discussed in Section 3.5.3, bound rubber can play an important role in efficiently slowing the flocculation rate of filler aggregates, but chain scission would facilitate filler agglomerate formation as it may reduce bound rubber content as well as introduce more chain ends in the polymer matrix, which is known to be one of the sources of mechanical hysteresis within polymeric materials. Accordingly, highly intensive mixing would not be favorable to the dynamic hysteresis of carbon black filled NR[114].

Bound rubber formation may also be associated with the breakdown of filler aggregates, another process occurring during mixing. This phenomenon would be more severe in intensive mixing. As mentioned before, high structure is more favorable for this process, which has been considered to be one of the reasons for the higher bound rubber of high structure carbon blacks. The same phenomenon can be observed with low structure blacks but to a lesser extent.

It should be mentioned that whereas the bound rubber content increases with intensive mixing, the viscosity of the compound does not. In fact, in most cases the viscosity decreases[130]. In view of the conflict between the viscosity expected from bound rubber and that measured, Wang[226] believed that the effect of filler agglomeration on the viscosity of uncured rubber can account for the low or comparable viscosity of an intensively mixed compound. It is believed that filler agglomeration can substantially increase the viscosity as the effective filler loading would be increased due to the trapped rubber in agglomerates[66]. Consequently, a less structured architectural complex of filler aggregates in a polymer matrix would over-compensate the effect of increased bound rubber. Thus, lower viscosity of the compound as a whole would not be unexpected. There is good evidence to support this inference. With a coupling agent, TESPT, the bound rubber content of a silica compound can increase by as much as 100% (21.1% *vs.* 41.5% for uncoupled compound and its coupled counterpart), but the viscosity measured with the DEFO plastometer decreases by 20%[67]. The Mooney viscosity of a compound containing 50 phr silica drops from 150 to 80 by adding 2.5 phr TESPT[144], which is certainly attributable to the depressed filler agglomeration.

Besides mixing temperature and time, the sequence of adding the ingredients has been found to be another critical parameter that influences rubber properties.

It has been reported that bound rubber formation varies significantly upon changing the mixing sequence of ingredients. This can be seen from Figure 5.22[33,145].

In fact, in order to obtain better dynamic properties (especially better temperature dependence of hysteresis), the mixing procedures of carbon black and carbon/silica dual phase filler filled compounds can be optimized in the way that the oil is added after the filler is incorporated into the polymer, followed by the addition of other ingredients[114].

When a coupling agent is used, the efficient way for its utilization is the premodification of the filler surface with the coupling agent. In fact, premodified silicas with various silanes have been commercialized[91] and premodified carbon blacks have also been patented[99]. Besides the higher efficiency of the coupling agents, the greater flexibility of the mixing procedure, the elimination of alcohol release (for silane coupling agents) during processing, and the reduction of moisture content are the additional advantages for premodified fillers, silica in particular[146]. However, when the modification of the filler surface has to be carried out in situ (i.e., reaction of coupling agent with filler takes place during mixing, which may be more economical than using premodified products), the mixing sequence of the coupling agent and other ingredients should be specifically considered. For example, in the case of a silica compound with TESPT, all compounding ingredients which might interfere with the ethoxy groups of the coupling agent have to be excluded[92]. Therefore, for a two-stage mixing procedure with the second stage only for mixing of curatives, the introduction of TESPT should be with the silica before other ingredients to make the best use of the reaction time independence to achieve the highest modification. In the case of multi-stage mixing, the silane is preferentially added in the first stage, which gives better dynamic hysteresis[147].

When the addition of oil is considered, it would be better to introduce the coupling agent before the oil since oil molecules may occupy the filler surface and interrupt the filler-coupling agent reaction[114]. This may not be very critical for a silica compound because of less interaction between the oil and the silica surface and higher surface polarity of the filler which may allow silanes, a polar material, to be driven to and react with silanols. This may, however, be very crucial for carbon black-filled rubber as the filler surface has a higher affinity with oil, blocking the active center from accepting the coupling agent. In this case, a better mixing procedure would be that the coupling agent goes with filler into the polymer first, followed by the oil and other ingredients.

### 8.1.4 Precrosslinking Effect

As discussed in Section 5.1.7.2.2, MRPRA[148] described a variety of polymer-carbon-black masterbatches that can be modified with some sulfur donors. They found that after the reactive mixing, a certain amount of sulfur donor is bonded on the polymer chains and the polymer-filler interaction is enhanced, indicated by a considerable increase in bound rubber and substantial reduction in dynamic hysteresis (characterized by tan $\delta$ at 23°C). Although the low hysteresis may be related to the higher crosslink density of the filled rubber since the sulfur donor would act as an additional crosslinking agent, the inhibition of filler flocculation by the high bound rubber would also be a key mechanism responsible for the improved hysteresis. The high bound rubber content may originate from polymer modification, some functional groups being grafted on the polymer chains. These functional groups, which generally have a higher polarity than the hydrocarbon polymer, may be strongly adsorbed on the filler surface. In addition, sulfur donors would be active for precrosslinking of the polymer at high temperature, preventing filler agglomeration during the post processes.

When epoxidized natural rubber (ENR) and 50 phr N330 carbon black were mixed with diamines in a Banbury, Terakawa and Muraoka[149] found that the bound rubber was considerably increased and Payne effect was markedly reduced when compared with a non-amine counterpart. The temperature dependence of tan $\delta$ was also improved with a clear crossover point shown at about 0°C (Figure 8.23). Terakawa and Muraoka attributed the improved dynamic properties to the increased crosslink density as well as to the improvement of filler-polymer interaction caused by the strong interaction between the amine and functional groups on the carbon black. However, low Payne effect and the crossover phenomenon of the temperature curves would strongly suggest the contribution of depressed filler agglomeration impacting the improved dynamic hysteresis, both at high and low temperatures. Although the filler-polymer interaction may be enhanced by the addition of amines, as cited before, the crosslinking of the epoxy groups with amines during mixing occurs with a high degree of certainty. It is well known that the crosslinking of epoxy materials with amines can take place at room temperature and the rate of this reaction can rise very rapidly with elevating temperatures[150]. Even though the temperature of mixing in Terakawa's study is not specified, the dump temperature is probably higher than 120°C so that gelation of the polymer would be expected.

Gerspacher and co-workers[151,152] were able to obtain vulcanizates with better macro-dispersion and micro-dispersion, hence, dynamic properties were improved by using dynamic vulcanization, i.e., adding all curatives, including sulfur and accelerators, into the mixer to allow certain degree of crosslinking generated. However, as the compound is relatively highly crosslinked, it cannot be processed in the downstream processing steps such as milling and extrusion with conventional equipment.

Another advantage of the early addition of crosslinkers is associated with the application of coupling agent. As cited in the Section 8.1.2.3.4, a coupling agent like TESPT has three functions: modifying the filler surface to reduce filler-filler interactions, introducing covalent bonds between filler surface and polymer chains to strengthen polymer-filler interactions, and generating higher bound rubber content (including polymer gel) to prevent filler flocculation. The last function can be compensated by adding additional sulfur in the early mixing stage while reducing the coupling agent dosage[153]. In this case, the crosslinking rate with sulfur alone is so low[154,155] that the compound viscosity does not develop very rapidly and the processability can easily be controlled in the accepted ranges by adjusting the temperature and time.

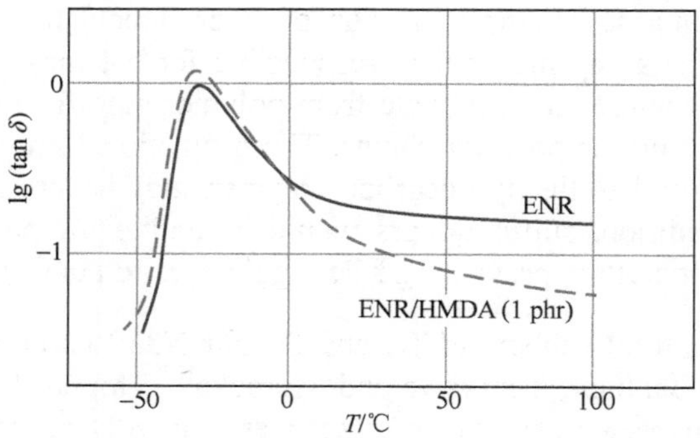


**Figure 8.23** Effect of hexa-methylenediamine on temperature dependence of N330 carbon black-filled ENR (NR epoxidized in 25% in mol)[149]

Since it has been recognized that filler flocculation in the compound takes place mainly in the vulcanization stage[123,156], it may be worthwhile to mention that improvement in cure characteristics of compound (i.e., in terms of reduction of unnecessary scorch time and increase in cure rate) would also kinetically inhibit filler agglomeration. It was reported[114] that for carbon/silica dual phase filler-filled compounds, tan $\delta$ at high temperatures and the Payne effect decrease monotonically with increasing ultra-accelerator tetrabenzyl thiuram disulfide (TBzTD) in the curative system over the range of dosages used (from 0 phr to 0.5 phr). Although this may influence the crosslink density and crosslink structure of the vulcanizate which impacts the dynamic properties, the short scorch time and high cure rate from the addition of ultra-accelerator would effectively slow down and finally arrest the filler agglomeration process, due to rapid rise of viscosity in the polymer matrix.

# ■ 8.2 Skid Resistance–Friction

It has been widely accepted that dynamic hysteresis, namely tan $\delta$ of tread compounds at low temperature and measurable frequencies, is of importance for wet skid resistance, due to the high-frequency nature of the dynamic strain involved[3,6,157–159]. Some workers have found a good correlation between viscous modulus and wet skid resistance[160,161] and many researchers have used different functions of dynamic moduli of the compounds[162–164], even in combination with abrasion loss during skid tests[165,166] to fit their test results. Most of the authors have accepted that the friction of rubber compounds on substrates consists of different components of energy dissipation, such as deformation hysteresis, adhesion, fracture (abrasion or/and tearing), and even viscous shear dissipation in water[162,167,168]. Almost all have agreed that the first two components play a dominant role in the determination of elastomer friction on rigid surfaces, and these are mainly governed by dynamic properties of the compounds, particularly hysteresis and dynamic modulus[162,167–170].

When the "green" tire was commercialized, it was found that, besides the low rolling resistance, this tire featured better-wet skid resistance. Several papers have addressed the mechanisms experimentally and all of the explanations are related to dynamic properties, such as higher tan $\delta$ at low temperature[158], high frequency[171], and higher $G''$ at higher speeds[160]. Each theory has been successful in explaining their experimental results. However, it was found that when these theories are applied to new fillers, the wet friction of the compounds cannot be fully explained by dynamic properties. For example, when a carbon-silica dual phase filler (CSDPF 2000) is compounded in a passenger tire tread compound, the dynamic properties can fully match those of silica-based compounds with similar formulations. This includes not only properties measured by temperature sweep (Figure 8.24) at constant strain, but

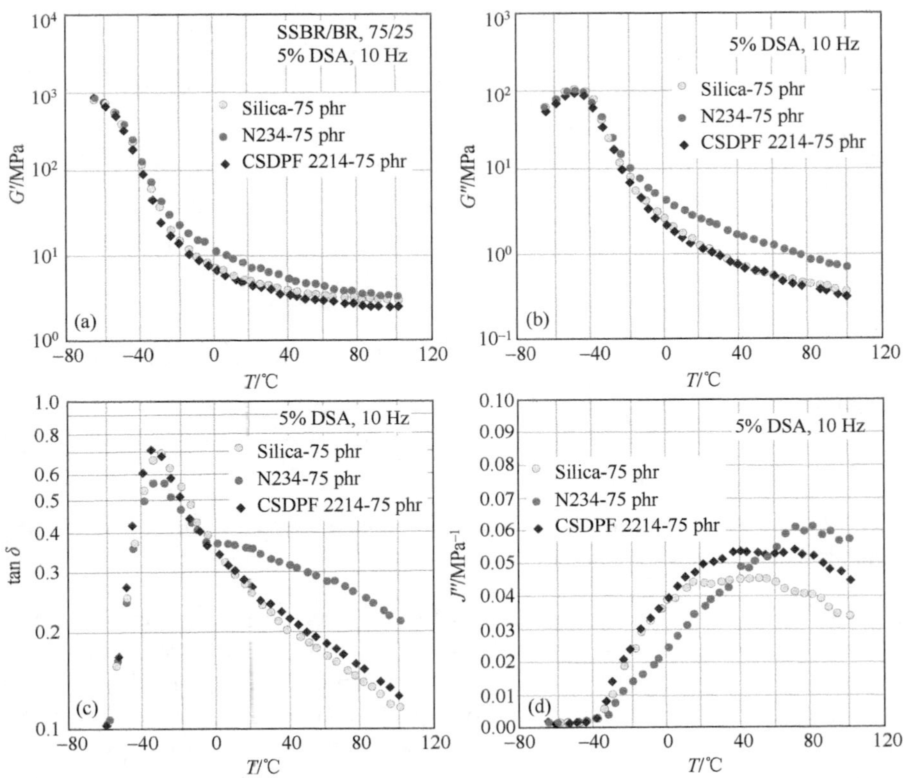


**Figure 8.24** Temperature dependence of dynamic properties at 5% DSA and 10 Hz for passenger tire tread compounds filled with a variety of fillers

also by strain sweeps at different temperatures (Figure 8.25 and Figure 8.26). The reduced dynamic property master curves, generated from the frequency dependence at different temperatures using the WLF principle, show that there is a similarity between CSDPF and silica compounds, which is quite different from those of carbon black[172]. However, while the wet skid resistance of CSDPF-filled tread compounds for passenger tire, measured with British Portable Skid Tester (BPST)[173] and Grosch Abrasion & Friction Tester (GAFT)[174], can be improved compared with their carbon black counterparts, they are still inferior to silica-based tread compounds.

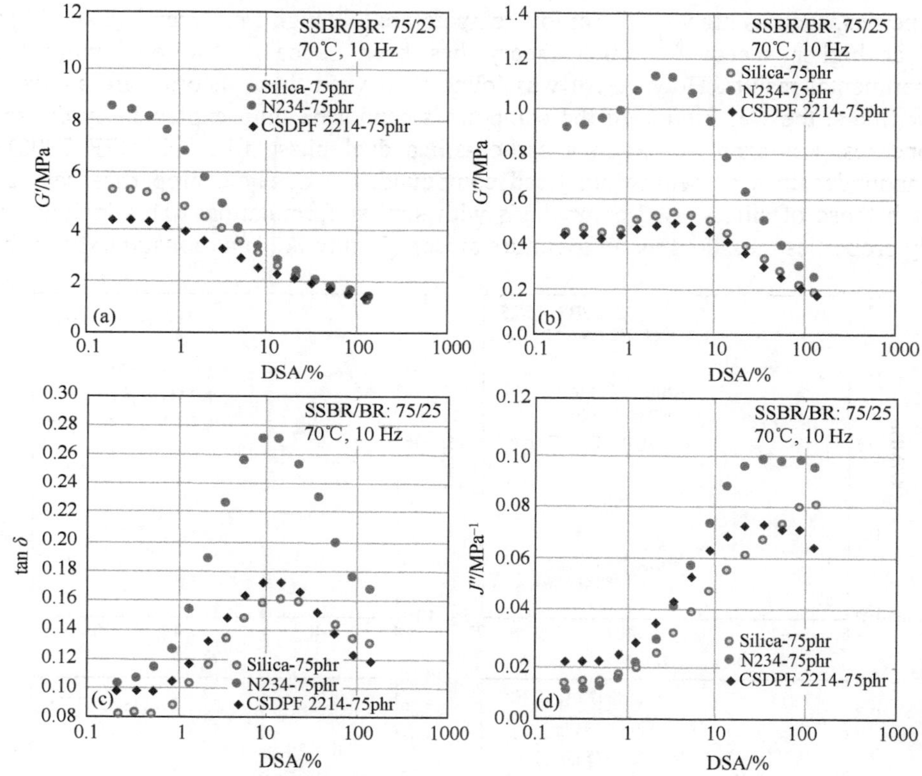


**Figure 8.25** Strain dependence of dynamic properties at 70°C for passenger tire tread compounds filled with a variety of fillers

In addition, a lot of work on wet skid resistance has been done and many results have been cumulated from both laboratory tests and tire tests since the green tire was commercialized. It has been found that skid resistance is not always better for silica compounds than for comparable carbon black compounds. For example, under conditions similar to those experienced by truck tires, silica compounds give poorer wet skid resistance[175]. For passenger car tires, the silica tire is not favored in dry skid resistance[176]. Even the wet skid resistance of passenger car tires is not always favorable to silica. In Table 8.3, the favorable conditions for silica and carbon black based on the input from tire and automobile manufactures and results from filler suppliers are summarized. These observations suggest that the mechanisms involved in wet skid resistance are more complicated than can be explained from dynamic properties of the compcunds. The need exists to explain the effect of all fillers on skid resistance in similar and relatively nonspecific terms, i.e., the observed phenomena related to all fillers should follow a general rule or principle.

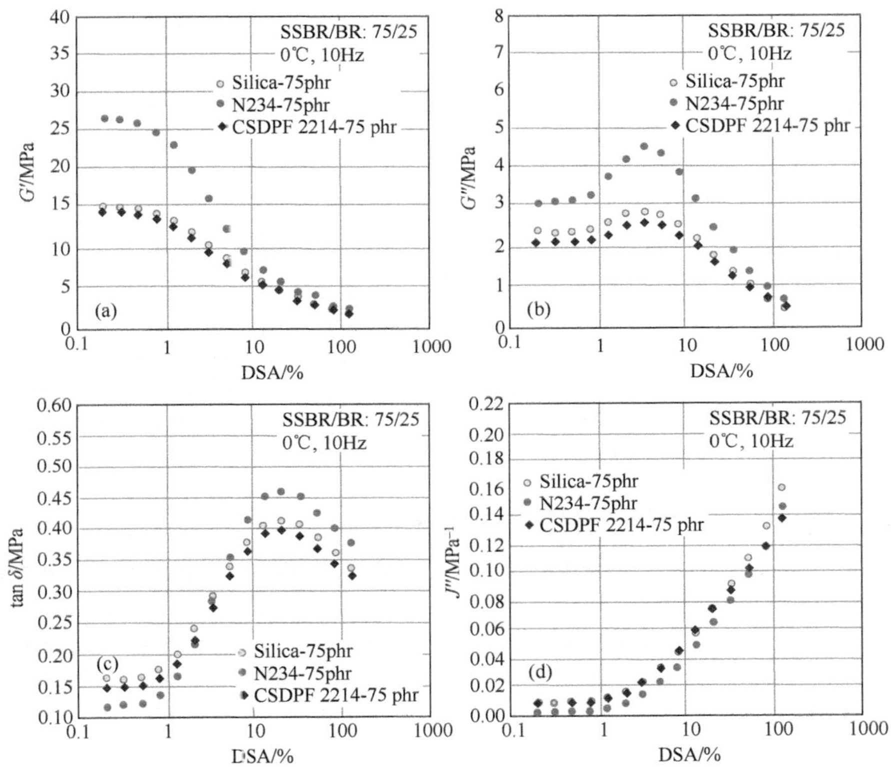


**Figure 8.26** Strain dependence of dynamic properties at 0°C for passenger tire tread compounds filled with a variety of fillers

**Table 8.3** Comparison of skid resistance of carbon black- and silica-filled tread compounds under different conditions[a]

| Conditions | | | Carbon black | Silica |
|---|---|---|---|---|
| Road surface | | Dry | + | – |
| | | Wet | + / – | – / + |
| Wet skid resistance | Road surface texture | Low $\mu$ (0.2-0.5) | – | + |
| | | High $\mu$ (> 0.6) | + | – |
| | Brake system | Locked wheel | + | – |
| | | ABS | – | + |

Continued

| Conditions | | | Carbon black | Silica |
|---|---|---|---|---|
| Wet skid resistance | Temperature | Low | − | + |
| | | Relatively high | + | − |
| | Speed | Lower | + | − |
| | | Higher | − | + |
| | Load | Low | − | + |
| | | High | + | − |

a.+: favorable, −: unfavorable.

It is now understood that while the dynamic properties of tread compounds, namely dynamic hysteresis and modulus at low temperature, are important, the hydrodynamic lubrication (HL), elasto-hydrodynamic lubrication (EHL), especially on the micro scale (MEHL), and boundary lubrication (BL) are also critical[175,177,178]. The relative role of MEHL in wet skid resistance under different test conditions is determined by the composition of the tread compound, such as the nature of the polymer and filler, and especially their interaction.

## 8.2.1 Mechanisms of Skid Resistance

### 8.2.1.1 Friction and Friction Coefficients–Static Friction and Dynamic Friction

Friction is a force of resistance to the relative motion of two contacting surfaces. This force, $F$, is largely independent of external influences and is proportional only to the normal reaction force, $N$, between the two surfaces. It follows a simple law:

$$F = \mu N \tag{8.2}$$

where $\mu$ is the friction coefficient. This coefficient is independent of the contact pressure and size of the contact area.

There are two friction coefficients. In a stationary specimen, the static friction coefficient applies: it is the force required to create motion divided by the force pressing mating surfaces together. However, with a specimen in motion, the dynamic coefficient of friction applies. This coefficient of friction is the ratio of the force needed to sustain motion at a specified surface velocity to the force pressing two surfaces together. The lower the friction coefficient, the easier the two surfaces slide over each other. The dynamic friction coefficient varies with applied load, velocity, and temperature[179].

### 8.2.1.2 Friction between Two Rigid Solid Surfaces

The frictional force between rigid surfaces is explained by assuming that the actual contact area is much smaller than the apparent one. Therefore, at the points of contact the pressure is very high. In the case of two metal surfaces sliding against each other, points of contact weld together and are sheared under the tangential friction force. Under these conditions, the friction coefficient is given by the ratio of the shear strength to the yielding pressure of the weaker of the two materials[160]. In the case of a hard solid in contact with a soft surface, the asperities of the hard solid penetrate until the contact area times the yielding pressure balance the local normal reaction. Under the tangential force, softer material is plowed out. This theory of friction suggests that the friction coefficient, $\mu$, is determined by the ratio $\tau_c/H$, where $\tau_c$ is the critical shear stress of the softer material and $H$ is its hardness[180]. When two rough surfaces intimately contact each other, the frictional energy is also dissipated when lifting the mating surfaces over their interlocking asperities[181]. Therefore, it is understood that the friction coefficient is also determined by other factors, such as surface roughness of the materials. In all these processes, plastic deformation of these materials plays an important role in the friction coefficient.

## 8.2.2 Friction of Rubber on Rigid Surface

Frictional energy dissipation in the sliding of viscoelastic materials still lies within the solid by virtue of its internal hysteresis. Unlike solid materials such as metals, which dissipate energy by plastic flow, the hysteresis in rubber involves very limited permanent deformation and is generally only associated in minor part with plastic flow[181]. However, in the rubbery state, the mobility of polymer chains is high, and this determines the general rules for friction in rubbery polymers even though the two main factors contributing to the friction are also adhesion and deformation or hysteresis.

### 8.2.2.1 Dry Friction

#### 8.2.2.1.1 Adhesion Friction

When a piece of rubber is loaded against a hard counter-face, pressure is generated at the true contact area. This area is determined by the normal force applied, elastic modulus of the rubber and the topography or micro- and macro-roughness of the counter-face. While the frictional force generated between rubber and rigid base also consists of adhesion and deformation terms, like those between two rigid bodies, a polymer, due to its easy deformability in the rubbery state, is able to form a true contact area comparable to the nominal one. The establishment of this true contact area gives rise to a finite amount of static friction, which is part of the resistance to motion under a tangential stress. This resistance is the force, $F_a$, due to adhesion of the polymer to the counter-face and is believed to be the result of interaction between surface molecules at the contact area. However, the hysteresis of the rubber also plays a significant role. This is due to the essential viscoelastic nature of the molecular adhesion in the friction process as described by the mixed theory (see below)[182]. Even in the case of weak adhesion of the polymer to the hard surface, tangential stresses arise at the contact boundary during shear and sometimes even surpass the maximum shear stresses throughout the bulk.

There are several theories of elastomeric friction concerning the adhesion component. Among others, the one proposed by Bulgin, Hubbard, and Walters[2] and so-called mixed theory[182] are simple but representative. Part of the mechanism in these theories considers molecular adhesion giving rise to stick-slip and the other part presents the information by using a mechanical model. The basic concept is schematically shown in Figure 8.27. When a piece of rubber is sliding across a hard surface with a velocity, $V$, suppose that "adhesion" takes place at point A and persists for a time during which the system moves a distance $\lambda$. Then release takes place at $A$ and adhesion at a new position $A'$ is established. During this process, the total energy loss due to the dynamic strain is equal to the external work of friction, $F_{ad}\lambda$, where $F_{ad}$ is the friction force. Accordingly, under normal load, $W$, the adhesional friction force is described by Eq.8.3:

$$F_{ad} = K\sigma_{ms}(W / H)\tan\delta \tag{8.3}$$

and the coefficient of adhesional friction, $\mu_{ad}$, defined by $F/W$, is:

$$\mu_{ad} = K(\sigma_{ms} / H)\tan\delta \tag{8.4}$$

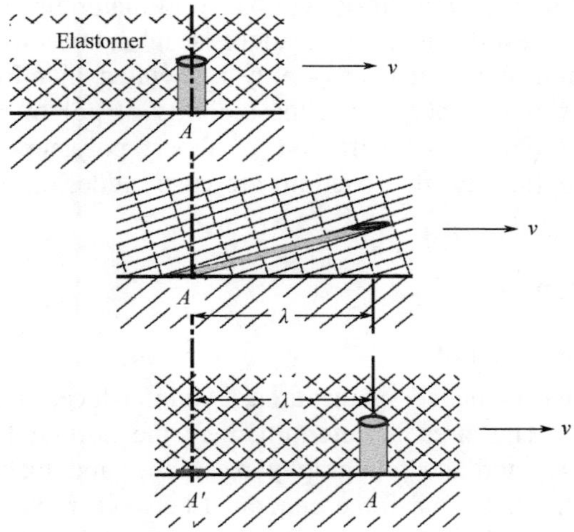


**Figure 8.27** Schematic adhesion mechanism of friction

where $K$ is a proportionality constant, $\sigma_{ms}$ is the maximum stress able to be generated on the elemental area, $H$ is hardness that, with $W$, determines the total true contact area between polymer and rigid surface, and tan $\delta$ is the loss factor which is related to the ratio of energy dissipated to the energy stored in this stick-slip process. Obviously, $\sigma_{ms}$ is determined by the adhesion strength at the contact area. On the other hand, while the adhesional friction resistance is due to adhesion $\mu_{ad}$ depends also on the tan $\delta$ of rubber, which is a viscoelastic property. This suggests that the principal mechanism of frictional energy dissipation from adhesion is through deformation loss, namely viscoelastic hysteresis. The role of adhesion is to increase the magnitude of the deformation loss[181].

With regard to adhesion, the concept is somewhat controversial, which is partly due to terminology. In a general sense, "adhesion" basically implies the sticking together of two surfaces without any external effects. However, solid surfaces, even when clean, do not normally adhere. Controversies also arise when the mechanism of energy dissipation is considered. The fundamental work expected in overcoming adhesion forces is the intrinsic surface energy, $\gamma$, which has a value of about 1.0 J/m$^2$ for metals and 0.03 J/m$^2$ for non-polar or low-polar solids such as rubber and polymer. This is far too small to account for observed friction forces[181]. Therefore, as described in Eq. 8.4, the source of energy dissipation lies also in deformation losses within the solid. Apparently, "tangential interaction" rather than "intermolecular interaction" might be a better description of this process and the term "sticktion" might be better used than the term "adhesion".

#### 8.2.2.1.2 Hysteresis Friction

While the adhesion component of friction is also dependent on deformation hysteresis of the material, it is confined to the surface, i.e., only a skin layer of the rubber undergoes deformation. With regard to deformation or hysteresis friction, the hysteresis effect is manifested by the deformation of the polymer bulk by macro and micro asperities such as encountered by the tires of automobiles on a road surface. A so-called "unified theory" can be used to better describe the principle of hysteresis friction[183]. Consider a piece of rubber pressed on an asperity, as schematically shown in Figure 8.28. In the static state, an elastic pressure distribution without net side force will be established. The hysteresis friction is zero. Under a horizontal pull, the rubber piles up at the leading edge of the asperity and the contact is broken at a high point on the downward slope. This creates a distorted pressure distribution, which gives rise to a net hysteresis friction force opposing the forward motion of the rubber. When further increasing sliding speed, the contact area takes up a new position and the degree of pressure asymmetry and the magnitude of hysteresis friction force also increase. Further increase in speed results in rubber contraction or stiffening in the contact area, hence, the friction force decreasing. The rubber hysteresis developed on a rough surface in such a way has been termed macro-hysteresis. Hysteresis may also be generated due to the micro-asperity interaction for which the term micro-hysteresis is used. The micro-hysteresis, however, decreases progressively as the surface finish of the opposing solid gets finer until a characteristic smoothness is reached. Then the frictional resistance can be attributed entirely to adhesion at the interface. Generally, when the sliding speed, $V$, varies, the coefficient of hysteresis friction, $\mu_{\text{hyst}}$, is given by[168]:

$$\mu_{\text{hyst}} = K'\left(p / E'\right)^n \tan\delta \tag{8.5}$$

where $n \geqslant 1$, $p$ is the mean pressure on each asperity, and $E'$ is the elastic modulus. In this equation the constant, $K'$, is related to the asymmetry of pressure distribution on asperities. This factor also shows a speed dependency[168,180,184].

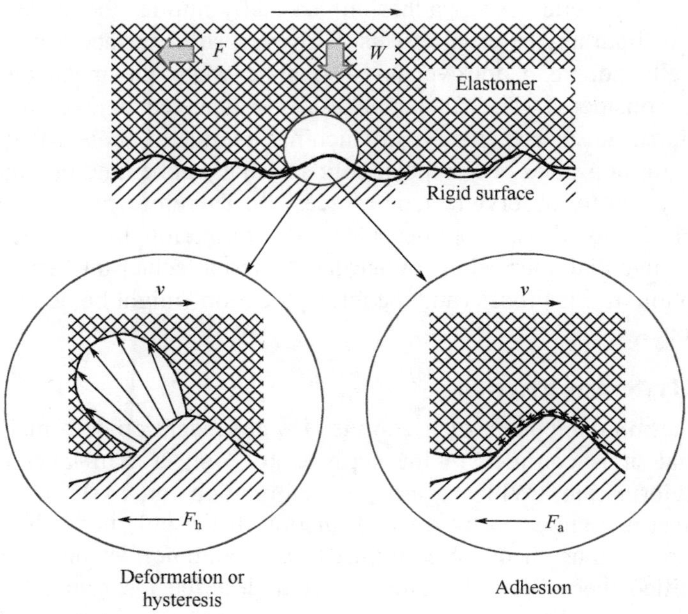


**Figure 8.28** Schematic unified friction mechanisms

In summary, with regards to the friction between rubber and a rigid solid surface, viscoelastic properties of the rubber play a dominant role in all the mechanisms involved.

### 8.2.2.2 Wet Friction

According to fluid mechanics, the introduction of a fluid film between two relatively moving solid surfaces will generate a pressure, i.e., hydrodynamic pressure, within the film, resulting in a lubrication effect. If the lubricant films are thick compared with asperity height of the solid surfaces, the friction resistance is determined by hydrodynamic mechanisms and by elastohydrodynamic (EHD) mechanisms in a high pressure non-conforming contact [see Figure 8.30(a)]. If the solid surfaces are so hard that the asperity deformation is largely elastic, the contribution to the friction force from the solid deformation losses is small. The friction then arises predominantly from the shear strength of the interfacial film[181]. The shear stress of the film, $\tau$ varies with pressure, $P$, at constant strain rate and temperature: $\tau = \tau_1 + CP$, where $\tau_1$ and $C$ are constants. This relationship seems to apply from film thicknesses of ~1.0 mm down to a few monolayers.

#### 8.2.2.2.1 Elastohydrodynamic Lubrication

Elastohydrodynamic lubrication is the situation in which the elastic deformation of the surrounding solid plays a significant role in the hydrodynamic lubrication process[185]. This results in two effects that are not accounted for in classical hydrodynamic lubrication: (1) the influence of high pressure on the viscosity of liquid lubricants and (2) significant local deformation of the elastic solid. These effects substantially change the geometry of the lubricating film, hence the pressure distribution at the contact area. In this case the hydrodynamic pressure generation must be matched with the elastic pressure in the contacting solids. The final elastohydrodynamic condition at the contact spots would be determined by basic lubrication theory and the elastic properties of the solids. Consequently, the lubricant film thickness due to elastohydrodynamic effect would be changed significantly, which would influence the transition from hydrodynamic to boundary lubrication.

#### 8.2.2.2.2 The Thickness of Lubricant Film for Rubber Sliding over Rigid Asperity

When an elastomer slides over a rigid asperity in the presence of lubricant, the elastohydrodynamic effect will occur. As shown by Moore[186], this can be represented by an elastomer sliding under a load on a lubricated two-dimensional surface having a sinusoidal asperity (Figure 8.29). Without tangential motion, the lubricant at the asperity tip will be squeezed out under the normal load, $W$. When a tangential motion with a velocity, $V$, is applied, a hydrodynamic upward pressure wedge in the inlet region of the lubricant film will be generated. As this pressure is resisted by the elasticity of the elastomer, a vertical elastohydrodynamic equilibrium is attained which produces a steady-state distortion in the surface of the elastomer during sliding. The center part of the film may be considered to have a constant thickness, and at the outlet region the pressure falls rapidly. This thickness is decreased to a value $h^*$ at point $B$ during sliding. When elastohydrodynamic equilibrium is reached, based on Reynolds' equation for hydrodynamic lubrication, Moore was able to give the following relationship between $h^*$ and the modulus of the elastomer, $E$[187]:

$$h^* = f_{\mathrm{ws}} \left[\eta V\right]^{1/2} \left[\frac{LR(1-v^2)}{\pi WE}\right]^{1/4} \tag{8.6}$$

where $\eta$ is the viscosity of the lubricant, $L$ is a length perpendicular to the paper, $R$ is the asperity radius, $W$ is the load, $V$ is the sliding speed, $v$ is Poisson's ratio of the elastomer, and $f_{\mathrm{ws}}$ is a wedge shape factor. It can be seen that besides the modulus of elastomer, the film thickness of the lubricant is dependent on sliding speed, surface roughness of the substrate, which is related to asperity radius, and viscosity of lubricant, which, in turn, is related to temperature and load.

#### 8.2.2.2.3 Boundary Lubrication

In practice, it is often impossible to obtain fluid lubrication, particularly if the sliding

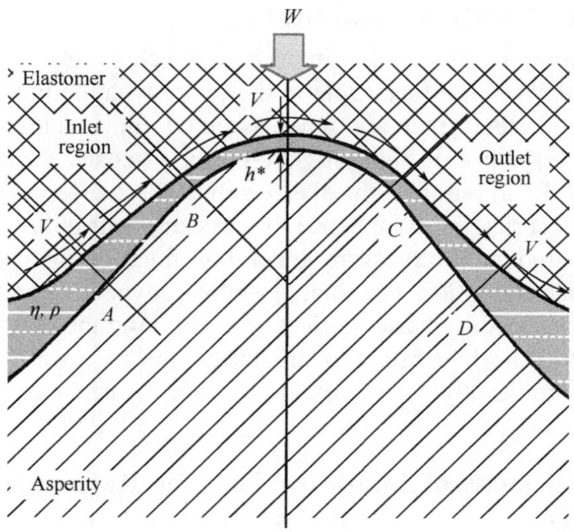


**Figure 8.29** Thickness of water film – elastohydrodynamic effect

speeds are low or loads are high. In such cases, the thick lubricant layer breaks down and the surfaces are separated by a lubricant film of only molecular dimensions. Under these conditions, which are referred to as boundary conditions, the very thin film of lubricant is bonded to the substrate by very strong molecular adhesion forces, and it has obviously lost its bulk fluid properties[187]. The bulk viscosity plays little or no part in the frictional behavior under these conditions. The friction is influenced by the nature of the underlying surface as well as by the chemical constitution of the lubricant. Boundary lubrication is of great importance in engineering and tire practice. The coefficient of friction for surfaces lubricated with boundary film is considerably higher than for fluid lubrication[188].

The transition from hydrodynamic to boundary lubrication is relatively gradual. As sliding speed is decreased or load is increased, the wedge of lubricant separating the surfaces becomes thinner and more surface asperities penetrate the film. At the tip of these asperities, the lubrication is of a boundary nature so that the amount of boundary lubrication increases gradually as fluid lubrication decreases[188]. Therefore, there exists a transition zone where this type of "mixed" lubrication is involved. However, the development of this transition zone is determined not only by the test conditions, including the types of lubricants, but also to a great extent by the properties of the solid surface.

#### 8.2.2.2.4 Difference in Boundary Lubrication between Rigid-Rigid and Rigid-Elastomer Surfaces

In general, under boundary lubrication conditions, the total friction coefficient, $\mu_{BL}$, can be written as a sum of the individual contributions of liquid, $\mu_{liq}$, solid, $\mu_{solid}$, and deformation components, $\mu_{deform}$:

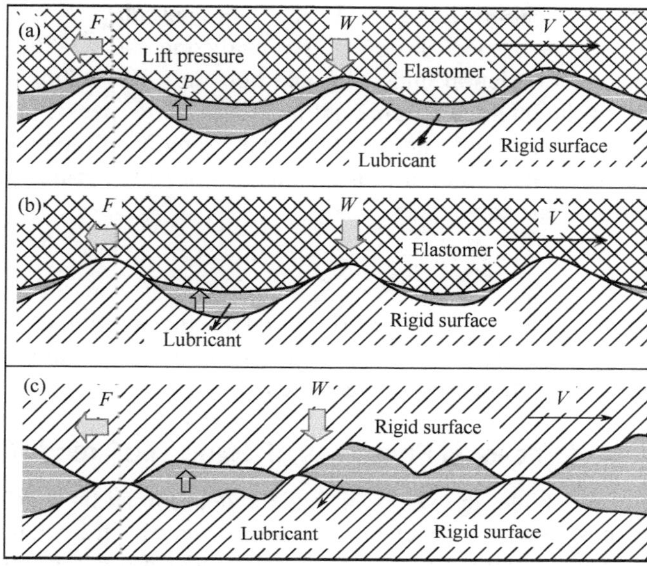


**Figure 8.30** (a) Elastohydrodynamic lubrication; (b) boundary lubrication – elastomer on rigid surface, and (c) boundary lubrication – rigid surface on rigid surface

$$\mu_{BL} = \mu_{liq} + \mu_{solid} + \mu_{deform} \tag{8.7}$$

For a rigid surface sliding on a rigid surface, like a metal sliding on metal as shown in Figure 8.30(c), the friction force, $F$, may be expressed as the sum of solid friction at the asperity peaks, liquid friction in the voids and plowing (deformation) contributions.

Based on the assumption that the term of plowing is negligible and the contribution from the liquid to the friction force is very small compared to that from the solid, Moore demonstrated the following for rigid surface-on-rigid surface sliding[189]:

$$\mu_{BL} \ll \mu_{dry} \tag{8.8}$$

where $\mu_{dry}$ is the friction coefficient under dry conditions. According to Eq. 8.8, the friction coefficient is reduced upon lubrication. However, the reduction in the friction coefficient is also determined by boundary properties of lubricant films, which in turn depend on the nature of the surfaces as well as on the composition of lubricant itself.

When an elastomer is sliding on a rough rigid substrate in the presence of a lubricant, the situation is quite different. If the sliding speed is low enough to neglect fluid entrainment over the asperity peaks but at the same time the sliding speed is large enough to establish hydrodynamic lift in the void spacing between neighboring asperities, the condition of boundary lubrication exists. However, in this case, unlike in Figure 8.30(c) (metal-on-metal), the liquid film is continuous between the surfaces [Figure 8.30(b)]. The film at the asperity peaks is extremely thinned and its properties are quite distinct from those of the bulk lubricant in the voids. Due to the physical-chemical interactions near asperity peaks, the lubricant may assume some of the properties related to the draped elastomer. By analyzing the relative magnitudes of hysteresis coefficients of friction under dry and boundary lubrication, and comparing the shear strength of the interacting lubricant layer at asperity peaks, the shear strength at asperity peaks under dry condition, and the shear strengths of the bulk liquid, Moore was able to conclude that for an elastomer sliding on a rigid surface under boundary lubrication conditions, one can expect[189]:

$$\mu_{\mathrm{BL}} \geqslant \mu_{\mathrm{dry}} \tag{8.9}$$

This suggests that in contrast to the rigid-rigid system, the addition of a lubricant to an elastomer-rigid sliding system in a boundary state does not imply a significant reduction in the resulting sliding friction coefficient.

### 8.2.2.3 Review of Frictional Properties of Some Tire Tread Materials

During braking of a vehicle, the skid resistance is determined by the friction level of the tread compound-pavement under dry conditions, and by the friction level of the water-lubricated tread compound-pavement surface under wet conditions. Therefore, it could be useful to briefly review some basic friction data of different components of tread compounds which could possibly have contact with the pavement. During the service of a tire, there are two possible situations of the worn surface of the tread element: (1) the surface entirely consists of rubber, implying that the filler aggregates are covered by a rubber film, or (2) a part of the tread surface is bare surface of filler. An investigation of the contribution of each component of the tread surface to the total friction would provide an estimate of the friction characteristics of tread compound on the road surface during service. However, such an investigation is difficult due to a lack of basic friction information on rubber, silica and carbon black on real road surfaces such as basalt slabs or asphalt cobblestone track. Therefore, some experimental friction data of analogous materials were reviewed, taking glass, carbon, and graphite materials as the analogues to silica and carbon black, and glass or metal as analogues of the road surface or skid track.

#### 8.2.2.3.1 Carbon and Graphite

Carbon black aggregates consist mainly of crystallites which have a so-called quasi-graphitic structure and a very small amount of unorganized (amorphous) carbon. Therefore, it is reasonable to take graphite and carbon materials to be model materials for carbon black. It is well known that the friction of graphite on a solid counter-face is low, even in the absence of a lubricant. The friction coefficient of steel on graphite is approximately $\mu = 0.1$ which can remain low up to very high temperatures. Such low friction has generally been considered to be due to the lamellar structure of graphite. This structure imposes marked anisotropic properties, showing very low shear strength in the direction parallel to the lamellae. For a hard non-graphitic carbon surface, it was reported that the friction coefficient, $\mu$, of steel on carbon and of carbon on carbon is approximately 0.16. This is reduced to about 0.13 when a lubricant is applied[190].

#### 8.2.2.3.2 Glass

Glass is chosen as the analogue of silica due to the similarity of its chemical nature. The typical friction coefficient of clean glass on glass was found in the literature to be 0.9, which is drastically higher than that of carbon materials on a rigid sliding track. It was also reported that the fragments of glass that were flaked off from the surface during sliding had been severely deformed. This suggests that strong glass-glass junctions are formed at the contact points and that these junctions are sheared during the sliding process[190].

Upon lubrication, the friction is reduced. The friction coefficient, $\mu$, of glass on glass is higher than that on steel if fatty acids, alcohols, and water are used as lubricant. It was also documented that glass (to glass) was poorly lubricated by these liquids. For example, Sameshima, Akamatu, and Isemura[191] reported that the friction coefficient of clean glass on clean glass is 1.04 which is reduced to 0.9 by pure water, showing only a small lubrication effect.

#### 8.2.2.3.3 Rubber

Due to the easy deformability of rubber, elastomers form a real contact area comparable to the nominal one during contact with a solid. This results in a high sticktion between rubber and skidding track. In combination with the high deformation of rubber under tangential stresses, the friction is much higher than that of a rigid material sliding on a rigid surface. Depending on the sliding velocity, temperature, and types of elastomers, the friction coefficients of rubber on glass surface can reach a level from 2.5 to 4[160,190]. The compounding ingredients appear to have little effect since it was found that friction depends mainly on the nature of the rubber matrix[190].

The effect of lubricants such as water is somewhat complicated. At the temperature and sliding speed in the regime of hydrodynamic lubrication, friction is mainly determined by the shear strength or absolute viscosity of the lubricant. In this case, the friction is extremely low. Due to the elastohydrodynamic effect, the sliding of the slider over the fluid creates hydrodynamic pressure that lifts the elastomer, resulting in a thicker liquid film compared with the rigid slider. At the same sliding velocity, the thicker film gives lower shear rate, thus lower friction force would be expected. Consequently, the friction coefficient of water-lubricated rubber on a rigid surface is lower than a water-lubricated rigid surface on the same track. When the liquid film is reduced to certain level at which the rigid sliding surface contacts the peak of asperities on the rigid track, a mixed lubrication starts. In this case hydrodynamic and boundary lubrications take place simultaneously under the same conditions of velocity, temperature and surface roughness of the track. For a water-lubricated rubber slider, in the area where the hydrodynamic lubrication occurs, the slider will be separated by lubricant, resulting from elastohydrodynamic effect. In this situation, rubber sliding on glass gives lower friction than glass on glass. In the regime of boundary lubrication, the friction of rubber on water-lubricated glass should be as high as that on un-lubricated glass (Eq. 8.9), and is much higher than that of glass on water-lubricated glass, as expected from Eq. 8.8.

**Table 8.4** Friction coefficients of different materials against steel and glass surfaces

| Items | Dry surface | Wet surface |
|---|---|---|
| Carbon black[a] | 0.1,[b] 0.16[c] | 0.13[c] |
| Silica[d] | 1.04 | 0.9 |
| Rubber[d] | 2.5–4 | Similar to dry surface – BL<br><< 0.1 –EHL |

a. On steel surface; b. graphite; c. hard non-graphite carbon; d. on glass surface.

In Table 8.4 the friction coefficients of the analogues of carbon black and silica on metal or glass are summarized[160,190,191].

#### 8.2.2.3.4 Prediction of Friction of Filled Rubbers on Dry and Wet Road Surfaces Based on Surface Characteristics of Different Materials

Based on the knowledge and observations mentioned above about friction of rubber and the analogues of silica and carbon black on a glass surface, the relative skid resistance under different conditions may be predicted based on surface characteristics of the tread compounds. Comparison of these predictions with road test results of tires under different conditions may provide some basic understanding of the mechanisms involved in wet skid resistance.

The comparisons of dry and wet skid resistance between silica- and carbon black-filled rubbers based on the basic friction data and observations are presented in Table 8.3. Disregarding the dynamic properties of the compounds, it would be expected that when both silica and carbon black are covered by polymer, there would be no significant difference in any skid resistance. If both carbon black and silica aggregates were exposed, silica compounds would be expected to give higher skid resistance on both wet and dry track surfaces. In the case that carbon black was exposed but silica has a rubber film, silica might give lower wet skid resistance but higher dry skid resistance. All these combinations do not appear to be in line with the observations obtained from lab and road tests. The experimental results show that carbon black exhibits a higher skid resistance on a dry surface, but a lower value on a wet surface. This is consistent with the hypothesis that during the skid test, carbon black aggregates would be covered by rubber while silica is exposed. Such an inference needs to be verified.

### 8.2.2.4 Morphology of the Worn Surface of Filled Vulcanizates

By reviewing the basic concepts of friction under dry and wet conditions, and existing data on friction coefficients of possible model materials that are at the worn surface of the tire tread compounds and road pavement surface, it is inferred that after skid testing under wet conditions, the top skin of the worn surface contains bare silica surface in silica-filled compounds, but carbon black aggregates remain covered by rubber. This inference is supported by measurements of the surface energies of the fillers, and adsorption energies of a variety of chemicals, analyzing the properties of filled vulcanizates, and direct SEM investigation of the worn surface of compounds after skid testing.

#### 8.2.2.4.1 Comparison of Polymer-Filler Interaction between Carbon Black and Silica

It is reasonable to believe that when a rubber is filled with rigid particles, it causes non-uniform stress distribution in the polymer matrix. In the case of extension[192], for a vulcanizate with strong polymer-filler interaction, the maximum stress is generated at a short distance from the filler surface into the polymer matrix. These regions may act as sites that favor the initiation and growth of internal voids and cracks. Accordingly, the rubber does not immediately detach itself from the filler surface but undergoes a rupture in the polymer matrix. For a vulcanizate having poor polymer-filler interaction, the dewetting stress at the direct interface between filler and polymer is lower. In this case, instead of matrix fracture, cavitation will occur at the filler surface by interfacial debonding or dewetting mechanisms[193–195]. Consequently, bare surface of filler may be exposed on the worn surface of vulcanizates.

**Table 8.5** $\gamma_s^d$ and specific components of adsorption energies of some polar chemicals on N234 carbon black, silica (ZeoSil 1165), and CSDPF 2214

| Items | Carbon black N234 | Silica | CSDPF 2214 |
|---|---|---|---|
| Test temperature/°C | 180 | 180 | 180 |
| $\gamma_s^d$ /(mJ/m$^2$) | 382 | 28 | 512 |
| $I^{sp}$($CH_3CN$)/(mJ/m$^2$) | 173 | 278 | 202 |
| $I^{sp}$(THF)/(mJ/m$^2$) | 96 | 271 | 86 |
| $I^{sp}$(acetone)/(mJ/m$^2$) | 86 | 264 | 90 |
| $I^{sp}$(ethyl acetate)/(mJ/m$^2$) | 48 | 206 | 53 |

It is well known that carbon black has stronger interaction with hydrocarbon polymer than silica. This is demonstrated not only by the energetic characteristics of the filler surface, but also by their in-rubber properties which are directly related to their reinforcing ability.

Energetically, carbon black is characterized by very high surface energy, especially the dispersive component $\gamma_s^d$. The $\gamma_s^d$ values for carbon black N234, silica, and CSDPF 2214 are listed in Table 8.5. The adsorption energies of different chemicals are plotted as a function of the area occupied by a molecule, $\sigma_m$ (Figure 5.21). While $\gamma_s^d$ of carbon black is about 13 times higher than that of silica typically used in tires, the surface energies of CSDPFs are even higher than those of commercial carbon blacks having similar surface areas. The polymer-filler interactions may be verified by the adsorption free energy of model chemicals of polymers. Generally speaking, the polymers used in tires are nonpolar or low-polar materials. The higher adsorption energy of the nonpolar adsorbents such as alkanes on carbon blacks, which is related to the high $\gamma_s^d$, would be indicative of stronger interaction with hydrocarbon rubbers. Here again the carbon blacks give high polymer-filler interaction.

When incorporated into rubber, the higher polymer-filler interaction of carbon black and CSDPF with polymers can also be demonstrated by higher bound rubber contents[66,196]. The bound rubber is the portion of the polymer in the filled uncured compounds which is unable to be extracted by good solvent for rubber and has been conventionally used as a measure of polymer-filler interaction. Such an observation is supported by the investigation of polymer-filler interfacial interaction by means of DSC. Compared to a silica-filled compound, the $T_g$ of the polymer is about 2°C higher for a carbon black-containing compound[197]. The $T_g$ of CSDPF-containing compound is still somewhat higher relative to carbon black. The increase in $T_g$ can be interpreted in terms of increased polymer-filler interaction due to the stronger physical adsorption, resulting in a reduction of the molecular mobility of the polymer at the interface. More evidence of the low polymer-filler interaction of silica compound is its substantially low stress at high elongation, such as 200% or 300%. At high elongation, the effect of filler agglomeration on the stress should disappear and polymer-filler interaction may now play an important role in the stress at high elongation. In the case of silica, the poor interaction with hydrocarbon polymer leads to slippage and detachment (dewetting) of the rubber molecules on the filler surface, resulting in relatively low stresses at high elongations (see Figure 6.26)[198].

While the poor polymer-filler interaction of silica is evident, a substantial complication is introduced due to addition of a coupling agent into the silica compound. One may argue that, in a practical tread formulation, coupling agents considerably increase the polymer-filler interaction. It has been observed that, with use of a coupling agent such as bis(3-triethoxysilylpropyl)-tetrasulfide (TESPT), bound rubber is higher in silica compounds compared to carbon black, and the modulus at 300% elongation can reach a level comparable to or even higher than that of carbon black-filled compounds. Therefore, the argument of the poor polymer-filler interaction of silica could not be valid for the coupled silica compound. However, the improvement of polymer-filler interaction is essentially due to the creation of covalent bonds between the polymer and filler surface. From the adsorption point of view, the application of coupling agents does not enhance the physical interaction between the polymer and silica. For carbon black-filled compounds, the filler-polymer interaction is essentially physical in nature[34,38]. On the other hand, for a silica having a surface area about 170 $m^2/g$, the practical TESPT dosage is about 8% of filler loading (by weight). If the entire coupling agent can react with the silica surface, the surface concentration of TESPT would be about 0.5 molecule/$nm^2$ (or 1 silane group attached per $nm^2$). For precipitated silica, the silanol concentration is 4–8 SiOH/$nm^2$. This implies that only a small portion of silanols react with coupling agents and small portion of silica surface is covered by the coupling agent. The majority of surface is in physical contact with polymer. This gives a lower polymer-filler interaction. This statement is supported by DSC results, showing that, by adding 8% TESPT, no effect of the coupling agent on the $T_g$ of the compound is observed[197]. It seems that a small amount of strong chemical bonds may be enough to provide necessary polymer-filler interaction for tensile modulus and strength. However, the fracture of the materials may still be initiated at the polymer-filler interface where the adhesion energy may be lower than the cohesive energy of the polymer matrix. Consequently, some bare surface of silica on the worn surface of a compound may be expected.

While intuitive reasoning leads one to the expectation that the silica would be exposed, this argument can be further reinforced if the tread compound is under a wet condition. It has been reported that for a highly silane modified silica, a small polar chemical such as acetonitrile is able to penetrate the graft layer and adsorb onto the silica surface[73,80]. It has also been demonstrated that, using toluene as a solvent, the bound rubber content of a silica-filled compound is drastically reduced and the swelling of its vulcanizate is significantly increased when the tests are conducted in an ammonia atmosphere[66,198]. This is due to the adsorption of ammonia molecules on the silica-polymer interface which separates the polymer from the silica surface. One may expect a similar effect of water due to its strong hydrogen bonding with the silica surface, especially uncoupled silanol groups. In the case of carbon black, ammonia does not show any significant effect on bound rubber and swelling behavior of the filled rubber, as would be expected for water as well. This suggests that under wet conditions, at the skin of the compound, the silica-polymer interaction would be further weakened by water adsorption whereas carbon black would not be affected.

The adsorption of water at the polymer-silica interface can be more safely demonstrated by the effect of water absorption on dynamic properties of filled rubber. Shown in Figure 8.31 are the strain dependence of the vulcanizates filled with silica, TESPT-coupled silica, and carbon black N234. As can be seen, after soaking the filled rubbers in water at 80°C for 10 days, a significant decrease in Payne effect (i.e., the difference in storage moduli between low strain amplitude and high amplitude) is found for TESPT-coupled silica, and to a much greater degree for uncoupled silica. No effect is observed on carbon black. The marked reduction in Payne effect by soaking the silica compound is a more reliable reflection of the adsorption of water on the silica surface in the vulcanizate. It is widely accepted that the Payne effect is a measure of filler agglomeration. The polymer trapped in the filler agglomerates behaves as filler in terms of stress-strain properties, resulting in increased filler effective volume, and hence a high modulus. The breakdown of the filler agglomerates by increasing strain amplitude would release the trapped rubber so that the apparent filler volume fraction and corresponding modulus would be substantially reduced[38]. Upon adsorption of water at the filler surface, the distance between aggregates increases and filler-filler interaction at the contact area decreases, thus resulting in reduced low-strain modulus and Payne effect. With regard to the carbon black, the unchanged Payne effect may suggest that no or a negligible amount of water molecules can be adsorbed at the filler-polymer interface due to the high polymer-filler interaction and low polarity of carbon black. Although these results were obtained at a relatively high temperature and longer soak time, the conclusion would be valid towards the wet skid test conditions. This is reasonable because, at the skin of the compound, the diffusion distance of the water in the compound is so small that the adsorption process may be quickly completed when the rubber compound contacts water.

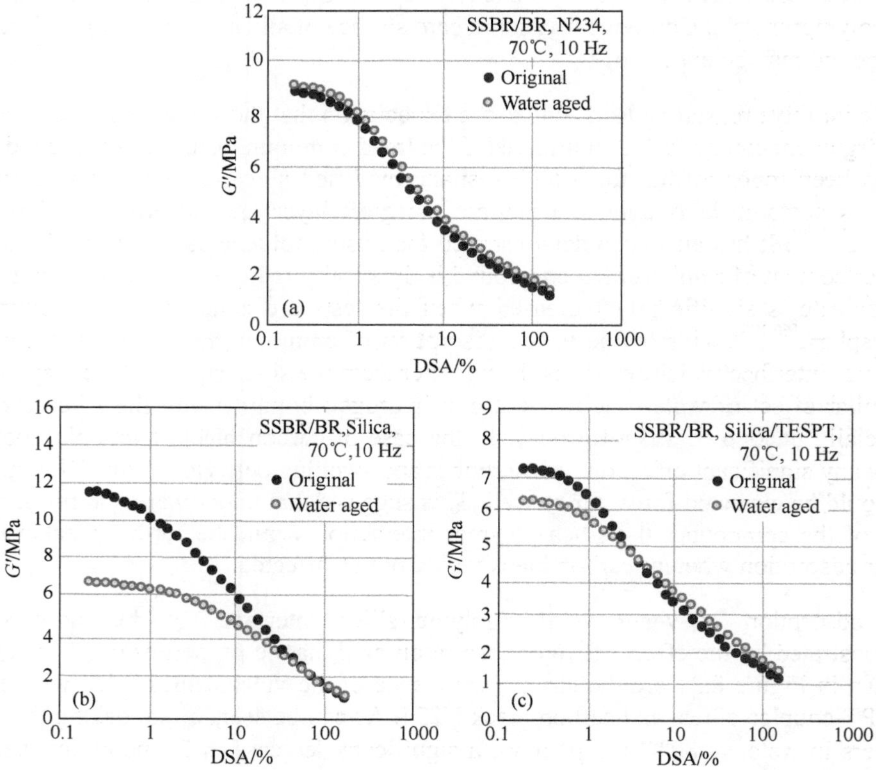


**Figure 8.31** Strain dependence of elastic moduli after soaking in water at 80°C for 10 days for the vulcanizates filled with a carbon black and silica. For carbon black (a) and silica with TESPT (c). For (b), the formulation is the same as (c) but without TESPT

In summary, it is proposed that the weaker polymer-filler interaction and high polarity of silica would allow the rubber on the filler surface to be peeled under the shear stress during skidding of the filled rubber on a track, resulting in the exposure of bare surface. This is especially true under wet conditions due to water adsorption. In contrast, under the same conditions, the surface of carbon black should remain covered by rubber due to its strong interaction with polymer and lower surface polarity. In many cases the surface characteristics of the vulcanizates filled with silica and carbon black can be identified by their effect on some test results.

#### 8.2.2.4.2 Effect of Break-in of Specimens under Wet Conditions on Friction Coefficients

It was observed that at the beginning of a skid test with the Grosch Abrasion and Friction Tester (GAFT) under wet conditions with a fresh test wheel (without run-in), there was no difference in friction coefficient between silica-and carbon black-filled compounds on a glass surface at 2°C, slip angle 14°, load 76.6 N, and high speeds (1.85 km/h). After break-in of the specimens under the same conditions for 50 minutes, the data had reached a steady state, at which point the side force of silica was about 15% higher than that of carbon black (Figure 8.32). One may argue that this difference could be related to the difference in the cure state of these two compounds because the specimen surface is generally over cured compared to the inside of the compounds due to heat transfer. The over-cure of the compound would influence the crosslink structure such as crosslink density and sulfur distribution in crosslinks, but this seems not to be the case for wet skid resistance. When the cure time of specimens of carbon black- and silica-filled rubbers were prolonged for an additional 10 minutes there were no changes in skid resistance measured by BPST. The changes in skid resistance during break-in may be readily interpreted by the change in the surfaces of these compounds. Before break-in, the fresh surfaces of both silica- and carbon black-filled vulcanizates may be covered by polymer, and thus no significant difference in skid resistance should be expected. Upon run-in of the specimens, silica surface may become exposed and carbon black surface may not. As a result, the silica would give higher skid resistance than carbon black, as discussed above. The rubber coverage of fresh surface of the cured rubber vulcanizates has been demonstrated by very low surface electrical conductivity. For carbon black compounds, after buffing the surface the conductivity is drastically increased. It was concluded from these results that the fresh surface of the filled compounds is rich with polymer compared to the bulk[199], probably due to filler-filler interaction, and the much lower surface energy of the polymer compared with that of fillers. The rich polymer on the surface would lead to a more stable state from the surface energy point of view of the compounds. This should be true for silica compounds as well.

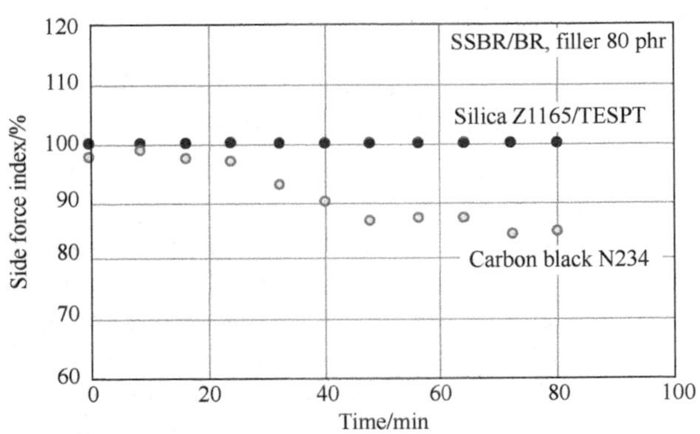


**Figure 8.32** Effect of run time of GAFT on side force index under wet conditions for vulcanizates filled with silica and carbon black

#### 8.2.2.4.3 Abrasion Resistance of Filled Vulcanizates under Wet and Dry Conditions

Abrasion results for carbon black and silica compounds are shown in Figure 8.33. The data under dry conditions were obtained from the Cabot Abrader[200] at 50°C and slip ratio of 14%. The wet abrasion resistance was determined by means of GAFT on a corundum surface with 180 grit at 14° and 25° slip angles, 5°C, a load of 50 N, and speed of 4.5 km/h On the dry surface, silica gives a little lower abrasion resistance than carbon black (88% *vs.* 100%). Under wet conditions, however, the abrasion index of silica compounds is much lower than that of carbon black (61% *vs.* 100% for 14° and 42% *vs.* 100% for 25° slip angles). Regardless of the difference in abrasion test methods which may result in a certain degree of uncertainty of the comparison, the relatively poorer abrasion resistance under wet conditions for silica compounds could be attributed to water adsorption at the interface which leads to weaker polymer-filler interaction.

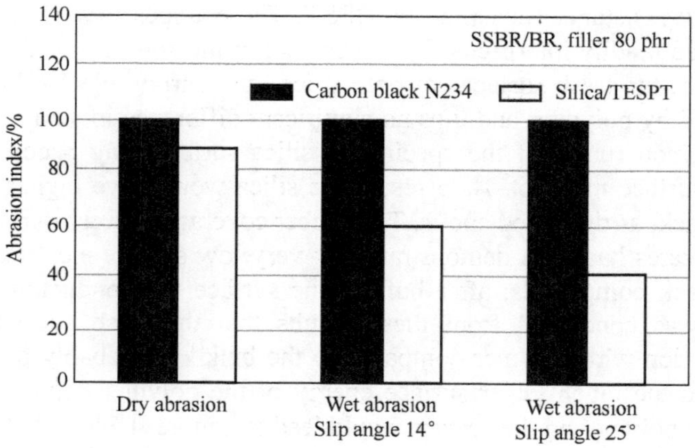


**Figure 8.33** Dry and wet abrasion resistance for carbon black-and silica-filled vulcanizates

#### 8.2.2.4.4 Observation of the Change in Friction Coefficients during Skid Test

During the friction test under wet conditions on BPST, the friction coefficient of the glass track surface is decreased. Obviously, the compound surfaces remove sharp protuberances initially, and then establish a smooth topography. In other words, the glass surface seems to be polished. While this phenomenon is minor for carbon black compounds, it seems to be critical for silica-filled vulcanizates. In fact, this process is so severe that after a certain period of testing silica compounds, reasonably reproducible results are unable to be obtained. Two reasons may account for these observations. One is the difference in abrasiveness between silica- and carbon black-filled vulcanizates. This may be attributed to their surface situation. If both carbon black aggregates are covered by polymer at the worn surfaces of the compounds, no significant difference in abrasiveness between silica and carbon black should be expected. Therefore, the highly effective reduction in friction for silica compounds can be interpreted in two ways: either both silica and carbon black aggregates are exposed at the surface, or silica is exposed and carbon black is shielded by rubber. It is common knowledge that due to its high hardness, silica is more abrasive than carbon black. Although this is true, it may not be the main reason. With this mechanism, the polishing effect would permanently remain. In fact, upon cleaning the substrates after the test, the friction coefficients can be substantially recovered. Another reason for the decrease in friction coefficient during skid test is that small fragments of rubber worn off during skidding stick on the tip of asperities, shielding sharp protuberances of the substrate. It may be reasonable that compared with carbon black, the poorer polymer-filler interaction of silica under wet conditions should lead to more rubber fragments being peeled off from the filler surface under shear stress during skidding.

#### 8.2.2.4.5 SEM Observation of Worn Surface

The assumption that bare silica exists, and rubber shields carbon black on the worn surface of compounds after wet skid testing has also been confirmed by direct observation with SEM. Shown in Figure 8.34 and Figure 8.35 are SEM images for passenger tire tread compounds filled with carbon black and silica (with coupling agent), respectively. After abrasion, the image of the silica aggregates in the compound is more evident at the sub-micron scale than that of its carbon black counterpart. It seems that carbon black aggregates on the specimen are certainly much more uniform over the surface range investigated while hard particles of silica are clearly seen.

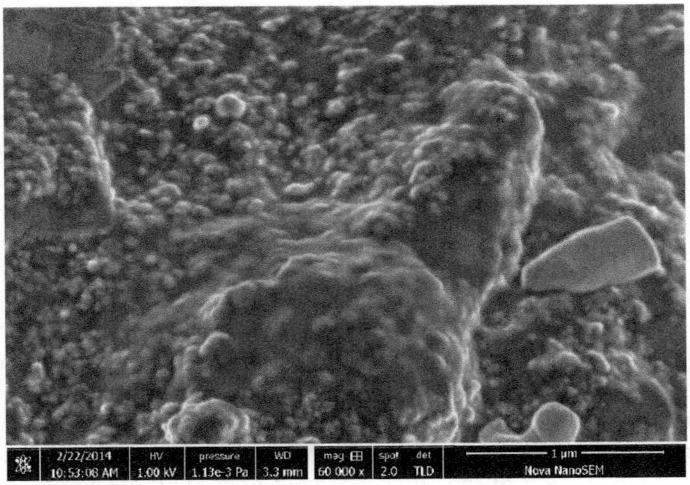


**Figure 8.34** SEM images of the surface of carbon black tread compound after wet skid test

The discussion above and the experimental observations presented are all consistent with the hypothesis that at the very top skin of the worn surface of filled compounds, filler aggregates are exposed in the case of silica and rubber-shielded for carbon black. This will be the basis of further discussion about the difference in wet skid resistance among different fillers.

### 8.2.3 Wet Skid Resistance of Tire

Through the narrative of last sections, we know that in addition to dynamic properties

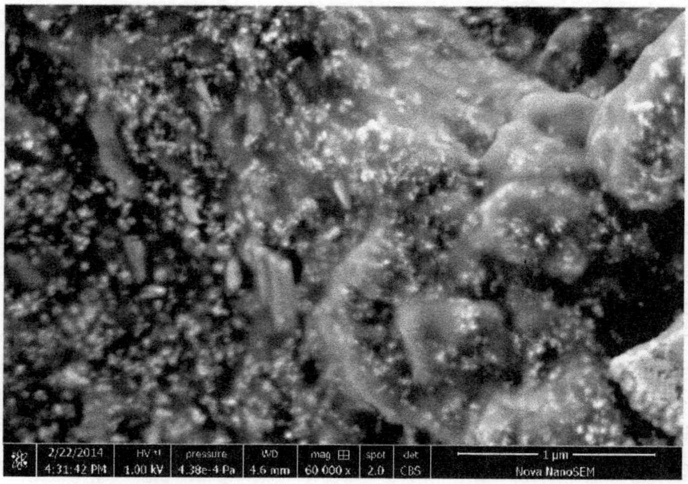


**Figure 8.35** SEM images of the surface of silica tread compound after wet skid test

of the filled vulcanizates, hydrodynamic lubrication (HL), elastohydro-dynamic lubrication (EHL), especially on the micro scale, and boundary lubrication (BL) mechanisms play a very important role in friction under wet conditions. It has been also demonstrated that after wet skid tests, the morphology of the worn surface of vulcanizates is quite different. While carbon black aggregates are covered by rubber, silica aggregates are exposed. This will substantially influence the lubrication mechanisms of the tire on the wet pavement, hence the wet skid resistance or grip.

#### 8.2.3.1 Three Zone Concept

It is well known that when a vehicle is braked on wet pavement surfaces, skidding of the vehicle occurs, which has been attributed to the existence of a thin film of water adhering to the road surface. If the thickness of the water film is over a few millimeters, dynamic hydroplaning takes place. This is characterized by a hydrodynamic upward thrust in the wedge region ahead of the contact area. This wedge region penetrates backwards from the leading edge of "contact"[201]. In normal wet weather and road conditions, however, the water film is generally much thinner. With regard to wet skid resistance, a very thin film of water (about 0.01 mm to 0.1 mm) on the road surface is just sufficient to destroy intimate tread-to-road contact by the mechanism of elastohydrodynamic separation. In this case, the resulting skidding phenomenon from such a thin film has been termed "viscous hydroplaning". Practically, the problem of viscous hydroplaning is much more serious because it is very frequently encountered. Further reduction in the water film to only few molecular layers, as in the case of the beginning of a rain shower, leads to water boundary lubrication. This type of lubrication, as discussed before, may not result in any reduction in skid resistance compared with a dry road surface[202]. In fact, all these situations may occur simultaneously at the tread-to-road contact area, and the different mechanisms may come into play to different degrees, depending on road surface, weather conditions, speed, vehicle, driver habits, and more importantly, tire design. Both the tire structure and the materials, especially in the tread compound, are important aspects of tire design.

For application of the HL theory for tires, the best model for the wet traction of tires is based on a "Three-Zone Concept". It was originally put forward by Gough[203] for sliding wheels and later extended by Moore[204] to cover the case of the rolling tire. This concept suggests that at speeds below the hydroplaning limit, the tire contact area can be divided longitudinally into three distinct regions (Figure 8.36), which are identified as follows[201]:

- Squeeze-film zone: In this forward region of contact area, a water wedge is formed due to the displacement inertia of the intercepted water film. The thickness of this film decreases progressively as under pressure, the individual tread elements pass through the contact area, squeezing out the water film. The key mechanism operating here is HL and EHL, but very little friction is generated in this zone.
- Transition zone: The transition zone begins when the tire elements, having penetrated the squeeze-film, start draping dynamically about the major asperities of the road surface, making contact with the lesser asperities. In this zone, a progressive breakdown of the water film occurs down to a thickness of a few layers of water molecules. Therefore, a mixed-lubrication regime exists, which is part EHL and part BL. The effective friction coefficient changes widely from a very low value of viscous hydroplaning at the leading edge of the transition zone, to the friction value of BL at the trailing edge of this zone, which is comparable to dry friction.
- Traction zone or actual contact zone: This region is the rear part of the contact area, beginning with the end of transition zone. It is the zone where most of the traction or skid resistance is developed. Here, the lubricating water film has been totally or substantially removed, and vertical equilibrium of the tread elements on pavement surface has been attained. In this zone, BL is the dominant mechanism.

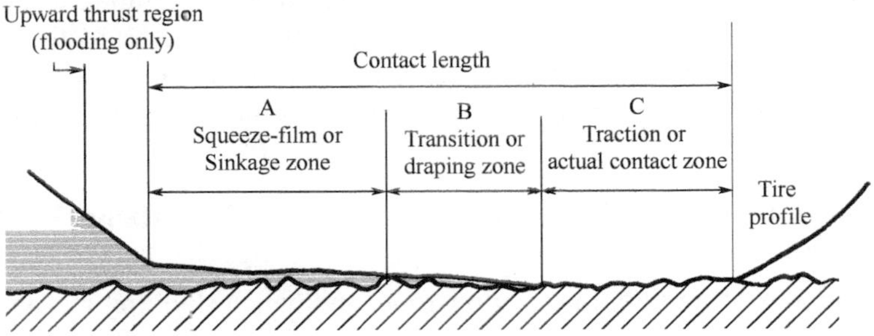


**Figure 8.36** The three-zone concept of tire contact on road in rolling under wet condition

The overall skid resistance or traction performance will be determined both by the relative size of the three zones, and the level of friction in each zone. Thus, to improve wet skid resistance, the goals should be to increase the traction zone relative to the squeeze-film and transition zones, and to increase the level of friction in the transition and traction zones.

### 8.2.3.2 Effect of Different Fillers in the Three Zones

#### 8.2.3.2.1 Minimization of Squeeze-Film Zone

The size of the squeeze-film zone is of primary importance in any investigation of hydroplaning or skidding phenomena. This is first determined by the rate of the squeezing, i.e., the time taken for a particular tread element to squeeze through the water film. According to Moore[205], the time of squeezing, $t_{sf}$, is given by:

$$t_{sf} = \frac{L_{sf}}{V} = \frac{K_{sf}}{W} \times \frac{\eta A^2}{f(\varepsilon / h) h_i^2} \left(1 - \frac{h_f^2}{h_i^2}\right) \tag{8.10}$$

where $L_{sf}$ is the length of squeeze-film zone, $V$ is speed, $W$ is the normal load on a tread element, $\eta$ is the water viscosity, $A$ is the area of tread element, $h_i$ is the initial film thickness, $h_f$ is the final film thickness, $K_{sf}$ is a constant related to the shape of tread element, and $f(\varepsilon/h)$ is the dimensionless surface roughness parameter expressed by a polynomial:

$$f(\varepsilon / h) = C_0 + C_1 (\varepsilon / h) + C_2 (\varepsilon / h)^2 + \ldots \tag{8.11}$$

where $\varepsilon$ is the average peak-to-trough measurement of asperity size in the rough surface, $h$ is the water film thickness, and $C_i$ ($i$ = 0, 1, 2, etc.) are constants for the particular pavement geometry.

For a given tire design, the rate of water film squeezing is dependent on the initial film thickness, load, water viscosity, and surface texture. If the time of squeezing is minimized, then the time remaining for the development of adequate skid resistance in the transition and actual contact zones is maximized. This is accomplished most readily by selecting a road texture having adequate drainage[206] and reasonably sharp asperities[201]. On the other hand, for a given pavement surface, reduction of speed is an efficient way to establish a high level of skid resistance. It can be seen that fillers in the tread compound have an influence in this zone only through its effect on water film thickness.

#### 8.2.3.2.2 Minimization of Transition Zone and Maximizing Its Boundary Lubrication Component

For a given squeeze-film rate, the traction zone, and hence skidding resistance, can be maximized by reduction of the transition zone, which is determined by draping time, $t_d$. It has been shown that the draping time can be expressed as follows[205]:

$$t_d = \frac{L_d}{V} = \left(\frac{\rho d}{g}\right)^{1/2} \left(\frac{A}{E^*}\right)^{1/3} (RP_i)^{-1/6} \tag{8.12}$$

where $L_d$ is the length of transition zone, $\rho$ is density of the tread element, $d$ is the tire crown thickness, $g$ is gravity, $R$ is radius of the asperity of road surface, $P_i$ is the inflation pressure of tire, and $E^*$ is the complex modulus of the tread compound. This equation suggests that in the transition zone, the compound modulus plays an important role in the draping process, in addition to road surface texture or roughness and inflation pressure. At the same speed, an increase in the compound stiffness would reduce the transition zone, giving a longer traction zone.

The road surface roughness can be characterized by its topography in terms of its fractal dimension. The road surface texture can be analyzed by moving an average of ordinates into components or scales ranging in pitch size from microscopic up to as large as one or two centimeters (see Figure 8.37)[207]. Moore emphasized the importance of small-scale- or micro-roughness. The existence of an adequate micro-roughness at the tips of road surface asperities is essential to provide grip between the tire tread and the road under wet conditions, enhancing wet skid resistance. In the absence of this micro-roughness, the slipping rubber in the contact area generates hydrodynamic pressure wedges on individual asperities of the road texture, tending to entrain a liquid film over asperity peaks. Such elastohydrodynamic action separates tire tread and surface locally and can produce a dangerous condition known as viscous hydroplaning.

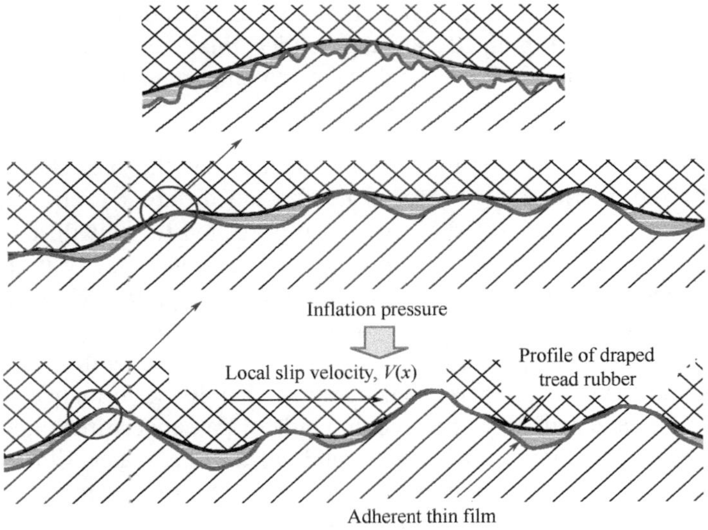


**Figure 8.37** Micro elasto-hydrodynamic lubrication *vs.* micro-surface roughness of pavement

When the EHL theory is applied to local micro-roughness, a silica compound will be more favorable towards wet skid resistance than a carbon black compound. This is due to the difference in surface characteristics between these two compounds. In these compounds, carbon black aggregates are covered by a rubber film while the silica surface is bare. The modulus of the rubber film is much lower than that of the bare surface of silica. Under the same conditions of speed, surface texture, and temperature, silica should give a thinner water film and a shorter draping time at the compound surface than carbon black according to Eq. 8.12 and the previous discussion on water film thickness[199,202]. This implies that the water film is more easily broken down by silica, and in the transition zone, the BL component is much more developed for the silica compound. Although under BL conditions the friction of silica on a road surface is somewhat lower than in the dry state[202], it is much higher than hydrodynamic lubrication (HL), which occurs for carbon black filled compounds under the same conditions. As a result, a higher friction coefficient in the transition zone would be expected for silica compounds.

#### 8.2.3.2.3 Maximization of Traction Zone

Any increase in the relative size of the traction zone will increase the wet skid resistance, since the majority of frictional force is developed in this region. The lubrication in the traction zone is of a boundary nature, so that the difference in BL between silica and carbon black compounds would be reflected in their wet skid resistance. At boundary conditions, silica, as a rigid material, gives lower friction than it does in the dry state. In the case of rubber, the friction is comparable to (if not higher than) dry friction. Due to the higher friction of rubber on dry road surfaces compared to silica, and the features of the worn surfaces of these compounds, the carbon black compound is favored in this zone relative to the silica compound even under wet skid conditions. It should be pointed out that under this condition, the deformation component of the frictional force would play an important role, i.e., the dynamic properties of the compounds would be equally important.

From the discussions above, it may be concluded that the wet skid resistance is determined by the distribution of three zones having different friction regimes. The effect of different fillers on wet skid resistance is related to their different effects in each of these zones, especially in the transition and traction zones. With regard to wet skid resistance ranking among fillers, from the lubrication point of view, it appears that increases in the squeezing water film and transition zones will favor the silica compound while the development of the traction zone will generally favor the carbon black compound. Based on this consideration, better performance in wet skid resistance can be expected from silica compounds of passenger tires under conditions of higher speeds, smother road surface, and where the water film squeeze and transition zones are dominant. If the traction zone is more developed, carbon black should be the preferred filler for wet skid resistance.

### 8.2.3.3 Influencing Factors on Wet Skid Resistance

Before discussing the factors that influence the wet skid resistance, it is better to introduce the method and equipment used for testing wet skid resistance, the British Portable Skid Tester (BPST), the Grosch Abrasion and Friction Tester (GAFT), as well as tire tests on a proving ground.

### Tests with BPST

BPST (shown in Figure 8.38) measurements were carried out at room temperature with a modified equipment[173]. It is basically a pendulum tester used to measure energy loss when a rubber pad slides over a flat wet surface under standard conditions. The energy loss, together with integral formula of contact forces, was used to obtain the friction coefficient.

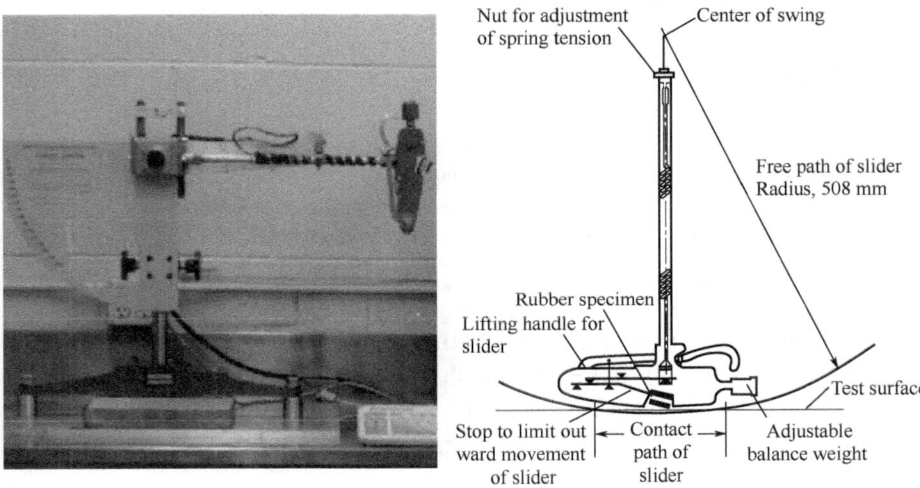


**Figure 8.38** Test apparatus of the British Portable Skid Tester (BPST)

### Tests with GAFT

Wet skid resistance was also measured by means of GAFT. It basically measures the side force when a specimen in the form of full rubber wheels is run on a rough disk at a run angle, under given load, velocity, and temperature.

Similar to the cornering of a car, as an element of rubber wheel impinges on the track surface as the wheel rotates, it is progressively distorted sideways along a path as shown in Figure 8.39. Towards the rear end of the contact area, as the load intensity decreases and the particular friction limits are reached, the rubber element returns quickly to the meridional (natural) position, involving a degree of local slippage. Consequently, side forces are derived from distortions within the wheel/substrate contact area. The distortion angle, which is the angle between the plane of the wheel and the traveling direction has been termed "slip angle" even though no overall "slip" or skidding is involved in the mechanism which generates side force. At constant slip angle the side force is determined by the friction coefficient, the load applied on the wheels, and slip angle as well. From the side force generated during running, the side force coefficient (side force/vertical load) can be calculated.

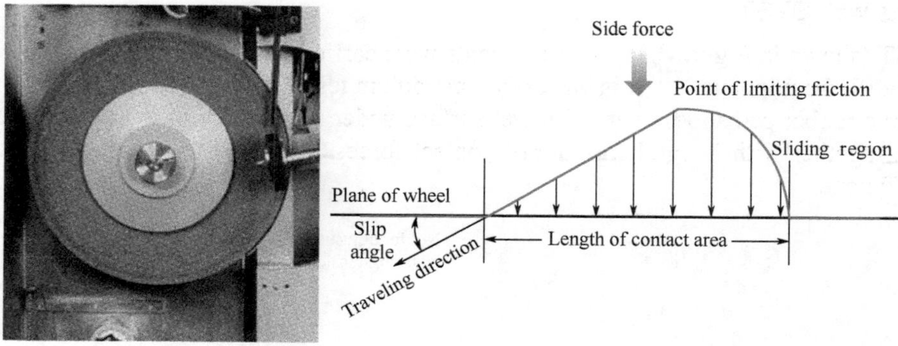


**Figure 8.39** Side force generation from distortions within the wheel/substrate contact area

### Tire Tests on a Proving Ground

In order to confirm the results from lab tests, additional tests using a car equipped with ABS were conducted on a proving ground. The experimental tires were new 185/70R14 passenger tires retreaded with different compounds. The wet skid resistance was measured by wet ABS braking distance at a speed from 40 km/h to 130 km/h on four types of road surfaces: ceramic slabs (0.1$\mu$), basalt slabs (0.3$\mu$), polished concrete (0.4$\mu$), and high friction asphalt (0.8$\mu$).

## 8.2.3.3.1 Effect of Test Conditions on Wet Skid Resistance

### Speed

Inspection of Eqs. 8.10 and 8.12 suggests that, for a given tread compound under given test conditions of temperature, load, water film thickness, and track surface, the lengths of the squeezing-film zone and the transition zone are proportional to speed. Consequently, if the contact length of the tread compound with track surface is kept constant, the traction zone will increase with decreasing speed. At a high speed, the wet skid resistance would be favored by silica compounds due to the early breakdown of water film and an increase in boundary component. With decreasing speed, the advantage that the silica has over the carbon black and CSDPF will diminish due to the reduction of the squeeze-film zone and transition zone and the increase in traction zone. This can be seen clearly from Figure 8.40 in which the rating of the friction coefficient is derived from the side force measured by means of GAFT at different slip angles and water temperatures, taking the silica compound as a reference. For example, at 2°C and speed of 1.87 km/h, the carbon black compound gives a side force coefficient about 30% lower than that of the silica compound. When the speed is reduced to 0.183 km/h, the difference is diminished to about 10%. A similar trend is also observed for the CSDPF 2214 containing compound.

### Temperature

The results obtained from two different substrate surfaces, namely ground glass and corundum (180 blunt), under the conditions of load 76.6 N, speed 1.89 km/h, and slip

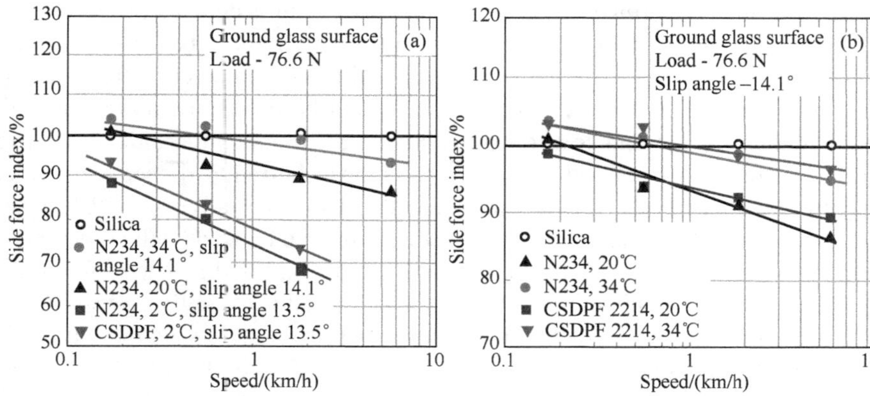


**Figure 8.40** Side force indexes measured with GAFT as a function of speed for vulcanizates filled with a variety of fillers

angle 14.1° are presented in Figure 8.41 as a function of temperature over the range from −2 (salt water) to 63°C. It can be seen that the advantage of silica in wet skid resistance is also diminished with rising temperature. Beyond 34°C, the skid resistance of silica compounds on ground glass surface is lower than carbon black- and CSDPF-filled rubber. This observation can be readily interpreted as resulting from the temperature dependence of different lubrication mechanisms involved in skidding.

Under wet conditions, two effects may be expected from a change in temperature. Upon an increase in temperature, the viscosity of water decreases along with the modulus of the filled rubber. As can be seen from Eq. 8.10, a decrease in water viscosity would result in a substantial reduction in squeezing film time, while a decrease in modulus simultaneously leads to an increased draping time (Eq. 8.12). These effects can affect the length of the traction zone in different ways. The accomplishment of a squeezing process in a short time at high temperature will maximize the length of the traction zone in which most skid resistance is developed. On the other hand, the extended transition zone at higher temperature, due to the decrease in modulus, will tend to minimize the traction zone. The net effect of temperature on the relative sizes of the transition and traction zones will be determined by the changes in the squeezing film time, $t_{sf}$, and draping time, $t_d$. Upon increasing temperature, for example, from 0°C to 40°C, the viscosity of water decreases by a factor of 2.75 (from 1793 μPa • s to 653 μPa • s), but the modulus of the compounds used in this test was reduced by only about 50%. However, while the squeezing film time is reduced linearly with decreasing water viscosity (Eq. 8.10), the draping time increases with a decrease in the cube root of the modulus (Eq. 8.12). This suggests that the traction zone will be appreciably extended by speeding up the water film squeeze process at higher temperatures, while the elevated temperature produces only a slightly longer transition zone. Consequently, the net effect of elevated temperature on the length of the traction zone should be significantly positive; thus, carbon black and CSDPF have an advantageous influence on wet skid resistance at higher temperatures.

An alternative interpretation of the speed and temperature dependencies of the skid resistance among the compounds filled with different fillers is based on viscoelastic mechanisms. Grosch[208] also found that, compared to carbon black, silica compounds give lower side force coefficients at low speeds and higher temperatures, and higher values at higher speeds and lower temperatures. By plotting the side force friction obtained at different speeds and temperatures against the reduced speeds obtained by multiplying the speed with shift factors calculated from the universal WLF equation, he was able to establish master curves. At higher reduced speeds, silica was better in wet skid resistance, whereas at lower reduced speeds, carbon black was favored. The higher side force coefficient of the silica compound at high speeds and low temperature was attributed to its higher dynamic hysteresis at low temperature relative to carbon black. However, when the skid resistance of CSDPF-filled compound is included for comparison, the similarity of its viscoelastic properties with those of silica compounds is not reflected in the wet skid resistance. Its similar side force coefficient with that of the carbon black compound is not in line with the large difference in their viscoelastic properties either (see Figure 8.24, Figure 8.25, and Figure 8.26). We therefore conclude that the differences in speed and temperature dependencies of the wet skid resistance among the compounds with different fillers are mainly determined by lubrication mechanisms, although the effect of their viscoelastic properties is also a factor. This is true at least for the speeds and temperatures covered in the test results presented in Figure 8.40 and Figure 8.41.

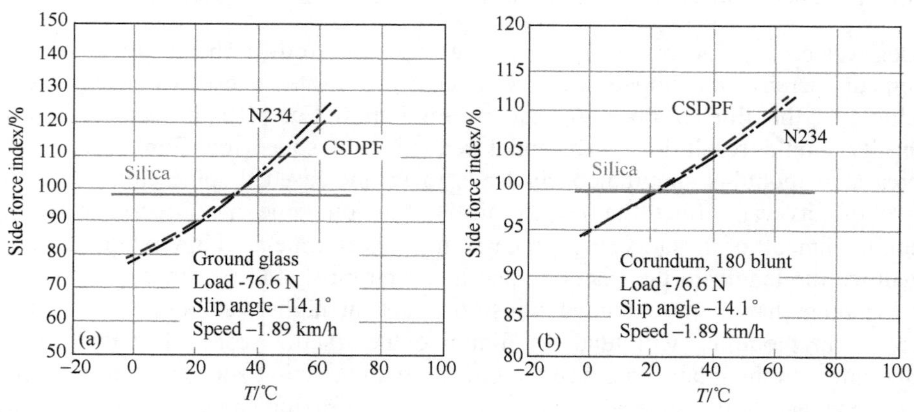


**Figure 8.41** Side force indexes measured with GAFT as a function of temperature for vulcanizates filled with a variety of fillers

## Surface Roughness

High roughness or high relative sharpness of asperities permits high localized pressures to be developed between the rubber and track surface, which ensures physical contact, hence optimum sticktion even under the most severely wetted conditions. However, the primary importance of surface roughness of a track is closely related to the squeeze-film time. When the squeeze term of Reynolds' equation is applied to a rough surface, squeezing-film rate can be expressed as[186]:

$$\frac{\mathrm{d}h}{\mathrm{d}t} = -2.37\frac{Wh^3}{\eta A^2}\left[1 + C_1\left(\frac{\varepsilon}{h}\right) + C_2\left(\frac{\varepsilon}{h}\right) + \ldots\right] = -C\frac{Wh^3}{\eta A^2} f\left(\frac{\varepsilon}{h}\right) \tag{8.13}$$

where $C$ is a constant. From this equation, it can be concluded that increasing surface roughness would significantly increase the drainage rate in the void spacing between roughness elements. As a consequence, the squeeze-film time would be lessened appreciably. On the other hand, upon increase in roughness, the draping time is affected only to a very small extent because $t_d$ changes with the 6$^{th}$ root of $R$, the asperity radius (Eq. 8.12). Therefore, the reduction of $t_{sf}$ will directly lead to an increase in the traction zone. Thus, the advantage of silica over carbon black should be diminished when the track surface becomes rougher. This has been confirmed experimentally. As can be seen from Figure 8.41, the difference in the side force coefficients between silica and carbon black measured by GAFT at high speed on ground glass is significantly reduced when a corundum track whose surface is much rougher than that of ground glass is used. For example, at 0°C, the side force coefficient of the carbon black-filled vulcanizate on a ground glass track is 21% lower than that of its silica counterpart, but on a corundum surface, the difference is reduced to 4.5%. In this test, the CSDPF shows very similar behavior to carbon black. In addition, on the side force coefficient/temperature curves of the three fillers, crossover points are observed beyond which the carbon black and CSDPF show higher friction than silica. This point is about 33°C with ground glass surface, and it drops to 20°C on corundum. Although there is some degree of uncertainty due to the difference in the physical properties between glass and corundum, the effect of the surface roughness should be mainly responsible for the observed phenomenon. On the rougher surface, the reductions in the squeezing-film zone (Eq. 8.10) and transition zone (Eq. 8.12) would maximize the traction zone, which is favorable for carbon black and CSDPF 2214. This is especially true at higher temperatures, in which the traction zone can be further increased.

The effect of road surface roughness on wet skid resistance was also investigated with a passenger vehicle equipped with ABS. The skid resistance was measured by the stopping distance on wet surfaces with water depth of 1 mm at 11–15°C. In Figure 8.42, the results of the stopping distance index measured at 50 km/h and 60 km/h are presented. It was found that tires having tread compounds with different fillers do not show a significant difference on the 0.1 $\mu$ ceramic slabs surface. However, on the 0.3 $\mu$ basalt slabs surface, silica shows a considerably higher wet skid resistance than carbon black and CSDPF. On the rougher 0.4 $\mu$ polished concrete surface, the difference in stopping distance between silica and other fillers is somewhat reduced. This difference is further diminished with increasing surface roughness. On the 0.8 $\mu$ high friction asphalt surface, the wet skid resistance of CSDPF 2214 is better than its silica counterparts.

### Load

With increasing load, the amount of BL increases gradually while EHL decreases. As shown by Eqs. 8.10 and 8.13, squeezing-film time is inversely related to load and squeezing-film rate increases with load. Upon increasing load, the increase in the traction zone, due to the reduction in lengths of both squeezing-film and transition zones, would result

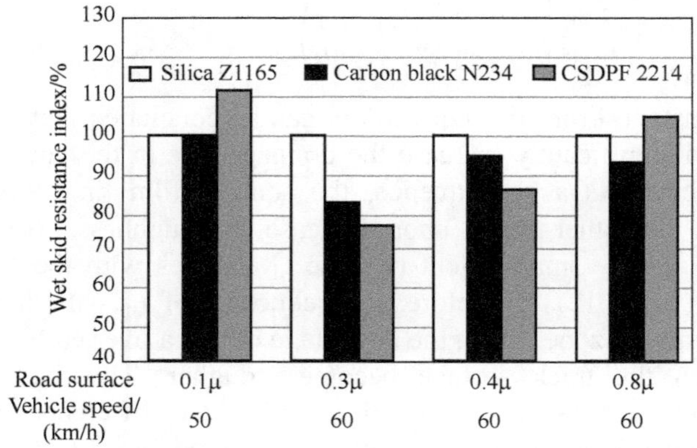


**Figure 8.42** Wet skid resistance index on difference track surfaces for passenger tire tread compounds filled with various fillers

in favorable ranking for carbon black and CSDPF, in relation to silica. This has been demonstrated experimentally with GAFT.

The effects of load on the side force coefficients obtained on ground glass with a speed of 1.44 km/h and a temperature of 3°C at different slip angles are shown in Figure 8.43. At lower loads, which correspond to the condition of passenger tires (around 40 N), silica gives the highest friction force. However, its advantage diminishes with increasing load. At high loads the CSDPF shows the best wet skid resistance and silica becomes the worst, with the crossover point occurring at around 60 N for a 14° slip angle. The same trend can be seen for carbon black, but the crossover point occurs at a higher load (around 75 N). The decrease in silica ranking with increasing loads can be explained based on a change in the relative dimensions of the three lubrication zones. However, the fact that the CSDPF compound generated significantly higher side forces compared with the carbon black compound, implies that other mechanisms influencing wet skid resistance are more important. Among others, the viscoelastic properties of the compounds may account for this observation. Although the dynamic properties of the compound play a role throughout the transition and traction zones, the role is more significant or dominant in the latter. As the traction zone is increased by increasing load, dynamic hysteresis becomes a major factor that governs the overall wet skid resistance. This favors CSDPF due to the higher dynamic hysteresis and lower modulus of CSDPF-filled compound at low temperatures.

If the better performance of CSDPF 2214 at higher load, in relation to carbon black, is attributed to its higher hysteresis and lower modulus at low temperature, this principle should be applicable to the silica compound since its viscoelastic properties are essentially similar to those of CSDPF. For the silica compound, the benefit from its dynamic properties at low temperature would be counter-acted by its deficiency in BL due to the lower friction of the exposed silica particles in the traction zone.

If the better performance of CSDPF 2214 at higher load, in relation to carbon black, is attributed to better dynamic properties at low temperature, the same argument could also be applied to higher temperature results, as the traction zone is also more developed due to the low viscosity of water. In fact, there is no significant difference between carbon black and CSDPF at higher temperatures (Figure 8.41). During the load study, the test temperature was 3°C while in the temperature study, the crossover points, beyond which the wet skid resistance of carbon black and CSDPF is higher than silica, is about 33°C. Over the temperature range from 33°C to 63°C, the hysteresis of the CSDPF 2214 compound is much lower compared to the carbon black compound, but the dynamic modulus is also lower. Consequently, the advantage of dynamic properties of the CSDPF 2214 compounds, is not observed in this particular test. This argument is supported by the tire test results (Figure 8.42). The higher wet skid resistance measured on the rough surface and relatively low temperature (11–15°C) for CSDPF 2214 compounds can be attributed to its developed traction zone and higher dynamic hysteresis at lower temperature.

### Slip Angle

The effect of slip angle on the side force coefficient as a function of load is shown in Figure 8.43 for angles of 10°, 14°, and 25° at constant temperature and speed. While

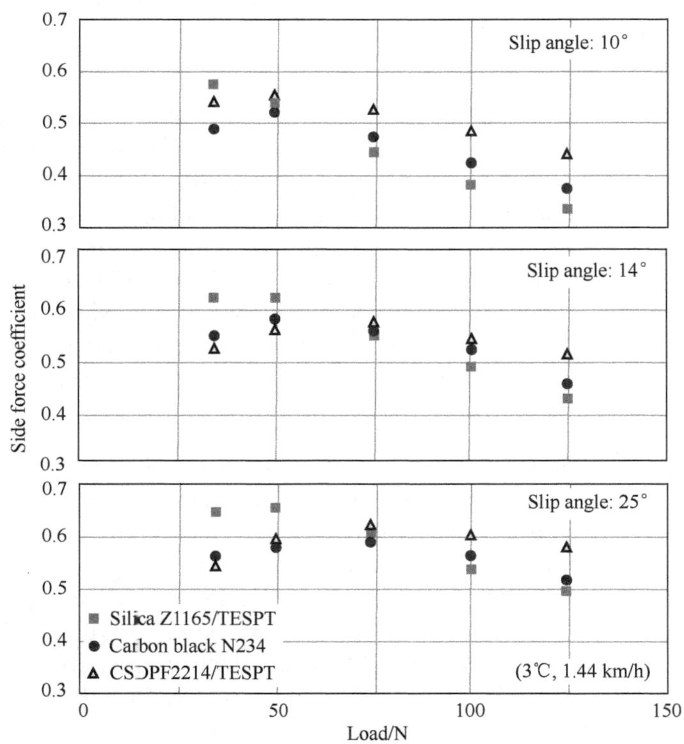


**Figure 8.43** Side force coefficients as a function of load at different slip angles for passenger tire tread compounds filled with various fillers

the overall side force coefficients increase with increasing slip angle, as a larger distortion of the tread rubber is generated, the advantage of the carbon black compound over its silica counterpart at high load diminishes gradually. In the side force coefficient/load plots, the crossover points of carbon black with silica are 64 N, 75 N, and 90 N for the slip angles 10°, 14°, and 25°, respectively. This increase in crossover point can be attributed to a reduction of the contact area of the wheel/substrate at high slip angles, caused by quickly reaching the overall friction limitation, accompanied by increased sliding at the rear of the contact area. In other words, with increasing slip angle, the traction zone in the contact area is reduced, hence the role of BL in the side force generation reduces. This is favorable for silica-filled compounds. Similar phenomena are observed for the CSDPF 2214-filled compound, but even with a high slip angle the skid resistance it is still significantly higher than the compounds filled with silica and carbon black. Accordingly, the crossover points of CSDPF 2214 with silica are 45 N, 60 N, and 75 N at slip angles of 10°, 14°, and 25°, respectively, which is significantly lower compared with those for silica/carbon black. The better performance of CSDPF over carbon black is certainly related to its higher tan $\delta$ and lower modulus at lower temperature, even though from lubrication point of view it may not be better due to its 20% silica coverage.

#### 8.2.3.3.2 Effect of Compound Properties and Test Methods on Wet Skid Resistance

##### Vulcanizate Hardness

Compared with the importance of hysteresis in wet skid resistance, the role of the hardness of the tread compounds has been much less discussed in the literature. The available reports on lab and tire tests regarding the effect of hardness on wet skid resistance are quite conflicting. With increasing compound hardness, some have shown an increase in wet friction[209–214], and some reported a decrease[209,215] or no effect[216,217]. This may be due to the fact that the hardness effect is masked by other compound properties, especially hysteresis. It may also be related to differences in test conditions, road/substrate surface, water depth, speed, load, test equipment, etc., as different conditions may change the lubrication mechanisms in which hardness plays different roles.

The effect of hardness of tread compounds on skid resistance measured with BPST and GAFT were investigated using carbon black N234 compounds. The compounds were varied in hardness by adjusting the carbon black loading (from 50 phr to 80 phr), oil level (from 0 phr to 30 phr), and dosage of sulfur (from 0.8 phr to 2 phr). The BPST tests were conducted at 23°C on sandblasted glass and GAFT tests were carried out on a corundum surface at 5°C with a load of 50 N, speed 0.2 km/h and slip angle 14°. Figure 8.44 shows clearly that compound hardness does have an effect on the ranking of the wet skid properties. While the BPST value decreases linearly with hardness, the GAFT value increases with hardness. Although the scatter of the data is quite large, the trends seem to hold up quite well over this large range of compound hardness. According to Veith[218], if tests are done under essentially BL, softer rubbers will yield higher friction due to increased true contact area. If wet tests are performed under the conditions where HL predominates, higher rubber hardness produces higher overall friction coefficient. The harder rubbers resist elastic deformation, which favors reduction of water squeeze zone, increasing friction performance. This can explain the GAFT results, for which HL may be dominant due to the relatively low load and low temperature. It is, however, very hard to explain the BPST results, even though the test wheel of GAFT rolls on substrate at certain slip angle while test pad of BPST slides over sanded glass. The reason for the different trend with hardness may be due principally to changes in the length of footprint in the direction of travel, although the effect of the hysteresis of the compounds with different hardness cannot be excluded. For GAFT wheels, the footprint at a load of 50 N is about 16 mm for a soft compound, but it decreases to 12 mm for the hard materials. In the case of BPST pads, the contact lengths in the sliding direction is about 1.2 mm for the hard materials, but the lengths for soft compounds is almost twice as long. It therefore appears that in the case of GAFT tests with increasing hardness, the effect of squeezing water film to increase the BL zone plays a dominant role. In this case, hard compounds would be preferred regardless of the lower friction coefficient in the traction zone. With regard to BPST tests, although the short footprint length in sliding direction may generally give a short traction zone, the significant increase in relative contact length for the soft compounds would result in an overall improvement of wet friction.

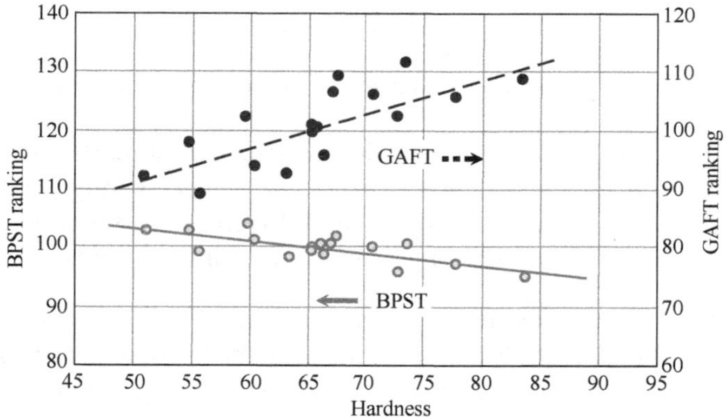


**Figure 8.44** Wet skid resistances measured with BPST and GAFT as a function of hardness of carbon black filled vulcanizates

## Braking Systems: Locked Wheel *vs.* ABS

At the start of braking, brake pressure is increased, generating a braking force at the wheels, which reduces vehicle speed. The percentage difference between the vehicle and the wheel speed is defined as brake "slip", which increases until the peak value on the friction coefficient $\mu$-slip curve is reached. Beyond this maximum, there is no further increase in braking force, even if the brake pressure is further increased. In fact, there is a significant reduction in braking force (Figure 8.45). In this process, a transition from static friction to kinetic friction takes place. The lowest coefficient of adhesional friction is attained for locked wheels. In this case, the braked wheels are no longer rotating and the brake slip has the value 100%. On the other hand, the corresponding lateral force is optimal near zero slip, and monotonically decreases to a negligible level at 100% slip when the wheel is locked. This is why the steering control and directional stability is lost for locked wheels. The purpose of an antilock braking system is to maintain the braking force near the peak adhesion, by employing cyclic release and re-apply techniques with a frequency from 3 Hz to 5 Hz. With this system, while providing a high level of braking force, sufficient steering control can also be achieved.

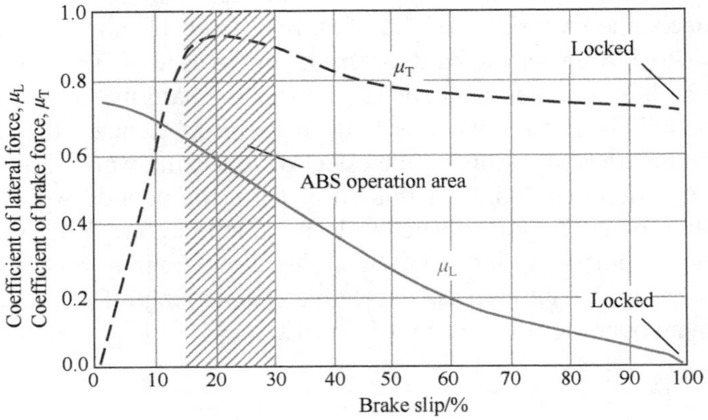


**Figure 8.45** Brake and lateral force coefficients as a function of brake slip

Soon after the commercialization of the silica passenger car tire, it was recognized that its advantage in wet skid resistance over carbon black is only noticeable if the passenger vehicles are equipped with ABS and driven on a smooth pavement. When the locked wheel brake system is used, carbon black tread compounds perform better.

According to Moore[219], below the hydroplaning limit and if the water film thickness is sufficiently small that hydrodynamic theory could be applied, the separating force in rolling would be twice that in sliding. This is because for rolling wheels, two surfaces, both the tire tread and pavement, entrain water into the wedge. For sliding wheels, however, only the pavement is performing this function. With ABS, as the tire intermittently rolls, the traction zone is smaller, which favors silica. For locked wheel, however, the tire tread slides on the pavement continuously, thus the traction zone is more developed. This suggests that carbon black would be the best filler, since the BL mechanism is dominant.

## BPST and GAFT *vs.* ABS

It has been reported that there is a good correlation of BPST results with the peak friction values from actual road testing of passenger tires[220]. Since ABS operates around the peak, the wet skid resistance of the vehicle with ABS should be predicted by BPST values. In fact, it was found that the ABS stopping distances measured on a low $\mu$ road surface correlate well with the BPST values obtained with the test procedures described previously on a sanded glass plate (Figure 8.46). This may result from their similar lubrication mechanisms: in both tests, MEHL plays an important role.

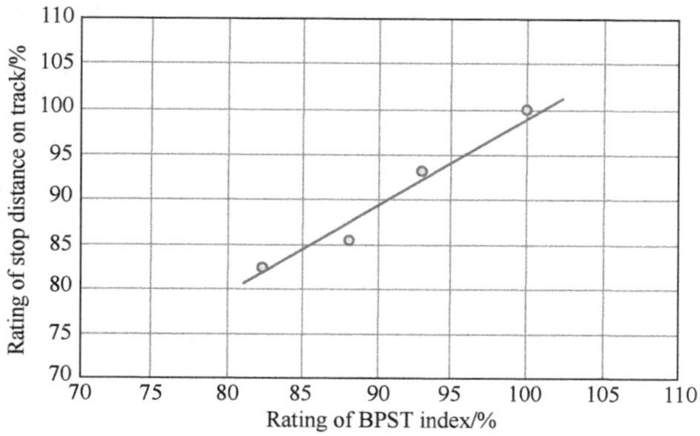


**Figure 8.46** ABS stopping distances of passenger car on wet asphalt surface (0.3 μ) *vs.* BPST index

Although hardness shows different effects on wet skid resistance when measured by BPST and GAFT, at normal hardnesses for passenger tire tread compounds, GAFT values can also be used to predict ABS stopping distance on a low $\mu$ road surface, provided that the tests are conducted at higher speed, low temperature, low load, and on a ground glass substrate. Under these conditions, the GAFT results are mainly determined by the MEHL mechanism.

#### 8.2.3.4 Development of a New Filler for Wet Skid Resistance

A group of dual phase fillers in which silica and carbon are *co*-formed in a carbon black-like reactor have been developed[196,221]. The first commercialized product was CSDPF 2000. This material could impart high wet skid resistance to truck tires, as at high load (about 100 N) corresponding to that of in a truck tire. The side force coefficient of this CSDPF is significantly higher than those of carbon black and silica (Figure 8.43). In addition, when this material is applied to NR-based truck tire tread compounds, the wet skid resistance measured by BPST is significantly higher than that of silica and carbon black. Therefore, CSDPF 2000 has been recommended for truck tire application to improve the global performance, including wet skid resistance[222]. However, as mentioned previously, although the dynamic properties of a passenger tire tread compound filled with CSDPF 2000 can fully match those of a silica compound, there is a deficiency in wet skid resistance on smooth surface for a car with ABS.

Using typical passenger tread compounds, comprehensive studies on the effect of different fillers on wet skid resistance have demonstrated that friction coefficients measured by means of BPST on wet and sandblasted glass surfaces are closely related to the silica surface area of the filler (Figure 8.47). Since the silica surface area is proportional to the silica-polymer interfacial areas in a unit volume of compound, this demonstrates that higher silica-polymer interfacial area gives better wet skid resistance. According to the correlation between BPST number and ABS stopping distance of passenger cars on wet smooth surface, fillers having higher silica-polymer interfacial areas in the compounds should give higher wet skid resistance for ABS.

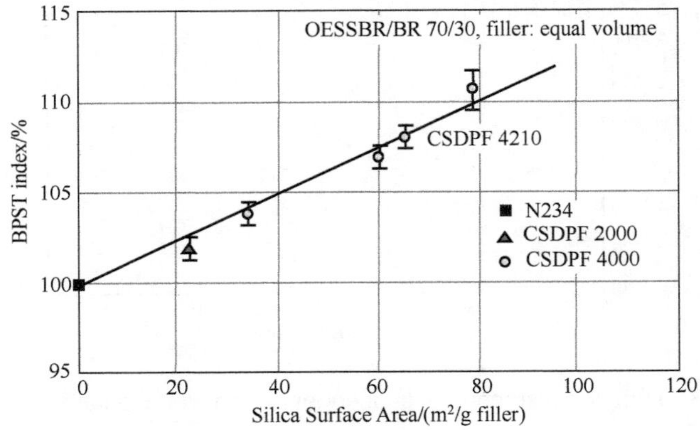


**Figure 8.47** Effect of silica surface area of fillers on wet skid resistance of vulcanizates filled with various fillers (OESSBR: oil-extended SSBR)

In order to improve wet skid resistance for passenger tires, a new carbon-silica dual phase filler has been developed, referred to as CSDPF 4000, with substantially increased levels of silica at the surface compared to earlier dual phase fillers[175,223]. As seen from Table 8.6, CSDPF 4210 has much higher silica-surface coverage relative to CSDPF 2214, due to the silica domain distribution and high silicon content. When incorporated in hydrocarbon polymers, CSDPF 4210 is characterized by higher filler-polymer interaction in relation to a physical blend of carbon black and silica, and lower filler-filler interaction compared to traditional carbon black and silica. Evidence for the high polymer-filler interaction can be obtained by comparing the adsorption energies of polymer analogs on the filler surfaces[175]. The lower filler-filler interaction of CSDPF 4210 can be demonstrated by the strain dependence of the elastic modulus, $G'$. For filled rubber, the elastic modulus decreases with strain amplitude, i.e., "Payne effect"[224]. This effect is generally used as a measure of filler agglomeration, controlled mainly by filler-filler interaction[224,225]. Although from the chemical composition point of view, CSDPF 4210 is in between carbon black and silica, it gives the lowest Payne effect of all compounds (Figure 8.48). It has been recognized that higher polymer-filler interaction and lower filler-filler interaction of CSDPF will allow a better compromise between abrasion resistance and temperature dependence of dynamic hysteresis (tan $\delta$, namely lower hysteresis at higher temperature and higher hysteresis at low temperature (Figure 8.49)[225,226]. This will result in a good wear resistance and lower rolling resistance of tires. In addition to this good combination of properties, the surface silica coverage of the new CSDPF 4000 materials should be very favorable for wet skid resistance. In fact, several road tests have demonstrated

**Table 8.6** Analytical properties of some typical carbon blacks and CSDPFs

| Items | Silicone content/% | | BET-surface area/($m^2$/g) | | STSA/($m^2$/g) | | CDBP/ (mL/100 g) | Silica coverage[a]/% |
|---|---|---|---|---|---|---|---|---|
| Filler | As is | HF[b] | As is | HF | As is | HF | | |
| N234 | N/A | N/A | 122 | 122 | 118 | 118 | 101 | NA |
| CSDPF CRX2214 | 4.8 | 0.72 | 154 | 295 | 121 | 150 | 101 | 21 |
| CSDPF CRX4210 | 10.0 | 0.01 | 154 | 167 | 123 | 155 | 108 | 55 |
| Silica Z1165 | 46.7 | N/A | 168 | N/A | 132 | N/A | NA | 100 |

a. The silica-surface coverage of CSDPF was estimated from iodine number and surface area that was the averaged value of BET-surface area (NSA) and statistic thickness surface area (STSA).

b. HF – hydrofluoric acid treated

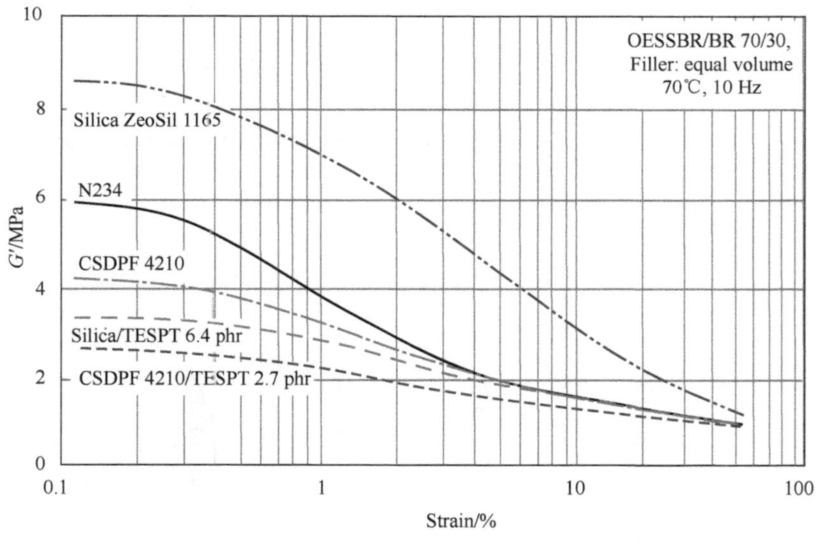


**Figure 8.48** Strain dependence of $G'$ for vulcanizates filled with various fillers

that for passenger vehicles with ABS, the wet skid resistance of CSDPF 4210 with silica surface coverage of about 55%, tested on smooth pavement surface, gives comparable results to silica (Figure 8.50). In the water squeeze and transition zones, where EHL and MEHL are predominant, CSDPF 4210 may not be as effective as silica but it is better than carbon black. In the traction zone where the friction coefficient is controlled by BL, CSDPF 4210 should be better than silica because at least 45% of the aggregate surface is covered by rubber due to the strong polymer-filler interaction, but it may not be as efficient as carbon black in this regard. However, the deformation component of the friction is also important in the traction zone, hence the dynamic properties of the compounds would be equally important. Relative to carbon black, the high hysteresis at low temperature (Figure 8.49) and low dynamic moduli of CSDPF 4210, which leads to higher energy dissipation and larger contact area between tread compound and road surface, would favor BL. Therefore, compared with conventional fillers, a better-balanced result would be expected in the road test for CSDPF 4210 due to its hybrid surface and dynamic properties.

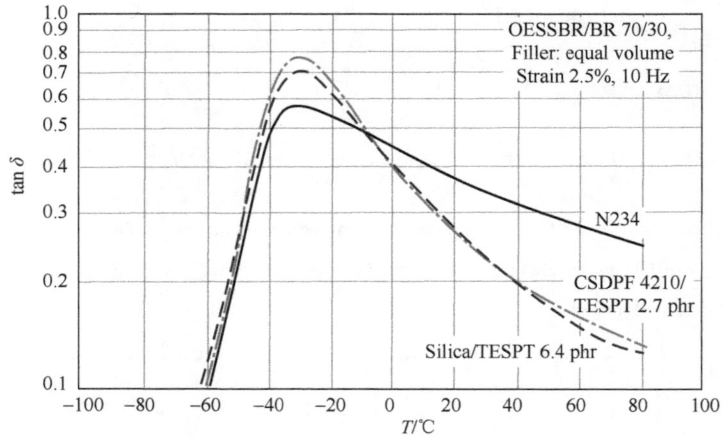


**Figure 8.49** Temperature dependences of tan $\delta$ for vulcanizates filled with various fillers

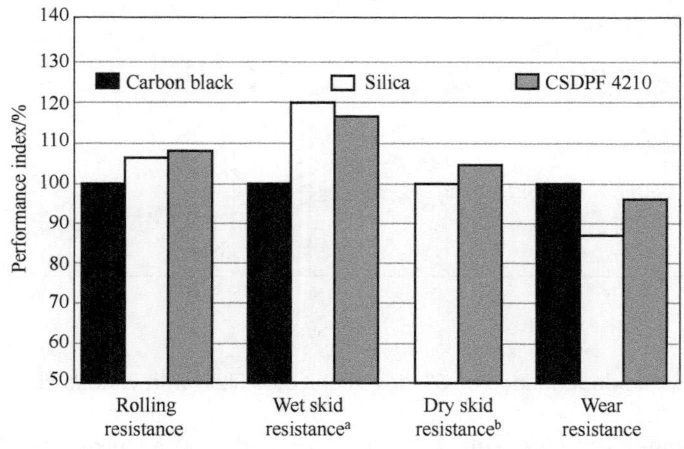


**Figure 8.50** Passenger tire performance of tread compounds filled with different fillers
a. Wet skid resistance was obtained with vehicles having ABS on asphalt road surface
b. Dry skid resistance is not available for carbon black tire

# 8.3 Abrasion Resistance

## 8.3.1 Abrasion Mechanisms

According to Medalia[252], the mechanisms of abrasion (loss of rubber) which have been identified include the following: abrasive or cutting abrasion, intrinsic or small-scale abrasion, pattern abrasion, abrasion by roll formation and mechanochemical degradation.

### Abrasive or Cutting Abrasion (ACA)

In this mechanism, it is considered that the scratches are typically made in the direction of abrasion[227]. According to Grosch and Schallamach[228], if the abrasion process considered is due to tensile failure, detachment of rubber particles from the bulk will be due to catastrophic tearing of the rubber surface behind the moving track asperities, similar to the tearing behind a needle. Incidence of tearing depends on whether the initiation or growth of a tear in a strained rubber can be sustained by the release of elastically stored energy. The energy which would become available by the opening of a cut of unit area is called the tearing energy $G$; its critical value is the catastrophic tearing energy $G_c$, which depends on the strength of the rubber and related to the energy density at break, $W_b$, and the diameter $d$ of the tip of the tear, thus:

$$G_c = W_b d \tag{8.14}$$

Therefore, the rate of abrasion should depend on the energy density at break rather than on the tensile strength itself.

On the other hand, continuous abrasion of rubber in the same direction tends to produce abrasion by removing large particles. A related effect, catastrophic tearing of the sample at the leading edge, is experienced with rubbers under high loads and results in a surface layer being peeled off. Up to certain pressures, the abraded volume is proportional to the normal load, and can therefore be expressed by an abrasion coefficient, $A$, defined as the volume abraded per unit normal load and unit sliding distance.

Rubber friction against the tracks employed in abrasion processes is, for a given temperature and velocity, described by the coefficient of friction $\mu$. It is therefore possible to characterize the abrasion behavior of rubber by the load-independent ratio $A/\mu$. This is the abraded volume per unit energy dissipation and has been called the abradibility. It is, therefore, assumed that the abraded volume multiplied by the energy density at break is proportional to the frictional energy dissipation, or:

$$A = k\mu / W_b \tag{8.15}$$

where $k$ is a constant.

The problem lies in the rate dependence of $W_b$ because tensile strength measurements should be carried out at a rate of extension similar to that on the rubber surface during abrasion, which was estimated to be about 10,000 percent/sec. Tensile experiments could be done at rates of this order of magnitude by using a specially constructed apparatus. The tensile failure at high extension rates can also be estimated with data measured at different rates and temperatures using the WLF transform.

### Intrinsic or Small-Scale Abrasion (ISS)

Abrasion of rubber, whether filled or unfilled, often appears to involve a competition between two different mechanisms: removal of microscopic particles of rubber by a fracture process (local mechanical rupture – tearing), which was termed as "intrinsic wear", and chemical deterioration of rubber by mechanoxidation (smearing).

According to Schallamach[229], due to the inhomogeneity in the physical properties of the rubber at the very small scale of the affected surface area, surface damage in real abrasion occurs with irregularity of traces. When a rubber surface moves over the track asperity, the damage of the rubber surface will occur in the weakest places, resulting in random nature of the damage. Also, the track asperities will differ in size and level, thus introducing an additional random element.

For this type of wear, the basic assumption is that in rubber abrasion, discrete particles of volume are detached from the bulk, where $a$ is a length proportional to the size of the area of contact between the rubber and a track asperity. The probability of a particle being torn off is taken to be proportional to the tensile stress near the rubber surface; this will be given by $F/a^2$, where $F$ is the frictional force produced by an asperity. Finally, it is assumed that the spacing between detachments along the path of an asperity is proportional to the contact length, $a$. The abrasion per unit distance of travel is then given by:

$$A = ka^3(F / a^2)(1 / a) = kF \qquad (8.16)$$

where the $k$ is a constant. Accordingly, the abrasion is proportional to the frictional energy dissipation and differences in the abrasion of different rubbers are contained in the constant. On the other hand, the frictional force on a rough track is roughly proportional to the normal load $W$, so that the following equation should also hold:

$$A = k'W / E \qquad (8.17)$$

where $k'$ is a constant and $E$ is the modulus of the rubber.

In intrinsic wear, while the rate of wear depends strongly upon the frictional force as shown by Eq. 8.16, the worn surface and the fineness of the particles also changes with frictional force. Gent and Pulford[230] demonstrated experimentally using a blade abrader that, at low levels of frictional force, the wear was light, the worn surface was relatively smooth, and the debris consisted of relatively small particles, such as 1–5 μm. At higher frictional forces, the wear was rapid, the surface was much rougher in texture, and the main wear particles were larger, frequently exceeding 100 μm in mean diameter. In their study, the experimentally determined relationships between wear *A* and frictional work *F* can be represented by the general result:

$$A = k''F^n \tag{8.18}$$

where the coefficient $k''$ and exponent $n$ are characteristic of the particular material being examined. For unfilled SBR and BR materials, the exponent $n$ was found to be about 2.9 and 3.5, respectively. In contrast, exponents for carbon black-filled materials SBR and BR were considerably smaller, 1.5 and 1.9, corresponding to a lesser dependence of the rate of wear upon the frictional force for these materials.

It was also found[230] that certain compounds, for example carbon black-filled SBR, NR, and EPM, were worn in a way that sticky and oily debris were formed. This was attributed to the degradation of the polymer chains. Carbon black-filled compounds of BR and EPDM, on the other hand, produced only dry, particulate debris with no signs of chemical deterioration.

It seems that the abrasion of rubber is governed by competition between two quite different processes: removal of microscopic particles from the rubber surface by fracture and mechanochemical decomposition of the rubber. Presumably the dominant process depends on the relative susceptibility of the material. Incorporation of carbon black in the compounds increases the stiffness and strength of rubber and thereby suppresses tearing failures. In this case, mechanochemical attack becomes the principal mode of wear in materials like NR and SBR that are susceptible to decomposition by this mechanism. For these polymers, the resonance-stabilized radicals formed react with oxygen (if present) and the initial chain fracture is rendered permanent. The presence of oily product appeared to decrease further wear, even when the frictional force was maintained constant. Apparently, oily product acted, on the one hand, as a viscous protective film, alleviating the local concentrations of tearing force that are presumably responsible for the detachment of wear particles. Also, oily debris stuck on the track of the abrader may reduce friction, resulting in lower abrasion rate. In this regard, oxygen and temperature play important roles for the decomposition of polymer molecules. For BR compounds, the radicals formed by main-chain rupture are known to react with the polymer itself, leading to further crosslinking rather than to decomposition. The presence or absence of oxygen has little effect on this type of wear. However, when the wear fragments were strongly adhering but not liquid-like, they accumulated into dry particles of debris.

### Pattern Abrasion (PST)

When rubber is abraded against abrasives, it is found that, under suitable conditions, an array of nearly parallel ridges is formed normal to the direction of abrasion. This array of ridges has been called the abrasion pattern. Similar ridges are found on the surface of tires, in particular on driving bands, where they are oriented at an angle to the direction of motion of the vehicle. The physical characteristics of the ridges on tires and of those produced in the laboratory are so similar that there can be no reasonable doubt[231].

Using a line contact (in practice a razor blade) as the abrading device, Southern and Thomas[232] were able to closely resemble the abrasion patterns from typical multi-asperity surfaces. Evidence for the essential correctness of the model and a further insight into the detailed mechanism of abrasion have been obtained.

Figure 8.51(a) shows schematically the formation of the abrasion pattern and the way it is deformed under the action of the blade[233]. As the blade passes over the pattern, the tongue of rubber is pulled back and then released as the blade passes on. The stress produced in this process is assumed to cause crack growth in the reentrant corners such as *P* and *Q*. This crack growth will have a component $\Delta x$ perpendicular to the surface for each pass of the blade. If, as is assumed to be the case, the surface has reached a steady state with the abrasion pattern maintaining a constant overall appearance, the surface of the rubber must be lowered on average by $\Delta x$ at each pass. Thus the volume of rubber removed per unit area of surface is $\Delta x$.

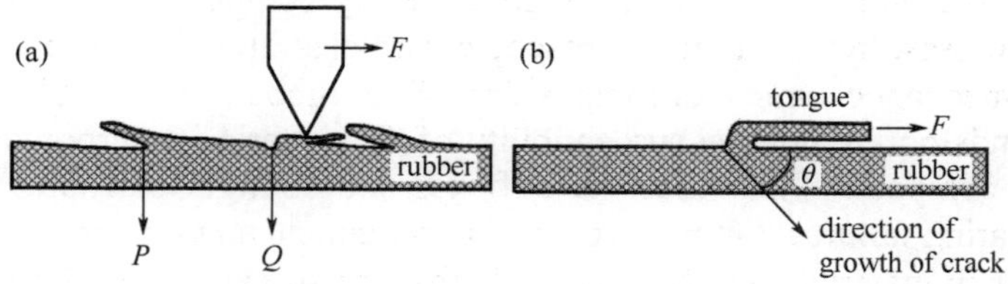


**Figure 8.51** (a) Diagram of abrasion pattern and its deformation by the blade; (b) model for crack growth under abrading force, *F*, showing crack propagation at an angle, *θ*, to the surface[233]

To apply the necessary fracture mechanics analysis, the deformation of the abrasion pattern is somewhat idealized as shown in Figure 8.51(b). The pattern is taken to be uniform across the surface of the sample, which observation indicates to be a fair approximation, and the frictional force, *F*, to be wholly sustained by the tongues of the pattern.

With the fracture mechanics approach, it is necessary to calculate the mechanical energy d$U$ released when the crack grows by an incremental amount d$c$. This energy release rate that has been termed the tearing energy and is denoted by $G$, defined as the energy released per unit area of new surface:

$$G = \frac{1}{w} \times \frac{\mathrm{d}U}{\mathrm{d}c} \tag{8.19}$$

where $w$ is the width of the specimen over which $F$ is applied. It has been found from extensive tear and crack growth studies that the $G$ value determines the rate of crack propagation, independently of the overall shape of the specimen and of the detailed way the forces are applied. Thus, if the tearing-energy/crack-growth relation is known for a particular material, the behavior in any type of specimen can be predicted, provided the $G$ value developed in that specimen can be calculated from the measurable applied forces.

Consideration of Figure 8.51(b) shows that, if the crack tip advances into the rubber a distance d$c$, the force $F$ will move by an amount d$c(l + \cos\theta)$ parallel to the surface of the rubber, provided the extension of the tongue is not large. Also, the component of crack growth perpendicular to the rubber surface, d$x$, is given by d$c$ sin $\theta$.

This is very similar to that existing for the tear tests of "simple extension" or "trouser test" pieces described by Rivlin and Thomas[234]. For the conditions stated above, it is easy to show, following a similar analysis, that the tearing energy is given by:

$$G = F / w(1 + \cos\theta) \tag{8.20}$$

The crack growth behavior of a number of rubbers under repeated stressing has been examined using the tearing energy approach, and the results have been expressed as the crack growth per cycle, $r$, as a function of the maximum $G$ value attained during each cycle[235,236]. For many materials over a fair range of $G$ values (approximately $0.1 < G < 10$ kN/m), it is found that this function is of the general form:

$$r = BG^{\alpha} \tag{8.21}$$

where $B$ is a constant. The exponent $\alpha$ varies from about 2 for natural rubber to 4 or more for noncrystallizing unfilled rubbers, such as SBR[237]. With the above model of the abrasion process, the volume of rubber abraded, $A$, per revolution of the wheel, will therefore be given by:

$$A = rsw\sin\theta \tag{8.22}$$

where $s$ is the circumference of the abrasion wheel. Combining Eqs. 8.20, 8.21, and 8.22, and taking logarithms gives:

$$\lg(A / sw\sin\theta) = \alpha\lg[F(1 + \cos\theta) / w] + \lg B \tag{8.23}$$

It should be noted that Eq. 8.23 contains no arbitrary constants and relates the quantities measured in an abrasion experiment to the crack growth constants, $\alpha$ and $B$, of the rubber. A direct test of the theory is obtained by plotting $A/(sw\sin\theta)$ against $F(1 + \cos\theta)/w$ on the same graph as crack growth data taken from the literature[236].

It is seen from Eq. 8.22 that the abrasion depth loss depends on two factors, the crack growth rate, $r$, and the angle at which the cracks grow into the rubber, $\theta$. The crack growth rate is related to the tearing energy which is the fundamental strength property of the rubber, but the factors which determine the angle at which cracks grow are less well understood. But it appears to be closely related to the geometry of the pattern, and it seems likely, from a study of the geometry, that much of the loss of rubber occurs from the steeply raked face of the pattern[233].

## Abrasion by Roll Formation (ARF)

In the wear by roll formation, a layer of rubber which has been mechano-chemically degraded (and may have absorbed dust) rolls up into small cylinders which are sloughed off. According to Reznikovskii and Brodskii[238], when a smooth abradant presses on a particular projection on the rubber surface with force $N$ and moves parallel to the surface of the rubber with velocity $\mu$ (Figure 8.52), a complex deformation of the projection in the contact zone takes place. During subsequent movement, an increase in the deformation of the projection leads to an increase in the force which opposes this deformation and consequently to an increase in the area of rubber making contact with abradant. General slipping begins when the tangential component of the elastic force in the contact zone is equal to the force of friction. If the rubber does not have the strength required and the friction force is fairly high, then it may happen that failure in the area of greatest deformation occurs before general slipping begins in the contact zone. The nature of this type of failure is such that with a complex stress condition in the rubber, failure normally occurs where the surface layer of materials is in a state of maximum strain. It is also known that failure begins with the appearance of a cut perpendicular to the direction of the action of the tensile force. When a cut appears, its subsequent growth takes place under the action of a fairly small force. The direction of cut growth depends in a complex manner on the nature of the stress condition and on a series of incidental factors related to the presence of micro heterogeneities in the structure of the material. Much more probable is the gradual tearing of the rubber, allowing relative movement in the contact zone without total slipping.

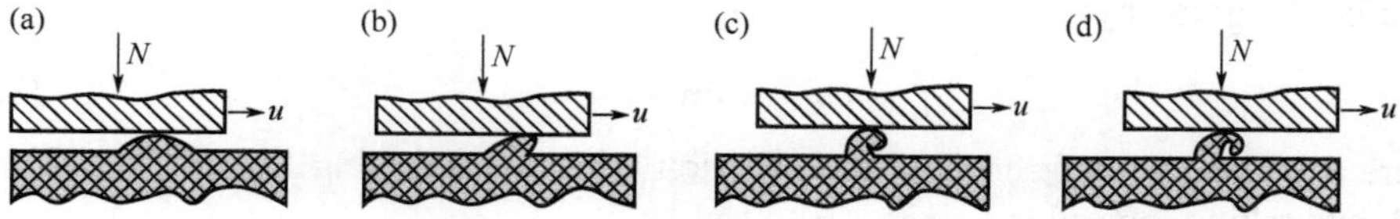


**Figure 8.52** Gradual stages [(a), (b), (c), (d)] in the formation of a "roll" in the friction of rubber against a smooth abradant[238]

Such movement is possible if the layer of rubber separated off during tearing winds into a roll. With such a mechanism, subsequent movement will take place under conditions of rolling friction accompanied by continuous tearing of the rubber and accumulation of the shred separated off to form the roll.

The shred is in a stressed condition. The force which causes elongation depends on the resistance of the rubber to tearing at the place where the shred separates off. The elongation of the shred depends on its cross-section, which is generally variable and connected in a complex manner with several factors determining the direction in which the tear grows. Failure of the shred at a portion where the elongation is critical causes detachment of the roll and thereby completes the elementary act of frictional wear.

It is therefore clear that ARF can only take place with a particular combination of external conditions and properties of the rubber. This type of abrasion is more likely in rubbers with a low tear strength and since the strength properties of the rubber depend considerably on temperature, the heating of the surface layer as a result of sliding friction may also be an extremely important factor. Under certain conditions this heating may cause resinification of the surface layer of rubber and appearance of a characteristic tackiness, which greatly increases the effective friction.

Based on the description above, Reznikovskii et al.[238,239] proposed an approximate quantitative analysis of the ARF process. In this analysis, the abrasion resistance is characterized by $\beta$, which is the ratio of the frictional work, $W_f$, to the volume abraded per unit time $dA/dt$. The frictional work consists of different components:

$$W_f = W_t + W_e + W_r \tag{8.24}$$

where $W_t$ is the work used in tearing the shred from the surface, $W_e$ is the work used in elongating the shred, and $W_r$ is the work consumed in hysteresis losses related to roll formation.

The main condition determining the possibility of roll formation is:

$$W_f \leqslant \mu N u \tag{8.25}$$

where $\mu$ is the coefficient of static friction between the rubber and abradant, $N$ is the load, and $u$ is the speed.

Assuming that $W_t$ can be expressed based on the theory of Rivlin, Thomas, and Greensmith[240–243], $W_e$ in terms of a linear law of elasticity, and $W_r$ as a somewhat simplified equation derived from the work of Bulgin et al.[244], one may have:

$$\beta = f(G, E, D, a, b, r) \tag{8.26}$$

where $G$ is the characteristic tear energy, $E$ is the elastic modulus, $D$ is the resilience of rubber, and $a$, $b$, and $r$ are the average values of the thickness and width of a separated shred, and the radius of a roll, respectively.

These equations show the relationship between the intensity of abrasion through roll formation and the main elastic-relaxation and strength properties of the rubber.

### Abrasion by Spalling (SPA)

The spalling (SPA) mechanism of abrasion proposed by Gent[245] was based on questioning the above four mechanisms as all those theories are related to the removal of particles of rubber by frictional forces and severe decomposition of the polymers. Among others, the questions are based on the following observations: (a) NR compounds are exceptionally resistant to crack growth because of their ability to crystallize at high strains, but they are not particularly resistant to abrasion, being similar to or somewhat worse than a comparable SBR compound, which does not crystallize under any circumstances and has markedly inferior crack-growth resistance;

(b) compounds filled with carbon black are strikingly more resistant to abrasion, and yet they are not much more resistant to crack growth under repeated stressing; and (c) crack-growth resistance diminishes rapidly as the test temperature is raised, whereas the resistance to abrasion is not so seriously affected.

According to Gent, for the spalling mechanism of abrasion, it is assumed that vulcanized rubber contains small, isolated voids that are too small to see. When a spherical cavity in rubber is subjected to internal pressure it will expand in a highly nonlinear way. For a material obeying the simple kinetic theory of rubber-like elasticity, the relation between inflating pressure $P$ and expansion ratio $\lambda$ of the cavity radius is given by[246,247]:

$$P / E = (5 - 4\lambda^{-1} - \lambda^{4}) / 6 \tag{8.27}$$

where $E$ is the tensile (Young's) modulus of elasticity of the rubber. This relation should also be a reasonable guide for materials that show somewhat more complex elastic behavior[247]. It predicts that, at a critical inflation pressure $P_c$, given by $5E/6$, the cavity will expand without limit. In practice, the cavity will tear open to form an internal crack when the maximum extensibility of the rubber is reached.

We should now examine the hypothesis that elastomers contain small voids and that such cavities can be the cause of internal failure as a result of unbounded elastic expansion. It seems clear that cavities, if they exist, must be relatively small, or otherwise they would be readily observed. But surface-energy terms will tend to make very small cavities collapse. Thus, the probable range of diameters for natural cavities in rubber is from about 0.1 to 1 μm[248]. The expansion relation for a pressurized cavity, Eq. 8.27, does not depend upon its original diameter, so that the actual size is rather unimportant; if cavities exist, they will inevitably become large internal cracks when an inflation pressure of $5E/6$ is developed.

Several experiments indicate that this hypothetical mechanism of internal rupture of rubber is valid. When rubber blocks are supersaturated with high pressure dissolved gases or liquids and the super-saturation pressure exceeds the critical value, $5E/6$, then visible bubbles appear in the interior of the blocks[249,250]. Each bubble corresponds to an internal crack. When rubber blocks are placed in a state of approximately uniform triaxial tension by bonding them between closely spaced rigid plates and pulling the plates apart, large internal cracks appear at stresses corresponding to the critical internal condition $P_c$[247]:

$$P_c = 5E / 6 \tag{8.28}$$

Gent believed that, in all of these cases, the critical pressure for internal rupture is found to be proportional to the elastic modulus $E$, rather than to a more conventional measure of strength, indicating that failure is governed by an elastic criterion.

It was, therefore, proposed by Gent that spalling of rubber by friction takes place as a result of fractures within the rubber, below the surface. This hypothesis has several interesting corollaries. First, it is clear that stiffer materials will be less prone to spalling, provided that they are sufficiently extensible for Eq. 8.28 to apply, at least approximately. Secondly, because resistance to spalling depends primarily on rubber stiffness, it will not show good correlation with normal measures of strength. For example, it should be rather similar for compounds with similar hardness, even if they have very different tear strengths. And it should depend to a much lesser degree upon test temperature, because the stiffness of compounds is usually less temperature-sensitive than their strength properties. Accordingly, all of these features are in accord with the observed abrasion resistance of rubber and lends some support to the hypothesis that abrasion is initiated by subsurface fractures that develop in accordance with an elastic criterion.

### Mechanochemical Degradation (MCD)

This is a pervasive phenomenon in rubber abrasion and probably involves a combination of mechanical action (stretching of the rubber molecules) and chemical degradation (including thermal and oxidative mechanisms). The effect of oxygen in promoting degradation and wear is well documented both by experiments in inert atmosphere and by varying the amount of antioxidant (in NR compounds). With carbon black N330 filled NR compounds with and without antioxidant (*N*-isopropyl-*N*-phenyl-*p*-phenylenediamine – IPPD), Schallamach[251] reported the effects of atmosphere and antioxidant on abrasion (Figure 8.53). He tested an unprotected and a protected compound which, after having been abraded in air, were abraded in nitrogen and then in air again. Each point gives the rate of abrasion during the preceding 250 revolutions of the grinding wheel. It is surprising to see that the first reading in air after the nitrogen experiment is below the level reached in nitrogen. It was explained by Schallamach that undegraded rubber in nitrogen is still abraded but smearing sets in instantaneously. The source of smearing must lie in the abrasion process itself and originate from local degradation in the presence of oxygen[251].

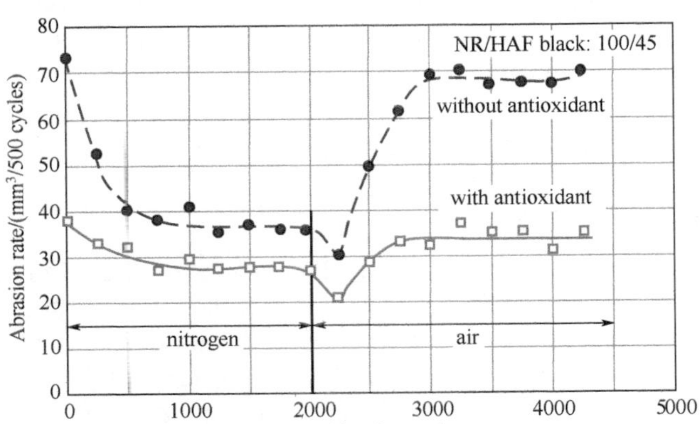


**Figure 8.53** Change in the rate of abrasion of two NR compounds in different atmospheres

As described by Medalia[252], the usual effect of MCD is to produce an oily sticky layer on the surface of the sample when abraded under fatigue wear conditions. This occurs with black-filled compounds of NR, SBR, and IR; with BR, however, it appears that the radicals formed by MCD lead to crosslinking and thus the compound remains dry. While some have contended that the oily layer represents exudation of process oil or extender oil in the compound, it has been shown that it is formed even in compounds without oil and thus must be due to degradation, as mentioned above. In the absence of oxygen, carbon black-filled NR forms a dry powder rather than an oily layer; this and other experiments demonstrate the formation of macroradicals by molecular fracture. It should be noted that during abrasion measurements, if an oily sticky layer is formed, the wheel of the abrader will be lubricated which leads to significant reduction of the abrasion rate, giving misleading results. However, for a tire on the road, the oily layer picks up road dust and becomes more or less dry, and is then removed by roll formation or other means.

To sum up, although a large number of mechanisms have been proposed in the literature, the above six may be the most representative. Based on these theories, the parameters of vulcanizates which influence the abrasion resistance and wear of the rubber products, tire in particular, are friction, tearing energy, tensile strength, and tensile (Young's) modulus. Lower friction coefficient, higher tearing energy, higher tensile strength, and higher modulus and hardness would be favorable for abrasion resistance.

Practically, abrasion is measured with different abraders, such as the DIN abrader, Akron angle abrader, Pico abrader, Lambourn abrader, and Cabot abrader. The abrasion of each abrader may involve one or more abrasion mechanisms, i.e., the mechanisms may be not mutually exclusive. For different abraders, one mechanism may be accompanied, to a greater or lesser degree, by others. Therefore, the rating of abrasion resistance of different compounds may be different for different machines.

In the case of treadwear of tires on the road, the results involve a mixture of mechanisms in which the relative importance of each depends on the road conditions including its microtexture, temperature, whether it is wet or dry, the slip and load on the tire, the tire construction, and the tread compound.

## 8.3.2 Effect of Filler Parameters on Abrasion

### 8.3.2.1 Effect of Filler Loading

The effect of filler loading on abrasion of filled vulcanizates used to be a hot topic in the rubber industry, especially in research of tire technology. With SBR vulcanizates filled with 17 different grades of carbon black, Wolff and Wang[253] were able to construct a master-curve which shows the abrasion (volume loss) measured with a DIN abrader as a function of $f\phi$, where $f$ is the shift factor, and $\phi$ is the filler volume fraction (Figure 8.54). As can be seen, over the range of filler loadings used, i.e., 0–70 phr for semi-reinforcing (soft) grades and 0–50 phr for reinforcing (hard) blacks, the abrasion decreases monotonically with increasing $f\phi$, which can be treated as the effective volume of the filler. Since the master curve is constructed only by horizontally shifting the abrasion-*vs.*-loading curves for all carbon blacks, all the individual filler-loading functions follow the same mode, i.e., the abrasion reduces with filler loading, but the reduction rate gradually becomes lower. It seems that if $f\phi$ is taken as the effective volume of the filler, some other factors influencing the filler effectiveness would be reflected in the shift factors $f$ for all carbon blacks.

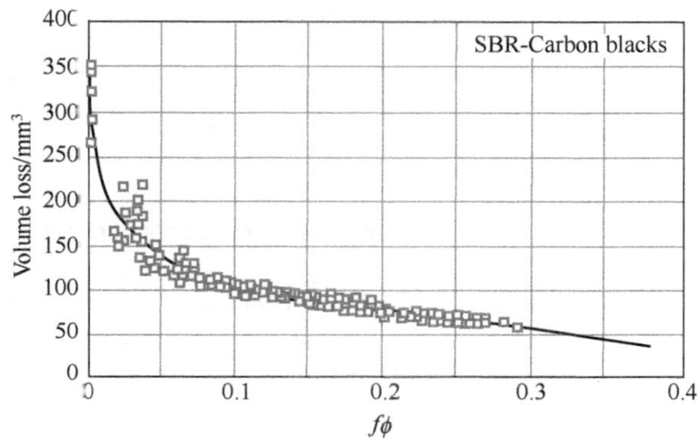


**Figure 8.54** Mastercurve – abrasion change in the rate of abrasion of two NR compounds in different atmospheres

The ultimate goal of most research on the abrasion of rubber is to improve the tread wear of tires on the road. Compared with abrasion measured using lab abraders, the wear of tires involves a mixture of complex mechanisms[252].

Using a trailer on a "harsh" road, the effect of filler loading on tread wear of tire was studied by Bulgin and Walters[254] under various conditions, showing a maximum in the wear resistance as a function of loading (Figure 8.55). Similar results were also reported by others[255]. Using a carbon black with surface area of 130 $m^2/g$ and DBPA of 124 mL/100 g, Veith and Chirico[256] reported that the optimum filler loadings increase linearly with wear severity. While the initial decrease in wear rate was attributed to

- reduction in slip angle at constant cornering force, and
- reduction in fatigue wear ("friction wear");

the decrease above the maximum was attributed to

- increase in ACA which became the dominant mechanism[254],
- decrease in tensile strength at high loadings, possibly due to "serious physical interference between neighboring particles"[257],
- poorer dispersion of the carbon black in the polymer matrix, and
- higher heat build-up due to the higher hysteresis.

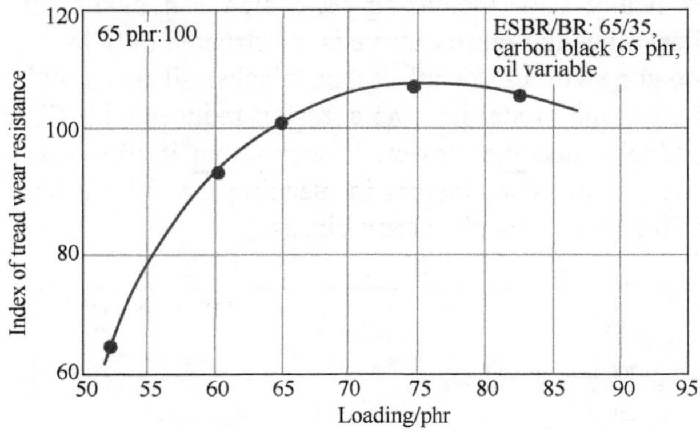


**Figure 8.55** Tread wear of passenger tire as a function of carbon black loading[255]

It was reported by Grosch and Shallamach[258] that the tread wear of the tire increases significantly with surface temperature (Figure 8.56). This is especially true for NR tread compounds as polymer degradation at high temperature is more severe.

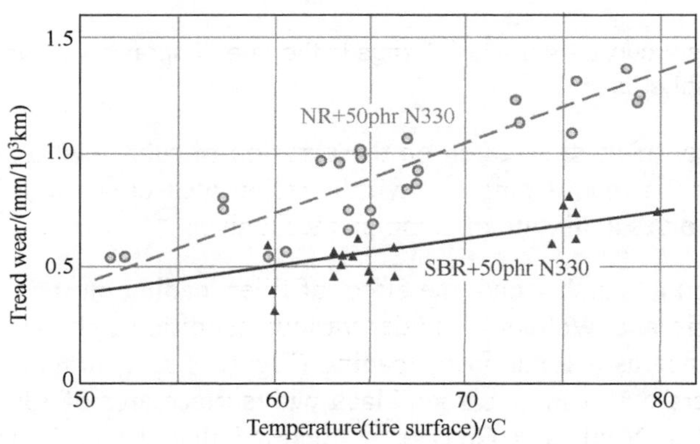


**Figure 8.56** Tread wear as a function of temperature at tire surface[258]

### 8.3.2.2 Effect of Filler Surface Area

Surface area is generally considered to be the most important parameter of fillers for abrasion[259–263]. Rubber filled with high-surface-area carbon black usually shows a higher abrasion resistance, although a practical limit is reached at higher surface area. As can be seen from Figure 8.57, at given structure of the blacks, the shift factor increases with surface area, passing a maximum and then decreases. Since the abrasion resistance of the filled SBR vulcanizates decreases monotonically with the effective volume, $f\phi$, the smaller of the shift factors should suggest that beyond the maximum shift factor, the abrasion resistance of vulcanizates should be reduced with surface area of the filler. The negative effect of high-surface-area blacks on abrasion resistance should be related to their difficulty in obtaining good dispersion. This should be due to stronger filler-filler interaction, since the surface energy of fillers increases with increasing surface area (see Section 8.3.2.4.1). On the other hand, the aggregate size of high-surface-area blacks is smaller (see Chapter 2), which leads to lower shear stress being applied to agglomerates during mixing. It should also be related to increased heat build-up, as the hysteresis of filled vulcanizates increases significantly with surface area of blacks. This argument can be confirmed by liquid phase mixing of NR/carbon black. With this mixing process, the filler dispersion can be considerably improved and the hysteresis can be reduced, which shifts the maximum to the side of higher surface area.

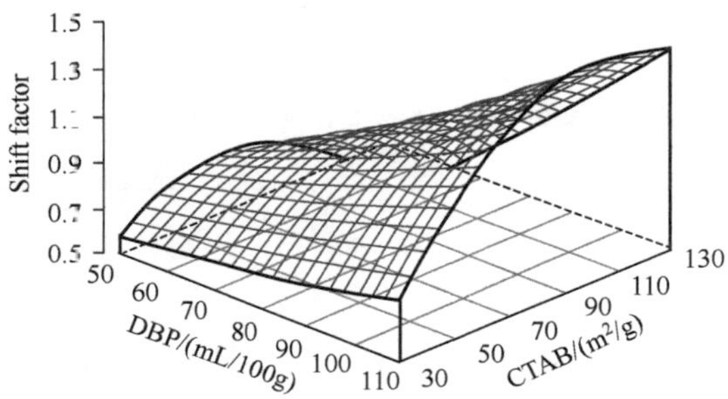


**Figure 8.57** Shift factors for abrasion as a function of CTAB surface area and DBP(24M4) for emulsion SBR vulcanizates

Similar results were obtained from road tests, i.e., higher surface area was associated with improved tread wear, although a practical limit was reached at which this improvement was counteracted. This can be seen in Figure 8.58. At a constant loading of 65 phr carbon black with 35 phr oil, the optimum surface area (EMSA) for an SBR/BR blend system lies between 130 and 150 $m^2/g$[255]. Also using ESBR/BR compounds filled with different carbon blacks, a more comprehensive study was carried out by Veith and Chirico[256] under different severities that were determined by the tread loss and test mileage. Figure 8.59 shows carbon black loading at the minimum-wear rate ($CBC_{min}$), obtained from the wear-loading plots, as a function of the log severity. Obviously, for carbon black having the same structure, in order to achieve a minimum wear rate at high general severity, lower-surface-area blacks require a higher carbon black level than higher-surface-area blacks, while no significant difference in wear resistance was found at lower test severities. However, with a limited surface area range of carbon blacks, Wilder and his coworkers[264] reported a positive effect of surface area at both high and low severities, with radial as well as bias tires. With a single supplier, Hess et al.[265] even showed that radial tire tread wear index is an excellent linear function of surface area.

### 8.3.2.3 Effect of Filler Structure

Besides surface area, the structure of fillers, carbon black in particular, is another important

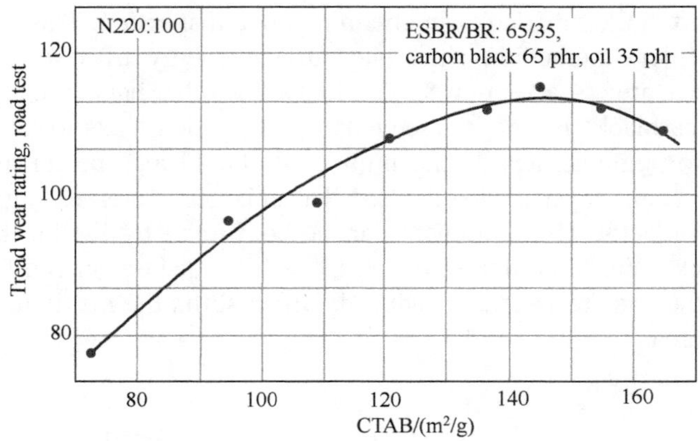


**Figure 8.58** Tread wear of passenger tire as a function of surface area of carbon black[255]

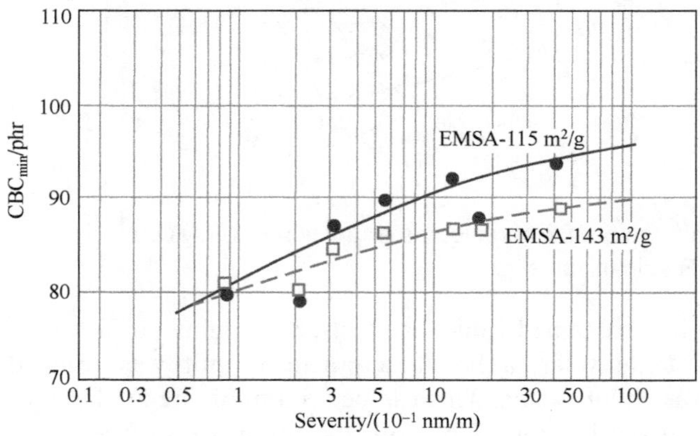


**Figure 8.59** Effect of severity on carbon black loading at minimum wear rate for low surface area (EMSA) and high surface area at DBP 124 mL/100 g

parameter to influence abrasion. As seen in Figure 8.57, the shift factor of abrasion resistance for black-filled ESBR increases with compressed DBPA at constant surface area. This is consistent with many studies showing that carbon blacks of higher structure give higher abrasion resistance. In most cases, as the tests were carried out at constant loading and similar surface area, the effect of higher structure might be due to the higher effective filler loading as more rubber is occluded, leading to increased modulus and hardness.

The general conclusion of the structure effect on abrasion resistance is also similar to the observation of tread wear tests. From the study of Veith and Chirico[256], it was found that, over the wear rate range studied, the $CBC_{min}$ plots are approximately linear with log severity, the curves for high and low DBP being essentially parallel (Figure 8.60). As DBP is decreased, more black must be added to maintain minimum wear at any severity.

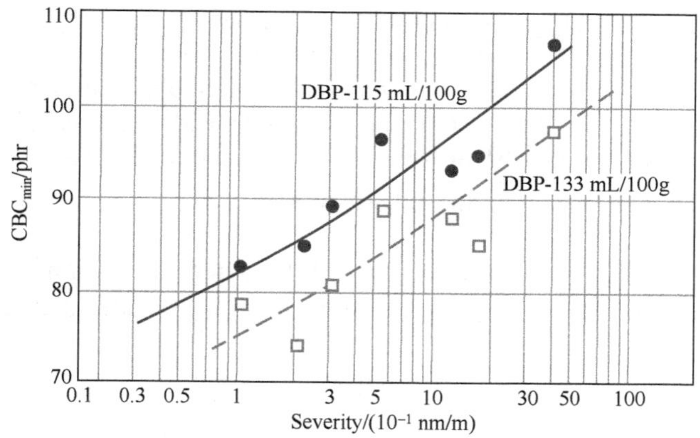


**Figure 8.60** Effect of severity on carbon black loading at minimum wear rate for low DBP black and high DBP black at EMSA 130 $m^2/g$

However, while high structure causes higher effective loading due to occlusion (see Eqs. 3.3, 3.4, and 3.5), rubber loading per se affects other properties beside modulus and tread wear: for example, higher loading gives higher heat build-up and rolling resistance, so for optimum performance it may be desirable to use a low loading of a high structure black[252].

### 8.3.2.4 Effect of Filler-Elastomer Interaction

While most of the studies on abrasion of filled vulcanizates have been focused on filler morphology, filler dispersion, and loading, very few studies have dealt with the effect of surface activity, even though its effect on abrasion has long been recognized. With the development of the IGC technique, the role of surface activity, in terms of polymer-filler interaction, in abrasion resistance has been comprehensively investigated. Some basic phenomena that have been observed in practice related to tire wear resistance can now be better understood. Most importantly, the abrasion resistance of filled compounds can be further optimized.

#### 8.3.2.4.1 Effect of Filler-Elastomer Interaction Related to Surface Area

As mentioned in Section 8.3.2.2, rubber filled with high-surface-area filler usually shows a higher abrasion resistance, but above a certain surface area limit, abrasion resistance cannot be increased further[30,266]. In fact, poor abrasion resistance was observed for carbon blacks with very high surface areas. As shown in Figure 8.58, at lower loading, the increase in abrasion resistance with increasing surface area may be attributed to several mechanisms related to increasing effective interfacial area between filler and polymer. It may also be related to polymer-filler interaction, as the dispersive component of the filler surface energy increases with surface area (Figure 2.56)[22]. The strong polymer-filler interaction of high-surface-area blacks has also been demonstrated by direct observation of aggregate separation stress from the polymer matrix in carbon black filled SBR vulcanizates by means of electron microscopy[267]. The lower abrasion resistance of rubber filled with very fine carbon blacks may be attributed to poor carbon black dispersion caused by the reduced inter-aggregate distance, which is inversely proportional to the surface area of carbon blacks[24]. The poorer dispersion of the high-surface-area blacks may also be associated with the increased filler-filler interaction. When considering the adhesion energies between different materials, the polymer-filler interaction determined by $W_a^{fp}$ varies with the square root of the $\gamma_f^d$ of the filler (Eq. 5.1).

According to the following equation, the filler-filler interaction characterized by cohesive energies between filler surfaces, $W_c^{ff}$, increases linearly with $\gamma_f^d$:

$$W_c^{ff} = 2\gamma_f^d \tag{8.29}$$

This equation suggests that, with increasing surface area of carbon black and hence $\gamma_f^d$, the filler-filler interaction increases more rapidly than polymer-filler interaction, making fine-particle carbon blacks more difficult to be dispersed, leading to increased abrasion rate.

The effect of surface energies on abrasion resistance for different surface area blacks has been difficult to prove because it is hard to separate from the effect of morphology. However, one approach to the problem is to investigate the effect of simple heat-treatment of blacks on abrasion resistance, where differences in morphology can be eliminated.

#### 8.3.2.4.2 Effect of Heat Treatment of Carbon Black

When carbon black was heated in an inert atmosphere (nitrogen in this case), the $\gamma_f^d$ of the carbon black increased with increasing temperature of heat treatment up to 900°C, which is well below the graphization temperature of carbon black [Figure 8.61(a)]. The changes in surface energies are related to changes in surface chemistry. It has been demonstrated[52] that the number of acid groups on carbon black starts to decrease above 200°C[268], and the oxygen-containing groups decompose during thermal treatment (between 200°C and 800°C), accompanied by the evolution of $CO_2$ and CO[52]. Therefore, it is logical to assume that after decomposition of the oxygen-containing groups during heat treatment, high-energy sites would be left, and/or that some original high-energy pots that were screened by the oxygen groups would be revealed. This suggests that with regard to polymer-filler interaction, the oxygen-containing groups are much less active, especially in less- or nonpolar polymers. When treated black N234 is compounded into emulsion SBR at 50 phr loading, the abrasion resistance is shown to improve with temperature up to a level around 600°C. Thereafter, a gradual deterioration of the property is observed, even though the $\gamma_f^d$ steadily goes up [Figure 8.61(a)]. Similar results have been reported by Brown et al.[269] Since, over the range of temperatures used, the morphologies of the carbon black cannot be changed, the change in abrasion resistance of the filled compounds upon heat treatment can only be interpreted in terms of their surface characteristics. As discussed before, the increase in abrasion resistance of the blacks treated at relatively low temperature is certainly attributed to the enhancement of polymer-filler interaction as shown by $\gamma_f^d$. The negative impact of further temperature increase appears to be related to the poorer dispersion of the filler due to the stronger filler-filler interaction originating from the very high $\gamma_f^d$. In fact, for carbon black treated at temperatures beyond 600°C, the undispersed area of black in the vulcanizates measured by dispersion analyzer increases with treatment temperature [Figure 8.61(b)]. It seems that the positive impact of the improved polymer-filler interaction on the abrasion resistance is more than offset by the deterioration of filler dispersion.

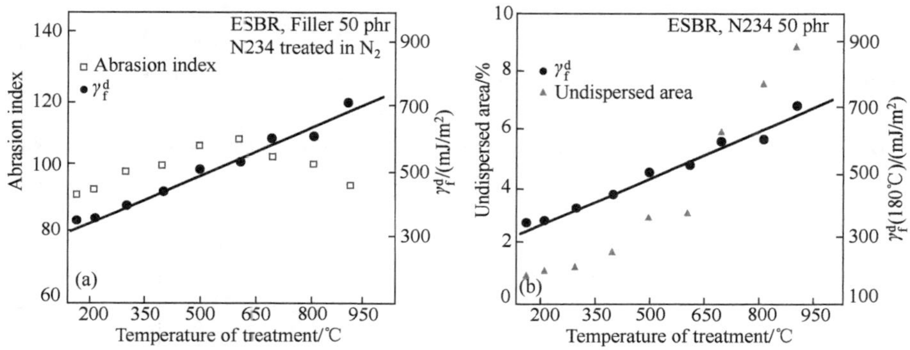


**Figure 8.61** Effect of heat-treatment of carbon black N234 on its $\gamma_f^d$ and abrasion resistance for filled ESBR vulcanization, and its $\gamma_f^d$ and its dispersion in the compounds

#### 8.3.2.4.3 Effect of Oxidation of Carbon Black

To further demonstrate the effect of oxygen-containing functional groups on carbon black on polymer-filler interaction, the surface energies of oxidized carbon blacks have been investigated. Upon oxidation of carbon black N234 with nitric acid (acid dosage: 30 parts of 65% acid per 100 parts carbon black), the $\gamma_f^d$ at 180°C is reduced by 26%. As a consequence, the abrasion resistance of an SSBR vulcanizate filled with 50 phr of this oxidized black is reduced by 38% at 14% slip ratio. As a matter of fact, the results of the changes in polymer-filler interaction by oxidation and heat treatment in inert atmosphere are in good agreement with the adhesion strength measured by electron microscopy[267]. The adhesion of carbon black GPF (N660) for BR was reduced by thermal and chemical oxidation. When the surface of carbon black was deoxidized, the adhesion between polymer and carbon black was improved.

#### 8.3.2.4.4 Effect of Physical Adsorption of Chemicals on Carbon Black Surface

The impact of filler surface characteristics on abrasion resistance may be further verified using physical adsorption of chemicals. In a comprehensive study, Dannenberg et. al[270] modified carbon black N339 with a surfactant, cetyltrimethyl ammonium bromide (CTAB), either by adding this chemical to the mixer or by pre-adsorption onto the carbon black surface. The modification resulted in a significant drop in abrasion resistance measured by the Akron angle abrader (Table 8.7) even though the dispersion of the filler was actually improved. Shown in Figure 8.62 is the dispersive component of the surface energies for carbon black N234 as a function of CTAB adsorption at 150°C.

**Table 8.7** Effect of CTAB adsorption on Akron angle abrasion resistance[a]

| Items | Carbon black N339 | | | |
|---|---|---|---|---|
| CTAB | 0 | 2.25 phr added to mix | 0.9% adsorbed | 4.3% adsorbed |
| Bound rubber/% | 32.6 | 31.3 | 24.5 | 18.0 |
| Abrasion index/% | 100.0 | 67.2 | 63.8 | 52.9 |

a. Formulation: SBR 1500:100, carbon black: 50, zinc oxide: 3, stearic acid: 1.5, antioxidant: 1, oil: 8, accelerator CBS: 1.25, sulfur: 1.25.

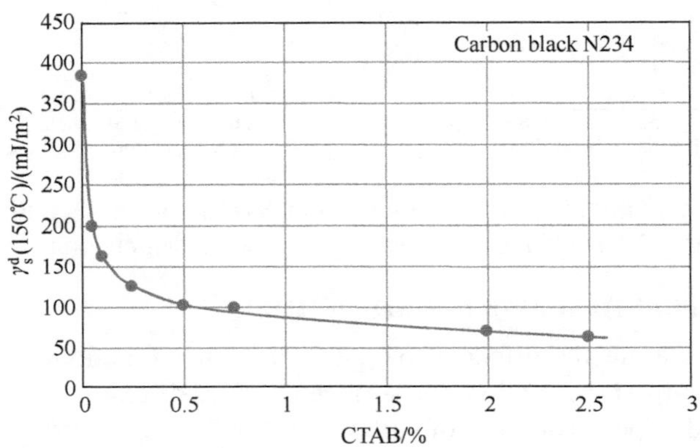


**Figure 8.62** $\gamma_s^d$ as a function of CTAB adsorption on carbon black N234

It can be seen that even a small amount of CTAB adsorbed on the surface, less than a monolayer, deteriorates the surface activity of carbon black significantly, reducing the abrasion resistance.

The correlation of surface characteristics obtained using a series of protein-treated blacks with abrasion resistance also confirms the effect. When carbon black was treated in aqueous solutions of bovine albumin, a model compound used for protein in natural rubber, the $\gamma_s^d$ decreased (Figure 8.62), and so did the polymer-filler interaction. As a result, significant loss in abrasion resistance is observed (Figure 8.63)[271].

#### 8.3.2.5 Effect of Carbon Black Mixing Procedure

For chemicals having the same cross-sectional area, polar chemicals give higher adsorption energies than alkanes, which can be taken as model compounds of hydrocarbon

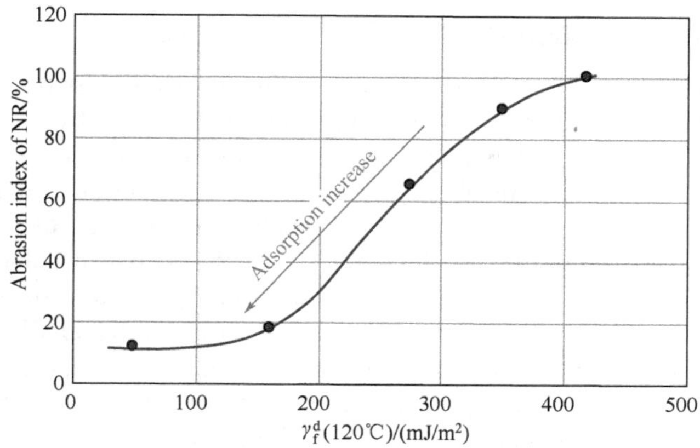


**Figure 8.63** Abrasion resistance *vs.* $\gamma_f^d$ of protein-treated N234 in NR compounds

polymers and oils (Figure 5.21). This suggests that when carbon black comes in contact with polar ingredients, such as antioxidants and stearic acid, the amount of available surface area and the number of filler active centers for polymer adsorption on the filler surface can be substantially reduced. Once molecules of these ingredients are adsorbed on the filler surface they are very difficult to be displaced by polymer chains. For oil, which has low polarity and better compatibility with polymer, the competition for the higher energetic sites on the filler surface between polymer molecules and oil would make the mixing sequence very important. From the point of view of polymer-filler interaction, the mixing procedure of carbon black-filled compounds should be optimized in such a way that the oil and other ingredients are added after the filler is incorporated into the polymer. For example, for a typical passenger tire tread compound using a blend of SSBR/BR (75/25), 75 phr carbon black N234 and 25 phr oil, when the carbon black was added to the polymer first and oil was added after carbon black had been incorporated, the abrasion resistance was significantly higher compared with that of the compound prepared by adding oil together with carbon black (Figure 8.64).

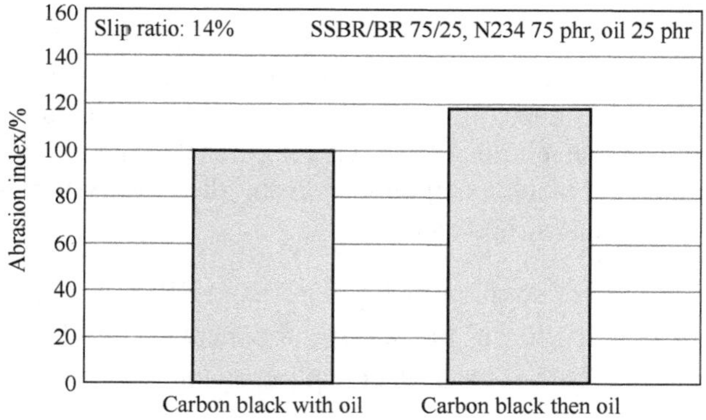


**Figure 8.64** Effect of mixing sequence on abrasion resistance of carbon black-filled vulcanizates[271]

#### 8.3.2.6 Silica *vs.* Carbon Black

Since the beginning of the industrial-scale production of fine-particle silicas, silica manufacturers have been seeking to have their products used in tires. The reason why it could not be used as the principal filler in tire tread compounds until recently is mainly related to the very poor abrasion resistance. This can be understood from its surface characteristics as shown by the adsorption energies of different chemicals (Figure 5.21). Compared to carbon black, the interaction between polar chemicals and silica is very high, while the interaction with alkanes is very low, indicating very weak interaction with hydrocarbon rubbers. The lack of strong polymer-filler interaction is the most important factor contributing to the lower abrasion resistance, as shown in Figure 8.65. Without a coupling agent, the abrasion resistance of silica in a typical passenger tire tread compound is reduced by about 70% compared to carbon black. These, in combination with poorer dynamic properties and poor processability, prevented the use of silica in tread compounds[68,272].

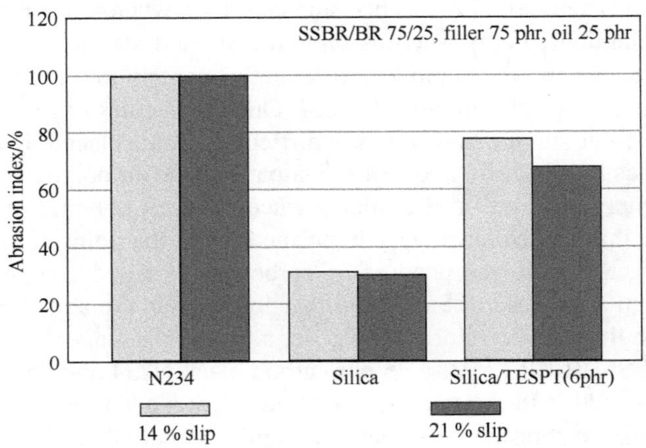


**Figure 8.65** Abrasion resistance of carbon black N234, silica, and silica/TESPT compounds

For the green tire tread compound patented by Michelin, besides low rolling resistance and better wet skid resistance, it was claimed that tread life and durability were comparable to those of conventional tires. This implies that the tread wear resistance is equal to that of carbon black compounds. From the material point of view, this compound is characterized as follows:

- the polymer is a blend of solution SBR (SSBR) and BR,
- the compound contains silica as the major component of the filler system, and
- bifunctional silane TESPT is used as a coupling agent between polymer and filler.

In the silica tire formulation, 6.4 phr of carbon black was used as the carrier for the coupling agent.

Among these, the application of a sulfur containing silane-coupling agent is the key technology. By organo-silane-surface modification, filler agglomeration can be effectively depressed via reduction of the specific component of the silica surface energy, leading to significantly improved compound processability and dynamic properties. The poorer polymer-filler interaction can also be considerably compensated for by chemical linkages between polymer and filler via sulfide groups. As a result, the abrasion resistance of the compound is drastically promoted (Figure 8.65), making it possible to use silica in passenger tire tread compounds as the principal filler[272].

#### 8.3.2.7 Silica in Emulsion SBR Compounds

One of the features of the silica-tread compound described in the Michelin patent is the polymer system: the principal rubber is SSBR. However, for typical passenger tread compounds, the dominant polymer was emulsion SBR (ESBR) even though SSBR had been commercialized for a few decades and its advantages in rolling resistance and skid resistance, though not significant, have been recognized[190]. Since the introduction of silica-based tread compounds, although great effort has been made by tire manufacturers to use ESBR in silica compounds due to its better processability and low cost, the solution polymerized rubbers (including BR) are still the exclusive polymers for silica tires. Among other reasons, poor abrasion resistance is the main deficiency of silica-ESBR compounds. While this inferior property is related to its weak polymer-filler interaction, unlike SSBR, it cannot be compensated for by a coupling reaction as illustrated in Figure 8.66. Even with a high dosage of coupling agent, the abrasion resistance is far from competing with that of carbon black compounds. Several factors may be responsible for the poor abrasion resistance of ESBR-silica compounds. Firstly it may be related to the efficiency of the coupling agent. Due to the stronger interaction with polar chemicals (Figure 5.21 and Table 8.8) in the SSBR-based passenger tire tread compound, the coupling agent can be driven to the filler surface by polar interaction. This facilitates the coupling reaction, and the poor polymer-filler interaction can be effectively offset by chemical bonds between polymer and filler via the coupling agent. However, in the case of ESBR-based tread compounds, the non-rubber impurities originating from the production process of emulsion SBR interfere with the coupling reaction. It is known that in ESBR there is about 5% – 8% non-rubbers, which are mainly surfactants. As the silica surface is highly polar in nature, it readily accepts the polar groups of these substances, leaving their non- or less-polar groups to interact with the polymer. In other words, the surfactants in ESBR can be easily adsorbed on the filler surface, weakening polymer-filler interaction. They also block the silanols on the silica surface, preventing chemical coupling reaction. In addition, the polarity of polysulfide groups in TESPT and/or their decomposition products is higher than that of the hydrocarbon rubber. In the compound they may preferably associate themselves with the non-rubber substances, reducing the reaction between coupling agent and polymer chains. Consequently, the poor polymer-silica interaction cannot be efficiently compensated for by the application of coupling agents, even at very high dosages. Such an argument can be further strengthened by the issues raised from development of silica compounds for truck tires.

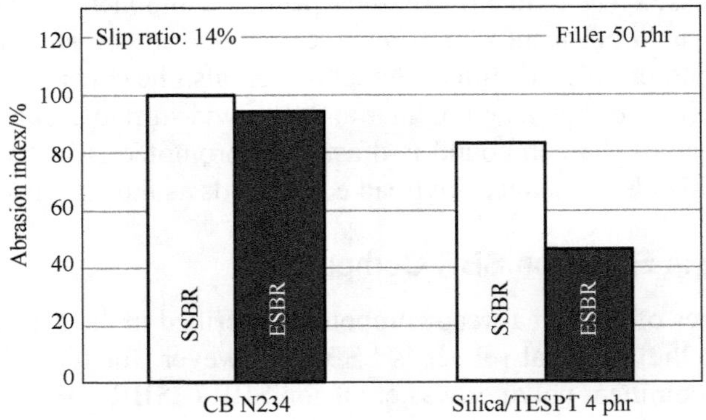


**Figure 8.66** Abrasion resistance of carbon black N234 and silica/TESPT in solution and emulsion SBRs

**Table 8.8** $\gamma_f^d$ and specific components of adsorption energies of some polar chemicals on carbon black N234 and silica (ZeoSil 1165) and TESPT-modified silica

| Items | Carbon black N234 | Silica | Silica | Silica/TESPT |
|---|---|---|---|---|
| Test temperature/°C | 180 | 180 | 130 | 130 |
| $\gamma_f^d$/(mJ/m$^2$) | 382 | 28 | 47 | 36 |
| $I^{sp}$(benzene)/(mJ/m$^2$) | 93 | 73 | 111 | 40 |
| $I^{sp}$($CH_3CN$)/(mJ/m$^2$) | 173 | 278 | | |
| $I^{sp}$(THF)/(mJ/m$^2$) | 74 | 271 | | |
| $I^{sp}$(acetone)/(mJ/m$^2$) | 86 | 264 | | |
| $I^{sp}$(ethyl acetate)/(mJ/m$^2$) | 48 | 206 | | |
| $I^{sp}$($CHCl_3$)/(mJ/m$^2$) | 72 | 39 | | |

### 8.3.2.8 Silica in NR Compounds

Although silica has been used with solution SBR in passenger tread compounds to improve rolling resistance and wet traction, it has been reported that the application of silica to truck tire compounds as a main filler has not been successful[273]. The key deficiency of silica in truck-tire-tread compounds is also poor abrasion resistance. In the last few decades, the radialization of truck tires has led to a higher proportion of natural rubber usage[274]. This is firstly due to its excellent green strength and tack, which are very desirable properties for radial truck tire building. Secondly, the low-angle steel cord breaker system of radial truck tires makes the crown area stiffer, leading to a deficiency in its ability to envelop sharp objects. This makes the tire more prone to cutting and chipping, especially under overloading and poor road conditions. This again drives higher usage of natural rubber for the tread component of the radial truck tire[275]. Multiple factors may account for the inferior properties of truck-tire-tread compounds containing silica, but a major one is related to the surface characteristics of the silica. In the case of NR-based truck tire tread compounds, the non-rubber substances originating from the natural rubber latex can interfere with the polymer-filler interaction, similar to the effect of surfactants in ESBR. In dry NR, there are more than 5% non-rubber substances, which are mainly proteins, fatty acids, phospholipids, other ester-like substances, and their degradation products. As they are easily adsorbed on the silica surface, these non-rubbers can deteriorate polymer-filler interaction and reduce the effectiveness of coupling agents. For example, even with a ratio of TESPT to silica as high as 15/100 by weight, the enhancement of abrasion resistance is inadequate[276,277]. Also, due to the surface characteristics of silica, more curatives, which are generally polar substances, can be adsorbed on the filler surface. Additionally, the surface acidity of the silica is much higher than that of carbon black. These effects result in poor cure characteristics, leading to a lower cure rate and lower yield of crosslinks during vulcanization. Therefore, increased amounts of curatives are necessary in order to be able to obtain reasonable crosslink density and cure kinetics. For NR compounds filled exclusively with silica, the dosage of sulfenamide accelerators used has been as high as 2.8−3.6 phr with 1.7−2.0 phr sulfur compared to 1−1.4 phr and 0.8−1.2 phr for carbon black-filled NR compounds[196,222]. This, along with the high level of TESPT, will lead to high levels of chain-modification of the polymer due to increased cyclic sulfides, pendant-S-accelerator groups, and pendant coupling agent moieties. This in turn will significantly reduce the flexibility of polymer chains, resulting in a deterioration of abrasion resistance. Moreover, due to the poorer processability of silica, especially during mixing, the filler incorporation time is increased. This leads to a drastic drop in molecular weight of polymer due to chain scission, as NR is very sensitive to mechano-oxidative degradation. This is particularly true when heat treatment is performed for the pre-coupling reaction. Consequently, the abrasion resistance is further impaired. As can be seen in Figure 8.67, the abrasion resistance of vulcanizates filled with 50 phr silica and 6 phr TESPT is about 40% lower than its carbon black counterpart even though a high dosage of coupling agent was applied[175].

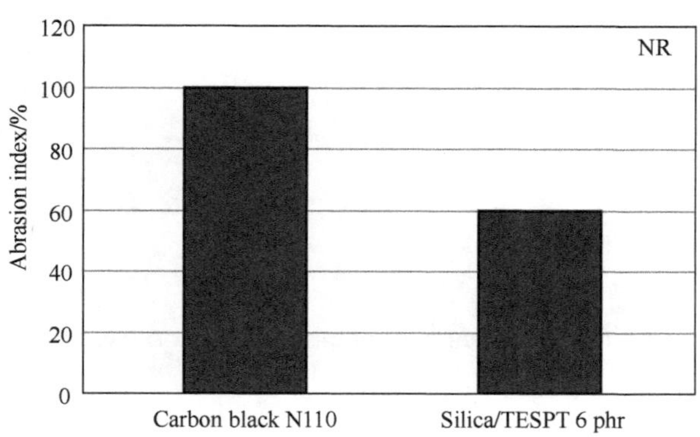


**Figure 8.67** Abrasion resistance of carbon black N110 and silica/TESPT in NR compounds

#### 8.3.2.9 Effect of CSDPF on Abrasion Resistance

When incorporated in hydrocarbon polymers, both CSDPF 2000 and 4000 are characterized not only by lower filler-filler interaction but also higher filler-polymer interaction compared to both carbon black and silica. Evidence of the higher filler-polymer interaction of dual phase fillers in hydrocarbon polymers can be obtained by comparing the adsorption energies of polymer analogs on the filler surfaces. Shown in Figure 8.68 are the adsorption free energies of heptane. The results for both CSDPF 2000 and 4000 are presented along with those of blends of carbon black N234 and silica. Regardless of the difference in surface areas, at the same silica contents the dual phase fillers give significantly higher adsorption energies. The higher surface activity has been attributed to the change in surface microstructure of the carbon black phase. Upon doping with a foreign substance in the carbon domains, more surface defects in the graphitic crystal lattice and/or smaller crystallite dimensions, which have been proposed as the active centers for rubber adsorption[22], may be formed. This consideration could find its qualitative evidence from scanning tunneling microscopic investigation. The statistically smaller area of well-organized carbon structure (quasi-graphitic) of CSDPF surface could be direct evidence of more active centers in the carbon phase[175]. When coupling agent is needed to strengthen the polymer-filler interaction on the silica domains, the dosage should be much lower than that used in silica compounds. Practically, CSDPF 4000, with higher surface coverage of silica, is recommended for passenger tire tread compounds to obtain comparable rolling resistance and wet skid resistance to silica compounds with better wear resistance (Figure 8.50). In the case of NR-based tread compounds for truck tire, the CSDPF 2000, with low silica coverage, is the most favorable filler that imparts superior wet friction coefficient[278], lower hysteresis, and comparable abrasion resistance relative to carbon black (Figure 8.69).

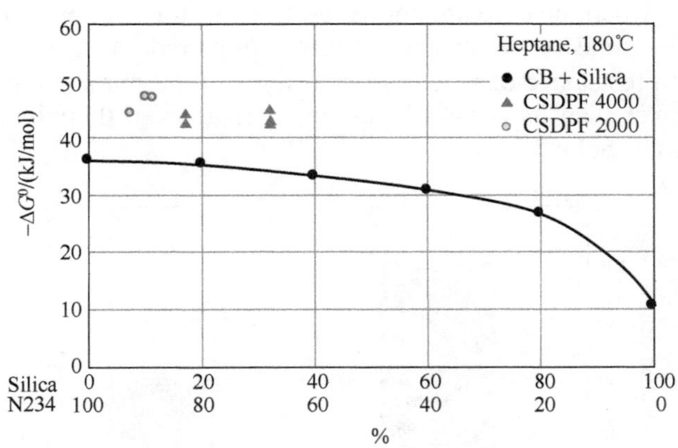


**Figure 8.68** Adsorption free energies of heptane at 180°C on a variety of fillers

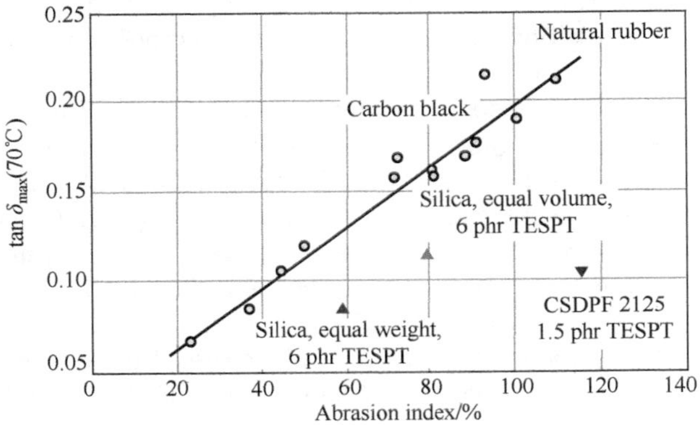


**Figure 8.69** Hysteresis and abrasion resistance of NR compounds filled with a variety of fillers

## References


[1] Medalia A I. Effect of Carbon Black on Dynamic Properties of Rubber Vulcanizates. *Rubber Chem. Technol.*, 1978, 51: 437.

[2] Bulgin D, Hubbard D G, Walters M H. *Proc. Rubber Technol. Conf. 4th*, London, 1962: 173.

[3] Saito Y. New Polymer Development for Low Rolling Resistance Tyres. *Kautsch. Gummi Kunstst.*, 1986, 39: 30.

[4] Medalia A I. Heat Generation in Elastomer Compounds Causes and Effects. *Rubber Chem. Technol.*, 1991, 64: 481.

[5] Duperray B, Leblanc J L. The Time-temperature Superposition Principle as Applied to Filled Elastomers. *Kautsch. Gummi Kunstst.*, 1982, 35: 298.

[6] Nordsiek K H. The Integral Rubber Concept-an Approach to an Ideal Tire Tread Rubber. *Kautsch. Gummi Kunstst.*, 1985, 38: 178.

[7] Wolff S, Wang M-J. Chapter 6 & Chapter 9. In: Donnet J-B, Bansal R C, Wang M-J (Editors). Carbon Black, Science and Technology. 2nd ed. New York: Marcel Dekker, 1993.

[8] Wang M-J. presented at a workshop "Praxis und Theorie der Verstärkung von Elastomeren," Hanover, Germany, Jun. 27-28, 1996.

[9] Grosch K A. The Rolling Resistance, Wear and Traction Properties of Tread Compounds. *Rubber Chem. Technol.*, 1996, 69: 495.

[10] Hess W M, P. C. Vegvari, and R. A. Swor, Carbon Black in NR/BR Blends for Truck Tires. *Rubber Chem. Technol.*, 1985, 58: 350.

[11] Ulmer J D, Chirico V E, Scott C E. The Effect of Carbon Black Type on the Dynamic Properties of Natural Rubber. *Rubber Chem. Technol.*, 1973, 46: 897.

[12] Ayala J A, Hess W M, Dotoson A O, et al. New Studies on the Surface Properties of Carbon Blacks. *Rubber Chem. Technol.*, 1990, 63: 747.

[13] Caruthers J M, Cohen R E, Medalia A I. Effect of Carbon Black on Hysteresis of Rubber Vulcanizates Equivalence of Surface Area and Loading. *Rubber Chem. Technol.,* 1976, 49: 1076.

[14] Patel A C, Jackson D C. Poster presented at IRC'91, Essen, Jun. 24-27, 1991.

[15] Patel A C, Lee K W. Characterizing Carbon Black Aggregate via Dynamic and Performance Properties. *Elastomerics,* 1990, 122(3), 14; 1990, 122(4): 22.

[16] Wang M-J, Etude du Renforcement des élastomères par les Charges: Effet Exercé par l'emploi de Silices Modifiées par Greffage de Chaines Hydrocarbonées. Sc.D. Dissertation, Université de Haute Alsace, Mulhouse, France, 1984.

[17] Staszezuk P. *Mater. Phys.,* 1980, 14: 279.

[18] Zettlemoyer A C. Hydrophobic Surfaces, F. M. Fowkes, Ed., Academic Press, 1969: 9.

[19] Cazeneuve C. Thesis de Docteur Ingénieur, Université de Haute Alsace, Mulhouse, France, 1980.

[20] Donnet J-B, Brendle M, Dhami T L, et al. Plasma Treatment Effect on the Surface Energy of Carbon and Carbon Fibers. *Carbon* 1986, 24: 757.

[21] Guilpain G. Etude de traitements superficiels de fibres de carbone par voie électrolytique. Ph.D. Thesis, Université de Haute Alsace, Mulhouse, France, 1988.

[22] Wang M-J, Wolff S, Donnet J-B. Filler-Elastomer Interactions. Part Ⅲ. Carbon-Black-Surface Energies and Interactions with Elastomer Analogs. *Rubber Chem. Technol.,* 1991, 64: 714.

[23] Donnet J-B, Lansiger C M. Characterization of Surface Energy of Carbon Black Surfaces and Relationship to Elastomer Reinforcement. *Kautsch. Gummi Kunstst.,* 1992, 45(6): 459.

[24] Wang M-J, Wolff S, Tan E-H. Filler-Elastomer Interactions. Part Ⅷ. The Role of the Distance between Filler Aggregates in the Dynamic Properties of Filled Vulcanizates. *Rubber Chem. Technol.,* 1993, 66: 178.

[25] Kraus G. *Proc. Int. Rubber Conf.*, Brighton, U.K., 1977, Paper 21.

[26] Janzen J, Kraus G. *Proc. Int. Rubber Conf.*, Brighton, U. K., 1972, G7-1.

[27] Stacy C J, Chirico V E, Kraus G. Effect of Carbon Black Structure Aggregate Size Distribution on Properties of Reinforced Rubber. *Rubber Chem. Technol.,* 1975, 48: 538.

[28] Hess W M, Chirico V E. Elastomer Blend Properties-Influence of Carbon Black Type and Location. *Rubber Chem. Technol.,* 1977, 50: 301.

[29] McDonald G C, Hess W M. Carbon Black Morphology in Rubber. *Rubber Chem. Technol.,* 1977, 50: 842.

[30] Shieh C-H, Mace M L, Ouyang G B, Branan J M, et al. Meeting of the Rubber Division, ACS, Toronto, Canada, May 21-24, 1991.

[31] Brennan J J, Jermyn T E, Boostra B B. Carbon Black-Polymer Interaction: a Measure of Reinforcement. *J. Appl. Polym. Sci.,* 1964, 8: 2687.

[32] Dannenberg E M. Bound Rubber and Carbon Black Reinforcement. *Rubber Chem. Technol.,* 1986, 59: 512.

[33] Leblanc J L, Hardy P, Leblanc J L, et al. Evolution of Bound Rubber During the Storage of Uncured Compounds. *Kautsch. Gummi Kunstst.,* 1991, 44: 1119.

[34] Wolff S, Wang M-J, Tan E-H. Filler-Elastomer Interactions. Part Ⅶ. Study on Bound Rubber. *Rubber Chem. Technol.,* 1993, 66: 163.

[35] Hess W M, Chirico V E, Burgess K A. Carbon-black Morphology in Rubber. *Kautsch. Gummi Kunstst.,* 1973, 26: 344.

[36] Heckman F A, Medalia A I. *J. Inst. Rubber Ind.* 1969, 3: 66.

[37] Gessler A M. Effect of Mechanical Shear on the Structure of Carbon Black in Reinforced Elastomers. *Rubber Chem. Technol.*, 1970, 43: 943.

[38] Ban L L, Hess W M. Current Progress in the Study of Carbon Black Microstructure and General Morphology. Interactions Entre les Elastomères et les Surfaces Solides Ayant Une Action Renforçant, *Colloques Int. C.N.R.S.*, Paris, France, 1975, No. 231: 81.

[39] Nakata T. The Theory and Application of Filler Reinforcement. *Nippon Gomu Kyokaishi,* 1985, 58: 713.

[40] Herd C R, Mcdonald G C, Hess W M. Morphology of Carbon-Black Aggregates: Fractal Versus Euclidean Geometry. *Rubber Chem. Technol.,* 1992, 65: 107.

[41] Wolff S, Wang M-J, Tan E-H, et al. Surface Energy of Fillers and its Effect on Rubber Reinforcement. *Kautsch. Gummi Kunstst.,* 1994, 47: 780.

[42] Wang W, Haidar B, Vidal A, et al. Study of Surface Activity of Carbon Black by Inverse Gas Chromatography. IV: Effect of Mechanical Action on Surface Activity of Carbon Black. *Kautsch. Gummi Kunstst.,* 1994, 47: 238.

[43] Gessler A M. *Proc. Int. Rubber Conf.*, Brighton, England, 1967: 249.

[44] Boonstra B B. Chapter 7. In: Blow C M, Hepburn C (Editors). Rubber Technology and Manufacture. 2nd ed. London: Butterworth Sci., 1982.

[45] Rivin D. Use of Lithium Aluminum Hydride in the Study of Surface Chemistry of Carbon Black. *Rubber Chem. Technol.*, 1963, 36: 729.

[46] Tan E-H. Ph.D. Thesis, Université de Haute Alsace, Mulhouse, France, 1992.

[47] Wolff S, Wang M-J, Tan E-H. Filler-Elastomer Interactions. Part X. The Effect of Filler-elastomer and Filler-filler Interaction on Rubber Reinforcement. *Kautsch. Gummi Kunstst.,* 1994, 47: 102.

[48] Dannenberg E M. *Rubber Age (NY),* 1966, 98 (9): 82.

[49] Bansal R C, Donnet J-B. Chapter 4. In: Donnet J-B, Bansal R C, Wang M-J (Editors). Carbon Black, Science and Technology. 2nd ed. New York: Marcel Dekker, 1993.

[50] Puri B R, Bansal R C. Iodine Adsorption Method for Measuring Surface Area of Carbon Blacks. *Carbon,* 1965, 3: 227.

[51] Donnet J-B, Voet A. Chapter 8. Carbon Black, New York: Marcel Dekker, 1976.

[52] Rivin D. Surface Properties of Carbon. *Rubber Chem. Technol.*, 1971, 44: 307.

[53] Shaeffer W D, Smith W R, Polley M H. Structure and Properties of Carbon Black-changes Induced by Heat Treatment. *Ind. Eng. Chem.,* 1953, 45: 1721.

[54] Franklin R E. *Acta Crystallogr.* 1950, 3: 1907; 1950, 4: 253.

[55] Heckman F A. Microstructure of Carbon Black. *Rubber Chem. Technol.*, 1964, 37: 1245.

[56] Hess W M, Herd C R. Chapter 3. In: Donnet J-B, Bansal R C, Wang M-J (Editors). Carbon Black, Science and Technology. 2nd ed. New York: Marcel Dekker, 1993.

[57] Wang M-J, Wolff S, Freund B. Filler-Elastomer Interactions. Part XI. Investigation of the Carbon-Black Surface by Scanning Tunneling Microscopy. *Rubber Chem. Technol.*, 1994, 67: 27.

[58] Zerda T W, Xu W, Yang H, et al. Meeting of the Rubber Division, ACS, Anaheim, California, May 6-9, 1997.

[59] Wolff S. Chemical Aspects of Rubber Reinforcement by Fillers. *Rubber Chem. Technol.*, 1996, 69: 325.

[60] Wang M-J, Wolff S. Filler-Elastomer Interactions. Part Ⅵ. Characterization of Carbon Blacks by Inverse Gas Chromatography at Finite Concentration. *Rubber Chem. Technol.*, 1992, 65: 890.

[61] Kraus G. Chapter 4. Reinforcement of Elastomers, New York: Interscience, 1965: 140.

[62] Funt J M. Dynamic Testing and Reinforcement of Rubber. *Rubber Chem. Technol.*, 1988, 61: 842.

[63] Medalia A I, Kraus G. Chapter 8. In: Mark J E, Erman B, Eirich F R (Editors). Science and Technology of Rubber. 2nd ed. San Diego: Academic Press, 1994.

[64] Wolff S, Görl U, Wang M-J, et al. Silica-based Tread Compounds. *Eur. Rubber J.*, 1994, 176 (11): 16.

[65] Wang M-J, Wolff S, Donnet J-B. Filler-Elastomer Interactions. Part Ⅰ. Silica Surface Energies and Interactions with Model Compounds. *Rubber Chem. Technol.*, 1991, 64: 559.

[66] Wolff S, Wang M-J. Filler-Elastomer Interactions. Part Ⅳ. The Effect of the Surface Energies of Fillers on Elastomer Reinforcement. *Rubber Chem. Technol.*, 1992, 65: 329.

[67] Wolff S, Wang M-J. *Proc. Int. Conf. Carbon Black,* Mulhouse, France, Sept. 27-30, 1993: 133.

[68] Wolff S, Wang M-J, Tan E-H. Surface Energy of Fillers and its Effect on Rubber Reinforcement. *Kautsch. Gummi Kunstst.*, 1994, 47: 873.

[69] Wang M-J, Patterson W J. *Proc. Int. Rubber Conf.* Manchester, June 17-20, 1996, paper no. 43.

[70] Tan E-H, Wolff S, Haddeman M, Grewatta H P, et al. Filler-Elastomer Interactions. Part IX. Performance of Silicas in Polar Elastomers. *Rubber Chem. Technol.*, 1993, 66: 594.

[71] Hofmann W. Chapter 4. Rubber Technology Handbook, Munich: Hanser Publishers, 1989.

[72] Iler P K. The Chemistry of Silica, New York: Interscience, 1979.

[73] Donnet J-B, Papirer E, Vidal A, et al. presented at Rubbercon'88, Oct. 10-14, 1988, Sydney, Australia.

[74] Vidal A, Shi Z H, Donnet J-B. Modification of Carbon Black Surfaces: Effects on Elastomer Reinforcement. *Kautsch. Gummi Kunstst.*, 1991, 44: 419.

[75] Vidal A, Papirer E, Wang M-J, et al. Modification of Silica Surfaces by Grafting of Alkyl Chains. I-Characterization of Silica Surfaces by Inverse Gas-solid Chromatography at Zero Surface Coverage. *Chromatographia,* 1987, 23: 121.

[76] Papirer E, Vidal A, Wang M-J, et al. Modification of Silica Surfaces by Grafting of Alkyl Chains. II-Characterization of Silica Surfaces by Inverse Gas-solid Chromatography at Finite Concentration. *Chromatographia,* 1987, 23: 279.

[77] Iler P K. Relation of Particle Size of Colloidal Silica to the Amount of a Cationic Polymer Required for Flocculation and Surface Coverage. *J. Colloid Interface Sci.*, 1971, 37: 364.

[78] Ribio J, Kitchener A J. The Mechanism of Adsorption of Poly (ethylene oxide) Flocculant on Silica. *J. Colloid Interface Sci.*, 1976, 57: 132.

[79] Mewis J. Rheology of Concentrated Dispersions. *Adv. Colloid Interface Sci.*, 1976, 6: 173.

[80] Wang M-J, Wolff S. Filler-Elastomer Interactions. Part Ⅴ. Investigation of the Surface Energies of Silane-Modified Silicas. *Rubber Chem. Technol.,* 1992, 65: 715.

[81] Monte S J, Sugerman G. Meeting of the Rubber Division, ACS, Chicago, Illinois, Oct. 5-7, 1982.

[82] Monte S J. Ken-React Preference Manual-Titanate, Zirconate and Aluminate Coupling Agents, Second rev sed edition, Kenrich Petrochemicals, Inc., 1993.

[83] Metal-Acid Esters-Chelates, Hüls, AG. 1992.

[84] Wang M-J, Vidal A, Papirer E, et al. Modification of Silica Surfaces by Grafting of Alkyl Chains. Part Ⅲ. Particle/particle Interactions: Rheology of Silica Suspensions in Low Molecular Weight Analogs of Elastomers. *Colloids Surf.,* 1989, 40: 279.

[85] Plueddemann E P. Silane Coupling Agents, New York: Plenum Press, 1982.

[86] Mittal K L. Silanes and Other Coupling Agents, *vs.* P, Utrecht, 1992.

[87] Donald G W. *Rubber Age* 1970, 102(4): 66.

[88] Grillo T A. *Rubber Age* 1971, 103(8): 37.

[89] Ranney M W, Pageno C A, Ziemiansky L P. *Rubber World,* 1970, 163: 54.

[90] Roland R. Rubber Compound and Tires Based on Such a Compound, European patent, EP0501227A1, 1992.

[91] Degussa A G, Silanized Silicas Coupsil, *Technical Information Bulletin*, No. 6031, Sept. 1, 1995.

[92] Wolff S. Optimization of Silane-Silica OTR Compounds. Part 1: Variations of Mixing Temperature and Time during the Modification of Silica with Bis-(3-Triethoxisilylpropyl)-Tetrasulfide. *Rubber Chem. Technol.*, 1982, 55: 967.

[93] Thurn F, Wolff S. *Proc. Int. Rubber Conf., Munich,* Sept. 2-5, 1974.

[94] Wolff S. Presented at the First Franco-German Rubber Symposium, Obernai, France, Nov. 14-16, 1985.

[95] Polmanteer K E, Lentz C W. Reinforcement Studies-Effect of Silica Structure on Properties and Crosslink Density. *Rubber Chem. Technol.*, 1975, 48: 795.

[96] Graves D F. Benzofuroxans as Rubber Additives. *Rubber Chem. Technol.*, 1993, 66: 61.

[97] Yamaguchi T, Kurimoto I, Ohashi K, et al. Novel Carbon Black/rubber Coupling Agent. *Kautsch. Gummi Kunstst.,* 1989, 42: 403.

[98] González L, Rodríguez A, de Benito J L, et al. A New Carbon Black-Rubber Coupling Agent to Improve Wet Grip and Rolling Resistance of Tires. *Rubber Chem. Technol.,* 1996, 69: 266.

[99] Wolff S, Görl U. Carbon Blacks Modified with Organosilicon Compounds, Method of Their Production and Their Use in Rubber Mixtures, US patent, US5159009, Oct. 27, 1992.

[100] Wolff S, Görl U. The Influence of Modified Carbon Blacks on Viscoelastic Compound Properties. *Kautsch. Gummi Kunstst.,* 1991, 44: 941.

[101] Swor R A, Taylor R L. Use of Silane Coupling Agent with Carbon Black to Enhance the Balance of Reinforcement Properties of Rubber Compounds, US patent, US5494955, Feb. 27, 1996.

[102] Boehm H P. Some Aspect of the Surface Chemistry of Carbon Black, Presented at The Second International Conference on Carbon Black, Mulhouse, France, Sept. 27-30, 1993.

[103] Wagner M P. Reinforcing Silicas and Silicates. *Rubber Chem. Technol.,* 1976, 49: 703.

[104] Austin A E. *Proc. Conf. Carbon*, 3rd. 1958: 389.
[105] Boehm H P. *Adv. Catal.,* 1968, 16: 161.
[106] Herd C R. Meeting of the Rubber Division, ACS, Montreal, Canada, May 5-8, 1996.
[107] Wang W, Vidal A, Donnet J-B, et al. Study of Surface Activity of Carbon Black by Inverse Gas Chromatography. Ⅲ: Superficial Plasma Treatment of Carbon Black and its Surface Activity. *Kautsch. Gummi Kunstst.,* 1993, 46: 933.
[108] Takeshita M. Mukai U, Sugawara T. (to Bridgestone), Rubber Composition for Tires, US patent, US4820751, Apr. 11, 1981.
[109] Kinoshita K. Chapter 4. Carbon, Electrochemical and Physico-Chemical Properties. New York: John Wiley & Sons, Inc., 1988.
[110] Donnet J-B, Bouland J C. *Rev. Gen. Caout.,* 1964, 41: 407.
[111] Donnet J-B, Shultz J, Eckhardt A. Etude de la microstructure d'un noir de carbone thermique. *Carbon,* 1968, 6: 781.
[112] Mahmud K, Wang M-J, Francis R A, Belmont J. Elastomeric Compounds Incorporating Silicon-treated Carbon Blacks, US patent, WO/1996/037547, 1996.
[113] Wang M-J, Mahmud K, Murphy L, et al. Meeting of the Rubber Division, ACS, Anaheim, California, May 6-9, 1997.
[114] Wang M-J, Patterson W J, Brown T A, et al. Meeting of the Rubber Division, ACS, Anaheim, California, May 6-9, 1997.
[115] Ohkita K. Grafting of Carbon Black, Japan: Rubber Dig. Co., 1982.
[116] Ohkita K. *Polymer Dig.*, 1993, No. 10, 3.
[117] Tsubokawa N. Functionalization of Carbon Black by Surface Grafting of Polymers. *Prog. Polym. Sci.,Chem.,* 1992, 17: 417.
[118] Kroker R, Schneider M, K. Hamann. Polymer Reactions on Powder Surfaces. *Prog. Org. Coat.,* 1972, 1: 23.
[119] Morton M, Healy J C, Denecour R L. *Proc. Int. Rubber. Conf.*, Brighton, 1967: 175.
[120] Schuster R H. Educ. Symp., Paper F., Meeting of the Rubber Division, ACS, Montreal, Canada, May 5-8, 1996.
[121] Nuyken O, Ko S, Voit B, et al. Core-shell Polymers as Reinforcing Polymeric Fillers for Elastomers. *Kautsch. Gummi Kunstst.,* 1995, 48: 784.
[122] Johnson P S. Rubber Products Manufacturing Technology, Bhowmick A K, Hall M M, Benarey H A, Eds., New York: Marcel Dekker, 1994.
[123] Nakajima N. Mixing and Viscoelasticity of Rubber (Part 1). *Int. Polym. Sci. Technol.* 1994, 21(11): T/47-67.
[124] Yoshida T. Foundations of Rubber Kneading. *Nippon Gomu Kyokaishi* 1992, 65: 325.
[125] Funt J M. Meeting of the Rubber Division, ACS, Cleveland, Ohio, Oct. 1-4,1985.
[126] Gerke R H, Ganzhorn G H, Howland L H, et al. Manufacture of Rubber, US patent, US2118601A, May 24, 1938.
[127] Dannenberg E M, Collyer H J, *Ind. Eng. Chem.,* 1949, 41: 1067.
[128] Dannenberg E M, Carbon Black Dispersion and Reinforcement. *Ind. Eng. Chem.,* 1952, 44: 813.
[129] Barton B C, Smallwood H M, Ganzhorn G H. Chemistry in Carbon Black Dispersion. *J. Polym. Sci.,* 1954, 13: 487.
[130] Welsh F E, Richmond B R, Keach C B, et al. Meeting of the Rubber Division, ACS, Philadelphia, Pennsylvania, May 2-5, 1995.
[131] Cotton G R. Mixing of Carbon Black with Rubber. II. Mechanism of Carbon Black Incorporation. *Rubber Chem. Technol.,* 1985, 58: 774.

[132] Boonstra B B. Mixing of Carbon Black and Polymer: Interaction and Reinforcement. *J. Appl. Polym. Sci.,* 1967, 11: 389.

[133] Gessler A M. *Rubber Age,* 1969, 101(12): 54.

[134] Berry J P, Cayré P J. The Interaction between Styrene-butadiene Rubber and Carbon Black on Heating. *J. Appl. Polym. Sci.,* 1960, 3: 213.

[135] Wang M-J. Studies on Basic Properties of Several Antioxidants. Technical Bulletin, BRDIRI, May 1975.

[136] Crowther B G, Edmondson H M. Chapter 8. In: Blow C M (Editor). Rubber Technology and Manufacture. Cleveland, Ohio: CRC Press, 1971.

[137] Kraus G, Gruver J T. Molecular Weight Effects in Adsorption of Rubbers on Carbon Black. *Rubber Chem. Technol.*, 1968, 41: 1256.

[138] Villars D S. Studies on Carbon Black. Ⅲ. Theory of Bound Rubber. *J. Polym. Sci.,* 1956, 21: 257.

[139] Stickney P B, McSweeney E E, Mueller W J. Bound-Rubber Formation in Diene Polymer Stocks. *Rubber Chem. Technol.,* 1958, 31: 369.

[140] Watson W F. Combination of Rubber and Carbon Black on Cold Milling. *Ind. Eng. Chem.,* 1955, 47: 1281.

[141] Ashida M, Abe K, Watanabe T. *Nippon Gomu. Kyokaishi*, 1976, 49: 11.

[142] Cotton G R. Influence of Carbon Black Activity on Processability of Rubber Stocks, Part I. *Cabot report*, 74-C-4, Sept., 1974.

[143] Cotten G R. Mixing of Carbon Black with Rubber, Ⅳ. Effect of Carbon Black Characteristics. *RUBBEREX 86 Proceeding, ARPMA*, Apr. 29-May 1, 1986.

[144] Degussa, Reforcing Agent Si 69, X50-S, X50.

[145] Donnet J B, Wang W, Vidal A, et al. Study of Surface Activity of Carbon Black by Inverse Gas Chromatography. II: Effect of Carbon Black Thermal Treatment on Its Surface Characteristics and Rubber Reinforcement. *Kautsch. Gummi Kunstst.*, 1993, 46: 866.

[146] Görl U, Panenka R. Silanisierte Kieselsäuren: eine neue Produktklasse für zeitgemässe Mischungsentwicklung. *Kautsch. Gummi Kunstst.*, 1993, 46: 538.

[147] Patkar S D, Evans L R, Waddel W H. presented at International Tire Exhibition and Conference, Akron Ohio, Sept. 10-12, 1996.

[148] MRPRA, Functionalization of Elastomers by Reactive Mixing," Research Disclosure, 308, Jun., 1994.

[149] Terakawa K, Muraoka K. *Proc. Int. Rubber Conf.*, Kobe, Japan, Oct. 23-27, 1995, Paper no. P24.

[150] Hamerton I. Recent Developments in Epoxy Resins. *RAPRA Rev. Rep.*, (Report 91), 1996, 8: 7.

[151] Wampler W A, Gerspacher M, Yang H H, et al. *Rubber Plast. News*, Apr. 24, 1995, 24(28): 45.

[152] Gerspacher M. *Proc. Int. Rubber Conf.* Manchester, June 17-20, 1996, Paper no. 44.

[153] Cruse R W, Hofstetter M H, Panzer L M, et al. Meeting of the Rubber Division, ACS, Louisville, KY, Oct. 8-11, 1996.

[154] Hofmann W. Chapter 2. Vulcanization and Vulcanizing Agents. London: Maclaran & Sons, 1967.

[155] Coran A Y. Chapter 7. In: Mark J E, Erman B, Eirich F R (Editors). Science and Technology of Rubber. 2nd ed. San Diego: Academic Press, 1994.

[156] Böhm G G A, Nguyen M N. Flocculation of Carbon Black in Filled Rubber Compounds. I. Flocculation Occurring in Unvulcanized Compounds During Annealing at Elevated Temperatures. *J. Appl. Polym. Sci.*, 1995, 55: 1041.

[157] Heinrich G. Dynamics of Carbon Black Filled Networks, Viscoelasticity, and Wet Skid Behavior. *Kauts, Gummi Kunsts.*, 1992, 45: 173.

[158] Heinrich G, Glave L, and Stanzel M. Material-und Reifenphysikalische Aspekte bei der Kraftschlußoptimierung von Nutzfahrzeugreifen *VDI Berichte*. 1995, 1188: 49.

[159] P. Roch. *Kauts, Gummi Kunsts*. 1995, 48: 430.

[160] Grosch K A. The Rolling Resistance, Wear and Traction Properties of Tread Compounds. *Rubber Chem. Technol.*, 1996, 69: 495.

[161] Nahmias M, Serra A. Correlation of Wet Traction with Viscoelastic Properties of Passenger Tread Compounds. *Rubber World*, 1997, 216(6): 38.

[162] Veith A G. Meeting of the Rubber Division, ACS, Cleveland, Ohio, Oct. 17-20, 1995.

[163] Futamura S. Effect of Material Properties on Tire Performance Characteristics-Part II, Tread Material. *Tire Sci. Technol.*, 1990, 18(1): 2.

[164] Kawakami S, Hirakawa H, Misawa M. Interpretation of Wet Friction Coefficient by the Viscoelastic Nature of Rubber. *J. Soc. Rubber. Ind. Jpn.*, 1988, 61: 722.

[165] Takino H, Nakayama R, Yamada Y, et al. Viscoelastic Properties of Elastomers and Tire Wet Skid Resistance. *Rubber Chem. Technol.*, 1997, 70: 584.

[166] Takino H, Takahashi H, Yamano K, et al. Effects of Carbon Black and Process Oil on Viscoelastic Properties and Tire Wet Skid Resistance. *Tire Sci. Technol.*, 1998, 26: 241.

[167] Veith A G. Tire Traction *vs.* Tread Compound Properties-How Pavement Texture and Test Conditions Influence the Relationship. *Rubber Chem. Technol.*, 1996, 69: 654.

[168] Moore D F. Chapter 2. The Friction and Lubrication of Elastomers. Oxford: Pergamon Press, 1972.

[169] A. Le Gal, Yang X, Kluppel M. Evaluation of Sliding Friction and Contact Mechanics of Elastomers Based on Dynamic-Mechanical Analysis. *J. Chem. Phys.*, 2005, 123(014704): 1.

[170] O. Le Maître, "La Résistance au Roulement des Pneumatiques: Une Concéquence du Comportement Viscoélastique de Ses Matériaux".

[171] Kabayashi N, Furuta I. Comparison between Silica and Carbon Black in Tire Tread Formulation. *J. Soc. Rubber. Ind. Jpn.*, 1997, 70: 147.

[172] Wang M-J, Lu S X, Mahmud K. Carbon-Silica Dual-Phase Filler, a New-generation Reinforcing Agent for Rubber. Part VI. Time-Temperature Superposition of Dynamic Properties of Carbon-Silica-Dual-Phase-Filler-Filled Vulcanizates. *J. Polym. Sci., Part B, Polymer Physics*, 2000, 38: 1240.

[173] Ouyang G B, Tokita N, Shieh C H. Carbon Black Effects on Friction Properties of Tread Compound Using A Modified ASTM-E303 Pendulum Skid Tester. Meeting of the Rubber Division, ACS, Denver, Colorado, May 18-21, 1993.

[174] Grosch K A. Abrasion of Rubber and Its Relation to Tire Wear. *Rubber Chem. Technol.*, 1992, 65: 78.

[175] Wang M-J, Kutsovsky Y, Zhang P, et al. New Generation Carbon-Silica Dual Phase Filler Part I. Characterization and Application to Passenger Tire. *Rubber Chem. Technol.*, 2002, 75: 247.

[176] Wang M-J. New Developments in Carbon Black Dispersion. *Kautsch, Gummi Kunstst.*, 2005, 58: 626.

[177] Wang M-J, Kutsovsky Y. Effect of Fillers on Wet Skid Resistance of Tires. Part I: Water Lubrication *vs*. Filler-Elastomer Interactions. *Rubber Chem. Technol.*, 2008, 81: 552.

[178] Wang M-J, Kutsovsky Y. Effect of Fillers on Wet Skid Resistance of Tires. Part II: Experimental Observations on Effect of Filler-Elastomer Interactions on Water Lubrication. *Rubber Chem. Technol.*, 2008, 81: 576.

[179] Fatigue and Tribological Properties of Plastics and Elastomers. Plastics Design Library, Morris, NY, 1995.

[180] Sarkar A D. Chapter 10. Friction and Wear. London: Academic Press, 1980: 244.

[181] Johnson K L. In: Dowson D, Taylor C M, Godet M, et al. (Editors). Friction and Traction. Guildford, UK: Westbury House, 1981: 3.

[182] Moore D F. Chapter 9. The Friction and Lubrication of Elastomers. Oxford: Pergamon Press, 1972.

[183] Kummer H W. Unfied Theory of Rubber and Tire Friction, Eng. Res. Bulletin B-94, The Pennsylvania State University, Jul., 1966.

[184] Moore D F. Chapter 10. The Friction and Lubrication of Elastomers. Oxford: Pergamon Press, 1972.

[185] Dowson D. Paper 1, Elastohydrodynamic Lubrication, *Proce. Inst. Mech. Engrs*, 1965-6, 180: Part 3B, P7.

[186] Moore D F. Chapter 7. The Friction and Lubrication of Elastomers. Oxford: Pergamon Press, 1972.

[187] Moore D F. Chapter 7. Principles and Applications of Tribology. Oxford: Pergamon Press, 1975.

[188] Bowden F P, Tabor D. Chapter 9. The Friction and Lubrication of Solids. Oxford, 1950: 176.

[189] Moore D F. Chapter 5. The Friction and Lubrication of Elastomers. Oxford: Pergamon Press, 1972.

[190] Bowden F P, Tabor D. Chapter 8. The Friction and Lubrication of Solids. Oxford, 1950: 163.

[191] Sameshima J, Akamatu H, Isemura T. *Rev. Physical Chem.*, 1940, 37: 90.

[192] Oberth A E, Bruenner R S. Tear Phenomena around Solid Inclusions in Castable Elastomers. *Trans. Soc. Rheol.*, 1965, 9: 165.

[193] Gent A N. Detachment of an Elastic Matrix from a Rigid Spherical Inclusion. *J. Mater. Sci.*, 1980, 15: 2884.

[194] Nicholson D W. On the Detachment of a Rigid Inclusion from an Elastic Matrix. *J. Adhesion*, 1979, 10: 255.

[195] Kinloch A J, Young R J. Chapter 10. Fracture Behavior of Polymers. London: Applied Science, 1983.

[196] Wang M-J, Mahmud K, Murphy L J, et al. Carbon-Silica Dual Phase Filler, a New Generation Reinforcing Agent for Rubber-Part I. Characterization. *Kautsch. Gummi Kunstst.*, 1998, 51: 348.

[197] Whitehouse R S. *Cabot report*, CIM- 99-27, Nov. 17, 1998.

[198] Wolff S, Wang M-J, Tan E H. Filler-Elastomer Interactions. Part X. The Effect of Filler-Elastomer and Filler-Filler Interaction on Rubber Reinforcement. *Kautsch. Gummi Kunsts.*, 1994, 47: 102.

[199] Tokita N. private communication.

[200] Shieh C H, Funt J M, Ouyang G B (to Cabot Corporation), Method of Abrading. US Patent 4,995,197, 1991.

[201] Moore D F. Tire Traction Under Elastohydrodynamic Conditions. "Friction and Traction", Dowson D, Taylor C M, Godet M and Berthe D, Eds., Guildford, UK: Westbury House, 1981: 221.

[202] Wang M-J, Kutsovsky Y. Effect of Fillers on Wet Skid Resistance of Tires. Part I: Water Lubrication *vs.* Filler-Elastomer Interactions. *Rubber Chem. Technol.*, 2008, 81 (4): 552; Effect of Fillers on Wet Skid Resistance of Tires. Part II: Experimental Observations on Effect of Filler-Elastomer Interactions on Water Lubrication. *Rubber Chem. Technol.*, 2008, 81 (4): 576.

[203] Gough V E. Friction of Rubber on Lubricated Surfaces. *Rev. Gen. Caoutch*, (Discussion of Paper by D. Tabor) 1959, 36: 1409.

[204] Moore D F. A review of squeeze films. *Wear,* 1965, 8: 245.

[205] Moore D F. The Logical Design of Optimum Skid-Resistant Surface. *Proc. Highway Res. Board Report*, 1968, 101: 39.

[206] Moore D F. The Measurement of Surface Texture and Drainage Capacity of Pavements. *Intemat. Colloq. Techn. Univ.* Berlin, 1968.

[207] Yandell W O, Taneerananon P, Zankin V. Frictional Interaction of Tire and Pavement.Walter J E and Meyer W E, Eds., Philadelphia: ASTM, 1983: 304.

[208] Grosch K A. Laborbestimmung der Abrieb-und Rutschfestigkeit von Laufflächenmischungen-Teil I: Rutschfestigkeit. *Kautsch. Gummi Kunstst.*, 1996, 49: 432.

[209] Veith A G. Measurement of Wet cornering Traction of Tires. *Rubber Chem. Technol.*, 1971, 44: 962.

[210] Giles C G, Sabey B E, Cardew K H F. Development and Performance of the Portable Skid-Resistance Tester. *Rubber Chem. Technol.*, 1965, 38: 840.

[211] Hallman R W, Brunot C A. A Proposed Method for Wet-Skid Evaluation. *Rubber Age*, 1964, 95: 886.

[212] Sarbach D V, Hallman R W, Brunot C A. Wet Skid-Laboratory vs Road Tests. *Rubber Age*, 1965, 97: 76.

[213] French T, Patton R G. *Proc. Fourth Rubber Technol. Conf.* London, 1963: 196.

[214] Grime G, Giles C G. *Proc. Inst. Mech. Engrs., (Automobile Division)*, (1954-55), 1: 19.

[215] Bevilacqua E M, Percarpio E P. Lubricated Friction of Rubber. I. Introduction. *Rubber Chem. Technol.,* 1968, 41: 832.

[216] Bassi A C. Measurements of Friction of Elastomers by the Skid Resistance Tester. *Rubber Chem. Technol.*, 1965, 38: 112.

[217] Sabey B E, Lupton G N. Friction on Wet Surfaces of Tire-Tread-Type Vulcanizates. *Rubber Chem. Technol.*, 1964, 37: 878.

[218] Veith A G. Tire-Road-Rainfall-Vehicles The Friction Connection. P3. In: Frictional Interaction of Tire and Pavement. In: Meyer W E and Walter J D (Editor). ASTM Special Technical Publication 793. ASTM, 1983.

[219] Moore D F. Chapter 5. The Friction of Pneumatic Tires. Amsterdam: Elsevier Scientific Publishing Co., 1975.

[220] Giustino J M, Emerson R J. Instrumentation of the British Portable Skid Tester. Meeting of the Rubber Division, ACS, Toronto, Canada, May 10-12, 1983.

[221] Mahmud K, Wang M-J, Francis R A. Elastomeric Compounds Incorporating Silicon-Treated Carbon Blacks. US Patent 5,830,930; Elastomeric Compounds Incorporating Silicon-Treated Carbon Blacks and Coupling Agents. US Patent 5,877,238 (to Cabot

Corp, 1998 and 1999); Mahmud K and Wang M-J. Method of Making a Multi-Phase Aggregate Using a Multi-Stage Process. US Patent 5,904,762; Method of Making a Multi-Phase Aggregate Using a Multi-Stage Process. US Patent 6,211,279 (to Cabot Corp., 1999 and 2001).

[222] Wang M-J, Zhang P, Mahmud K. Carbon-Silica Dual Phase Filler, a New Generation Reinforcing Agent for Rubber: Part IX. Application to Truck Tire Tread Compound. *Rubber Chem. Technol.*, 2001, 74: 124.

[223] Mahmud K, Wang M-J, Kutsovsky Y. Method of Making a Multi-Phase Aggregate Using a Multi-Stage Process. US Patent 6,364,944 (to Cabot Corp., 2002).

[224] Payne A R. The Dynamic Properties of Carbon Black-Loaded Natural Rubber Vulcanizates. Part I. *J. Polym. Sci.*, 1962, 6: 57; Payne A R, Whittaker R E. Low Strain Dynamic Properties of Filled Rubbers. *Rubber Chem. Technol.*, 1971, 44: 440.

[225] Wang M-J. Effect of Polymer-Filler and Filler-Filler Interactions on Dynamic Properties of Filled Vulcanizates. *Rubber Chem. Technol.*, 1998, 71: 520.

[226] Wang M-J. Application of Inverse Gas Chromatography to the study of Rubber Reinforcement. Chapter 3. In: Hardin M and Papirer E (Editor). Powders and Fibers: Interfacial Science and Applications. Boca Raton, FL: CRC Press, 2006.

[227] Bulgin D and Walters M H. The Abrasion of Elastomer Under Laboratory and Service Conditions. *Proc. 5th Intern. Rubber Conf.*, Brighton, UK, 1967: 445.

[228] Grosch K A and Schallamach A. Relation between Abrasion and Strength of Rubber. Inst. Rubber Ind. *Trans. IRI,* 1965, 41: 80.

[229] Schallamach A. Chapter 13. In: Bateman L(Editor). The Chemistry and Physics of Rubber-Like Substance. London: Mclaren, 1963.

[230] Gent A N, Pulford C T R. Mechanisms of Rubber Abrasion. J. *Appl. Polym. Sci.*, 1983, 28: 943.

[231] Schallamach A. *Trans. IRI,* 1952, 28, 256. Abrasion pattern on rubber. Rubber chemistry and technology, 1953.

[232] Southern E, Thomas A G. Studies of Rubber Abrasion. *Plast. Rubber: Mater. Appl.*, 1978, 3: 133.

[233] Southern E, Thomas A G. Studies of Rubber Abrasion. *Rubber Chem. Technol.*, 1979, 52: 1008.

[234] Rivlin R S, Thomas A G. Rupture of Rubber. I. Characteristic Energy for Tearing. *J. Polym. Sci.*, 1953, 10: 291.

[235] Thomas A G. Rupture of Rubber. V. Cut Growth in Natural Rubber Vulcanizates. *J. Polym. Sci.*, 1958, 31: 467.

[236] Lake G J, Lindley P B. Ozone Cracking, Flex Cracking and Fatigue of Rubber. *Rubber J.*, 1964, 146(11): 30.

[237] Lindley P B, Thomas A G. *Proc. 4th Rubber Technol. Conf.* IRI, London, 1963: 428.

[238] Reznikovskii M M, Brodskii G I. Abrasion of Rubber. D. I. James, Ed., Palmerton Publishing, Co., 1964: 14.

[239] M. M. Reznikovskii. Abrasion of Rubber, D. I. James, Ed., Palmerton Publishing, Co., 1964: 119.

[240] Rivlin R S, Thomas A G. Rupture of Rubber. I. Characteristic Energy for Tearing. *J. Polym. Sci.*, 1953, 10: 291.

[241] Thomas A G. Rupture of Rubber. II. The Strain Concentration at an Incision. *J. Polym. Sci.*, 1955, 18: 177.

[242] Greensmith H W, Thomas A G. Rupture of Rubber. III. Determination of Tear Properties. *J. Polym. Sci.*, 1955, 18: 189.

[243] Greensmith H W. Rupture of Rubber. IV. Tear Properties of Vulcanizates Containing Carbon Black. *J. Polym. Sci.*, 1956, 21(98): 175.

[244] Bulgin D, Hubbard G. Rotary Power Loss Machine. *Trans. IRI,* 1958, 34: 201.

[245] Gent A N. A Hypothetical Mechanism for Rubber Abrasion. *Rubber Chem. Technol.*, 1989, 62: 750.

[246] Green A E, Zerna W. Theoretical Elasticity. London: Oxford University Press, 1960, Section 3.10.

[247] Gent A N, Lindley P B. Internal Rupture of Bonded Rubber Cylinders in Tension. *Proc. R. Soc. London, Ser.*, 1958, A249: 195.

[248] Gent A N, Tompkins D A. Surface Energy Effects for Small Holes or Particles in Elastomers. *J. Polym. Sci.,* 1969, Part A, 27: 1483.

[249] Gent A N, Tompkins D A. Nucleation and Growth of Gas Bubbles in Elastomers. *J. Appl. Phys.*, 1969, 40: 2520.

[250] R. L. Denecour, A. N. Gent, Bubble Formation in Vulcanized Rubbers. *J. Polym. Sci.*, 1968, Part A, 27: 1853.

[251] Schallamach A. Abrasion, Fatigue, and Smearing of Rubber. *J. Appl. Polym. Sci.,* 1968, 12: 281.

[252] Medalia A I. Effects of Carbon Black on Abrasion and Treadwear. *Kautsch. Gummi Kunstst.*, 1994, 47: 364.

[253] Wolff S, Wang M-J. Physical Properties of Vulcanizates and Shift Factors. *Kautsch. Gummi Kunstst.,* 1994, 47: 17.

[254] Bulgin D, Walters M H. *Proc. 5th Intern. Rubber Conf.*, Brighton, UK, 1967: 445.

[255] Shien C H, Mace M L, Ouyang G B, et al. Meeting of the Rubber Division, ACS, Toronto, May 21-24, 1991.

[256] Veith A G, Chirico A E. A Quantitative Study of the Carbon Black Reinforcement System for Tire Tread Compounds. *Rubber Chem. Technol,* 1979, 52: 748.

[257] Mullins L H. The Chemistry and Physics of Rubber-Like Substances. Bateman L, Ed., London: McLaren, 1963: 322.

[258] Grosch K A, Shallamach A. Tyre Wear at Controlled Slip. *Wear*, 1961, 4: 356.

[259] Studebaker M L. Chapter 12. In: Kraus G (Editor). Reinforcement of Elastomers. New York: Wiley Intersci., 1965.

[260] Wolff S. *Kautschukchemikalien und Fūllstoffe in der modernen Kautschuktechnologi.,* Eiterbildungsstudium Kautschuktechnologie, Universität Hannover, 1989/1990.

[261] Bulgin D. Reinforcement of Rubbers and Plastics by Particulate Fillers. *Composite*, 1971, 2(3): 165.

[262] Dizon E S. Processing in an Internal Mixer as Affected by Carbon Black Properties. *Rubber Chem. Technol.*, 1976, 49: 12.

[263] Dannenberg E M. *Rubber Age,* 1966, 98: 82.

[264] Wilder C R, Haws J R, Cooper W T. Effects of Carbon Black Types on Treadwear of Radial and Bias Tires at Variable Test Severities. *Rubber Chem. Technol.*, 1981, 54: 427.

[265] Hess W M, Ayala J A, Vegvari P C, et al. The Influence of Carbon Black Properties on Tread Performance. *Kautsch. Gummi Kunstst.*, 1988, 41: 1215.

[266] Cotton G R, Dannenberg E M. A Method for Evaluation of Carbon Blacks and Correlation with Road Wear Ratings. *Tire Sci. Technol.*, TSTCA, 1974, 2(3): 211.

[267] Hess W M, Lyon F, Burgess K A. Einfluss der Adhäsion zwischen Ruß und Kautschuk auf die Eigenschaften der Vulkanisate. *Kautsch. Gummi Kunstst.*, 1967, 20: 135.

[268] Garten V A and Weiss D E. A New Interpretation of the Acidic and Basic Structures in Carbons. II. The Chromene-Carbonium Ion Couple in Carbon. *Aust. J. Chem.*, 1957, 10: 309.

[269] Brown W A, Patel A C. *Proc. Intern. Rubber Conf., G5*, 1981.

[270] Dannenberg E M, Papirer E, Donnet J B. in "Interactions Entre les Elastomeres et les Surfaces Solides Ayant Une Action Renforcante" Collogues Internationaux du C. N. R. S., 1975.

[271] Wang M-J. Effect of Filler-Elastomer Interaction on Tire Tread Performance, Part III. Effect on Abrasion. *Kautsch. Gummi Kunstst.*, 2008, 61: 159.

[272] Wolff S, Görl U, Wang M-J. Silica-based Tread Compounds: Background and Performance. Paper presented at the TYRETECH'93 Conference, Basal, Switzerland, Oct. 28-29, 1993.

[273] Freund B. *Eur. Rubber J.*, Sept. 34, 1998.

[274] Watson P J. TireTech'99, 1999.

[275] Knill R B, Shepherd D J, Urbon J P, et al. *Proc. Int. Rubb. Conf.*, 1975, Volume V, RRIM, Kuala Lumpur, 1976.

[276] Wolff S. Meeting of the Rubber Division, ACS, New York, Apr. 8-11, 1986.

[277] Wolff S, Panenka R. presented at IRC'85, Kyoto, Japan, Oct. 15-18, 1985.

[278] Bomal Y, Cochet P, Dejean B, Gelling I, Newell R. Influence of Precipitated Silica Characteristics on the Properties of a Truck Tyre Tread, II, *Kautsch. Gummi Kunstst.*, 1998, 51: 259.

# 9 Development of New Materials for Tire Application

## 9.1 Chemical Modified Carbon Black

Reinforcement of rubber by extremely fine filler such as carbon black and silica is most dominant requirement for tread compounds in tires. Filler manufacturers have continuously strived to meet ever increasing reinforcement demands in this area. Starting from the urgent requirement for the reduction of fuel consumption of automobiles at the middle of 1970s, emphasis has been placed on reducing the rolling resistance while increasing abrasion resistance. Up until the late 1980s, the largest contribution to reduction of rolling resistance from materials has been made by the rubber manufacturers. Low hysteresis polymers such as functional polymers like Sn-coupled and chemically modified SBR and BR have proved to be very effective for carbon black compounds. This was due to the improvement of filler dispersion and agglomeration by introducing chemically reacting functional polymer molecules with some active sites of fillers during mixing. On the other hand, during almost the same period, the invention of some coupling agents enabled fine silica to be dispersible in hydrocarbon polymers. The surface modification with organo silane, such as bis[3-(triethoxysilyl)propyl] tetrasulfide (TESPT), effectively promotes polymer-filler interaction through chemical linkages and depresses filler agglomeration, which gives substantially lower rolling resistance with acceptable abrasion resistance, compared with carbon black.

The success of the application of silica modification encouraged the application of carbon black modification in the same approach to improve its performance in the filled compounds. In fact, since the 1950s, attempts at surface modification of carbon black have been reported, including physical adsorption of chemicals on the surface, heat treatment, plasma treatment, frequently oxidation,as well as chemical and polymer grafting. Several new approaches have been published in the literature for carbon black modification[1-5]. Among others, a number of patents have been issued to Cabot[6,7] which describe that the decomposition of a diazonium compound derived from a substituted aromatic or aliphatic amine precursor resulting in the attachment of a substituted aromatic ring or chain onto the surface of the carbon black. Examples show attachment of amines, anionic and cationic moieties, alkyl, polyethoxyl, and vinyl groups, and polysulfide moieties. One of the precursors is 4-aminophenyl disulfide (APDS), which yields a modified carbon black that can be used in elastomers.

With APDS-modified carbon black, the dispersive component of surface energy of the carbon black can be drastically reduced[8], depressing the driving force of black agglomeration, resulting in a better micro dispersion. On the other hand, the sulfide group is able to react with polymer molecules upon heating, so that the coupling reaction between polymer and filler can be achieved during vulcanization. The effectiveness of surface modification in suppression of filler agglomeration due to both lowering filler-filler interaction, and increasing polymer-filler interaction via chemical bridges, has been verified by dynamic rheological testing[9]. These fillers were tested in typical passenger tire tread compounds alongside the virgin carbon black and silane modified silica. In filled rubbers, the elastic modulus decreases with strain amplitude, which has been termed the "Payne effect"[10,11]. As mentioned before (see Section 7.2.1), this effect is generally used as a measure of filler agglomeration, controlled mainly by filler-filler interactions. The strain dependence of elastic modulus results (Figure 9.1) show that the APDS-modified carbon black exhibits a lower Payne effect compared to bis[3-(triethoxysilyl)propyl]tetrasulfide (TESPT)-modified silica, which is itself much lower than that found for unmodified carbon black. This is a clear indication that filler agglomeration is substantially reduced. Accordingly, a striking similarity between the modified carbon black and coupled silica is observed in the tan $\delta$ at 70°C measured in strain sweep (Figure 9.2), and the temperature sweep results. Thus, the APDS-modified carbon black provides viscoelastic properties comparable to TESPT-modified silica, which is now well known to be better in terms of hysteresis and hence rolling resistance compared to conventional carbon black. Additionally, the ability of the chemically modified carbon black to form chemical bonds with the polymer through its disulfide groups provides improved polymer-filler interaction, leading to increased abrasion resistance (Figure 9.3).

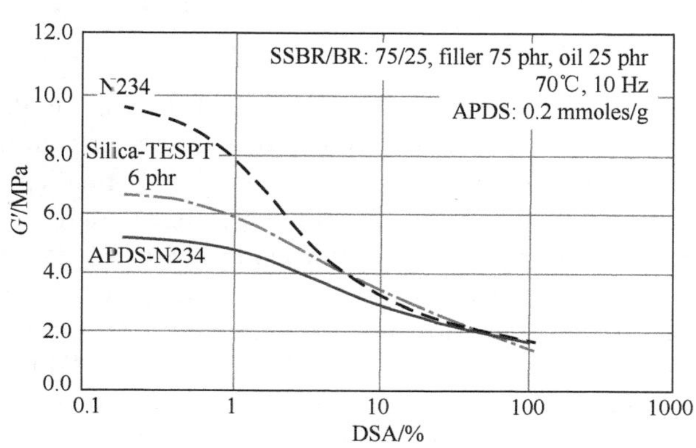


**Figure 9.1** Strain dependence of *G'* for vulcanizates filled with a variety of fillers

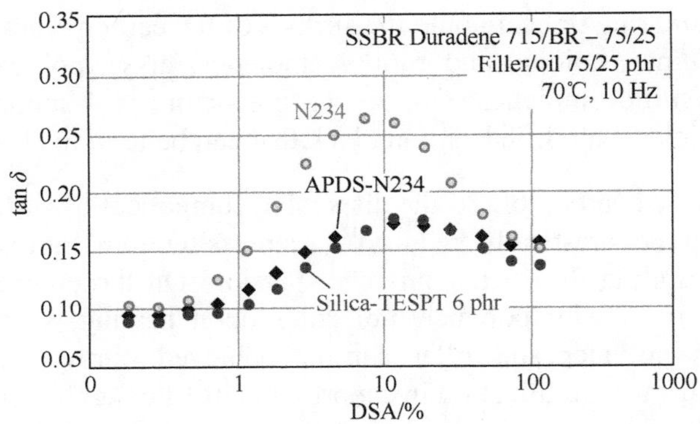


**Figure 9.2** Strain dependence of tan $\delta$ at 70°C for vulcanizates filled with a variety of fillers

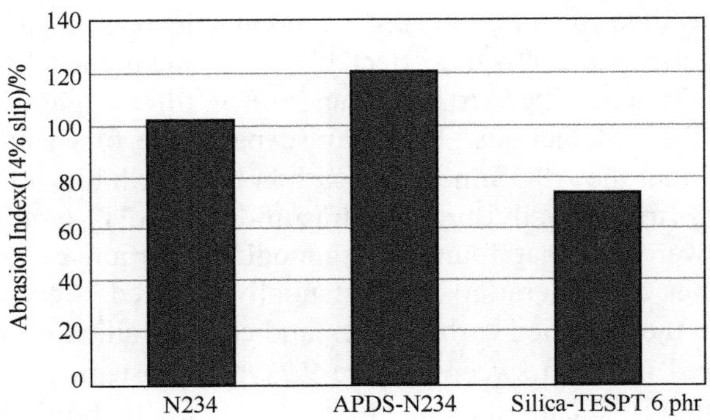


**Figure 9.3** Abrasion resistance for vulcanizates filled with a variety of fillers

## ■ 9.2 Carbon-Silica Dual Phase Filler (CSDPF)

During the period from the 1970s to 1990s, a group of dual and multiple phase fillers in which silica or/and other metal oxides and carbon are co-formed in a carbon black-like reactor were developed. In particular, two types of carbon-silica dual phase fillers (CSDPF) (see Figure 9.4) have been commercialized under the trade name ECOBLACK™ XXXX. The two series, identified by the numbers 2XXX and 4XXX, differ in the silica domain distribution within the aggregates. In the aggregates of CSDPF 2XXX, the carbon black and silica are intimately intermixed on a scale that is about the same size as the carbon black crystallite. For the CSDPF 4XXX, the silica location is more on the exterior of the aggregates. In each type of material, there are several grades having different morphologies and silica contents.

The classification system of CSDPF is based on the type, particle size, structure, and

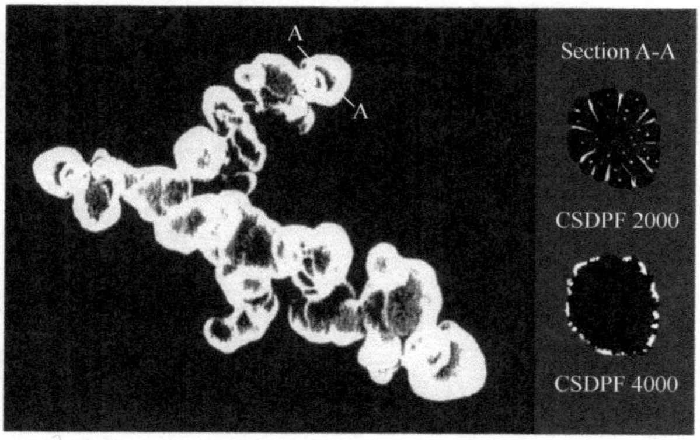


**Figure 9.4** Carbon-silica dual phase fillers (CSDPF) schematic cartoon

silicon content, and is composed of a prefix CSDPF followed by four numerals. The first numeral represents the technology: 2 for the materials produced with a single stage process[12–15] and 4 for those produced with a multi-stage process[16,17]. The second numeral represents the particle size of the filler, which correspond to the ASTM designation for carbon blacks. For CSDPF 2XXX, the third letter indicates the structure, with 0 = low, 1 = normal, and 2 = high, and the last latter represents the silicon content as shown in Table 9.1. In the case of CSDPF 4XXX, the silica content is expressed by the last two numbers.

**Table 9.1** Numbers representing silica contents in CSDPF nomenclature

| CSDPF 2XXX | | CSDPF 4XXX | |
|---|---|---|---|
| Last number | Silicon content (by weight)/% | Last two numbers | Silicon content (by weight)/% |
| 0 | 0.1–0.9 | 00 | 0.1–0.9 |
| 1 | 1.0–1.9 | 01 | 1.0–1.9 |
| 2 | 2.0–2.9 | 02 | 2.0–2.9 |
| 3 | 3.0–3.9 | 03 | 3.0–3.9 |
| 4 | 4.0–4.9 | 04 | 4.0–4.9 |
| 5 | 5.0–5.9 | 05 | 5.0–5.9 |
| 6 | 6.0–6.9 | 06 | 6.0–6.9 |
| 7 | 7.0–7.9 | 07 | 7.0–7.9 |
| 8 | 8.0–8.9 | 08 | 8.0–8.9 |
| 9 | 9.0–9.9 | 09 | 9.0–9.9 |
| | | 10 | 10.0–10.9 |
| | | 11 | 11.0–11.9 |

Continued

| CSDPF 2XXX | | CSDPF 4XXX | |
|---|---|---|---|
| Last number | Silicon content (by weight)/% | Last two numbers | Silicon content (by weight)/% |
| | | 12 | 12.0–12.9 |
| | | ... | ... |
| | | 19 | 19.0–19.9 |
| | | 20 | 20.0–20.9 |

**Table 9.2** Analytical properties of some typical carbon blacks and CSDPFs

| Filler | Silica content/% | | NSA-surface area/($m^2$/g) | | STSA/($m^2$/g) | | CDBP/(mL/100 g) | Silica coverage/% |
|---|---|---|---|---|---|---|---|---|
| | As is | HF | As is | HF | As is | HF | | |
| N134 | N/A | N/A | 146 | N/A | 134 | NA | 104 | N/A |
| N234 | N/A | N/A | 122 | 122 | 118 | 118 | 100 | N/A |
| CSDPF 2124 | 4.1 | 0.85 | 171 | 251 | 133 | 146 | 115 | 20 |
| CSDPF 4210 | 10.0 | 0.01 | 154 | 167 | 123 | 155 | 108 | 55 |
| Silica Z1165 | 46.7 | N/A | 168 | N/A | 132 | N/A | N/A | 100 |

### 9.2.1 Characteristics of Chemistry

Table 9.2 shows the basic properties of typical CSDPF 2XXX and 4XXX materials compared with those of conventional fillers. CSDPF 4XXX has much higher silica-surface coverage relative to CSDPF 2XXX, due to the silica domain distribution and higher silicon content of the 4000 series. This is demonstrated by the changes in silica content and surface area upon hydrofluoric acid (HF) extraction. During HF extraction, the silica will be dissolved with the carbon domain remaining unchanged (Table 9.2).

The fact that significant portions of silica still remain, and surface areas drastically increase upon HF treatment, suggests that the silica of CSDPF 2XXX is distributed throughout the aggregates. By contrast, CSDPF 4XXX shows a small change in surface area, and the fact that there is almost no silica left after HF extraction indicates that the silica in the CSDPF 4XXX aggregates is located predominantly on the surface. This observation is supported by the images and the carbon- and silicon-maps obtained using scanning transmission electron microscope equipped with an energy dispersive X-ray analyzer (STEM/EDX). In the aggregates of CSDPF 4XXX, the silica domains are distributed within the aggregates[18,19].

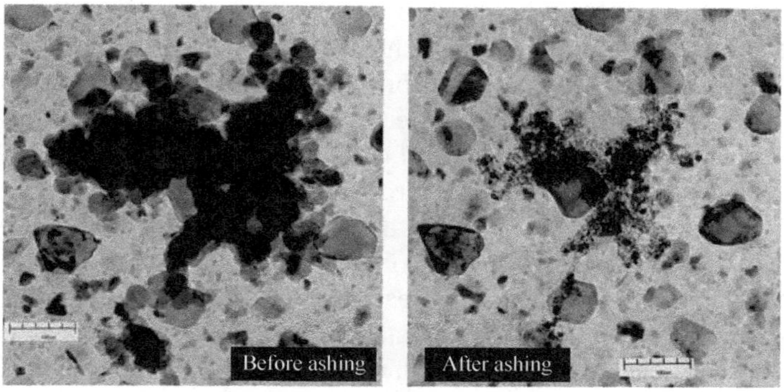


**Figure 9.5** TEM images cf CSDPF 2XXX aggregate before and after ashing

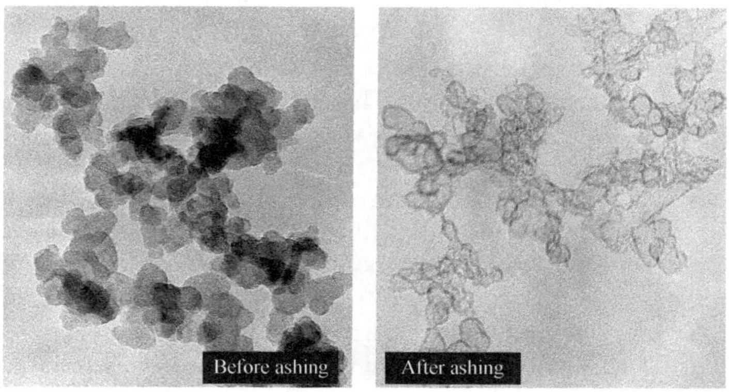


**Figure 9.6** TEM images of CSDPF 4XXX aggregate before and after ashing

The silica distributions can be also visualized from transmission electron microscope (TEM) images of the filler particles before and after ashing. Figure 9.5 shows a TEM image of a CSDPF 2XXX aggregate and an image of the same aggregate that was ashed, leaving behind the silica phase. The silica domains can be seen as having been finely distributed throughout the aggregate. Ashing of the CSDPF 4XXX aggregates, however, resulted in a quite different picture (Figure 9.6): silica is left as a shell-like structure whose shape is similar to the contour of the unashed particles. It is apparent that the silica domain was intimately attached on the carbon black surface with a high surface coverage.

### 9.2.2 Characteristics of Compounding

When incorporated in hydrocarbon polymers, both CSDPF 2XXX and 4XXX are characterized by higher filler-polymer interaction in relation to a physical blend of carbon black and silica, and lower filler-filler interaction compared to either traditional carbon black or silica. Evidence of the higher filler-polymer interaction of the silica-containing fillers in hydrocarbon polymers can be obtained by comparing the adsorption energies of polymer analogs on the filler surfaces[20]. The adsorption free energies of heptane, a model compound for hydrocarbon polymers, were measured by inverse gas chromatography (IGC)[21]. The results for both CSDPF 2XXX and 4XXX are presented along with those of blends of carbon black N234 and silica (Figure 5.15). Regardless of the difference in surface areas, at the same silica contents, the silica-containing fillers give significantly higher adsorption energies than the physical blend. The high surface activity of CSDPF is attributed to the presence of heteroatoms in the carbon domains. This leads to an increased population of crystal defects in the graphitic crystal lattice and/or smaller crystallite dimensions, leading to more edges of carbon basal planes. The edges of basal planes and crystal defects have been proposed as the active centers for rubber adsorption[22].

The weaker filler-filler interaction of CSDPF 2XXX and 4XXX is demonstrated by the strain dependence of the elastic modulus, *G'* (Figure 9.7). Although from the chemical composition point of view, both CSDPF 2XXX and 4XXX are between carbon black and silica, what is actually observed here is that the two new fillers give a lower Payne effect than either the black or the silica. This unique behavior is readily explained by the hybrid surfaces of the new materials, which minimizes the contact between like surfaces. It has been established from surface energy studies that interaction between unlike surfaces is lower than that between the same type of surface[23]. The high polymer-filler interaction and low filler-filler interaction of CSDPF should allow a better compromise between abrasion resistance and dynamic hysteresis at high temperature. This will result in a good balance between wear resistance and rolling resistance in tires. Furthermore, the lower surface silica content and high surface activity of the carbon domains means that these fillers will require significantly lower silane coupling agent compared to silica. Although both types of CSDPFs can be used advantageously in passenger tire tread compounds, considering the global performance requirements of different tires, the CSDPF 4XXX is recommended for passenger tires, and the CSDPF 2XXX for truck tire tread formulations. (See Sections 8.1.2.3.6, 8.2.3.4, and 8.3.2.9.)

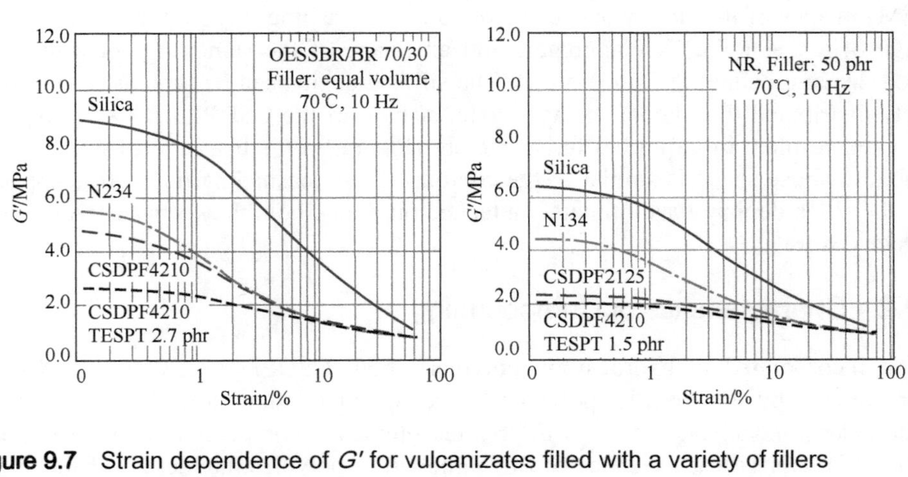


**Figure 9.7** Strain dependence of *G'* for vulcanizates filled with a variety of fillers

### 9.2.3 Application of CSDPF 4000 in Passenger Tires

The recommendation of CSDPF 4XXX for passenger tires is based on its high silica surface coverage, which is very favorable to wet skid resistance. During service, a passenger tire has a smaller footprint and lower load compared with a truck tire. Therefore, under wet conditions, the water squeezing and transition zones are relatively large, especially for vehicles equipped ABS on smoother roads. The high silica coverage of CSDPF 4XXX results in shorter water squeezing and transition zones compared to carbon black and CSDPF 2XXX, similar to the behavior of silica. This behavior also results in an increase in the traction zone, where BL is dominant and carbon fillers have an advantageous effect. Therefore, in relation to silica, CSDPF 4XXX has a disadvantage in the water squeezing and transition zones due to its lower silica surface coverage, but this is offset to a certain degree by an increase in the traction zone. In the traction zone, the BL friction is actually higher than that of carbon black compounds due to the improved dynamic properties of the compounds, i.e., high hysteresis and lower modulus at low temperature. As a result, the overall wet skid resistance of the tires filled with CSDPF 4XXX can be improved. On the other hand, it has been recognized that the abrasion resistance of the silica compound is still significantly poor relative to a carbon black compound. This is explained by the relatively weak polymer-filler interaction between silica and hydrocarbon rubber, even with solution polymers and a high dosage of coupling agent. However, this deficiency would be significantly compensated for by the high surface activity of the carbon domain of CSDPF. In Figure 8.50, the average performances of passenger car tires with different fillers in tread compounds are shown. Compared with a silica tire, CSDPF 4210 gave similar rolling resistance, close wet skid resistance for the ABS vehicle, higher dry skid resistance, and significantly improved tread wear resistance. When CSDPF is used to replace carbon black, the rolling resistance and wet skid resistance will be significantly improved without significant trade-off in tread wear and dry traction[18].

### 9.2.4 Application of CSDPF 2000 in Truck Tires

In truck tires, the principal requirements of higher wear resistance, lower rolling resistance, and improved wet skid resistance are the same as those for passenger cars. However, unlike passenger car tires, very little silica has been used in heavy truck tires. The main reason for this arises from the different polymers used. In car tires, solution SBR is the dominant polymer in tread compounds, whilst heavy truck tires use mainly NR for its low heat build-up and resistance to tearing[24,25]. Silica gives relatively poor tread wear in these compounds due to the high non-rubber content of the natural rubber and high polarity of the silica surface. The weak polymer-filler interaction is not well compensated for by coupling agents, so that there is a significant deficiency in abrasion resistance[26]. The CSDPF 2000 series, with its lower silica surface coverage and high surface activity of carbon domains of, is very favorable in this system. As the silica coverage is smaller, the low polymer-filler interaction on the silica domains due to the lower efficiency of the coupling agent is largely offset by the high surface activity of the carbon domains. Therefore, tread wear is not significantly deteriorated when carbon black is replaced by CSDPF 2XXX, especially its higher surface area and high structure grades CSDPF 2124 and 2125 (Figure 8.69). On the other hand, the lower filler-filler interaction of CSDPF leads to an improved balance of dynamic properties at different temperatures, even at very low dosages of coupling agents. Consequently, while the low hysteresis at high temperature imparts substantially reduced rolling resistance, the wet skid resistance is significantly increased. The reason for this is that, in truck tires, the length of the traction zone is dominant over the water squeezing and transition zones due to the bigger size and high load. Under these circumstances, the wet skid resistance of the tire is governed by BL, which is determined by the dynamic properties of the compounds. Under BL conditions, the high hysteresis at low temperatures of CSDPF 2XXX compounds[26] results in higher energy dissipation, and low dynamic modulus leads to a larger contact area between road surface and tread. Both of these phenomena lead to higher wet skid resistance, as demonstrated by the load dependence of the friction coefficient (Figure 9.8). Accordingly, CSDPF 2XXX is an excellent filler for NR-based truck tire tread compounds to ensure the best global performance.

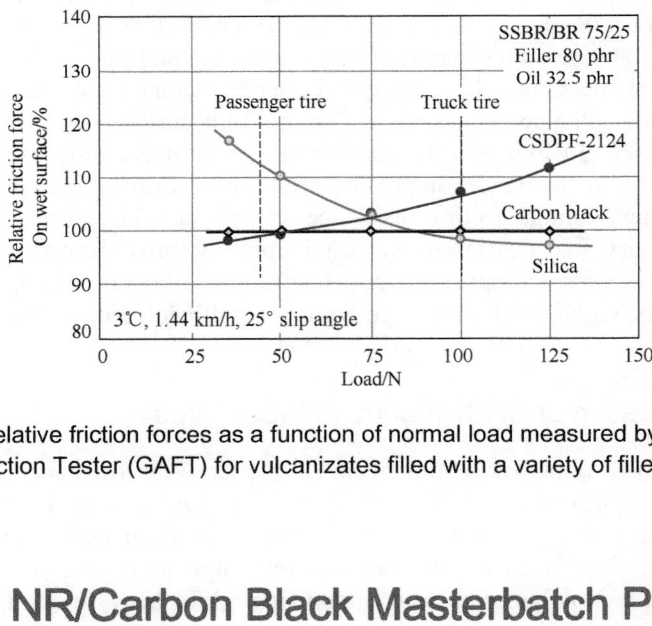


**Figure 9.8** Relative friction forces as a function of normal load measured by Grosch Abrasion & Friction Tester (GAFT) for vulcanizates filled with a variety of fillers

## ■ 9.3 NR/Carbon Black Masterbatch Produced by Liquid Phase Mixing

From a processing point of view, mixing is the most critical process for rubber compounding. Along with the basic physical changes of the materials, in some cases chemical reactions, the primary functions of mixing are incorporation, dispersion, and distribution of the filler and other ingredients in the polymers. Traditionally, this is achieved by using batch mixing and continuous mixing of fillers and solid rubber or pellets, referred to as dry mixing[27].

During the last few decades, a great effort has been made to produce carbon black-polymer masterbatches by mixing polymer latex with filler slurry and then coagulating the mixture chemically. The commercial products are exclusively made by a batch process. With this process, generally the filler dispersion can be improved relative to dry mixing. However, the long mixing and coagulation time reduces the productivity. For natural rubber, some non-rubber substances in NR latex, protein in particular, can be adsorbed on the filler surface and interfere with polymer-filler interaction.

Cabot Elastomer Composite, known as CEC or $E^2C$, is a natural rubber-carbon black masterbatch produced with a unique continuous liquid phase mixing/coagulation process. During this process, carbon black incorporation, dispersion, and distribution are completed in a very short period of time. It has been found that this liquid phase mixing offers the following benefits over the conventional dry mixing and wet batch processing:

- Simplified mixing procedure;
- Reduced mixing costs due to reduced mixing equipment, energy, and labor;
- Elimination of free carbon black handling and reduced dust emission;
- Excellent dispersion of filler independent of filler morphology;
- Improved vulcanizate properties;
- Improved capital efficiency; and
- Facilitation of continuous mixing.

In this part, after introduction of the production processes of CEC, the processability and physical properties of this material will be discussed in comparison with dry mixing.

### 9.3.1 Mechanisms of Mixing, Coagulation, and Dewatering

The process of CEC production, as presented in Figure 9.9, consists of carbon black slurry make-up, NR latex storage, mixing and coagulation of carbon black slurry and latex, dewatering of the coagulum, drying, finishing, and packaging.

The carbon black slurry is prepared by finely dispersing carbon black in water mechanically without any surfactant. The slurry is injected into the mixer at very high speed and mixes continuously with NR latex stream. Under highly energetic and turbulent conditions, the mixing and coagulation of polymer with filler is completed mechanically at room temperature in less than 0.1 second, without the aid of chemicals.

After dewatering of the coagulum in an extruder, the material is continuously fed into the dryer to further reduce the moisture to less than 1%. The residence time in the dryer is 30–60 seconds. Over the entire drying process, only for a very short period,

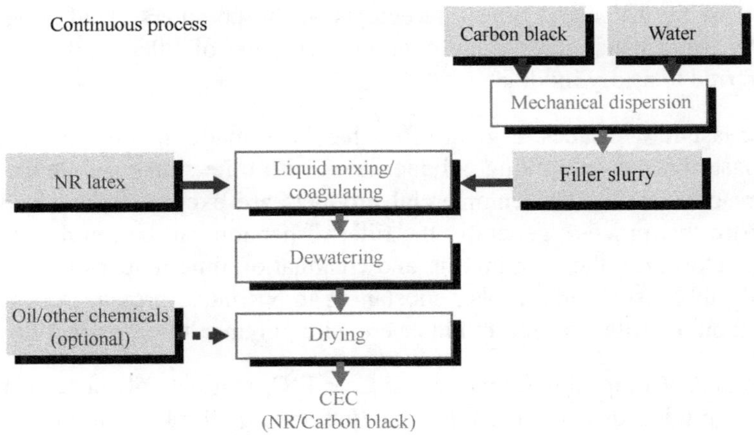


**Figure 9.9** CEC process flow diagram

typically 5–10 seconds, will the compound temperature reach 140–150°C. This is to say that, during drying, the thermo-oxidative degradation of natural rubber can be essentially prevented. In the dryer, small amounts of antioxidants can also be introduced as a stabilizer for storage. Optionally, the small ingredients in the compounds, such as zinc oxide, stearic acid, antiozonants, antioxidants, and wax can be added in this stage.

The dried material can then be slabbed, cut, or pelletized. Currently, CEC is packaged into highly friable bale form consisting of compressed small strips.

The key feature of the CEC process is the fast mixing and coagulation, and a short drying time at high temperature. This achieves excellent performance for the material as polymer-filler interaction can be better preserved and polymer degradation can be well eliminated.

## 9.3.2 Compounding Characteristics

One of the features of CEC is its superior filler dispersion. The dispersion of carbon black in the polymer is, therefore, not a concern for compounders using CEC. Mixing other ingredients, such as antioxidants, wax, oil, vulcanization activators, and curatives, requires distinct procedures to take full advantage of CEC's unique properties due to the different rheological behaviors and different requirements between conventional rubber and CEC. The exact mixing specifications will be dependent on the type of compounds, mixing equipment, desired downstream processability, product performance requirements, and other factors.

CEC has a higher viscosity relative to pure polymer and the dry-mixed masterbatch. The high viscosity of CEC is mainly caused by the hardening of the material during storage. This is especially the case for highly loaded compounds with high surface area carbon blacks, such as those used for tire treads. Besides the hydrodynamic effect of the filler, three mechanisms are responsible for the hardening effect of CEC: polymer gelation, bound rubber formation, and carbon black flocculation. However, equilibrium is reached in a short period of time.

**Polymer Gelation:** This is the same mechanism as that for pure NR hardening during storage[28]. In this mechanism, it has been proposed that the condensation of the biochemically formed aldehyde groups along the polymer chains, probably via association with non-rubbers, gives rise to increased viscosity of the rubber. A similar effect should be evident in CEC.

**Bound Rubber Formation:** This mechanism is related to the addition of carbon black in the polymer resulting in adsorption of polymer chains on the filler surface. This process may continue during storage as in the case of dry mixing where the bound rubber increases rapidly at the beginning of storage and reaches equilibrium in about 1 month[29]. The formation of bound rubber will significantly raise the compound viscosity as the effective filler volume increases[29–31] due to the immobilization of the polymer segments on the filler surface. It is also related to the entanglement between the adsorbed polymer chains and polymer molecules in the matrix (sol).

**Carbon Black Flocculation:** It has been recognized that, through filler-filler interaction, filler flocculation or agglomeration can take place during storage. This is especially true at the very early stage of storage when the bound rubber has not been fully developed. As a consequence, the rubber trapped in the agglomerates will lose, at least partially, its identity as rubber and behave as a filler due to its immobilization as far as the agglomerates cannot be broken down under an applied stress[23]. This leads to a substantial increase in apparent volume fraction of filler, hence viscosity of the compound.

The storage hardening effect of pure NR can be efficiently inhibited by using some chemicals, such as hydroxylamine, which can react with aldehyde groups on the polymer chains via condensation mechanism[32]. However, due to the filler effect, effectiveness of these chemicals for preventing storage hardening of CEC is considerably reduced.

Bound rubber formation is an indicator of polymer-filler interaction, which is an important parameter that governs rubber reinforcement. Any attempt to reduce bound rubber formation will result in deterioration in rubber properties, abrasion resistance in particular.

It should be point out that when some small chemicals, such as stearic acid, antioxidants, oil, and wax, are added in the CEC process, the viscosity of CEC will be, to a certain extent, reduced, which facilitates the mixing process.

#### 9.3.2.1 Mastication Efficiency

A favorable feature of CEC related to mixing is that, compared with dry-mixed carbon black masterbatch, the Mooney viscosity of CEC drops more rapidly upon mastication and during downstream processing. This may be associated with the high viscosity of CEC, which gives rise to high shear stress during mixing, facilitating the mechano-oxidative mastication and breakdown of agglomerates to release the trapped rubber. After mixing, while the compound viscosity can easily reach the same level as that for dry-mixed compounds, the viscosity will also be stabilized, showing that the Mooney viscosity of the compound does not change with storage (Figure 9.10). As will be shown below, the rapid breakdown may lead to over-mixing unless the mixing time is considerably reduced.

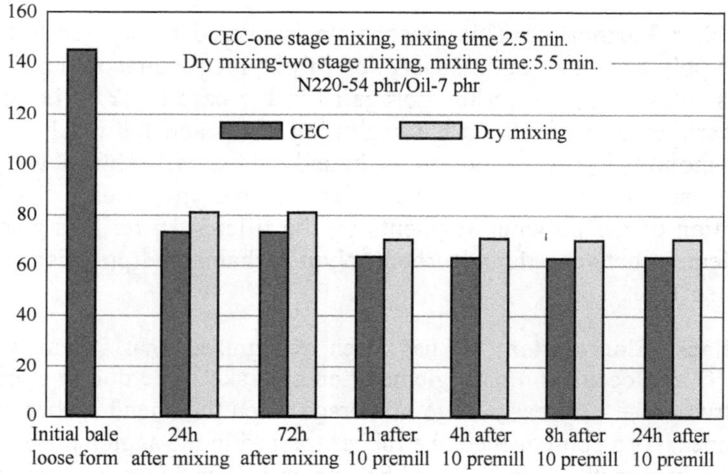


**Figure 9.10** Effect of mixing in Banbury-1.6 L and remilling (6″ roll mill) on Mooney viscosity

### 9.3.2.2 CEC Product Form

Generally, high torque may be generated when high viscosity materials are directly charged into the Banbury. Practically, to avoid excess power or unacceptable peak torque, the materials are either warmed up in a hot room or masticated on an open mill to reduce the viscosity. In order to simplify mixing procedures and minimize torque, CEC compounds are made into a "loose" bale form. In these loose bales, besides that the bale can be easily broken down, there is a certain amount of void volume, which substantially reduces the peak power when the CEC bale is engaged between the rotors and wall of the mixer chamber in the early stage of breaking down the compound. This allows the loose bale to be fed directly into the Banbury without prewarming and premastication. Commercially, CEC is packaged in easy-to-handle 16 kg loose bales in low melt bags. Product form alternatives, such as slabs or pellets, can also be used.

### 9.3.2.3 Mixing Equipment

There are broadly two types of batch internal mixers: tangential and intermeshing mixers. Both can be used for CEC mixing. From the point of view of structure design, the main difference between these two types of mixers is whether the rotation paths of the major diameters of the two rotors intercept or not. The features of the tangential

Banbury are a large available mixing chamber volume, high fill factor, quick feeding and discharging of material, and high productivity per energy input[33]. The intermeshing mixer has the following advantages: efficient mastication, temperature control, incorporation of oily material, and dispersion of ingredients. Therefore, similar to dry mixing, when different mixers are used for CEC mixing, the operational parameters, such as the fill factor, material feeding, rotor speed, ram pressure, and mixing cycle, must be adjusted accordingly.

#### 9.3.2.4 Mixing Procedures

Similar to conventional mixing, the mixing of CEC consists of several phases. After feeding in the internal mixer, the CEC material is broken down and masticated for a short period of time. This is followed by the addition of the small chemical ingredients such as cure activators, antioxidants, wax, and oil. Generally the batch is dumped after the first stage addition of the small chemicals and the curatives are added in the following or second stage. This is referred to as two-stage mixing. In certain instances the curatives can be added with, or just after the addition of the small ingredients for single-stage mixing.

##### 9.3.2.4.1 Two-Stage Mixing

*CEC mastication:* After charging of CEC into the mixer, the mastication of CEC prior to the incorporation of other ingredients is critical. In this step, the material viscosity drops substantially, which is essential for chemical addition and downstream processing. The time of mastication is mainly determined by the target Mooney viscosity, which should target down-stream mixing and processing. For tire tread compounds, for example, when a cold-feed extruder is used, a slightly longer mastication period is favorable for the extrudate quality. The mastication time is also related to the incorporation period and dispersion of other ingredients. Depending on the type of equipment, rotor speed, and compound, the mastication time can vary from 30 seconds to 90 seconds and small adjustments may be necessary when the batch size and ram pressure are changed.

*Addition of small ingredients and oil:* Following the initial mastication period, for CEC only having rubber and carbon black, the small ingredients, such as zinc oxide, stearic acid, antioxidants, and wax can be added. Generally, there is no significant difference in the incorporation and dispersion of unmeltable powder chemicals between dry mixing and CEC mixing. However, the behavior of oily materials is somewhat different. This includes oil, liquid chemicals, and solid materials that melt at the mixing temperature such as stearic acid, wax, and some antioxidants. When oily materials are added, the stretching and tearing of the batch by the tips of wings no longer occurs. Therefore, no material transfer takes place between the rotors and chamber side surfaces, and hence no energy is imparted into the polymer matrix. The oily materials essentially behave as a lubricant. The mixing action does not start until the oily materials are absorbed into the rubber. Therefore, the lubrication time or incorporation time of oily materials is determined by the rate of their absorption in rubber.

In practice, an increase in mastication time leads to shorter lubrication time. In a limited range, an increase in batch size also results in speeding up the incorporation of oily materials. It has also been observed that the lubrication time decreases with increasing temperature when the mastication temperature is relatively higher and an increase in rotor speed, if allowed, can effectively reduce the lubrication time. Shown in Figure 9.11 is a mixing chart for CEC with 50 phr N234 in a F270 Banbury.

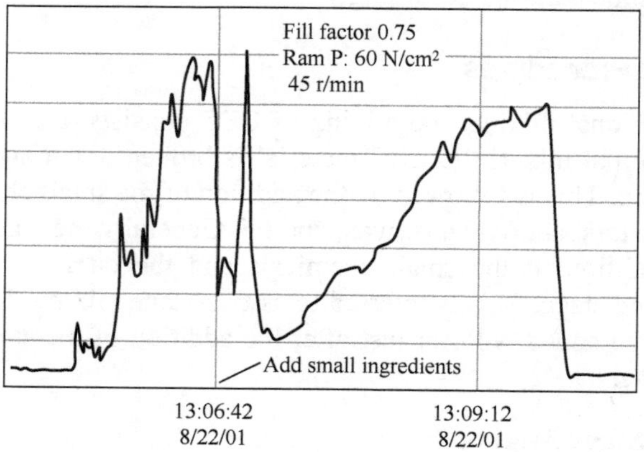


**Figure 9.11** Typical CEC mixing power in F270 Banbury

After the incorporation of the ingredients, the batch can be dumped either onto a roll mill or into a single-or twin-screw extruder for further blending, cooling, and sheeting or pelletizing, which are similar operations to those of dry mixing.

When the oily chemicals are added in CEC production process, the mixing is facilitated due to reduced viscosity and elimination of lubrication. For CEC containing all small ingredients, except curatives, this stage of mixing is only for mastication. Therefore, the mixing productivity can be significantly increased by pre-addition of oily chemicals. In a mixing trial, the lubrication time is reduced by about 0.5 minute to 1 minute for the CEC with pre-addition of stearic acid. This results in a reduction in the mixing cycle by about 25%—30% with Banbury F50-4WST and intermeshing mixer Intermix K-2A Mark 5 for the first stage compared with the CEC without stearic acid.

*Second stage of mixing:* For two-stage mixing, the second stage is used mainly to add curatives. The mixing procedure is the same as that for dry mixing.

#### 9.3.2.4.2 Single-Stage Mixing

Single-stage mixing enhances productivity as the mastication, incorporation, and dispersion of all ingredients are completed in a single pass of the internal mixer with limited roll-mill operation. However, this is only feasible where sufficient mixing time is allowed to achieve the required dispersion of the ingredients and adequate downstream processability. With single-stage mixing, curatives can be added either with or after other ingredient addition. The batch should generally be dumped below a temperature limit of 125°C to prevent precure or scorch.

The key parameters to determine the feasibility of single-stage mixing are, among others, the energy input and temperature profile. While enough energy is needed for ingredient dispersion and for bringing down the compound viscosity, the temperature has to be well controlled to a narrow range during and after curative addition for the safety of downstream processing, such as extrusion, calendering, or molding.

For mixers whose rotor speed cannot be adjusted during mixing and whose cooling system is not efficient enough, the mixer should run at lower rotor speed in order to keep the batch temperature in the appropriate range. For example, the rotor speed should be similar to that used in the second stage of a two-stage mixing procedure. Consequently, the mixing cycle should be longer than that of the first stage of the two-stage mixing, but shorter than a two-stage mix.

The mixing efficiency can be substantially improved and the mixing cycle can be significantly reduced for single-stage mixing when using a mixer with a variable speed drive and an intensive cooling system. In this case, the mastication and addition of small ingredients, except curatives, can be performed at high rotor speed. When the rotor speed is reduced the mixer can be used as a cooling unit to bring the compound temperature down to a limit in an acceptable time frame to allow the curative addition and dispersion. Intensively cooled machines can reach excellent cooling rates. In this regard, the intermeshing mixer exhibits a significant advantage over a tangential Banbury due to its high mastication efficiency and high contact area between the compound and the mixer temperature-controlled surface.

In addition, pre-addition of small chemicals in the CEC compound facilitates single-stage mixing. Additionally, this results from lower compound viscosity, hence easier processability and lower temperature.

#### 9.3.2.5 Total Mixing Cycle

The total mixing cycle, as mentioned before, is determined by the dispersion of the ingredients and target viscosity of the compound. Too short of a mixing cycle may lead to a high compound viscosity, which in the worst case will leave some of the materials in the compound not fully masticated, giving a rough extrudate appearance. This is especially critical when a cold feed extruder is utilized. In the case of hot feed extrusion, the compound can be further masticated on the roll mills during warming before extrusion so that the mixing cycle in the mixer may be reduced.

Over-mixing of the compound will result in a very low Mooney viscosity. However, this will have a significantly negative effect on compound properties, viscoelastic properties in particular. More severe mechano-oxidation of the polymer may influence the aging resistance. Also, a low compound viscosity can facilitate the flocculation of the filler during vulcanization, resulting in higher hysteresis[34]. This will increase the rolling resistance when the compounds are used for tires, especially in tread compounds. On the other hand, where low viscosity is required, such as in wire skim compounds, CEC can be particularly advantageous.

Generally, the CEC mixing cycle can be 30% to 70% shorter than the traditional mixing time, depending on the mixing procedures of dry mixing.

In summary, CEC is a unique material that can be easily mixed in conventional rubber mixing equipment. To make maximum use of its inherent qualities to improve productivity and to realize the performance benefits that result, the different mixing behaviors of CEC have to be recognized.

### 9.3.3 Cure Characteristics

Generally, the cure characteristics of CEC compounds are similar to those of dry-mixed materials. However, somewhat shorter scorch time and higher cure rate may be observed with CEC compounds. This is generally due to the moisture content. It has been found that moisture can catalyze the decomposition of the accelerators into intermediate active products, for example, amine and 2-mercaptobenzothiazole from sulfenamide, so the cure onset and cure kinetics can be significantly sped up. The effect of moisture on cure characteristics is similar for both CEC and dry-mixed compounds. Presented in Figure 9.12 is the effect of moisture content on scorch time of N234-filled compounds. The moisture contents in the final compounds are varied either by adding moisture in the compounds or by drying the NR and CEC in an oven before mixing, and are measured just before the scorch test. Obviously, the scorch time of both CEC and dry-mixed compounds follow almost the same function as moisture content. CEC's moisture content is 0.5% to 1%.

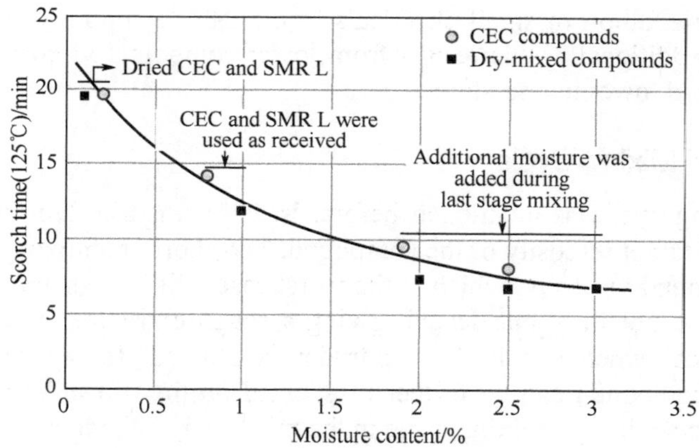


**Figure 9.12** Effect of moisture on scorch time of CEC and dry-mixed compounds

### 9.3.4 Physical Properties of CEC Vulcanizates

#### 9.3.4.1 Stress-Strain Properties

It is generally observed that, with the same formulation, the static moduli measured at high elongations (100% and 300%) of CEC vulcanizates are similar to those of their dry-mixed counterparts. Statistically, the CEC gives slightly higher tensile strength, longer elongation at break, and 1.5 points to 2.5 points lower hardness at the practical loading.

#### 9.3.4.2 Abrasion Resistance

Generally speaking, for carbon blacks that are easily dispersed, the abrasion resistance of CEC is comparable to that of dry-mixed counterparts at lower loading. For the high-surface-area and/or low-structure carbon blacks that have poor dispersibility, the CEC gives advantages over dry mixing in abrasion resistance, especially at higher filler loading.

Presented in Figure 9.13 is the effect of loading (carbon black N134) on abrasion resistance measured by means of Cabot Abrader (a Lambourn-type machine) at 7% slip ratio. For both CEC and dry-mixed compounds with SMR 20 that is generally used in tire tread, the abrasion resistance increases with filler loading, passing through a maximum, and then drops. Several mechanisms may be involved in the reduction in abrasion resistance at high filler loading. Among others such as reduction in rubber content, rapid increase in hardness, and deterioration of fatigue resistance, the poor dispersion of carbon black at high loading plays an important role for the decline of abrasion resistance. The excellent carbon black dispersion, even at high loading, accounts for the significantly better abrasion resistance of CEC, compared with dry-mixed compounds at high loading. Also due to the improvement of filler dispersion, the CEC technology is able to move the maximum abrasion resistance and optimum filler loading to higher levels.

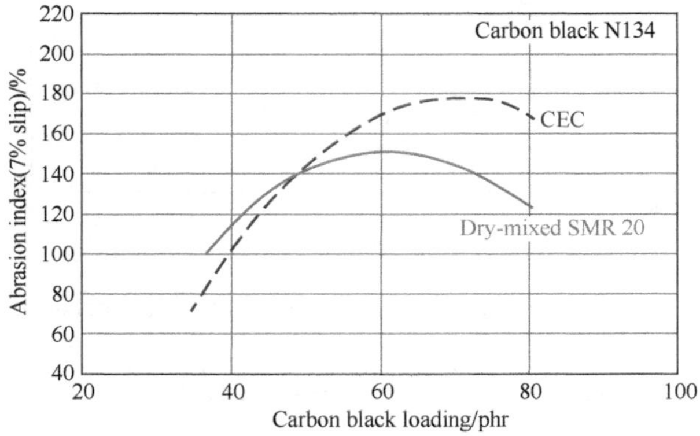


**Figure 9.13** Effect of carbon black loading on abrasion resistance

Practically, for certain applications, such as tire tread compounds, the hardness of the materials has to be limited in a relatively narrow range, due to the balance of wear resistance, skid resistance, and other performance. For a given carbon black, this is generally achieved by adjusting carbon black loading and oil contents. It has been found that, at constant hardness, the abrasion resistance follows different curves as a function of filler and oil loading. Figure 9.14 illustrates the effect of carbon black

N234 and oil on abrasion resistance at 7% slip ratio for the compounds with hardness of 65. For dry-mixed compounds, abrasion resistance decreases monotonically with increasing carbon black and oil loading. In contrast, for CEC, the abrasion resistance of vulcanizates passes through a maximum. This is because of the effect of oil adsorption on polymer-filler interaction, one of the most critical parameters controlling abrasion resistance. In dry-mixed materials, the oil adsorption interferes with the polymer-filler interaction. In contrast, the polymer-filler interaction in CEC is much less affected by the addition of oil during mixing, as the adsorption of polymer chains on carbon black surface has already been completed. This can be demonstrated by evaluating the effects of oil addition on the bound rubber content. Figure 5.23 shows the differences in bound rubber content between CEC and dry-mixed compounds with different levels of oil. The results are taken from a statistical analysis of about 200 pairs, each pair having one CEC and one dry-mixed compound with the same formulation. For all compounds with and without oil, CEC gives higher bound rubber. The difference increases as the oil content increases for the compounds. The drop in abrasion resistance of CEC compounds at high loading of oil and carbon black is mainly caused by the drastic reduction of the polymer content, the basic component in the vulcanizate.

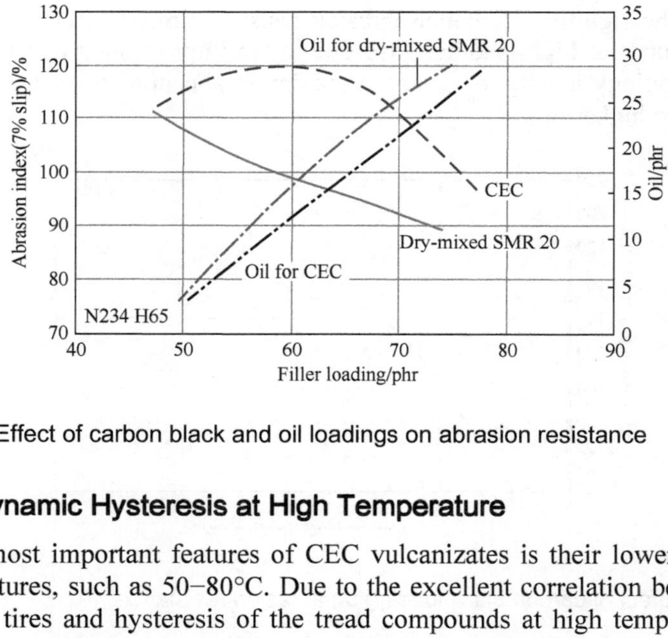


**Figure 9.14** Effect of carbon black and oil loadings on abrasion resistance

### 9.3.4.3 Dynamic Hysteresis at High Temperature

One of the most important features of CEC vulcanizates is their lower hysteresis at high temperatures, such as 50–80°C. Due to the excellent correlation between rolling resistance of tires and hysteresis of the tread compounds at high temperature, lower rolling resistance and lower heat build-up are expected with CEC.

For eight carbon blacks including one with surface area from 110 $m^2/g$ to 200 $m^2/g$ and CDBP from 52 mL/100 g to 116 mL/100 g, compounds having oil loading from 0 phr to 30 phr and filler loading from 30 phr to 75 phr, the reductions in maximum loss tangent obtained from strain sweeps, tan $\delta_{max}$, at 60°C from 340 compounds are presented in Figure 9.15. When the hysteresis of the compounds is compared pair by pair between CEC and their dry-mixed counterparts with the same formulations, the hysteresis of CEC vulcanizates is on average 8% lower. This number increases to 10.2% and 13.4% for the compounds containing oil (from 5 phr to 30 phr) and high oil (from 10 phr to 30 phr), respectively. The lower hysteresis for CEC compounds is mainly due to improved micro-dispersion of carbon blacks, i.e., depressed filler agglomeration. This can be demonstrated by the Payne effect, i.e., the difference in dynamic moduli measured at low strain amplitude and high strain amplitude[33]. For all 340 compounds mentioned above, the average Payne effect measured at strain amplitude at 0.1% and 60% is about 9% lower for CEC vulcanizates (Figure 9.16) and this difference increases with increasing oil contents in the compounds. Similar results were also observed when the compounds were adjusted to the same hardness. As can be seen from Figure 9.17, taking carbon black N234-filled NR as an example, the hysteresis is always lower for CEC and it is even more favorable at higher filler and oil loading.

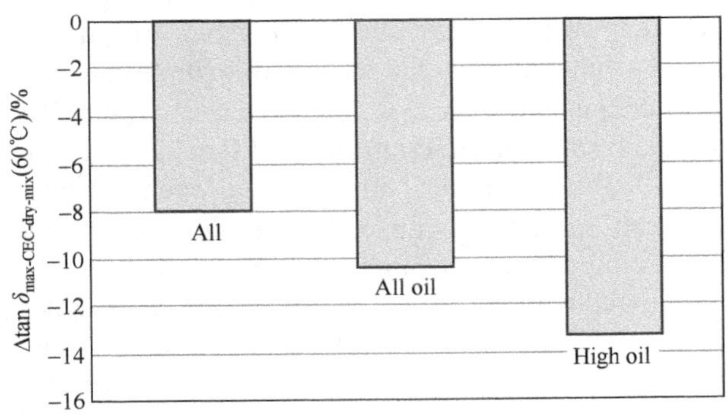


**Figure 9.15** Effect of oil on hysteresis at 60°C

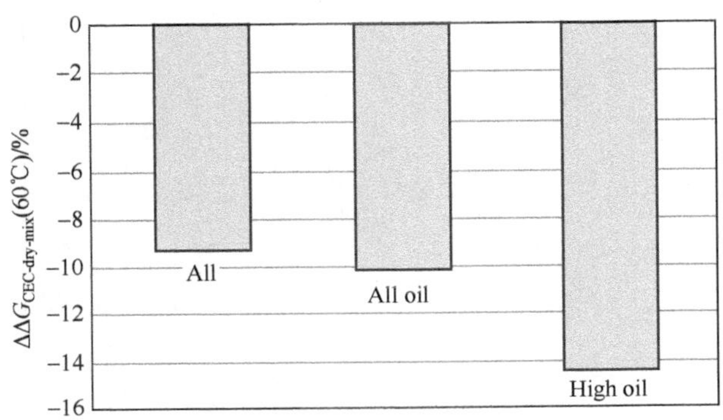


**Figure 9.16** Effect of oil loading on Payne effects at 60°C

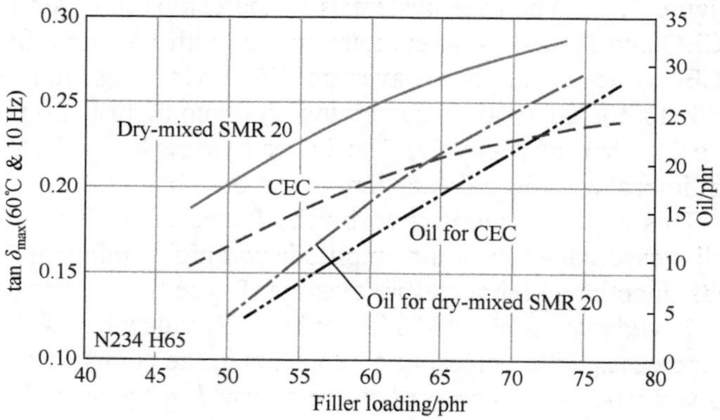


**Figure 9.17** Effect of carbon black and oil loadings on hysteresis

The rebound results are consistent with the observations on hysteresis. In Figure 9.18, the rebounds measured with Zwick tester for a series of carbon blacks and filler loading are plotted as a function of $\psi\phi$, a so-called loading-interfacial-area parameter. Here, $\psi$ is interfacial area equaling to $\rho S\phi$. $\rho$ is the density of the filler, $S$ is the specific surface area, and $\phi$ is the volume fraction of the filler in the compounds. Caruthers, Cohen, and Medalia[35] found a good correlation between $\psi\phi$ and hysteresis, hence rebound. As can be seen, for the compounds having identical $\psi\phi$, the average rebound of CEC compounds is improved by 3% absolute points (or 5% to 10% in total) with CEC technology compared with conventional dry mixing. The higher rebound or lower hysteresis of CEC compounds results in lower heat build-up. In a typical endurance test for CEC truck tires, compared to an existing commercial tire, the running temperature was reduced by 10°C and endurance life increased by 17% for CEC-tread compounds prepared with a comparable formulation as the dry mix.

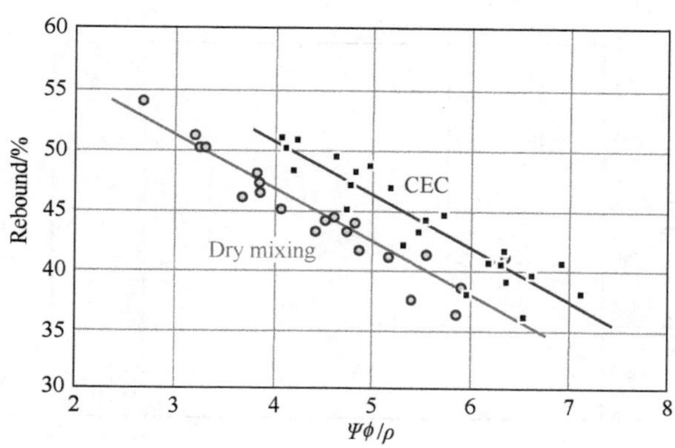


**Figure 9.18** Rebound measured with Zwick tester for a series of carbon blacks and filler loading

### 9.3.4.4 Cut-Chip Resistance

The cut-chip resistance of CEC compounds was investigated by means of the Cabot OTR-Service Simulator with nylon 6 ply rated 6.90 × 9 inch diameter tires. The tires were retreaded with compounds containing carbon black N234, N220, N231, and Regal 660, which is a low-structure member of the N200 group of carbon blacks with surface area 112 $m^2/g$ and CDBP 52 mL/100 g. The cut-chip ratings were determined by counting and totaling the number of defects (3.2 mm or larger) in the tread area which result from cutting, chipping, chunking, and abrasive action of the tire traveling over the Simulator track. The results are presented in Figure 9.19. Compared to the traditional materials, the CEC tread compounds give significantly improved cut-chip resistance. This is especially true for the low-structure carbon blacks. It is generally observed that low-structure carbon blacks give higher tearing strength, but this advantage is offset by their poorer dispersibility in dry mixing. For carbon black Regal 660, while cut-chip properties of dry-mixed compounds is even poorer than that of carbon black N231, its potential for improvement of cutting and chipping properties is fully brought out by CEC technology.

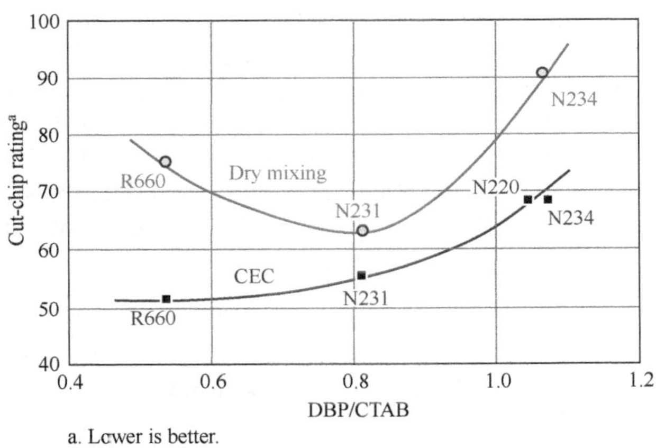


**Figure 9.19** Cut-chip resistance of CEC compounds containing different carbon blacks (N234, N220, N231, and Regal 660)

### 9.3.4.5 Flex Fatigue

Mainly thanks to the excellent dispersibility of carbon blacks, and maybe also due to lower heat build-up, a great benefit for fatigue resistance has been obtained from CEC. Compared with the traditional compounds, the average fatigue life in compression mode increases by well over 90%, which will impart a significant improvement to service life of some rubber goods, such as anti-vibration products, wipers, and belts; even sidewall of tires are significantly improved over their dry-mixed counterparts.

## ■ 9.4 Synthetic Rubber/Silica Masterbatch Produced with Liquid Phase Mixing

"Green tire" technology has been widely considered as a major breakthrough in the history of the tire industry. With acceptable abrasion resistance, it greatly improved tire rolling resistance and wet skid resistance for societal benefits[36,37].

The basic features of the tread compound of a green tire are as follows: the reinforcing system is silica with silane coupling agent; a blend of SSBR (solution styrene-butadiene rubber) and BR (butadiene rubber) is the main polymer system, and DPG (diphenylguanidine) is used as secondary accelerator besides sulfenamides.

The function of the silane coupling agent is for silica surface modification, reducing its polarity, and increasing filler-polymer interaction, and depressing filler agglomeration in rubber. As a result, the hysteresis of the vulcanizates at high temperature (such as 60°C) decreases, leading to lower rolling resistance of the tire. The less filler agglomeration also leads to increased tan $\delta$ at low temperature, which is beneficial for wet skid resistance[23]. This is especially true in the traction zone of the contact patch, where higher hysteresis at low temperature would increase friction (see Section 8.2.3)[38,39]. Silane coupling reaction will generate chemical linkage between filler and rubber which improves the polymer-filler interactions and thus abrasion resistance (see Section 8.3.2)[40]. Compared with carbon black, the interaction of silica with polymer is much lower without silane coupling agent[41,42].

The application of SSBR in green tire compounds[43] is also related to the improvement of tread wear resistance, as the polymer-filler interaction of solution polymerized polymer is better than that of its emulsion polymerized counterpart, which was discussed in Section 8.3.2.

Since the polar silanols (isolated, vicinal, and geminal) on the silica surface are acidic, the cure kinetics can be drastically affected, resulting in much lower crosslink density. By adding some diphenylguanidine (DPG), which is basic, the cure rate can be significantly increased, and so also the crosslink density.

Silica-filled compound are conventionally mixed in a Banbury-type mixer in several stages. The disadvantages of this dry-mixing technique are as follows:

- severe Banbury wear,
- poorer filler dispersion, even for highly dispersible silica,
- higher Mooney viscosity,
- poorer processability – large die swell and rough extrudate, and
- lower abrasion resistance than that of carbon black-filled compound.

More recently a state-of-the-art technology has been developed by the EVE Rubber Institute which produces rubber masterbatch by continuously mixing synthetic rubbers and fillers in the liquid phase (EVE process for short)[44–50]. The resulting products are known as EVEC® (Eco-Visco-Elastomer Composite) in the industry. Compared with silica-filled compounds mixed in a dry state, the EVEC is characterized by the following:

- excellent filler dispersion,
- high efficiency of silanization, leading to stronger filler-polymer interaction and weaker filler-filler interaction,
- improved compound processability, and
- better strain-stress properties, dynamic properties, friction and abrasion, and tire performance.

### 9.4.1 Production Process of EVEC

A simplified flow diagram of EVEC production is illustrated in Figure 9.20. In the production process of EVEC, the rubber solution may be received directly from upstream polymerization process. Alternatively, the pulverized synthetic rubber is loaded in a rubber solution preparation tank via a dosing device. After dissolving by solvent, it becomes a viscous and uniform rubber solution. Meanwhile, silica is pre-mixed with solvent and finely dispersed by mechanical means.

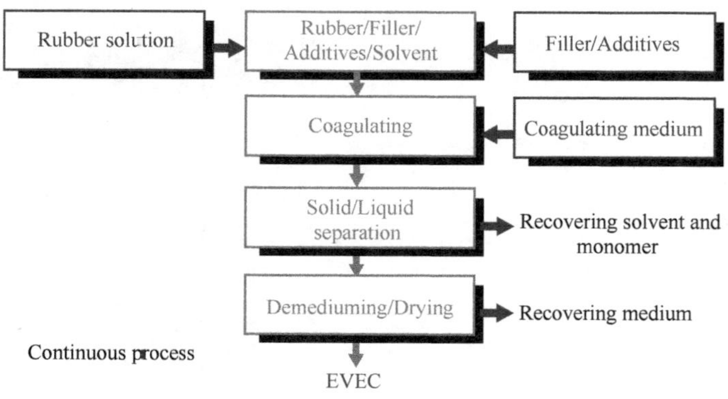


**Figure 9.20** Simplified flow diagram of EVE process

Both rubber solution and silica slurry are mixed instantaneously under highly turbulent conditions though a special device at high pressure. Subsequently, the silica, coupling agent, other additives, and polymers are fully dispersed together. The mixture is then sent to a stripping device to remove the solvent at elevated temperature and low pressure. Meanwhile, the solvent is recovered for reuse. Through this process, not only are the filler-polymer interactions greatly enhanced but also filler-filler interaction is drastically reduced.

The continuous liquid phase mixing technology is a dust-free process that delivers high-performance masterbatch with low energy consumption.

## 9.4.2 Compound Properties

A typical silica-tread formulation of passenger tire was used for evaluation of EVEC, using polymer blends of solution OE-SBR and BR with a weight ratio of 96.3/30. The detailed formulations are listed in Table 9.3. A carbon black compound was introduced as a reference. The formulation of EVEC-L is exactly the same as the dry-mixed silica compound but compound EVEC-H has 6 phr more silica in the formulation[51]. The dry-mixed compounds were prepared in a Banbury with a three-stage mixing procedure. The properties of compounds and vulcanizates are shown in Table 9.4.

**Table 9.3** Typical tread formulation of passenger tire

| Formulation | CB N234 | Silica-dry mixing | EVEC-L | EVEC-H |
|---|---|---|---|---|
| SSBR OIL extended | 96.3 | 96.3 | 96.3 | 96.3 |
| BR | 30 | 30 | 30 | 30 |
| Silica NS165MP | – | 78 | 78 | 84 |
| Carbon black N234 | 78 | – | – | – |
| TESPT | – | 6.4 | 6.4 | 6.4 |
| Oil | 1.8 | 1.8 | 1.8 | 1.8 |
| ZnO | 3.5 | 3.5 | 3.5 | 3.5 |
| Stearic acid | 2 | 2 | 2 | 2 |
| Wax | 1 | 1 | 1 | 1 |
| Antioxidant RD | 1.5 | 1.5 | 1.5 | 1.5 |
| Antioxidant 6PPD | 2 | 2 | 2 | 2 |
| Accelerator DPG | – | 2.1 | 2.1 | 2.1 |
| Accelerator CBS | 1.35 | 2 | 2 | 2 |
| Sulfur | 1.4 | 1.4 | 1.4 | 1.4 |

**Table 9.4** Physical properties of rubber compounds

| Property | CB N234 | Silica-dry mixing | EVEC-L | EVEC-H |
|---|---|---|---|---|
| Bound rubber/% | 44.5 | 75.3 | 67.5 | 68.2 |
| ML1+4' (100°C) | 66 | 74 | 68 | 68 |
| Scorch time (t5)/min | 25.4 | 24.0 | 18.6 | 18.2 |
| Hardness (RT, Shore A) | 68 | 66 | 57 | 62 |

Continued

| Property | CB N234 | Silica-dry mixing | EVEC-L | EVEC-H |
|---|---|---|---|---|
| Hardness (60°C, Shore A) | 66 | 65 | 56 | 59 |
| T100/MPa | 2.2 | 2.6 | 2.2 | 2.8 |
| T300/MPa | 10.3 | 12.2 | 14.3 | 17.3 |
| T300/T100 | 4.7 | 4.7 | 6.5 | 6.2 |
| Tensile strength/MPa | 16.7 | 16.4 | 19.8 | 19.5 |
| Elongation at break/% | 452 | 409 | 348 | 324 |
| Tear strength/(N/mm) | 40 | 46 | 36 | 35 |
| Rebound (RT)/% | 30.2 | 36.3 | 42.1 | 40 |
| Rebound (60°C)/% | 42 | 55 | 60 | 60 |
| Heat build-up, $\Delta T$/°C | 62.2 | 32.2 | 24.5 | 28.2 |
| SS-tan$\delta_{max}$ (60°C) | 0.318 | 0.164 | 0.103 | 0.111 |
| Akron abrasion index/% | 100 | 83 | 105 | 117 |

### 9.4.2.1 Bound Rubber Content

The results in Table 9.4 show that the highest bound rubber content is for the dry-mixed silica compound, followed by EVEC-H and EVEC-L. The bound rubber of the dry-mixed compound filled with carbon black is the lowest. It was reported[52] that silica-filled SBR compounds without coupling agent showed lower bound rubber content than carbon black-filled SBR compounds at the same filler loading and similar filler surface area. This was mainly attributed to their different polymer-filler interactions, as the dispersive component of surface energy ( $\gamma_f^d$ ) of silica is substantially lower than that of carbon black, and so is the polymer-filler interaction. The fact that bound rubber content of dry-mixed silica compound in Table 9.4 is considerably higher than that of carbon black-filled compound is evidently attributed to the application of bifunctional silane coupling agent. The coupled silica-polymer system would increase the polymer-filler interaction caused by chemical reaction. In addition, in dry-mixed silica compound, some coupling agent may remain in the polymer matrix after mixing so that the polysulfide in the coupling agent may generate some pre-crosslinks between polymer molecules, forming some gel. This gel becomes part of the bound rubber as the gel cannot be extracted out by the solvent during testing. Obviously, the lower bound rubber contents for the EVECs is related to the fact that much less polymer gel is formed in the polymer matrix as most of the coupling agent (if not all) is adsorbed on silica surface during the production process.

### 9.4.2.2 Mooney Viscosity

The rating of Mooney viscosity for the four compounds is similar to that of bound rubber. The Mooney of the dry-mixed silica compound is the highest, then EVEC, and carbon black gives the lowest value. However, differences in viscosity between carbon black and EVEC are relatively small.

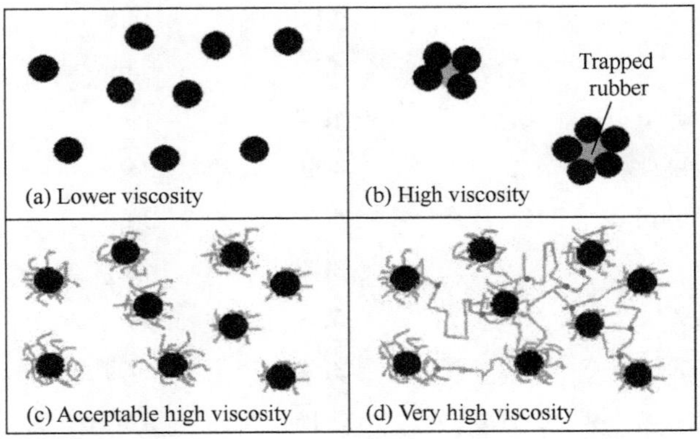


Figure 9.21 Illustration of factors influencing the Mooney viscosity for different compounds

In addition to loading and filler morphology, there are three factors influencing Mooney viscosity: filler agglomeration, the bound rubber content, and the polymer gel content. If the filler aggregates are fully dispersed without agglomeration, the Mooney viscosity of rubber compound should be the lowest [Figure 9.21(a)]. Once the aggregates are agglomerated, the rubber trapped in the agglomerates will increase the effective volume fraction, and hence the viscosity of the compound would be increased [Figure 9.21(b)]. The compound viscosity would also increase if bound rubber forms [Figure 9.21(c)]. When pre-crosslinking of the polymer matrix and bound rubber take place simultaneously, the viscosity will be augmented drastically [Figure 9.21(d)]. It seems that the lower viscosity of the carbon black-filled compound may be caused by the mechanisms shown in Figure 9.21(b) and (c), while the relatively higher viscosities of EVEC compounds are probably more related to the case illustrated in Figure 9.21(c) rather than (b). For the dry-mixed silica compound, the highest viscosity likely results from the combination of the mechanisms of Figure 9.21(b), (c), and (d).

### 9.4.2.3 Extrusion

Compared with the dry-mixed compounds, the die swell of EVEC-L is somewhat smaller and its surface appearance is smoother compared with its dry-mixed counterpart (Figure 5.42). This is certainly related to the low bound rubber, low viscosity, and less crosslinking of rubber molecules in the compound. For these reasons, the stress relaxation of rubber molecules is faster in the extrusion process.

#### 9.4.2.4 Cure Characteristics

The cure characteristics of the compounds are shown in Figure 9.22. Although the curing systems of the carbon black compound and EVEC-L are different, their cure curves are similar, except for minimum torque and scorch time. However, the cure characteristics of the dry silica compound are quite different, even though it used the same curing system as the EVEC-L compound.

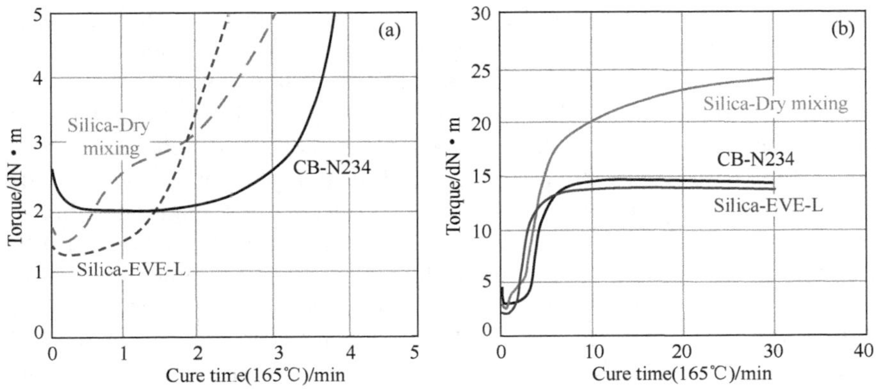


**Figure 9.22** Comparison of the curing properties of different tread rubbers

In the induction period of crosslinking, the cure curves are quite different [Figure 9.22(a)]. The minimum torque (ML) is often taken to be a measure of compound viscosity before crosslinking. However, while the Mooney viscosity is the highest for dry-mixed silica compound, its ML falls in between the two other compounds; the value of ML is the lowest for EVEC, and highest for the carbon black-filled compound.

The different ranking of the two measurements of viscosity can be attributed to the different experimental conditions. For the Mooney viscosity measurement, the rotor rotates uni-directionally in the cavity at a rate of 2 r/min. In the case of the rheometer, the torsional vibration angle is 0.5° and the frequency is 1.7 Hz. The shear amplitude of the former is infinite, while the latter is very small. Moreover, the testing temperature of the Mooney was 100°C while the rheometer was run at 165°C. The higher ML of the carbon black-filled compound, which also gives the lowest Mooney viscosity, could be related to filler agglomeration during tests. Filler agglomeration could lead the trapped rubber in agglomerates to be shielded from external stress, acting like filler. As a consequence, the effective filler volume would increase, resulting in an increased viscosity. In the carbon black compound, the higher surface energy of the filler and lower bound rubber would cause the filler aggregates to become agglomerated more easily according to both thermodynamic and kinetic mechanisms, especially at higher temperature. In the rheometer test, the shear strain is so low that developed agglomerates of carbon black cannot be broken down, so the ML is relatively high. In the Mooney viscosity measurement, since the shear strain is infinite, the agglomerates are almost fully broken down, so that the carbon black compound gives the lowest viscosity. The low ML of EVEC, which is similar to that of the carbon black compound, also demonstrates less agglomeration of silica in the compounds.

Generally, the cure profile of rubber compounds shows several distinct phases: induction of cure, onset of cure, undercure, optimum cure, and overcure. The cure profile is typically an "*S*" shape[53], as seen for the carbon black and EVEC compounds here. The cure curve of the dry-mixed silica-filled compound, however, rises rapidly with time and a bulge appears in the induction period (Figure 9.22). The presence of the bulge seems to suggest that, in addition to crosslinking, another process may be occurring which may reasonably be related to filler agglomeration. The effect of this can be superposed on the crosslinking of the polymer, resulting in a bulged curve in the induction period. This phenomenon does not appear in the cure of EVEC, which could be attributed to the excellent surface modification of fillers caused by sufficient adsorption of coupling agents on the silica surface[43]. As mentioned before, the effective modification of silica in EVEC increases the compatibility between filler and rubber and greatly reduces the driving force of filler agglomeration. Since the compound formulations were the same for dry-mixed compound and EVEC-L, the more developed agglomeration in the dry-mixed material can be explained by its inefficient surface modification. Earlier studies have found that agglomeration of filler also occurs in the induction period of high-temperature vulcanization for carbon black-filled compounds[54]. However, due to less bound rubber content and no or less gel formation in the compounds during mixing, at high cure temperature, filler agglomeration may form much earlier and faster, resulting in a high minimum torque. Of course, this phenomenon should also occur during the test process of Mooney viscosity. But due to the low test temperature and high shear rate, the filler agglomerates may be formed slowly and be broken down easily. Consequently, filler agglomeration is not able to be formed significantly during the test (preheating for 1 minute and then rotating for 4 minutes)[54].

In the cure curves of the carbon black-filled compound and EVEC-L, overcure does not result in a significant change in torque. The crosslinking rate of dry-mixed silica compounds increases after onset, reaches a certain level, and then decreases gradually; however, the torque is still rising over the time of the test (marching modulus). This is probably related to a certain amount of reactive chemical derived from silane coupling agent, i.e., polysulfide-containing chemical, still remaining in the rubber matrix. With prolonging cure time, the polymer will be continuously but slowly crosslinked with polysulfide, making the torque rise slightly. Silane coupling agents in EVEC-L compound are basically fully adsorbed on the filler surface, and thus there is no sulfur-donor in the rubber matrix. Once the crosslinking of the polymer by curatives is completed, the torque is stabilized.

If the above argument is reasonable, it seems that coupling agent adsorbed on the filler surface has no contribution to the crosslinking of rubber molecules during cure, or the sulfur-donor adsorbed on the filler surface has no effect on the torque.

### 9.4.3 Vulcanizate Properties

#### 9.4.3.1 Hardness of Vulcanizates

The hardness of EVEC-L is significantly lower than that of the equivalent dry-mixed vulcanizates and much lower than that of the carbon black-filled vulcanizate. As expected, the hardness of EVEC vulcanizate increases with increasing silica loading, but even with 6 phr more silica, the hardness is still lower compared with its dry-mixed counterpart (Table 9.4).

As mentioned in Section 6.2.2, hardness is related to Young's modulus, which is the slope at the origin of the tensile stress-strain curve (see Figure 6.7) and the slope is mainly determined by the filler agglomeration when the filler loading and filler morphology are similar. The implication of this is that the degree of filler agglomeration in carbon black-filled vulcanizate is the highest, followed by dry-mixed silica-filled vulcanizate, and that of EVEC is lowest. This is consistent with result of dynamic elastic-modulus which will be discussed later.

#### 9.4.3.2 Static Stress–Strain Properties

Apart from crosslink density and filler loading, the factors influencing the tensile stress of filled vulcanizates include filler morphology, filler agglomeration in the rubber matrix and polymer-filler interaction. When vulcanizates are stretched, filler agglomeration will be broken. As a result, the trapped rubber will be released and take part in the deformation of rubber matrix, leading to lower modulus. In other words, at very low elongations, the tensile stresses are much higher for vulcanizates with more developed filler agglomeration. This is the case at elongations lower than 50%, where the highest stress is observed for the dry-mixed silica vulcanizate without coupling agent, then carbon black and the dry-mixed silica compound with coupling agent. The tensile stress of EVEC-L is the lowest. Beyond 50% elongation, the rate of increase of stress with increasing elongation decreases as the breakdown of agglomerates occurs. As can be seen in Figure 6.7, when the elongation increases further to levels beyond 125%, the ranking of tensile stress at any given elongation is totally reversed: the tensile stress decreases in the order EVEC > dry-mixed silica vulcanizate with coupling agent > carbon black vulcanizate.

While filler agglomeration plays an important role in the tensile stress at low elongation, at high elongation where the agglomerates are almost totally broken down and the trapped rubber in the agglomerates is released, the main factor influencing stress of compounds, besides crosslink density, filler loading, and filler structure (concerning the occluded rubber in the filler aggregates), is polymer-filler interaction.

According to a model proposed by Dannenberg[55,56], when crosslinks are introduced in gum compounds, a random distribution of the chain lengths between crosslinks is formed. As a consequence, when the sample is subjected to the tensile test, the shortest or most strained, chains will break early in the test, followed by the next most strained ones, until at the moment just before break only a few chains effectively carry the load. In this case, the tensile strength should be lower. If the chains which are most strained are given a chance of slipping to relieve the tensions caused by stretching, added to those built in already, they will not break prematurely but will survive to the very moment before rupture. More chains then effectively carry the load, and a higher stress results.

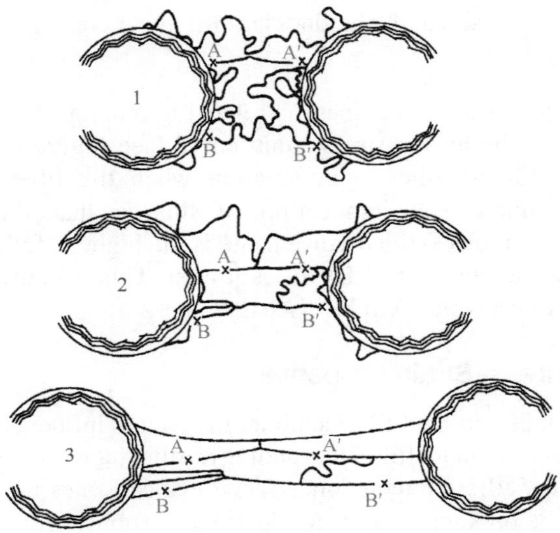


**Figure 9.23** Molecular orientation and slippage model during the tensile process of filled vulcanizates raised by Dannenberg[55,56]

For filled vulcanizates, the situation is somewhat complicated. A schematic picture of the slippage process is shown in Figure 9.23, which models three different distances between two filler particles in the direction of stress. As the stretching process proceeds from stage 1, the first chain slips at the points of connection A and A′ until the second chain is also taut between B and B′ (stage 2). Then, on continuing elongation, B and B′ start to slip there so that finally a stage 3 is reached in which all three chains are stretched to their maximum and share the imposed load equally.

It is obvious from this model that the high stress is mainly due to the energy necessary for slippage. Therefore, with increasing elongation the stress increase will be determined by the sites of polymer-filler interaction. For the silica compound without coupling agent the polymer-filler interaction is the lowest; therefore, the tensile stress should be low. The stress at a given elongation would increase more rapidly for coupled silica, especially EVEC, due to the higher polymer-filler interaction (Figure 6.10).

When elongation continues to increase, the high stress can also lead to slippage and even detachment of rubber chains from the filler surface. This may result in vacuole formation[57] and elimination of local stress concentration in the rubber network. On the other hand, the detachment of rubber chains from the filler surface should give lower stress at high elongation. This is why the carbon black vulcanizate has significantly higher stress at higher elongations, such as 300%. In the case of TESPT-modified silica vulcanizates, a certain amount of rubber chains are fixed chemically onto the filler surface and could not slip, thus the tensile stress is even higher than that of carbon black vulcanizate (Figure 6.10). On the other hand, the number of stress concentration points in the rubber matrix and on the filler surface increases extensively at high elongation. Compared with the dry-mixed silica compound containing TESPT, EVEC shows much higher stress at high elongations, such as 300%, which is certainly related to the higher efficiency of the coupling reaction. It may also be related to the even distribution of covalent bonds between polymer chains and filler over the filler surface. In addition, the excellent micro-dispersion of the filler aggregates in the polymer matrix leads to more chains effectively carrying the load without rupture and/or detachment from the filler surface. For the dry-mixed silica compound, due to the imperfection of coupling and filler agglomeration, detachments and vacuolation would take place easily, resulting in lower stress at higher elongation. This is demonstrated by the SEM observation of the stretched specimens (Figure 9.24). Almost no vacuoles were observed at a stress of 17.3 MPa and elongation of 330% for the EVEC vulcanizate, a significant number of holes were observed in the dry-mixed silica counterpart even at a stress of 12.2 MPa and elongation of 300%.

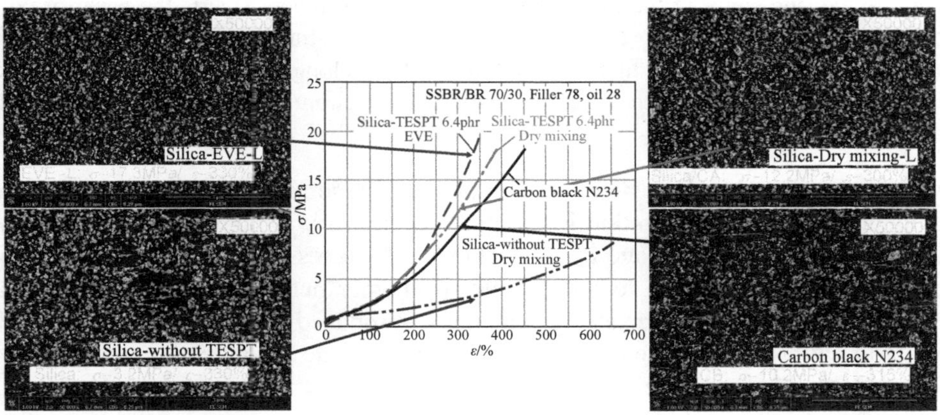


**Figure 9.24** SEM of the tread vulcanizates containing silica at different strains from 130% to 480%

As discussed in Section 6.2.3, the factors that affect the tensile stress at high elongations, such as crosslink density, filler loading, and filler morphology, also have the same effects on the tensile stress at low and elongations. Since the effect of filler agglomeration on stress-strain behavior has been excluded at low elongations, such as 100%, the ratio of tensile stress at 300% elongation (T300) to that at 100% elongation (T100) has been taken as a measure of the polymer-filler interaction (also see Section 6.2.3). Accordingly, we can conclude that the polymer-filler interactions in the filled vulcanizates increase in the order of silica without TESPT < carbon black < dry-mixed silica with TESPT < EVEC.

#### 9.4.3.3 Tensile Strength and Elongation at Break

The tensile strength of the EVEC vulcanizate is higher than those of the other vulcanizates (Table 9.4). This should be firstly attributed to the excellent macro-dispersion of the silica (see Figure 4.12) as the tensile strength is an ultimate property in which failure is initiated from small flaws such as undispersed fillers. Secondly, the excellent micro-dispersion of the filler, illustrated by TEM (Figure 9.27), also provides homogeneous stress distribution throughout the polymer which allows more polymer chains to effectively carry the load, giving the high strength.

Similar to the mechanism described for stress at high elongations, when the vulcanizate is stretched to a certain extent, rubber chains align along the stress direction in the rubber matrix. If the filler-polymer interaction is weak, slippage or detachment of polymer chains from the filler surface will release the stress concentration. This would lead to a more even stress distribution in the polymer network, avoiding rubber rupture at lower elongations. As the elongation continuously increases, more and more detached polymer chains from the filler surface can occur, resulting in vacuolation and fiber-like alignment of the polymer network. Under this condition, while the elongation may reach a very high level without break, the tensile stress at the given elongation is much lower. This is the case for silica compounds without coupling agent which can be demonstrated by SEM images (Figure 9.24). This could also account for the relatively high elongation at break of the carbon black-filled vulcanizate, since the polymer-filler interaction of carbon black is generally stronger than that of silica.

Due to chemical bonding between polymer and filler for TESPT-modified silica compounds, stress concentration in the polymer network at higher elongation cannot be easily released by slippage and detachment. As a result, the stress in the rubber network rises steeply. Due to the restrained polymer network, along with a certain degree of flaws or undispersed agglomerates introduced through silica mixing, the elongation at break is of the dry-mixed silica vulcanizate is lower compared with the carbon black-filled vulcanizate.

Unlike tensile strength, the EVEC vulcanizate gives the lowest elongation at break. The root cause may also be its stronger filler-polymer interaction resulting from the high efficiency of coupling reaction and better macro-dispersion and micro-dispersion of the fillers. For this vulcanizate, the slippage and detachment of the polymer chains is significantly reduced and the stress concentration in the polymer network cannot be released, resulting in lower elongation at break, even though the tensile stress at high elongation is considerably higher.

#### 9.4.3.4 Tear Strength

Compared with NR and IR, the SBR and BR polymers used in passenger car tires have inherently lower tear strength. As stress is applied to a crack or defect, an orientated structure (crystallization for strain-crystallizing rubber like NR and IR, or alignment of molecules for amorphous rubber such as SBR and BR) is created in the region of the tear tip, which is associated with large local deformations. For carbon black-filled rubber, the alignment of polymer chains can be facilitated in the tear tip by slippage and detachment of the polymer chains from filler surface, due to the relatively low polymer-filler interaction, which is physical in nature. This oriented structure is formed parallel to the direction of applied stress, and perpendicular to the direction of tear growth, leading to higher tear energy. For the silica-silane coupled vulcanizates, the polymer-filler interaction is mainly from covalent bonds that fix the polymer chains on the filler surface chemically. This would limit the orientation of polymer chains in the tear tip area, leading to a lower tear strength. The effect is more developed for the EVEC vulcanizates since the coupling reaction is more efficient (Table 9.4).

### 9.4.3.5 Dynamic Properties

#### 9.4.3.5.1 Strain Dependence of Dynamic Properties

#### (a) Elastic Modulus

Figure 9.25 the "Payne effect"[58]. Compared with the carbon black-filled vulcanizate, the silica compounds show lower Payne effects due to the coupling reaction of silica with polymers, which reduces the difference in surface energy between filler and polymer, and increases polymer-filler interaction. The higher efficiency of surface modification of silica in the EVE process further depresses filler agglomeration, which can be demonstrated by TEM images (Figure 9.26 and Figure 9.27). Obviously, while silica aggregates are dispersed uniformly in the EVEC, agglomeration is more developed in the dry-mixed silica vulcanizate. This is consistent with the Young's modulus, which is also determined by filler agglomeration.

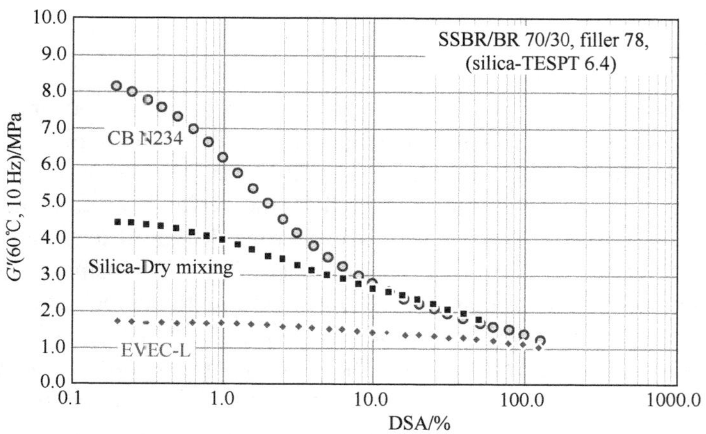


**Figure 9.25** Elastic Modulus *G'* and Payne effect of different tread vulcanizates

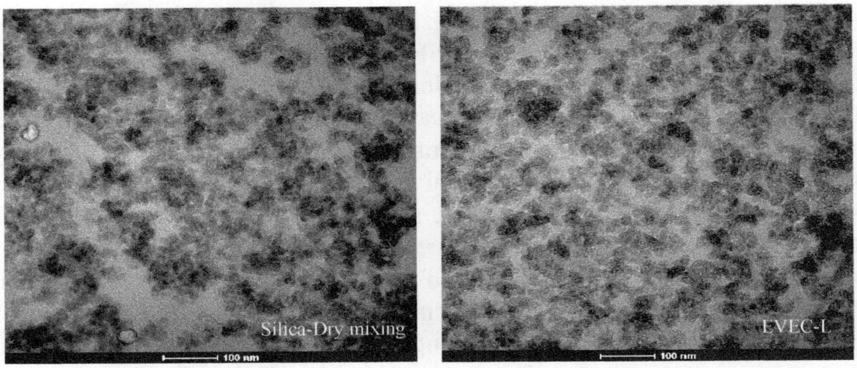


**Figure 9.26** HR-TEM of silica-containing vulcanizates by dry mixing and EVEC-L mixing (50,000-fold)

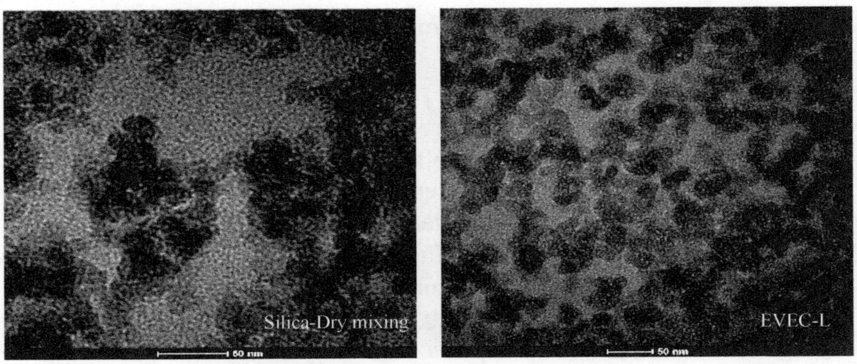


**Figure 9.27** HR-TEM of silica-containing vulcanizates by dry mixing and EVEC-L mixing (100,000-fold)

## (b) Viscous Modulus

Strain dependence of $G''$ for several vulcanizates is illustrated in Figure 9.28. Generally, $G''$ increases with strain at low amplitude, goes through the maximum in the region of 2%–3% strain, and then declines rapidly. It is believed that the loss modulus is dependent on energy consumption arising from the cyclic disruption and reformation of filler agglomerates[58,59]. Less breakdown and reformation of agglomerates at small strain will result in low energy loss, thus lower $G''$. With increasing strain amplitude, more agglomerates will be broken and reformed, leading to augmentation of $G''$. When the strain amplitude further increases, the disruption of agglomerates is so severe that there is not enough time for reformation. Therefore, $G''$ will decrease due to the decline in energy dissipation. Accordingly, vulcanizates having more developed filler agglomeration would give higher $G''$ over the dynamic strain amplitudes used. This explains why the carbon black vulcanizate gives highest $G''$ at 60°C and the $G''$ of EVEC is the lowest.

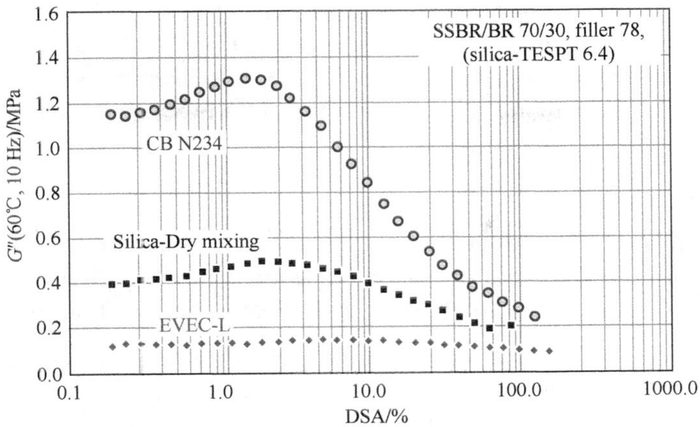


**Figure 9.28** Viscous Moculus $G''$ of different tread vulcanizates

### (c) tan $\delta$

By definition, tan $\delta$ is the ratio of the viscous modulus to the elastic modulus. As discussed in Section 8.1.1.1, there is a good correlation between tan $\delta$ at 60°C and rolling resistance of tire. At this temperature, the polymer is in its rubbery state with higher entropic elasticity and lower hysteresis, so energy dissipation is mainly related to the change of filler agglomerate structure during periodic strain. It can be expected, therefore, that under these conditions, the level of filler agglomeration determines the level of tan $\delta$[43]. Shown in Figure 9.29 is the strain dependence of tan $\delta$ for several compounds measured at 60°C and 10 Hz. According to the degree of filler agglomeration discussed above, carbon black vulcanizates should therefore have the highest tan $\delta$, and then the conventionally mixed silica compound with TESPT. The EVEC vulcanizates give the lowest hysteresis. This is consistent with the Payne effect.

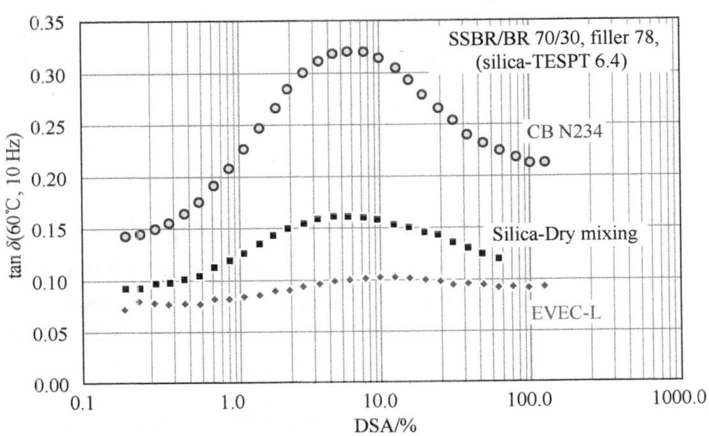


**Figure 9.29** Loss factor tan $\delta$ of different tread vulcanizates

Practically, the rolling resistance of tire can be estimated using the maximum values of tan $\delta$ (tan $\delta_{max}$) in the strain sweep. The comparison of tan $\delta_{max}$ is shown in Table 9.4 for these vulcanizates. It can be seen that tan $\delta_{max}$ of dry-mixed silica compound is 48% lower than that of the carbon black compound, and EVEC-L gives 20% lower tan $\delta_{max}$ compared to its dry-mixed counterpart.

### (d) Loss Compliance

The loss compliance $J''$ is defined as $J'' = G''/G^{*2}$. The effect of material stiffness on the energy loss is emphasized in loss compliance, i.e., the higher the stiffness, the lower the energy loss under constant stress in a dynamic deformation (see Eq. 7.12).

Figure 9.30 shows that $J''$ measured at 60°C and a deformation frequency of 10 Hz generally increases with increasing strain. The carbon black compound possesses the lowest $J''$, while that of EVEC is the highest under very low strain (<1%). However, with increasing strain, $J''$ of the carbon black compound rises faster and reaches a higher level than that of EVEC-L at high strain, showing a crossover point at a strain amplitude around 2%. The EVEC vulcanizate, on the other hand, gives higher $J''$ compared with its dry-mixed silica counterpart, mainly due to lower stiffness, but the difference in the $J''$ between the two silica vulcanizates gets smaller with increasing strain amplitude.

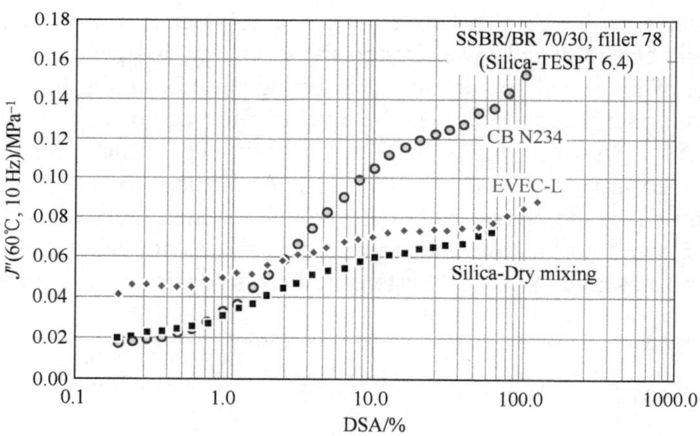


**Figure 9.30** Loss compliance $J''$ of different tread vulcanizates

#### 9.4.3.5.2 Temperature Dependence of Dynamic Properties

Temperature dependence of $G'$, $G''$, tan $\delta$, and $J''$ was measured at a frequency of 10 Hz. It can be seen from Figure 9.31 and Figure 9.32 that the relative ratings of $G'$ and $G''$ of the three vulcanizates are similar: the carbon black compound gives the highest value, followed by the dry-mixed silica compound, and the EVEC-L has the lowest value.

With regard to tan $\delta$, EVEC-L gives the highest values at low temperature, while the carbon black vulcanizate has the lowest. The reverse is true at higher temperature, with a crossover point around 8°C for this particular polymer system (Figure 9.33). This phenomenon has been attributed to filler agglomeration[43] and has been discussed in detail in Section 7.2.5. It is generally believed that the wet skid resistance property of tires is related to tan $\delta$ of the tread compound at low temperature (around 0°C). The higher the tan $\delta$, the better is the performance of wet skid resistance. From this point of view, a silica compound, especially EVEC, has some advantages, although the factors that influence the wet skid resistance are more complicated (see Section 8.2). The loss compliances $J''$ of all the vulcanizates increase gradually with temperature and reach inflexion points at temperatures around −3°C to −10°C (Figure 9.34). Above these temperatures, $J''$ of carbon black compound still rises slowly, while that of EVEC drops slightly with the dry-mixed silica vulcanizate being in the middle. The loss compliance is highest for the EVEC compound over the whole range of temperature tested, but above 30°C, the $J''$ of the carbon black-filled vulcanizate is higher than that of the conventional silica vulcanizate.

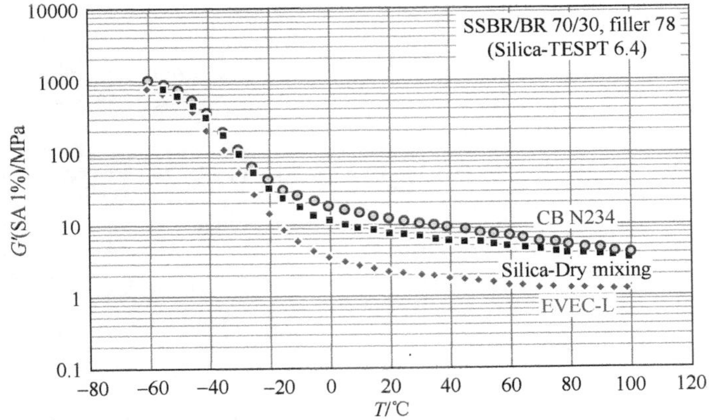


**Figure 9.31** Temperature dependence of $G'$ for different tread vulcanizates

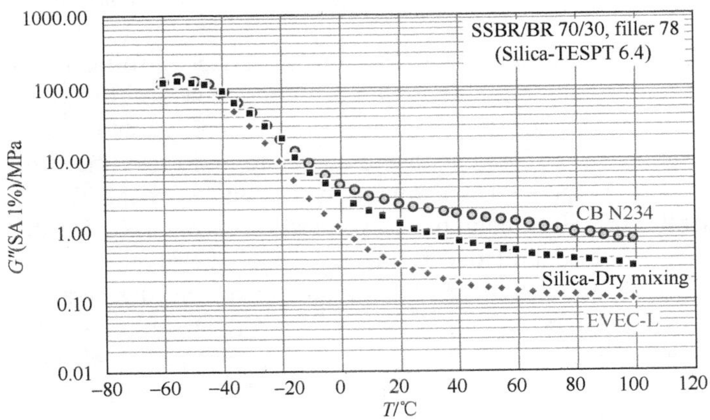


**Figure 9.32** Temperature dependence of $G''$ for different tread vulcanizates

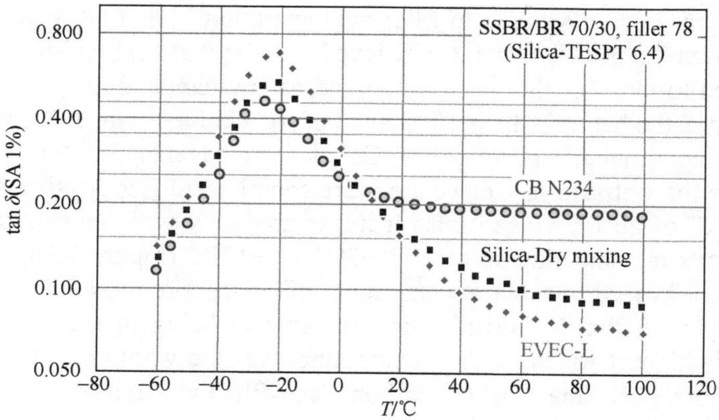


**Figure 9.33** Temperature dependence of tan $\delta$ for different tread vulcanizates

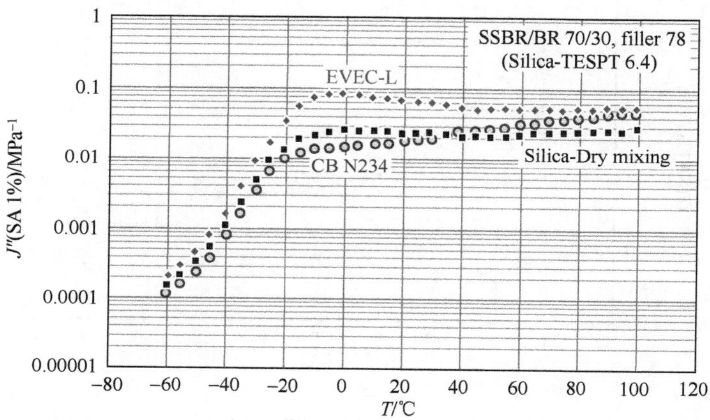


**Figure 9.34** Temperature dependence of $J''$ for different tread vulcanizates

The differences in $J''$ between the vulcanizates is probably also mainly related to filler agglomeration. According to the definition of $J''$ and the effect of filler agglomeration on the elastic and viscous moduli, hence tan $\delta$ of rubber, $J''$ should be also related to filler agglomeration. At moderate to low temperature in the transition zone, while tan $\delta$ of the carbon black compound is the lowest, the $G'$ and $G''$ are the highest, and so is the $G^*$. Consequently, the black compound possesses the lowest $J''$. On the contrary, the tan $\delta$ of EVEC is the highest due to the least agglomeration, while $G'$ and $G''$ are relatively low. This leads to the EVEC compound having the highest $J''$. At a high temperature in the rubbery region, $G^*$ declines at a relatively high rate in the carbon black vulcanizate because of its higher hysteresis and higher break rate of agglomerates. As a result, $J''$ keeps rising slowly. With regard to EVEC, $G^*$ declines somewhat slowly due to the lower level of agglomeration, and so does $J''$. Another reason for the slight drop of $J''$ for EVEC maybe that, according to the entropic elasticity theory, due to less rubber being trapped in filler agglomerates, the modulus of un-trapped rubber will rise with temperature, resulting in lower $J''$.

The relatively high $J''$ of the EVEC compound at low temperature may be beneficial to the skid resistance and handling performance of the tire. Generally, the skid resistance is determined by the friction between the tire and pavement. As discussed in Section 8.2.2, the friction force $F$ or the friction coefficient $\mu$ between rubber and the dry solid surface is mainly determined by two components: the hysteresis component and the adhesion component. According to Moore's theory[60], the hysteresis component of friction coefficient $\mu_h$ is expressed by Eq. 8.5, which may be roughly re-written as:

$$\mu_h \cong K\left(p^n / E'^{n-1}\right) J'' \tag{9.1}$$

where $n \geqslant 1$, $K$ is a constant, $p$ is the average pressure on every convex of the solid surface, and $E'$ is the elastic modulus. The complex modulus in Eq. 8.5 is replaced by the elastic modulus.

The adhesion component of friction coefficient $\mu_a$ is expressed by Eq. 8.4[61], which may also be approximately given by:

$$\mu_a \cong K' \sigma_{ms} J'' \tag{9.2}$$

where $K'$ is a constant and $\sigma_{ms}$ is the maximum stress able to be generated on the elemental area. In this equation, the hardness of rubber in Eq. 8.4 is replaced by complex modulus in Eq. 9.2. It can be seen that both components of friction coefficient increase with the loss compliance. Since Eqs. 8.4 and 8.5 were derived based on a dry solid surface, it should be noted that friction involves a very high frequency of deformation from about $10^5$ Hz to $10^8$ Hz. Therefore, it is possible to estimate the dry skid resistance and handling performance of a tire in the laboratory at low temperature (such as 0–20°C) according to WLF time-temperature equivalent principle of rubber. The higher the $J''$, the better is the dry skid resistance and handling performance. In other words, EVEC tread compound would be expected to give better performances on a dry road surface.

With regard to wet skid resistance, in addition to the hysteresis of the tread compound at low temperature, other factors also exert an influence, such as hydrodynamic lubrication of water, especially in the micro-elasto-hydrodynamic lubrication scale[38,39].

As discussed in Section 8.2.3, when a vehicle brakes on a wet road surface, the tire tread contact area can be divided into three regions: the front is a squeeze-film zone, the rear is the traction zone, with the transition zone in between. In the squeeze-film zone, the friction mechanism is elasto-hydrodynamic lubrication, which gives a very low friction coefficient. In the rear traction zone, where boundary lubrication is dominant, the main wet traction of the tire is provided. The friction coefficient in the transition zone is in the range from squeeze-film zone to traction zone. Therefore, the most effective way to improve the wet skid resistance of a tire is to reduce the length of squeeze-water film and transition zones, and hence to increase the traction zone where most frictional force is generated. In addition, an increase of friction coefficient in the traction zone is also beneficial. It has been known that, in the traction zone, the friction coefficient of boundary lubrication is the same as, or even higher than, that of the dry road surface (Eq. 8.9). According to Eqs. 9.1 and 9.2, $J''$ of vulcanizates is also one of the most important parameters that impact wet skid resistance. On the other hand, the ambient temperature involved in wet skid resistance is relatively low compared to dry skid resistance. Due to the very high frequencies of strain involved in the wet skid process, if $J''$ is taken as a measure of wet skid resistance, the test temperature of the materials should be lower, such as below 0°C. As can be seen from Figure 9.34, $J''$ of EVEC is the highest in low-temperature region, and that of conventional black compound is the lowest. Therefore, considering only the dynamic properties, the best wet skid resistance would be expected for EVEC tread compound.

#### 9.4.3.5.3 Rebound and Heat Build-up

The carbon black vulcanizate has the lowest rebound at both room temperature and 60°C, followed by the dry-mixed silica vulcanizate with TESPT (Table 9.4). EVEC gives the highest rebound, especially when the filler loading is lower.

The heat build-up test is carried out on a Goodrich flexometer with a load 1 MPa, stroke 4.45 mm, and frequency 30 Hz. The bottom $\Delta T$ are shown in Table 9.4. Obviously, the heat build-up of the carbon black vulcanizates is the highest while EVEC gives the lowest, with dry-mixed silica in between.

### 9.4.3.6 Abrasion Resistance

The abrasion resistance index from an Akron abrader is also presented in Table 9.4. At the same filler loading, while abrasion resistance of the dry-mixed silica vulcanizate is much lower than that of the carbon black-filled compound, EVEC-L shows the highest abrasion resistance. Abrasion resistance is further improved by increasing filler content.

It is generally accepted that, besides the polymer composition, the abrasion resistance of tread compounds is mainly affected by filler loading, dispersion, morphology, and polymer-filler interaction[40]. Mainly due to the difference in their surface nature, carbon black-filled vulcanizates generally perform much better in abrasion than silica-filled vulcanizates. Although the abrasion resistance could be improved by usage of silane coupling agent, it is still inferior to its carbon black counterpart when mixed using dry mixing technology. With the same filler content, EVEC vulcanizates show much higher abrasion resistance than chemically modified silica compounds, and even better than conventional carbon black vulcanizates. Besides the excellent filler dispersion mentioned above, it is also related to the improvement of polymer-filler interactions. Since interactions between polymer and silica surface mainly depend upon covalent bonds generated by the silane coupling reaction, this interaction in EVEC is greatly improved due to the highly effective silanization caused by liquid phase mixing. This can be demonstrated by the SEM observation on stretched samples (see Figure 9.24), and verified by T300/T100. Compared with traditional silica-filled vulcanizates, T300/T100 of EVEC-L is increased by 38%. It should be pointed out that, because of different conditions and severities involved in laboratory abrasion tests and tire wear tests, there is no obvious correlation between laboratory data and result of tread wear on the road. However, several road tests have shown that the order of tread wear resistance of the compounds shown in the Table 9.3. It is the same as that of the abrasion resistance measured with Akron abrader (see Table 9.4), but the differences among the three types of compounds are even larger.

# ■ 9.5 Powdered Rubber

The development of filler/rubber masterbatches in a free-flowing powder form was first carried out in the 1970s with the goal of simplifying the production of rubber compounds and reducing compounding cost[62]. This technology has been further developed in the 1990s, and like CEC, it has enabled some new levels of performance to be achieved with conventional types of fillers. Powdered rubber technology is also well suited to continuous mixing as well as batch mixing processes.

## 9.5.1 Production of Powdered Rubber

Production process is basically a batch process, which is presented in Figure 9.35[63]. First, a stable polymer latex (either NR latex, emulsion polymerized polymer latexes, or emulsionized solution-polymerized polymers) is prepared. A filler suspension that has previously been exactly adjusted according to the particular particle size of the filler is then added, together with various additives. The mixture of the latex and filler suspension is homogenized under intensive stirring in an agitator tank. Co-precipitation of the mixture is achieved using various chemicals that influence the pH of the suspension, causing the coagulating latex to surround each individual filler particle. A separate filler fraction is added to form an efficient separating layer around each powdered rubber particle so that, after precipitation, the particles do not stick together. If necessary, the product in water can be subjected to a maturing process of several hours in a homogenization tank. Then, most of the water is removed in a centrifuge, after which the finely divided batch is further dried down to a moisture content of under 1%. The products end up in a free-flowing form after drying.

## 9.5.2 Mixing of Powdered Rubber

Powdered rubber can be mixed with other compounding ingredients using typical batch mixing processes (internal mixer or open mill) or by continuous mixing processes. In the internal mixer, it is suggested that the other ingredients are best added in a modified upside-down process. Because a high quality of dispersion can be obtained after a very short mixing time, it is suggested that the mixing stages and time can be reduced.

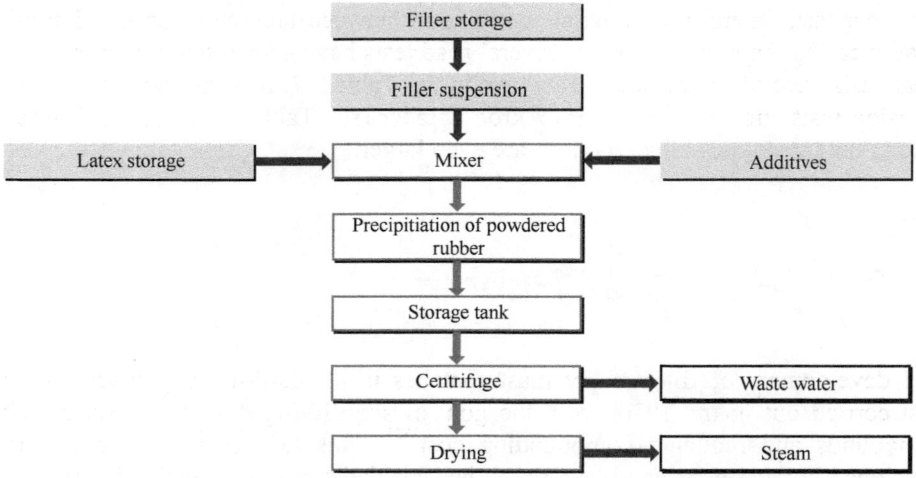


**Figure 9.35** Flow diagram of powdered rubber production

Due to the free-flowing form of the masterbatch, the powdered rubbers are especially suitable for the continuous mixing processes that have long been applied to thermoplastic materials. With powdered rubber, all compound ingredients can be fed into a continuously operating extruder mixer that can achieve enough shear force to equal the shear condition of a mill or internal mixer over a relatively short time.

## 9.5.3 Properties of Powdered Rubber Compounds

### Carbon Black-Filled Powdered Rubber

Compared with conventionally mixed compounds, the properties of this compound are as follows: The emulsion SBR/carbon black powdered rubber gives better filler dispersion and higher bound rubber. The die swell at different shear rates is also lower. With regard to the physical properties of vulcanizates, slightly better or comparable ultimate properties were observed in the carbon black powdered rubber compounds, with almost equal rebound, hysteresis, and hardness. The DIN abrasion was somewhat improved, which is consistent with good filler dispersion[63,64].

In a natural rubber formulation, comparable compound viscosity was achieved at 3 phr lower filler loading compared to the standard mixed counterpart. The powdered rubber material gave similar vulcanizate stress-strain properties, higher ball rebound, and lower hysteresis, but its abrasion resistance has not been reported.

### Silica-Filled ESBR Powdered Rubber

When all-silica tread compounds were introduced to passenger car tires, one of the disadvantages was their poor processability, especially related to mixing. Another feature of these compounds was the application of solution-polymerized polymers, i.e.,

SSBR and BR[65]. Emulsion SBR has not been used much in all-silica tread compounds due to the high concentration of surfactants and other polar substances remaining in the polymer from its production. These small chemicals may interfere with the coupling reactions between silane and polymer chains, resulting in a poor polymer-filler interaction, hence unsatisfactory tread wear resistance[66,67]. By using the powdered rubber process, it is possible to complete the reaction of the silane coupling agent with the silica before it is exposed to the emulsion polymer, thus overcoming this problem. In practice, silica and silane are brought together during the production of the silica suspension in water. Since the two materials have very different polarities, a homogeneous phase that is essential for the silane attachment on the silica surface is not formed. Therefore, so-called phase mediators are also added in the suspension. After the rubber emulsion is added, the silica/silane/rubber system is coagulated at room temperature using acids. This process takes place exclusively in the aqueous phase, i.e., in a single-phase system. During coagulation, the silica treated with silane is enveloped by the polymer. After coagulation, the process water is separated mechanically and the product is dried. During the drying process, the organosilane is chemically reacted on the silica surface with ethanol being split off[68]. The silica/silane/ESBR powdered rubbers can be mixed with either conventional internal mixers or a continuous mixer. In both cases, the mixing procedures can be significantly simplified. When the materials are compared with their conventionally mixed counterparts, filler dispersion is improved, the Mooney viscosities of the compounds are higher, and die swell is smaller. For the physical properties of the vulcanizates, the powdered rubber is characterized by higher tensile strength and stress at 300% elongation, somewhat lower hysteresis, and significantly improved DIN abrasion. However, no comparison was reported with dry-mixed SSBR compounds, which are commonly used in silica tires.

## ■ 9.6 Masterbatches with Other Fillers

### 9.6.1 Starch

In the late 1990s, Goodyear introduced a new line of tires, whose tread formulation contained a certain amount of a starch-based filler material called BioTred[69]. The starch filler, in the form of a starch/plasticizer composite, is applied with a fairly high dosage of organosilane coupling agent to improve polymer-filler interaction.

#### Structure and Features

Starches, $(C_6H_{10}O_5)_n$, are natural products made of a mixture of two polymers: amylose, composed of long straight chains containing 200–1000 glucose units linked by $\alpha$-1,4-glycoside links, and amylopectin, consisting of comparatively short-chained molecules of about 20 glucose units cross-linked by $\alpha$-1,6-glycoside links (Figure

9.36). Most starch materials have about 25% linear and 75% branched/cross-linked molecules[70].

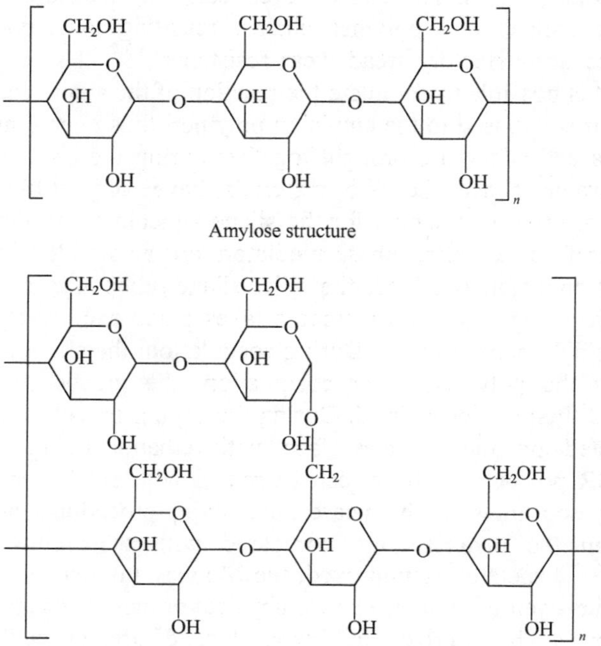


**Figure 9.36** Starch structure of amylose and amylopectin

Starches exist in nature in the form of granules with a quasi-crystalline structure. The size of starch granules varies in diameters from 1 μm to 150 μm, depending on the sources from which the starch is produced. Cornstarch, the main commercially available starch material, which was used in the new tires, has granules with a size from 5 μm to 26 μm with an average diameter of 15 μm.

Due to a very high softening point (200°C or above) and a large granule size, starch by itself cannot be mixed directly into rubber compounds using a conventional mixing facility, since the typical temperature for compound mixing is in the range of about 150°C to 170°C. The mixing temperature is not sufficient to cause starches to melt and to blend uniformly with the rubber composition[71]. A procedure was therefore developed to mix starch first with polymeric plasticizers and other ingredients in a separate operation to form a starch composite so that the starch filler can be incorporated with other rubber compound ingredients using conventional mixers[72]. The polymeric plasticizers used include poly(ethylene-vinyl alcohol), ethylene-glycidyl acrylate, vinyl acetate, and cellulose that is compatible with the starch. It is believed that the strong chemical and/or physical interactions between the starch and the plasticizer result in the reduction of the softening point of the starch composite such that it can be well mixed with polymer and other ingredients for rubber compounding[71,73].

It is should be pointed out that in tire tread compounds, starch cannot be used as a principal filler. It was reported that a small quantity of a starch composite silane coupling agent does not bring about any significant performance advantages over carbon black-based compounds. The use of the silane coupling agent is believed to give better dispersion of the fillers and to reduce the filler agglomeration of both starch and base filler in the compounds. With a high level of the silane coupling agent, the reduction in tan $\delta$ at high temperature is about 21% relative to the compound filled with pure carbon black.

It was claimed that tire rolling resistance was significantly reduced when part of carbon black or silica, for example 30%, was replaced with the starch/plasticizer composite in tread compounds. Results were explained in terms of lowering compound hysteresis and reduction of tire weight. Some enhanced wet skid resistance, without compromise in tread wear, was also claimed in comparison to conventional carbon black-based tires[73].

### 9.6.2 Organo-Clays

There has been considerable recent research interest in clays[74–78], particularly the montmorillonite type, which have been cation exchanged with organic ammonium salts, including ammonium terminated liquid rubbers[79,80]. These materials are sometimes termed nano-clays. The cation exchange reaction makes the clay organophilic, enhancing dispersion in organic polymers, and also facilitates exfoliation of the silicate layers. The result is a potentially very high aspect ratio filler. Filler agglomeration can therefore be achieved at much lower loading levels than are required for carbon black or silica. In rubber reinforcement applications, increased tensile properties at equivalent loadings have been observed compared to conventional fillers. Another advantage of this type of filler is the low gas permeability, which could be useful for tire inner liners, for example. On the negative side, the organo-clays tend to give high hysteresis and loss tangent, and often high permanent set. Research is continuing, particularly through the use of coupling agents, and blending with other fillers, with the aim of maintaining the advantages of the organo-clays and overcoming their disadvantages.

## References


[1] Wang M-J, Morris M. Rubber Technologist's Handbook, Volume 2. UK: Smithers Rapra, 2009.

[2] Wolff S, Görl U. Carbon Blacks Modified with Organosilicon Compounds, Method of Their Production and Their Use in Rubber Mixtures. US Patent, 5159009, 1992.

[3] Keoshkerian B, Georges M K, Drappel S V. Ink Jettable Toner Compositions and Processes for Making and Using. US Patent, 5545504, 1996.

[4] Catia B, Vittorio B, Gianfranco D T. Polymer Composition Including Destructured Starch and an Ethylene Copolymer. US Patent, 5409973, 1995

[5] Joyce G A, Little E L. Thermoplastic Composition Comprising Chemically Modified Carbon Black and Their Applications. US Patent, 5708055, 1998.

[6] Belmont J A. Process for Preparing Carbon Materials with Diazonium Salts and Resultant Carbon Products. US Patent, 5554739, 1996.

[7] Belmont J A, Amici R M, Galloway C P. Reaction of Carbon Black with Diazonium Salts,Resultant Carbon Black Products and Their Uses. US Patent, 5851280, 1998.

[8] Wang M-J. New Developments in Carbon Black Dispersion. *Kautsch. Gummi Kunstst,* 2005, 58(12): 626.

[9] Tokita N, Wang M-J, Chung B, et al. Future Carbon Blacks and New Concept of Advanced Filler Dispersion. *J. Soc. Rubber Ind.*, 1998, 71(9): 522.

[10] Payne A R. The Dynamic Properties of Carbon Black-loaded Natural Rubber Vulcanizates. Part I. *J. Polym. Sci.*, 1962, 6(19): 57.

[11] Payne A R, Whittaker R E. Low Strain Dynamic Properties of Filled Rubbers. *Rubber Chem. Technol.*, 1971, 44(2): 440.

[12] Mahmud K, Wang M-J, Francis R A. Elastomeric Compounds Incorporating Silicon-Treated Carbon Blacks. US Patent, 5830930, 1998.

[13] Mahmud K, Wang M-J, Francis R A. Elastomeric Compounds Incorporating Silicon-treated Carbon Blacks and Coupling Agents. US Patent, 5877238, 1999.

[14] Mahmud K, Wang M-J. Method of Making a Multi-phase Aggregate Using a Multi-stage Process. US Patent, 6211279, 2000.

[15] Mahmud K, Wang M-J. Method to Improve Traction Using Silicon-treated Carbon Blacks. US Patent, 5869550, 1999.

[16] Mahmud K, Wang M-J. Method of Making a Multi-phase Aggregate Using a Multi-stage Process. US Patent, 6364944, 2002.

[17] Mahmud K, Wang M-J, Francis R A. Method of Making a Multi-phase Aggregate Using a Multi-stage Process. US Patent, 5904762, 1999.

[18] Wang M-J, Kutsovsky Y, Zhang P, et al. New Generation Carbon-silica Dual Phase Filler Part I. Characterization and Application to Passenger Tire. *Rubber Chem. Technol.*, 2002, 75(2): 247.

[19] Murphy L J, Wang M-J, Mahmud K. Carbon-Silica Dual Phase Filler: Part V. Nano-Morphology. *Rubber. Chem. Technol.*, 2000, 73(1): 25.

[20] Wang M-J, Kutsovsky Y, Zhang P, et al. Using Carbon-silica Dual Phase Filler-Improve Global Compromise Between Rolling Resistance, Wear Resistance and Wet Skid Resistance for Tires. *Kautsch. Gummi Kunstst.*, 2002, 55(1-2): 33.

[21] Wang M-J, Tu H, Kutsovsky Y, et al. Carbon-Silica Dual Phase Filler, A New Generation Reinforcing Agent for Rubber: Part VIII. Surface Characterization by IGC. *Rubber. Chem. Technol.*, 2000, 73(4): 666.

[22] Wang M-J, Mahmud K, Murphy L J, et al. Carbon-silica dual phase filler, a new generation reinforcing agent for rubber-Part I. Characterization. *Kautsch. Gummi Kunstst.*, 1998, 51(5): 348.

[23] Wang M-J. Effect of polymer-filler and filler-filler interactions on dynamic properties of filled vulcanizates. *Rubber Chem. Technol.*, 1998, 71(3): 520.

[24] Knill R B, Shepherd D J, Urbon J P, et al. *Proceedings of the Intemational Rubber Conference*, 1975.

[25] Mahmud K, Wang M-J. Reznek S R, et al. Elastomeric Compounds Incorporating Partially Coated Carbon Blacks. US Patent, 5916934, 1999.

[26] Wang M-J, Zhang P, Mahmud K. Carbon-silica Dual Phase Filler, a New Generation Reinforcing Agent for Rubber: Part IX. Application to Truck Tire Tread Compound. *Rubber. Chem. Technol.*, 2001, 74(1): 124.

[27] Wang M-J, Wang T, Wong Y L, et al. NR/Carbon Black Masterbatch Produced with Continuous Liquid Phase Mixing. *Kautsch. Gummi Kunstst.*, 2002, 55(7-8): 388.

[28] Gregiry M J, Tan A S. Proc. Int. Rubber Conf. 1975, 4:28.

[29] Leblanc J L. *IRC'98*, 1998.

[30] Westlinning H, Kautschuk V F. *D.K.G*, 1962.

[31] Smit P P A. Le Renforcement des Élastomères et les Surfaces Solides Ayant Une Action Reforçante. *Colloques Internationaux du la CNRS*, 1975: 231.

[32] Chin P S. Viscosity Stabilized Heveaerumb., *J. Rubber Res. Inst.* 1969, 22(1):56.

[33] Nortey N O. Enhanced Mixing in the Intermeshing Batch Mixer. *Meeting of Rubber Division*, 1998: 70.

[34] Wang T, Wang M-J, Shell J, et al. The Effect of Compound Processing on Filler Flocculation[J]. *Kautsch. Gummi Kunstst.*, 2000, 53: 497.

[35] Caruthers J M, Cohen R E. Effect of Carbon Black on Hysteresis of Rubber Vulcanizates: Equivalence of Surface Area and Loading. *Rubber Chem. Technol.*, 1976, 49: 1076.

[36] Roland R. Rubber Compound and Tires Based on such a Compound. Europe Patent, 0501227A1, 1992.

[37] Roland R. Copolymer Rubber Composition with Silica Filler, Tires Having a Base of Said Composition and Method of Preparing Same. US Patent, 5227425,1993.

[38] Wang M-J, Kutsovsky Y. Effect of Fillers on Wet Skid Resistance of Tires. Part Ⅰ: Water Lubrication *vs.* Filler-Elastomer Interactions. *Rubber Chem. Technol.*, 2008, 81:552.

[39] Wang M-J, Kutsovsky Y. Effect of Fillers on Wet Skid Resistance of Tires. Part Ⅱ: Experimental Observations on Effect of Filler-Elastomer Interactions on Water Lubrication. *Rubber Chem. Technol.*, 2008, 81:576.

[40] Wang M-J. Effect of Filler-Elastomer Interaction on tire Tread Performance Part Ⅲ. *Kautsch. Gummi Kunstst.*, 2008, 61: 159.

[41] Wang M-J, Wolff S. Filler-Elastomer interactions. Part Ⅲ. Carbon-Black-Surface Energies and Interactions with Elastomer Analogs. *Rubber Chem. Technol.*, 1991, 64: 714.

[42] Wolff S, Wang M-J. Filler-Elastomer Interactions. Part Ⅳ. The Effect of the Surface Energies of Fillers on Elastomer Reinforcement. *Rubber Chem. Technol.*, 1992, 65: 329.

[43] Wollf S. *The International Conference of the Deutsche Kautschuk-Gesellschaft*, 1974.

[44] Wang M-J, Song J J, Dai D Y. Continuous Manufacturing Method for Rubber Masterbatch, Rubber Masterbatch Prepared by Using Continuous Manufacturing Method and Rubber Product. China Patent, 201310027024.1, 2013.

[45] Wang M-J, Song J J, Dai D Y. Continuous Manufacturing Process for Rubber Mastrbatch and Rubber Masterbatch Prepared Therefrom. China Patent, 201310338268.1, 2013.

[46] Wang M-J, Song J J, Dai D Y. Continuous Manufacturing Process for Rubber Mastrbatch and Rubber Masterbatch Prepared Therefrom. China Patent, 201310337560.1, 2013.

[47] Wang M-J, Song J J, Dai D Y. Continuous Manufacturing Process for Rubber Mastrbatch and Rubber Masterbatch Prepared Therefrom. China Patent, 201310337779.1, 2013.

[48] Wang M-J, Song J J, Dai D Y. Continuous Manufacturing Process for Rubber Mastrbatch and Rubber Masterbatch Prepared Therefrom. China Patent, 201310337559.9, 2013.

[49] Wang M-J, Song J J, Dai D Y. Continuous Manufacturing Process for Rubber Mastrbatch and Rubber Masterbatch Prepared Therefrom. China Patent, 201310337578.1, 2013.

[50] Wang M-J. *Tire Technology EXPO*, 2016.

[51] Wang M-J, Song J, Wang Z, et al. *The Fall 2019 196th Technical Meeting of the Rubber Division. ACS*, 2019.

[52] Wolff S, Wang M-J, Tan E-H. Surface Energy of Fillers and Its Effect on Rubber Reinforcement. Part 2, *Kautsch. Gummi Kunstst.*, 1994, 47(12): 873.

[53] Hofman W. Vulcanization and Vulcanizing Agents. London: Maclaren & Sons, 1967.

[54] Wang M-J. Filled Rubber *vs.* Gum. *12th Fall Rubber Colloquium, Hannover*, 2016:22.

[55] Dannenberg E M. Molecular Slippage Mechanism of Reinforcement., *Trans. Inst. Rubber Ind.*, 1966, 42: 26.

[56] Dannenberg E M. The Effects of Surface Chemical Interactions on the Properties of Filler-Reinforced Rubbers. *Rubber Chem. Technol.*, 1975, 48(3): 410.

[57] Bryant K C, Bisset D C. *Rubber Technol. Conf., 3rd Conf*, 1954: 655.

[58] Wang M-J. The Role of Filler Networking in Dynamic Properties of Filled Rubber. *Rubber Chem. Technol.*, 1999, 72(2): 430.

[59] G. Kraus, Mechanical Losses in Carbon Black Filled Rubbers.*, Appl. Polym, Appl. Polym. Symp.*, 39, 75 (1984).

[60] Moore D F. The Friction and Lubrication of Elastomers. Oxford: Pergamon Press, 1972.

[61] Bulgin D, Hubbard D G, Walters M H. *Proc. Rubber Technol. Conf. 4th*, 1962: 173.

[62] Nordseik K H Berg G. *Kautsch. Gummi Kunstst.*, 1975, 28: 397.

[63] Görl U, Nordseik, K H, Berg G. Rubber/Filler Batches in Powder Form- Contribution to the Simplified Production of Rubber Compounds., *Kautsch. Gummi Kunstst.*, 1998, 51: 250.

[64] Görl U, Schmitt M. *Rubber World*, Powder Rubber-a New Raw Material Generation for Simplifying Production., 2001, 224: 1.

[65] Roland R. Rubber Compound and Tires Based on such a Compound. Europe Patent, 0501227, 1992.

[66] Wang M-J. In Powders and Fibers: Interfacial Science and Applications, ed. M. Nardin and E. Papirer, Marcel Dekker Inc., 2006.

[67] Wang M-J, Zhang P, Mahmud K. Carbon-Silica Dual Phase Filler, a new Generation Reinforcing Agent for Rubber- Part IX. Application to Truck Tire Tread Compound. *Rubber Chem. Technol.,* 2001, 74(1): 124.

[68] Görl U. *Meeting of the Rubber Division*, ACS, 2002: 11.

[69] Trade Publication, Eur. Rubber J., 183, 34 (2001).

[70] Whistler R L,Daniel J R in Encyclopedia of Chemical Technology, 4th Ed. Vol. 22, p. 699, John Wiley and Sons, New York, 1997.

[71] Corvasce F G, Linster T D, Thielen G. Europe Patent: 795581B1, 2001.

[72] C. Bastioli, V. Bellotti, and G. Del Tredici, inventors; Butterfly S.R.L., assignee; WO91/02025, 1991.

[73] Sandstrom P H. Rubber Containing Starch Reinforcement and Tire Having Component Thereof. Europe Patent, 1074582A1, 2001.

[74] Zhang L, Wang Y, Wang Y, et al. Morphology and Mechanical Properties of Clay/styrene-Butadiene Rubber Nanocomposites. *J. Appl. Polym. Sci.*, 2000, 78(11): 1873-1878.

[75] Du M, Guo B, Lei Y, et al. Carboxylated Butadiene-Styrene Rubber/halloysite Nanotube Nanocomposites: Interfacial Interaction and Performance. *Polymer*, 2008, 49(22): 4871.

[76] Lvov Y, Wang W, Zhang L, et al. Halloysite Clay Nanotubes for Loading and Sustained Release of Functional Compounds. *Adv. Mater.*, 2016, 28(6): 1227.

[77] Du M, Guo B, Jia D. Newly Emerging Applications of Halloysite Nanotubes: a Review. *Polym. Int.*, 2010, 59(5): 574.

[78] Guo B, Chen F, Lei Y, et al. Styrene-butadiene Rubber/halloysite Nanotubes Nanocomposites Modified by Sorbic Acid. *Appl. Surf. Sci.*, 2009, 255(16): 7329.

[79] Ganter M, Gronski W, Semke H, et al. Surface-compatibilized Layered Silicates-A Novel Class of Nanofillers for Rubbers with Improved Mechanical Properties. *Kautsch. Gummi Kunstst.*, 2001, 54(4): 166.

[80] Gronski W, Schon F. *Proceedings of the International Rubber Conference*, 2003: 297.

# 10 Reinforcement of Silicone Rubber

Silicone rubber is a class of elastomers based on poly(dimethylsiloxane) (PDMS). The polymer has no carbon atoms in its main chain, which distinguishes it from all other widely used elastomers. It has outstanding resistance to extreme temperatures and it possesses high chemical resistance. Its combination of extremely low glass transition temperature and the high chemical inertness of its main chain means that silicone rubber can operate normally at temperatures from −100°C to +300°C. Silicone elastomers are resistant to attack by oxygen, ozone, and many other chemicals due to of the absence of double bonds in the main chain and the high bond energy of the Si–O bond. A number of variants of the basic PDMS polymer exist, based on molecular weight, molecular weight distribution, terminal functional groups, and partial replacement of the dimethyl siloxane monomer with phenyl or vinyl-substituted analogs. The phenyl groups provide better low-temperature performance and a low concentration of vinyl groups facilitates crosslinking by peroxides. Apart from peroxide curing, the main type of crosslinking that is used for heat-cured silicone rubbers is a platinum-catalyzed addition reaction. Whilst silicone elastomers have outstanding chemical and temperature resistance, the mechanical properties of the neat polymer are poor, especially tensile strength, tear strength, and abrasion resistance. Particulate reinforcement is therefore essential in order to provide useful rubber products.

Although the principles of particulate reinforcement of silicone rubbers are the same as for other elastomeric polymers, there are some significant differences which warrant a separate discussion. Silicone rubber is unique among commonly used elastomers in that silica is the predominant particulate filler used for reinforcement. Carbon black, when used, is usually present as a secondary filler to provide color, UV absorption, electrical conductivity, or some other specific property. From the first commercial introduction of silicone rubbers in the 1940s, it was realized that silica was an ideal reinforcing agent. The reason can be easily attributed to the natural affinity of Si–OH (silanol) groups, which are invariably present on the surface of silica, for the siloxane backbone of the PDMS polymer.

One consequence of using silica rather than carbon black as the main reinforcing filler for silicone rubber is that products with colors other than black can easily be produced. In fact, because of the similarity in refractive index between amorphous silica and PDMS polymer, it is possible to produce reinforced rubber articles which are completely transparent if a well-dispersed, fine-particle silica reinforcing agent is used. This has come to be one of the defining characteristics of silicone rubber, especially in the medical field.

## 10.1 Fumed *vs.* Precipitated Silica

The two types of silica that are suitable for rubber reinforcement are fumed and precipitated silicas. These two types of synthetic silica are produced by different manufacturing processes, as discussed in Chapter 1, but both provide sufficiently small primary particle size and an aggregated particle structure that makes them suitable for rubber reinforcement. Traditionally, fumed silica has been the main reinforcing particle used in silicone rubber, even though its production cost is significantly higher than that of precipitated silica. Nevertheless, precipitated silicas are also used in large quantities for reinforcements of silicone rubber, when a balance of cost and performance is required.

The preference for fumed silica in this application is due to certain silica properties, which can be related to the different manufacturing processes used. These manufacturing processes are discussed in Chapter 1.

- Fumed silica generally has a lower overall concentration of silanol groups on the surface due to the fact that it is produced at very high temperature, where condensation reactions leading to elimination of water from adjacent silanol groups is facilitated. As a consequence of the overall reduction in silanol groups, the proportion of isolated silanol groups, or silanol groups which cannot hydrogen-bond to adjacent silanol groups on the silica surface, is higher. The isolated silanol groups are thought to be important for interaction with the silicone polymer. Infrared measurements have shown that isolated silanols interact more strongly with the oxygen in the siloxane backbone than do vicinal or geminal silanols[1].
- A second consequence of the lower overall silanol concentration in fumed silica is that its affinity for water is lower than that of precipitated silica. Any water that remains on the silica surface after mixing can significantly interfere with the silica-polymer interaction, and removal of the strongly adsorbed water from precipitated silica can be problematic.
- Precipitated silicas are produced in an aqueous medium, and so the last step in the production process is drying of the silica from a water-based medium. This can lead to strong capillary forces between particles and strong particle-particle interactions, which can cause the silica to be more difficult to disperse in polymers. The introduction of spray-drying processes and so-called "highly dispersible" silicas has alleviated this disadvantage, but precipitated silicas are still considered to be less dispersible than fumed silicas, which have never been exposed to liquid water.

- Fumed silicas are chemically purer. Precipitated silicas contain significant amounts of sodium, other metals, and ionic species such as sulfate and chloride, which can have negative effects on PDMS polymers under some conditions.

## 10.2 Interaction between Silica and Silicone Polymers

The interaction between polymer and filler surface has been shown to generally play a critical role in rubber reinforcement. In the case of silica, and its interaction with silicone polymers, two approaches to characterizing the silica surface have been taken. The first is based on chemical characterization of the types and concentrations of various functional groups on the silica surface. The second approach is based on surface energy analysis. The chemical analysis of functional groups on silica surface has mostly been done using IR and NMR spectroscopy. Four types of functional group have been identified: siloxane, isolated silanol, geminal silanol, and vicinal silanol groups (Figure 10.1). Whilst it has been shown that precipitated silicas have higher concentrations of geminal and vicinal silanol groups and fumed silicas have higher concentrations of siloxane and isolated silanol groups on the surface[2], no correlation has been shown between these concentrations and PDMS polymer interactions for a given type of silica.

**Figure 10.1** Schematic illustration of silica surface showing isolated, geminal, and vicinal silanols and siloxane groups

### 10.2.1 Surface Energy Characterization by Inverse Gas Chromatography

As discussed in Chapter 2, surface activity can be determined, in a physical sense, by measuring the surface energy of fillers. Variations in surface energy determine the adsorptive capacity and energy of adsorption. Physically, the interaction between two materials is determined by their surface free energies. It is known that several types of interactions exist between molecules close to each other, such as dispersive, dipole-dipole, induced dipole-dipole, hydrogen bonding, and acid-base. These result in different types of cohesive forces, which are the origin of the surface free energy. The surface free energy of a material can be expressed as the sum of several components, each corresponding to a type of molecular interaction (dispersive, polar, hydrogen bonding, etc.). Since the effect of the dispersive force is universal, the dispersive component of the surface free energy, $\gamma^{d}$, is particularly important. Inverse gas chromatography (IGC) using a homologous series of hydrocarbons can be used to determine the dispersive component of surface energy. The method and the theory behind it have been described in Chapter 2. Results in Figure 10.2 show the dispersive surface energy, measured in $mJ/m^2$, of a series of untreated fumed silicas with surface area from 90–420 $m^2/g$[3]. The specific grades of fumed silica that were tested are described in Table 2.9. It can be seen that the specific surface energy increases with increasing surface area. A similar effect has been observed for a series of carbon blacks.

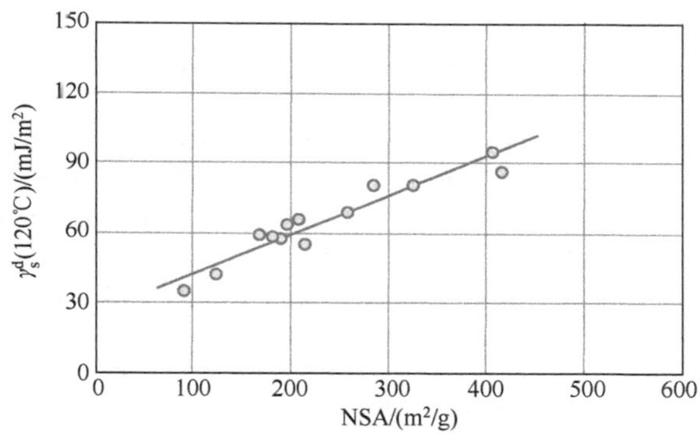


**Figure 10.2** Dispersive surface energy of a series of fumed silicas as a function of the specific surface area of the silica, measured by inverse gas chromatography (IGC)

However, unlike carbon black, it can be expected that specific polar interactions, particularly hydrogen bonding, would play an important role in the overall interaction between silica and PDMS polymer. In order to measure this effect more directly, the free energy of adsorption of hexamethyldisiloxane (HMDS) was determined by IGC. HMDS can be considered a model compound for PDMS polymer; it is essentially a methyl-terminated dimer of the dimethyl(silyl) repeat unit. It can be seen in Figure 10.3 that this measure of adsorption energy also increases with increasing surface area. Polar components of adsorption free energies were also measured using acetonitrile and THF, and these parameters also increased with silica surface area[3]. Thus, it has been clearly shown that the energy of interaction between silica and PDMS should increase with higher surface area fumed silica, although the specific type of interaction has not been demonstrated by IGC.

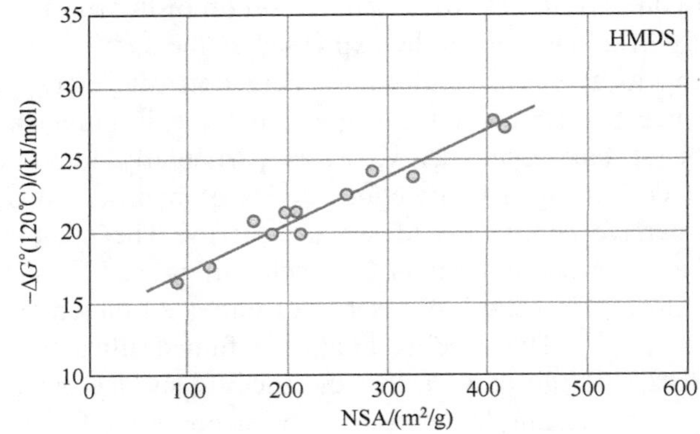


**Figure 10.3** Free energy of adsorption for a series of untreated fumed silicas

### 10.2.2 Bound Rubber in Silica-PDMS Systems

Bound rubber in silicone rubber filled with silica has been studied considerably. A discussion of the significance and meaning of bound rubber can be found in Chapter 5. In PDMS, the increase in bound rubber content after mixing has been found to increase very slowly, and reach saturation only after very long times[4]. The kinetics of this process has been thoroughly studied and found to fit well with a model based on diffusion and random polymer adsorption[5].

In the silica-silicone rubber system, hydrogen bonding between silanol groups on the silica surface and the siloxane bonds of the polymer is thought to be the cause of the strong adsorption of this polymer. If ammonia, which is a strongly hydrogen-bonding chemical, is added during bound rubber measurement, it can preferentially adsorb and displace silicone polymer from the silica surface. This method was used to demonstrate that bound silicone rubber on silica is mainly physisorbed through hydrogen bonding[6]. However, if the silica-PDMS mixture is heated to temperatures between 130°C and 280°C, the polymer becomes progressively chemically attached through a condensation reaction[7].

From a study on a series of fumed silicas with different surface areas in vinyl-terminated PDMS, bound rubber results showed that there was a critical surface area of fumed silica around 170 $m^2$/g, below which no coherent mass could be obtained during solvent extraction[3]. Beyond the critical surface area, the bound rubber content seemed to decrease with increasing silica surface area (Figure 5.25). These observations are qualitatively similar to those observed for carbon black in hydrocarbon rubbers. The mechanistic explanation for this has been discussed in detail in Section 5.1.7.4.

# ■ 10.3 Crepe Hardening

The increase in viscosity or modulus of a particle-filled elastomer compound immediately after mixing is well known (see Section 5.2.4). When applied to organic elastomers, this phenomenon is generally referred to as storage hardening, and lasts for several days before reaching a plateau. For silica-filled silicone elastomers, the hardening phenomenon is often much more severe. It is known as crepe hardening and can continue for a period of several weeks after mixing. Results in Table 10.1 show the magnitude of the increase, measured by plasticity, of typical high consistency rubber (HCR) compounds. Increasing the silica loading causes an increase in plasticity, but the effect of loading is overshadowed by the large increases that occur on storage of the freshly mixed compounds. This can cause an increase in plasticity of 4–7 times over a 1-month period of storage at room temperature. It can be seen from Table 10.1 that most of the increase occurs over the first 14 days.

**Table 10.1** Williams plasticity of mixtures of high MW, methyl-terminated PDMS with various amounts of Cabo-O-Sil MS 75-D fumed silica as a function of time after mixing

| Silica loading/phr | 22 | 24 | 26 | 28 | 30 | 32 | 34 | 36 | 38 |
|---|---|---|---|---|---|---|---|---|---|
| Plasticity: fresh | 19 | 24 | 51 | 52 | 49 | 39 | 55 | 59 | 56 |
| Plasticity: 1 day | 34 | 38 | 64 | 64 | 73 | 73 | 70 | 71 | 72 |
| Plasticity: 7 days | 79 | 83 | 147 | 148 | 159 | 166 | 181 | 184 | 217 |
| Plasticity: 14 days | 146 | 153 | 155 | 161 | 179 | 189 | 206 | 213 | 245 |
| Plasticity: 28 days | 149 | 156 | 171 | 176 | 195 | 206 | 227 | 234 | 275 |

The crepe-hardening phenomenon has been studied as a function of temperature and polymer molecular weight[8]. An activation energy for the crepe hardening process of 16.8 kJ/mol has been reported[9], which is consistent with a physical rather than a chemical interaction. Beyond the filler agglomeration, some studies have concluded that the hardening is caused by hydrogen bonding between the polymer and silanol groups on the silica surface[10,11]. When the distance between particles is small enough, adsorption of polymer chains on adjacent particles, in combination with normal polymer entanglement, can lead to a large increase in stiffness, manifested by increases in Mooney viscosity and plasticity of the uncured compounds. The magnitude of the hardening effect is determined by multiple factors, notably:

- polymer properties such as molecular weight and terminal groups,
- silica properties, especially surface area and structure, and
- availability of silanol groups on the silica surface.

Substances which block the silanol groups on the silica surface can significantly inhibit the crepe hardening. Even water is effective as a short-term anti-crepe-hardening agent[2].

For practical applications of silicone rubbers, it is found to be necessary to mitigate crepe hardening in order to make silicone compounds that are processable. Two general approaches are used to achieve this. The first is to use processing aids during mixing, and the second is to use pre-treated silicas.

## 10.4 Silica Surface Modification

In organic elastomers, when silica surface modification is employed, there are usually two purposes. The first is to facilitate dispersion by reducing particle-particle interactions, and the second is to promote bonding between polymer and filler, especially during vulcanization. This second effect has led to the term coupling agents being applied to the surface modifiers. In silicone rubber, due to the natural strong interaction between silica and the siloxane backbone, there is no need for chemical coupling of the filler. The purpose of the silica surface treatment is solely to reduce particle-particle interaction and minimize crepe hardening. For this reason, different chemicals are used, and the surface modifiers are usually referred to as plasticizers, processing aids, or anti-crepe hardening agents. Like coupling agents in organic elastomers, surface modifying agents may be added in situ, that is, during the mixing stage of silica with polymer, or the silica may be pre-treated.

A number of chemicals have been found to be effective as in situ plasticizers for PDMS-silica systems, including various surfactants. The most widely used plasticizer, however, is low molecular weight, hydroxy-terminated PDMS. During mixing with high molecular weight PDMS, the smaller plasticizer molecules can diffuse more rapidly, and quickly come into contact with the silica surface. When they make contact, the hydroxyl terminal groups ensure that the plasticizer remains strongly adsorbed, occupying silanol sites on the silica surface and thus limiting the ability of the high molecular weight polymer to adsorb. This also results in lower filler-filler interaction, hence less filler agglomeration, which is also essential for the reducing crepe hardening. Although the adsorbed molecules are essentially the same as the bulk polymer, the low molecular weight means that bridging of polymer chains between adjacent particles is hardly possible, and entanglement of polymer molecules attached to adjacent silica particles is also greatly reduced[8,12]. Since these are the three mechanisms that lead to the substantial crepe hardening effect in silicone rubbers, the low molecular weight plasticizers are able to have a strong inhibiting effect on hardening.

If too much processing aid is used in a formulation, it can diminish the mechanical properties of the cured vulcanizate. On the other hand, if not enough is used, crepe hardening can become a serious practical problem. It is therefore common to determine the amount of plasticizer or processing aid based on the interfacial area in the compound. This obviously depends on the loading of silica and also its specific surface area.

Pretreatment of silica is a viable alternative to silica surface treatment in situ. Generally, it has the advantage of enabling more uniform and controlled surface treatment, and more efficient use of treating agents. However, it usually adds to total cost compared with in situ treatment. Quite a large number of surface-treated silicas have been made experimentally and produced commercially. Some of these have been reviewed elsewhere[2]. Most of the effective surface treatments involve silanes, and attachment is through siloxane bonds using silazanes or alkoxysilanes, for example. The simplest type of treated silica is one in which surface hydroxyl groups are replaced with trimethylsilyl groups (Figure 10.4). The trimethylsilyl groups cause the silica to become hydrophobic if the treatment level is high enough. In a practical sense, this chemical modification of the surface can greatly reduce the filler agglomeration and the interaction with siloxane polymers, leading to lower viscosity and less crepe hardening. However, if the treatment level is too high, not enough silanol groups remain to interact with the polymer, leading to a reduction in mechanical properties like tensile strength[1,13]. There is thus an optimum level of surface treatment to enable satisfactory processability of silicone compounds, yet also maintain high mechanical properties of the vulcanizates.

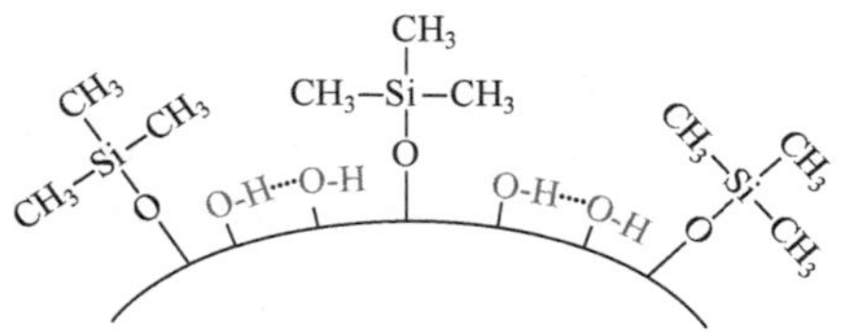


**Figure 10.4** Representation of a chemically modified silica surface where a portion of the silanol groups have been replaced with trimethylsilyl groups

# ■ 10.5 Morphological Properties of Silica

The morphology of silica particles, both fumed and precipitated, is not very different from that of carbon black; however, the methods used to measure morphological properties are somewhat different.

## 10.5.1 Surface Area

Specific surface area is the single most important property for reinforcing fillers because firstly, it provides a measure of the interface between polymer and particles, which is important for many reinforcement properties. One of the consequences of the incorporation of silica into a polymer is the creation of an interface between a rigid solid phase and a soft elastomer phase. The total area of the interface depends on the filler loading and the specific surface area of the filler (see Eq. 3.1). Thus, at constant loading, silica with higher specific surface area will result in higher total interfacial area in the compound. Some well-documented phenomena, such as adsorption of polymer and other compound ingredients, occur at the rubber-filler interface. In addition, surface area, to a large degree, is related to the average size of primary particles in the filler, which is also important in many reinforcement mechanisms. Although these statements are true for carbon black and for silica, the methods used are not exactly the same.

BET surface area measured by nitrogen adsorption is the most commonly used measure of surface area for silica. It is considered to be a measure of the total surface area accessible to nitrogen. The method and theory behind this method has been described in Chapter 2. The BET surface areas of silicas used in silicone rubber reinforcement is in the range of 90–400 $m^2/g$, whether fumed or precipitated[3,11]. On an order of magnitude basis, this range is similar to that of carbon blacks used in rubber, especially when compared at surface area per volume instead of surface area per weight (silica has a density about 20% higher than that of carbon black). In carbon black characterization for rubber applications, the STSA surface area is more often used because it is considered a more accurate measure of the surface area available to rubber polymer. For silica, however, STSA is rarely cited. The reason is that the STSA calculation is dependent on a certain surface property, and for silica, differences in surface chemistry and activity would cause this test to be unreliable. It has been reported that a modified version of the CTAB test can be used to measure the "external" surface area of silica, but this has not been widely adopted (see Chapter 2).

Although external surface area is not generally measured for silica, the concept remains significant. It seems, from several studies, that certain rubber compound properties such as modulus increase with increasing silica surface area, up to around 250 $m^2/g$, after which a further increase in surface area makes little or no difference[3,11]. An example is shown in Figure 10.5. Independently, studies of primary particle size of fumed

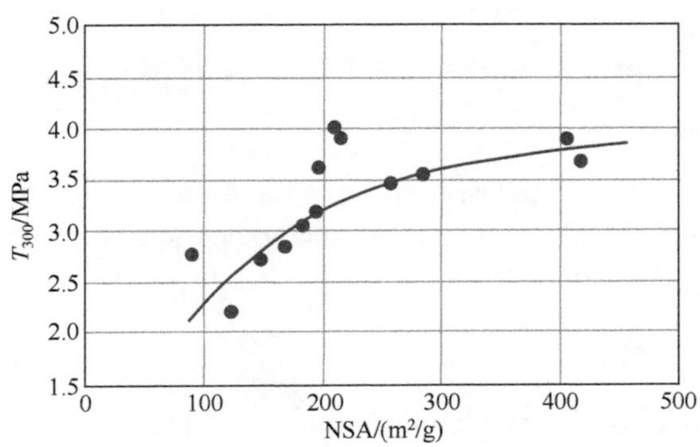


**Figure 10.5** Tensile stress at 300% elongation (M300) as a function of silica surface area at constant loading (38 phr) in a silicone compound

silica by TEM have shown that primary particle size has an inverse relationship with specific surface area up to around 250 $m^2$/g, beyond which increase in surface area seems to occur without any change in primary particle size. The change in surface area was attributed to surface roughness, although direct evidence for that was weak.

## 10.5.2 Structure Properties of Silica

Many mechanical properties of filled elastomers have been explained in terms of the occlusion of rubber by filler aggregates. When structured fillers are dispersed in rubber, the polymer portion filling the internal voids of the filler aggregates is unable to participate fully in the macro-deformation. The immobilization (at least partially) in the form of occluded rubber, causes this portion of rubber to behave like filler rather than like the polymer matrix. As a result, the effective volume of the filler, with regard to rheological and stress-strain behavior, is increased considerably.

Knowledge of the effect of structure of silica on silicone rubber properties is still lacking, for several reasons. First, there is no simple and reliable test method for directly characterizing structure of fumed silica, even though the particles, both fumed and precipitated, are clearly structured aggregates, as seen by transmission electron microscopy. The method used for carbon black structure determination, namely oil absorption, is not reliable for silica particles. Void volume measurements are frequently used to characterize precipitated and other wet-process silicas. Carman surface area, which measures gas transport through a packed bed[13], has been used as an indirect measure of structure for fumed silica[14]. One study, in which silica particles were synthesized with extremely low and high structures, but the same primary particle size, demonstrates the importance of particle structure, especially for tensile modulus (Figure 10.6)[15]. In this study, all particles were surface-modified with trimethylsilyl groups, and loading was fixed at 20 phr to accentuate the difference.

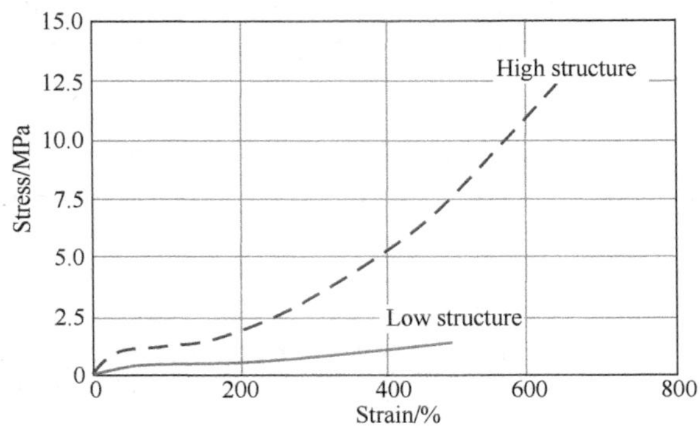


**Figure 10.6** Stress-strain curves for silicone elastomers filled with low and high structure silica. The primary particle size for both silicas was 5.8 nm, and pore volumes were 0.6 $mm^3$/kg and 5.8 $mm^3$/kg (from ref [15])

Another reason for the relatively scarce information on silica structure is that, in the fumed silica manufacturing process, there is limited ability to control structure independently of surface area. This is one of the major differentiators between fumed silica and carbon black, in terms of morphology space.

## 10.6 Mixing and Processing of Silicone Compounds

Mixing conditions for silicone compounds are quite different from those used to mix organic elastomer compounds. Typically, much longer mixing times are employed for silicone compounds and the mixers are at low shear rates, compared to Banbury mixing, for example. Mixing times can be upwards of one hour, although shear can be very low or zero for a portion of that time. This is in contrast with typical mixing times for organic rubber compounds, which are of the order of minutes. There are a number of reasons for this difference:

- Fumed silica is often used as a reinforcing agent. The fluffy nature of fumed silica means that incorporation times are relatively long.
- Wetting of silica surface by PDMS polymer is relatively slow. It can be accelerated by increasing temperature, but still takes longer than most organic polymers.
- Shear-generated frictional heating of compounds is much lower than in typical Banbury mixing of commodity elastomers. This means that heating occurs primarily by transfer of heat from the mixer body, which is relatively slow.
- Silicone compounds generally require complete removal of air bubbles and moisture. This is usually achieved by applying reduced pressure to the mixer at a temperature above 100°C.

The extrudate appearance of fumed silica-filled silicone rubber compounds was investigated at different extrusion rates at constant temperature (25°C) by varying screw speeds over a range from 5 rpm to 70 rpm (rotations per minute)[3]. The surface roughness of the extrudates cured was measured by a portable glossmeter (Micro-TRI-Gloss, from Byk Gardner) at 60°. Gloss results are from the specular reflection of light on a surface. As the surface becomes rougher, the diffuse reflection from the surface increases and the gloss is reduced. Figure 10.7 shows the results of gloss as a function of the silica surface area for extrudates extruded at a speed of 35 rpm and cured in an oven at 177°C for 5 minutes. With increasing surface area, the surface roughness of the extrudates increases rapidly between 170 $m^2/g$ and about 300 $m^2/g$, beyond which the smoothness of the materials improves slightly.

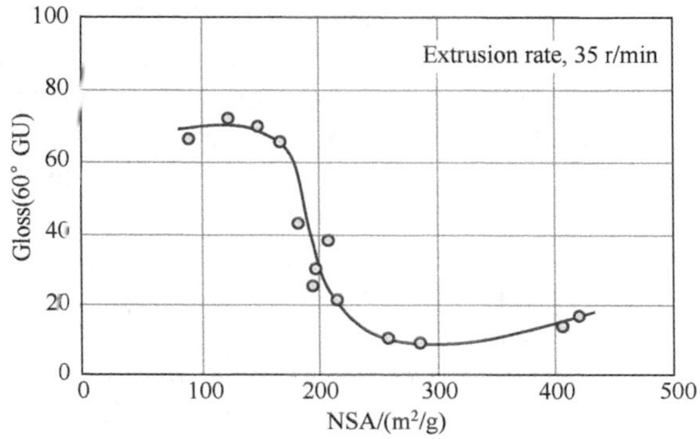


**Figure 10.7** Surface roughness of the extrudate as a function of surface area of silica

At low surface area, the roughness does not seem to change significantly. The results obtained at other extrusion rates [Figure 10.8(a)] show that gloss becomes lower as extrusion rate increases. The overall surface appearances were also estimated visually [Figure 10.8(b)], and there seems to be a good general correlation between these ratings and the surface gloss measured using a glossmeter.

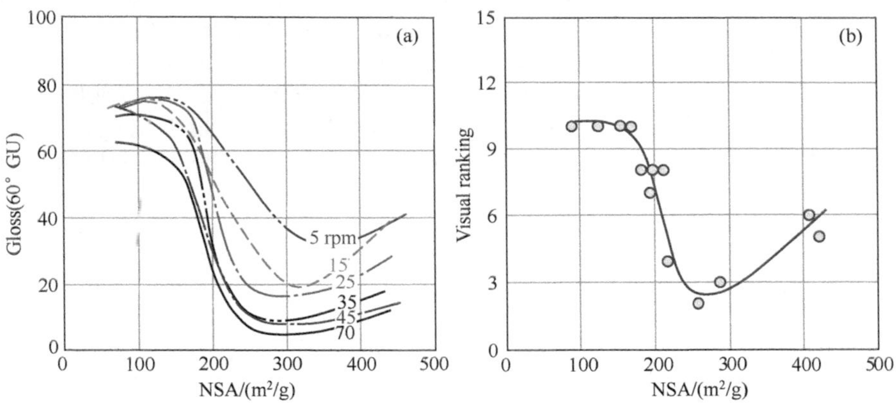


**Figure 10.8** Surface roughness of the extrudate as a function of surface area of silica

At low surface area, where polymer bridging between filler particles is minimal, increasing surface area causes an increase in bound rubber, causing the elastic memory to increase somewhat, but this appears to be largely offset by the increased filler agglomeration, which tends to reduce the elastic memory. Above the critical surface area where polymer bridging becomes significant, the effect of filler particles acting as extra crosslinks becomes dominant, leading to poorer extrudate appearance. At higher surface area, the effect of highly developed filler agglomeration becomes predominant, which, along with the reduction of bound rubber content, leads to an improved appearance of the extrudate.

Since the surface roughness of the extrudate is strongly affected by the extrusion rate, the critical rates of the compounds were also measured. The surface is considered rough below 30 gloss units at 60°. The critical rates were also estimated visually. The results shown in Figure 10.9 include some that were extrapolated from the results obtained at lower extrusion rates because even at the highest rate used in the study (70 rpm) the roughness of the extrudate was still below the criteria that were set. As expected, both the measured and visually estimated results generally show the same pattern: the critical rate decreases with surface area, passing a minimum and then going up slightly. It is more evident in the visual ratings. The correlation between the measured data and those estimated visually was generally good (Figure 10.10), but the compound filled with silica S17D is a clear exception. For this silica, the critical extrusion rate measured by glossmeter is extremely low and that obtained visually is even higher than 70 rpm, i.e., all extrudates looked smooth. It was also found that at low extrusion rate, the gloss of the S17D-compound was the lowest. The reduction in gloss is very small for the S17D-filled compound with an increased extrusion rate, while the gloss of other compounds drops rapidly. The conflict in the ratings of the S17D compound between different methods appears to be related to the scale of the surface roughness. The glossmeter can measure surface roughness at the micron scale, whereas visual rating measures larger scale distortions. This effect is illustrated by the images shown in Figure 10.11. The extrudate of silica S17D is characterized by higher roughness on the micro scale at low extrusion rates, but less distortion at high extrusion rate. The reasons for this unique performance are not fully understood, but it may be that the particle architecture leads to a highly developed filler agglomerate. It was demonstrated[16] that the S17D compound showed a high degree of filler agglomeration compared to other silicas, but that it seems to be easily broken down upon strain.

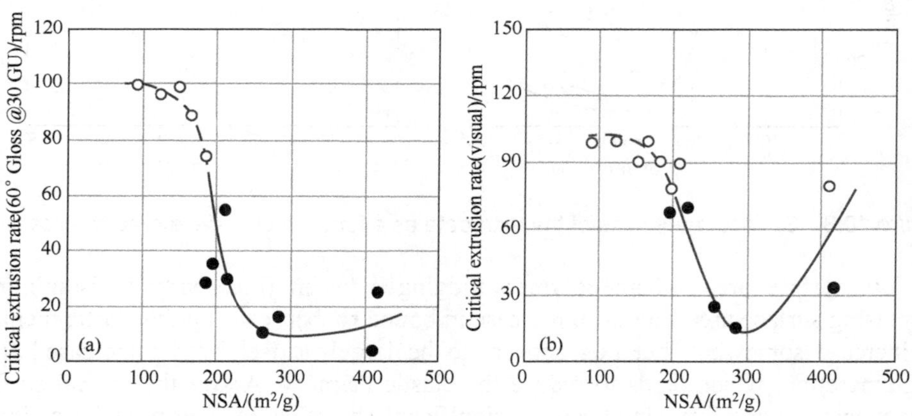


**Figure 10.9** Critical extrusion rates for surface roughness as a function of surface area of silica. The empty symbols represent the values extrapolated from the result obtained at the measured extrusion rates

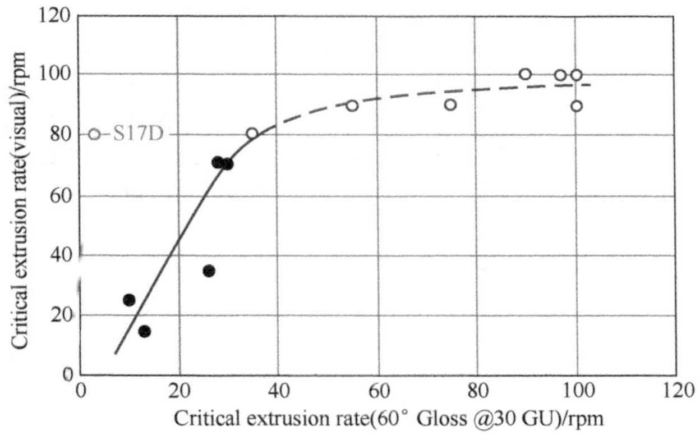


**Figure 10.10** Critical extrusion rates: visual *vs.* measured with glossmeter

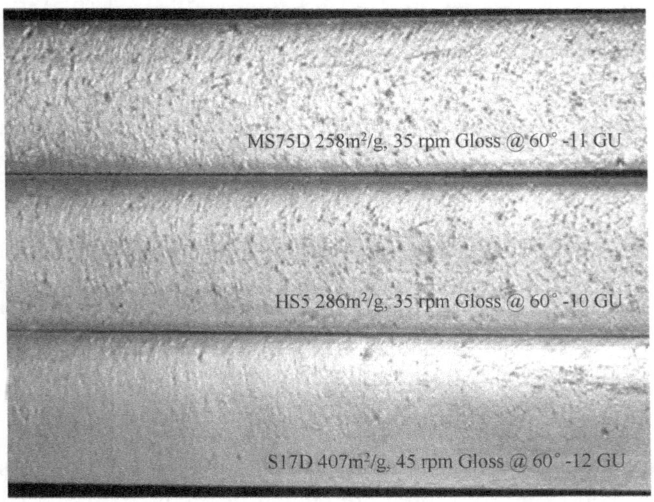


**Figure 10.11** Surface roughness of extrudates having similar gloss at 60°

The effects of filler surface area and extrusion rate on the extrudate roughness can be explained on the basis of their effects on the elastic response or elastic memory of the compounds, as has been discussed in Section 5.3.3. It has been recognized that the surface appearance of an extrudate is associated with elastic recovery due to incomplete release of stress caused when long-chain molecules are orientated by shear in the die. The distortion of the extrudate increases with increasing unreleased stress and may, in some cases, even lead to the complete fracture of the extrudate (melt fracture). For a given compound, high temperature, which effectively reduces the relaxation time of the polymer, and lower extrusion rate, which allows the polymer enough time to relax in the die, both work to reduce the elastic recovery of the extrudate, giving a smoother surface. In terms of the polymer, the primary factor influencing the surface roughness of unfilled elastomers is the entanglement of elastomer molecules, which, in turn, is determined by their molecular weight and its distribution. Since the elastic memory occurs in the rubber phase alone, the surface roughness of the extrudate for any given polymer is generally improved by the addition of filler. On the other hand, filler aggregates may serve as multi-functional crosslinks, due to the adsorption of polymer molecules, causing an increase in elasticity of the compounds.

## ■ 10.7 Silica Dispersion in Silicone Rubber

Achieving a good level of dispersion is important in silicone rubber compounds as it is in all filled rubber compounds. Optical dispersion methods do not always work because of the similarity in refractive index between silica and PDMS. However, it is still possible to measure dispersion with optical reflection methods such as that used by the Dispergrader, provided the illumination and measurement angles are chosen appropriately. The reason is that the Dispergrader test does not exactly "see" undispersed filler particles as an optical transmission method would. What is measured is actually surface roughness or surface irregularities caused by undispersed filler particles. Other methods have also been developed to measure roughness in uncured silicone compounds as a way to characterize the state of dispersion. One such method is gloss measurement from an extruded surface[3]. It should be noted that if silicone vulcanizates that have been cured with peroxides are tested, some crystalline decomposition products from the peroxide curatives can remain in the compound and distort the measured dispersion quality. Normally, those residues are removed by a fairly extensive post-curing at a temperature around 200°C.

Dispersion ratings, determined by visual observation, for a series of silicone rubber vulcanizates that had been post-cured are shown in Figure 10.12. It can be seen that filler dispersion quality deteriorates with increasing surface area of the silica up to around 250 $m^2/g$, beyond which no further deterioration occurs. The general decrease in dispersion quality with increasing filler surface area is expected based on experience from carbon black, as discussed Chapter 4. The leveling off of dispersion quality at high surface area is consistent with the observation that primary particle size does not change significantly above a surface area around 250 $m^2/g$. The higher surface area at that level may be related to surface nano-roughness, which would not affect dispersibility.

Generally, there appears to be little concern regarding macrodispersion of silica in silicone rubber in the literature. This is partly because of the long mixing times that are usually used, and the fact that silica is usually surface-modified. These two factors, which are necessary in order to make commercially useful products, lead to a higher

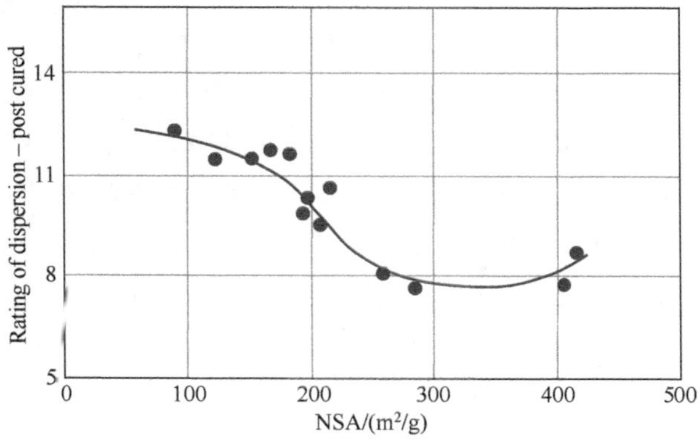


**Figure 10.12** Dispersion rating of post-cured silicone vulcanizates containing 38 phr of fumed silica

degree of macrodispersion than is usually achieved in other elastomers filled with carbon black. In addition, silica that is less than well-dispersed is not usually visible because of the similarity in refractive index with the polymer. Microdispersion, on the other hand, is important, but, as in other polymers, it is usually measured indirectly, such as by through dynamic mechanical properties (see Section 10.9).

# 10.8 Static Mechanical Properties

## 10.8.1 Tensile Modulus

The primary effect of adding hard, reinforcing particles to elastomers is to increase stiffness, which can be observed by hardness and static modulus. The results in Table 10.2 show the effect of increasing loading of the fumed silica in a typical HCR (high consistency rubber) silicone rubber. All the measures of static modulus show a steady, monotonic increase with increasing silica loading, as does hardness (Shore A). In this example, the increase in modulus is accompanied by a continuous increase in tensile strength and tear strength, with little diminishing of elongation at break. This is a clear indication of the reinforcing nature of silica particles in silicone elastomers. At some higher silica loading, it can be expected that tensile strength would go down as modulus continues to increase with loading. In other words, the basic reinforcement mechanism parallels the effect of carbon black reinforcement of organic elastomers.

**Table 10.2** Vulcanizate properties of silicone rubber compounds filled with varying amounts of Cab-O-Sil MS-75 D fumed silica

| Silica loading/phr | 22 | 24 | 26 | 28 | 30 | 32 | 34 | 36 | 38 |
|---|---|---|---|---|---|---|---|---|---|
| Tear strength/(kN/m) | 7.88 | 8.93 | 9.63 | 10.86 | 11.21 | 12.08 | 12.96 | 13.66 | 15.06 |
| T50/MPa | 0.47 | 0.47 | 0.56 | 0.57 | 0.59 | 0.64 | 0.68 | 0.7 | 0.77 |
| T100/MPa | 0.6 | 0.61 | 0.71 | 0.72 | 0.74 | 0.78 | 0.83 | 0.84 | 0.89 |
| T200/MPa | 0.92 | 0.9 | 1.15 | 1.14 | 1.17 | 1.24 | 1.34 | 1.32 | 1.37 |
| T300/MPa | 1.46 | 1.43 | 1.92 | 1.92 | 1.94 | 2.12 | 2.28 | 2.2 | 2.31 |
| T400/MPa | 2.26 | 2.28 | 3.09 | 3.1 | 3.08 | 3.48 | 3.63 | 3.53 | 3.76 |
| T500/MPa | 3.31 | 3.47 | 4.59 | 4.66 | 4.69 | 5.24 | 5.39 | 5.16 | 5.74 |
| Tensile strength/MPa | 4.36 | 4.79 | 5.14 | 5.52 | 6.27 | 6.45 | 6.68 | 6.73 | 7.11 |
| Elongation@break/% | 592 | 590 | 532 | 546 | 587 | 567 | 570 | 591 | 564 |
| Shore A hardness | 41 | 42 | 45 | 45 | 48 | 49 | 51 | 53 | 55 |
| Compression set/% | 8 | 9 | 8 | 8 | 11 | 11 | 10 | 10 | 11 |

The effect of fumed silica surface area at constant loading also mirrors quite well the effect of carbon black in other polymers, as has been discussed above (see Figure 10.5). A similar study on a series of precipitated silicas with surface areas ranging from 50 $m^2/g$ to 250 $m^2/g$ found a strong correlation between surface area and tensile modulus[11]. For carbon black, however, structure, determined by COAN (compressed oil absorption number), is also an important determinant of tensile modulus. COAN and OAN (oil absorption number) are not often reported for silica; however, in one study on precipitated silica reinforcement of silicone, a positive correlation was found between OAN number and tensile modulus[11]. It is not clear whether the effect was confounded with surface area. In a separate study, the reinforcement effects of three fumed silicas with similar BET surface area but different Carman surface area were compared. The results showed increasing hardness, modulus, and tensile strength with increasing Carman surface area, which is an indirect measure of aggregate structure[12]. Note also Figure 10.6 showing results from an experimental study in which particles with extremely different structures were produced. So, it seems that morphological effects are similar in principle to those of carbon black. However, for fumed silica in particular, the range of structure available at any given surface area is more limited compared to carbon black.

The effect of silica surface chemistry of tensile modulus has been reported in numerous studies. Maxson and Lee first showed how large the effect of silica surface modification could be on tensile properties[17]. As discussed previously, the chemical modification of the very active silica surface, whether in situ or by pretreating the silica, can have a dramatic effect on composite properties. This is illustrated in Figure 10.13, from reference 18, which shows the stress-strain curves for PDMS filled with 20 phr of fumed silica with different types of surface treatment. The key to achieving higher reinforcement seems to be formation of a strong bond with the polymer during vulcanization. In the example in Figure 10.13, modified silica 3 was treated with divinyl tetramethyldisilazane, so that vinyl groups became chemically bonded to the silica surface. These groups have the ability to become bonded to the polymer during peroxide vulcanization. The effect of the new groups was much larger than simply increasing crosslink density by a similar amount[18]. In fact, it was shown on modulus provided by such treated silica was similar to that provided by untreated silica, which has a high concentration of hydrogen bonding silanol groups. Both untreated silica and vinyl-group-treated silica can form strong polymer-filler bonds, by different mechanisms. However, the vinyl treated silica has much weaker filler-filler interactions, which enables easier dispersion and much lower viscosity in the uncured compound, which is a big practical advantage[18].

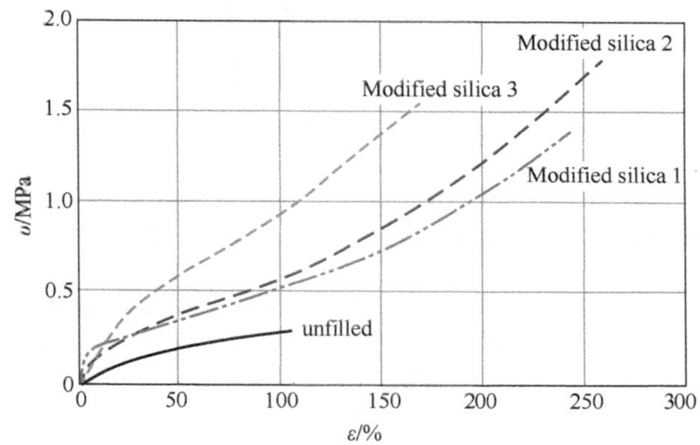


**Figure 10.13** Stress-strain curves of 118,000 MW PDMS filled with 20 phr of silica with different surface treatments (from ref [18]). Modified silica 1 treatment: hexamethyldisilazane; Modified silica 2 treatment: dichlorodimethylsilane; Modified silica 3 treatment: divinyl tetramethyldisilazane

Filled rubbers have been shown to undergo stress softening, so that for a second and subsequent stretch the stress at any strain up to the maximum strain reached in the first stretch is lower than the stress obtained at the same strain in the first stretch. This is known as the Mullins effect, and has been discussed in Section 6.3[19]. The basic characteristics of the effect are the same as in other elastomers filled with carbon black. The model that was found to best fit the results obtained at different temperatures in the PDMS system was one based on the limited extensibility of polymer chains. When chains approach the limit of their extensibility, they become non-Gaussian, causing an increase in modulus. When the limit of extensibility is exceeded, slippage or detachment from the filler surface can occur, leading to a decrease in modulus. These effects have been discussed previously. Results from the PDMS system were found to fit well with the model of Bueche[20], from which an estimate, $b$, of the inter-particle distance can be derived. The value of $b$ derived from the Bueche model was found to be smaller than the average inter-particle distance determined by AFM analysis. However, it was close the shortest inter-particle distances. Clement et al. concluded that regions of high local volume fraction of silica predominantly contributed to the Mullins effect. In their model, there are silica-rich and silica-poor regions within the vulcanizate. In the silica-rich regions, the silica structure deforms less under applied strain, but the polymer is much more deformed due to strain amplification. As the applied strain increases, regions of decreasing volume fraction become affected (Figure 10.14)[20]. This model can easily explain why a poorer dispersion of silica leads to a higher Mullins effect.

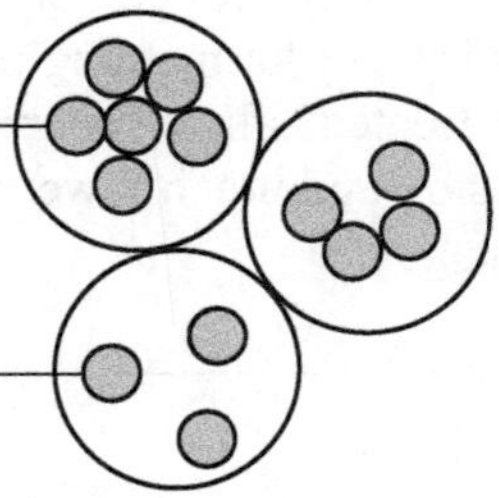


**Figure 10.14** Schematic view of variations in local volume fraction of silica. Gray circles represent silica particles, white is matrix

### 10.8.2 Tensile Strength and Elongation Properties

The surface area dependence of tensile strength and breaking elongation (Figure 10.15), measured at constant temperature and normal strain rate, show that while the elongation at break decreases monotonically with surface area, the effect on tensile strength is more complicated. Breaking stress does not change significantly until 170 $m^2/g$, beyond which it increases sharply to a maximum at around 200 $m^2/g$, followed by a reduction at higher surface area. The lower tensile strength in vulcanizates filled with lower surface area silicas can be attributed to their lower polymer-filler interaction, which in turn can be related to lower interfacial area between filler and polymer, higher surface coverage with the processing aid, and lower surface energies of the fillers. When the vulcanizate is subjected to high tensile stresses, polymer chains may be more easily detached from the surface, thus providing less of a barrier to crack propagation and coalescence. With increasing surface area, both the polymer-filler interaction and the interfacial area between filler and polymer increase. This will significantly reduce the possibility of vacuole formation and favor slippage of chains along the surface to relieve stress concentration[21,22]. The reduction of breaking stress at higher surface area is probably caused by poorer dispersion. In this case, the poorly dispersed filler may furnish flaws that can initiate cracks and then develop into catastrophic failure under lower stresses. The increase in size or concentration of flaws appears to outweigh the effects of finer particle silica beyond about 250 $m^2/g$.

### 10.8.3 Compression Set

Compression set is an important property for any rubber product used in sealing applications. For silicone rubbers, it is usually tested after compression for 22 hours at

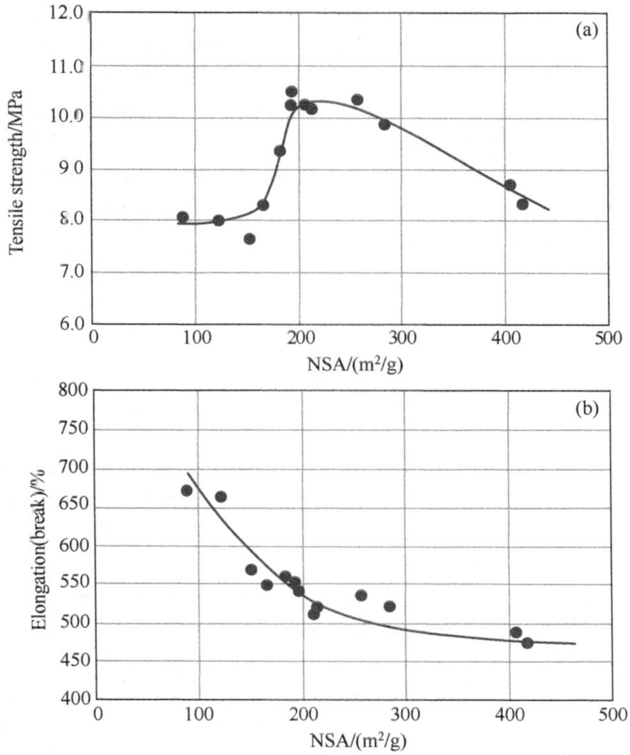


**Figure 10.15** (a) Tensile strength and (b) elongation at break for a series of silicone rubber compounds filled with 38 phr of silica with different surface areas

a temperature around 175°C. At this temperature, any non-permanent, non-covalent crosslinks can rearrange, leading to permanent set. In general, compression set of silicone rubber compounds can vary over a wide range, and, in the worst case, can be large enough to make the vulcanizate useless as a sealing material. Cochrane and Lin reported compression set of up to 80% for silicone compounds filled with untreated fumed silica[12]. The compression set had a strong dependence on silica surface area, increasing from 45% to 80% as surface area was increased from 100 $m^2/g$ to 400 $m^2/g$. Increasing structure, determined by Carman surface area, also led to higher compression set. The lowest set values were obtained from compounds filled with pre-treated silica. These results can all be explained in terms of the contribution of filler-mediated effective crosslinks to the overall crosslink network. With high surface area untreated silica, bridging of polymer chains between filler particles and entanglement of chains attached to adjacent particles by hydrogen bonding can make a large contribution to the crosslink network. On the other hand, if the surface area is lower, the particles have larger average spacing distance, and if they are surface treated, the amount of hydrogen bonding between polymer and silica is reduced. Both these effects tend to reduce the contribution of filler-effective crosslinks relative to the carbon-carbon crosslinks introduced by peroxide vulcanization.

In a similar study using precipitated silicas instead of fumed, a similar strong dependence of compression set on silica surface area and structure was found[11]. In addition, it was found that pH of the silica was also an important factor for precipitated silicas. At acidic pH (below 5), there was a strong correlation between compression set and pH, lower pH leading to higher compression set. This can be explained by a chemical effect. At low pH, the silica surface can cause an acid-catalyzed scission of the polymer chain, leading to rearrangement and a reduction in molecular weight, both of which lead to increased compression set.

## ■ 10.9 Dynamic Mechanical Properties

The decrease in elastic modulus with increasing dynamic strain, known as the Payne effect, is a common phenomenon in filled rubber materials. As has been discussed in detail in Chapter 7, the Payne effect is usually associated with filler agglomeration or networking, leading to trapped or occluded rubber. It has practical importance for dynamic rubber applications because the maximum in loss or viscous modulus, $G''$ is correlated with the change in $G'$ from low to high strains[23,24]. This means that materials with a high Payne effect will have higher viscous modulus in relation to storage modulus and therefore higher energy is lost to heat during cyclic deformation. In a series of studies on silica-filled PDMS networks, Clement et al. showed that exactly the same phenomenon occurred in silicone rubbers[25]. A series of silica-filled PDMS compounds loaded from 10–60 phr, showed the classical Payne effect. A linear relationship was found between the maximum in $G''$ and $\Delta G'$, the change in elastic modulus from low strain to its plateau level at high strain (Figure 10.16). It was noted that a Payne effect was observed even at the lowest filler loading, corresponding to a volume fraction of 0.04, which is well below the percolation point reported elsewhere[26].

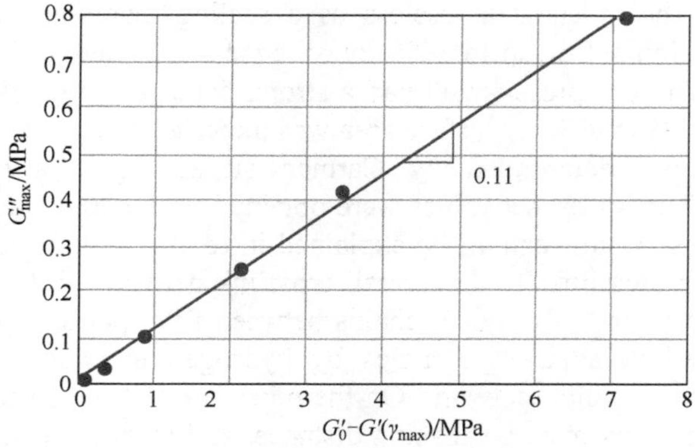


**Figure 10.16** Relationship between the $G''$maximum and the $G'$drop for networks filled with various A300-t silica loadings, at room temperature (from Ref [23])

A series of silica-filled PDMS compounds with different silica surface areas at equal loading show the typical change in elastic modulus with strain amplitude at constant frequency (1.7 Hz) and temperature (30°C), characteristic of the Payne effect (Figure 10.17)[3]. The magnitude of the Payne effect, measured as the difference between the lowest-strain $G'$ and the $G'$ at 98% strain, shows a strong correlation with the surface area of the silica (see Table 2.9 and Figure 10.17). Both the surface area effect at constant loading and the loading effect at constant surface area discussed above indicate that interfacial area between filler and polymer is a dominant factor for determining the magnitude of the Payne effect.

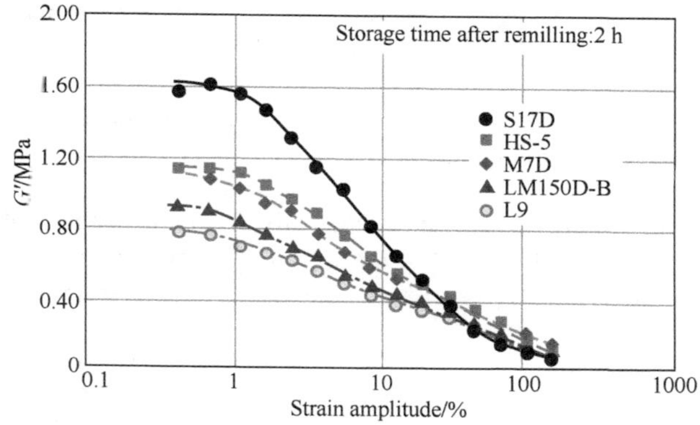


**Figure 10.17** Storage modulus as a function of dynamic strain amplitude for a series of silicone rubber compounds filled with 38 phr of fumed silica with different surface areas. Surface areas of fumed silica grades are given in Table 2.9

It has widely been accepted that the Payne effect is mainly, if not only, related to filler agglomeration in the polymer matrix. As mentioned before, trapped rubber within agglomerates can increase the effective filler volume fraction, increasing modulus of the filled vulcanizates. At moderate and high strains, the agglomerates are, to a certain degree, broken down and the rubber trapped within the agglomerates is released, lowering modulus. Thus, the Payne effect can serve as a measure of filler agglomeration or micro-dispersion. It can be seen from Figure 10.18 that for the two silicas with the highest surface areas, the Payne effects are significantly different, even though the surface areas are similar, indicating more developed filler agglomerates in the S17D-filled vulcanizate compared to HS-5. The S17D silica also gives weaker agglomerates, indicated by the more rapid drop in $G'$ with strain. It seems that some mechanisms other than those associated with the surface area play a considerable role in filler agglomeration in this case.

As has been discussed earlier, filler surface chemistry is another important factor. It has been shown that surface-modified silica shows much reduced Payne effect in PDMS compared to untreated silica[25], as has been observed in other polymers. This can be explained by the strong filler-filler interaction due to hydrogen bonding, which is greatly reduced when the silica is surface-modified. A similar effect was observed

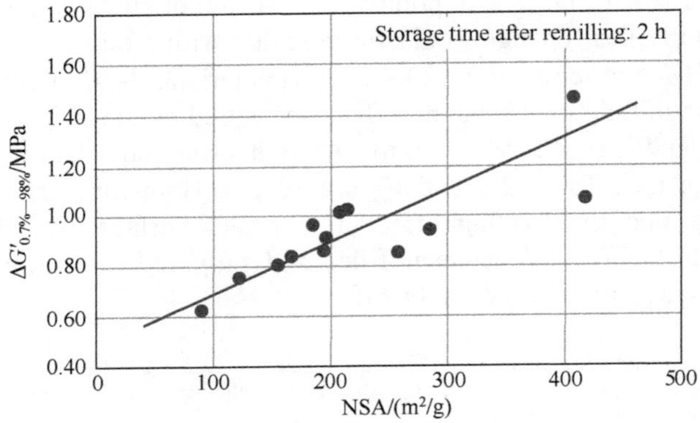


**Figure 10.18** Payne effect as a function of the surface area of the silica filler in silicone rubber compounds at constant loading

when processing aid was used with untreated silica (in situ treatment). However, the results indicate that pre-treatment of silica is much more effective at reducing filler-filler interaction, and thus Payne effect, than is in situ treatment with processing aid[25].

Temperature effects on dynamic modulus have also been measured in the silica-PDMS system. It was found that low stain $G'$ and the plateau-level $G'$ both decreased as temperature increased, consistent with what has been observed in carbon-black filled systems. It shows that an enthalpic mechanism is operating, because the entropic effect on elasticity would have the opposite effect, namely an increase in modulus with increasing temperature. An analysis of results from PDMS-silica systems with respect to various mechanistic models that have been proposed in the literature to explain the Payne effect[27] concluded that the Payne effect does not originate neither completely from the breakdown and reformation of the filler network as proposed by Kraus[28] and Huber and Vilgis[29], nor from the adsorption and reformation of rubber chains at the filler surface, as proposed by Maier and Goritz[30]. The Payne effect in silica-filled silicone elastomers was best explained by the existence of a gradient in the mobility of the polymeric chains from the PDMS/silica interface to the bulk, in agreement with Wang's approach[31]. This is the only interpretation that was able to explain both the temperature effect and the filler volume fraction effect, as well as the recovery of the Payne effect over short times[25,27].

## References

[1] Boonstra B, Cochrane H, Dannenberg E. Reinforcement of Silicone Rubber by Particulate Silica. *Rubber Chem. Technol.*, 1975, 48: 558.

[2] Warrick E L, Pierce O R, Polmanteer K E, Saam J C. Silicone Elastomer Developments 1967-1977. *Rubber Chem. Technol.*, 1979, 52: 437.

[3] Morris M D, Wang M-J, Kutsovsky Y. Effect of Surface Area of Fumed Silica on Silicone Rubber Reinforcement. Paper #49 Presented at the spring 167th Technical Meeting ACS Rubber Division, San Antonio, TX, 2005.

[4] Cohen-Addad J P, Huchot P, Jost P, Pouchelon A. Hydroxyl or Methyl Terminated Poly(dimethylsiloxane) Chains: Kinetics of Adsorption on Silica in Mechanical Mixtures. *Polymer*, 1989, 30: 143.

[5] Levresse P, Feke D L, Manas-Zloczower I. Analysis of the Formation of Bound Poly(dimethylsiloxane) on Silica. *Polymer*, 1998, 39: 3919.

[6] Wolf S, Wang M-J, and Tan E-H. Filler-Elastomer Interactions. Part VII. Study on Bound Rubber. *Rubber Chem. Technol.,* 1993, 66: 163.

[7] Li Y-F, Xia Y, Xu D, Li G. *J. of Appl.Polym.Sci.*, 1981, 19: 3069.

[8] DeGroot J V, Macosko C W. Aging Phenomena in Silica-Filled Polydimethylsiloxane. *J. Colloid Interface Sci.,* 1999, 217: 86.

[9] Vondracek P, Schatz M. Bound Rubber and "Crepe Hardening" in Silicone Rubber. *J. Appl. Polym. Sci.,* 1977, 21: 3211.

[10] Aranguren M I, Mora E, DeGroot J V, Macosko C W. Effect of Reinforcing Fillers on the Rheology of Polymer Melts. *J. Rheol.*, 1992, 36: 1165.

[11] Okel T A, Waddell W H. Effect of Precipitated Silica Physical Properties on Silicone Rubber Performance. *Rubber Chem. Technol.,* 1995, 68: 59.

[12] Cochrane H, Lin C S. The Influence of Fumed Silica Properties on the Processing, Curing, and Reinforcement Properties of Silicone Rubber. *Rubber Chem. Technol.* 1993, 66: 48.

[13] Carman P C, Malherbe P., *J. Soc. Chem. Ind.*(*London*), 1950, 69: 134.

[14] DeGroot J V. *ACS Rubber Division Meeting, Orlando,* 1999, Paper No. 94.

[15] Polmanteer K E, Lentz C W. Reinforcement Studies-Effect of Silica Structure on Properties and Crosslink Density. *Rubber Chem. Technol.,* 1975, 48: 795.

[16] Wang M-J, Morris M D, Kutsovsky Y. Effect of Fumed Silica Surface Area on Silicone Rubber Reinforcement. *Kautsch. Gummi Kunstst.*, 2008, 61: 107.

[17] Maxson M T, Lee C L. Effects of Fumed Silica Treated with Functional Disilazanes on Silicone Elastomers Properties. *Rubber Chem. Technol.,* 1982, 55: 233.

[18] Aranguren M I, Mora E, Macosco C W, Saam J. Rheological and Mechanical Properties of Filled Rubber: Silica-Silicone. *Rubber Chem. Technol.,* 1994, 67: 820.

[19] Clement F, Bokobza L, Monnerie L. On the Mullins Effect in Silica-Filled Polydimethylsiloxane Networks. *Rubber Chem. Technol.,* 2001, 74: 847.

[20] Bueche F. Mullins Effect and Rubber-Filler Interaction. *J. Appl. Polym. Sci.,* 1961, 5: 271.

[21] Dannenberg E M. *Trans. Inst. Rubber Ind.* 1966, 42: T26.

[22] Dannenberg E M. The Effects of Surface Chemical Interactions on the Properties of Filler-Reinforced Rubbers. *Rubber Chem. Technol.,* 1975, 48: 410.

[23] Payne A R, *J. Polym. Sci.,* 1962, 6: 57.

[24] Payne A R, Whittaker R E. Low Strain Dynamic Properties of Filled Rubbers. *Rubber Chem. Technol.,* 1971, 44: 440.

[25] Clement F, Bokobza L, Monnerie L. Investigation of the Payne Effect and its Temperature Dependence on Silica-Filled Polydimethylsiloxane Networks. Part I: Experimental Results. *Rubber Chem. Technol.*, 2005, 78: 211.

[26] Pouchelon A, and Vondracek P. Semiempirical Relationships Between Properties and Loading in Filled Elastomers. *Rubber Chem. Technol.,* 1989, 62: 788.

[27] Clement F, Bokobza L, Monnerie L. Investigation of the Payne Effect and its Temperature Dependence on Silica-Filled Polydimethylsiloxane Networks. Part II: Test of Quantitative Models. *Rubber Chem. Technol.*, 2005, 78: 232.

[28] Kraus G. *J. Appl. Polym. Sci. – Appl. Polym. Symp.*, 1984, 39: 75.

[29] Huber G, Vilgis T A, Heinrich G. Universal Properties in the Dynamical Deformation of Filled Rubbers. *J. Phys: Condens. Matter,* 1996, 8(29): L409.

[30] Maier P G, Goritz D. *Kautsch. Gummi Kunstst.,* 1996, 49: 18.

[31] Wang M-J, Effect of Polymer-Filler and Filler-Filler Interactions on Dynamic Properties of Filled Vulcanizates. *Rubber Chem. Technol.,* 1998, 71: 520.

# Index